AF886589

Resource Allocation and Performance Optimization in Communication Networks and the Internet

Resource Allocation and Performance Optimization in Communication Networks and the Internet

Liansheng Tan

CRC Press
Taylor & Francis Group
Boca Raton London New York

CRC Press is an imprint of the
Taylor & Francis Group, an **informa** business

MATLAB® is a trademark of The MathWorks, Inc. and is used with permission. The MathWorks does not warrant the accuracy of the text or exercises in this book. This book's use or discussion of MATLAB® software or related products does not constitute endorsement or sponsorship by The MathWorks of a particular pedagogical approach or particular use of the MATLAB® software.

CRC Press
Taylor & Francis Group
6000 Broken Sound Parkway NW, Suite 300
Boca Raton, FL 33487-2742

Printed on acid-free paper

International Standard Book Number-13: 978-1-4987-6944-0 (Hardback)

Visit the Taylor & Francis Web site at
http://www.taylorandfrancis.com

and the CRC Press Web site at
http://www.crcpress.com

Dedication

This book is dedicated to my dear son Wenlin and my lovely daughter Rani.

Contents

Preface

Being formed by wire-lined and wireless communication systems, communication networks including the Internet have demonstrated their importance in the past decade as a fundamental driver of economic growth. Over the years, they have not only expanded in their sizes, such as geographical area and number of terminals, but also in the variety of services, users, and deployment environments. With the dramatic developments and fast evolution of communication networks and the Internet, resource allocation continues to be the fundamental challenge because better quality of service (QoS) is required with the increasing demand for bandwidth-hungry and/or computation-intensive services. The purpose of resource allocation in such environments is to intelligently assign the limited available resources among terminals/clients in an effective and efficient way to satisfy end-users service requirements. The central aim of resource allocation is to optimize network performance.

Regarding resource allocation and performance optimization issues in communication networks, the key challenge is how to cope with various new emerging system architectures, such as wireless sensor networks (WSNs), cognitive networks, mesh networks, multihop networks, peer-to-peer networks, cloud computing systems, and data centers. Traditional yet successful centralized allocation mechanisms have been widely used in transmission control protocol/Internet protocol (TCP/IP) networks (namely, the Internet) while fully distributed solutions are applicable to wireless communication networks. In recent years, many approaches and tools, including optimization theory, control theory, game theory, and auction theory, have been employed to model and solve a variety of practical resource allocation problems. Therefore, resource allocation in communication networks is a pressing research topic that has huge applications. Despite the significant advancements in this area, it is still imperative to develop advanced resource allocation techniques for ensuring the optimal performance of these systems and networks.

As a network of networks, the Internet is the global system of interconnected computer networks that use the IP suite to link devices worldwide. It consists of private, public, academic, business, and government networks of local to global scope, linked by a broad array of electronic, wireless, and optical networking technologies. The Internet carries an extensive range of information resources and services, such as the inter-linked hypertext documents and applications of the World Wide Web (WWW), electronic mail, telephony, real-time multimedia applications, and peer-to-peer networks for file sharing. Resource allocation and performance optimization for such a complex communication system will remain as a challenge and a long-standing interesting topic.

Ubiquitous access to information, anywhere, anyplace, and anytime is a remarkable feature of all sorts of information systems in the future. This will be realized by rapidly emerging wireless communications systems, which are based on radio and infrared transmission mechanisms, and utilizing such technologies as cellular telephony, personal communications systems, wireless Internet Protocol Private Branch Exchange (IP PBX), wireless local area networks, WSNs, Internet of things (IoT), and cloud computing systems. Wireless communication is a truly revolutionary technology, enabling every form of communications including multimedia communications between people and devices from any location in a wireless and/or wire-lined network including the Internet. It also underpins exciting applications such as sensor networks, smart homes, telemedicine, and automated highways.

A central element of the design philosophy that shaped the wireless communication systems is the QoS guaranteed for users. QoS is the overall performance of a network, particularly the performance seen by the users of the network. In wireless networks, resource allocation is a key factor that affects network performance due to scarce resource and variable channel conditions and users demands in all architectures of wireless communications. Hence, the problem of allocating resources with QoS guarantee attracts much research interest in recent years. While there is a rich body of publications associated with wireless communication engineering, there lacks a focused and dedicated reference to summarize the recently developed theoretical and technical findings on resource allocation and performance optimization in wireless communication networks. This needed expertise must build upon to encompass network management, including network resource allocation and resource management, integration of wireless and wire-lined networks, system support for mobility, computing system architectures for wireless nodes/base stations/servers, user interfaces appropriate for small-handled portable devices, and new applications that can exploit various wireless communication resources, including WSNs and the IoT systems.

As the field of communication networks and the Internet continues to spread and evolve, students, engineers, practitioners, and researchers face enormous literature. This book offers an up-to-date, comprehensive, and valuable alternative for the readers to access knowledge in the specific areas of resource allocation and performance optimization within the relevant topics. This book is presented also as a broad reference for individuals working in control and automation, communications, computing, and networking communities.

The present book provides a comprehensive introduction to the underlying theory, design techniques, and analytical results of communication networks and the Internet, focusing primarily on the core principles of network design. The main presented elements are models, protocols, and algorithms, which are keys in resource allocation and performance optimization. The key components of the current book include the following:

- Within the framework of resource allocation and performance optimization, an overview of the general background, including the relevant protocols, architectures, methods, and algorithms in communication networks and the Internet, is outlined.
- We present a number of approaches, which are based on control theory, within the framework of resource allocation and optimization of communication networks and the Internet. These control theory-oriented approaches are found to be successful in modeling and analyzing the dynamics of TCP/IP network (the Internet) that is run by high-speed protocol such as Fast TCP. Further, the stability analyzing results with fairness considerations are helpful for establishing novel transmission mechanisms, which lead to new insights into inventing new transmission protocols such as the generalized Fast TCP.
- We elaborate on the network utility maximization (NUM) theory with applications in resource allocation of wire-lined networks, wireless networks, and optical networks, with a central aim of design, i.e., the QoS guarantee. The state of the art of developments in this line presented and discussed.
- A couple of novel theories, for example, the resource allocation approach in the aggregate flow level and the NUM approach in the two-way flow scenario, are firstly presented.

The book is organized as follows. We start with a detailed introduction to the evolving communication technologies in wireless networks in Chapter 1. In this chapter, we highlight the recently and dynamically developed wireless communication technologies, although an introduction to the wire-lined networking technologies is not given here as they have been mature relatively.

Chapter 2 gives a survey on utility, utility function, and NUM theory. Herein, we give an application of NUM theory: the reverse engineering design of TCP Reno from the flow dynamical model to the packet-level implementation, which can be guided and understood by the NUM theory.

The notable approaches to address the congestion control and resource allocation issues of communication networks and the Internet from a macroscopic perspective are the primal algorithm, the dual algorithm, and the primal-dual algorithm. Chapter 3 focuses on introduction to these algorithms. Features of this chapter include a detailed analysis of the above three algorithms from stability theoretical point of view, the workable methods of how to obtain the individual flow rate, the individual flow price correspondingly from the aggregate flow rate, and the aggregate link price using the generalized inverse of matrix.

In Chapter 4, we present a novel virtual price FAST TCP congestion control approach termed as REM-FAST, a new algorithm termed as the Generalized FAST TCP, and a new FAST TCP flow level model, leading to a new FAST TCP variant to deal with packet loss in wireless links.

In communication networks, the notion of fairness characterizes how competing users should share the bottleneck resources. In Chapter 5, we review and analyze various fairness concepts and outline the state-of-the-art approaches in achieving various fairness criteria in the context of resource allocation in communication networks and the Internet. Taking into account to the fact that in wireless networks, resources (e.g., bandwidth, time slot, and frequency) are usually very limited and expensive, this chapter pays particular attention to resource allocation of wireless networks to satisfy certain fairness criterion.

Regarding the networks with a ring topology, the key performance objectives are to simultaneously achieve fairness, high utilization, and spatial reuse for fairness eligible traffics over the ring, and as such designing efficient fair bandwidth allocation algorithms to achieve such goals is an important task. Considering all these objectives and the special feature of fairness criterion particularly for the ring metropolitan area networks, Chapter 6 presents discussions on the existing fair bandwidth allocation approaches that are applied to these networks in a surveying manner but with technical expositions.

Further to Chapter 6, Chapter 7 continues on discussions on the designing issues with relation to efficiency, capability to meet QoS requirements and fairness in wide and metropolitan area networks. The approaches presented herein are based on control theory, which successfully fulfill the aim of meeting efficiency, fairness, and spatial reuse requirements of a general network with a ring topology.

Trade-off between efficiency (utilization, throughput, or revenue) and fairness in communication networks and the Internet with relation to any fairness criterion, that is, compromising between the fairness among users and the utilization of resources is interesting and important within the context of resource allocation and performance optimization. Chapter 8 is dedicated to the trade-off approaches between the efficiency and fairness for communication networks and the Internet.

Chapter 9 presents the theoretical and experimental results of fairness comparison among the leading high-speed protocols. This chapter mainly covers the following discussions and analyses: (1) fairness comparison between FAST TCP and TCP Reno; (2) fairness comparison between FAST TCP and TCP Vegas; and finally (3) experimental fairness comparison among FAST TCP, H-STCP, STCP, and TCP Reno.

Chapter 10 focuses on the bidirectional data transmissions of a connection, with flows of TCP packets going with flows of ACK packets together but in opposite directions. We discuss the performance of FAST TCP in a bidirectional transmission scenario. We further take the first step to investigate the NUM theory in the two-way flow scenario. The modeling and solution lead to some straightforward discoveries of a critical problem with network bandwidth allocation and performance optimization in the NUM framework to account for the impact of ACK packet flow.

Directly relevant to the task of resource allocation and performance optimization in communication networks and the Internet, it is very useful to estimate traffic matrix (TM) from link measurements and routing information, especially for the tasks of capacity planning, traffic engineering, and network reliability analysis. Chapter 11 presents some novel methods to estimate IP TM and provides a brief survey on this subject.

In Chapter 12, we study utility maximization problems for communication networks but from a router-level (aggregate flow-level) bandwidth allocation standpoint. Note that, when talking about the router-level bandwidth allocation, we also mean the aggregate flow-level bandwidth allocation for wire-lined networks, including the Internet, as in this case flows are aggregated at the routers. However, once wireless networks are concerned, it is more precise to use the term, aggregate flow-level bandwidth allocation. By using the generalized matrix inverse, we propose a general model of utility-optimized router-level bandwidth allocation and its solution.

To meet the increasing bandwidth demands and to reduce costs, optical burst switching (OBS) network is a promising candidate for the next-generation optical Internet. Chapter 13 presents novel methods and algorithms of bandwidth allocation in OBS networks, using the aggregate flow-level NUM approach.

Power saving is one of the key issues in mobile ad hoc networks (MANETs) which can be realized both in MAC and network layers. Attacking this issue, Chapter 14 presents power-adjusting algorithm on mobility control for MANETs.

In Chapter 15, we propose a routing algorithm termed as PCA-guided routing algorithm (PCA-RA) by exploring the PCA approach. The algorithm remarkably reduces energy consumption and prolongs network lifetime by realizing the objective of minimizing the sum of distances between the nodes and the cluster centers in a WSN network. Both theoretical analyses and simulation results are presented to support the above claim.

Chapter 16 presents the following approaches for optimally configuring and clustering a WSN: (1) to optimally configure a WSN by time matrix and (2) to optimally cluster a WSN into a multi-tier architecture.

One of the most widespread and important applications in WSNs is continuous data collection. However, how to efficiently and effectively collect and process big data in WSNs still remains a challenge, which is crucial to network lifetime, and computing and communication performance. Subsequently, Chapter 17 is dedicated to reporting some novel methods and algorithms of big data collection in WSNs.

In Chapter 18, we study the challenge to simultaneously achieve maximal lifetime and maximal total utility in a multihop WSN communication system. Given the fact that there is usually a contradiction between these two objectives and that a trade-off exists between them, we provide insightful analyses, theoretical models, algorithms, and simulation results on this trade-off.

Resource allocation among multimedia users to meet the various QoS requirements in a wireless network is still a challenge. Chapter 19 provides an approximate NUM model and its solution for resource allocation among real-time multimedia users in wireless networks.

Chapter 20 studies the wireless resource allocation problem in the downlink of a wireless network with a central control system, such as a cellular base station. We attempt to maximize the total utility of all users. We consider two common types of traffic: hard QoS traffic and best-effort traffic. Three allocation algorithms are presented for these two types of traffic, namely, the HQ allocation for hard QoS traffic, the elastic allocation for best-effort traffic, and the mixed allocation for the coexistence of both types of traffic. A number of key theorems as the general design guidelines for utility-based resource allocation in wireless networks are given.

For wireless networks, how to allocate resources among the soft QoS traffic remains an important design challenge. Mathematically, the most difficulty comes from the nonconcave utility function of soft QoS traffic in the NUM problem. Facing this challenge, Chapter 21 establishes some key theorems to find the optimal solution and then presents a complete algorithm called utility-based allocation for soft QoS (USQ) to obtain the desired optimal solution. The proposed theorems and algorithm act as designing guidelines for resource allocation of soft QoS traffic in a wireless network, which take into account the total available network resources, the users' traffic characteristics, and the users' channel qualities. By numerical examples, we illustrate the explicit solution procedures.

Acknowledgments

This book was supported by the National Natural Science Foundation of China (No. 61370107 and No. 61672258).

1 Introduction to Wireless Networks: Evolving Communication Technology

Over the years, wireless technologies have advanced from theoretical establishments to readily applied engineering techniques that play a major role in many aspects of modern life. The first-generation (1G) wireless network has fulfilled the basic mobile voice, while the second-generation (2G) wireless network has introduced capacity and coverage. This is followed by the third-generation (3G) wireless network, which has quest for data at higher speeds to open the door for truly "mobile broadband" experience, which was further realized by the fourth-generation (4G) wireless network. The 4G wireless network provides access to a wide range of telecommunication services, including advanced mobile services, supported by mobile and fixed networks, which are increasingly packet based, along with a support for low- to high-mobility applications and a wide range of data rates, in accordance with service demands in multiuser environment. Fifth-generation (5G) should be a more intelligent technology that interconnects the entire world.

In the context of service provider (SP), with the exponential growth of mobile data traffic, wireless operators are currently interested heavily in infrastructure, including base stations, distributed antenna system (DAS), small cells, and others, to satisfy the demand of the users. They are looking at radio access network (RAN), visualization and software defined networking (SDN) to simplify network management, enable rapid reconfiguration of the networks to remain agile and flexible, and to reduce cost. What this means is that much of the hardware-based topology and the control and monitoring functionality of the network elements will be visualized in web-scale data centers to eliminate much of the hardware. With this, wireless operators can move from being hardware centric to software centric and become digital wireless content providers and (SPs).

This chapter gives a comprehensive survey on the evolving map of wireless communication networks covering the evolution of their basic principles of operation, and their architectures, from 1G wireless networks to 5G wireless networks and beyond. In this chapter, we not only summarize the evolution of the wireless generations, indicating the great advances that have so far been achieved in wireless communications in general, but also discuss the future developing trend of wireless communication networks.

1.1 EVOLUTION OF WIRELESS NETWORKS

Broadly speaking, a wireless network is a form of communication network, which uses wireless data connections for performing data communications in a wireless manner to realize various applications and services. Examples of wireless networks include cell phone networks [4], Wi-Fi local networks [5,6], wireless mesh networks (WMN) [7], wireless sensor networks [8–10], wireless ad hoc, networks [11–13], and terrestrial microwave networks [14,15].

Wireless communication is the transfer of information between two or more points that are not connected by an electrical conductor. Therefore, wireless telecommunication networks are generally implemented, functioned, and administered by using radio communications. The radio waves distance have a large range, which can be a few meters for television or as far as thousands or even millions of kilometers, for instance, the deep-space radio communications. It includes various types of fixed, mobile, and portable applications, including two-way radios, cellular telephones, personal digital assistants (PDAs), and wireless networking. Key applications of radio wireless technology

Table 1.1
Various Layers of Wireless Networking Technologies

Layer	Technologies
Device	Mobile devices: PDAs, notebooks, cellular phones, pagers, handheld PCs, wearable computers, smart sensors, Global Positioning System (GPS) tracking device, network wireless storage, wireless power bank
Application and service	Wireless applications: Wireless Application Protocol (WAP), i-mode, messaging, Voice over IP (VoIP), location-based services, microphone for input of user speech utterances, audible sound signal processing, data describing data resources
Physical	Wireless standards: 802.11a, 802.11b, 802.11g, AX.25, 3G, 4G, 5G, CDPD, CDMA, WCDMA, TSCDMA, WCTP, GSM, General Packet Radio Service (GPRS), radio, microwave, laser, Bluetooth, 802.15, 802.16, IrDA, XHTML, ZigBee

can be seen in garage door openers, GPS units, wireless computer mice, keyboards and headsets, headphones, satellite television, radio receivers, broadcast television, and cordless telephones. The implementations of wireless telecommunications usually take place at the physical level (layer) of the OSI model in network architecture [3].

Being strikingly different from the wire-lined networking, wireless networking is a remarkable approach by which homes, telecommunication networks, and enterprise (business) installations are able to avoid the costly process of introducing cables into a building, or as a connection between various equipment locations (for the features of wireless networking see, e.g., [2]). When performing the wireless networking, a user can have a notebook computer, PDA, Pocket PC, Tablet PC, or just a cell phone and stay online anywhere a wireless signal is available.

The basic mechanism behind wireless networking is that signals can be carried by electromagnetic waves that are further transmitted to a signal receiver. However, to make two wireless devices understand each other, one needs to design protocols for communication. If we categorize the wireless networking technologies into three layers (see [20]), it is easy to understand their working mechanism, as shown in Table 1.1. These three layers are device layer, physical layer, and application and service (protocol) layer, respectively.

In the device layer, mobile devices are gadgets ranging from the smallest cell phone to PDAs and notebook computers. These devices use wireless technologies to communicate with each other. The physical layer inherits various physical encoding mechanisms for wireless communications. Bluetooth, 802.11x, CDMA, GSM, 3G, 4G, 5G, and beyond are various standards that define different methods to physically encode the data for transmission across the airwaves. The application and service layer, also referred to as ISO layers 2–7, contains the protocols that enable wireless devices to process transmission of data in an end-to-end way. Protocols like WAP, VoIP, and i-mode are performed in this layer.

Since 1947, Bell Laboratories pioneered the innovative approaches of using cells for wireless communications [1], the wireless communication networks have been rapidly developed and evolved, which enabled the diversity of wireless applications to be enriched greatly. From late 1970s, when the wireless communications epoch started, till now significant changes and enormous growth have been seen in wireless networks. Figure 1.1 displays the evolution of wireless networks [25] in terms of data rates, services, and technologies.

1.2 1G WIRELESS NETWORKS

Following the mobile radio telephone, namely the 0G, is the 1G technology. The term of 1G (or 1-G) refers to the first generation of wireless telephone technology (mobile telecommunications).

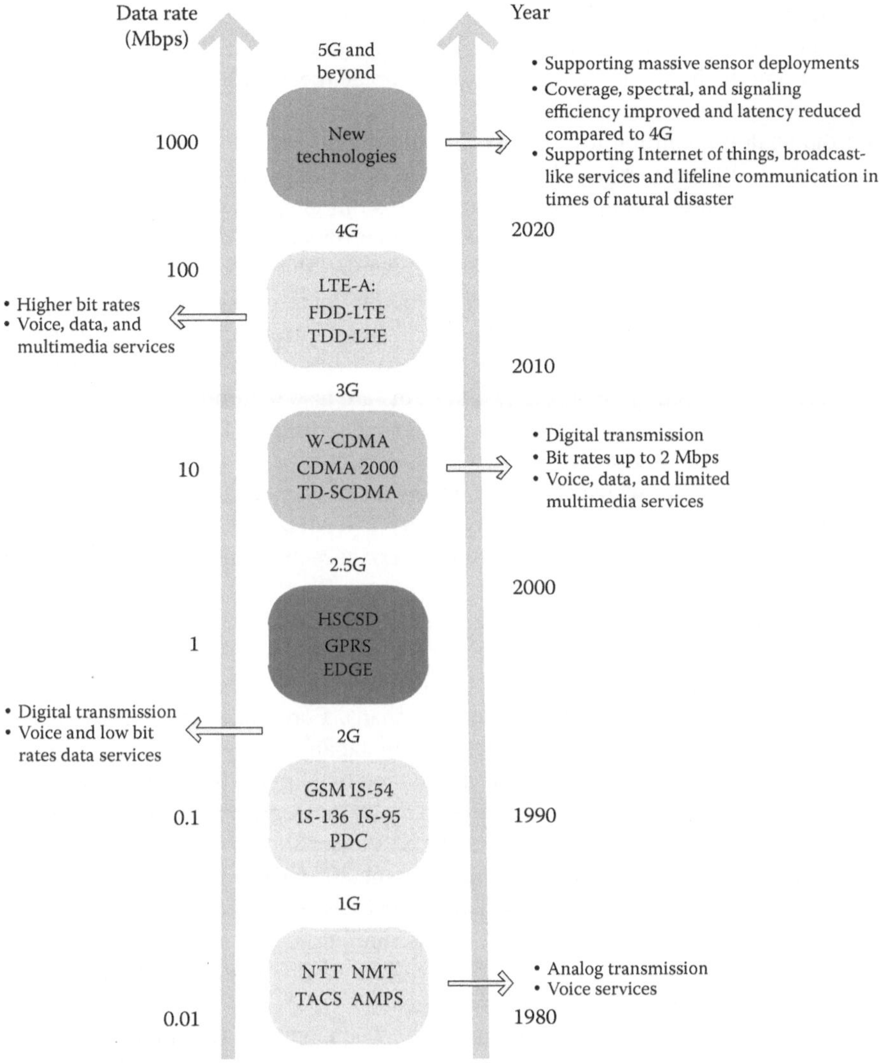

Figure 1.1 Evolution of wireless networks: a road map from 1G to 5G.

1G is governed by the analog telecommunications standards that were launched in the 1980s and continued until being replaced by 2G digital telecommunications. The main difference between the two mobile telephone systems (1G and 2G) is that the radio signals used by 1G networks are analog, while 2G networks use digital signals.

In analog recording technology, a property or characteristic of a physical recording medium is brought to change in a manner analogous to the variations in air pressure of the original sound. Usually, the air pressure variations are first converted (by a transducer such as a microphone) into an electrical analog signal, in which either the instantaneous voltage or current is directly proportional to the instantaneous air pressure (or is a function of the pressure). The variations of the electrical signal are then converted to variations in the recording medium by a recording machine, for instance, a tape recorder.

In digital recording technology, the physical properties of the original sound are converted into a sequence of numbers, which can then be stored and read back for reproduction. Normally, the sound is transduced (as by a microphone) to an analog signal in the same way as for analog recording, and then the analog signal is digitized or converted to a digital signal through an analog-to-digital converter and finally recorded onto a digital storage medium such as a compact disk or hard disk.

Between the analog recording technology and the digital recording technology, two main differences in functionality are bandwidth and signal-to-noise ratio (S/N). However, both digital and analog systems have inherent strengths and weaknesses of themselves. In terms of the Nyquist frequency [17], the bandwidth of the digital system is determined by the sample rate it uses. On the contrary, the bandwidth of an analog system is dependent on the physical capabilities of the analog circuits. The S/N of a digital system is limited by the bit depth of the digitization process, but the electronic implementation of the digital audio circuit may introduce additional noise. In an analog system, other natural analog noise sources exist such as flicker noise and imperfections in the recording medium. Some functions of the two systems are also naturally exclusive to either one or the other, such as the ability for more transparent filtering algorithms [16] in digital systems and the harmonic saturation of analog systems.

With comparison to 2G, in which the voice during a call is encoded to digital signals, in 1G the voice is only modulated to higher frequency, typically 150 MHz and up. The common thing in 1G and 2G is that both of them use digital signaling to connect the radio towers (which listen to the handsets) to the rest of the telephone system.

One of 1G standards is Nordic Mobile Telephone 450 (NMT-450), used in Nordic countries, Switzerland, the Netherlands, Eastern Europe, and Russia. Others include Advanced Mobile Phone System (AMPS) used in North America and Australia, Total Access Communications System (TACS) in the United Kingdom, C-450 in West Germany, Portugal, and South Africa, Radiocom 2000 in France, TMA in Spain, and RTMI in Italy. In Japan, there were multiple systems. Three standards, TZ-801, TZ-802, and TZ-803, were developed by Nippon Telegraph and Telephone Corporation (NTT), while a competing system operated by Daini Denden Planning, Inc. (DDI) used the Japan Total Access Communications System (JTACS) standard.

The NTT systems used 600 FM duplex channels in the 800 MHz band, with a channel separation of 25 kHz. The NMT-450 wireless communication systems were operated in the 450 MHz range with a total bandwidth of 10 MHz. The JTACS was running at 900 MHz with a band of 25 MHz for each path and a channel bandwidth of 25 kHz. The AMPS system was allocated a 40 MHz bandwidth within the 800–900 MHz frequency range.

All aforementioned 1G wireless networks used the frequency division multiple access (FDMA) approach [26–29] to attain spectrum sharing among multiple users. Although 1G wireless networks inherit handover and roaming capabilities, having used different frequencies and communication protocols, they were not able to interoperate between countries. This constituted of the disadvantage of 1G wireless networks.

More details of the 1G wireless network technologies mentioned earlier can be found in, for example, [18].

1.3 2G WIRELESS NETWORKS

Motivated by the global roaming and the comprehensive needs for advanced wireless communications, 2G wireless networks come to a reality, which then obsoleted the 1G wireless networks. Thanks to the advances in large-scale integrated circuits (ICs) technology, digital communications became much more practical and economical than analogous communications. Therefore, 2G wireless networks used digital transmission rather than analog transmission and the multiple access techniques, including the time division multiple access (TDMA) [21,22] and code division multiple access (CDMA) [1,19,23,24]. Based on the above technologies, 2G systems were significantly more efficient on the spectrum allowing for far greater mobile phone penetration levels. 2G networks introduced data services for mobile, starting with SMS text messages. The 2G networking technologies realized various advanced services such as text messages, picture messages, and multimedia messages by allowing for the transfer of data in a way that only the intended receiver can receive and read it, as all text messages are digitally encrypted.

There are several 2G technologies that have been implemented across the world. The most widespread deployment is the TDMA-based Global System for Mobile Communications (GMS) and the CDMA-based IS-95 system. Other 2G technologies that have been deployed include Digital European Cordless Telephone (DECT), IS-136, and the PDC-based personal handy-phone system (PHS).

The TDMA-based GSM is a wireless communications system, which started in the 1980s in Europe with the aim to get rid of the capacity limitation being experienced by analog networks.

The GSM 900 system utilizes two 25-MHz bands for the uplink and downlink, and within this spectrum 200 kHz channels are allocated. The uplink and downlink are separated by a 45-MHz spacing. GSM 1800 uses two 75-MHz bands for the uplink and downlink. Again 200 kHz channels are allocated within those bands and are separated by a 95-MHz spacing. The 1900-MHz systems use two 60-MHz bands for the uplink and downlink using 200-MHz channels within those bands and separated by 80-MHz spacing.

CDMA is a relatively new technology in the mobile cellular industry. The CDMA commercial networks were first deployed in the mid-1990s. However, they have been growing very quickly and they have occupied about 25% of the wireless networks globally.

In a wireless network, the wireless spectrum is a scarce resource, with tight regulations in terms of usage and power radiated along with licenses required to operate. Interference is an issue that wireless networks must confront. Spread spectrum is one of the employed techniques, being inherently less sensitive to interference. Spread spectrum techniques typically use more bandwidth than necessary to transmit and receive bits.

Spread spectrum techniques inherently offer more privacy than narrow-band techniques, as they are more difficult to intercept or spy on. They use a code, which is known only to the transmitter and the receiver. Spread spectrum techniques can also coexist with other technologies as the transmitter and receiver can obtain information as long as they are decoding with the same code.

There are two popular spread spectrum techniques. The first is frequency hopping spread spectrum, where the transmitter and the receiver hop in a predetermined sequence through a wide band of frequencies. The second one is direct sequence spread spectrum (DSSS), in which data bits are transformed by codes, which in turn occupy a wide band of frequencies. This technique is already in use in public wireless networks such as IS-95, i.e., CDMA, as it is popularly known. DSSS is a technique wherein the carrier is modulated by a digital code in which the code bit rate is much larger than the information signal bit rate.

The DSSS system is a wide-band system, in which the entire bandwidth of the system is totally available to the users. The user's data are spread using a spreading signal referred to as the code signal. The code signal or the spreading signal has a much higher data rate than the user data rates. For instance, the 1.2288 MCPS in CDMA vs. user data rates that are much lower. At the receiving end, de-spreading is achieved by the cross correlation of the signal with a synchronized replica of the same signal used to spread the data. In CDMA systems, pseudo-random noise (PN) sequences are used to spread the bandwidth and distinguish among various users' signals. PN sequences, as the name suggests, are not random but rather deterministic.

IS-41 is standardized by the Telecommunications Industry Association (TIA). Revision C is the latest version of the protocol and is called IS-41C. IS-41 is the core networking protocol that supports mobility, authentication, and roaming. IS-41 allows network equipment to be multivendor. Since the equipment has to conform to the standard interface, it is possible to have an environment wherein Mobile Station Switching Centers (MSCs) are from vendor A and the Base Station Controller (BSC) radio network is from vendor B. Roaming between networks that use GSM MAP and IS-41 requires the use of gateway functions that convert messages from one protocol to another. Such gateways can be considered protocol translators. IS-41C is an application-layer protocol. IS-41 is normally operated over SS7 networks, which provide the reliability required for signaling. The CDMA standard IS-95 is specified by Telecommunications Industry Association (TIA)/Electronic Industries Alliance (EIA).

Europe firstly identified the need for a common mobile telephony standard since different countries had differing analog networks, and consequently roaming of subscribers between these networks was not possible. The Conference European des Postes et Telecommunications (CEPT) is a standardization arena in Europe, within which a new group called Groupe Special Mobile (GSM) was formed in 1982 with the specific task to specify a unique radio communication system for Europe at 900 MHz frequency band with a total bandwidth of 50 MHz. The 2G cellular telecommunication networks were commercially launched on the GSM standard in Finland by Radiolinja in 1991.

GSM solved the incompatibility issue, which was originally staying with the 1G wireless communication systems, by offering a single unified standard, which made seamless roaming possible through the Europe. In the United States, there were three 2G standards. Firstly, Digital AMPS (D-AMPS), which used TDMA as air interface, was introduced in 1991 as the first 2G system Interim Standard 54 (IS-54). Secondly, in 1996, a new D-AMPS version IS-136 was created. A number of appealing features had been added to the original IS-54 specification, including text messaging, circuit switched data (CSD), and an enhanced compression protocol. Thirdly, the first CDMA-based digital IS-95 was implemented in 1993 by the CDMA Development Group (CDG) and TIA. The American Federal Communications Commission (FCC) also auctioned a new block of spectrum in the 1900 MHz band, which allowed GSM1900 to enter the US market. In Japan, the Personal Digital Cellular (PDC) system was initially introduced in 1991, NTT DoCoMo launched its service in 1993. PDC used TDMA air interface and was functioning in 800 MHz and 1.5 GHz band. One can find more details of the 2G wireless networks, for example, in Ref. [18].

Compared to 1G, 2G networks essentially had higher spectrum efficiency, better quality data services, and more advanced roaming ability [1,18]. Besides the traditional voice services, low bit rates data services were supported and more advanced mobility was designed to get rid of the disadvantages of the 1G wireless communication system. Till today, 2G wireless networks still occupy the communication markets throughout the whole world. However, being driven by the dramatically increasing mobile subscribers and the evolution of new kinds of services, the data rates of plain GSM were not sufficient. Some more advanced technologies were brought to be existed, such as High Speed Circuit Switched Data (HSCSD), GPRS, and Enhanced Data Rates for Global Evolution (EDGE), commonly termed as 2.5 generation (2.5G) wireless networks [30,31].

In order to obtain higher data transmission rates, the GSM air interface was first enhanced to have yielded HSCSD [32]. In HSCSD, multiple time slots were multiplexed together. Thus, the total data rate of HSCSD comes out to be the product of the number of the time slots timing the data rate per time slot. In practical scenario, the maximal number of time slots is usually 4, thus the theoretical maximal data rate of HSCSD is 4×14.4 kbps $= 57.6$ kbps full-rate time slots. Therefore, the high-speed data rate is upgraded, in this way, without any hardware modification being involved. The only thing to do is to upgrade its software.

The next modification to GSM air interface is GPRS [33,34], in which four new channel coding schemes were proposed. In GPRS, the data rates can be lifted up to 160 kbp. Being packet-switched, the GPRS system does not allocate the radio resource continuously but rather than when needed. Therefore, GPRS is particularly suitable for non-real-time services such as E-mail and Web surfing.

Finally, a notable refinement to the former system is the so-called EDGE [35], which was designed as an add-on of the existing digital systems to obtain higher data rates. In EDGE, a new modulation scheme termed as eight-phase shift keying (8PSK) was implemented on the basis of the GSM radio structure and TDMA framing. EDGE thereby increased the data rate of the normal GSM systems by three times. EDGE had been able to conduct wireless multimedia services, for instance, video phone and video conference at the data rate of 384 kbps.

1.4 3G WIRELESS NETWORKS

At the initial stage of wireless networks, wireless communication technology had offered the nomadic user a diversity of data services, but higher data rates would provide far more sophisticated applications to meet the increasingly demands from users. The limitation in providing these new services is the transmission data rate. It has taken a decade to bring reasonable speeds to the desktop using Integrated Services Data Network (ISDN) [36,37], Digital Subscriber Link (DSL) [38], and cable modems. To provide similar speeds to the mobile user is even more difficult. Two factors are the limited bandwidth and the harsh radio frequency (RF) environment [39]. Further, the complex wireless communication environment plagued by deep fades, diversity, and high bit error rates held back the advancement of new technology.

The huge customer demand for digital services in wide diversity and high quality is the major stimulus for the 3G wireless networks. However, the new technologies in 3G and standards for 3G took a number of years to develop and to deploy as there is a too big technical jump from 2G to 3G. 2.5G (a.k.a. "two and a half G") is essentially bridging in terms of technology that allowed SPs to transition from 2G to 3G systems smoothly. Customers could also start receiving limited 3G features before 3G is fully available.

2.5G systems uses improved digital radio and packet-based technology with new modulation techniques to increase data rates, system efficiency, and system performance. Their advantage is that despite changes they remain compatible with 2G systems. An added benefit is the low cost of the changes as compared to moving to totally new 3G systems. From a users point of view, the transition from 2G to 3G seems to be straightforward as 2G services are also provided by 3G services. However, from the SPs viewpoint, the switching from 2G to 3G is not easy, but it is risky and costly as most 2G equipment must be replaced to support 3G. Furthermore, 3G also requires more frequency spectrum, which is particularly scarce and costly. For this reason, deploying interim 2.5G systems did help. It provides significant improvements over 2G but does not require large investments.

In 1992, the World Administrative Radio Conference (WARC) defined the frequency spectrum for 3G wireless communication to support advanced services, particularly data services. International Mobile Telecommunications-2000 (IMT-2000) is the International Telecommunications Unions (ITU) effort to develop a rigorous definition. The IMT-2000 technologies [1,40–42] have the following features:

- *High-speed data transmissions*: 3G brings an order of magnitude improvement to data communications over 2G, that enables bandwidth hungry multimedia applications. Although the use of e-mail and web browsing is dominating, they are not crucial applications that ask for a huge investment in 3G. Multimedia and video conferences are the exact services that are expected to justify the new infrastructure. 3G data rates fall into three categories: 2 Mbps to stationary users (i.e., fixed location), 384 kbps to pedestrian users, and finally 144 kbps to vehicular users.
- *Higher voice quality and high Quality of Service (QoS)*: 3G is trying to provide voice quality comparable to that of wire-line telephony. 3G is also trying to provide high QoS to a diversity of service.
- *Greater capacity*: With the explosion in usage, the need to efficiently use frequency spectrum is extremely important.
- *Multiple simultaneous services*: This feature enables a user to download an MP3 audio file at the same time talking on the same cell phone.
- *Symmetrical and asymmetrical data transmission support*: E-mail and web browsing present asymmetrical nature. For instance, data transmitted by the user is much less than that transmitted to the user. However, services such as video conferences are symmetrical resulting in equal data transmission in both directions. This requires 3G to have the capability in supporting both the symmetrical and asymmetrical data transmission.

- *Global roaming across networks*: 3G is expected to support global roaming. Currently, roaming between international networks requires a different cellular phone for each network. 3G is facing this challenge.
- *Enhanced security*: In 3G, users will be able to communicate and conduct business in a secure environment. This will ultimately foster additional commercial services.
- *High level of service flexibility*: In 3G wireless networks, both circuits witched (e.g., voice, real-time video) and packet-switched services (e.g., Internet Protocol) shall be supported.

Looking into the above features of 3G networks more closely and taking, for example, the security issue, one sees that 3G networks have followed the rule that any new security architecture must be based on an evolution of GSM and must adopt four basic principles:

- It will take into account the additional features needed for actual or predicted change in the operating environment;
- It will maintain compatibility with GSM wherever possible;
- It will retain those features of GSM that have proved to be robust and useful to the user and network operator;
- It will add or enhance features to overcome actual or perceived weaknesses in 2G system.

Therefore, one can see that, compared to the previous generations of wireless network (0G, 1G, and 2G), the following enhancements on the aspects of security were seen as a priority for 3G [46]:

1. Mutual authentication,
2. Assurance that authentication information and keys are not being reused (key freshness),
3. Integrity protection of signaling messages, specifically the secured encryption algorithm negotiation process,
4. Use of stronger encryption (a combination of key length and algorithm design),
5. Termination of the encryption further into the core network to encompass microwave links.

Among the above mentioned features of 3G network, one notable characteristic is the global roaming capability. In 3G, Mobile IP technology [47] envisions the transfer of data and voice traffic in a packet-switched fashion. Given the fact that, in the Internet, each host has an IP address using which data are transmitted to it. With the hosts being stationary, the physical location of the host does not change. Therefore, the routing tables (updated in relatively large time intervals) in routers function very well, guiding data to the destination address. However, in mobile communication systems, nodes keep changing their location frequently. It is not practical to change the routing tables every time when the mobile node moves. To attack this difficulty, a protocol called Mobile IP has been proposed by IETF as a solution, particularly for wireless networks. Mobile IP is only suitable for best effort traffic. But 3G wireless systems must provide multimedia and other real-time, delay constraint services. Hence, reservation of resources has to be carried on. Although Resource Reservation Protocol (RSVP) (see, e.g., [50]) is a popular protocol for resource reservation on the Internet, it lacks the ability to deal with the mobility issue. Therefore, an enhancement of RSVP, termed as Mobile RSVP (see, e.g., [49]), which works on Mobile IP, has been proposed. The main characteristic of this protocol is that, the base station supporting the Mobile Terminal (MT) is responsible for ensuring that the QoS guarantees [274] are maintained, when the MT moves to a nearby cell. This aim is realized by passively reserving resources in all the neighboring cells [48].

The 3G specifications required participation from companies and agencies worldwide to assure that the standard is globally accepted. The 3G partnership project was tasked with the overall coordination responsibilities. Applications of 3G include GPS, Location-based services, Mobile TV, Telemedicine, Video Conferences, and Video on demand.

At the beginning, IMT-2000 was supposed to be a single, unified, worldwide standard, but later on, it has been split into three categories: Wideband CDMA (WCDMA), CDMA2000, and Time

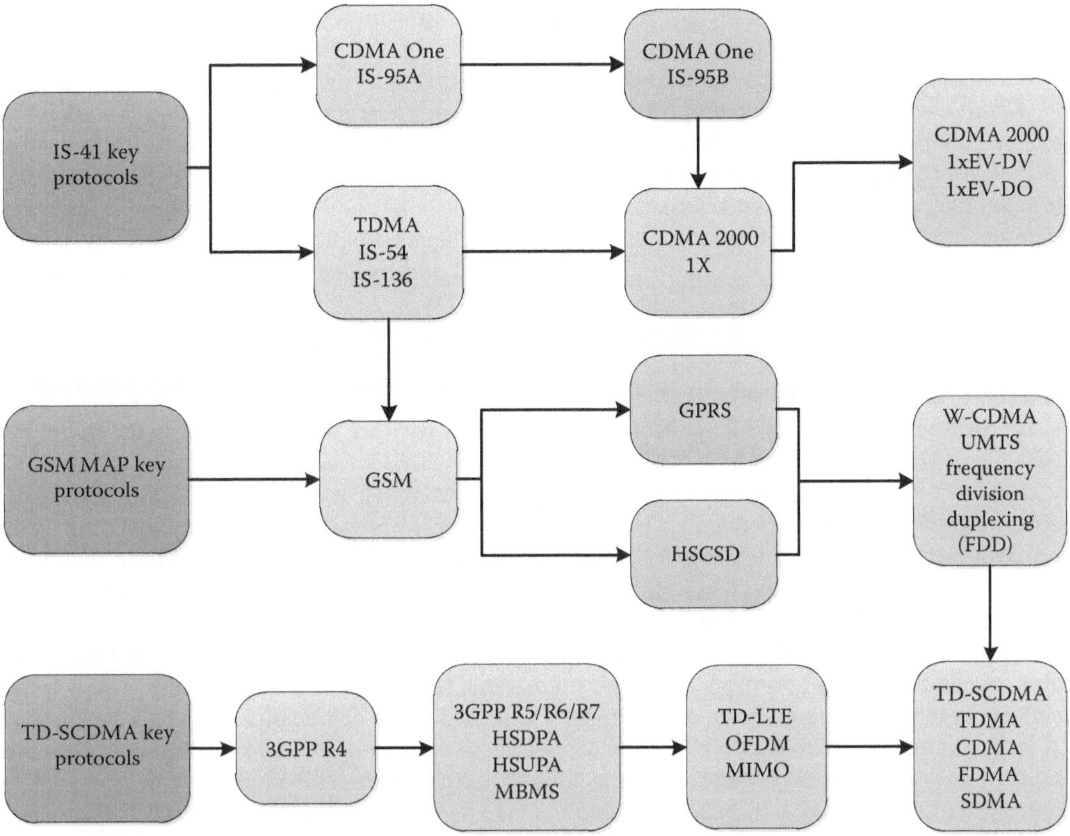

Figure 1.2 A road map of transition from 2G to 3G and beyond in technologies.

Division Synchronous CDMA (TD-SCDMA). A road map of 3G development logically in terms of technology evolution has been shown in Figure 1.2, which is modified from Figure 2 in Ref. [1].

WCDMA (see, e.g., [45]) is a 3G standard evolved from GSM networks. It is led by the 3GPP organization in Europe to provide multimedia communications and integrated services. Most major mobile network operators (MNOs) in Europe have chosen WCDMA as their 3G solutions.

CDMA2000 (see, e.g., [43]) is a following up and refined version of 2G CDMA standard IS-95. It is managed by 3GPP2, which is separated and independent from 3GPP. The mobile operators in North America and Asia Pacific are committed to CDMA2000 for 3G wireless communications.

TD-SCDMA (see, e.g., [44]) is a standard developed by Chinese telecommunications company Datang. It was proposed by the China Wireless Technology Standard (CWTS) Group in 1998, approved as one of the 3G standards by ITU in May 2000, and joined 3GPP in March 2001. China, the world's largest market for wireless communications, is using TD-SCDMA as the 3G standard.

1.5 4G WIRELESS NETWORKS

Following 0G, 1G, 2G, and 3G, wireless networks have further been driven to the 4G by the dramatically increasing mobile subscribers and the huge demands of new multimedia applications. In recent years, more and more interests are directed to 4G wireless networks in study, research, and commercial activities. More generous bandwidth, smoother and quicker hand off, wider coverage area, and more services are obtained in 4G wireless networks by tightly following up and overtaking the existed wireless communication technologies. In particular, 4G supports a wide variety of new services with high data rates, including the real-time multimedia services. The services offered by 4G include mobile multimedia [52], anytime anywhere [53], global mobility support [54], integrated wireless solution [55], and customized personal service [51,56].

The 4G wireless networks adopt flexible and adaptive integration of network technologies to enable mobile station to seamlessly roam between access networks. They should allow more advanced broadband multimedia services with varying QoS requirements, as they should not only provide the commonly known Internet services but also should transport the traditional voice service and other real-time applications. Those QoS requirements should include preferred low handoff delay, high bandwidth utilization, low new call and hand off blocking probability, low loss packets, and high packet rate.

The QoS mechanisms in 4G should include admission control, and resource scheduling and allocation to allow multimedia applications to get certain QoS guarantees on bandwidth, loss rate, and delay for its packets delivery. However, providing QoS guarantees in 4G networks is a nontrivial issue. Subsequently, QoS challenges for 4G wireless networks are well addressed in the literatures (see, e.g., [57]).

Recently, a novel medium access control, called MACW4G, was developed [58]. MACW4G covers radio link control, medium access control, and part of physical layers corresponding to the protocol stack for wireless Internet. MACW4G takes into account the constraints imposed by the heterogeneity of access networks. Reference [59] addresses the QoS degradation problem of wireless relay node failures in multihop Worldwide inter-operable Microwave Access (WiMAX) and Long-Term Evolution (LTE) networks. In 4G networks, hot-spot cells can be occurred when available wireless resources at some location are not enough to sustain the needs of users. The hot-spot cell can potentially lead to blocked and dropped calls and cause degradation of the service quality. A large amount of users enjoying multimedia services can move around in a 4G mobile network, which may generate heavy flows of traffic load in the network. This situation can generate the hot-spot cell, which has a short length of life at most a few minutes. Reference [60] proposes a handover-based scheme, which can effectively manage hot-spot cells in 4G mobile networks. The proposed scheme adopts hard handoff mechanism and dynamically controls the time point of handover based on the amount of load of the current serving cell and the target cell.

In 4G wireless networks, there have been significant research initiatives to explore potential solutions to better optimize network resources and maintain a high quality of experience (QoE) for subscribers. Reference [61] provides a survey on one specific emerging solution, called offload. The main aim of offloading is to avoid transporting low-priority traffic in costly networks in order to avoid degrading the perceived QoE. Offloading supports traffic redirection from parts of the network where congestion could occur to other low-cost parts of the network where capacity is available and less expensive. Reference surveys various offload approaches at different parts of the global network (access, core, and gateway). The aim of these approaches is to help mobile operators to face challenges such as capacity crunch, radio access saturation, average revenues per user decrease, and lack of indoor coverage.

4G wireless networks are described by the idea of heterogeneous network infrastructures integrating both existing wire-line and wireless access systems through advanced technologies. The wireless access systems include not only 2G/3G wireless networks but also broadband wireless networks, including satellite, digital audio broadcasting (DAB), digital video broadcasting (DVB), wireless local area network (WLAN), etc. In the form of high-speed ubiquitous connectivity, 4G encompasses all systems of various networks, ranging from public to private, from operator-driven to ad hoc, from broadband to personal area networks [62] using IP as the integrating mechanism. Being in such a converged ubiquitous environment, 4G actually envisioned that mobile users can roam between a broad range of communication systems and access information anywhere, anytime with seamless connection to any network through mobile phones, PDA, and laptops.

IEEE 802.16m and 3GPP LTE-Advanced are the two evolving standards targeting 4G wireless systems. In both standards, multiple-input multiple-output (MIMO) antenna technologies play an important role in meeting the 4G requirements. The application of MIMO technologies is one of the most critical differences between 3G and 4G. It not only enhances the conventional point-to-point link but also enables new types of links such as downlink multiuser MIMO. A large family of MIMO techniques have been developed for various links and with various amounts of available channel state information in both IEEE 802.16e/m and 3GPP LTE/LTE-Advanced [65].

IEEE 802.16/WiMAX-based broadband and mobile wireless access play an important in 4G wireless systems. IEEE 802.16/WiMAX standard [66] is designed to provide broadband wireless capability using a well-defined QoS framework, by incorporating several advanced radio transmission technologies such as orthogonal-frequency division multiplexing (OFDM), adaptive modulation and coding, and adaptive forward error correction (FEC). Therefore, it is a promising technology for wireless services requiring high-rate transmission (in the range of tens of Mbps) and strict QoS requirements in both indoor and outdoor environments. This technology is currently receiving big interests to researchers and practitioners in the design, analysis, installation, and management of 4G wireless access systems and networks. Several applications, including wireless telemedicine and e-health services, are brought into being by IEEE 802.16/WiMAX technology in 4G wireless networks.

Application adaptability to resources is a key feature of 4G communication services. From the view of mobile users, this means that services can be delivered automatically according to their personal preferences, while in view of terminals, this means various terminals are able to run one application with various formats depending on their capabilities. In connection with networks, applications can be transformed into various forms and levels in order to adapt to various networks resource availability [63].

Like in any wireless networking context, in 4G, resources are considered to be scarce and insufficient. Appropriate management of resources in 4G is crucial for efficient networks' usage and users' satisfaction. 4G wireless networks present the resource management issues, especially in wireless heterogeneous networks. Two network architectures, i.e., the RIWCoS and the ARAGORN architecture, were designed [67] to deal with resource management under various access technologies. The former one is specifically tailored to meet the demands of emergency situations, while the latter one introduces the notion of cognitive resource management and policy-regulated networking.

Due to budget limitations, network providers have to dynamically plan and deploy the 4G networks through multiple stages of time. By considering one-time deployment cost, daily operational cost, and 3G network congestion, the issue of how an operator financially manages the cash flow and plans the 4G deployment in a finite time horizon to maximize his final-stage profit was discussed in Ref. [64]. Although the operator provides both the traditional 3G services and the new 4G services, it is suggested that users start to use the 4G services only when they reach a sizable coverage. At each time stage, the operator should first decide an additional 4G deployment size by predicting users' responses in choosing between the 3G and 4G services. The problem is formulated as a dynamic programming model. An optimal threshold-based 4G deployment policy is suggested in Ref. [64]. It is suggested that the operators not deploy a full 4G coverage in an area with low user density or high deployment/operational cost. During the 4G deployment process, it is believed that the 4G subscriber number first increases and then decreases, as the 4G service helps mitigate 3G network congestion and increases its QoS.

1.6 5G WIRELESS NETWORKS

So far, 4G wireless networks can support data rates of up to 1 Gbps for low mobility, such as nomadic or local wireless access, and up to 100 Mbps for high mobility, such as mobile access. LTE and its extension, LTE-Advanced systems, as practical 4G systems, have recently been deployed or soon will be deployed around the world. However, with an explosion of wireless mobile devices and services, 4G is still far from being expected.

Taking China for example, more and more people are now accessing the Internet from a mobile device than a PC. China has been going through an explosive Internet accessing period, with mobile playing a key role in getting people online. The most recent data of Internet accessors and Internet coverage in China, which were released by the state-affiliated research organization China Internet Network Information Center (CNNIC), are shown in Figure 1.3. The latest report[1] published by CNNIC shows that the percentage of Chinese users accessing the Internet via mobile grew to 83.4% as of June 2014, for the first time surpassing the percentage of users who access the Internet via PCs (80.9%). It shows also that, the percentage of Chinese users accessing the Internet via mobile grew to 88.9% as of June 2015.

There had been a surprising surge in 4G/LTE services as China Mobile rapidly expanded its TD-LTE network and busily signed up subscribers. It claimed more than 50 million subscribers on its TD-LTE service by late 2014. The operator was making the most of its early lead on rivals China Unicom and China Telecom, both of which had been far more conservative in their deployment of TD-LTE networks, relying instead on Frequency Division (FD)-LTE technology C full licenses for which had yet to be allocated C over the home-grown TD standard favored by China Mobile and, importantly, the MIIT.

Confronted with a continuous decline in the fixed-line market, the government has again intervened to set aggressive targets for broadband services. Globally, China became the leading country in the number of Digital Subscriber Line (DSL) users in 2003 and by 2010 was the leading country in terms of FttX deployment. In the meantime, there has been a rapid development of Internet businesses in China as the digital economy takes shape. The pace of development has escalated as more and more of the population gain access to the Internet. The number of Internet users had passed the 650 million mark by mid-2014, increasingly accessing websites via mobile phones.

China continues to build a substantial world-class telecommunications infrastructure and the investments show no sign of abating. As data traffic grows, the major operators are keeping pace by increasing both domestic and international connectivity through submarine and terrestrial cable links. The country also has high aspirations with its space program and has developed a local industry to develop, build, and deploy communications satellites.

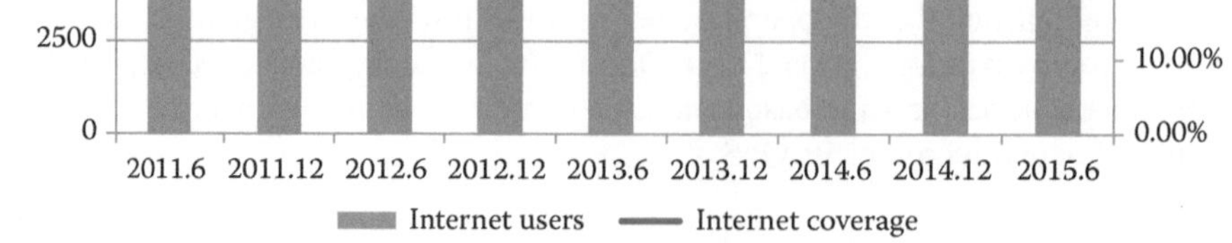

Figure 1.3 Internet accessors and Internet coverage in China. (From China Internet Network Information Center (CINIC), The 36rd report on the statistics of internet developing status in China, July 2015, http://cnnic.cn/hlwfzyj/hlwxzbg/hlwtjbg/201507/P020150723549500667087.pdf.)

[1] http://www.cnnic.net.cn/.

China is at the forefront of technology development, strongly supported by all levels of government. Among these initiatives are Cloud Computing and Smart Grid deployments and the building of Smart Cities that support the governments climate change targets set out in the 12th Five Year Plan.

As of June 2014, China had 632 million Internet users and 527 million mobile Internet users, according to CNNIC. This means the country tacked on 14 million new Internet users and 27 million new mobile Internet users in just 6 months. The Internet penetration rate stood at 46.9%, up 1.1% from the end of 2013. As of June 2015, China had 593.57 million mobile Internet users already. The statistics of mobile Internet accessors and mobile Internet coverage in China in recent years has been shown in Figure 1.4.

The amount of time each Internet user spends online on average a week is about 25.9 h, up 0.9 h from the end of 2013. CNNIC credits this to more Wi-Fi access across the country, as well as the maturity of 3G coverage and the onset of 4G access. CNNIC notes that there has been a rapid growth in the use of mobile payment services, with the percentage of users increasing from 25.1% to 38.9%. In turn, this has also led to the growth of m-commerce, group buying, and lifestyle services such as online travel booking.

In particular, mobile shoppers numbered 205 million in China as of June 2014, growing 42% in just 6 months. The portion of mobile Internet users who shopped online also rose from 28.9% to 38.9%.[2]

As more and more devices run wireless, one of the most crucial challenges in 4G wireless networks is the physical scarcity of RF spectra allocated for cellular communications. Cellular frequencies use ultra-high frequency bands for cellular phones, normally ranging from several hundred megahertz to several gigahertz. These frequency spectra have nearly been used up, making it difficult for operators to acquire more.

Another challenge is that the deployment of advanced wireless technologies comes at the cost of high-energy consumption. The increase of energy consumption in wireless communication systems

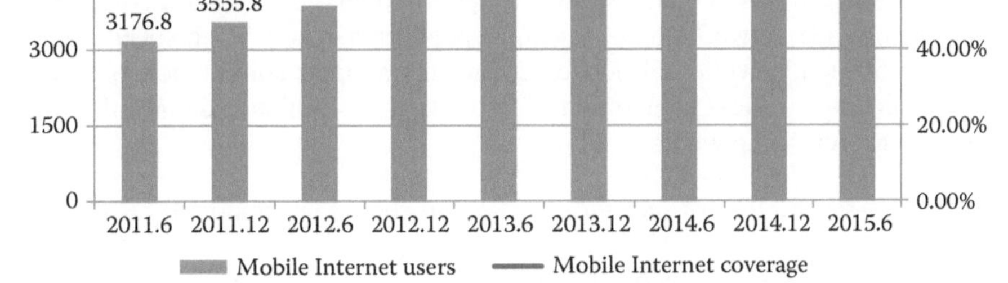

Figure 1.4 Mobile Internet accessors and mobile Internet coverage in China. (From China Internet Network Information Center (CINIC), The 36rd report on the statistics of internet developing status in China, July 2015, http://cnnic.cn/hlwfzyj/hlwxzbg/hlwtjbg/201507/P020150723549500667087.pdf.)

[2]http://www.chinatravelnews.com/article/82135.

causes an increase of CO_2 emission indirectly, which is currently considered to be a major threat for the environment. Moreover, it has been reported by cellular operators that the energy consumption of base stations contributes to over 70% of their electricity bill. In fact, energy-efficient communication was not one of the initial concerns in 4G wireless systems, but it came up as an issue at a later stage.

Other challenges are, for example, average spectral efficiency, high data rate and high mobility, seamless coverage, diverse QoS requirements, and fragmented user experience (incompatibility of different wireless devices/interfaces and heterogeneous networks), to mention only a few. 4G wireless networks still have constraints in facing the above challenges.

To face the above challenges and to respond to the continuously increasing demand for high data rates and high-mobility requirement raised by new wireless applications, wireless communication community and industry have started research on 5G wireless systems that are expected to be deployed beyond 2020.

We need groundbreaking wireless technologies to solve the difficulty caused by trillions of wireless devices in 4G wireless networks. Researchers have already started to investigate beyond 4G (B4G) or 5G wireless techniques. The project UK-China Science Bridges: (B)4G Wireless Mobile Communications[3] is perhaps one of the first projects in the world to start B4G research, where some potential B4G technologies were identified. Europe and China have also initiated some 5G projects, such as METIS 2020,[4] supported by EU and National 863 Key Project in 5G supported by the Ministry of Science and Technology (MOST) in China. Nokia Siemens Networks described how the underlying radio access technologies can be developed further to support up to 1000 times higher traffic volumes compared to 2010 travel levels over the next 10 years. Samsung demonstrated a wireless system using millimeter (mm) wave technologies with data rates faster than 1 Gbps over 2 km.

It is too early to define 5G with any certainty at the present. However, it is widely agreed that compared to the 4G network, the 5G network should achieve 1000 times the system capacity, 10 times the spectral efficiency, energy efficiency, and data rate (i.e., peak data rate of 10 Gbps for low mobility and peak data rate of 1 Gbps for high mobility), and 25 times the average cell throughput. The aim is to connect the entire world and achieve seamless and ubiquitous communications between

- Anybody (people to people),
- Anything (people to machine and machine to machine),
- Anywhere (wherever they are),
- Anytime (whenever they need),
- Anyhow (by whatever electronic devices/services/networks they wish).

This means that 5G networks should be able to support communications for some special scenarios not supported by 4G networks (e.g., for high-speed train users). High-speed trains can easily reach 350 up to 500 km/h, while 4G networks can only support communication scenarios up to 250 km/h. Specifically, the Next Generation Mobile Networks Alliance defines the following particular requirements for 5G networks:

- Data rates of several tens of megabits per second should be supported for tens of thousands of users;
- 1 Gbps to be offered simultaneously to many workers on the same office floor;
- Several hundreds of thousands of simultaneous connections to be supported for massive sensor deployments to support the Internet of things (IoT);

---

[3] http://www.ukchinab4g.ac.uk/.

[4] https://www.metis2020.com/.

- Spectral efficiency should be significantly enhanced compared to 4G with the emphasis on average and cell-edge spectrum efficiency (relation to throughput and capacity);
- Coverage should be improved to support high density of connections where needed;
- Very large traffic volumes and variations should be able to be handled;
- Significant improvement in power efficiency;
- Security and protection in pervasive and highly heterogeneous environments;
- High resiliency for public safety and emergency communication use cases;
- Signaling efficiency should be enhanced so that signaling does not represent a higher volume than data, and the associated overhead should be justified by the application needs;
- Seamless connectivity, inter-system mobility, and seamless inter-system authentication should be supported;
- Energy efficiency: the 5G system has to support the traffic increase of the next decade but with a reduction of the energy consumption of the whole network;
- Latency should be reduced significantly compared to LTE: 10 ms end-to-end user plane latency, and 1 ms for ultra-high reliability, ultra-low latency services.

A conceptual 5G wireless networks, in terms of service rates, delays, and frequencies, are displayed in Figure 1.5. In addition to providing faster speeds, it is predicted that 5G networks will also need to meet the needs of new use cases such as the IoT as well as broadcast-like services and lifeline communication in times of natural disaster.

In 5G wireless networks, various potential technologies include massive MIMO, energy-efficient communications, cognitive radio networks, and visible light communications. Regarding the above

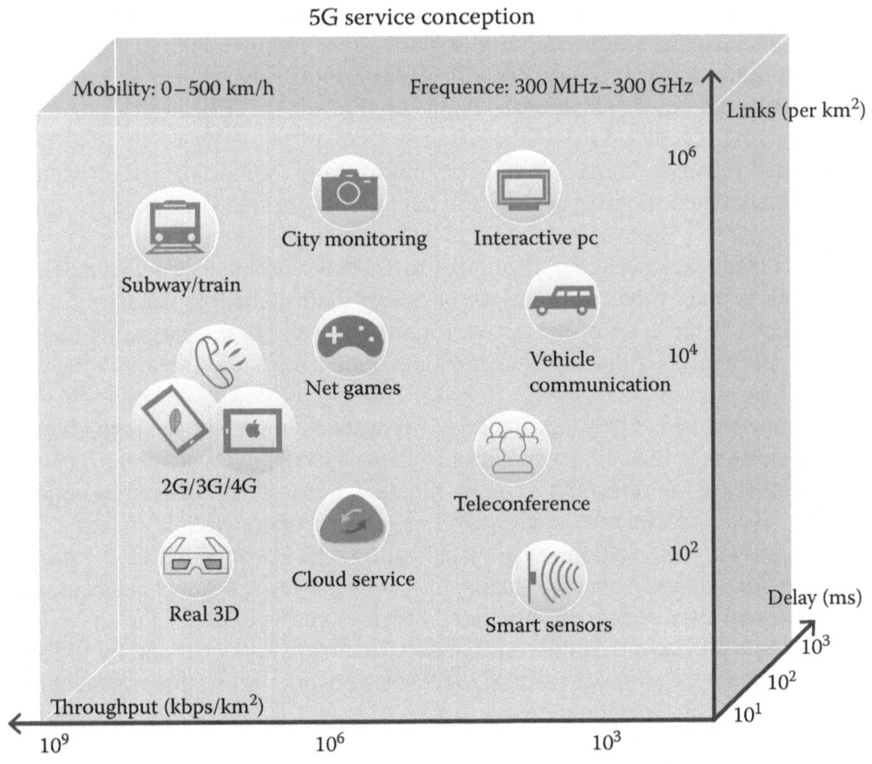

Figure 1.5 Networking architecture, service rates, delays, and frequencies in 5G wireless networks.

potential technologies, so far comprehensive research interests have been directed on the following aspects[5]:

1. Worldwide wireless web (WWWW) [86], i.e., comprehensive wireless-based web applications that include full multimedia capability beyond 4G speeds.
2. Pervasive networks providing IoT, wireless sensor networks, and ubiquitous computing [82]: the user can be connected simultaneously by using several wireless access technologies and can move seamlessly between them. These access technologies including 2.5G, 3G, 4G, or 5G mobile networks, Wi-Fi, WPAN, or any other future access technology.
3. Radio propagation and channel models for millimeter-wave wireless communication in relation to potential 5G technologies, including standards for major global 60 GHz WLAN and personal local-area networks (WPAN) (see [69–71]).
4. Massive Dense Networks, also known as Massive Distributed MIMO, which provide green flexible small cells 5G Green Dense Small Cells, with massive MIMO multiple messages for several terminals being transmitted on the same time-frequency resource, to maximize beam-forming gain while to minimize interference (see, e.g., [72–75]).
5. Advanced interference and mobility management [79–81], which should be attaining the cooperation of different transmission points with overlapped coverage, to realize a flexible use of resources for uplink and downlink transmission in each cell, and to perform the direct device-to-device (D2D) transmission by utilizing advanced interference cancellation techniques.
6. Li-Fi[6]: a portmanteau of light and Wi-Fi is a massive MIMO visible light communication network to advance 5G. Li-Fi uses light-emitting diodes to transmit data, rather than radio waves like Wi-Fi.
7. Cognitive radio technology [84], also known as smart radio. This technology allows various radio technologies to share the same spectrum efficiently by adaptively finding unused spectrum and adapting the transmission scheme to the requirements of the technologies currently sharing the spectrum. This dynamic radio resource management is achieved in a distributed manner and relies on software-defined radio (see the IEEE 802.22 standard for Wireless Regional Area Networks).
8. Vandermonde-subspace frequency division multiplexing (VFDM) [85]: a modulation scheme to allow the coexistence of macro cells and cognitive radio small cells in a two-tiered LTE/4G network.
9. Proactive content caching at the edge aims to solve the problem of reducing the end-to-end delay, which is one of the requirements of 5G. If the backhaul is capacity limited, caching users' contents at the edge of the network (namely at the base stations and user terminals) provides a solution to offload the back haul and reduce the access delays to the contents (see, e.g., [76–78]).
10. Multiple-hop networks: a major difficulty in 5G systems is to make the high bit rates available in a larger portion of the cell, particularly to those users being a position in between several base stations. By use of direct D2D communication, this issue is attacked by cellular repeaters and macro diversity techniques, also known as group cooperative relay.
11. Wireless network visualization [83]: Visualization will be extended to 5G mobile wireless networks. With wireless network visualization, network infrastructure can be decoupled from the services that it provides, differentiated services can coexist on the same infrastructure to maximize its utilization. Multiple wireless virtual networks operated by different SPs can dynamically share the physical substrate wireless networks, which are operated by MNOs. In

[5] https://en.wikipedia.org/wiki/5G.

[6] https://en.wikipedia.org/wiki/Li-Fi.

5G, mobile virtual network operators (MVNOs) may provide some specific telecom services (e.g., VoIP, video call, and over-the-top services).

12. IPv6, wearable devices with AI capabilities such as smart watches and optical head-mounted displays for augmented reality.
13. Dynamic ad hoc Wireless Networks, essentially combining Mobile ad hoc network, WMN or wireless grids, combined with smart antennas, cooperative diversity, and flexible modulation.

Driven by the applications of mobile Internet and IoT, the 5G mobile communication system and network strive to meet the forthcoming challenging requirements under a diverse set of extreme use cases in the 2020s, characterized by extremely high data rates, high mobility, low latency, massive number of connections, and high traffic density, among others. Compared with the previous generations, 5G demands tremendously much higher energy efficiency and spectrum efficiency to provide immersed and tactile service to its users.

In order to fulfill the purpose of energy efficient and flexible spectrum utilization in 5G, innovative network architecture, spectrum access regimes, and physical technologies have recently been studied in various aspects where evolutionary or even revolutionary solutions for 5G are under research and development. For example, large-scale MIMO, full duplex, and ultra-dense small cells may ameliorate spectrum efficiency through refining usage of the limited spectrum resource.

To obtain further spectrum efficiency gain, cognitive radio and license-assisted access concepts contribute a lot in bringing more spectrum opportunities in underutilized licensed/unlicensed frequency bands. With an aim to explore the much higher frequency bands, mmWave and THz communications benefit 5G with unprecedented potentials in extremely large bandwidth. In other words, refining the spectrum utilization with novel technologies, cultivating new access mechanism over the licensed or unlicensed frequency bands, and exploring the decolonized radio spectrum play critical roles in paving the way to 5G era with effective energy efficiency, effective spectrum efficiency, and effective cost attributes.

Challenges within the area of energy efficient and flexible spectrum utilization in 5G system and networks include

- Network or system level architecture design,
- Ultra-dense network-based spectrum access technologies,
- Physical layer waveform and unified air interface design,
- Nonorthogonal multiple access technologies,
- Full duplex spectrum access technologies,
- Massive MIMO technologies,
- Spatial spectrum usage technologies,
- Flexible spectrum access mechanism and technologies,
- Cognitive radio and LAA-based spectrum access regime and technologies,
- Licensed/Unlicensed spectrum aggregation technologies,
- Device to device communication technologies,
- mmWave and THz communication.

In summary, 5G supports a fully mobile and connected society, which is featured by tremendous level of connectivity and density, the huge volume of traffic, the required multilayer denazification, the broad range of use cases, and a huge diversity of business models. For 5G, there is a need to push the integration of performance to provide, where needed, for example, much greater throughput, much lower latency, ultra-high reliability, much higher connectivity density, and higher mobility range. This dramatically improved performance is expected to be provided along with the capability to control a highly heterogeneous environment, and capability to, among others, ensure security and trust, identity, and privacy. For 5G, it is anticipated [98] that the need for new radio interface(s) driven by use of higher frequencies, specific use cases such as IoT or specific capabilities (e.g., lower latency), will go beyond what 4G and its enhancements can support. 5G will operate in a highly

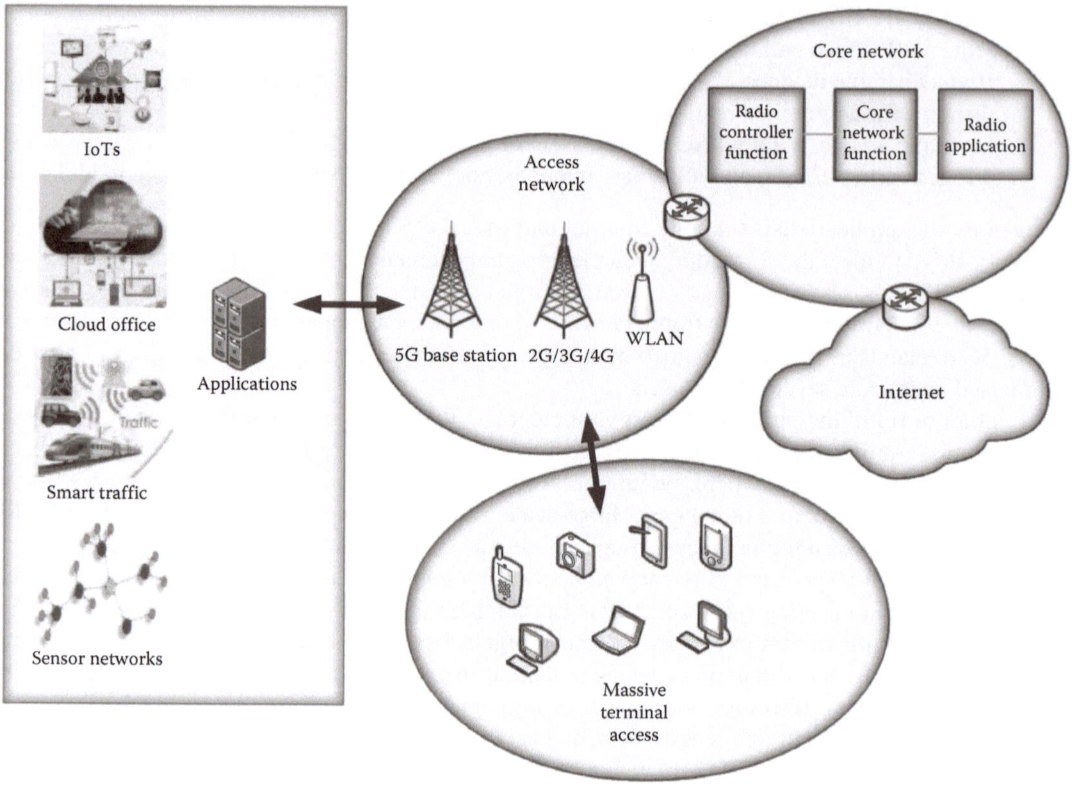

Figure 1.6 A vision of 5G wireless networks.

heterogeneous environment, where there exists multiple types of access technologies, multilayer networks, multiple types of devices, multiple types of user interactions, etc. In such an environment, there is a fundamental need for 5G to achieve seamless and consistent user experience across time and space. 5G will require the allocation of additional spectrum for mobile broadband by being supported by flexible spectrum management capabilities. In addition, an IPR eco system is needed to be developed in 5G. Figure 1.6 depicts a simple vision of 5G wireless networks.

1.7 FUTURE TRENDS ON WIRELESS COMMUNICATION NETWORKS

1.7.1 SUPPORTING CLOUD COMPUTING

Big data, IoT, and content delivery services are dramatically increasing the Internet traffic, which introduces the cloud computing approach to enable the computing process and data storage being moved away from the end users. The cloud computing systems call for not only outstanding, collecting, processing, and computing capabilities but also efficient communication approaches. From the perspective of transmission efficiency, the infrastructure of optical networks should be used within the cloud computing systems, while the ubiquitous access and end user mobility introduce the need for wireless networks, which should be seamlessly integrated with the optical DCs. Therefore, future wireless communication networks should be in converged infrastructure to support cloud and mobile cloud computing services [87]. The infrastructure will interconnect both fixed and mobile end users with DCs through a heterogeneous networking structure by integrating optical and wireless domains. Virtualization [88,89] across all technologies will be adopted, and the virtual infrastructure should be designed [90] to maximize service rates of users, to minimize end-to-end delays, and energy consumption. The future technologies that comprise the converged network infrastructure are [87] as follows:

- Wireless access and backhauling solution: A heterogeneous topology consisting of a cellular LTE system for the wireless access components and a set of wireless microwave links for the interconnection of the LTE-enabled based stations and the edge nodes of the optical metro network solution [91]. Using this approach, it is supported by a wide range of services and performance requirements, for instance, real-time and non-real-time streaming, conversational and interactive services with low or high delays in a wide range of environments, including indoor, urban, and rural.
- Optical metro networking approach: Time Shared Optical Network (TSON) [92] is proposed to be implemented by utilizing the wavelength division multiplexed (WDM) optical network technology. TSON has the capacity to support short-lived connections and facilitate fast service delivery (as low as 300 μs) by using the tool of subwavelength granularity and by applying the concept of virtual machine migration.
- Converged infrastructure: An extremely important function in the converged infrastructure is that, the map and aggregation/deaggregation of the traffic from one domain to the other should be established.

1.7.2 SUPPORTING IoT

With the dramatically increase of desktop applications and web applications, the mobile computing and pervasive/ubiquitous computing have triggered the IoT over the last decade. Thanks to the advances in sensor technology, sensors are getting more powerful in data collecting, data processing, and transmission, as well as they are cheaper and smaller in size. This has stimulated large-scale deployments. As a result, today we have a large number of sensors already deployed and it is predicted [94] that this number will grow even more rapidly over the next decade. Eventually, these sensors will generate big data [95]. One is forced to analyze, interpret, and understand the big data in an efficient and effective way. This calls for the development of context-aware computing [93], which is playing an important role in the IoT paradigm. Therefore, future wireless communication technology will serve in the IoT platform to perform machine to machine communication and to integrate with the context-aware computing technology.

When large numbers of sensors are deployed, and start generating data, significant amounts of middle ware solutions need to be introduced to focus on various aspects in the IoT such as device management, interoperability, platform portability, context awareness, security and privacy, and others. Besides, a huge amount of mobile devices are connected to the Internet and formed the Mobile Internet [96]. With the evolution of social networking, users started to become connected together over the Internet. Our next step in the IoT is where objects around us will be able to connect to each other (e.g., machine to machine) and communicate via the Internet [97]. By 2020, it is predicted [94] there should be 50–100 billion devices connected to the Internet. Therefore, the future wireless networking technology should support all these sorts of communications in one way or another.

The essential component of IoT is sensor networks. Despite the following facts, there have been several major wireless technologies used to build wireless sensor networks: wireless personal area network (WPAN) (e.g., Bluetooth), WLAN (e.g., Wi-Fi), wireless metropolitan area network (e.g., WiMAX), wireless wide area network (e.g., 2G and 3G networks), and satellite network (e.g., GPS). Sensor networks also have used two kinds of protocols for communication: non-IP based (e.g., Zigbee and Sensor-Net) and IP-based protocols (NanoStack, PhyNet, and IPv6). In addition, there have already been three categories of sensor networks comprising the IoT: body sensor networks (BSN), object sensor networks (OSN), and environment sensor networks (ESN). Such challenges

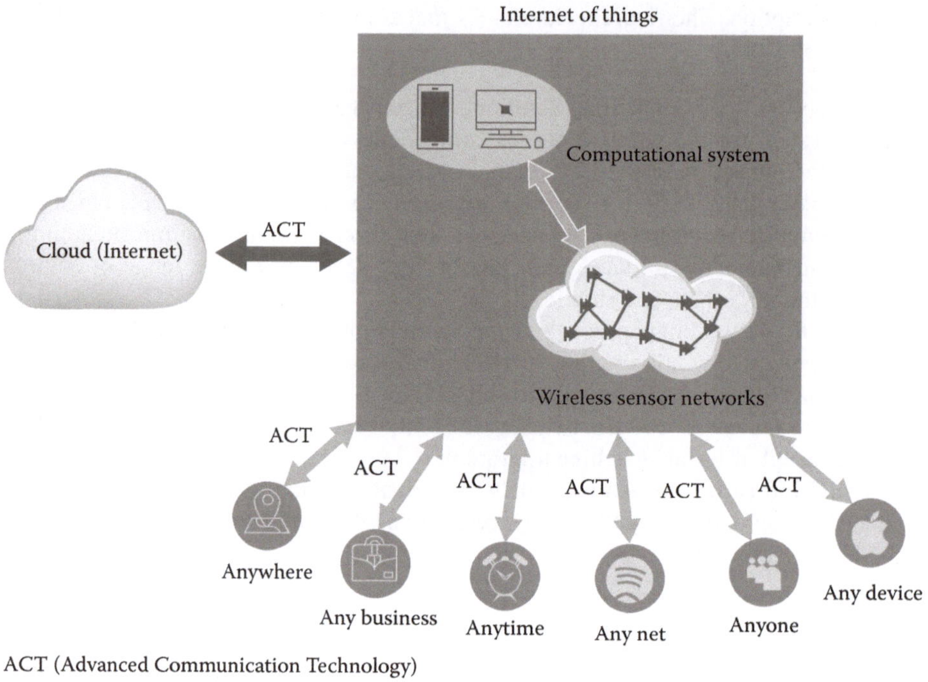

Figure 1.7 Future wireless communication networks supporting IoT and cloud computing.

are still facing us that need to be considered when developing sensor network middle ware solutions: abstraction support, data fusion, resource constraints, dynamic topology, application knowledge, programming paradigm, adaptability, scalability, security, and QoS support.

In summary, future wireless communication networks (5G and beyond) should support IoT and cloud computing in a more efficient and more effective way, which is illustrated by Figure 1.7.

2 Network Utility Maximization (NUM) Theory

Kelly et al. [104,105] presented a fundamental approach of network utility maximization (NUM) for bandwidth allocation and congestion control in computer networks including the Internet. Following Kelly's pioneering work, there have been numerous researches (see, e.g., [100,112,115,119,121,122, 124,127,128,132,190,199]) to provide utility-aware network bandwidth allocation schemes based on optimization theory [99].

Protocol design for various functionalities within a communication network architecture, either for wire-lined networks or for wireless networks, can be viewed as a distributed or centralized resource allocation problem. This involves understanding what resources are, how to allocate them fairly, and perhaps most importantly, how to achieve this goal in a distributed or a centralized and stable fashion. A NUM model leads us to find the centralized or distributed solution for congestion control, routing, and scheduling in wired and wireless networks. These controllers are the analytical equivalent of protocols in use on the Internet today and some existing protocols as realizations of such controllers.

NUM theory has comprehensively motivated the current flow control approaches, which establish the methodologies of congestion controlling and resource competing mechanism, both for the wire-lined and wireless network architectures. These methodologies can be designed to allocate bandwidth resource fairly among competing users with the total utility of a network being maximized.

In particular, the flow control approaches in protocol design consist of two components: a source algorithm that dynamically adjusts rate (or window size) in response to congestion in its path and a link algorithm that updates, implicitly or explicitly, a congestion measure and sends it back to sources that use that link. These algorithms, well known as the primal-dual algorithm [111], are all resulted from the conventional flow-level optimization model, i.e., NUM model. They have explained the congestion control and pricing mechanism in networks in a systematic way.

2.1 UTILITY AND UTILITY FUNCTION

In terms of economics, utility is a measurement of usefulness or preferences over some set of goods and services. It represents a satisfaction level experienced by consuming a good. Due to the fact that one cannot directly measure benefit, satisfaction, or happiness from a good or service, a way to represent and measure them with relation to economic choices has been chosen instead. Economists have attempted to propose highly abstract methods of comparing utilities by observing and calculating the economic choices. In the simplest manner, economists consider that utility can reflect people's willingness to pay different amounts for different goods.

Individual utility and social utility can be constructed as the value of a utility function and a social welfare function, respectively. When coupled with production or commodity constraints, under some assumptions, these functions can be used to analyze Pareto efficiency, such as illustrated by Edgeworth boxes in contract curves. Such efficiency is a central concept in welfare economics.

In finance, utility is applied to generate an individual's price for an asset called the indifference price. Utility functions are also related to risk measures, with the most common example being the entropy risk measure.[1]

---

[1] See https://en.wikipedia.org/wiki/Utility.

In the domain of networking, a utility provides a measure of user-perceived quality of service (QoS), whose function can be related to such QoS metrics as user's sending rate, packet loss, frequency, signal-to-noise ratio (SNR), time slots, or power. In the sequel, let us take a multiple radio access network, as an example, to explain the utility function and the indifference curve of utility function. In this instance, we are going to just relate the utility function with the variable bandwidth.

Multiple radio access technologies (multi-RATs) have been considered as one of the promising features in the next-generation wireless networks, and users are capable of communicating through any of these RATs (e.g., wireless local area network or WLAN, wideband code division multiple access WCDMA, and long-term evolution LTE in a smartphone), in what is called multi-RAT heterogeneous networks (multi-RAT Hetnets) [134]. Due to the scarce spectrum and time-variant wireless channel capacity, resource allocation mechanisms are expected to play a key role in providing QoS guarantees and in improving the resource utilization in multi-RAT Hetnets [133].

Figure 2.1 is an illustration of a multi-RAT wireless network in "consuming" the total resource (bandwidth) at the base station: two WLANs are competing for the total bandwidth. The component x stands for the total bandwidth that is allocated to the first group of mobile devices, while y stands for the total bandwidth that is allocated to the second group of mobile devices. Note that, herein, the term "resource" may also refer to other metrics, e.g., packet loss, frequency, SNR, time slots, power, etc.

For the multi-RAT wireless network displayed by Figure 2.1, the utility function on its bandwidth allocation among the two WLANs can be defined as a function on the two-dimensional vector (x, y), where the component x stands for the total bandwidth that is allocated to the first WLAN, while y stands for the total that is allocated to the second WLAN. There are, for example, three typical utility functions as follows:

- The Cobb–Douglas utility function $U(x,y) = x^{\alpha}y^{1-\alpha}, \ 0 \le \alpha \le 1$,
- The linear utility function $U(x,y) = \alpha x + \beta y$,
- The Leontief function $U(x,y) = \min\{\alpha x, \beta y\}$.

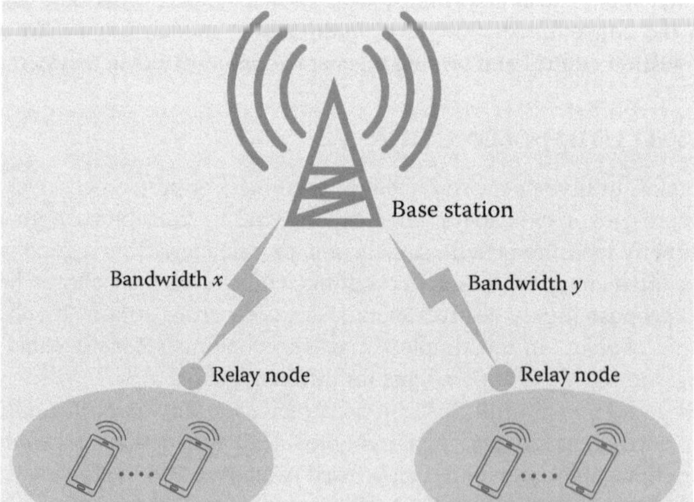

Figure 2.1 An illustration of a radio access network in "consuming" the total resource; two wireless local area networks are competing for the total resource of the base station. The component x stands for the total resource that is allocated to the first group of mobile devices, while y stands for the total resource that is allocated to the second group of mobile devices.

To take a close look into the Cobb–Douglas utility function on the increasing and concave property on its individual variable, for $\alpha = 1/2$, we have the utility function to be expressed as

$$U(x,y) = (xy)^{1/2}.$$

Next, we fix the value of bandwidth allocation to the second WLAN y, and let $y = 1, 2^2, 3^2, 4^2, 5^2$, we then plot all the curves of the following utility functions:

$$\begin{aligned} U(x,1^2) &= x^{1/2}, \\ U(x,2^2) &= 2x^{1/2}, \\ U(x,3^2) &= 3x^{1/2}, \\ U(x,4^2) &= 4x^{1/2}, \\ U(x,5^2) &= 5x^{1/2} \end{aligned}$$

into Figure 2.2. Each curve in this figure shows how utility increases as the base station gives the bandwidth x to the first WLAN while holding the bandwidth allocation y to the second WLAN fixed. One can draw a similar figure in the u, y plane, in which one fixes x at various levels.

Now to take a look at the concave property of all curves plotted in Figure 2.2, we take the partial derivative of the utility function with respect to x, which gives us

$$\frac{\partial U}{\partial x} = \frac{1}{2}\frac{1}{x^{1/2}}y^{1/2}, \forall y.$$

The slope of the curves in Figure 2.2 at a point is determined by the above partial derivative. The value of the partial derivative is decreasing in x, which suggests the slope of the utility curves becomes flatter at high levels of x, i.e., the curves are concave. This is known as diminishing

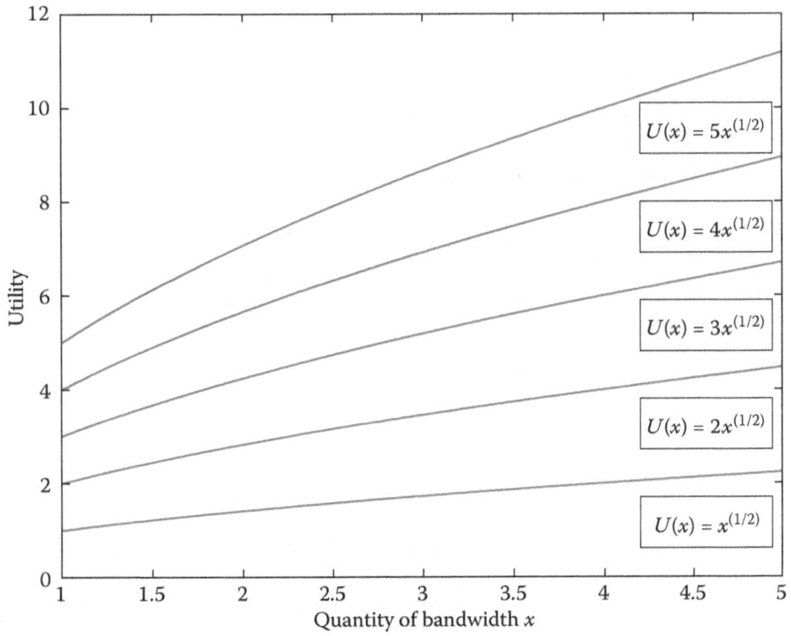

Figure 2.2 The curves of the Cobb–Douglas utility function with y being fixed at the values of 1, 4, 9, 16, and 25.

marginal utility, which tells us each unit of bandwidth x gives the base station less additional utility than it did for the last one.

By taking these curves and plotting them in the three-dimensional x, y, U space gives us a three-dimensional surface, as shown in Figures 2.3 and 2.4. Given any combination of x and y, one can determine from these figures the corresponding level of utility for the base station. Higher points on the surface represent higher levels of utility. The base station will normally seek to allocate the total bandwidth among the two competing WLANs to obtain the highest utility.

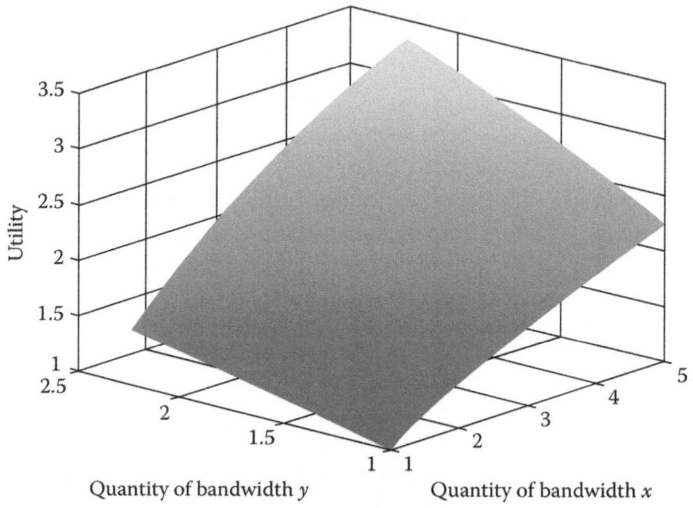

Figure 2.3 A three-dimensional plot of the Cobb–Douglas utility function in the x, y, U space.

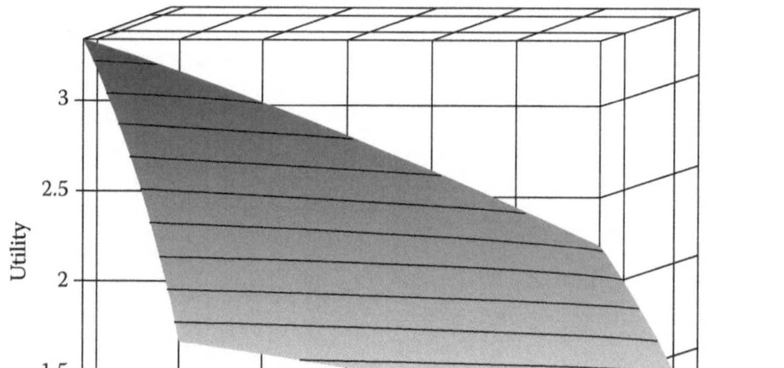

Figure 2.4 A three-dimensional plot of x, y, and U. Given any combination of bandwidth x and y, one can read the corresponding level of utility. Higher points on the surface represent higher levels of utility. The base station will normally seek to allocate the total bandwidth among the two competing WLANs to obtain the highest utility.

In microeconomic theory, an indifference curve [135] is a graph showing different bundles of goods between which a consumer is indifferent in terms of utility. That is, at each point on the curve, the consumer has no preference for one bundle over another. One can equivalently refer to each point on the indifference curve as outputting the same degree of utility (satisfaction) for the consumer. In other words, an indifference curve is the locus of various points showing different combinations of two goods providing equal utility to the consumer. The main use of indifference curves is in the representation of potentially observable demand patterns for individual consumers over commodity bundles [136].

Economists usually apply utility to construct the indifference curve, which plots the combination of commodities that an individual or a society would accept to maintain a given level of satisfaction. Therefore, utility and indifference curves are used in understanding the underpinnings of demand curves, which are crucial in the supply and demand analysis that is used to study the goods markets.

Indifference curves, generally, have the following properties[2]:

- Being defined only in the nonnegative quadrant of commodity quantities. In the present example of the multi-RAT wireless network, all bandwidth allocations (x, y) should be positive.
- Being sloped negatively. That is, as the bandwidth allocation (x) to the first WLAN increases, the total satisfaction level at the base station would increase if not offset by a decrease in the quantity of the other portion of bandwidth (y), which is taken by the second WLAN. The negative slope of the indifference curve reflects the assumption of the monotonicity of base station's preferences, which generates monotonically increasing utility functions, and the assumption of nonsatiation (marginal utility for all bandwidth allocation is always positive). Because of monotonicity of preferences and nonsatiation, a bundle with more of both bandwidth allocations must be preferred to one with less of both; thus, the first bundle bandwidth must yield a higher utility and lie on a different indifference curve at a higher utility level. The negative slope of the indifference curve implies that the marginal rate of substitution is always positive.
- Being complete. All points on an indifference curve are ranked equally preferred and ranked either more or less preferred than every other point not on the curve.
- Being transitive with respect to points on distinct indifference curves. That is, if each point on I_2 is (strictly) preferred to each point on I_1, and each point on I_3 is preferred to each point on I_2, thus each point on I_3 is preferred to each point on I_1 (it is seen from Figure 2.5). A negative slope and transitivity exclude indifference curves crossing since straight lines from the origin on both sides of where they crossed would give opposite and intransitive preference rankings.
- Being strictly convex. If f is the function that produces the indifference curve, it must satisfy $\forall x_1 \neq x_2 \in X, \forall t \in (0,1): \qquad f(tx_1 + (1-t)x_2) < tf(x_1) + (1-t)f(x_2).$

In terms of the utility theory (see, e.g., [137]), the utility function of the base station is a function that ranks all pairs of bandwidth allocations by order of preference (completeness) such that any set of three or more bandwidth combinations forms a transitive relation. This means that for each combination (x,y) there is a unique relation, $U(x,y)$, representing the utility (satisfaction) relation associated with (x,y). The relation $(x,y) \rightarrow U(x,y)$ is called the utility function. The range of the function is a set of real numbers. The actual values of the function have no importance. Only the ranking of those values has implications in the theory. More specifically, if $U(x,y) \geq U(x',y')$, then the bandwidth combination (x,y) is described as at least as good as the the combination (x',y'). If $U(x,y) > U(x',y')$, the combination (x,y) is referred to be strictly preferred to the combination (x',y').

[2]https://en.wikipedia.org/wiki/Indifference_curve.

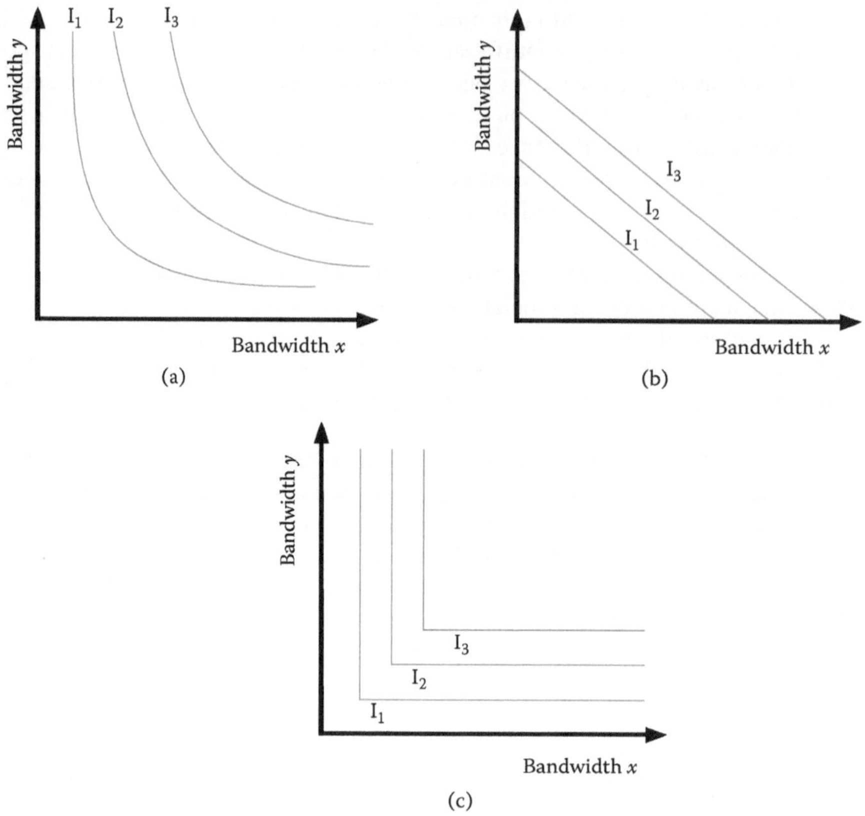

Figure 2.5 Three indifference curves of utility: (a) The Cobb–Douglas utility function $U(x,y) = x^{\alpha}y^{1-\alpha},\ 0 \le \alpha \le 1$, (b) a linear utility function $U(x,y) = \alpha x + \beta y$, and (c) the Leontief function $U(x,y) = \min\{\alpha x, \beta y\}$.

Let us consider a particular bandwidth allocation $(x_0,\ y_0)$ and take the total derivative of $U(x,y)$ at this point:

$$dU(x_0,\ y_0) = U_1(x_0,\ y_0)\,dx + U_2(x_0,\ y_0)\,dy.$$

This is then reduced to

$$\frac{dU(x_0,\ y_0)}{dx} = U_1(x_0,\ y_0).1 + U_2(x_0,\ y_0)\frac{dy}{dx}, \tag{2.1}$$

where $U_1(x,y)$ is the partial derivative of $U(x,\ y)$ with respect to its first argument, evaluated at $(x,\ y)$. Similarly, we can obtain the formulation for $U_2(x,\ y)$ as follows:

$$\frac{dU(x_0,\ y_0)}{dy} = U_1(x_0,y_0)\frac{dy}{dx} + U_2(x_0,y_0).1.$$

The indifference curve through $(x_0,\ y_0)$ has at each combination on the curve the same utility level as combination $(x_0,\ y_0)$. That is, when preferences are represented by a utility function, the indifference curves are the level curves of the utility function. Therefore, if one needs to change the quantity of x, by dx, without moving off the indifference curve, one must also change the quantity of y, by an amount dy, such that, in the end, there is no change in U:

$$\frac{dU(x_0,\ y_0)}{dx} = 0,$$

or substituting 0 into (2.1) to solve for dy/dx:

$$\frac{dU(x_0, y_0)}{dx} = 0 \Leftrightarrow \frac{dy}{dx} = -\frac{U_1(x_0,y_0)}{U_2(x_0,y_0)}.$$

Thus, the ratio of marginal utilities gives the absolute value of the slope of the indifference curve at point (x_0, y_0). This ratio is called the marginal rate of substitution between x and y.

For the case, where the utility function is in the form of the Cobb–Douglas utility:

$$U(x,y) = x^{\alpha}y^{1-\alpha},$$

the marginal utility of x is

$$U_1(x,y) = \alpha\left(\frac{x}{y}\right)^{\alpha-1},$$

and the marginal utility of y is

$$U_2(x,y) = (1-\alpha)\left(\frac{x}{y}\right)^{\alpha}.$$

In the above $\alpha < 1$. The slope of the indifference curve is then

$$\frac{dx}{dy} = -\frac{1-\alpha}{\alpha}\left(\frac{x}{y}\right).$$

This curve is plotted in Figure 2.5a.

If the utility function is of the form $U(x,y) = \alpha x + \beta y$, then the marginal utility of x is $U_1(x,y) = \alpha$, and the marginal utility of y is $U_2(x,y) = \beta$. Therefore, the slope of the indifference curve is

$$\frac{dx}{dy} = -\frac{\beta}{\alpha}.$$

Note that the slope does not depend on x or y; the indifference curves are straight lines (refer to Figure 2.5b). For the case, where the utility function is of the form of the Leontief function,

$$U(x,y) = \min\{\alpha x, \beta y\},$$

the marginal utility of x is

$$\frac{\partial U(x,y)}{\partial x}\Big|_{(x_0,y_0)} = U_1(x_0,y_0) = \begin{cases} \alpha, & \alpha x \le \beta y, \\ 0, & \alpha x > \beta y. \end{cases}$$

The marginal utility of y is

$$\frac{\partial U(x,y)}{\partial y}\Big|_{(x_0,y_0)} = U_2(x_0,y_0) = \begin{cases} 0, & \alpha x \le \beta y, \\ \beta, & \alpha x > \beta y. \end{cases}$$

Therefore, the slope of the indifference curve is given by

$$\frac{dx}{dy} = -\frac{U_2(x_0,y_0)}{U_1(x_0,y_0)} = \begin{cases} 0, & \alpha x \le \beta y, \\ +\infty, & \alpha x > \beta y, \end{cases}$$

which is plotted in Figure 2.5c.

2.2 NUM THEORY

Data networks have traditionally been designed in architecture as a set of layers with each layer being independent of the others. This layering technique has contributed greatly to the success of the Internet at present. However, even though the layers are treated individually during design, the parameter choices at one layer have impact on other parameters at other layers. NUM, as an approach of optimized layering decomposition, addresses this issue by posing the entire layered stack as a unified global utility optimization problem. The network is, therefore, designed as a distributed solution to this optimization problem, which is an approach utilizing a systematic, rather than an ad hoc, process of designing layered protocol stack for wired and wireless networks. This approach is to incorporate the relevant protocol layers into a unified theory by regarding them as carrying out an asynchronous distributed computation over the whole network to, implicitly or explicitly, solve a global optimization problem, that is, modeling the network. Different layers perform interactions on different subsets of the targeted variables using local information to achieve individual optimality. Being unified from those individual algorithms, the main algorithm attempts to achieve a global objective. Such a design process can be understood through the mathematical language of decomposition theory for constrained optimization (see the framework of layering as optimization decomposition approach [184–187]). This framework enables us to break the complex system down into simpler modules, and it allows us to systematically carry out the layering process and explicitly trade-off design objectives.

The NUM theory started from the pioneering work by Kelly et al. [104,105], which proposes an approach of optimization modeling and decomposition solutions to study the complex interactions of network congestion control. Since then, this approach has been substantially and extensively extended in both wire-lined and wireless areas and now forms a promising direction toward a mathematical theory of network architectures. The paper by Chiang et al. [184] provides a survey on the current status of the approaches "layering as optimization decomposition." They comprehensively discuss the decomposition approaches, which lead to distributed computation, and related protocols and algorithms on the aspects of congestion control, routing, scheduling, random access, power control, and channel coding.

The basic NUM model can be expressed in terms of a utility function $U_i(x_i)$ in the sense that the desired bandwidth allocation $x^* = (x_i^*$, all sources $i)$ solves the utility maximization problem:

$$\max_x \sum_i U_i(x_i) \text{ subject to link capacity constraints.} \tag{2.2}$$

Utility functions are often assumed to be smooth, increasing, concave, and associated with source rate only, although recent investigations have lifted some of these assumptions for certain service applications where they are needed.

There are three key steps for applying the above model in cross-layer design, resource allocation, and congestion control in wireless and wire-lined networks. They are

- To formulate a specific NUM problem by choosing the appropriate utility function,
- To seek the modularized and distributed solution following a particular decomposition by using the primal algorithm, or the dual algorithm or the primal-dual algorithm,
- To explore the space of alternative decompositions that provide a choice of layered protocol stacks.

There are two important concerns: how to choose the utility functions and how to ensure the resulting resource allocation schemes to guarantee QoS to users? The utility functions are either implicitly determined by the given protocols already or are to be chosen rather than designed. Utility functions can be chosen by considering the following factors:

- First, various traffic types have different utility functions. Elastic traffic is characterized by a logarithmic utility function, whereas inelastic traffic uses a step utility function (hard QoS traffic) or a sigmoid utility function (soft QoS traffic).
- Second, utility function can be determined by human psychological and behavioral models such as mean opinion score (quality of experience), in voice over IP applications.
- Third, utility functions can be used as a metric to define optimality of resource allocation efficiency.
- Fourth, different shapes of utility functions lead to optimal resource allocations that satisfy well-established definitions of various fairness.

There are two types of objective functions: the summation of all individual utility functions of individual users and a network-wide cost function. These individual utility functions can be associated with users' rate, reliability, delay, delay jitter, power level, and SNR. The network-wide cost function is specified by network operators, which can be functions of congestion degree, energy efficiency, network lifetime, collective estimation error, packet loss, etc. Utility functions can be coupled across the users and may not have an additive structure (e.g., network lifetime).

The utility function is usually formulated in terms of individual source rate; however, in the recent study by Tan et al. [164], there will be benefits if the utility function is formulated by the aggregate flow rate. For instance, the approach proposed by Tan et al. [164] has direct applications in the optical burst switching (OBS) networks, where the burst assembly is used as the procedure of aggregating packets from various IP routers into bursts at the edge of an OBS network.

Regarding the objective functions, there are formulations to maximize a weighted sum of all utility functions, to characterize the Pareto optimal trade-off (multiobjective optimization) between the user objective and the operator objective, and to formulate the game-theoretic issues between users and operators, or among users or operators themselves. There are two types of constraints: the collection of physical, technological, and economic requirements within the communication infrastructure and the set of per-user, hard, inelastic QoS restrictions that need to be followed at the equilibrium.

Having the NUM formulation at hand, we are not interested in solving it through centralized computation. Rather, we are interested in modularizing the solution method through decomposition approach to obtain the explicitly iterating procedure. Motivated by this, the original problem is therefore decomposed into a number of subproblems. In the decomposed problem, each subproblem involves only a subset of variables and applies only a subset of constant parameters. The decomposition techniques use primal decomposition and dual-decomposition methods. The former is to decompose the original primal problem, while the latter is to decompose the Lagrange dual of the problem. After decomposition, the interactions are presented in these subproblems to have variables and parameters to be solved and determined in a layering manner.

The main methods and their resulting algorithms or protocols developed in recent publications for the study and application of the NUM model by using the decomposition approaches are summarized in Table 2 in [184]. These methods, algorithms, or protocols and their references include

- Stability conditions from Lyapunov function construction [112]
- Cooperative protocol resulted from an optimization model in a reverse engineering manner by using Kakutanis fixed point theorem [115]
- Dual-decomposition descent algorithm [105]
- Stability results deducted from singular perturbation theory [160,194]
- Stability conditions by passivity argument [160,194]
- Stability conditions by using the time-delayed control theory [127]
- Stability conditions by using the classical control theory [175,182]
- Equilibrium properties via vector field representation [195–197]
- Noncooperative protocol as a game playing result [198]

- Contraction function by restricting Jacobian's norm [217]
- Cross-layer interactions by generalized NUM model [200]
- End-user generating price methods [184]
- A general NUM model formulated by the aggregate flow rate [164]
- Various timescales of protocol stack interactions through different decompositions [201]
- Maximum differential congestion pricing methods for node-based back-pressure scheduling [202]
- Incorporate routing functions into congestion control and scheduling [184]
- Primal and dual decomposition for coupling constraints [184]
- Partial and hierarchical decomposition for architectural alternatives [203,204]
- Budget-balanced distribution rule resulted from Nash equilibrium [189,317]
- Stable utility design for distributed resource allocation [188]
- Perron–Frobenius theorem to power control in wireless networks [191–193].

We now give a list of some of the recent publications that use layering approaches in optimization decomposition. Key presentations in these papers are generalized NUM formulations, dual decomposition or primal penalty function approach, and modularized and distributed solution algorithm with results in optimality, stability, and fairness. The presented results both in theory and practice range from adaptive routing, distributed theoretic source coding, video signal processing, interactions of appropriate layering coordinating prices. A list of the recent publications in layering approaches are as follows[3]:

- Joint congestion and contention control [205–210,217];
- Joint congestion control and adaptive coding or power control [211–213];
- Joint congestion control and scheduling [202,214,216,217];
- Joint routing and power control [218–220];
- Joint congestion control, routing, and scheduling [202,216,221–225];
- Joint routing, scheduling, and power control [226,227];
- Joint routing, resource allocation, and source coding [228,229];
- Transmission Control Protocol/Internet Protocol (TCP/IP) interactions [200,230,231];
- Hyper-Text Transfer Protocol (HTTP)/TCP interactions [232];
- Joint congestion control and routing [233–237];
- Network lifetime maximization [238];
- Lagrange dual decomposition leading to the trade-off between energy consumption and application performance in wireless sensor networks [239];
- Distributed utility maximizing rate allocation in cyber-physical multihop wireless networks [240];
- Cross-layer interaction between the transport protocols such as TCP and the underlying Media Access Control (MAC) level rate adaptation [241];
- Maximization of network utility function via joint design of rate control, rate scheduling, power control, and secure coding over wireless broadcast networks [242];
- Utility-based resource allocation in backbone wireless mesh networks [243].

2.3 APPLICATION OF NUM: REVERSE ENGINEERING OF TCP RENO

The most famous application of NUM is the reverse engineering of TCP Reno as an optimization problem. The NUM model is posed as the maximization problem of the summation of all users' utility functions, in which the transmission rates are the optimization variables and the constraints

[3]Note that there are a huge number of publications, that are broadly in this area. What we have listed here is some selected representatives, far from a complete list. We apologize for any missed paper that we should include here.

are the basic capacity constraints. It can be shown that a distributed iterative solution to the NUM is TCP Reno. This finding is important for two reasons: First, it shows that TCP Reno is an optimal rate allocation algorithm. Second, it gives insight into the mathematical equations that govern the congestion and bandwidth allocation of a network.

The difficult question in all NUM problems is how to distributively solve the equation in such a way that the result is meaningful. Unfortunately, there is no simple answer, since an approach that works well for one network formulation may be infeasible for another. One common approach is to optimize the dual problem, rather than the primal. In this case, the dual variables can often be deemed as the price of the constraint. For example, in the case of TCP Reno, the dual variable is multiplied by the sum of the data rates over a link minus the capacity of that link. Ideally, we would like this difference to be zero so that the data rates can reach the maximum without causing congestion. As this difference deviates from zero, the dual variable can be seen as the price for that deviation.

Next, we present the reverse engineering technique of TCP Reno [168] by using the NUM theory [166]. The predominant TCP implementations are TCP Tahoe and TCP Reno. The basic mechanism of these protocols is to linearly increase the source's sending rate for utilizing the spare capacity, but to exponentially reduce the source sending rate when there is congestion at the link. Congestion is detected when the source detects a packet loss. A connection begins transmission with a small window size of one packet or less than four packets, and the source will increase its window size by one if it receives an acknowledgment. This is the so-called slow-start phase, under which the window size is doubled every round-trip time. When the window reaches a threshold, the source enters the congestion avoidance phase, where it increases its window by the reciprocal of the current window size every time it receives an acknowledgment. This increases the window by one in each round-trip time, which is the so-called additive increase. Upon detecting a packet loss, the source first reduces the slow start threshold to half of the current window size, and then retransmits the lost packet, and finally re-enters the slow-start phase by resetting its window to one.

A network is modeled as a set of L links with finite capacities $c = (c_l, l \in Ł)$. They are shared by a set of N sources indexed by i in set I. Each source i uses a set $L_i \subseteq L$ of links. The sets L_i define an $L \times N$ routing matrix, which is described by

$$R_{li} = \begin{cases} 1, & \text{if} l \in L_i, \\ 0, & \text{otherwise.} \end{cases}$$

If each source is associated with a utility function $U_i(x_i)$ in terms of the source sending rate x_i, then the well-known utility maximization problem is described by

$$\max_{x_i \geq 0} \sum_i U_i(x_i), \quad \text{subject to} \quad Rx \leq c. \tag{2.3}$$

The above model is to maximize the aggregate utility of all sources, subject to the link capacity constraints. To solve it in a distributed manner for a large network, we now consider the following Lagrange function

$$\begin{aligned} L(x,p) &= \sum_i U_i(x_i) - \sum_l p_l(y_l - c_l) \\ &= \sum_i U_i(x_i) - q_i x_i + \sum_l p_l c_l, \end{aligned} \tag{2.4}$$

where q_i is the congestion price and p_l is the link price. We then obtain its following duality problem

$$\min_{p \geq 0} \sum_i B_i(q_i) + \sum_l p_l(c_l), \tag{2.5}$$

where

$$B_i(q_i) = \max_{x_i \geqslant 0} U_i(x_i) - x_i q_i. \tag{2.6}$$

By applying the Karush–Kuhn–Tucker condition to the above duality problem, one has the following optimum condition

$$U_i^{'}(x_i^*) = q_i^*. \tag{2.7}$$

We only consider the congestion avoidance phase of TCP Reno, in which an elephant typically spends most of its time. In our treatment, the source rates are acting as the primal variable x, whereas the link packet loss probabilities act as the prices p. We further assume that the round-trip time τ_i of source i is kept to be a constant and that rate x_i is related to window size w_i by

$$x_i(t) = \frac{w_i(t)}{\tau_i}. \tag{2.8}$$

At time t, $x_i(t)$ is the rate at which packets are sent and acknowledgments are received. A fraction $(1 - q_i(t))$ of these acknowledgments are positive, each incrementing the window size $w_i(t)$ by $1/w_i(t)$. Therefore, the window size $w_i(t)$ increases, if averaged, at the rate of $x_i(t)(1 - q_i(t))/w_i(t)$. In a similar way, negative acknowledgments are resulting in an averaged rate of $x_i(t)q_i(t)$, each halving the window size, and hence the window $w_i(t)$ decreases at a rate of $x_i(t)q_i(t)w_i(t)/2$. Since $x_i(t) = w_i(t)/\tau_i$, for Reno, we have the following average model to express its source throughput

$$\frac{dx_i(t)}{dt} = \frac{1 - q_i(t)}{\tau_i^2} - \frac{1}{2} q_i(t) x_i^2(t). \tag{2.9}$$

The equilibrium point of the above is given by

$$q_i^* = \frac{2}{2 + \tau_i^2 (x_i^*)^2}. \tag{2.10}$$

Based on the above one can obtain the utility function of TCP Reno by identifying the above with the condition 2.7. We then deduce the utility function

$$U_i(x_i) = \frac{\sqrt{2}}{\tau_i} \tan^{-1}\left(\frac{\tau_i x_i}{\sqrt{2}}\right). \tag{2.11}$$

The above results can be seen in Refs. [244,245]. From Equation 2.10, we have the following formulation to relate the equilibrium source rate with the packet loss probability:

$$x_i^* = \frac{\sqrt{2 - 2q_i^*}}{\tau_i \sqrt{q_i^*}}. \tag{2.12}$$

This verifies the throughput formula given by [246,247]. A three-dimensional plot of the above relation between the source rate and the round-trip delay, the packet loss probability comes out as Figure 2.6.

By substituting Equations 2.12 into 2.11, we obtain the following utility function at the equilibrium point (x_i^*, τ_i, q_i^*)

$$U_i(x_i^*) = \frac{\sqrt{2}}{\tau_i} \tan^{-1}\left(\sqrt{\frac{1}{q_i^*} - 1}\right). \tag{2.13}$$

A three-dimensional plot of the above relation between the source utility function and the round-trip delay, the packet loss probability, is shown in Figure 2.7.

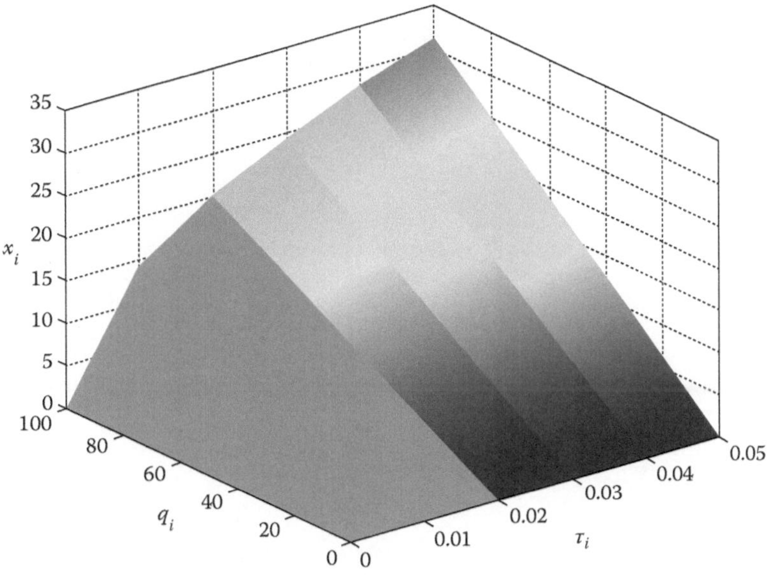

Figure 2.6 A three-dimensional plot of the relation between the source rate and the round-trip delay, the packet loss probability in TCP Reno.

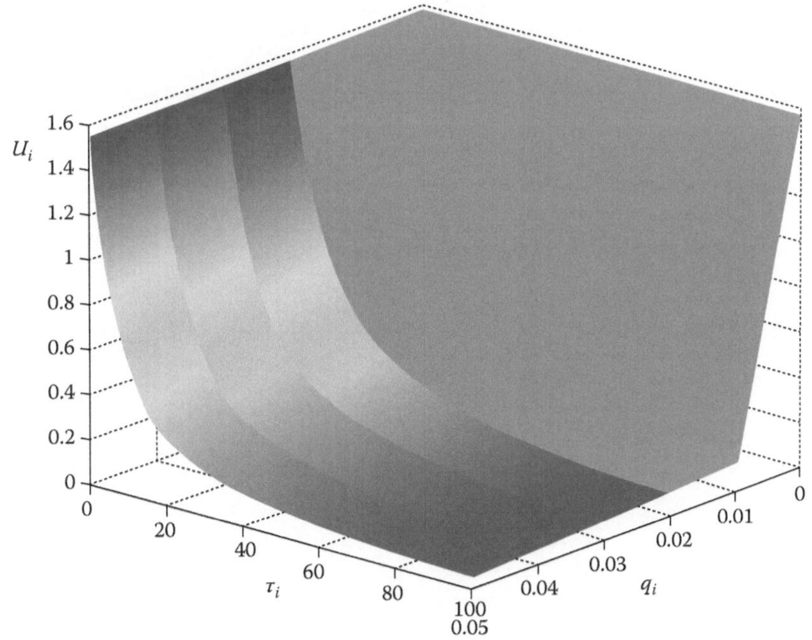

Figure 2.7 A three-dimensional plot of the relation between the source utility function and the round-trip delay, the packet loss probability in TCP Reno.

3 Congestion Control Approaches: Primal Algorithm, Dual Algorithm, and Primal-Dual Algorithm

Congestion control [215,319,322] and resource allocation mechanisms [562,574–600,613–615] in today's wireless and wire-lined networks, including the Internet, have already presented many challenges in design as they continue to expand in size, diversity, reaching scope, integration, and convergence. Having a deep understanding of how their fundamental resource is controlled and how the allocated resource is contributing to congestion control and service quality is extremely important. Networks, including the Internet, are initially designed microscopically on a heuristic, intricate basis of many control mechanisms in the packet level; they are understood afterward macroscopically in the flow level.

The notable approaches to address the congestion control and resource allocation issues from a macroscopical perspective are the primal algorithm, the dual algorithm, and the primal-dual algorithm. All these algorithms result from the network utility maximization model. The significance of these algorithms is the modeling of the congestion measuring and source actions. It is presented in the following:

- Links can feed back to data sources about the information of congestion in resources being used. It is assumed that each network link measures its congestion by a scalar variable (termed price) and that sources have information of the aggregate price of links in their path.
- With the congestion information, flows can take actions by using their control mechanisms to react to the congestion along the path to adjust their sending rate.

These assumptions are implicitly presented in many variants of the transmission control protocols (TCPs) of today. The price signal being used in these protocols can be, for example, loss probability and queueing delay. Moreover, it is possible to design alternative protocols based on other price metrics, for example, the random marking and random dropping actions of packets.

In the primal-dual congestion control algorithm, there are two key blocks: the source rates are adjusted to be adapted to aggregate prices in the TCP algorithm and the link prices are updated in terms of link utilization. Mathematically, we can model the TCP/AQM (active queue management) systems using the following general forms:

$$\dot{X}(t) = F(X(t), Q(t)) \times Q(t), \tag{3.1}$$

$$\dot{P}(t) = G(P(t), Y(t)) \times Y(t). \tag{3.2}$$

Equation 3.1 is the mechanism of source rate vector X updating in terms of the aggregate price vector Q, while Equation (3.2) is responsible for generating the link price vector P in terms of the aggregate flow rate Y. Note that $F(X(t), Q(t))$ and $G(P(t), Y(t))$ are matrix functions that are involved in generating the source rate vector and the link price vector, respectively. Deviating from the model presented by [165,166], herein, we consider a more general communication model where there are possibly multicast communications.

A multicast (one-to-many or many-to-many distribution [167]) is group communication, where information is addressed to a group of destination computers simultaneously. Group communication has two forms: application layer multicast and network-assisted multicast. Network-assisted multicast makes it possible for the source to efficiently send information to a specific group in a single transmission. Copies are automatically created in other network elements such as routers, switches, and cellular network base stations, but only to network segments that currently contain members of the group. A typical example of multicast communications is the Internet protocol (IP) multicast. IP multicast is a technique for one-to-many communication over an IP infrastructure in a network. The destination nodes send join and leave messages, for example, in the case of Internet television when the user changes from one TV channel to another. IP multicast scales to a larger receiver population by not requiring prior knowledge of who or how many receivers there are. Multicast uses network infrastructure efficiently by requiring the source to send a packet only once, even if it needs to be delivered to a large number of receivers. The nodes in the network take care of replicating the packet to reach multiple receivers only when it is necessary.

In a multicast, there are interactions (coupled components) in the source rate vector X and in the the link price vector P, as one source may issue more than one flow and the link may set more than one price value for a flow. In the usual model presented in, for example, [165,166], an inherited assumption is that any source only issues one flow. This usual model does not consider multicast flows and multiple pricing mechanism. Therefore, in our model presented in Equations 3.1 and 3.2, there may be interactions among the components of X and P. This is the reason why we have written the source rate updating and the link price updating mechanisms in the matrix equation forms rather than in the individual and independent equations.

The complete system, which is coupled and interacted by Equations 3.1 and 3.2, determines both the equilibrium and dynamic characteristics of the TCP/AQM network [165]. However, the equilibrium point is not necessarily unique; it will depend on the initial settings of the system. We will see this later in the next section. Therefore, the dynamic characteristics of the network may have a diversity. This issue is, however, very complex and deserves further study.

The mechanism of the source rate updating and link price updating, the relation between the individual source rate and link aggregate rate, which is related to the routing matrix, and the relation between the individual link price and the aggregated flow price, which is also related to the routing matrix, are demonstrated in Figure 3.1.

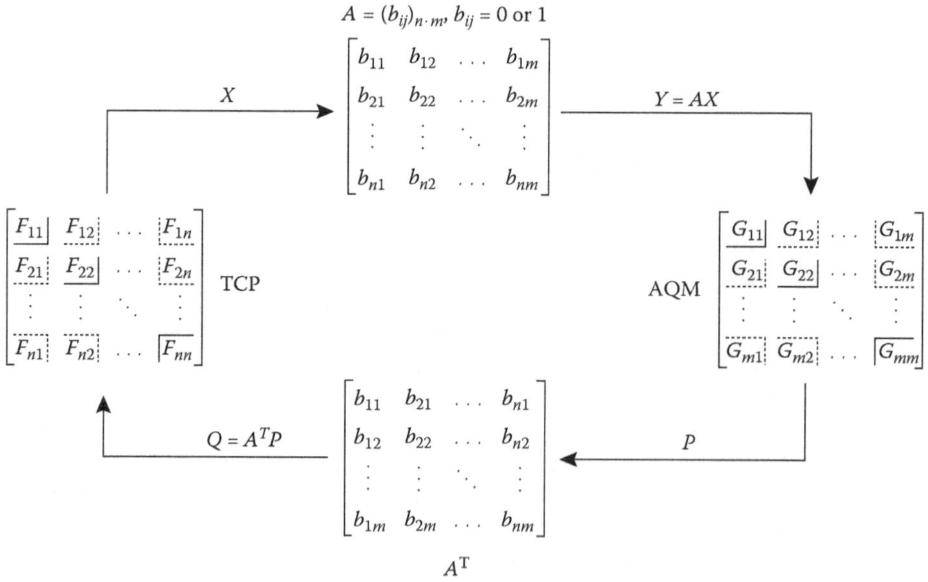

Figure 3.1 The mechanism of the source rate updating and link price updating in wireless and wire-lined networks.

3.1 BASIC MODEL FOR CONGESTION CONTROL IN THE INTERNET

Now, let us consider the following general communication network, which can be a wire-lined communication network, including the Internet, a wireless communication network, or a heterogenous network composed of both wire-lined and wireless networks. The terminologies, for example, users (flows), link, bandwidth, and others, that are used in all the aforementioned networks are the same except that, for the case of a wireless communication network, a router may refer to a relay node or a base station.

We now use the basic model introduced in [147,161]. Suppose that we have a set of users (flows), denoted by R, and a set of links (routes), denoted by L. The component L_r denotes the set of links through which flow $r \in R$ is transiting. One has $L_r \subseteq L$. Each link $l \in L$ may be used by a number of flows. For a specific link $l \in L$, we denote the set of all flows using this link for data transmission as R_l. Each flow $r \in R$ has an allocated flow bandwidth (rate) x_r and a utility function $U_r(x_r)$. Furthermore, the capacity of each link $l \in L$ is assumed to be fixed at the value c_l during the time period of our interest. We then take the summation of the individual flow rate to obtain the aggregate rate at this link:

$$y_l = \sum_{r \in R_l} x_r, \quad \forall l \in L.$$

By defining the routing matrix,

$$A \triangleq (A_{lr}, l \in L, r \in R)_{|L| \times |R|} = \begin{cases} 1, & \text{if flow } r \text{ uses link } l, \\ 0, & \text{otherwise,} \end{cases}$$

the source sending rates (column) vector $X \triangleq (x_r,\ r \in R)$ and the link rates (column) vector $Y \triangleq (y_l,\ l \in L)$. Therefore, we are able to formulate the link aggregate rate vector in terms of the individual flow rate vector through the routing matrix in the following equation:

$$Y = AX.$$

If we associate each user (flow) with a utility function $U_r(x_r)$, which is increasing, strictly concave, and continuously differentiable in x_r, over the range $x_r \geq 0$, then the so-called network utility maximization system problem, first introduced by Kelly [104], is described by

$$\max_X U(X) = \max_X \sum_{r \in R} U_r(x_r), \quad \text{subject to } y_l \leq c_l, \quad \text{over } x_r \geq 0, \tag{3.3}$$

where $C = (c_l, l \in L)$ is a column vector. The Lagrangian for the system problem is

$$L(x,p) = \sum_{r \in R} U_r(x_r) + \sum_{l \in L} p_l(c_l - y_l), \tag{3.4}$$

where p_l is the so-called Lagrange multiplier with relation to link l, $l \in L$. As the shadow price of the link, it also refers to the link congestion information.

Regarding each user (flow) r, we define an aggregate flow price q_r as

$$q_r \triangleq \sum_{l \in L} p_l,$$

which is the total price for this specific flow summated from each individual price of every link it transits.

Now, one has two price vectors: the link price (column) vector $P = (p_l, l \in L)$ and the flow price (column) vector $Q = (q_r, r \in R)$. Lagrangian equation 3.4 can be rewritten as follows:

$$L(x,p) = \sum_{r \in R} (U_r(x_r) - x_r q_r) + \sum_{l \in L} p_l c_l, \tag{3.5}$$

where we have used the equation

$$P^T Y = P^T AX = X^T A^T P = X^T Q,$$

which is true because $Q = A^T P$.

We have the following very important theorem of the well-known Karush–Kuhn–Tucker (KKT) conditions (also known as the Kuhn–Tucker conditions).

Theorem 3.1.1

[162,163] For the following optimization problem,

$$max_x F(x)$$

subject to

$$g_i(x) \leq 0, \ \ \forall i = 1,2,\ldots,m$$

and

$$h_j(x) = 0, \ \ \forall j = 1,2,\ldots,l,$$

where $x \in R^n$, f is a concave function, g_i are convex functions, and h_j are affine functions. Let x^* be a feasible point that satisfies all the constraints. Suppose that there is constant $\lambda_i \geq 0$ and μ_j such that the following equations hold

$$\frac{\partial f}{\partial x_k}(x^*) - \sum_i \lambda_i \frac{\partial g_i}{\partial x_k}(x^*) + \sum_j \mu_j \frac{\partial h_j}{\partial x_k}(x^*) = 0, \ \ \forall k, \tag{3.6}$$

$$\lambda_i g_i(x^*) = 0, \ \ \forall i. \tag{3.7}$$

Then, x^* is a global maximum. If f is strictly concave, then x^* is the unique global maximum. ■

From the Karush–Kuhn–Tucker conditions in Theorem 3.1.1, $\hat{x}$ solves Equation 3.3 if and only if there exists a $\hat{p}$ such that the pair $(\hat{x}, \hat{p})$ constitutes a saddle point for the function $L(x, p)$ expressed in Equation 3.5, where x is the maximizer and p is the minimizer, with $x \geq 0$ and $p \geq 0$.

Without loss of generality, we make an assumption that the constraint set for x is not empty and is bounded. The aggregate utility function (the summation) is concave as each individual utility function is concave. Therefore, there is a unique optimal solution $\hat{x}$ to Equation 3.3. Subsequently, there exists $\hat{p}$, and thus $\hat{q}$. Omitting the *hat*, a pair $(x, \ p)$ is in saddle-point equilibrium if and only if [104]

$$U_r'(x_r) = q_r, \ \forall r \in R, \tag{3.8}$$

$$p_l(c_l - y_l) = 0 \text{ and } p_l \geq 0, \text{ and } c_l \geq y_l, \ \forall l \in L. \tag{3.9}$$

For convenience, let us term the preceding equations as the saddle equilibrium point equations.

3.2 SOLVING THE SADDLE EQUILIBRIUM POINT EQUATIONS TO OBTAIN THE INDIVIDUAL FLOW RATE VECTOR AND THE LINK PRICE VECTOR

Next, let us discuss how to solve Equations 3.8 and 3.9 to obtain the source rate vector and link price vector pair. Observing Equation 3.9, we have the following two situations:

- For a certain link $\forall l \in L$, if $p_l = 0$, then this link is not congested. We thus can have $y_l < c_l$, which suggests that this link is not fully occupied.

- If all links are saturated (congested or bottlenecked), then we must have

$$c_l - y_l = 0, \quad \forall l \in L.$$

In the first case, it is impossible to determine the aggregate flow rate y_l. Consequently, it is impossible to determine the individual flow rate x_r. Let us consider the latter case. That is, we assume that all links are bottlenecked. By grouping all equations in Equation 3.9, we have the following linear matrix equation:

$$C = Y = A \times X. \tag{3.10}$$

Note that in real networks, the routing matrix A in Equation 3.10 is not usually a square one as there are usually much more flows (users) than links. This suggests that matrix A is not invertible. To solve matrix equation 3.10, we have to solicit the generalized inverse of matrix A.

Definition 3.2.1 [120] *For the routing matrix $A \in R^{|L|\times|R|}$, a unique matrix $A^{\dagger} \in R^{|R|\times|L|}$, which is called the Moore–Penrose inverse, exists satisfying*

$$AA^{\dagger}A = A,$$
$$A^{\dagger}AA^{\dagger} = A^{\dagger},$$
$$(AA^{\dagger})^T = AA^{\dagger},$$
$$(A^{\dagger}A)^T = A^{\dagger}A,$$

where A^T denotes the transpose of A. In the special case that A is a square non-singular matrix, the Moore–Penrose inverse of A is simply its inverse, i.e., $A^{\dagger} = A^{-1}$. In case a matrix A^{-} satisfies only the first condition, it is called a $\{1\}$-inverse. $\{1\}$*-inverses are not unique but play an important role in solving linear matrix equations.*

Lemma 3.2.1

[120] The general form of any $\{1\}$-inverse of a matrix M is

$$M^G = M^- + K_M - M^- M K_M M M^-,$$

where M^- is a particular $\{1\}$-inverse of M such as its Moore–Penrose inverse, and K_M is an arbitrary rational matrix with appropriate dimensions. ■

Based on Definition 3.2.1 and Lemma 3.2.1, we now have the following theorem regarding the solution of the linear matrix equation 3.10:

Theorem 3.2.2

The matrix equation $C = Y = A \times X$, with $A \in R^{|L|\times|R|}$, is consistent (can be solved) for X if and only if

$$(I - AA^-)C = 0, \tag{3.11}$$

for any $\{1\}$-inverse A^-. If the equation is consistent, its general solution can be written in the following form:

$$X = A^- C + (I_{|R|} - A^- A)H, \tag{3.12}$$

where $I_{|R|}$ is the $|R| \times |R|$ identity matrix, and H is an arbitrary $|R| \times 1$ vector. ■

In real networks, there are usually much more flows than links, rendering the routing matrix rectangular. However, it can frequently be of full row rank. The routing matrix A ($\in R^{|L|\times|R|}$) is seen to be a full row rank matrix if there are $|L|$ paths of source–destination (OD) pairs (flows) having passed only one link and these links are different. This appears in practice frequently, for example, most peer-to-peer applications generate one-link flows between a data source and a receiver (downloader). Overlay communication sessions, such as overlay multicast, correspond to a single-hop flow in topology (single tree, multiple trees, or mesh) [131].

In the earlier case, by interchanging [164], the relevant columns of A, we can represent A as

$$A = [I_{|L|\times|L|}, A'],$$

where $I_{|L|\times|L|}$ is an $|L|\times|L|$ identical matrix, and A' is an $|L|\times(|L|-|R|)$ matrix. For this full row rank matrix A, it can be seen that $A\times A^T$ is invertible. Subsequently, the Moore–Penrose inverse A^+ turns out to be its "right pseudo-inverse," which is denoted as A_R^+ [120]. We further have the following equation:

$$A^+ = A_R^+ = A^T(AA^T)^{-1}. \tag{3.13}$$

For the case where the routing matrix is of full row rank, we have the following important result.

Theorem 3.2.3

If the routing matrix A is of full row rank, the unique solution to the matrix equation $C = Y = A\times X$, with $A \in R^{|L|\times|R|}$, is given by

$$X = A^T(AA^T)^{-1}C. \tag{3.14}$$

■

The computation of A^+ is relatively easy if the routing matrix is a full row rank matrix. However, a more general method to compute A^+ is the singular value decomposition (SVD) [110]. This method can be used even if A is not a full row rank matrix, which is implemented, for example, in the "pinv" function in MATLAB®.

The SVD of the real matrix A can be performed as follows:

$$A = U\begin{pmatrix} \Sigma_{m\times m} & O \\ O & O \end{pmatrix}V^T, \tag{3.15}$$

where

$$U^TU = V^TV = VV^T = I_{|R|\times|R|},$$

$$UU^T = I_{|L|\times|L|},$$

$\Sigma = \mathrm{diag}(\sigma_1\ \sigma_2\ \cdots\ \sigma_m)$, the matrix U consists of $|L|$ ortho-normalized eigenvectors associated with the largest $|L|$ eigenvalues of AA^T, the matrix V consists of the $|R|$ ortho-normalized eigenvectors A^TA, the diagonal elements of Σ (called singular values) are non-negative square roots of the eigenvalues of A^TA, and $m = \mathrm{rank}(A)$. For the full rank matrix A, m is equal to its full rank value.

Then, we find that the following matrix

$$A^+ = V\begin{pmatrix} \Sigma_{m\times m} & O \\ O & O \end{pmatrix}^+ U^T \tag{3.16}$$

satisfies all four Moore–Penrose inverse's properties.

Further, we can simplify Equation 3.16 by the following equation:

$$\begin{pmatrix} \Sigma_{m\times m} & O \\ O & O \end{pmatrix}^{+} = \begin{pmatrix} \Sigma_{m\times m}^{-1} & O \\ O & O \end{pmatrix}. \tag{3.17}$$

However, for large-scale routing matrix A, SVD cannot be used to achieve the computation because it is highly time consuming. However, the computation can be achieved by a recurrent neural network if A is a full row rank matrix, and the computation time is satisfactory because of the parallel and distributed computation model. What is more, this method is even capable of being implemented in real-time computation.

Having solved Equation(s) 3.9 and obtained the individual flow rate vector X, we now go back to solve Equation(s) 3.8. To this end, let us group all the equations in 3.8 together and formulate them in the following way:

$$\begin{aligned} \frac{DU}{DX} &\triangleq \begin{pmatrix} U_1'(x_1) \\ U_2'(x_2) \\ \vdots \\ U_{|R|}'(x_{|R|}) \end{pmatrix} \\ &= Q = A^T \times P. \end{aligned} \tag{3.18}$$

Given that we have already obtained the individual flow rate vector X by the aforementioned solving approach, we are then able to substitute this vector into Equation 3.18. By a similar method, we have the following theorems.

Theorem 3.2.4

Matrix equation 3.18, with $A \in R^{|L|\times|R|}$, is consistent (can be solved) for P if and only if

$$(I - A^T(A^T)^-)\frac{DU}{DX} = 0, \tag{3.19}$$

For any $\{1\}$-inverse $(A^T)^-$. If the equation is consistent, its general solution can be written in the following form:

$$P = (A^T)^-\frac{DU}{DX} + (I_{|L|} - (A^T)^-(A^T))H_1, \tag{3.20}$$

where $I_{|L|}$ is the $|L| \times |L|$ identity matrix, and H_1 is an arbitrary $|L| \times 1$ vector. ■

When the routing matrix A is of full column rank, that is, A^T is of full row rank, we have the following important result to calculate the link price vector P.

Theorem 3.2.5

If the routing matrix A is of full column rank, the unique solution to matrix equation 3.18 is given by

$$P = A(A^TA)^{-1}\frac{DU}{DX}. \tag{3.21}$$

■

In summary, we have the following observation.

Theorem 3.2.6

The two sets of Equations 3.8 and 3.9, which result from the optimal solution to the network utility maximization model, have their unique solutions on the individual flow rate vector and the link price vector if and only if the routing matrix $A \in R^{|L|\times|R|}$ has full row rank $|L|$ and full column rank $|R|$. ■

3.3 EXAMPLE OF SOLVING THE SADDLE EQUILIBRIUM POINT EQUATIONS

In this section, we present an example to show how to solve saddle equilibrium point Equations 3.8 and 3.9.

We use a clustered wireless sensor network (WSN) as an example. A WSN [9], occasionally termed as a wireless sensor and actuator network (WSAN), is a network that is formed by a group of spatially distributed and linked autonomous sensors. WSNs are usually used to monitor physical or environmental conditions, for example, temperature, sound, and pressure through cooperatively passing the collected data through the whole network to a main data center for processing and knowledge.

In a sensor network, there are usually a large number of sensor nodes that are densely and randomly deployed in a certain area. The sensor network protocols and algorithms implemented in the network must have a high level of self-organizing capability. For instance, a WSN must have the capability of forming a new routing topology very quickly for smooth data transmissions in order to cope with the frequently broken links caused by the temporal death of some nodes. There are cooperative activities among the individual sensor nodes. Sensor nodes are usually fitted with an on-board processor. Therefore, they can apply their processing abilities to perform simple computations locally and only transmit the necessary data and partially processed data to the nodes responsible for the fusion or to the data center (base station) [180].

Clustering plays an important role for energy saving in WSNs. Using hierarchical routing or cluster-based routing [181], the clustering task is conducted to involve the sensor nodes in multi-hop communication within a particular cluster. Cluster formation is usually carried out considering the energy reserve of sensors and the cluster head (CHs). Clustering in WSNs improves energy consumption, lifetime of the network, and scalability. Because only the cluster head node in every cluster is required to perform transmission task, the other sensor nodes just forward their data to their corresponding cluster head. Clustering now has very important applications in high-density sensor networks because it is much easier to manage a set of cluster heads from each cluster than to manage the whole set of sensor nodes. In a clustered WSN, sensor nodes mainly undergo broadcast communications [180] in a cluster. The data gathered at the cluster heads in all the clusters are then transmitted to the sink through a hop or multiple hops.

Based on the aforementioned features of WSN, in order to illustrate how to solve saddle equilibrium point Equations 3.8 and 3.9, we now design an example of a clustered WSN, which is shown in Figure 3.2. We assume that this WSN is clustered with three cluster head nodes and one base station. We also assume that all the local data in each cluster have been gathered by the cluster head. Thus, we only consider flows between cluster heads and base station; we do not consider the broadcasting flows among the sensor nodes and the cluster head in the cluster. In this WSN, there are three links: Link1, Link2, and Link3, with capacity c_1, c_2, and c_3, correspondingly. There are four flows: three one-hop flows: f_1, f_2, and f_3 transiting Link1, Link2, and Link3, respectively, and one three-hop flow: f_4 transiting all the three links. By denoting the flow rate $X \triangleq (x_1, x_2, x_3, x_4)^T$, where x_i is the

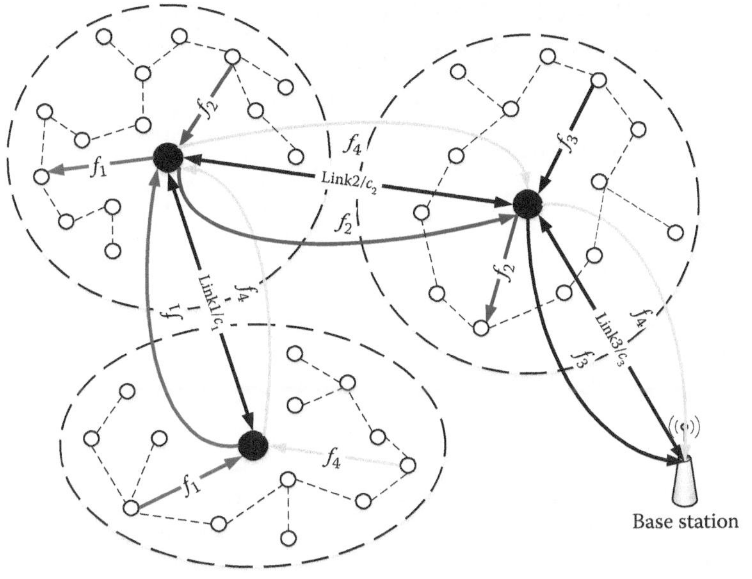

Figure 3.2 An example of a wireless sensor network: three links and four flows.

source rate for flow f_i, and the capacity vector $C = (c_1, c_2, c_3)^T$, we have

$$C = AX, \tag{3.22}$$

where the routing matrix A is given by

$$A = \begin{bmatrix} 1 & 0 & 0 & 1 \\ 0 & 1 & 0 & 1 \\ 0 & 0 & 1 & 1 \end{bmatrix}.$$

It can be seen that the routing matrix A is of full row rank and of full column rank, so the solutions to saddle equilibrium point Equations 3.8 and 3.9 uniquely exist. That is, we have

$$\begin{bmatrix} c_1 \\ c_2 \\ c_3 \end{bmatrix} = \begin{bmatrix} 1 & 0 & 0 & 1 \\ 0 & 1 & 0 & 1 \\ 0 & 0 & 1 & 1 \end{bmatrix} \begin{bmatrix} x_1 \\ x_2 \\ x_3 \\ x_4 \end{bmatrix}.$$

Now, we assume that all the flows are elastic traffic, whose utility functions are described by

$$U_i(x_i) = \log(1 + x_i), \quad i = 1, 2, 3, 4.$$

Formulating the saddle equilibrium point from Equations 3.8 and 3.9, we have the following seven equations:

$$U_i'(x_i) = \frac{1}{1 + x_i} = q_i, \quad i = 1, 2, 3, 4, \tag{3.23}$$

$$p_1(c_1 - (x_1 + x_4)) = 0, \quad x_1 + x_4 \leq c_1, \tag{3.24}$$

$$p_2(c_2 - (x_2 + x_4)) = 0, \quad x_2 + x_4 \leq c_2, \tag{3.25}$$

$$p_3(c_3 - (x_3 + x_4)) = 0, \quad x_3 + x_4 \leq c_3. \tag{3.26}$$

Assume that the prices of Link1, Link2, and Link3 are denoted by p_1, p_2, and p_3, respectively. The flow aggregate price vector is denoted by $Q = [q_1, q_2, q_3, q_4]^T$. Using the equation

$$Q = A^T P,$$

and Equation(s) 3.23, we further have

$$p_1 = q_1 = \frac{1}{1+x_1}, \tag{3.27}$$

$$p_2 = q_2 = \frac{1}{1+x_2}, \tag{3.28}$$

$$p_3 = q_3 = \frac{1}{1+x_3}, \tag{3.29}$$

$$p_1 + p_2 + p_3 = q_4 = \frac{1}{1+x_4}. \tag{3.30}$$

From Equations 3.24 through 3.26, we have $x_1 + x_4 = c_1$, $x_2 + x_4 = c_2$, and $x_3 + x_4 = c_3$. By substituting the preceding equations into Equations 3.27 through 3.29 and putting all the obtained equations into 3.30, we obtain

$$\frac{1}{1+c_1-x_1} = \frac{1}{1+x_1} + \frac{1}{1+x_1+c_2-c_1} + \frac{1}{1+x_1+c_3-c_1}. \tag{3.31}$$

By letting $w = 1 + x_1 - c_1$, Equation 3.31 is equivalent to

$$\frac{1}{w+c_1} + \frac{1}{w+c_2} + \frac{1}{w+c_3} + \frac{1}{w-2} = 0, \tag{3.32}$$

which is further reduced to

$$\begin{aligned} 4w^3 + 3(c_1+c_2+c_3-2)w^2 + 2[c_1c_2 + c_1c_3 + c_2c_3 - 2(c_1+c_2+c_3)]w + \\ c_1c_2c_3 - 2(c_1c_2 + c_1c_3 + c_2c_3) = 0. \end{aligned} \tag{3.33}$$

We solve Equation 3.33 to yield

$$\begin{aligned} w = -\frac{J}{12} + \sqrt[3]{\frac{432D-36JB+2J^3}{3456} + \sqrt{\left(\frac{432D-36JB+2J^3}{3456}\right)^2 + \left(\frac{12B-J^2}{144}\right)^3}} \\ + \sqrt[3]{\frac{432D-36JB+2J^3}{3456} - \sqrt{\left(\frac{432D-36JB+2J^3}{3456}\right)^2 + \left(\frac{12B-J^2}{144}\right)^3}}, \end{aligned} \tag{3.34}$$

where J, B, and D are given by

$$J = 3(c_1+c_2+c_3-2), \tag{3.35}$$

$$B = 2[c_1c_2 + c_1c_3 + c_2c_3 - 2(c_1+c_2+c_3)], \tag{3.36}$$

$$D = c_1c_2c_3 - 2(c_1c_2 + c_1c_3 + c_2c_3). \tag{3.37}$$

Finally, we obtain the flow rate

$$x_1 = w + c_1 - 1,$$

$$x_2 = w + c_2 - 1,$$

$$x_3 = w + c_3 - 1,$$

$$x_4 = 1 - w,$$

where w is given by Equation 3.34.

3.4 PRIMAL ALGORITHM

It is very important, both in theory and practice, to seek a distributed computing method for obtaining the source rate and price pair X and P, which uses only the information that is decentralized and available to individual users. The congestion control problem, which we had posed in the previous section, opens the possibility of finding such algorithms that are decentralized. From these algorithms, one is able to find one with satisfactory performance, for example, fast convergence and good fairness level.

For TCP/IP networks, the congestion control works in the following way: the users adapt x_r with respect to q_r, and the links adapt p_l with respect to y_l. TCP algorithms are among the user adaptation laws, and AQM algorithms are among the link adaptation laws [161]. TCP algorithms include TCP Reno [168], TCP Vegas [169,174], High-speed TCP [170], Scalable-TCP [171] and FAST TCP [172,334,352], and Generalized FAST TCP [175,176,642–644], while AQM algorithms include DropTail [177], RED [178], BLUE [179], REM/PI [100], and adaptive virtual queue (AVQ) [173, 352]. The working mechanism of TCP algorithms and AQM algorithms is displayed in Figure 3.3.

Taking the logarithmic utility function $U_r(x_r) = w_r \log x_r$, Kelly introduced the following primal algorithm [105]:

- User: each user r is updated by the following algorithm:

$$\dot{x}_r = k_r(w_r - x_r q_r), \tag{3.38}$$

  where k_r is a step size.
- Link: each link l computes its price as

$$p_l(t) = f_l(y_l(t)), \text{ where } f_l \geq 0,\ f_l' > 0. \tag{3.39}$$

The aforementioned primal algorithm globally converges [105] to a point that uniquely maximizes the following function:

$$\tilde{U}(x) = \sum_{r \in R} w_r \log x_r - \sum_{l \in L} \int_0^{y_l} f_l(y)\mathrm{d}y. \tag{3.40}$$

It is noted that delay is not considered in the aforementioned algorithms. If we take the propagation delay into account and neglect queueing delay and processing delay, thus delay is kept as a constant,

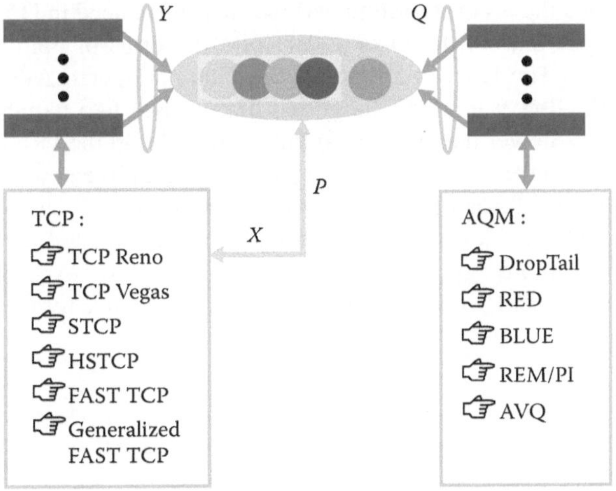

Figure 3.3 The interactions between TCP algorithms and AQM algorithms.

then we have the following equations:

$$y_l(t) = \sum_{r,l\in\tau} x_r(t-\tau_{lr}^f), \tag{3.41}$$

$$q_r(t) = \sum_{l\in\tau} p_l(t-\tau_{lr}^b), \tag{3.42}$$

where τ_{lr}^f and τ_{lr}^b are the forward delay from source r to link l and the backward delay from link l back to source r, respectively. The round trip delays (RTDs) are calculated as

$$\mathrm{RTD}_r \triangleq \tau_{lr}^f + \tau_{lr}^b, \quad \text{for any } l \in r.$$

Denote the Laplace transforms of $z(t)$, $y(t)$, $p(t)$, and $q(t)$, respectively, by $X(s)$, $Y(s)$, $P(s)$, and $Q(s)$, and define the routing matrix with delay in the frequency domain as $A(s) = (A_{lr}(s), l \in L, r \in R)$, where

$$\begin{aligned} A_{lr}(s) &= \exp(-s\tau_{lr}^f) && \text{if } l \in r, \\ &= 0 && \text{otherwise.} \end{aligned}$$

Then, we have the relationships [149]

$$Y(s) = A(s)X(s), \tag{3.43}$$

$$Q(s) = \mathrm{diag}(e^{-sRTD_r})A^T(-s)P(s), \tag{3.44}$$

and the primal algorithm with delay is computed by Ref. [150]:
Flow updates

$$\dot{x}_r(t) = k_r(w_r - x_r(t-RTD_r)q_r(t)). \tag{3.45}$$

Link updates

$$p_l(t) = f_l(y_l(t)). \tag{3.46}$$

With delays being considered, it is not easy to find the global stability condition for this general framework. However, conditions for local stability were first discussed in [150], and then further studied in [149,151,152]. For an explicit TCP/AQM system, with delays being considered, the paper by Tan et al. [127] provids results on stability conditions by using the time-delayed control theory.

Note that the link price computation in the aforementioned primal algorithm is performed statically. An algorithm called the AVQ algorithm was recently introduced in [153], where users choose the same dynamic adaptation, but the links update $p_l(t)$ in terms of both $y_l(t)$ and the so-called virtual queue capacity $\tilde{c}_l$. The latter is a time-varying link parameter. AVQ is a rate-based AQM scheme that controls the flow rate at the congested link. It uses two variables to update, namely, the virtual (link) capacity (lower than the actual link capacity) and the virtual queue (whose buffer capacity equals the actual buffer size). Upon each packet arrival, if the new packet overflows the virtual buffer, then the packet is discarded from the virtual buffer and the real packet is either dropped or marked. When a packet arrives, the virtual link capacity is updated according to its updating equation. The marking is more aggressive when the link utilization exceeds the target utilization and less aggressive otherwise. Particularly, in AVQ, the virtual capacity $\tilde{C}$ of each link is updated according to the following equation:

$$\frac{d}{dt}\tilde{C}(t) = \alpha(\gamma C - \lambda(t)),$$

where

$$\lambda(t) = \sum_{j=1}^{N} x_j(t)$$

is the total flow into the link, C is the link capacity, and $\alpha > 0$ is the control parameter, also termed as damping factor. The preceding equation acts as the main function of the AVQ, so that by modifying the virtual capacity of a link, the total flow entering the link attains the desired utilization of the link. The parameter α plays an important role in AVQ in the sense that it determines how fast one adapts the marking probability at the link to the changing networking conditions; and therefore, the performance is dependent on the choice of this parameter. This mechanism performs well. However, the rule for setting the AVQ control parameter is a challenge, particularly as link capacity scales upward. The paper [182] obtained some stability conditions on AVQ that decoupled the control parameter from the parameters related to network conditions, such as, number of sources, link capacity, and propagation delay, so that one could tune the control parameter to maintain stability with good scalability of capacity.

The global stability of the primal algorithm is presented in [154] without delay, and local stability has been given in [155,156] with feedback delays.

3.5 DUAL ALGORITHM

Dual algorithm was introduced in Ref. [112]. It applies the following algorithm at the link and user ends:

Links:

$$\frac{dp_l(t)}{dt} = \gamma_l(c_l - y_l(t)), \quad l = 1,2,\ldots,|L|, \tag{3.47}$$

where $\gamma_l > 0$ is a step size. γ_l is chosen in Ref. [112] to be l/c_l.

Users:

$$x_r(t) = x_r(q_r(t)) = \left(\frac{dU_r(t)}{dt}\right)^{-1}(q_r(t)), \quad r = 1,2,\ldots,|R|. \tag{3.48}$$

This dual algorithm is seen to be a special case of the dual algorithm in [105,148]. It was also given in Ref. [157] for the case where the users take logarithmic utility functions.

Considering Equation 3.47, for a link l, if $c_l - y_l(t) > 0$, which means that the link is not fully occupied, then $dp_l(t)/dt > 0$. In this case, the link price keeps increasing so that the congestion situation gets better. When the link is saturated, that is, $c_l = y_l(t)$, then $dp_l(t)/dt = 0$, which means that the link price is kept at a constant.

One can actually solve differential Equation 3.47 to obtain

$$p_l(t) = \gamma_l \int_0^t (c_l - y_l(s))ds + \varepsilon_l(0), \quad l = 1,2,\ldots,|L|, \tag{3.49}$$

where $\varepsilon_l(0)$ is a given initial value. Now, we have the following link price vector:

$$P(t) = \begin{pmatrix} p_1(t) \\ \vdots \\ p_{|L|}(t) \end{pmatrix} = \begin{pmatrix} \gamma_1 \int_0^t (c_1 - y_1(s))ds + \varepsilon_1(0) \\ \vdots \\ \gamma_{|L|} \int_0^t (c_{|L|} - y_{|L|}(s))ds + \varepsilon_{|L|}(0) \end{pmatrix}. \tag{3.50}$$

Now, we use the relation

$$Q(t) = A^T P(t) \tag{3.51}$$

to obtain the aggregate flow price vector $Q(t)$,

$$q_r(t) = \sum_{l=1}^{|L|} a_{lr}\left\{\gamma_l \int_0^t (c_l - y_l(s))ds + \varepsilon_l(0)\right\}, \ r = 1,2,\ldots,|R|, \tag{3.52}$$

where a_{lr} is the (l,r) position element of the routing matrix A, being either 0 or 1. By substituting Equation 3.52 in 3.48, we finally obtain the source rate vector $X(t)$ as follows:

$$x_r(t) = \left(\frac{dU_r(t)}{dt}\right)^{-1} \left(\sum_{l=1}^{|L|} a_{lr} \left\{\gamma_l \int_0^t (c_l - y_l(s))ds + \varepsilon_l(0)\right\}\right), \; r = 1,2,\ldots,|R|. \quad (3.53)$$

So far, we have actually yielded a novel source updating rule (algorithm), which is given by Equation 3.53. We summarize this novel dual algorithm as follows:

- Links: Calculate the aggregate flow rate vector Y, set the step size $0 < \gamma_l < c_l$, and communicate them to the sources.
- Sources: Update the flow rates according to Equation 3.53.

In this algorithm, note that the aggregate flow rate vector is easily available in the current TCP/IP network by using the existing simple network management protocol (SNMP). SNMP contains the following important operations:

- GET, to retrieve data from a network node;
- GETNEXT, to retrieve the next element from a network node;
- SET, to send configuration or control commands to a network node;
- TRAP, through this operation, a network node can send a notification to the management station;
- INFORM, through this operation, network nodes can try and send it again if no acknowledgment is received.

In particular, Net-SNMP is a suite of software for using and deploying the SNMP protocol (v1, v2c, and v3 and the AgentX subagent protocol). It supports IPv4, IPv6, IPX, AAL5, Unix domain sockets, and other transports. Net-SNMP contains a generic client library, a suite of command line applications, a highly extensible SNMP agent, Perl modules, and Python modules.

Various network protocols have been developed to support communication between computers and other types of electronic devices. The so-called routing protocols are the family of network protocols that enable computer routers to communicate with each other and in turn to intelligently transit traffic between their respective networks.

In the aforementioned novel dual algorithm, the routing matrix and the aggregate flow rate vector can be communicated to the sources using the existing network routing protocol. Every network routing protocol performs three basic functions:

- To identify other routers on the network,
- To keep track of all possible destinations (for network messages) along with some data describing the pathway of each,
- To make dynamic decisions regarding where to send each network message.

Particularly, the link state protocols enable a router to organize and follow a full map of all links in a region.

Routing information protocol (RIP)-enabled routers will discover the network by first sending a message requesting router tables from neighboring devices; then, neighbor routers send the full routing tables back to the requestor in response, and thereby, the requestor follows an algorithm to organize, merge, and transform all of these updates into its own table. If routers are implemented by RIP, they will then periodically send out their router tables to their neighbors so that any changes can be communicated across the whole network. Both IP4 and IP6 networks (through the newer routing information protocol next-generation (RIPng) standard) are supported by RIP. Based on the RIP protocol, we have the enhanced open shortest path first (OSPF) protocol, which overcomes some limitations of RIP, for example, by the features of a routing hierarchy, and lifting the network traffic load to avoid repeatedly resending full router tables at scheduled intervals.

OSPF is an open public standard with widespread adoption across many industry vendors. OSPF-enabled routers discover the network by sending identification messages to each other followed by messages that capture specific routing items rather than the entire routing table.

Cisco developed the Internet gateway routing protocol (IGRP) as another alternative to RIP. The newer enhanced IGRP (EIGRP) made IGRP obsolete in the 1990s. EIGRP supports classless IP subnets and improves the efficiency of the routing algorithms compared to older IGRP.

The intermediate system to intermediate system (IS-IS) protocol performs functions similar to OSPF. Even though OSPF became the more popular choice overall, IS-IS remains in widespread use among service providers. IS-IS does not run over IP but uses its own addressing scheme instead.

The border gateway protocol (BGP) is the Internet standard external gateway protocol (EGP). BGP detects modifications to routing tables and selectively communicates those changes to other routers over TCP/IP.

Internet providers commonly use BGP to join their networks together. In addition, one uses BGP to join together multiple internal networks. Among all routing protocols, BGP is perhaps the most challenging one to master due to its configuration complexity.

Compared to the original dual algorithm given by Equations 3.47 and 3.48, our novel dual algorithm does not bother the source (flow) aggregate price $q_r(t)$. The computing of the components of $q_r(t)$ is actually not an easy task. To understand the difficulty in obtaining the flow price vector, one needs the full information of the routing matrix and all the link prices of those links through which the flow is transiting. All the information is needed to be generated from the relevant links before the source can update its sending rate. This will bring a huge computing overhead to the network. Given this fact, our novel dual algorithm has the advantage over the dual algorithm given by Equations 3.47 and 3.48.

Regarding the stability issue, for the dual algorithm given by Equations 3.47 and 3.48, global stability was discussed in Ref. [112], while local stability was proved for a particular choice of utility functions in Ref. [158]. The constraint on the choice of utility functions means that certain fairness among the users cannot be achieved using the algorithm in Ref. [158]. Following the timescale decomposition idea in Refs. [154–156], to target an arbitrary utility function, the dual algorithm was modified by introducing slow timescale dynamics at the user end [159]. In that dual algorithm, $x_r = x_r(\xi_r, q_r)$, where ξ_r is adjusted slowly, just like the virtual queue capacity $\tilde{C}_l$ in the AVQ algorithms. As such, the dual algorithm can also allocate network resources in a reasonably fair basis.

3.6 PRIMAL-DUAL ALGORITHM

As we saw in the previous sections, the primal and the dual algorithms only update the dynamic variables at one end; they perform static adaptation at the other end. Primal algorithm updates its variables at the user end, while the dual algorithm updates its variables at the link end. However, in the current Internet, it is more practical to directly link $\dot{x}_r$ to q_r and $\dot{p}_l$ to y_l to have a primal-dual algorithm. This necessity is seen from the architecture of the current Internet, where TCP algorithms like TCP Vegas, TCP Reno, and Fast TCP, is a dynamic source algorithm, and most AQM algorithms, like RED and AVQ, associate the price to the queue length or queueing delay; thus, they all are dynamic link algorithms.

As discussed before, the variable $\hat{x}$ solves the system problem iff there exists a variable $\hat{p}$ such that $(\hat{x}, \hat{p})$ is the saddle point of $L(x, p)$. One thus can approach this optimization problem using the game theory. $L(x, p)$ is the gain for player a (Pl) who controls x, and loss for player b (P2) who controls p. So, P1 is the maximizer, while P2 is the minimizer, and we have a two-player zero-sum game with $(\hat{x}, \hat{p})$, the unique saddle point satisfying the following relation:

$$L(x, \hat{p}) \leq L(\hat{x}, \hat{p}) \leq L(\hat{x}, p), \ \forall x \geq 0, p \geq 0.$$

To attain the unique saddle point equilibrium, we can choose the following gradient algorithm:

$$\dot{x}_r = k_r \frac{\partial L(x,p)}{\partial x_r} = k_r (U_r'(x_r) - q_r), \tag{3.54}$$

$$\dot{p}_l = -\gamma_l \frac{\partial L(x,p)}{\partial p_l} = \gamma_l (y_l - c_l). \tag{3.55}$$

Such algorithms are studied in Refs. [160,161,183]; they can be regarded as primal-dual algorithms in the sense that $\dot{x}_r$ is associated with q_r and $\dot{p}_l$ is associated with y_l. From Equations 3.54 and 3.55, we have the following equations to formulate x_r and p_l explicitly:

$$x_r(t) = k_r \int_0^t \frac{\partial L(x_r(s), p_l(s))}{\partial x_r(s)} dx_r(s) = k_r \int_0^t ((U_r'(x_r(s)) - q_r(s))ds, \tag{3.56}$$

$$p_l(t) = -\gamma_l \int_0^t \frac{\partial L(x_r(s), p_l(s))}{\partial p_l(s)} dp_l(s) = \gamma_l \int_0^t (y_l(s) - c_l)ds. \tag{3.57}$$

In these two equations, we have assumed that all initial condition values of $x_r(t)$ and $p_l(t)$ are zero. For this algorithm, at equilibrium, we obtain

$$x_r(t) = U_r'^{-1}(q_r(t)), \ r = 1, 2, \ldots, |R| \tag{3.58}$$

and

$$y_l(t) = \begin{cases} = c_l & \text{if} \quad p_l > 0, \\ \leq c_l & \text{if} \quad p_l = 0, \end{cases} \qquad l = 1, 2, \ldots, |L|. \tag{3.59}$$

Hence, the equilibrium point solves system problem 3.3 exactly.

3.7 STABILITY OF PRIMAL-DUAL ALGORITHMS TOWARD SATISFACTORY PERFORMANCE

As a distributed algorithm to allocate network resources among competing users, congestion control consists of two components: a source algorithm that dynamically adjusts the rate in response to the congestion in its path and a link algorithm that implicitly or explicitly updates a congestion measure and sends it back to the sources that use that link. In the current Internet, the source algorithm is carried out by TCP and the link algorithm by AQM schemes such as DropTail or RED [178]. Different protocols use different metrics to measure congestion. For example, TCP Reno [168] and its variants use loss probability as a congestion measure, while TCP Vegas [169,174] and FAST TCP [172] use queueing delay.

Among all the congestion control algorithms, the primal-dual algorithm has attracted much attention recently (see [105,161,187,194,313,330]). In Ref. [105], the process of congestion control is regarded by Kelly and his coworkers as carrying out a distributed computation by sources and links over a network in real time to solve an optimization problem, and the objective is to maximize the aggregate source utility subject to capacity constraints. It has been shown that two classes of rate control algorithm are naturally associated with the objective functions appearing in, respectively, the primal and the dual formulation of the network's problem.

Then, in the so-called primal-dual algorithm, source rates are interpreted as primal variables, and congestion measures (link price) are interpreted as dual variables. In Ref. [111], TCP/AQM protocols are interpreted as distributed primal-dual algorithms to solve this optimization problem and its associated dual problem. In Ref. [172], FAST TCP/FIFO protocols are another implementation of the primal-dual algorithm. Different protocols such as Reno, Vegas, FAST, DropTail, RED, and REM [100] all solve the same prototypical problem with different utility functions. Moreover, all these protocols generate congestion measures (Lagrange multipliers) that solve the dual problem in equilibrium. A detailed introduction to the recent significant progress in the mathematical modeling of congestion control following the work of Kelly can be found in Ref. [187].

To optimize network performance, several improvements have been made on the primal-dual algorithm [105,112]. An optimization approach has been developed in Ref. [112] for reactive flow control, and a simple synchronous distributed algorithm has been derived, which is applied to solve the primal and dual problem. However, delay has not been considered in the analyses of both the synchronous distributed algorithm and its asynchronous version. Delays can cause instability of these congestion control algorithms and impair the performance of the algorithms in real networks. Stability of network congestion control plays a key role in ensuring that the system operates at the intended equilibrium point with the desired efficiency and fairness.

Earlier relevant discussions about the dynamics of the primal-dual algorithm are reported in Refs. [187,194,313]. Recently, in [320], the stability criteria for the primal-dual algorithm are discussed. However, only a sufficient condition on stability has been obtained in [320], and that condition (Theorem 1, [320]) is, however, specified by an additional condition (inequality (25), [320]). No guideline on how to apply the theoretical results to real networks with specific structure and with a specific protocol is given. Therefore, it is still an open challenge to obtain a sufficient and necessary stability condition on the primal-dual algorithm that facilitates network protocol design and parameter setting.

It is established in Ref. [127] that the overall operations of TCP congestion control algorithms work very similarly to a time-delayed closed-loop control system. Motivated by this, we use feedback time-delayed control theory to attack the stability issue of primal-dual congestion control in this section. We firstly describe the primal-dual congestion control as a time-delayed multivariable control system for a general network. We then discuss stability of the system by using time-delayed control techniques. Based on the analysis of the characteristic equation of the congestion control system in the frequency domain, the sufficient and necessary condition for the stability of the primal-dual algorithm is established for a dumbbell network. Further, the stability condition is applied to the design of the new protocol FAST TCP. Simulation results are presented to validate the theoretical results and to display their applications in real networks.

Next, we briefly review the mathematical model of the primal-dual congestion control algorithm with heterogeneous delays and obtain the linearized equations of the algorithm. In order to optimize the measurement of network performance, the vector that consists of source rates (primal variable) and link prices (dual variable) is viewed as the state variable. By applying multivariable control theory, we establish the multivariable control form of the primal-dual algorithm in the frequency domain and deduce the system characteristic equation. We analyze the stability of the primal-dual algorithm under a dumbbell network model. By using the techniques introduced in Refs. [315,316] to analyze the Hermite matrix of the time-delayed control system, we propose our main results, which give the stability criteria for the primal-dual algorithm in the presence of heterogeneous RTDs, in terms of both source rates and link prices. Further, we explore the techniques that apply our stability condition to the parameter setting of FAST TCP. By using the proposed stability conditions, we provide a guideline for choosing the control gains of FAST TCP to achieve satisfactory performance. We then present a set of simulations to validate the proposed results and also to demonstrate the use of the stability conditions to achieve satisfactory performance.

3.7.1 ANALYSES TO THE PRIMAL-DUAL ALGORITHM

In this section, by describing the congestion control network model as a time-delayed multivariable control system, we propose a method to establish the characteristic equation for networks with arbitrary topologies. The characteristic equation can further be utilized to analyze important properties, such as stability, of the network systems.

Consider a network that consists of a set $L = \{1,\cdots,M\}$ of unidirectional links of capacity $c_l, l \in L$. The network is shared by a set $I = \{1,\cdots,N\}$ of sources. Let x_i be the rate of flow i, and let $X(t) = \{x_i(t),\ i \in I\}$ be the rate vector. Let $R = (R_{li}, i \in I, l \in L)$ be the routing matrix, where $R_{li} = 1$ if flow i traverses link l, and 0 otherwise. By introducing the link rate vector $Y(t) = \{y_l(t), l \in L\}$, the link price vector $P(t) = \{p_l(t), l \in L\}$, and the source price vector $Q(t) = \{q_i(t), i \in I\}$, we then have $Q(t) = R^T P(t)$, $Y(t) = RX(t)$. We assume that source i attains a utility $U_i(x_i)$ and U_i is increasing and strictly concave in its argument.

Now, consider the following primal-dual algorithm [105,112]:

$$\dot{x}_i(t) = \kappa\left(\alpha_i - x_i(t)q_i(t)\right), \tag{3.60}$$

$$\dot{p}_l(t) = \gamma(y_l(t) - c_l), \tag{3.61}$$

where p_l is the price of link l, and the constants κ and γ are source control gain and link control gain, respectively. The positive constant α_i is introduced in the utility function $U_i(x_i) = \alpha_i \log x_i$. At time t, we assume that link l observes the aggregate source rate:

$$y_l(t) = \sum_{i=1}^{N} R_{li} x_i(t - \tau_{li}^f), \tag{3.62}$$

and source i observes the aggregate price in its path:

$$q_i(t) = \sum_{l=1}^{M} R_{li} p_l(t - \tau_{li}^b), \tag{3.63}$$

where τ_{li}^f is the forward feedback delay from source i to link l, and τ_{li}^b is the backward feedback delay from link l to source i. For simplicity, we assume that the feedback delays τ_{li}^f and τ_{li}^b are constants. In this section, we consider propagation delay and neglect the queueing delay; thus, treating RTD for any link l on the path of source i, that is, $\tau_i = \tau_{li}^f + \tau_{li}^b$, as a constant.

The network's optimization problem can be cast in primal or dual form; this leads to primal-dual algorithms that may be interpreted in terms of either explicit rates based on shadow prices or congestion feedback signal (link price) [105]. Equations 3.60 and 3.61 then reflect the dynamics of the primal algorithm implemented at sources and dual algorithm at links, respectively. This primal-dual algorithm represented by Equations 3.60 and 3.61 determines the system transient and steady properties; it henceforth governs the network performance like throughput, fairness, and delay. Stability of the algorithm is especially important when it is implemented within a network. For example, to have an efficient throughput, to bound the delay, and to minimize packet loss, the rate of data flow through its links should tend toward an equilibrium value rather than continually oscillating between having bandwidth spare and being completely overloaded.

Equations 3.60 through 3.63 present a closed-loop feedback system with a unique equilibrium $(x_i^*, p_l^*, y_l^*, q_i^*)$ satisfying

$$y_l^* = \sum_{i=1}^{N} R_{li} x_i^* = c_l,$$

$$q_i^* = \sum_{l=1}^{M} R_{li} p_l^*.$$

Define

$$x_i(t) = x_i^* + \delta x_i(t),\ y_l(t) = y_l^* + \delta y_l(t),$$
$$q_i(t) = q_i^* + \delta q_i(t),\ p_l(t) = p_l^* + \delta p_l(t).$$

We linearize Equations 3.60 and 3.61 around the equilibrium point and thus obtain the following time-delayed feedback control system:

$$\delta\dot{x}_i(t) = -\kappa\left(q_i^*\delta x_i(t) + x_i^*\delta q_i(t)\right), \tag{3.64}$$

$$\delta\dot{p}_l(t) = \gamma\delta y_l(t). \tag{3.65}$$

We also rewrite Equations 3.62 and 3.63 as follows:

$$\delta y_l(t) = \sum_{i=1}^{N} R_{li}\delta x_i(t-\tau_{li}^f), \tag{3.66}$$

$$\delta q_i(t) = \sum_{l=1}^{M} R_{li}\delta p_l(t-\tau_{li}^b). \tag{3.67}$$

Now, we define the following perturbed vectors: $\delta X(t) = \{\delta x_i(t), i \in I\}$, $\delta Y(t) = \{\delta y_l(t), l \in L\}$, $\delta P(t) = \{\delta p_l(t), l \in L\}$, and $\delta Q(t) = \{\delta q_i(t), i \in I\}$. We also assume that the forward delay routing matrix in the frequency domain is given by

$$R(s) = (R_{li}(s), i \in I, l \in L),$$

where

$$R_{li}(s) = \begin{cases} e^{-\tau_{li}^f s} & \text{if } R_{li} = 1, \\ 0 & \text{otherwise.} \end{cases}$$

Taking the Laplace transform of Equations 3.64 through 3.67, respectively, and rewriting them in matrix form

$$(sE_N + \kappa\hat{Q})X(s) = -\kappa\hat{X}Q(s), \tag{3.68}$$

$$sP(s) = \gamma Y(s), \tag{3.69}$$

$$Y(s) = R(s)X(s), \tag{3.70}$$

$$Q(s) = D(s)R^T(-s)P(s), \tag{3.71}$$

where $X(s)$, $Y(s)$, $P(s)$, and $Q(s)$ denote the Laplace transform of the perturbed vectors $\delta X(t)$, $\delta Y(t)$, $\delta P(t)$, and $\delta Q(t)$, respectively. $D(s) = \mathrm{diag}\{e^{-\tau_i s}, i \in I\}$, $\hat{Q} = \mathrm{diag}\{q_i^*, i \in I\}$, and $\hat{X} = \mathrm{diag}\{x_i^*, i \in I\}$.

From Equations 3.68 through 3.71, we have

$$\begin{bmatrix} sE_N + \kappa\hat{Q} & \kappa\hat{X}D(s)R^T(-s) \\ -\gamma R(s) & sE_M \end{bmatrix} \times \begin{bmatrix} X(s) \\ P(s) \end{bmatrix} = 0, \tag{3.72}$$

where E_N and E_M denote the N-order identity matrix and M-order identity matrix, respectively.

Therefore, the characteristic polynomial of closed-loop system Equations 3.60 through 3.63 is obtained from

$$\Delta(s) = \begin{vmatrix} sE_N + \kappa\hat{Q} & \kappa\hat{X}D(s)R^T(-s) \\ -\gamma R(s) & sE_M \end{vmatrix}. \tag{3.73}$$

Let

$$A(s) = (sE_N + \kappa\hat{Q}) - \kappa\hat{X}D(s)R^T(-s)(sE_M)^{-1}(-\gamma R(s))$$

be the Schur complement of the matrix sE_M in the block matrix

$$\begin{bmatrix} sE_N + \kappa\hat{Q} & \kappa\hat{X}D(s)R^T(-s) \\ -\gamma R(s) & sE_M \end{bmatrix}.$$

The characteristic equation is then given by

$$\det(sE_M)\det A(s) = 0. \tag{3.74}$$

Now, denote $z = e^{-s}$ and

$$\Delta(s) = B(s,z) = \det(sE_M)\det A(s,z),$$

where

$$A(s,z) = (sE_N + \kappa\hat{Q}) - \kappa\hat{X}D(z)R^T(1/z)(sE_M)^{-1}(-\gamma R(z)),$$

and

$$D(z) = \text{diag}\{z^{\tau_i}, i \in I\},$$
$$R(z) = (R_{li}(z), i \in I, l \in L),$$

where

$$R_{li}(z) = \begin{cases} z^{\tau_{li}^f} & \text{if } R_{li} = 1, \\ 0 & \text{otherwise.} \end{cases}$$

Then, based on the results on stability of time-delayed control systems [315], we have the following lemmas on the stability of the primal-dual system for a general network.

Lemma 3.7.2

For a network with the set $L = \{1,\ldots,M\}$ of links and the set $I = \{1,\ldots,N\}$ of sources that share the links, the primal-dual system described by Equations 3.64 through 3.67 is asymptotically stable if and only if

$$B(s,z) \neq 0, \text{ when Re } s \geq 0 \text{ and } |z| = 1, \tag{3.75}$$

where $B(s,z)$ is the characteristic polynomial of the primal-dual system. ■

According to $|z| = 1$, let $z = e^{j\lambda}$. The polynomial $B(s,z)$ is then viewed as a polynomial in s whose coefficients are polynomials in $e^{j\lambda}$. One can construct the $(N+M)\times(N+M)$ Hermite matrix $H(e^{j\lambda})$ [316] associated with $B(s,e^{j\lambda})$. Using Marden's recursive scheme [317], we have the following lemma.

Lemma 3.7.3

Stability condition 3.75 in Lemma 3.7.2 is equivalent to

$$H(1) = H(e^{j0}) \text{ is positive definite}$$

and

$$H(e^{j\lambda}) \neq 0 \text{ for all } \lambda \in [0, 2\pi],$$

where $H(e^{j\lambda})$ is the Hermite matrix that is associated with the system characteristic polynomial $B(s,z)$. ■

This general condition is applied to a dumbbell network in the next section.

3.7.2 STABILITY RESULTS FOR A DUMBBELL NETWORK

In this section, we discuss the stability of the primal-dual congestion control system for a dumbbell network model. We establish the explicit conditions under which the dumbbell network is stable in terms of the price and rate vector.

For a dumbbell topology with N sources and only one link, $q_i(t) = p(t - \tau_i^b)$, and then $q_i^* = p^*$, $i = 1, 2, \cdots, N$. The closed-loop system determined by Equations 3.64 through 3.67 is reduced to

$$\delta \dot{x}_i(t) = -\kappa \left(p^* \delta x_i(t) + x_i^* \delta p(t - \tau_i^b) \right), \tag{3.76}$$

$$\delta \dot{p}(t) = \gamma \sum_{i=1}^{N} \delta x_i(t - \tau_i^f). \tag{3.77}$$

Using the method developed in the previous section, the characteristic equation of 3.76 and 3.77 is obtained as follows:

$$\Delta(s) = (s + \kappa p^*)^{N-1} \left(s^2 + \kappa p^* s + \kappa\gamma \sum_{i=1}^{N} x_i^* e^{-\tau_i s} \right) = 0. \tag{3.78}$$

Now, we can determine that when $\gamma = 0$, characteristic Equation 3.78 has a repeated root $-\kappa p^*$ of multiplicity N and only one zero root. Therefore, the system determined by Equations 3.76 and 3.77 is marginally stable.

Next, we focus on the analysis in the case of $\gamma \neq 0$. We use the time-delayed control system theory [315,316] to investigate the system stability. Define

$$B(s, e^{-s}) = s^2 + \kappa p^* s + \kappa\gamma \sum_{i=1}^{N} x_i^* e^{-\tau_i s}. \tag{3.79}$$

Let $e^{-s} = z$, and $a_0 = 1$, $a_1 = \kappa p^*$, $a_2 = \kappa\gamma \sum_{i=1}^{N} x_i^* z^{\tau_i}$. According to [315], we can associate the function $B(s, e^{-s})$ with a polynomial $B(s, z)$, which is in terms of two independent complex variables s and z, and formulate Equation 3.79 as

$$B(s, z) = a_0 s^2 + a_1 s + a_2.$$

On the basis of this polynomial and by using the technique in Ref. [316], one constructs a Hermite matrix

$$H = \begin{bmatrix} (0,1) & (0,2) \\ (0,2) & (1,2) \end{bmatrix},$$

where the four elements are given by

$$\begin{aligned}
(0,1) &= j^{0+1+1}\left((-1)a_0\bar{a}_1 + (-1)\bar{a}_0 a_1\right) = 2a_1, \\
(0,2) &= j^{0+2+1}\left((-1)^2 a_0\bar{a}_2 + (-1)\bar{a}_0 a_2\right) = -2\mathrm{Im}a_2, \\
(1,2) &= j^{1+2+1}\left((-1)^2 a_1\bar{a}_2 + (-1)^{1+1}\bar{a}_1 a_2\right) = 2a_1\mathrm{Re}a_2.
\end{aligned}$$

Furthermore, let $z = e^{j\lambda}$. Then,

$$a_2 = \kappa\gamma \sum_{i=1}^{N} x_i^* e^{j\lambda\tau_i}.$$

One thus obtains the Hermite matrix

$$\begin{bmatrix} H(e^{j\lambda}) = 2\kappa p^* & -2\kappa\gamma \sum_{i=1}^{N} x_i^* \sin \tau_i\lambda \\ -2\kappa\gamma \sum_{i=1}^{N} x_i^* \sin \tau_i\lambda & 2\kappa^2\gamma p^* \sum_{i=1}^{N} x_i^* \cos \tau_i\lambda \end{bmatrix}.$$

Theorem 3.7.7

For a dumbbell network, the primal-dual system given by Equations 3.76 and 3.77 is asymptotically stable in terms of its state variable $\delta x_i(t)$ and $\delta p(t)$ if and only if

$$\kappa(p^*)^3 \sum_{i=1}^{N} \alpha_i \cos \tau_i \lambda - \gamma \left(\sum_{i=1}^{N} \alpha_i \sin \tau_i \lambda \right)^2 \neq 0, \tag{3.80}$$

where $\kappa > 0$, $\gamma > 0$, and $\lambda \in [0, 2\pi]$ is an arbitrary constant.

Proof: According to Lemma 3.7.3, system Equations 3.76 and 3.77 is asymptotically stable if and only if $H(1) = H(e^{j0})$ is positive definite and $\det H(e^{j\lambda}) \neq 0$ for all $\lambda \in [0, 2\pi]$. This then leads to the following two conditions.

Condition 1. $H(1) = H(e^{j0})$ is positive definite. In other words, the sequential principal minor of $H(1)$ are all positive. Since

$$H(1) = \begin{bmatrix} 2\kappa p^* & 0 \\ 0 & 2\kappa^2 \gamma p^* \sum_{i=1}^{N} x_i^* \end{bmatrix},$$

one has

$$\begin{cases} 2\kappa p^* > 0 \\ 4\kappa^3 \gamma (p^*)^2 \sum_{i=1}^{N} x_i^* > 0, \end{cases}$$

which is equivalent to $\kappa > 0$ and $\gamma > 0$ since $p^*, x_i^* > 0$.

Condition 2. $\det H(e^{j\lambda}) \neq 0$ for all $\lambda \in [0, 2\pi]$, where

$$\begin{aligned} \det H(e^{j\lambda}) = {} & 4\kappa^3 \gamma (p^*)^2 \sum_{i=1}^{N} x_i^* \cos \tau_i \lambda \\ & -4\kappa^2 \gamma^2 \left(\sum_{i=1}^{N} x_i^* \sin \tau_i \lambda \right)^2 \neq 0. \end{aligned}$$

Combining Conditions 1 and 2, one obtains that

$$\kappa(p^*)^2 \sum_{i=1}^{N} x_i^* \cos \tau_i \lambda - \gamma \left(\sum_{i=1}^{N} x_i^* \sin \tau_i \lambda \right)^2 \neq 0. \tag{3.81}$$

From Equation 3.60, we have

$$\alpha_i = x_i^* q_i^*. \tag{3.82}$$

For a network with a dumbbell topology, Equation 3.82 is changed into

$$\alpha_i = x_i^* p^*. \tag{3.83}$$

Substituting Equation 3.83 in 4.44, we have

$$\kappa(p^*)^3 \sum_{i=1}^{N} \alpha_i \cos \tau_i \lambda - \gamma \left(\sum_{i=1}^{N} \alpha_i \sin \tau_i \lambda \right)^2 \neq 0.$$

This concludes the proof.

Note that in equilibrium, we have the following relationship:

$$\sum_{i=1}^{N} x_i^* = c \text{ and } x_i^* = \frac{\alpha_i}{p^*}, \; i = 1, 2, \cdots, N.$$

Therefore, we further have

$$p^* = \sum_{i=1}^{N} \frac{\alpha_i}{c}. \tag{3.84}$$

We define

$$\begin{aligned} &F(\kappa, \gamma, \alpha_i, \lambda, \tau_i, N, c) \\ &= \kappa (p^*)^3 \sum_{i=1}^{N} \alpha_i \cos \tau_i \lambda - \gamma \left(\sum_{i=1}^{N} \alpha_i \sin \tau_i \lambda \right)^2 \\ &= \frac{\kappa}{c^3} \left(\sum_{i=1}^{N} \alpha_i \right)^3 \sum_{i=1}^{N} \alpha_i \cos \tau_i \lambda - \gamma \left(\sum_{i=1}^{N} \alpha_i \sin \tau_i \lambda \right)^2. \end{aligned} \tag{3.85}$$

In the rest of this section, we use F to denote the function $F(\kappa, \gamma, \alpha_i, \lambda, \tau_i, N, c)$ for simplicity. ■

There are a number of parameters coupled in this function that determine the value of the function. These parameters include the control gains in the primal-dual algorithm and the network parameters like the capacity, the delay, etc. To have a basic idea of how these parameters determine the stability of the primal-dual system within a given network structure, we next take a close look at the relationships between the function F and the variables, such as the parameters κ, γ, α_i, and λ. This sort of analyses will be beneficial to network design by providing a guideline of parameter settings.

Example 1:

Let us consider a dumbbell network with a link capacity of 100 Mb/s and with each packet size being 1000 bytes. Let the parameters $\gamma = 0.1$, the number of flows $N = 5$, and let $\alpha_1 = 50$, $\alpha_2 = 56.5$, $\alpha_3 = 62.5$, $\alpha_4 = 68.5$, $\alpha_5 = 75$, $\tau_1 = 10$ ms, $\tau_2 = 33$ ms, $\tau_3 = 55$ ms, $\tau_4 = 77$ ms, and $\tau_5 = 100$ ms.

Now, we study the relationship between the value of the function $F(\kappa, \lambda)$ and the source control gain κ and the parameter λ, which is plotted in Figure 3.4. As can be seen, F is equal to zero only when κ and λ take values in the square area marked in Figure 3.4, while F is always less than zero outside the area, and decreases as λ increases. Next, we put $\kappa = 0.5$, let γ change from 0 to 0.5, and keep other parameter settings. Figure 3.5 then shows the relationship between the value of the function $F(\gamma, \lambda)$ and the link control gain γ and the parameter λ. Observation suggests that F is equal to zero only when γ and λ take values in the narrow area marked in Figure 3.5, and F decreases as γ or λ increases. Finally, we assume that $\alpha_i = \alpha$ $(i = 1, 2, 3, 4, 5)$. Let $\gamma = 0.1$, α change from 50 to 75, and keep other parameter settings. We analyze the relationship between the value of the function $F(\alpha, \lambda)$ and the weighted parameters α of the utility function and the parameter λ. This result is depicted in Figure 3.6. One observes from Figure 3.6 that again F is equal to zero only when parameters α and λ take values in the narrow area marked in the figure, and F decreases as α or λ increases.

Because the multivariable function F is the compound function of the trigonometric function sine and cosine, we now consider some special angles. When $\tau_i \lambda = n\pi$ (n is an arbitrary integer),

$$F = \frac{\kappa}{c^3} \left(\sum_{i=1}^{N} \alpha_i \right)^4.$$

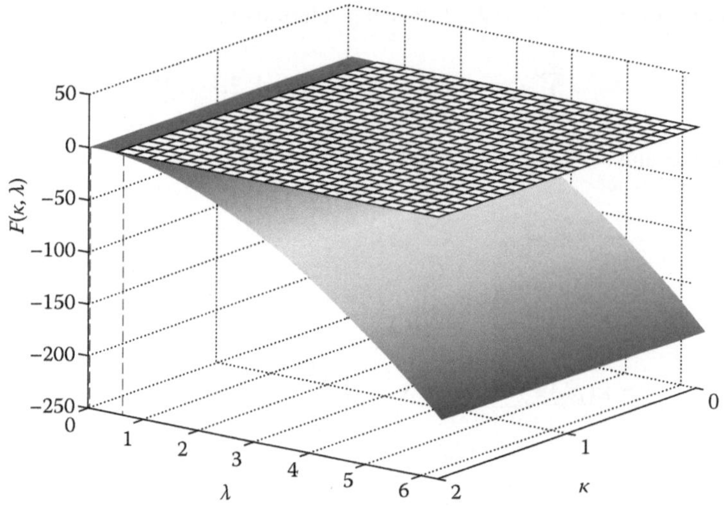

Figure 3.4 The relationship between $F(\kappa, \lambda)$ and κ, λ.

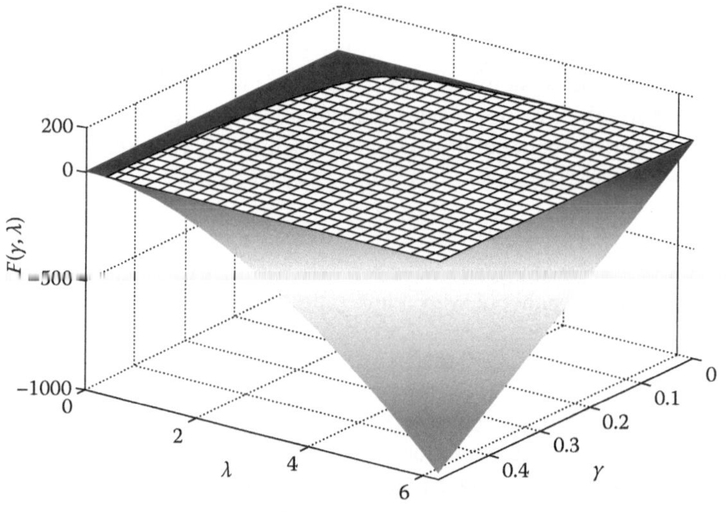

Figure 3.5 The relationship between $F(\gamma, \lambda)$ and γ, λ.

From Theorem 3.7.7, we know $\kappa > 0$. Then, F is nonzero.

When $\tau_i \lambda = n\pi/2$,

$$F = -\gamma \left(\sum_{i=1}^{N} \alpha_i \right)^2 .$$

For $\gamma > 0$, F is nonzero again.

In the next section, we will discuss the application of the obtained results.

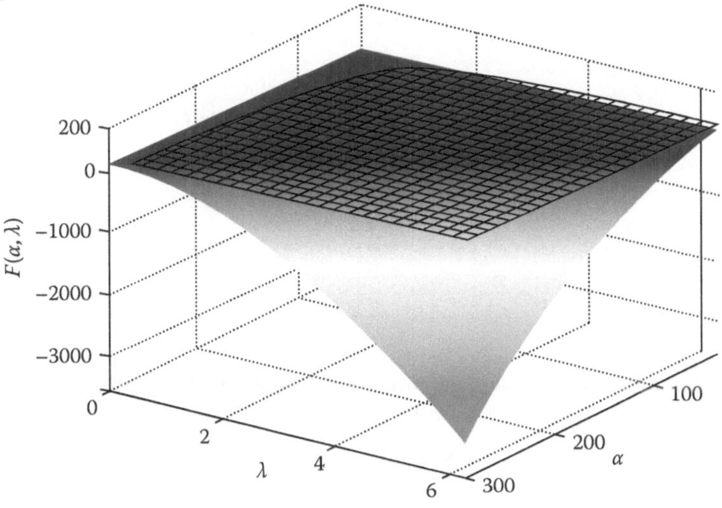

Figure 3.6 The relationship between $F(\alpha,\lambda)$ and α,λ.

3.7.3 APPLICATIONS TO PROTOCOL DESIGN AND PARAMETER SETTING OF FAST TCP

It is a challenging task to properly set the parameters of the TCP and AQM system in order to support efficient and stable operation of end-to-end congestion control [323]. As described earlier, many TCP/AQM systems can be considered as primal-dual systems. The system where FAST TCP acts as the source algorithm and DropTail acts as the link algorithm is such an example. To demonstrate the applications of the proposed stability conditions, we only focus on the stability issue of FAST TCP and its parameter setting in the present section. The obtained stability condition can be applied to other TCP/AQM systems. We use the recently developed FAST TCP link model [318], where both the self-clocking and link integrator effects are taken into account.

FAST TCP [172] is a new version of TCP, and it is proposed for high-speed networks. As a protocol that has applied the primal-dual algorithms, FAST TCP does not use packet loss but queueing delay as the "link price." By defining its utility function as $U_i(x_i) = \alpha_i \log x_i$, FAST TCP allocates the link bandwidth among flows to achieve proportional fairness under the framework of the primal-dual algorithm. The parameter α_i, which represents the total packets that flow i buffers in the link queue at equilibrium, is important in network performance. Too large a value will lead to a large queue delay, and too small a value may result in network instability. Although FAST TCP has shown promising results of achieving high performance, there are still no guidelines for choosing the parameter α_i. Moreover, FAST TCP can be unstable in the presence of heterogeneity of round-trip time (RTT) of the flows [328] in a single bottleneck link network.

In this section, we try to provide a rule of thumb to set the value of α according to the obtained stability condition.

The window update equation of FAST is given by Ref. [318]

$$\dot{w}_i(t+1) = \kappa\left(\frac{\alpha_i}{d_i+q_i(t)} - \frac{q_i(t)}{(d_i+q_i(t))^2}w_i(t)\right), \tag{3.86}$$

where $\kappa \in (0,1]$, d_i is the propagation delay, α_i is the weighting parameter, and $q_i(t)$ is the sum of queueing delay for all links in flow i's path, given by

$$q_i(t) = \sum_{l=1}^{M} R_{li}p_l(t-\tau_{li}^b).$$

The queueing delay is understood as the "link price," and it has been traditionally modeled as follows:

$$\dot{q}_i(t) = \frac{1}{c_l}\left(\sum_{i=1}^{N} R_{li}x_i(t-\tau_{li}^f) - c_l\right). \tag{3.87}$$

From Equation 3.86, we have the following relationship in steady state:

$$\alpha_i = \frac{q_i^*}{R_i^*} w_i^*. \tag{3.88}$$

Considering $w_i^* = x_i^* R_i^*$, we rewrite Equation 3.88 as follows:

$$\alpha_i = x_i^* q_i^*.$$

Therefore, Equation 3.86 has the same equilibrium point (x_i^*, q_i^*) with as Equation 3.60 in the primal-dual algorithm, which solves the utility optimization problem:

$$\max_{x \geq 0} \sum_i \alpha_i \log x_i \;\; \text{s.t. } Rx \leq c.$$

In steady state, we have

$$\alpha_i = x_i^* q_i^*, \; \sum_{i=1}^{N} x_i^* = c \;\text{ and }\; q_i^* = \sum_{l=1}^{M} R_{li} p_l^*.$$

Comparing Equation 3.87 with 3.61, we can see that they are very similar, and the difference between these two equations is only the corresponding control gain parameters. For the dumbbell network model, one has $q_i(t) = p(t - \tau_i^b)$ and $q_i^* = p^*$. We associate the control gain parameter of Equation 3.61 with that of Equation 3.87, and obtain

$$\gamma = \frac{1}{c}. \tag{3.89}$$

Combining Equation 3.89 with condition 3.80 derived in the previous section, we obtain

$$F = \kappa (p^*)^3 \sum_{i=1}^{N} \alpha_i \cos \tau_i \lambda - \frac{1}{c}\left(\sum_{i=1}^{N} \alpha_i \sin \tau_i \lambda\right)^2 \neq 0. \tag{3.90}$$

If all the sources choose the same weighted parameter α, Equation 3.90 is very useful to set the parameter α in their utility functions. Let $\alpha_i = \alpha$. From Equation 3.84, one obtains

$$p^* = \frac{N\alpha}{c}. \tag{3.91}$$

Substituting $\alpha_i = \alpha$ and Equation 3.91 into 3.90, by a direct computation, we get the following result:

$$F = \kappa N^3 \alpha^2 \sum_{i=1}^{N} \cos \tau_i \lambda - c^2 \left(\sum_{i=1}^{N} \sin \tau_i \lambda\right)^2 \neq 0, \tag{3.92}$$

where $\kappa > 0$, λ is an arbitrary constant, and $\lambda \in [0, 2\pi]$. Then, the inequality (Equation 3.92) is ready to be used in setting the parameters in FAST TCP. We will run some ns2 simulations [321] to verify this condition in the next section.

3.7.4 SIMULATION RESULTS

In this section, we provide simulation results to show how the conditions developed in the previous sections can be used in the network environment. In simulations, we use the single bottleneck link model with the capacity of 100 Mb/s and with N homogeneous sources. This model is depicted in Figure 3.7. We run the simulation by using *ns* [321] to validate the packet-level implementations and by using MATLAB to obtain the flow-level dynamics as well.

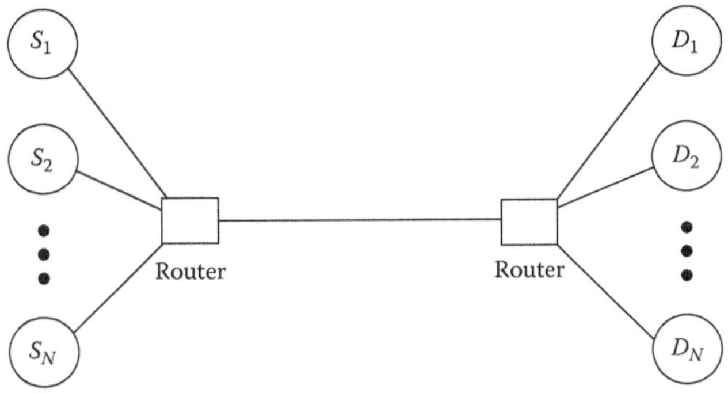

Figure 3.7 The dumbbell topology.

SIMULATION 1

In this simulation, we assume that the RTT τ_i for each source is equal to τ. Let all the sources have the same parameter $\alpha_i = \alpha$. Then, all the sources have the same sending rates in steady state.

We set the parameters as $N = 10$, $\alpha = 250$, $\tau = 10$ ms, source control gain $\kappa = 0.5$, and link control gain $\gamma = 1 \times 10^{-4}$. From Equation 3.84, one has

$$p^* = \frac{N\alpha}{c}.$$

Then, the link price at the equilibrium point is calculated to be

$$p^* = 0.2.$$

By using these set parameters, we perform some calculations in Equation 3.80 to obtain

$$F = N^2\alpha^2\left(\frac{\kappa}{c^3}N^2\alpha^2 \cos\tau\lambda - \gamma\sin^2\tau\lambda\right) = 9.7237 \neq 0,$$

which verifies that under the aforementioned parameter setting, the time-delayed multivariable control system for the primal-dual algorithm is stable about the source rates and link prices. Figures 3.8 and 3.9 illustrate the source sending rates and link prices for the aforementioned data. One observes that when the parameter settings satisfy condition 3.80, the source rates and link price can converge to the equilibrium point very quickly (rates converge at 1250 packets/s and prices at 0.2).

SIMULATION 2

Now, let us keep all the parameter settings in Simulation 1 but reduce the flow number to 5. Then, let us assume that they have different α_i. Let $\alpha_1 = 230$, $\alpha_2 = 240$, $\alpha_3 = 250$, $\alpha_4 = 260$, and $\alpha_5 = 270$. From Equation 3.84, the value of p in equilibrium is calculated to be $p^* = 0.1$. Using Equation 3.80, one obtains

$$F = \left(\sum_{i=1}^{N}\alpha_i\right)^2\left(\frac{\kappa}{c^3}\cos\tau\lambda\left(\sum_{i=1}^{N}\alpha_i\right)^2 - \gamma\sin^2\tau\lambda\right)$$
$$= 0.5563 \neq 0.$$

According to Theorem 3.7.7, the control system of the primal-dual algorithm is stable about the source rates and link prices under the aforementioned parameters settings. Figures 3.10 and 3.11

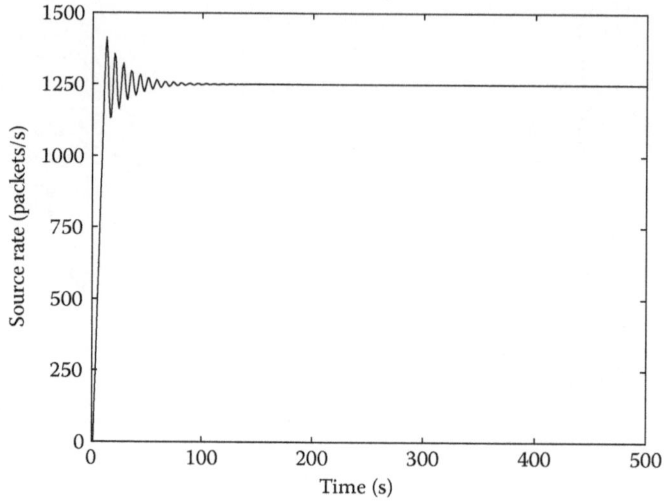

Figure 3.8 Source rates with the same value of α_i in the stable primal-dual case.

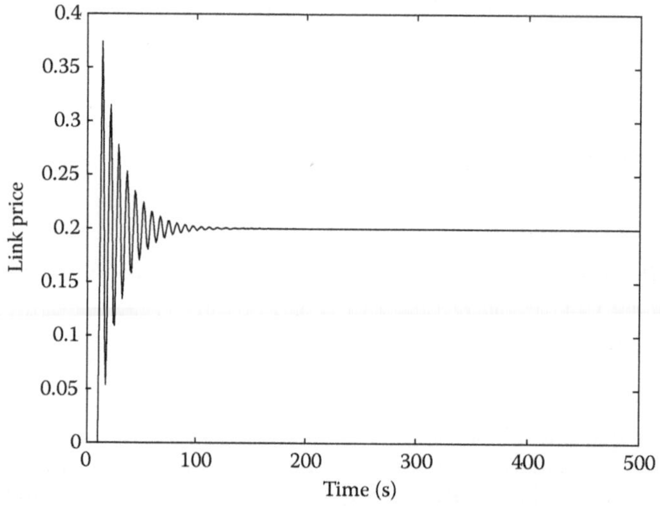

Figure 3.9 Link price with the same value of α_i in the stable primal-dual case.

show the plots of the simulation results. Figure 3.10 shows that all the five heterogeneous flow rates converge to their equilibrium points, while Figure 3.11 illustrates the convergence of the link price.

Furthermore, we use a stability index [172] to validate our result on the system's stability. Define

$$S_i = \frac{1}{\bar{x}_i}\sqrt{\frac{1}{m-1}\sum_{k=1}^{m}(x_i(k) - \bar{x}_i)},$$

where $\bar{x}_i$ is the average throughput of flow i over the interval $[1, m]$ and m is the upper bound of the sampling interval. Then, the stability index is the average over the N active sources:

$$S = \frac{1}{N}\sum_{i=1}^{N} S_i.$$

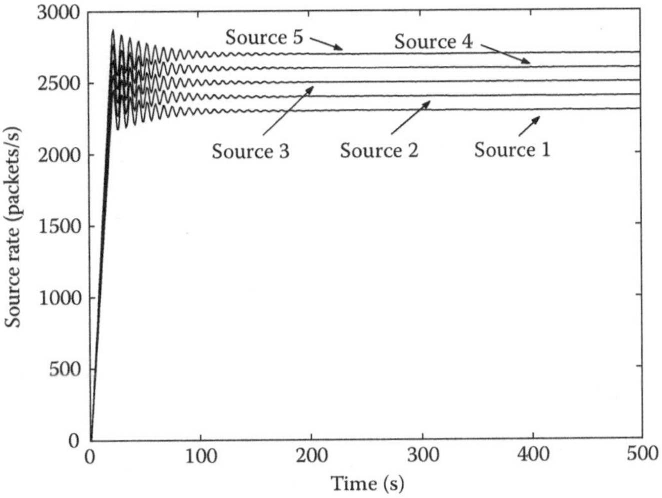

Figure 3.10 Source rates with different values of α_i in the stable primal-dual case.

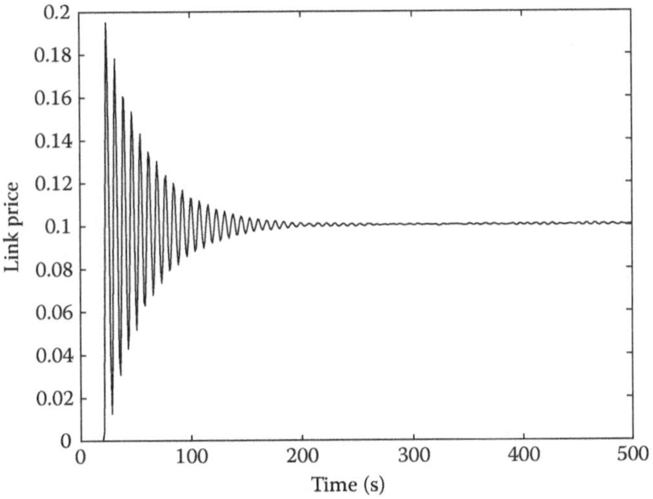

Figure 3.11 Link price with different values of α_i in the stable primal-dual case.

The smaller the stability index, the less oscillation a source experiences. Let $m = 5$ and keep other parameter settings the same. The curve of the aforementioned index is depicted in Figure 3.12, which suggests the stability about the source rates of system Equations 3.64 and 3.65.

We also use the following fairness index [338] to investigate the fairness of the five flows:

$$F = \frac{\left(\sum_{i=1}^{N}\right)^2}{N \sum_{i=1}^{N} x_i^2}.$$

The fairness index is plotted in Figure 3.13. As shown, the fairness index is approximately equal to 0.997 along the whole simulation. Therefore, choosing the parameters for the primal-dual algorithm according to our condition 3.80 can achieve good performance on both stability and fairness.

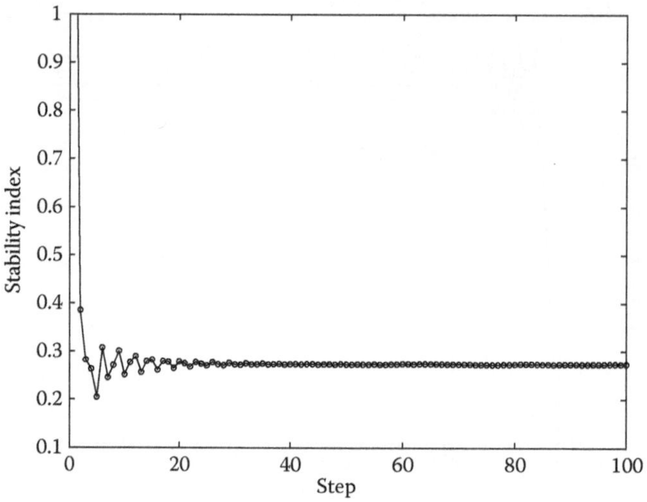

Figure 3.12 Stability index with different values of α_i in the a stable primal-dual case.

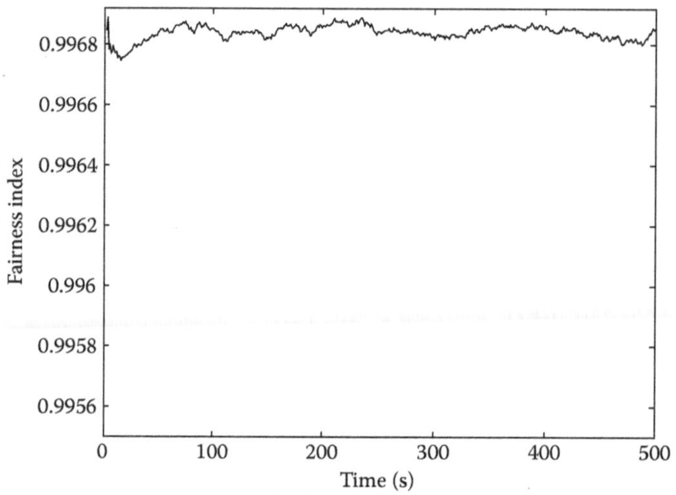

Figure 3.13 Fairness index with different values of α_i in the stable primal-dual case.

SIMULATION 3

For the same network topology presented in Simulation 1, one can set the parameters for FAST TCP by using Equation 3.92. Here, the link price p denotes the queueing delay.

Let $\tau_i = \tau$. From Equation 3.92, one obtains

$$F = \kappa N^2 \alpha^2 \cos \tau\lambda - c^2 \sin^2 \tau\lambda \neq 0. \tag{3.93}$$

Now, we run ns2 simulation [321] for the dumbbell network model with a link capacity of 1 Gb/s and each packet size being 1000 bytes. Let flow number $N = 3$, delay $\tau = 130$ ms, $\alpha = 1000$, source control gain $\kappa = 0.5$, and the parameter $\lambda = 0.015$. According to Equation 3.93, we calculate that

$$F = 4.4406 \times 10^6 \neq 0.$$

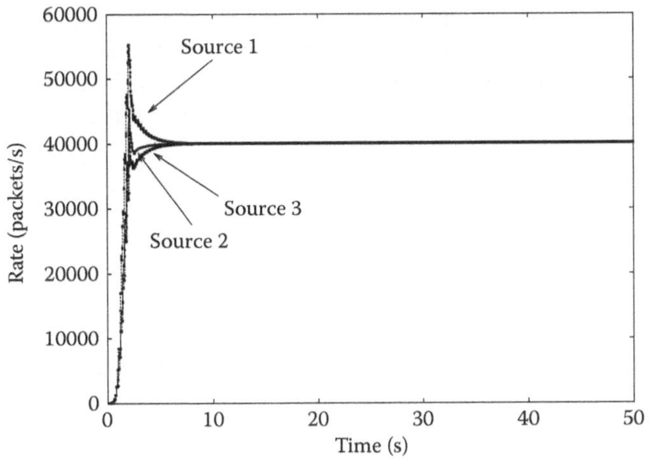

Figure 3.14 Source rate of FAST TCP (α = 1000).

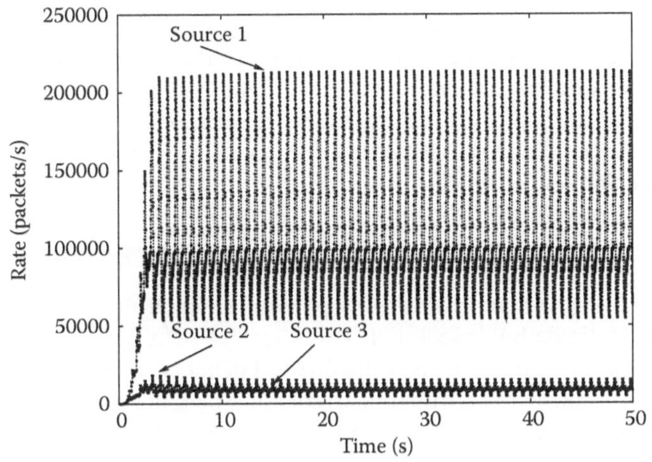

Figure 3.15 Source rate of FAST TCP ($\alpha = 115$).

From Theorem 3.7.7, the FAST TCP/DropTail system is stable. Figure 3.14 illustrates that when $\alpha = 1000$, the source rates converge very quickly and there is little oscillation.

Next, we change the value of α to 115 and keep other parameters settings as such. Then, we have

$$F = \kappa N^2 \alpha^2 \cos \tau\lambda - c^2 \sin^2 \tau\lambda = 0.$$

From Theorem 3.7.7, the FAST TCP/DropTail system is then unstable. The source rate in this setting is plotted in Figure 3.15. From Figure 3.15, we can observe that when $F = 0$, the sources have large oscillations on rates.

3.7.5 MORE RESULTS ON PARAMETER TUNING OF FAST TCP

FAST TCP [172,334] outperforms its predecessors in networks with large bandwidth-delay product as queueing delay is used as the congestion indication. However, parameter setting of FAST TCP still remains a challenging task [325,352] as there may be no single set of parameters that works well under different network scenarios and usually a dozen parameters are coupled together.

Recently, stability of the TCP dynamic has attracted much attention (see, e.g., [127,291,327]). Particularly, local and global stability on FAST TCP has been studied in [324,328–330]. The stability of the protocol dynamic is seen to play a key role in parameter setting to obtain satisfactory performance. The basic requirements for a stability condition to serve as a guideline of parameter setting are:

- It should relate all the relevant parameters in a comprehensive manner;
- The key parameter should be decoupled from other parameters in the condition.

From this point of view, the existing stability results on FAST TCP are not straightforward for application of parameter setting. There is still no guideline available on FAST TCP parameter setting that can obtain satisfactory performance under various network environments.

In FAST TCP, the parameter α is the number of the packets each source attempts to maintain in the network buffers at equilibrium [172]. This key parameter together with other parameters, including the link capacity c, the propagation delay d, and the source control gain γ, are responsible for the network performance under FAST TCP. In Ref. [325], on a simulation basis, it has been suggested that the value of α should be at least 0.0075 times the link capacity in a network with a single bottleneck link. However, no theoretical basis is given to support their claim. How this parameter setting is related to other parameters, like the propagation delay, is still unknown.

In this section, which discusses the stability of FAST TCP, a sufficient condition on asymptotical stability of FAST TCP congestion window is obtained. This condition relates all the relevant parameters in FAST TCP and decouples the key parameter α from others. Based on this, a guideline on choosing this parameter to obtain satisfactory performance is provided. We provide ns2 simulation results to validate the theoretical results.

3.7.6 ANALYSES OF THE FAST TCP MODEL AND RESULTS ON STABILITY

Consider a single-link single-source network topology in which the source and destination node is connected through a single bottleneck link with capacity c. Let $p(t)$ be the queueing delay and $x(t)$ the source sending rate. Let τ^f denotes the forward delay between source and link, and τ^b, the backward delay in the feedback path from link to source. Then, the source rate observed by the link at time t is $x(t-\tau^f)$, and $q(t)=p(t-\tau^b)$ is the queueing delay observed by the source at time t. Let $w(t)$ be the source congestion window size at time t. The source sending rate $x(t)$ is then defined as

$$x(t)=\frac{w(t)}{d+q(t)}, \tag{3.94}$$

where d denotes the constant propagation delay of the source, and the RTT at time t is given by $R(t)=d+q(t)$. Here, we interpret $\tau^f+\tau^b=R_0$ as the equilibrium value of $R(t)$.

The original FAST TCP adapts the congestion window periodically according to the discrete time model [172]

$$w[k+1]=(1-\gamma)w[k]+\gamma\left(\alpha+\frac{d}{d+q[k]}w[k]\right), \tag{3.95}$$

where α and γ are protocol parameters. This update is performed once every RTT. We follow [328] to obtain a continuous-time approximation of the window control. We then rewrite Equation 3.95 as

$$\frac{w[k+1]-w[k]}{R[k]}=-\gamma\frac{w[k]}{R[k]}+\gamma\left(\frac{\alpha}{R[k]}+\frac{d}{(R[k])^2}w[k]\right).$$

Using a first-order Euler approximation of the derivative and applying the identity

$$R[k]=d+q[k],$$

we derive the continuous-time window update equation

$$\dot{w}(t) = \gamma\left(\frac{\alpha}{d+q(t)} - \frac{q(t)}{(d+q(t))^2}w(t)\right). \tag{3.96}$$

Queueing delay has been traditionally modeled by

$$\dot{p}(t) = \frac{1}{c}\left(x(t-\tau^f)-c\right) = \frac{1}{c}\left(\frac{w(t-\tau^f)}{d+p(t-R_0)} - c\right). \tag{3.97}$$

Noting that

$$q(t) = p(t-\tau^b)$$

and in equilibrium

$$\frac{\alpha}{p_0} = x_0 = \frac{w_0}{d+p_0},\ x_0 = c,\ d+p_0 = R_0,$$

we linearize Equations 3.96 and 3.97 around (w_0, p_0) to obtain

$$\begin{cases} \delta\dot{w}(t) = -\frac{\gamma p_0}{R_0^2}\delta w(t) - \frac{\gamma\alpha d}{p_0 R_0^2}\delta p(t-\tau^b), \\ \delta\dot{p}(t) = \frac{1}{R_0 c}\delta w(t-\tau^f) - \frac{1}{R_0}\delta p(t-R_0), \end{cases} \tag{3.98}$$

where

$$\delta w(t) = w(t) - w_0,$$

$$\delta p(t) = p(t) - p_0.$$

$\delta w(t)$ and $\delta p(t)$ are both perturbed variables.

By taking the Laplace transformation of Equation 3.98, we have

$$\begin{cases} s\delta W(s) - \delta w(0) = -\frac{\gamma p_0}{R_0^2}\delta W(s) - \frac{\gamma\alpha d}{p_0 R_0^2}\delta P(s)e^{-\tau^b s} \\ s\delta P(s) - \delta p(0) = \frac{1}{R_0 c}\delta W(s)e^{-\tau^f s} - \frac{1}{R_0}\delta P(s)e^{-R_0 s}, \end{cases} \tag{3.99}$$

where $\delta W(s)$ and $\delta P(s)$ denote the Laplace transform of $\delta w(t)$ and $\delta p(t)$, respectively. By direct computing, one has

$$\begin{aligned} &\left(s+\tfrac{1}{R_0}e^{-R_0 s}\right)\left(s+\tfrac{\gamma p_0}{R_0^2}\right)\delta W(s) = \tfrac{1}{R_0}e^{-R_0 s}\delta w(0) \\ &-\tfrac{\gamma\alpha d}{p_0 R_0^2}e^{-\tau^b s}\left(\delta p_0 + \tfrac{1}{R_0 c}\delta W(s)e^{-\tau^f s}\right) + s\delta w(0). \end{aligned} \tag{3.100}$$

Then, we obtain the so-called *characteristic polynomial* $\Delta(s)$

$$\Delta(s) = s^2 + \left(\frac{\gamma p_0}{R_0^2} + \frac{1}{R_0}e^{-R_0 s}\right)s + \gamma\frac{1}{R_0^2}e^{-R_0 s}. \tag{3.101}$$

Characteristic polynomial Equation 3.101 determines the stability in terms of the congestion window size of linearized closed-loop, time-delayed system Equation 3.98. We use the Routh–Hurwitz stability criteria [326] to study its stability. To analyze the system stability, it is adequate to have the Padé approximation [326]

$$e^{-R_0 s} = \frac{2-R_0 s}{2+R_0 s}.$$

The approximated characteristic equation of linearized system Equation 3.98 is then reached at

$$\Delta(s) = R_0 s^3 + \left(\frac{\gamma p_0}{R_0} + 1\right) s^2 + \left(\frac{2}{R_0} + \frac{\gamma(2p_0 - R_0)}{R_0^2}\right) s + \frac{2\gamma}{R_0^2} = 0. \quad (3.102)$$

By denoting

$$a_3 = R_0,$$
$$a_2 = 1 + \frac{\gamma p_0}{R_0},$$
$$a_1 = \frac{2}{R_0} + \frac{\gamma(2p_0 - R_0)}{R_0^2},$$
$$a_0 = \frac{2\gamma}{R_0^2},$$

characteristic Equation 3.102 becomes

$$\Delta(s) = a_3 s^3 + a_2 s^2 + a_1 s + a_0 = 0. \quad (3.103)$$

According to the Routh–Hurwitz criteria [326], all the coefficients of Equation 3.103 on s have to be positive, which gives

$$\frac{2}{R_0} + \frac{\gamma(2p_0 - R_0)}{R_0^2} > 0.$$

Substituting $R_0 = d + p_0$ and $p_0 = \alpha/c$ into this inequality, we have the following conditions:

1. γ can take any positive value if $\alpha > cd$.
2. $\frac{\gamma < 2(cd+\alpha)}{(cd-\alpha)} = \frac{2+4\alpha}{(cd-\alpha)}$ if $\alpha < cd$.

Combining these two conditions, we focus on the case of $\gamma < 2$ in this section. We now compute and construct the Routh table by using these coefficients of characteristic equation (Table 3.1).

According to the Routh–Hurwitz stability test, linearized system Equation 3.98 is stable if and only if all the values of the first column of the Routh table are positive, i.e.,

$$\begin{cases} R_0 > 0, \frac{\gamma p_0}{R_0} + 1 > 0 \\ \frac{2}{R_0} + \frac{\gamma(2p_0 - R_0)}{R_0^2} - \frac{R_0}{\frac{\gamma p_0}{R_0} + 1} \cdot \frac{2\gamma}{R_0^2} > 0 \\ \frac{2\gamma}{R_0^2} > 0. \end{cases} \quad (3.104)$$

Considering $R_0 = d + p_0$ and $p_0 = \alpha/c$, the inequalities in Equation 3.104 are equivalent to the following inequality on α

$$\frac{\gamma^2 + \gamma + 2}{c^2}\alpha^2 - \frac{\gamma^2 + 2\gamma - 4}{c} d \cdot \alpha + (2 - 3\gamma) d^2 > 0. \quad (3.105)$$

Table 3.1
Routh Table

a_3	a_1	0
a_2	a_0	0
$a_1 - a_3 a_0 / a_2$	0	0
a_0	0	0

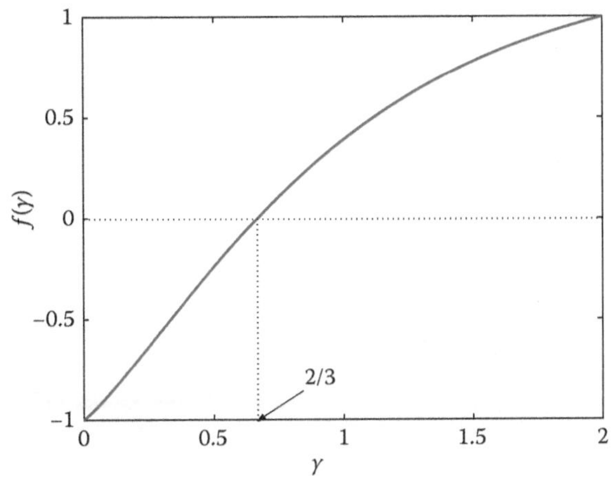

Figure 3.16 Relationship between $f(\gamma)$ and γ.

Solving Equation 3.105, one obtains

$$\alpha < \frac{(\gamma^2+2\gamma-4)-\gamma\sqrt{16\gamma+\gamma^2}}{2(\gamma^2+\gamma+2)} d \cdot c, \tag{3.106}$$

or

$$\alpha > \frac{(\gamma^2+2\gamma-4)+\gamma\sqrt{16\gamma+\gamma^2}}{2(\gamma^2+\gamma+2)} d \cdot c. \tag{3.107}$$

Let

$$f(\gamma) = \frac{(\gamma^2+2\gamma-4)+\gamma\sqrt{16\gamma+\gamma^2}}{2(\gamma^2+\gamma+2)}. \tag{3.108}$$

Solving $f(\gamma) > 0$, one then has $\gamma > 2/3$.

Figure 3.16 illustrates that $f(\gamma) < 1$ if $\gamma < 2$. Considering $\alpha > 0$, we thus obtain the following theorem.

Theorem 3.7.8

Given the network parameters c, d, and γ, FAST TCP described by Equation 3.98 is asymptotically stable in terms of the variable $\delta w(t)$ if

$$\alpha > f(\gamma) \cdot d \cdot c, \tag{3.109}$$

when $2/3 < \gamma < 2$. Here, $f(\gamma)$ is given by Equation 3.108. ■

Remark 1: In ns2 implementation of FAST TCP [325], it is intuitively suggested that on a path with capacity c, α/c must be five times the TCL variable *mi_threshold*, which defaults to 1.5 ms. Theorem 3.7.8 gives the condition $\alpha/c > f(\gamma) \cdot d$, which provides a theoretic support for this intuition. Further, it can be applied for different settings of source control gain γ, which has suggested the scalability of the result.

Remark 2: Unlike the existing stability conditions of FAST TCP [328–330], the result in Theorem 3.7.8 actually relates all the relevant parameters and decouples the key parameter α from all the other parameters (c, d, and γ). Therefore, this result is in a particular form that facilities parameter setting for FAST TCP.

Remark 3: Note that in the usual FAST TCP setting [172], the source control gain γ is set to be within $(0,1]$ and a default value of 0.5 is used. Theorem 3.7.8 also explains why FAST TCP has been stable for all experimental cases studied so far when $\gamma = 0.5$. Besides, the guideline on setting the parameter given by Theorem 3.7.8 allows a larger source control gain range.

3.7.7 SIMULATION VERIFICATION

We provide ns2 [321] simulations to show how the obtained stability condition can serve as a guideline on parameter setting in various network environments. Consider a single-link single-source network, with a link capacity of 1 Gbps and each packet size being 1000 bytes. The propagation delay is 50 ms. First, we assume the source control gain $\gamma = 0.8$. From Theorem 3.7.8, the stability range of α is calculated to be within $(1066, \infty)$. Setting $\alpha = 1250$, the simulation results are given in Figures 3.17 and 3.18, which show that both the congestion window and queue size are stable, and that they converge to their equilibrium points very fast. Next, we change the value of γ to 1.2. Then, the stability range of α is $(3564, \infty)$, obtained from condition 3.109. Setting $\alpha = 3750$, the simula-

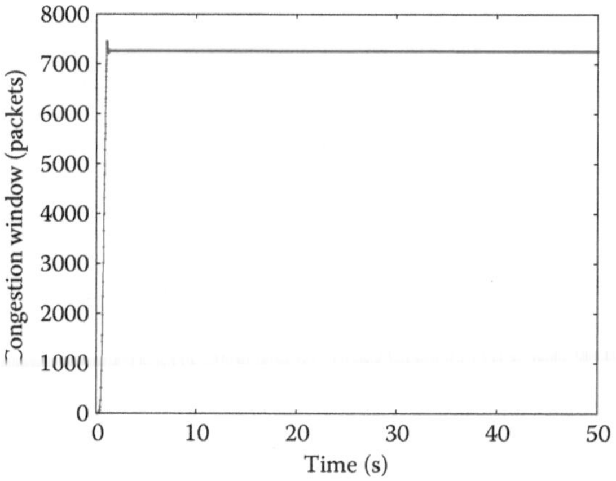

Figure 3.17 Congestion window when $\gamma = 0.8$ and $\alpha = 1250$.

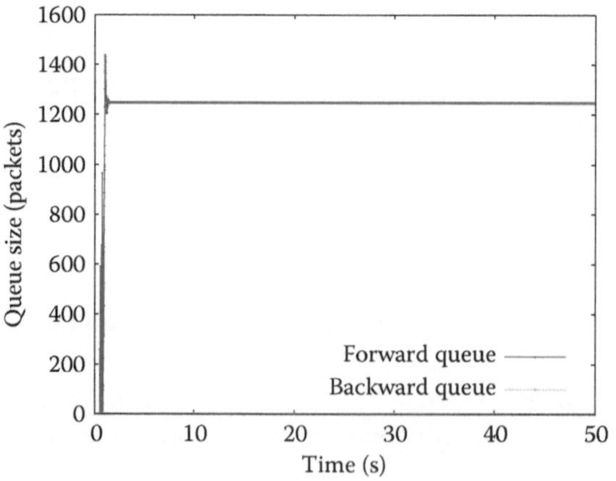

Figure 3.18 Queue size when $\gamma = 0.8$ and $\alpha = 1250$.

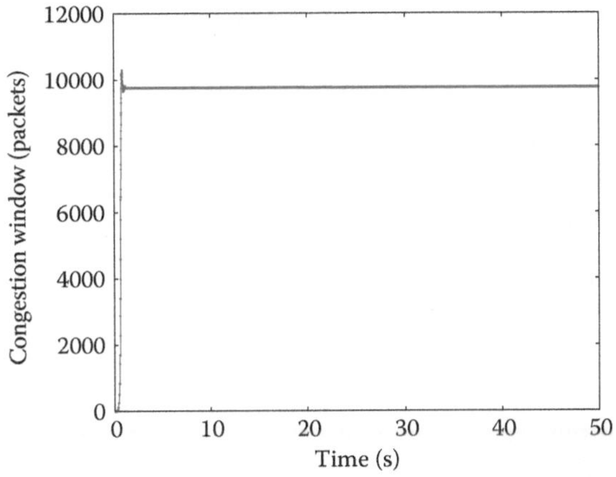

Figure 3.19 Congestion window when $\gamma = 1.2$ and $\alpha = 3750$.

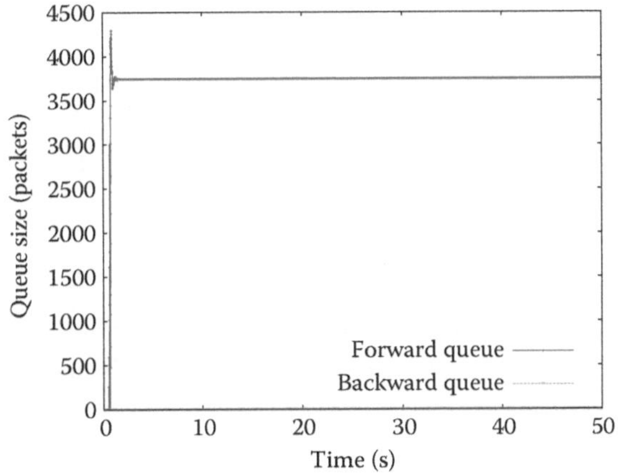

Figure 3.20 Queue size when $\gamma = 1.2$ and $\alpha = 3750$.

tion results can be plotted as in Figures 3.19 and 3.20. From Figures 3.19 and 3.20, one observes that the stability of FAST TCP can be achieved once again by setting appropriate value of α, even though the value of γ is greater than 1.

3.7.8 CONCLUDING REMARKS

In this section, we have analyzed the primal-dual algorithm from the multivariable time-delay control theory standpoint. Based on the characteristic equation of the multivariable control description, we have investigated the stability of the primal-dual congestion control system and obtained a sufficient and necessary stability condition. Furthermore, we have applied the stability condition to the parameter setting of FAST TCP protocol. Both theoretical analyses and simulation results have demonstrated that the stability condition can provide a guideline for choosing the parameters of FAST TCP to achieve satisfactory performance in communication networks.

Based on a continuous-time model of FAST TCP and by using control theory, we have obtained the sufficient stability condition for FAST TCP. The condition has given a guideline to select FAST TCP's parameter α. The stability condition can also be applied to environments with different source control gain settings of FAST TCP. We have validated the theoretical result by simulations.

3.8 STABILITY ANALYSES OF A PRIMAL-DUAL ALGORITHM: FAST TCP OVER RED

FAST TCP [172,334] is a new TCP congestion control protocol scheme for high-speed, long-distance networks. Unlike the additive increase multiplicative decrease (AIMD)-based TCP schemes, it uses queueing delay as a congestion indication [334], which allows FAST TCP to outperform its predecessors such as TCP Reno [292] in networks with large bandwidth-delay products.

As an important property, stability of FAST TCP has been studied in Refs. [324,329,330]. Extensive experiments on FAST TCP have also been conducted, and the results are promising [172]. In Ref. [172], local stability of FAST TCP in the absence of feedback delay was proved for the case of a single link.

Random early detection (RED) [178] is an AQM scheme [333] that complements TCP to improve its stability. RED is implemented in routers. It has been considered a challenge to optimize adaptively the parameters of RED (see [327,331,332], and references therein).

Given the importance of FAST TCP and RED and given the open question of optimizing RED parameters, it is natural that optimizing the operation of FAST TCP over RED (denoted from now on as FAST TCP/RED) is likely to become an important and challenging research topic.

In this section, we propose a new flow-level model for FAST TCP that efficiently reacts to packet loss, and then develop a closed-loop time-delayed feedback system for FAST TCP/RED that applies to the single-link, single-source case. By using the Routh–Hurwitz stability criteria [326], we derive a sufficient and necessary condition for stability in terms of the sending rate. Using this condition, we select the appropriate value of RED control parameter $\max_p$, and demonstrate, by ns2 simulations [321], significant stability improvement.

3.8.1 PRIMAL-DUAL MODEL: DEFINITIONS AND NOTATIONS

Consider a single-link, single-source network topology in which the source and destination node is connected through a single bottleneck link of capacity c. Let $q(t)$ be the queueing delay, and let $x(t)$ be the source sending rate.

Let τ^f denotes the forward delay between source and link and τ^b the backward delay in the feedback path from link to source. Then, let the source rate observed by the link at time t be $x(t-\tau^f)$, and let $q(t-\tau^b)$ be the queueing delay observed by the source. Let $w(t)$ be the source congestion window size at time t. The source sending rate $x(t)$ is then defined as follows:

$$x(t) = \frac{w(t)}{d+q(t-\tau^b)}, \tag{3.110}$$

where d denotes the constant propagation delay of the source, and the RTT is given by $R(t) = d+q(t)$. Here, we interpret $\tau^f + \tau^b = R_0$ as the equilibrium value of $R(t)$.

3.8.2 MODIFIED MODEL OF FAST TCP

We adopt the modified flow-level FAST TCP model and write it in the following continuous-time form:

$$\dot{w}(t) = \gamma\left(\alpha - x(t)q\left(t-\tau^b\right) - x(t)p\left(t-\tau^b\right)w(t)\right), \tag{3.111}$$

$$\dot{q}(t) = \frac{x\left(t-\tau^f\right)}{c-1}, \tag{3.112}$$

where $p(t)$ is the end-to-end packet loss probability, the constants α and γ are the congestion control parameters of FAST algorithm, and $\gamma \in (0,1]$. This flow-level FAST TCP is an improvement on its predecessor [169] as it considers packet loss.

3.8.3 CLOSED-LOOP FEEDBACK SYSTEM OF FAST TCP/RED

We consider the RED AQM [178] as the link algorithm. Under RED, the average queue size at the nth interval, denoted by $\text{avg}q_L(n)$, is defined as a measure of congestion of the network. When $\text{avg}q_L(n)$ is less than the minimum threshold min_{th}, no packets are dropped. When it exceeds the maximum threshold max_{th}, all incoming packets are dropped. When it is in between, a packet is dropped with a probability $p(n)$, an increasing function of $\text{avg}q_L(n)$. The dropping probability at the nth interval, $p(n)$, is therefore calculated as follows:

$$p(n) = \text{max}_p \times \frac{\text{avg}q_L(n) - \text{min}_{\text{th}}}{\text{max}_{\text{th}} - \text{min}_{\text{th}}}, \tag{3.113}$$

where max_p is the maximum value of $p(n)$. In short, we denote $k = \text{max}_p > 0$, $B = \text{max}_{\text{th}} - \text{min}_{\text{th}}$. Now, we consider the following continuous-time form of Equation 3.113:

$$p(t) = \frac{k \cdot (\text{avg}q_L(t) - \text{min}_{\text{th}})}{B}. \tag{3.114}$$

As the original FAST periodically updates its congestion window based on the average RTT and average queueing delay [172], using $q(t)$ to denote the average queueing delay at the tth RTT sample, $\text{avg}q_L(t) = c \cdot q(t)$. Therefore, the FAST TCP/RED closed-loop time-delayed feedback control system is described by

$$\begin{cases} \dot{w}(t) = \gamma\left(\alpha - x(t)q\left(t-\tau^b\right) - x(t)p\left(t-\tau^b\right)w(t)\right) \\ \dot{q}(t) = x\left(t-\tau^f\right)/c - 1 \\ p(t) = k \cdot (c \cdot q(t) - \text{min}_{\text{th}})/B. \end{cases} \tag{3.115}$$

3.8.4 STABILITY ANALYSIS

Having developed the closed-loop time-delayed feedback system of FAST TCP/RED, we will derive here, using the Routh–Hurwitz stability criteria [326], a sufficient and necessary condition for stability. By Equation 3.110, we have

$$w(t) = x(t)d + x(t)q\left(t-\tau^b\right). \tag{3.116}$$

By differentiating both sides of Equation (3.116), we obtain

$$\dot{w}(t) = d\dot{x}(t) + q\left(t-\tau^b\right)\dot{x}(t) + x(t)\dot{q}\left(t-\tau^b\right). \tag{3.117}$$

Substituting Equation 3.117 into 3.115, we reach the following updating equation of the source sending rate $x(t)$:

$$\begin{aligned} \dot{x}(t) = & \frac{\gamma\alpha}{d+q(t-\tau^b)} - \frac{\gamma x(t)q(t-\tau^b)}{d+q(t-\tau^b)} \\ & -\gamma x^2(t)p(t-\tau^b) - \frac{x(t)\dot{q}(t-\tau^b)}{d+q(t-\tau^b)}. \end{aligned} \tag{3.118}$$

Let

$$\dot{x}(t) = f\left(x(t), q\left(t-\tau^b\right), \dot{q}\left(t-\tau^b\right), p\left(t-\tau^b\right)\right),$$

and linearize it around the equilibrium point (x_0, q_0, p_0),

$$\begin{aligned} x(t) &= \delta x(t) + x_0, \\ q(t) &= \delta q(t) + q_0, \\ p(t) &= \delta p(t) + p_0, \end{aligned} \tag{3.119}$$

where the operating point (x_0, q_0, p_0) is defined by $\dot{w}(t) = 0$ and $\dot{q}(t) = 0$, given by

$$p_0 = \frac{\alpha - x_0 q_0}{x_0^2(d+q_0)},\ x_0 = c. \tag{3.120}$$

Then, to the first order,

$$\begin{aligned}\delta\dot{x}(t) = &\ \dot{x}(t) = \left.\frac{\partial f}{\partial x}\right|_* \delta x(t) + \left.\frac{\partial f}{\partial q}\right|_* \delta q(t-\tau^b) \\ &+ \left.\frac{\partial f}{\partial \dot{q}}\right|_* \delta\dot{q}(t-\tau^b) + \left.\frac{\partial f}{\partial p}\right|_* \delta p(t-\tau^b),\end{aligned} \tag{3.121}$$

where

$$\begin{aligned}\left.\frac{\partial f}{\partial x}\right|_* &= -\frac{\gamma q_0}{d+q_0} - 2\gamma x_0 p_0, \\ \left.\frac{\partial f}{\partial \dot{q}}\right|_* &= -\frac{x_0}{d+q_0}, \\ \left.\frac{\partial f}{\partial q}\right|_* &= -\frac{\gamma(\alpha + x_0 d)}{(d+q_0)^2}, \\ \left.\frac{\partial f}{\partial p}\right|_* &= -\gamma x_0^2.\end{aligned} \tag{3.122}$$

Considering the case $x_0 = c$, $d + q_0 = R_0$, we rewrite Equation 3.115 as follows:

$$\begin{cases}\delta\dot{x}(t) = -\gamma\left(\frac{q_0}{R_0} + 2cp_0\right)\delta x(t) - \frac{c}{R_0}\delta\dot{q}(t-\tau^b) \\ \qquad -\frac{\gamma(\alpha+cd)}{R_0^2}\delta q(t-\tau^b) - \gamma c^2 \delta p(t-\tau^b), \\ \delta\dot{q}(t) = \frac{\delta x\left(t-\tau^f\right)}{c}, \\ \delta p(t) = \frac{kc\cdot\delta q(t)}{B} + \frac{k(c\cdot q_0 - \min_{\text{th}})}{B-p_0}.\end{cases} \tag{3.123}$$

By taking the Laplace transformation of the above, we have

$$\begin{cases}s\delta x(s) = -\gamma\left(\frac{q_0}{R_0} + 2cp_0\right)\delta x(s) - \frac{c}{R_0}s\delta q(s)e^{-\tau^b s} \\ \qquad -\frac{\gamma(\alpha+cd)}{R_0^2}\delta q(s)e^{-\tau^b s} - \gamma c^2\delta p(s)e^{-\tau^b s} \\ \qquad + \delta x(0), \\ s\delta q(s) = \frac{\delta x(s)e^{-\tau^f s}}{c + \delta q(0)}, \\ \delta p(s) = \frac{kc}{B}\delta q(s) + \frac{1}{s}\left(\frac{k}{B}(cq_0 - min_{th}) - p_0\right),\end{cases} \tag{3.124}$$

where $\delta x(s)$, $\delta q(s)$, and $\delta p(s)$ denote the Laplace transform of $x(s)$, $q(s)$, and $p(s)$, respectively. By direct computing, one has

$$\begin{aligned}\Delta(s)\delta x(s) = &\ s\delta x(0) - \gamma c^2 e^{-\tau^b s}\left(\frac{k}{B}(cq_0 - \min_{\text{th}}) - p_0\right) \\ &- \left(\frac{\gamma(\alpha+cd)}{R_0^2} + \frac{c}{R_0}s + \gamma\frac{kc^3}{B}\right)e^{-\tau^b s}\delta q(0),\end{aligned} \tag{3.125}$$

where $\Delta(s)$ is the so-called characteristic polynomial given as follows:

$$\Delta(s) = s^2 + \left(\frac{\gamma q_0}{R_0} + 2\gamma c p_0 + \frac{1}{R_0} e^{-R_0 s}\right) s + \gamma\left(\frac{\alpha + cd}{R_0^2 c} + \frac{kc^2}{B}\right) e^{-R_0 s}. \quad (3.126)$$

Characteristic polynomial Equation 3.126 determines the stability (in terms of the source sending rate) of linearized closed-loop time-delayed FAST TCP/RED system (Equation 3.124). We use the Routh–Hurwitz stability criteria [326] to investigate its stability. To analyze the system stability, it is adequate to have the approximation: $e^{-R_0 s} = 1/(1+R_0 s)$ [326]. Let $\Delta(s) = 0$. The approximated characteristic equation of linearized FAST TCP/RED system (Equation 3.124) is then reached:

$$\Delta(s) = R_0 s^3 + \left(\frac{\gamma q_0 + 1}{R_0} + 2\gamma c p_0\right) R_0 s^2 + \left(\frac{\gamma q_0 + 1}{R_0} + 2\gamma c p_0\right) s + \frac{\gamma(\alpha + cd)}{R_0^2 c} + \gamma\frac{kc^2}{B} = 0. \quad (3.127)$$

We will now find a range of $\max_p$, where the linearized FAST TCP/RED system described by Equation 3.123 is asymptotically stable. Denoting

$$a_3 = R_0,$$

$$a_2 = \gamma q_0 + 2\gamma c p_0 R_0 + 1,$$

$$a_1 = \frac{(\gamma q_0 + 1)}{R_0} + 2\gamma c p_0,$$

$$a_0 = \frac{\gamma(\alpha + cd)}{(R_0^2 c)} + \frac{\gamma kc^2}{B},$$

the characteristic Equation 3.127 becomes

$$\Delta(s) = a_3 s^3 + a_2 s^2 + a_1 s + a_0 = 0.$$

We now compute and construct [326] the Routh table by using these coefficients of characteristic equation.

According to the Routh–Hurwitz stability test [326], linearized system (Equation 3.124) is stable if and only if all the values of the first column of the aforementioned Routh table are positive, i.e.,

$$\begin{cases} R_0 > 0,\ \gamma q_0 + 2\gamma c p_0 R_0 + 1 > 0, \\ \left(\dfrac{\gamma q_0 + 1}{R_0} + 2\gamma c p_0\right)^2 - \dfrac{\gamma(\alpha + cd)}{R_0^2 c} - \gamma\dfrac{kc^2}{B} > 0, \\ \dfrac{\gamma(\alpha + cd)}{R_0^2 c} + \gamma\dfrac{kc^2}{B} > 0. \end{cases} \quad (3.128)$$

Solving the inequalities in Equation (3.128) and considering $\max_p > 0$, one obtains

$$0 < \max_p < \frac{B}{\gamma c^2}\left[\left(\frac{\gamma q_0 + 1}{R_0} + 2\gamma c p_0\right)^2 - \frac{\gamma(\alpha + cd)}{R_0^2 c}\right]. \quad (3.129)$$

This stability condition relates the upper bound of $\max_p$'s stability range with both the RED and the FAST TCP parameters.

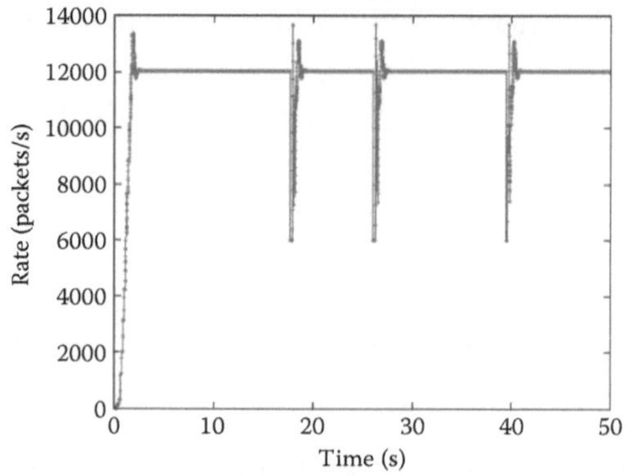

Figure 3.21 The sending rate of the stable FAST TCP/RED.

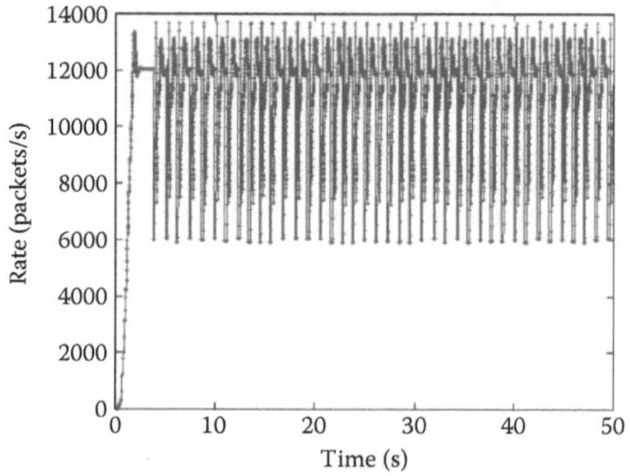

Figure 3.22 The sending rate of the unstable FAST TCP/RED.

3.8.5 SIMULATION RESULTS

Using the ns2 simulator [321], we now demonstrate that the choice of $\max_p$ based on Equation 3.129 can indeed yield satisfactory performance for a FAST TCP/RED system.

We again consider a single-source, single-bottleneck link of 100 Mb/s capacity and set each packet size to 1000 bytes. The source has a propagation delay of 100 ms with $\alpha = 200$ packets and $\gamma = 0.5$. We assume that $R_0 = 110$ ms, then we calculate $p_0 = 4.36 \times 10^{-6}$. The basic RED parameters are set to $\min_{th} = 125$ packets and $\max_{th} = 275$ packets. Then, $B = \max_{th} - \min_{th} = 150$ packets. For this data, the stability range is calculated by Equation (3.129) to be $0 < \max_p < 1.53 \times 10^{-4}$. We consider two values: $\max_p = 2 \times 10^{-5}$ and $\max_p = 1 \times 10^{-3}$. The former value is within the stability range, while the latter is not. Figures 3.21 and 3.22 illustrate the sending rate for this data and the two values of $\max_p$. We observe that when the control gain $\max_p$ is chosen within the stability range, the FAST TCP/RED system is stable, while as shown in Figure 3.22, there are many oscillations for the sending rate that the FAST TCP/RED system becomes unstable when the value of $\max_p$ is out of the stability range.

3.8.6 CONCLUSIONS

Based on a modified flow-level model of FAST TCP, we have developed a model of a closed-loop system of FAST TCP/RED for a single-link, single-source network. Using feedback control theory, we have obtained the sufficient and necessary stability condition for our FAST TCP/RED system, in terms of sending rate. The condition has led to appropriate choices of $\max_p$ value, and we have demonstrated, by simulation, the performance benefit achieved by our choice of the $\max_p$ value.

4 FAST TCP and Extensions

This chapter presents the following:

- A novel virtual price FAST TCP congestion control approach termed REM-FAST;
- A new algorithm termed as Generalized FAST TCP;
- A new FAST TCP flow-level model, leading to a new FAST TCP variant to deal with packet loss.

4.1 NOVEL VIRTUAL PRICE FAST TCP CONGESTION CONTROL APPROACH

The major objectives of congestion control algorithms in Internet include high link utilization, good fairness in bandwidth allocation, low packet loss, low queuing delay, and freedom from oscillation. In recent years, many congestion control schemes have been proposed to achieve these goals. However, none of them achieve these goals perfectly. FAST TCP [172,334,642–644] is a new TCP protocol proposed for high-speed networks, which is promising in terms of system stability, throughput. However, there are still challenges to face in the design of its technology to achieve satisfactory network performance.

One challenge comes from the so-called persistent congestion [174,352]. Just like Vegas, FAST TCP also produces persistent congestion because they rely on queuing delay as a congestion measurement, which makes backlog indispensable in conveying congestion to the sources. The persistent congestion will impair the fairness of FAST TCP and introduce extra backlog in the network.

In a general prim-dual network system, the link price is produced only when the aggregate input rate is larger than the link capacity, and the sum of the links' price will increase as the number of the "demand" increases. If this usual link pricing method is used, FAST TCP makes the backlog to increase linearly with the number of flows resulting in large backlog in the network which will affect the network performance of delay-sensitive applications such as online game. The second challenge is, therefore, how to choose the proper link pricing method to corporate with FAST TCP to achieve good performance.

The third challenge is how to design the window-updating equation of FAST TCP such that its equilibrium bandwidth allocations reach good fairness. Usually, in a multi-bottleneck link network, the long FAST TCP flows that pass three or more bottleneck links will suffer very low bandwidth allocation due to the limited fairness of the current FAST TCP.

There are two related works that motivate our present approach: REM-Vegas [174] provides a mechanism that conveys the sum of link price to source by using ECN marking. Under this scheme, it is possible to eliminate the backlog in the network and thus to avoid persistent congestion. Mo and Walrand propose the concept of (α, n)-proportional fairness [115] (note that we have also termed it as (p, β)-proportional fairness in later chapters to avoid notation confusion), which allows the bandwidth allocation to be close to the max–min fairness allocation by setting the parameters.

In this section, we propose a novel congestion control scheme termed REM-FAST. Based on the REM pricing mechanism, this scheme still works under the framework of FAST TCP. In REM-FAST, we suggest

- The link algorithm uses a virtual price method to develop the link price to avoid large backlog in the network;
- The aggregate link price be used in the source window updating as congestion indication;
- The source algorithm adopts a modified FAST TCP window-updating mechanism, at which equilibrium flows satisfy (α, n)-proportional fairness.

We validate this new scheme by ns2 simulations [321]. Simulation results confirm that in both single-bottleneck and multi-bottleneck link networks, good fairness is achieved and backlog is reduced.

4.1.1 LINK ALGORITHM AND SOURCE ALGORITHM

A general network can be modeled as a set $L = \{1, \ldots, M\}$ of links, shared by a set $I = \{1, \ldots, N\}$ of flows. Each link $l \in L$ has capacity c_l. Flow $i \in I$ traveled a router L_i consisting of a subset of links, i.e., $L_i = \{l | i \text{ traverses } l\}$. A link l is shared by a subset I_l of flows where $I_l = \{i \in I | i \text{ traverses } l\}$. Let x_i be the rate of flow i, and let $x = \{x_i, i \in I\}$ be the rate vector. Let $A = \{A_{li}, i \in I, l \in L\}$ be the routing matrix, where $A_{li} = 1$ if flow i traverses link l, and 0 otherwise. Throughout this section, the terms "flow" and "sources" are used synonymously. We next propose link and source algorithms under the above model.

4.1.1.1 Link Algorithm: Virtual Pricing Approach

In prim-dual algorithms, the source is supposed to know the sum of the passed links price, which is usually in the form of queuing delay or packet loss. However, there are some disadvantages for these two forms. By using packet loss as link price, the source only gets a two-bit congestion information and cannot receive the accurate sum of each links price because one packet can be lost only once. If queuing delay is used as the link price, the source can easily obtain the sum of link prices because the total queuing delay is equal to the sum of the queuing delay produced in each passed link. Unfortunately, queuing delay indication will result in persistent congestion and impair the performance of network. REM [1] and virtual queue (VQ) [153] support a mechanism that can avoid over-large queuing delay, and in this section, we implement it in link algorithms.

Usually, the link price p_l is updated according to [112]

$$p_l(t+1) = [p_l(t) + \gamma(x^l(t) - c_l)]^+, \tag{4.1}$$

where $\gamma > 0$ is a small constant, and if $\gamma = 1/c_l$, then p_l can be recognized as queuing delay, $x_l(t)$ is the aggregate input rate at time t, and c_l is the link capacity. Similarly, we propose the following link price-updating equation:

$$p_l^v(t+1) = [p_l^v(t) + \gamma(x^l(t) - c_l^*)]^+, \tag{4.2}$$

where c_l^* is the virtual capacity which is less than the link capacity c_l (in general, $c_l^* = 0.95 * c_l$). We call the price produced by real capacity Equation 4.1 as real capacity price, and the price produced by virtual capacity Equation 4.2 as virtual capacity price, abbreviated to RC price and VC price, respectively.

Consider that the equilibrium point of $x^l(t)$ is equal to c_l^* in (4.2). This implies that the input rate is less than the link capacity c_l in equilibrium. Therefore, the backlog will be drained more quickly by using the virtual price updating.

In order to convey the value of p_l^v to source, link l marks each packet with a probability $m_l(t)$ according to

$$m_l(t) = 1 - \phi^{-p_l^v(t)}, \tag{4.3}$$

where ϕ is a constant. Once a packet is marked, its mark is carried to the destination and then conveyed back to the source via acknowledgment. The exponential form is crucial for multi-bottleneck links, because after traversing a set $L(s)$ of links, the end-to-end marked probability is

$$m^s(t) = 1 - \prod_{l \in L(s)} (1 - m_l(t)) = 1 - \phi^{-p_s^v(t)}, \tag{4.4}$$

where

$$p_s^v(t) = \sum_{l \in L(s)} p_l^v(t)$$

is the total virtual price of the link set $L(s)$. The source calculates it by using Equation 4.4.

4.1.1.2 Source Algorithm: Modified Form of FAST TCP

The source algorithm can determine the fairness of congestion control scheme using its utility function. For example, FAST TCP can achieve the so-called proportional fairness by using $\log x$ as its utility function. However, this kind of fairness leads to the "long flow" suffers very low bandwidth allocation. As a generalization of proportional fairness and max–min fairness, the definition of (α, n)-proportional fairness is given by Mo and Walrand [115], and is described as follows. A rate vector x^* is (α, n)-proportionally fair if it is feasible, and for any other feasible vector x,

$$\sum_{i \in I_l} \alpha_i \frac{x_i - x_i^*}{(x_i^*)^n} \leq 0,$$

where α_i is positive numbers, for $i \in I$. When $n = 1$, it is proportional fairness; when n becomes large, the (α, n)-proportional fair rate vector approaches the max–min fair rate vector. Achieving (α, n)-proportional fairness corresponds to maximizing the sum of users' utilities of the form [115]:

$$U_i(x_i) = \begin{cases} \alpha_i \log x_i, & n = 1, \\ \alpha_i (1-n)^{-1} x_i^{1-n}, & \text{otherwise.} \end{cases}$$

From [105], the resulted optimal rates satisfy

$$x_i^* = \frac{\alpha_i^{1/n}}{(p_i^*)^{1/n}}, \tag{4.5}$$

where p_i is the link price.

REM-FAST achieves (α, n)-proportional fairness by using the following window update equation:

$$w_i(t+1) = w_i(t) + \gamma_i(\alpha_i^{1/n} - x_i(t)(p_s^v(t))^{1/n}), \tag{4.6}$$

where $x_i(t)$ is equal to $w_i(t)/RTT$, and $p_s^v(t)$ can be calculated by the following equation:

$$p_s^v(t) = -\log_\phi(1 - m^s(t)), \tag{4.7}$$

where ϕ is the parameter given in the link algorithm. $m^s(t)$ is the end-to-end marking probability and is updated periodically by source. In each period, the source counts the number of received ECN message N_e and the number of all the ACK packets N, and then it estimates $m^s(t)$ by the fraction $m^s(t) = N_e/N$. It is noted that window update Equation 4.6 is still in the spirit of the usual FAST TCP, but at equilibrium, its rate achieves the (α, n)-proportional fairness. Such achievement helps to mitigate the bias against long flows by adjusting the fairness parameters. This will be verified by simulations.

4.1.2 SIMULATION RESULTS

We design two simulations to show two main advantages of the scheme REM-FAST: small backlog and fair bandwidth allocation in multi-bottleneck network. In the simulations, the values of parameters are set to be $\phi = 1.15$, $\alpha_i^{1/n} = 100$, and $n = 3$. Figures 4.1 and 4.2 show the simulation scenarios we used.

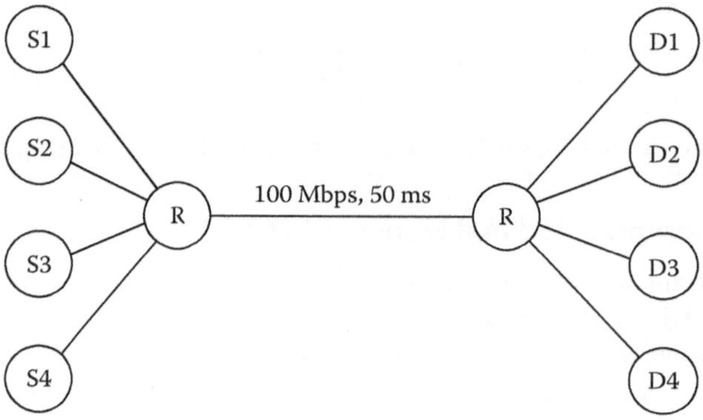

Figure 4.1 Simulation model 1: a single-bottleneck link network.

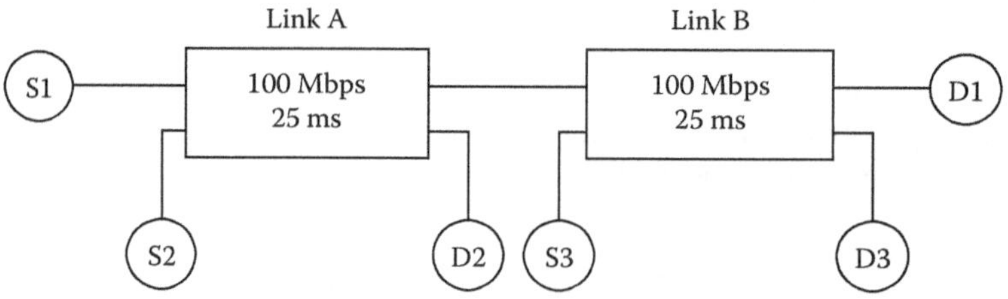

Figure 4.2 Simulation model 2: a linear network with two bottleneck links.

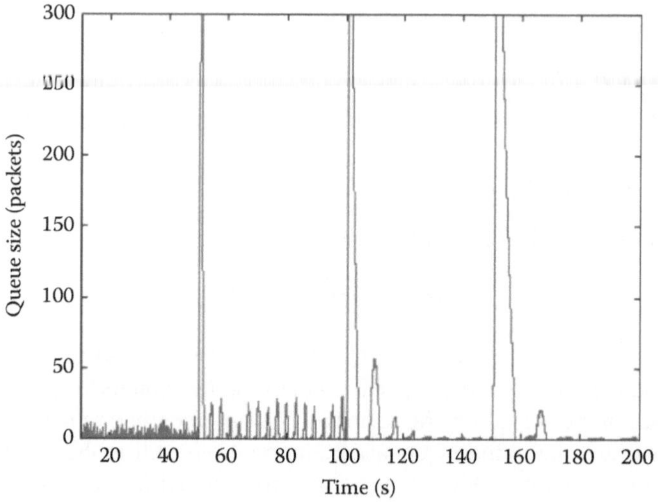

Figure 4.3 The sending rate for single-bottleneck link.

In the first simulation, there are four send–receiver pairs, and the time scheme for these four flows to be active is set as follows: S1–D1 (0–100 s), S2–D2 (25–100 s), S3–D3 (50–100 s), and S4–D4 (75–100 s), and the capacity of bottleneck link is 100 Mbps. Figure 4.3 plots the source sending rates under REM-FAST. In this figure, we can see that sources share the bandwidth fairly. Figure 4.4 plots the queuing delay of the link for the case of updating link price by Equation 4.2 in

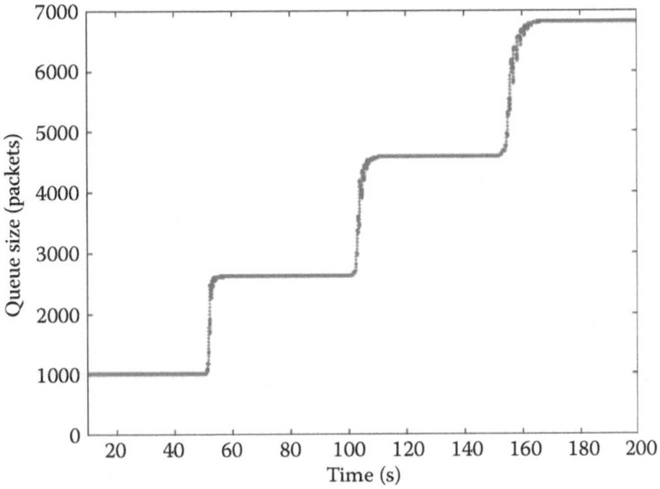

Figure 4.4 The queuing delay with VC price.

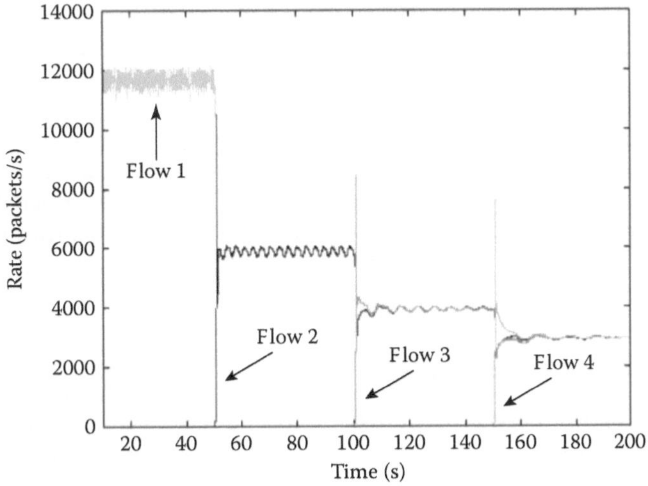

Figure 4.5 The queuing delay with RC price.

a virtual pricing manner. As we expected, the buffer occupancy is close to zero. Figure 4.5 plots the queuing delay of the link for the case of updating the link price by Equation 4.1 in a real pricing manner. In the real pricing case, as shown in Figure 4.5, the backlog is linearly increasing with the number of flow. However, such increasing tendency is not displayed in Figure 4.4, which implies that the deficiency of persistent congestion is remedied by using the virtual pricing mechanism.

The second simulation model is a simple linear network with two bottleneck links: one long flow (Flow 1) passes links A and B, and two short flows (Flows 2 and 3) pass links A and B separately. The time scheme for these three flows to be active is S1–D1 (0–100 s), S2–D2 (25–100 s), and S3–D3 (50–100 s). In the usual FAST TCP, the long flow, Flow 1, will suffer from low rate allocation as the short flows, Flows 2 and 3, join one by one. However, within the new scheme REM FAST, as

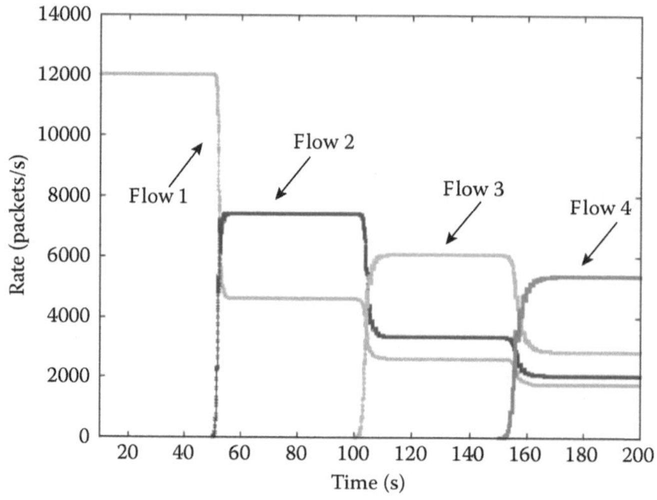

Figure 4.6 The sending rate for multi-bottleneck link.

suggested in Figure 4.6, the bandwidth allocation for Flow 1 is very close to that for Flows 2 and 3. This simulation result shows that REM FAST improves the fairness of the network.

4.1.3 CONCLUDING REMARKS

In this section, we propose a new congestion control scheme REM FAST, which is based on VC price and the general proportional fairness concepts. We design the link algorithm to generate a virtual price and a modified FAST TCP window-updating equation that uses the aggregate virtual price as the congestion indication. Simulation results are provided to show two main advantages of REM FAST: small backlog and fair bandwidth allocation.

4.2 GENERALIZED FAST TCP SCHEME

FAST TCP has been shown to be promising in terms of system stability, throughput, and fairness. However, it requires buffering that increases linearly with the number of flows bottlenecked at a link.

We propose herein a new TCP algorithm that extends FAST TCP to achieve (α, n)-proportional fairness in steady state, yielding buffer requirements that grow as the nth power of the number of flows. We call the new algorithm Generalized FAST TCP. We then provide a stability analysis and prove stability for the case of a single-bottleneck link with homogeneous sources in the absence of feedback delay. Simulation results verify that the new scheme is stable and that its buffering requirements can be made to scale significantly better than standard FAST TCP.

All congestion control algorithms can be understood as algorithms to solve an optimization problem, in which the network seeks to maximize the sum of the users' "utilities" subject to link capacity constraints. A user's utility is the benefit it derives from transmitting at a given rate. The equilibrium rates are determined by the objective of the optimization, while the dynamics are determined by the optimization procedure. In this framework, users pay a "price" for transmitting data on a congested link, typically either in terms of loss or in terms of queuing delay, and the equilibrium value of this price depends on the users' utility functions. As both of these price mechanisms impact adversely on users, it is desirable to use a utility function, which both achieves a fair rate allocation and imposes low (and fair) prices on users. We suggest to adapt the dynamics of FAST [172] to allow it to optimize a more general form of utility function. This allows a trade-off to be made between fairness and low queuing delay.

Unlike AIMD-based TCP schemes, FAST TCP uses queuing delay as the congestion indication, or price. Users' utilities are logarithmic, making the solution to the optimization problem satisfies the proportional fairness criteria [105]. If all users use FAST, the unique equilibrium rate vector is the unique solution of the utility maximization problem. One drawback of this approach is that the queuing delay (and hence buffer requirements) at a node increases in proportion to the number of flows bottlenecked there.

To allow a trade-off between fairness and network utilization, Mo and Walrand [115] proposed the concept of (α, n)-proportional fairness, which generalizes max–min fairness [139], proportional fairness [105], and minimum potential delay [278]. This corresponds to a simple family of power-law utility functions. We propose an extended version of FAST TCP, termed *Generalized FAST TCP*, in which equilibrium rates are (α, n)-proportional fair. This is achieved by making a slight change to the window update equation, which implicitly optimizes a suitable utility function. As well as allowing increased fairness, corresponding to $n > 1$, Generalized FAST TCP allows the queuing delay to be reduced at nodes carrying many flows by setting $n < 1$.

Our proposed scheme is a generalization of the existing FAST TCP [172]. It is a supplement of FAST TCP by adding a new parameter n. Specifically, by setting $n = 1$, our Generalized FAST TCP scheme is reduced to the existing FAST TCP. We will show that the new scheme inherits the merits of the current FAST TCP regarding stability and throughput for any value of n and not just for $n = 1$. We also provide stability analysis and prove that Generalized FAST TCP achieves (α, n)-proportional fairness.

4.2.1 GENERALIZED FAST TCP

A general network can be described as a set $L = \{1, \ldots, M\}$ of links shared by a set $I = \{1, \ldots, N\}$ of flows. Each link $l \in L$ has capacity c_l. Flow $i \in I$ traveled a router L_i consisting of a subset of links, i.e., $L_i = \{l \in L \,|\, i \text{ traverses } l\}$. A link l is shared by a subset I_l of flows where $I_l = \{i \in I \,|\, i \text{ traverses } l\}$. Let x_i be the rate of flow i, and $x = \{x_i, i \in I\}$ be the rate vector. Let $A = (A_{li}, i \in I, l \in L)$ be the routing matrix, where $A_{li} = 1$ if flow i traverses link l, and 0 otherwise. Throughout this section, the terms "flow", "sources" and "users" are used synonymously.

A rate vector $x \geq 0$ is called *feasible* if

$$\sum_{i \in I_l} x_i \leq c_l, \ \forall l \in L. \tag{4.8}$$

A rate vector x^* is α_i-weighted proportional fair if it is feasible, and for any other feasible vector x_i, the aggregate of proportional change is non-positive:

$$\sum_{i \in I} \alpha_i \frac{x_i - x_i^*}{x_i^*} \leq 0, \tag{4.9}$$

where α_i is positive numbers, $i = 1, 2, \ldots$.

Consider the following optimization problem (**P**):

$$\max_{x \geq 0} \sum_{i \in I} U_i(x_i), \tag{4.10}$$

subject to the constraint given by Equation 4.8, where U_i is the utility function of user i. We follow the standard approach of taking the Lagrangian

$$L(x; p) = \sum_i (U_i(x_i) - x_i q_i) - \sum_l p_l c_l, \tag{4.11}$$

where p_l, called the *price* of link l, is the Lagrange multiplier of the constraint due to the capacity of link l. We assume that

$$q_i(t) = \sum_{l=1}^{M} A_{li} p_l(t - \tau_{li}^b) \tag{4.12}$$

is the aggregate price observed by source i in its path, and link l observes the aggregate source rate

$$y_l(t) = \sum_{i=1}^{N} A_{li} x_i(t - \tau_{li}^f), \tag{4.13}$$

where τ_{li}^f is the forward feedback delay from source i to link l, and τ_{li}^b is the backward feedback delay from link l to source i. For simplicity, we assume that the feedback delays τ_{li}^f and τ_{li}^b are constants.

For given link prices, each source i determines its optimal rate as

$$x_i(p) = \arg\max_{x_i} U_i(x_i) - x_i q_i = (U_i')^{-1}(q_i). \tag{4.14}$$

The primal optimization (**P**) can then be replaced by its dual (**D**) given by

$$\min_{p \geq 0} \sum_i (U_i(x_i(p)) - q_i x_i(p)) + \sum_l c_l p_l. \tag{4.15}$$

According to [105], α_i-weighted proportional fairness is achieved within a system of social welfare maximization if all users have utility functions of the following form:

$$f_i(x_i) = \alpha_i \log x_i. \tag{4.16}$$

That is, a α_i-weighted proportional fair vector solves the above optimization problem (**P**) by maximizing the sum of all the logarithmic utility functions. In this case, Equation 5.20 becomes

$$x_i = \frac{\alpha_i}{q_i}. \tag{4.17}$$

For the existing version of FAST TCP, it is known [172] that the following source window-updating equation has a unique equilibrium point (x_i^*, q_i^*) satisfying Equation 4.17:

$$w_i(t+1) = \gamma \left(\frac{d_i w_i(t)}{d_i + q_i(t)} + \alpha_i(w_i(t), q_i(t)) \right) + (1-\gamma) w_i(t), \tag{4.18}$$

where

$$\alpha_i(w_i, q_i) = \begin{cases} \alpha_i w_i & \text{if } q_i = 0 \\ \alpha_i & \text{otherwise.} \end{cases}$$

Since this equilibrium point is known ([172], Theorem 1) to be the unique optimal solution of the above problem (**P**) with the specific utility functions given by Equation 4.16, FAST TCP maximizes the sum of logarithmic utility functions. This implies in particular that the current FAST TCP achieves α_i-weighted proportional fairness. Note that the fairness parameter α_i is also the number of flow i's packets that are buffered in the routers in its path at equilibrium. If there are N flows, the total number of packets buffered in the routers at equilibrium is $\sum_{i=1}^{N} \alpha_i$ (see [334]). From this, it is seen that the buffer occupancy increases linearly with the number of flows.

As a generalization of proportional fairness and max–min fairness, the definition of (α, n)-proportional fairness is given by Mo and Walrand [115], which is described in the following. Note that our notation differs slightly from that of [115], so that it corresponds to its usual meaning in the FAST algorithm. A rate vector x^* is (α, n)-proportionally fair, if it is feasible, and for any other feasible vector x,

$$\sum_{i \in I_l} \alpha_i \frac{x_i - x_i^*}{(x_i^*)^n} \leq 0, \tag{4.19}$$

where α_i are positive numbers, for $i \in I$. Note that Equation 4.19 reduces to Equation 4.9 when $n = 1$. It is also seen that when n becomes large, the (α, n)-proportional fair rate vector approaches the max–min fair rate vector. Achieving (α, n)-proportional fairness corresponds to maximizing the sum of users' utilities of the following form [115]:

$$U_i(x_i) = \begin{cases} \alpha_i \log x_i; & n = 1 \\ \alpha_i (1-n)^{-1} x_i^{1-n}; & \text{otherwise.} \end{cases} \tag{4.20}$$

Thus, from Equation 4.14, the optimal rates satisfy

$$x_i^* = \frac{\alpha_i^{1/n}}{(q_i^*)^{1/n}}. \tag{4.21}$$

Generalized FAST TCP seeks to achieve (α, n)-proportional fairness. This is achieved by modifying the window update equation as

$$w_i(t+1) = w_i(t) + \gamma_i \left(\alpha_i^{1/n} - \frac{(q_i(t))^{1/n}}{d_i + q_i(t)} w_i(t) \right), \tag{4.22}$$

where $\gamma_i \in (0, 1]$. It can easily be seen that the equilibrium point (x_i^*, q_i^*) of Equation 4.22 exactly satisfies (4.21).

In fact, most of the parameters of Generalized FAST TCP have already existed in the original FAST TCP; hence, we do not need to modify the original codes significantly. $\alpha_i^{1/n}$ is a constant, and all we need to do is to choose the power of $q_i(t)$. The queuing delay $q_i(t)$ can be calculated by *RTT-baseRTT*. The FAST algorithm is usually expressed in terms of an estimate, denoted *baseRTT*, of the pure propagation delay of flow i, d_i, and of the measured round-trip time $d_i + q_i$, denoted by *RTT*. Using that notation and the fact that

$$x_i(t) = \frac{w_i(t)}{(d_i + q_i(t))}, \tag{4.23}$$

the window update rule for Generalized FAST can be written as the pseudo-code

$$w \leftarrow w + \gamma \left(\alpha^{1/n} - \frac{w}{RTT} (RTT - baseRTT)^{1/n} \right),$$

where $\gamma \in (0, 1]$. Note that the equilibrium window size, w_i^*, queuing delay, q_i^* and rate, x_i^* of source i are associated by $x_i^* = w_i^* / (d_i + q_i^*)$, where d_i is the propagation delay that flow i experienced. Analogously to the analysis of FAST [172], by using the notion of (α, n)-proportional fairness, it is straightforward to prove the following theorem.

Theorem 4.2.1

Consider the case that the routing matrix, A, has full row rank and given identities (4.12) and (4.23), the unique equilibrium point (x^*, q^*) of the system of window-updating equations (4.22) is such that the rate vector $x^* = (x_1^*, \ldots, x_N^*)^T$ is the unique maximizer of the problem P with the utility function given by (4.20) and the queuing delay vector $q^* = (q_1^*, \ldots, q_N^*)$ is such that $p^* = (p_1^*, \ldots, p_M^*)$ is the unique minimizer of (4.15). ■

The above theorem implies in particular that the equilibrium rate vector determined by the new Generalized FAST TCP proposal achieves (α, n)-proportional fairness.

Now, we study the queue size at equilibrium. According to Equation 4.21,

$$q_i^* = \frac{\alpha_i}{(x_i^*)^n},$$

the backlog of each source kept in the links is

$$b_i = x_i^* \times q_i^* = \frac{\alpha_i}{(x_i^*)^{n-1}}. \tag{4.24}$$

Under the new Generalized FAST TCP scheme, we consider a dumbbell network with N flows sharing one bottleneck link with capacity C, the total buffer occupancy at equilibrium denoted by B is

$$B = \sum_{i=1}^{N} b_i = \sum_{i=1}^{N} \frac{\alpha_i}{(x_i^*)^{n-1}}. \tag{4.25}$$

By setting the same value of $\alpha^{1/n}$ for each flow and noting $x^* = C/N$, we can rewrite Equation 4.24 as follows:

$$B = \frac{N^n}{C^{n-1}} \cdot \left(\alpha^{\frac{1}{n}}\right)^n. \tag{4.26}$$

When $n = 1$, the summation of Equation 4.25 is the total number of packets buffered in the router at equilibrium under the original FAST TCP scheme. However, this can be reduced by altering n, which will be discussed further later on.

First, let us fix the value of $\alpha_i^{1/n}$ so as to study the relationship between the buffer occupancy and the number of flows under Generalized FAST TCP. We illustrate this relationship in Figure 4.7, by setting $\alpha_i^{1/n} = 100$ and the router capacity to be 2500 packet/s in a dumbbell network. In the figure, we plot four curves of buffer occupancy changing with the number of flows, which is corresponding to $n = 1/3, 1/2, 1, 2$.

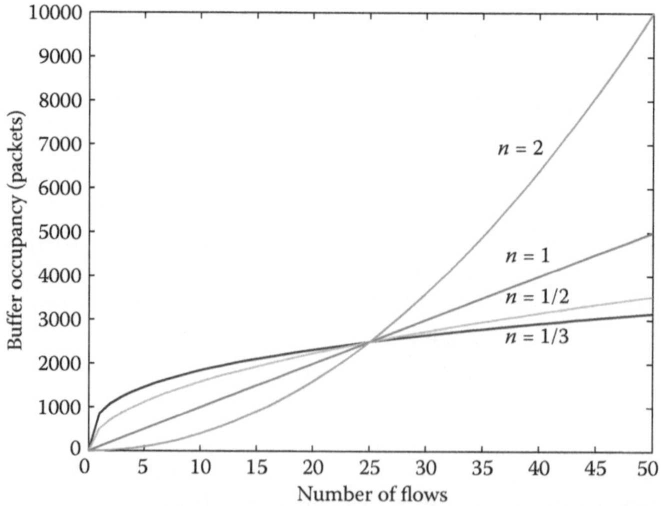

Figure 4.7 The relationship between the buffer occupancy and the number of flows with $\alpha_i^{1/n} = 100$.

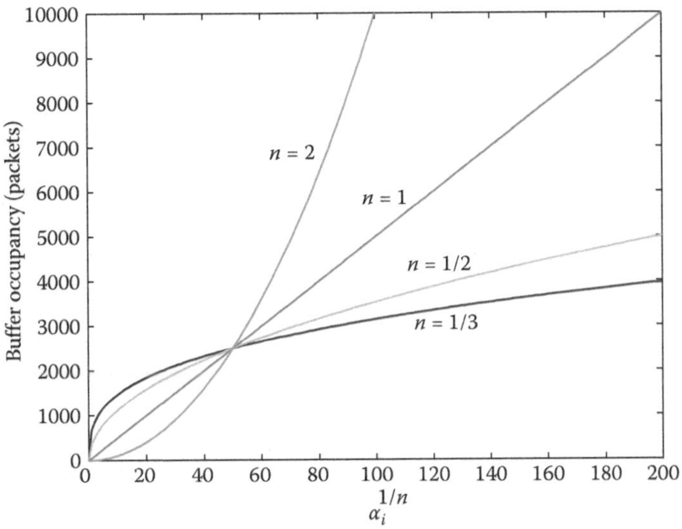

Figure 4.8 The relationship between the buffer occupancy and parameter $\alpha_i^{1/n}$ with the number of flows being 50.

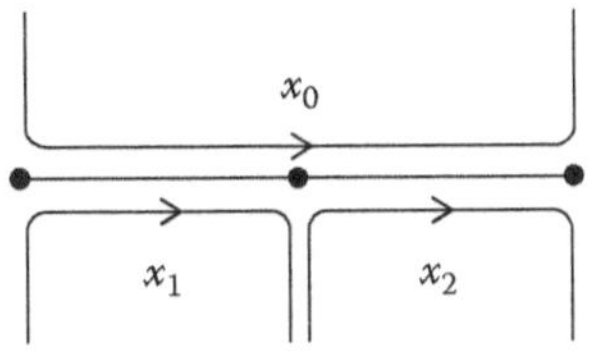

Figure 4.9 A linear network.

Observing Figure 4.7, one finds that the smaller the n values, the slower the increase in buffer occupancy as the number of connection increases.

Next, let us fix the number of flows so as to study the relationship between the buffer occupancy and parameter $\alpha_i^{1/n}$. We illustrate this relationship in Figure 4.8, by setting the number of flows as 50 and the router capacity as 2500 packet/s in a dumbbell network. In the figure, we plot four curves of buffer occupancy changing with the parameter $\alpha_i^{1/n}$, which is corresponding to $n = 1/3, 1/2, 1, 2$.

That is to say, buffer occupancy is related with the value of n and is affected by the parameter $\alpha^{1/n}$. However, a certain threshold of $\alpha^{1/n}$ value is necessary for keeping the stability of network [334]. We cannot reduce the buffer occupancy by setting $\alpha^{1/n}$ to 1. Too large value will lead to large buffer occupancy, and too small value will result in network instability. Thus far, there is no clear rule for choosing a reasonable value for the parameter $\alpha^{1/n}$.

Although small n can alleviate the increase in buffer occupancy, it also results in unfairness in a multi-bottleneck network. We study the relationship between bandwidth allocation and the value of n using a network with the simplest multi-bottleneck topology as shown in Figure 4.9. We assume that both multi-bottleneck link capacities are 1. We plot the (α,n) fair bandwidth allocation for this network into Figure 4.10.

As shown in Figure 4.10, as n increases, the bandwidth allocation converges to max–min fairness; when $n = 1$, it satisfies the so-called proportional fairness; and as n approaches zero, it becomes the maximum throughput allocation. We will revisit this issue in Section 4.5 and provide a more detailed quantitative discussion on the relationship between the fairness and value of n.

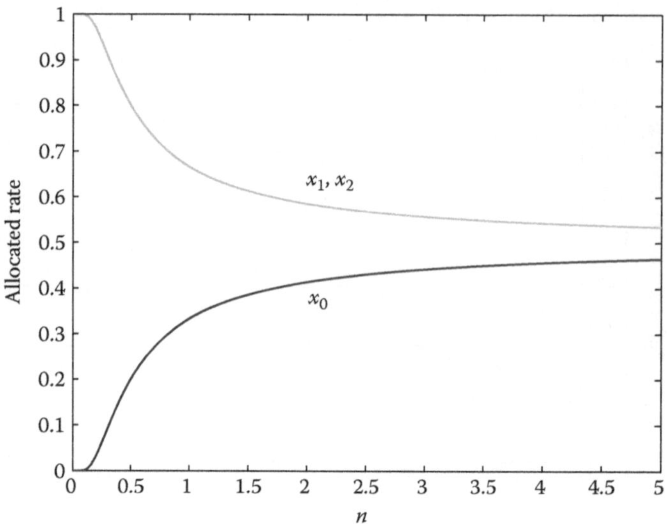

Figure 4.10 Bandwidth allocation of a linear network.

4.2.2 STABILITY ANALYSES

We now analyze the stability of the Generalized FAST TCP under the dumbbell (single-bottleneck) topology with N greedy sources, and with the link capacity of c. Following [318], we obtain the continuous form of Equation 4.22, which is given by

$$\dot{w}_i(t) = \gamma\left(\frac{\alpha_i^{1/n}}{d_i + q_i(t)} - \frac{(q_i(t))^{1/n}}{d_i + q_i(t)}x_i(t)\right), \; i = 1,\ldots,N. \tag{4.27}$$

The round-trip time of flow i is given by $R_i(t) = d_i + q_i(t)$, and in steady state, we have $R_i^* = d_i + q_i^*$, where R_i^* can be interpreted as the sum of τ_{li}^f and τ_{li}^b. From $w_i(t) = x_i(t)R_i(t)$, we have

$$\dot{w}_i(t) = \dot{x}_i(t)R_i(t) + x_i(t)\dot{q}(t).$$

Therefore, we have

$$\dot{x}_i(t) = \frac{\gamma\alpha_i^{1/n}}{R_i^2(t)} - \frac{\gamma x_i(t)(q_i(t))^{1/n}}{R_i^2(t)} - \frac{x_i(t)\dot{q}_i(t)}{R_i(t)}. \tag{4.28}$$

We now linearize the above equation around the equilibrium point

$$\begin{cases} x_i(t) = x_i^* + \delta x_i(t) \\ q_i(t) = q_i^* + \delta q_i(t). \end{cases}$$

Thus, the linearized system is given by

$$\delta\dot{x}_i(t) = -\frac{\gamma(q_i^*)^{1/n}}{R_i^{*2}}\delta x_i(t) - \frac{x_i^*}{R_i^*}\delta\dot{q}_i(t) - \frac{\gamma x_i^*}{nR_i^{*2}}(q_i^*)^{1/n-1}\delta q_i(t). \tag{4.29}$$

For the network with a single-bottleneck link, the queuing delay is modeled with

$$\dot{p}(t) = \frac{1}{c}(y(t) - c), \tag{4.30}$$

and the equilibrium points (y^*, p^*) satisfy $y^* = c$. Linearize Equation 4.30 around

$$\begin{cases} y(t) = y^* + \delta y(t) \\ p(t) = p^* + \delta p(t), \end{cases}$$

giving

$$\delta \dot{p}(t) = \frac{1}{c}\delta y(t). \tag{4.31}$$

Considering the number of links $M = 1$, from Equations 4.12 and 4.13, we also obtain their error equations as follows:

$$\delta q_i(t) = \delta p(t - \tau_i^b), \tag{4.32}$$

$$\delta y(t) = \sum_{i=1}^{N} \delta x_i(t - \tau_i^f). \tag{4.33}$$

We express the matrix form of Equations 4.29, 4.31, through 4.33, in Laplace domain, respectively,

$$E(s)\delta X(s) = -F(s)\delta Q(s), \tag{4.34}$$

$$\delta P(s) = \frac{1}{c \cdot s}\delta Y(s), \tag{4.35}$$

$$\delta Q(s) = D(s)A^T(-s)\delta P(s), \tag{4.36}$$

$$\delta Y(s) = A(s)\delta X(s), \tag{4.37}$$

where

$$E(s) = \text{diag}\left(s + \tfrac{\gamma}{R_i^{*2}}(q_i^*)^{1/n}\right),$$
$$F(s) = \text{diag}\left(\tfrac{x_i^*}{R_i^*}s + \tfrac{\gamma x_i^*}{nR_i^{*2}}(q_i^*)^{1/n-1}\right),$$
$$D(s) = \text{diag}(e^{-R_i^* s}),$$
$$A(s) = (e^{-s\tau_1^f}, e^{-s\tau_2^f}, \ldots, e^{-s\tau_N^f})^T,$$

where T denotes the transposition.

Theorem 4.2.2

Suppose for all i, $k_0 R_i^* \geq \max_i R_i^*$ for some k_0. When $n < R_i^*/q_i^*$, the Generalized FAST TCP is locally stable for a single-bottleneck link topology if

$$\gamma < \frac{n\phi}{k_0}(q_i^*)^{1-1/n}\sqrt{\frac{\phi^2 - k_0^2}{k_0^2 - \phi^2 (nq_i^*/R_i^*)^2}}, \tag{4.38}$$

where $(nq_i^*/R_i^*)\phi < k_0 < \phi$, and

$$\phi = \tan^{-1}\frac{2\sqrt{nq_i^*/R_i^*}}{1 - nq_i^*/R_i^*}.$$

Proof: Using Equations 4.34 through 4.37, the return ration seen at the source is described as

$$\text{diag}\left(\frac{\frac{x_i^*}{R_i^*}s + \frac{\gamma x_i^*}{nR_i^{*2}}(q_i^*)^{1/n-1}}{s + \frac{\gamma}{R_i^{*2}}(q_i^*)^{1/n}} e^{-R_i^* s}\right) A^T(-s)\frac{1}{c \cdot s}A(s). \tag{4.39}$$

For stability, it is sufficient that the eigenvalues of function (4.39) does not encircle -1 in the complex plane for $s = j\omega$, $\omega \geq 0$. The set of eigenvalues is identical to that of

$$L(j\omega) = \text{diag}(H(j\omega R_i^*))A^T(-j\omega)\text{diag}(x_i^*/c)A(j\omega), \tag{4.40}$$

where

$$H(j\omega R_i^*) = \frac{e^{-j\omega R_i^*}}{j\omega R_i^*} \cdot \frac{j\omega R_i^* + \frac{\gamma}{n}(q_i^*)^{1/n-1}}{j\omega R_i^* + \frac{\gamma}{R_i^*}(q_i^*)^{1/n}}.$$

From the lemma of [151], the spectrum of $L(j\omega)$ satisfies

$$\begin{aligned}\sigma(L(j\omega)) &= \sigma\left(\text{diag}(H(j\omega R_i^*))A^T(-j\omega)\text{diag}(x_i^*/c)A(j\omega)\right)\\ &\subseteq \rho\left(A^T(-j\omega)\text{diag}(x_i^*/c)A(j\omega)\right)\cdot\\ &\quad \text{co}\left(0\bigcup\left\{\frac{e^{-j\omega R_i^*}}{j\omega R_i^*}\cdot\frac{j\omega R_i^*+\frac{\gamma}{n}(q_i^*)^{1/n-1}}{j\omega R_i^*+\frac{\gamma}{R_i^*}(q_i^*)^{1/n}}, i=1,\ldots,N\right\}\right),\end{aligned}$$

where co$(\cdot)$ denotes the convex hull of the N eigentrajectories and the origin. ■

Since

$$\begin{aligned}&\rho\left(A^T(-j\omega)\text{diag}(x_i^*/c)A(j\omega)\right)\\ &\leq \left\|A^T(-j\omega)\text{diag}(x_i^*/c)A(j\omega)\right\|_\infty\\ &\leq 1,\end{aligned}$$

all the absolute row sums are equal to 1. Hence,

$$\begin{aligned}&\sigma(L(j\omega))\\ &\subseteq \text{co}\left(0\bigcup\left\{\frac{e^{-j\omega R_i^*}}{j\omega R_i^*}\cdot\frac{j\omega R_i^*+\frac{\gamma}{n}(q_i^*)^{1/n-1}}{j\omega R_i^*+\frac{\gamma}{R_i^*}(q_i^*)^{1/n}}, i=1,\ldots,N\right\}\right).\end{aligned} \tag{4.41}$$

By the generalized Nyquist stability criterion [335], systems (4.29), (4.31) through (4.33) are stable if the set in (4.41) does not encircle -1.

Let ω_i be the critical frequency at which the phase $\angle H(j\omega_i R_i^*)$ is $-\pi$. Then,

$$\angle H(j\omega_i R_i^*) = -\omega_i R_i^* - \frac{\pi}{2} + \angle\frac{j\omega R_i^* + \frac{\gamma}{n}(q_i^*)^{1/n-1}}{j\omega R_i^* + \frac{\gamma}{R_i^*}(q_i^*)^{1/n}} = -\pi.$$

Hence,

$$\omega_i R_i^* = \frac{\pi}{2} + \angle\frac{j\omega R_i^* + \frac{\gamma}{n}(q_i^*)^{1/n-1}}{j\omega R_i^* + \frac{\gamma}{R_i^*}(q_i^*)^{1/n}}. \tag{4.42}$$

From Equation (4.42) and Lemma 4.2.1, we have the lower boundary of $\omega_i R_i^*$, for all i, and when $n < R_i^*/q_i^*$,

$$\omega_i R_i^* \geq \frac{\pi}{2} - \tan^{-1}\frac{1 - nq_i^*/R_i^*}{2\sqrt{nq_i^*/R_i^*}} = \tan^{-1}\frac{2\sqrt{nq_i^*/R_i^*}}{1 - nq_i^*/R_i^*} = \phi. \tag{4.43}$$

Without loss of generality, we assume $R_1^* \geq R_i^*$ for all i, and suppose for all i, $k_0 R_i^* \geq \max_i R_i^*$ for some k_0. Then, $\omega_1 \leq \omega_i$ for all i since $\omega_i R_i^* = \omega_1 R_1^*$ for all i. Thus, at $\omega \leq \omega_1$, the convex hull of Equation 4.41 cannot encircle -1. At $\omega \geq \omega_1$, the set in Equation 4.41 does not encircle -1 if, for all i,

$$|H(j\omega R_i^*)| < 1. \tag{4.44}$$

For $\omega \geq \omega_1$, we have $\omega R_i^* \geq \omega_1 R_i^* \geq \omega_1 R_1^*/k_0$. Note that

$$|H(j\omega R_i^*)| = \frac{1}{\omega R_i^*} \cdot \sqrt{\frac{(\omega R_i^*)^2 + \left((\gamma(q_i^*)^{1/n-1})/n\right)^2}{(\omega R_i^*)^2 + \left((\gamma(q_i^*)^{1/n-1})/n\right)^2 \cdot (nq_i^*/R_i^*)^2}} \tag{4.45}$$

is a strictly decreasing function of ωR_i^*.

Therefore, for all i,

$$\begin{aligned} |H(j\omega R_i^*)| &\leq \left|H\left(j\frac{\omega_1 R_1^*}{k_0}\right)\right| \\ &= \frac{k_0}{\omega_1 R_1^*} \cdot \sqrt{\frac{(\omega_1 R_1^*)^2 + \left((k_0\gamma(q_i^*)^{1/n-1})/n\right)^2}{(\omega_1 R_1^*)^2 + \left((k_0\gamma(q_i^*)^{1/n-1})/n\right)^2 \cdot (nq_i^*/R_i^*)^2}} \\ &\leq \frac{k_0}{\phi} \cdot \sqrt{\frac{\phi^2 + \gamma^2\left((k_0(q_i^*)^{1/n-1})/n\right)^2}{\phi^2 + \gamma^2\left((k_0(q_i^*)^{1/n-1})/n\right)^2 \cdot (nq_i^*/R_i^*)^2}}. \end{aligned}$$

Then, from Equation 4.44, we obtain

$$\gamma < \frac{n\phi}{k_0}(q_i^*)^{1-1/n}\sqrt{\frac{\phi^2 - k_0^2}{k_0^2 - \phi^2 (nq_i^*/R_i^*)^2}},$$

where $(nq_i^*/R_i^*)\phi < k_0 < \phi$. Hence, the proof is completed with the following lemma.

Lemma 4.2.1

Let

$$h_i(\omega) = \angle \frac{j\omega R_i^* + \frac{\gamma}{n}(q_i^*)^{1/n-1}}{j\omega R_i^* + \frac{\gamma}{R_i^*}(q_i^*)^{1/n}}. \tag{4.46}$$

Then, for all i, $\omega \geq 0$, when $R_i^*/q_i^* > n$,

$$h_i(\omega) \geq -\tan^{-1}\frac{1 - nq_i^*/R_i^*}{2\sqrt{nq_i^*/R_i^*}}. \tag{4.47}$$

■

Proof:

$$h_i(\omega) = \tan^{-1}\left(\frac{n\omega R_i^*}{\gamma(q_i^*)^{1/n-1}}\right) - \tan^{-1}\left(\frac{\omega R_i^{*2}}{\gamma(q_i^*)^{1/n}}\right).$$

Then,

$$h_i'(\omega) = \frac{\gamma R_i^*(q_i^*)^{1/n} \cdot (R_i^*/q_i^* - n) \cdot [n\omega^2 R_i^{*3} - (\gamma(q_i^*)^{1/n})^2/q_i^*]}{[(n\omega R_i^*)^2 + (\gamma(q_i^*)^{1/n-1})^2] \cdot [(\omega R_i^{*2})^2 + (\gamma(q_i^*)^{1/n})^2]}. \tag{4.48}$$

When $R_i^*/q_i^* > n$, we can check that the solution,

$$\omega_i^* = \frac{\gamma(q_i^*)^{1/n}}{R_i^*}\sqrt{\frac{1}{nq_i^* R_i^*}},$$

of $h_i'(\omega) = 0$ minimizes the phase $h_i(\omega)$. Hence,

$$h_i(\omega) \geq \tan^{-1}\sqrt{\frac{nq_i^*}{R_i^*}} - \tan^{-1}\sqrt{\frac{R_i^*}{nq_i^*}} =: \varphi. \tag{4.49}$$

Moreover,

$$\tan\varphi = \frac{\sqrt{\frac{nq_i^*}{R_i^*}} - \sqrt{\frac{R_i^*}{nq_i^*}}}{2} = -\frac{1-(nq_i^*)/R_i^*}{2\sqrt{nq_i^*/R_i^*}}. \tag{4.50}$$

Therefore,

$$h_i(\omega) \geq -\tan^{-1}\frac{1-(nq_i^*)/R_i^*}{2\sqrt{nq_i^*/R_i^*}}.$$

Then, the lemma follows.

Interestedly, the above stability is independent of the fairness parameter n. When $n = 1$, it reduces to the case of the usual FAST TCP, the stability of which was discussed in [330]. The simulation results presented in the next section also demonstrate that there is almost no oscillation in the dynamics of the queue size and rate allocations under Generalized FAST TCP.

4.2.3 FAIRNESS-SCALABILITY TRADE-OFF

The concept of (α, n)-proportional fairness has often been used to investigate the trade-off between fairness and total throughput (see, e.g., [328]). In the context of Generalized FAST, it also provides a trade-off between fairness and scalability.

In a network with N flows and a single-bottleneck running standard FAST, the mean queue size scales linearly with N. Under Generalized FAST, if all flows have the same α, the mean queue size scales as N^n. Specifically, Equation 4.25 shows that the mean queue size is $(\alpha/C^{n-1})N^n$ for a capacity C, since $x_i^* = C/n$. Thus, setting $n < 1$ causes the queue size to scale better as the number of flows increases.

However, intuition says that reducing n pushes the equilibrium further from max–min fairness, since max–min fairness is the limiting case for large n. In particular, the demand function, which maps the route price to the flow rate, becomes $D(q_i) = (U')^{-1}(q_i) = q_i^{-1/n}$. To see the impact of this, consider a linear network with $N = M+1$ flows, in which flow 1 traverses all M links, and for $i = 2, \ldots, M+1$, flow i traverses only link $i-1$. Thus $q_q = Mq_i$ for all $i \neq 1$, giving $x_1 = M^{-1/n}x_i$. Setting increases the disparity between flow 1 and the remaining flows.

Let us first consider a suitable way of measuring the fairness of a set of rates. A common measure, advocated by Jain [141], is

$$J(x_1, \ldots, x_N) = \frac{\left(\sum_{i=1}^N x_i\right)^2}{N\sum_{i=1}^N x_i^2} = \left(\frac{\mathscr{M}_1}{\mathscr{M}_2}\right)^2,$$

where $\mathscr{M}_i = (x_1^i + \ldots + x_N^i)^{1/i}$ denotes the ith power mean of the values x_i. By the power mean inequality ([336], Theorem 16), this maps vectors into the interval (0,1], with a value of 1 denoting all values being equal, and lower values corresponding to less fairness. Note that users do not observe rate directly; rather, they observe the time required to transfer a given amount of data. The least fair allocation is one in which a user receives zero rate, requiring an infinite download time. The severity of this is significantly understated by J.

In particular, $J(0, x, \ldots, x) = (N-1)/N$, suggesting that starving one user from a large group is quite fair. It could be argued that starving one user is analogous to a form of admission control and hence is not necessarily unfair. However, the unfairness in the above scenario is based on the topology of the path, with the longer flow consistently receiving a lower rate. Thus, the user cannot simply give up and expect to be treated fairly on a new attempt. Such consistent discrimination is unfair.

A more appropriate measure may be obtained by applying Jain's measure to the download times instead of the rates, giving $J(1/x_1,\dots,1/x_N)$. If one rate tends to zero, this measure tends to $1/N$, which better indicates the unfairness, but still does not reflect the complete starvation of one source.

An alternative quantitative measure of fairness of rates is the ratio of the harmonic mean to the arithmetic mean:

$$F(x_1,\dots,x_N) = \frac{N^2}{\left(\sum_{i=1}^N 1/x_i\right)\left(\sum_{i=1}^N x_i\right)} = \frac{\mathscr{M}_{-1}}{\mathscr{M}_1}.$$

By the harmonic-arithmetic mean inequality (see Theorem 16 in [336]), this is again a number in the interval [0,1], and equal to 1 only if all rates are equal. It is more suitable than J because it is 0 if and only if a flow is entirely starved of bandwidth, $x_i = 0$ for some i. Note that this measure is independent of the network topology. For complex topologies, it may be desirable to consider a more sophisticated measure, such as the ratio of the harmonic mean of the rate vector to the harmonic mean of the max–min fair rate vector. However, the function F is sufficient for this example.

In the linear network scenario, the fairness ratio F of the equilibrium rates can be shown to be

$$\frac{M^2+2M+1}{M^2+M^{1+1/n}+M^{1-1/n}+1}.$$

This is $1-O\left((\log M)^2/Mn^2\right)$ for large M and large n, but $O(M^{1-1/n})$ for large M and small n. This shows that the fairness goes to 0 for small n, indicating that the long flow would be starved of bandwidth if too small a value of n were used.

Generalized FAST allows the protocol designer to trade fairness (large n) against queue scalability (small n). This may be particularly useful for private networks where knowledge of the topology and expected load is available; if the expected number of flows at any router is expected to be small, then scalability can be traded for increased fairness, while if the diameter of the network is small, then scalability can be improved.

4.2.4 SIMULATION RESULTS

We perform two sets of ns2 simulations [321,325]. The main objective of the first set of simulations is to show the relationship between the buffer occupancy increase and the value of n, and that of the second set is to verify that the fairness is related to the value of n.

The first set is for the dumbbell network topology, as shown in Figure 4.11, involving the five sender–receiver pairs. The link capacity is set to 1250 packets/s, and propagation delay is 50 ms. The size of every packet is set at 1000 bytes in all the simulations. As shown in Figure 4.12, we add a

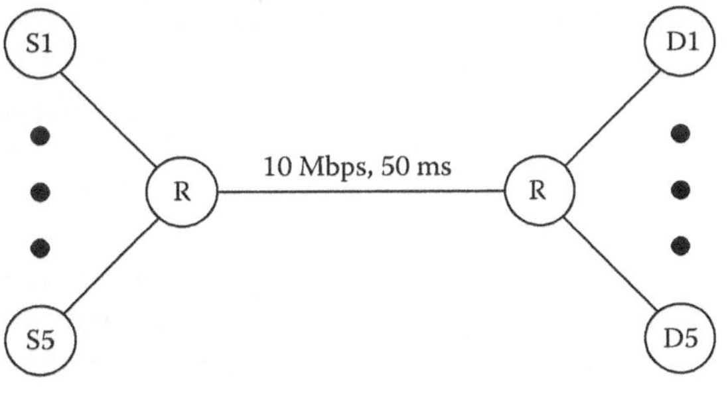

Figure 4.11 The simulation model of a dumbbell topology.

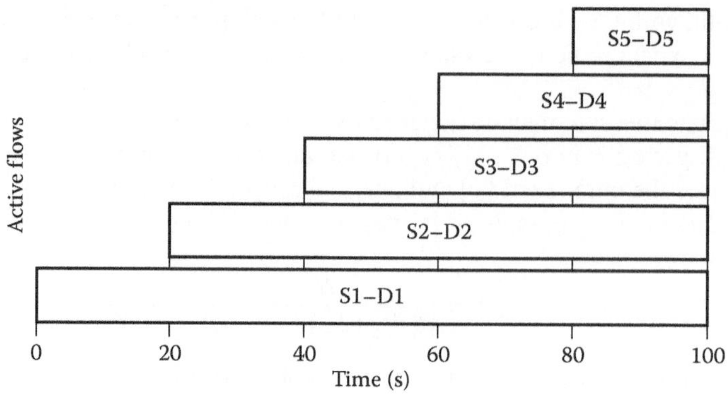

Figure 4.12 The active periods of the five flows.

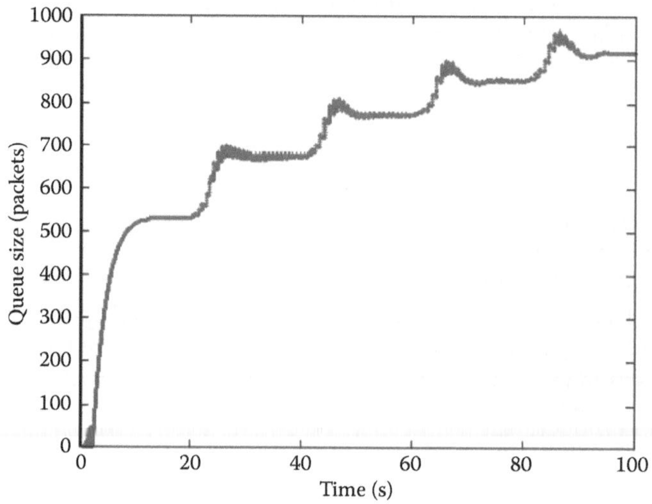

Figure 4.13 Queue size for the Generalized FAST TCP with $n = 1/3$ and $\alpha^3 = 100$.

new flow into the network after each 20 s. To avoid the persistent congestion [174] (the appearance of persistent congestion is also been observed in real network test of FAST TCP [172]), instead, we use the priority queue (PQ) method [352] in this scheme. Each flow has the same values of $\alpha^{1/n} = 100$ in the individual simulation. The parameter n for each simulation is set to 1/3, 1/2, and 1, respectively.

From the simulation results (Figures 4.13 through 4.18), we can see that there are various ways the buffer increases for different values of n. As shown in Figures 4.13, 4.15 and 4.17, buffer requirements grow as the nth power of the number of flows. For $n = 1$, buffer increases linearly with the number of flows bottlenecked at a link, while buffer increases as the 1/2th power of the number of flows when $n = 1/2$, similar relationship is observed for $n = 1/3$. Therefore, we can reduce the trend of buffer increment by setting small value of n. Figures 4.14, 4.16 and 4.18 illustrate that there is no significant oscillation in the dynamic source sending rate, which validates the stability.

The second set of simulations is for the multi-bottleneck network, and its topology is shown in Figure 4.19 with the three sender–receiver pairs. We assume that the bandwidth of each link is 12500 packets/s, and propagation delay is 50 ms. There are two short flows traveled one link and a long flow traveled two links.

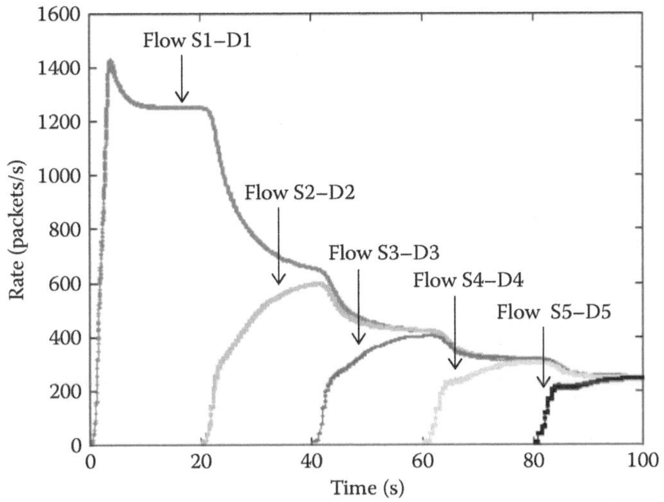

Figure 4.14 Sending rate for the Generalized FAST TCP with $n = 1/3$ and $\alpha^3 = 100$.

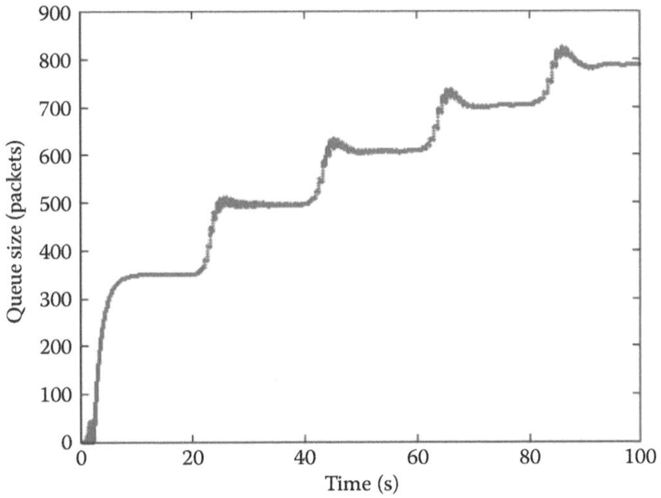

Figure 4.15 Queue size for the Generalized FAST TCP with $n = 1/2$ and $\alpha^2 = 100$.

The simulation results plotted in Figures 4.20 through 4.22 suggest that bandwidth allocation in multi-bottleneck link depends on the value of n. In the case of $n = 1$, the so-called proportion fairness, the sending rate of the short flow is double that of the long flow. While for $n = 1/2$, the sending rate of the short flow is four times that of the long flow, and for $n = 1/3$, it is eight times. From Figures 4.20 through 4.22, one can find that a smaller value of n will lead to larger difference of rate allocation between the short flow and the long flow, while as n increases, the bandwidth allocation converges to max–min fairness: the sending rate of the short flow equals to that of the long flow.

4.2.5 CONCLUDING REMARKS

We generalize the current FAST TCP scheme in such a way that the parameters n and $\alpha^{1/n}$ in the new window-updating equation can be set to satisfy the (α, n)-proportional fairness and control the

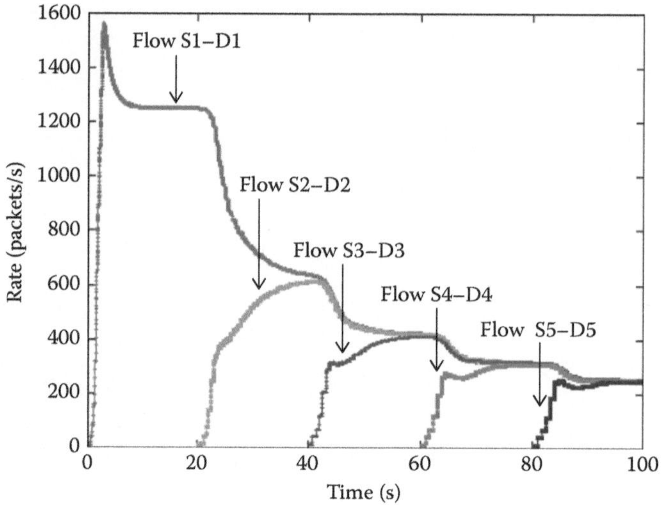

Figure 4.16 Sending rate for the Generalized FAST TCP with $n = 1/2$ and $\alpha^2 = 100$.

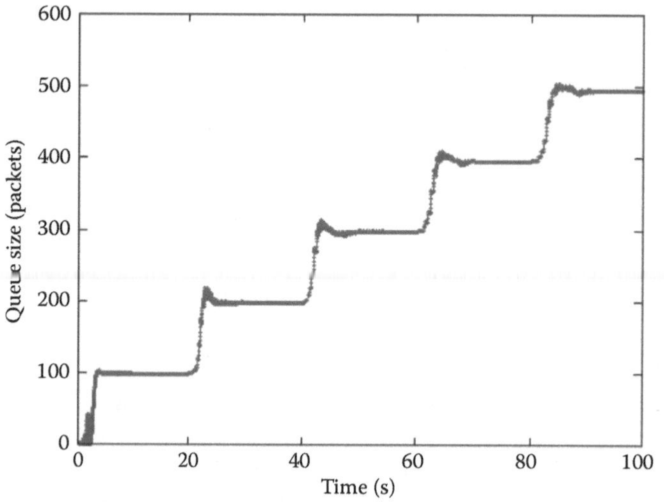

Figure 4.17 Queue size for the Generalized FAST TCP with $n = 1$ and $\alpha = 100$.

manner of buffer increase. We prove that the window-updating system in this general scheme is stable for the scenario of a single-bottleneck link with flows of equal propagation delay, ignoring feedback delay, and we have discussed the trade-off between fairness and buffer increment.

Simulation results have verified the stability of Generalized FAST and its ability to allocate rates in accordance with (α, n)-proportional fairness at equilibrium. Future research will investigate the performance of networks with general topology and a variety of traffic scenarios, including other TCP versions and UDP traffic, under the Generalized FAST TCP with different fairness parameters from design to implementation, and the impact of feedback delay. Furthermore, we can use virtual-queue-based active queue management (AQM) approach [173] and REM [100] to achieve fair rate allocation without dramatically increasing buffer occupancy.

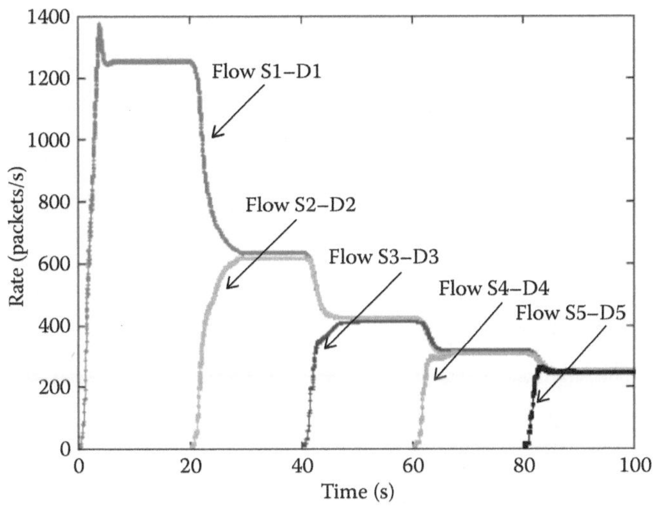

Figure 4.18 Sending rate for the Generalized FAST TCP with $n = 1$ and $\alpha = 100$.

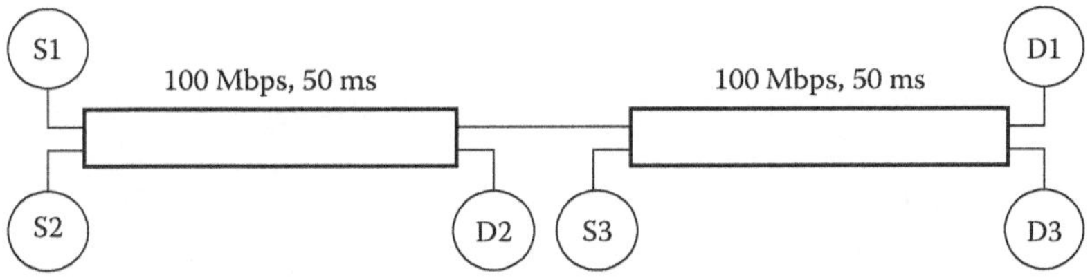

Figure 4.19 The simulation model of multi-bottleneck topology.

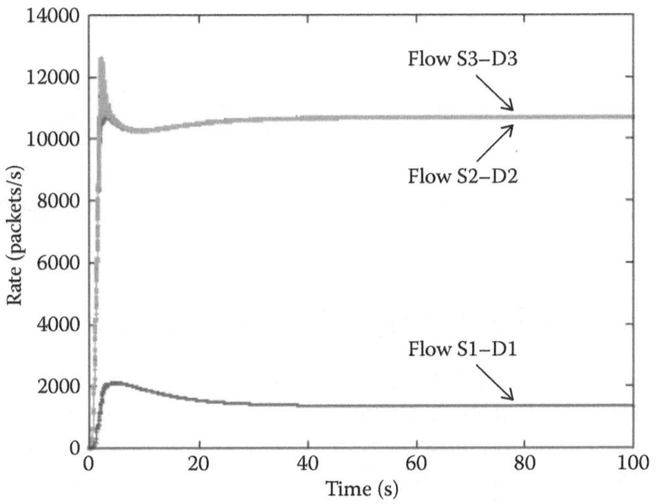

Figure 4.20 Sending rate for the Generalized FAST TCP with $n = 1/3$ under multi-bottleneck network.

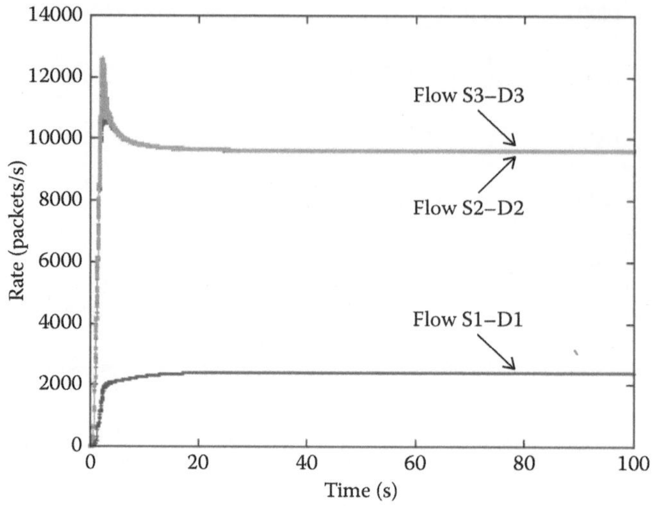

Figure 4.21 Sending rate for the Generalized FAST TCP with $n = 1/2$ under multi-bottleneck network.

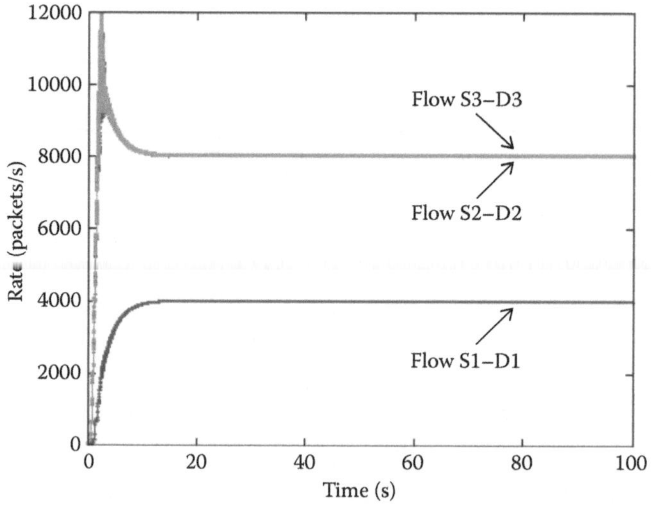

Figure 4.22 Sending rate for the Generalized FAST TCP with $n = 1$ in a multi-bottleneck network.

4.3 NEW FAST TCP VARIANT CONSIDERING PACKET LOSS

Packet loss events are not explicitly reflected in the flow-level model of FAST TCP. Considering the usual utility maximizing problem and a new utility function that takes into account packet loss, this section proposes a new FAST TCP flow-level model that results in a new equilibrium point, which is exactly the unique solution of utility maximizing problem. The new model performs better in real networks with unavoidable packet loss, which is particularly suitable for implementation in wireless communication networks.

Unlike AIMD-based TCP schemes, FAST TCP uses queuing delay, in addition to packet loss, as congestion indication. It yields the proportional fairness proposed by Kelly [105]. Like other congestion control algorithms, FAST TCP can also be designed at two levels, namely, the packet level and the flow level. The flow-level (macroscopic) algorithm aims to achieve high utilization, low queuing delay and loss, fairness, and stability, while the packet-level (microscopic) algorithms support its flow-level counterpart in achieving these goals. The current main components of packet-level algorithm of FAST TCP do not take packet loss into account. However, in addition, the packet-level algorithm includes a "patch" that does consider packet loss [334]. Both the macroscopic and microscopic attributes of FAST TCP have been widely studied [324,329,330,352]. However, all these studies on the macroscopic properties of FAST TCP are based on the existing flow-level model, which does not take the packet loss into account. Packet loss is inevitable in real networks. For instance, both buffer overflowing and the adoption of AQM schemes [178] may lead to loss of packets. Therefore, it is desirable to have a flow-level model that considers packet loss directly.

In this section, we present a new flow-level model of FAST TCP that accounts for packet loss such that FAST TCP window dynamics relate to both queuing delay and packet loss probability. This is achieved by a new utility function that in turn leads to a new utility optimization problem. We show that the solution of this new optimization problem is the new equilibrium point of the modified flow-level model of FAST TCP. We present simulation results that demonstrate the improvement achieved by the modified over the original FAST TCP flow-level model.

4.3.1 OPTIMIZATION PROBLEM

Here, we provide as background the optimization problem of [105] that sets the scene for our new flow-level FAST TCP model.

Consider a network that consists of a set $L = \{1, \ldots, L\}$ of unidirectional links of capacities c_l, $l \in L$. The network is shared by a set $I = \{1, \ldots, I\}$ of sources, where source i is characterized by a utility function $U_i(x_i)$ that is concave increasing in its transmission rate x_i. We interpret $\hat{q}_l$ and $\hat{p}_l$ as the queuing delay and packet loss probability at link l, respectively. Every source is assumed to generate one flow. Let the path $L(i) \subseteq L$ be the set of links that the flow generated by source i, called flow i, traverses. For each link l, let $I(l) = \{i \in I | l \in L(i)\}$ be the set of sources that use link l. Note that $l \in L(i)$ if and only if $i \in I(l)$.

Consider the following network optimization problem:

$$\mathbf{P}\text{: } \max_{x \geq 0} \sum_i U_i(x_i), \tag{4.51}$$

$$\text{subject to } \sum_{i \in I(l)} x_i \leq c_l, \ \forall l \in L. \tag{4.52}$$

Constraint Equation 4.52 implies that the aggregate source rate at any link l does not exceed the capacity. Since the objective function is concave and increasing, a unique equilibrium vector $x = \{x_i, i \in I\}$ exists, which is the unique solution to the above utility optimization problem.

Let $\sum_{l \in L(i)} \hat{q}_l$ be the sum of queuing delay for all links in $L(i)$. Assuming that packet loss probabilities on different links are independent, then

$$1 - \prod_{l \in L(i)} (1 - \hat{p}_l)$$

is the probability that a packet successfully traverses through $L(i)$. Because the current Internet is a SumNet, we obtain [337]

$$1 - \prod_{l \in L(i)} (1 - \hat{p}_l) \approx \sum_{l \in L(i)} \hat{p}_l. \tag{4.53}$$

For simplicity, we denote

$$q_i = \sum_{l \in L(i)} \hat{q}_l, \ p_i = \sum_{l \in L(i)} \hat{p}_l.$$

Following the approach of [105], the Lagrangian is

$$\begin{aligned} L(x,q) &= \sum_i U_i(x_i) + \sum_l \left(\hat{q}_l(c_l - \sum_{i \in I(l)} x_i) \right) \\ &= \sum_i \left(U_i(x_i) - x_i \sum_{l \in L(i)} \hat{q}_l \right) + \sum_l c_l \hat{q}_l \\ &= \sum_i (U_i(x_i) - x_i q_i) + \sum_l c_l \hat{q}_l, \end{aligned} \tag{4.54}$$

where the term $\hat{q}_l$ in the first row is the Lagrange multiplier. One obtains the unique maximizer to the above optimization problem by solving

$$\frac{\partial L}{\partial x_i} = 0.$$

4.3.2 NEW FLOW-LEVEL MODEL OF FAST TCP

Let w_i be the source congestion window size (packets). Unlike the original utility function of FAST TCP [172], which depends only on source rate x_i (windows/RTT), we propose the following new utility function which is associated with both source rate and loss probability:

$$U_i(x_i) = \alpha_i \log x_i - \frac{1}{2} x_i p_i w_i, \tag{4.55}$$

where α_i is a positive constant.

Note that

$$w_i(t) = x_i(t) \cdot (d_i + q_i(t)), \tag{4.56}$$

where d_i is the pure propagation delay of flow i, $R_i(t) = d_i + q_i(t)$ is the round-trip time (RTT) of flow i. The time unit is of the order of several RTTs, and source rate $x_i(t)$ should be interpreted as the average rate over this timescale. Dynamics smaller than several RTTs is not captured by the fluid model [111].

By setting the following parameters: $\alpha = 50$ packets, $p = 0.001$, RTT$= 0.1$ s, we plot the curves of the original utility function of FAST TCP and our new utility functions into Figure 4.23. The figure compares between FAST TCPs and our new utility function. Both of these two curves are concave, while the main difference is in the slope. For the same value U_0 of utility function, the source rate x_0 allocated to FAST TCP is smaller than the new utility function resulted rate x'_0. As the utility function indicates the user's want-satisfaction degree, Figure 4.23 suggests that to achieve the same want-satisfaction degree, users have to be allocated larger source rate amid packet loss.

$U_i(x_i)$ defined by Equation 4.55 is strictly concave, and when

$$p_i < \frac{\alpha_i \cdot \text{RTT}}{w_i^2}, \tag{4.57}$$

$U_i(x_i)$ is increasing.

Note that the relationship between loss probability and window size in equilibrium for TCP Reno has been given by [172]

$$p_i^* = \frac{3}{2 w_i^{*2}}.$$

It is reasonable to assume that

$$p_i^* = \frac{3}{2 w_i^{*2}} < \frac{\alpha_i \cdot \text{RTT}}{w_i^2}.$$

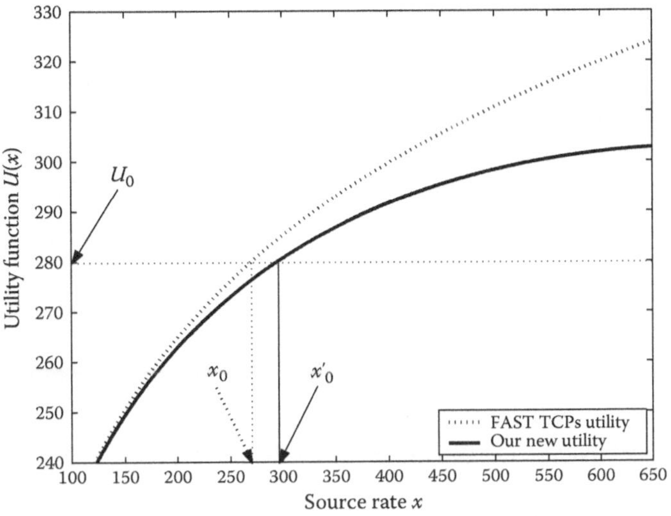

Figure 4.23 The comparison of FAST TCPs and our utility function.

Therefore, condition Equation 4.57 can be assumed to be satisfied.

Taking partial derivative of $L(x,q)$ with respect to x_i, we have

$$\frac{\partial L}{\partial x_i} = U_i'(x_i) - q_i.$$

Then, the optimal solution to the primal problem is given by

$$\alpha_i = x_i^* q_i^* + x_i^* p_i^* w_i^*, \tag{4.58}$$

where

$$q_i^* = \sum_{l \in L(i)} \hat{q}_l^*, \; p_i^* = \sum_{l \in L(i)} \hat{p}_l^*. \tag{4.59}$$

Equation 4.58 implies that in equilibrium, source i buffers $x_i^* q_i^*$ packets and drops $x_i^* p_i^* w_i^*$ packets per RTT. This equilibrium property is different from the original FAST TCP [172] in the sense that the original FAST TCP only considers the back off packets of flows at equilibrium but does not account for the lost packets.

We propose the following flow-level model of FAST TCP:

$$\dot{w}_i(t) = \gamma_i \left(\alpha_i - x_i(t) q_i(t) - x_i(t) p_i(t) w_i(t)\right), \tag{4.60}$$

where

$$q_i(t) = \sum_{l \in L(i)} \hat{q}_l(t - \tau_{li}^b), \; p_i(t) = \sum_{l \in L(i)} \hat{p}_l(t - \tau_{li}^b).$$

The constant $\gamma_i \in (0,\ 1]$ $(i = 1, 2, \ldots, I)$, and τ_{li}^b denotes the backward delay from link l to source i. Note that Equation 4.60 reduces to the original flow-level model of FAST TCP when $p_i(t) = 0$.

Letting $\dot{w}_i(t) = 0$ in Equation 4.60, we obtain the equilibrium point $(x_i^*, q_i^*, p_i^*, w_i^*)$ of Equation 4.60 that exactly satisfies (4.58). By Equation 4.56, the equilibrium values of the congestion window, the source sending rate, and the queuing delay of the source i are associated by

$$x_i^* = \frac{w_i^*}{d_i + q_i^*}. \tag{4.61}$$

Now, we propose the following theorem.

Theorem 4.3.3

Assume $p_i < \alpha_i \cdot \mathrm{RTT}/w_i^2$. The unique equilibrium point (x^*, q^*, p^*, w^*) of the system of FAST TCPs flow-level model (4.60) exists, and the rate vector $x^* = \{x_1^*, \ldots, x_I^*\}^T$ is the unique maximizer of the problem **P** (4.51) and (4.52) with the utility function given by (4.55).

Proof: An optimal rate vector x^* exists since the objective function in Equation 4.51 is continuous and the feasible solution set is compact. It is unique if U_i is strictly concave.

The four-tuple (x^*, q^*, p^*, w^*) is optimal if and only if x^* is feasible, and

$$x^* = \arg\max_{x \geq 0} L(x, q^*), \tag{4.62}$$

where L is the Lagrangian given in Equation 4.54.

Hence, to prove Theorem 4.3.3, we only need to establish Equation 4.62. Now,

$$\begin{aligned}\max_{x \geq 0} L(x, q^*) &= \max_{x \geq 0} \left(\sum_i (U_i(x_i) - x_i q_i) + \sum_l c_l \hat{q}_l \right) \\ &= \sum_i \max_{x_i \geq 0} (U_i(x_i) - x_i q_i) + \sum_l c_l \hat{q}_l.\end{aligned}$$

By construction of U_i in (4.55), we have

$$U_i'(x_i^*) = \frac{\alpha_i}{x_i^*} - p_i^* w_i^* = q_i^*,$$

which implies that

$$\left. \frac{\partial L}{\partial x_i} \right|_{(x_i^*, q_i^*)} = 0,$$

if $x_i^* > 0$. Since $L(x, q^*)$ is concave in x, this is the necessary and sufficient Karush–Kuhn–Tucker condition [144] for x^* to maximize $L(x, q^*)$ over $x \geq 0$. ∎

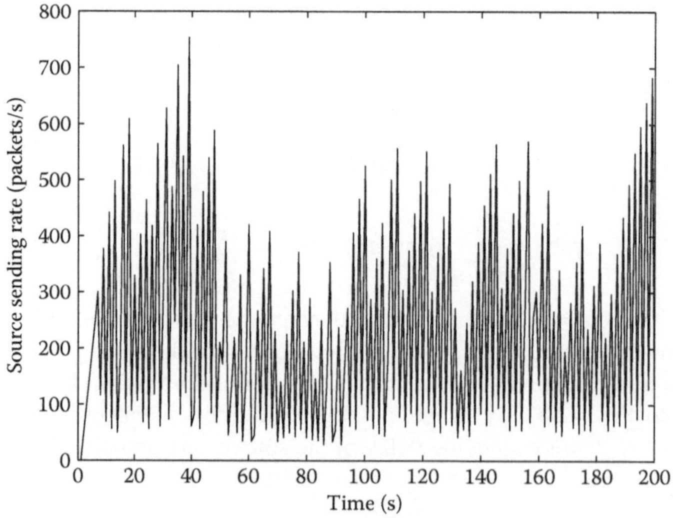

Figure 4.24 The source sending rate under the original flow-level FAST TCP.

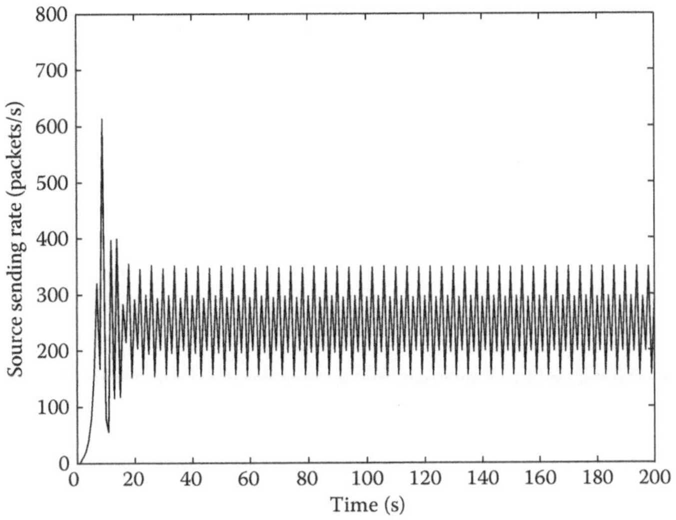

Figure 4.25 The source sending rate under the modified flow-level FAST TCP.

4.3.3 SIMULATION RESULTS

We now present simulation results based on a dumbbell network topology with five FAST TCP sender–receiver flows. For each flow i, we set the parameters $\gamma_i = 0.01$, $\alpha_i = 50$ packets, and $d_i = 0.1$ s. The capacity of the common link is set to 10 Mbps, and every packet size is set to 1 kbyte. We consider the random early detection (RED) [178] algorithm as the AQM in the router. The basic RED parameters are set to: $\text{min}_{th} = 150$ packets, $\text{max}_{th} = 350$ packets, $\text{max}_p = 0.1$, and the target queue length $q_T = 300$ packets. The simulation results for the sending rate of one of the five flows are shown in Figures 4.24 and 4.25. The other four flows exhibited similar behavior. We can observe that the modified flow-level model of FAST TCP reduces the oscillations of the sending rates.

4.3.4 CONCLUDING REMARKS

In this section, we have proposed a new FAST TCP variant and proved that it results in an equilibrium point which is the unique solution of the utility maximizing problem. We have demonstrated that this new variant can reduce oscillations of the sending rates.

5 Fairness and Bandwidth Allocation

In a social science context, fairness is defined as Justice. In [145], John Rawls proposes a social contract argument to define that justice, and especially distributive justice, is a form of fairness: an impartial distribution of goods. It is argued in [145] that each individual would not reject the utilitarian theory of justice. The total welfare of the society can be maximized in the way that every one may deduce his (or her) good for the greater benefit of others. There are two principles of justice according to Rawls's theory [145]:

- Each person is to have an equal right to the most extensive total system of equal basic liberties compatible with a similar system of liberty for all.
- Social and economic inequalities are to be arranged so that they are both to the greatest benefit of the least advantaged, consistent with the just savings principle, and attached to offices and positions open to all under conditions of fair equality of opportunity.

As we will see later, the above two principles are just the basis of the definition of fairness for a communication network with relation to, for example, the bandwidth allocation.

According to the utilitarian [146], justice requires the maximization of the total or average welfare across all relevant individuals. This may require sacrifice of some for the good of others, so long as each individual's good is taken impartially into account [271,283]. Utilitarianism, in general, then argues that the standard of justification for actions, institutions, or the whole world is impartial welfare consequentialism, and only indirectly, if at all, to do with rights, property, need, or any other nonutilitarian criterion. These other criteria might be indirectly important to the extent that human welfare involves them. But even then, such demands as human rights would only be elements in the calculation of overall welfare, not uncrossable barriers to action. This theory is then seen to be the theoretical basis of the network utility maximization theory, which has been widely used in congestion control and bandwidth allocation for both wire-lined and wireless communication networks. We will visit the network utility maximization theory and its applications in a number of places in this book.

Fairness has been defined in a number of different ways so far. The notion of fairness characterizes how competing users should share the bottleneck resources. In this section, we review and compare the standard definitions of fairness for a communication network.

5.1 GENERAL DESCRIPTION OF MAX–MIN FAIRNESS

One of the most common fairness definitions is the so-called max–min or bottleneck optimality criterion [138–142,248]. A feasible flow rate x is defined to be max–min fair if any rate x_i cannot be increased without decreasing some x_j, which is smaller than or equal to x_i [139]. A max–min fair allocation is achieved when bandwidth is allocated equally and in infinitesimal increments to all flows until one is satisfied, then among the remainder of the flows and so on until all flows are satisfied or the bandwidth is exhausted. The max–min fairness is seen to be exactly a realization of the aforementioned two principles of Rawls's Justice theory [145].

Many researchers have developed algorithms achieving max–min fair rates [139,141,142]. But a max–min fair vector needs global information [143] and most of those algorithms require exchange of information between networks and hosts. In [140], Hahne suggested a simple round-robin way of control, but it requires all the links perform round-robin scheduling and it needs to be guaranteed that packets of users are ready for all links.

In best-effort statistical multiplexing, a first-come first-served (FCFS) scheduling policy is often used. The advantage with max–min fairness over FCFS is that it results in traffic shaping, meaning that an ill-behaved flow, consisting of large data packets or bursts of many packets, will only punish itself and not other flows. Network congestion is consequently to some extent avoided.

Fair queuing is an example of a max–min fair packet scheduling algorithm for statistical multiplexing and best-effort packet-switched networks because it gives scheduling priority to users who have achieved lowest data rate since they became active. In case of equally sized data packets, round-robin scheduling is max–min fair.

Max–min fairness is used widely in networking areas for the purpose of bandwidth allocation, with the aims at allocating as much as possible to poor users, while not unnecessarily wasting resources. Almost all algorithms for computing the max–min fair bandwidth allocation are set upon the notion of bottlenecks. However, for mobile or peer-to-peer networks, the existing theories that rely on the bottleneck concept do not apply directly. In the paper [248], the authors give a unifying treatment of max–min fairness that covers the existing results within a framework and extend the applicability of the current max–min concept to mobile and peer-to-peer networks. It is observed that the existence of max–min fairness is actually a geometric property of the set of feasible allocations. Even though there exist sets on which max–min fairness does not exist, one can find a large class of sets, within which a max–min fair allocation does exist. The paper [248] presents a general and centralized algorithm with linear programming computing complexity, termed as max–min programming (MP), for computing the max–min fair allocation in all cases where it exists. If the set of feasible allocations has the free disposal property, then this algorithm reduces to the well-known simpler algorithm: water filling (WF). The findings are based on the relation between max–min fairness and leximin ordering. The results in [248] apply to weighted, unweighted max–min and min–max fairness.

5.2 MAX–MIN AND MIN–MAX FAIRNESS IN EUCLIDEAN SPACES

5.2.1 DEFINITIONS AND UNIQUENESS

Consider a set $\chi \subset \Re^N$. We define the max–min and min–max fair vectors with respect to set χ as follows:

Definition 5.2.1 [139] *A vector $\vec{x}$ is "max–min fair on set χ" if and only if*

$$(\forall \vec{z} \in \chi)(\exists s \in \{1,\ldots,N\})z_s > x_s \Rightarrow (\exists t \in \{1,\ldots,N\})z_t < x_t \leq x_s. \tag{5.1}$$

This means that increasing some component x_s must be at the expense of decreasing some already smaller component x_t.

Definition 5.2.2 *A vector $\vec{x}$ is "min–max fair on set χ" if and only if*

$$(\forall \vec{z} \in \chi)(\exists s \in \{1,\ldots,N\})z_s < x_s \Rightarrow (\exists t \in \{1,\ldots,N\})z_t > x_t \geq x_s. \tag{5.2}$$

This means that to decrease some component x_s must be at the expense of increasing some already larger component x_t.

One sees that if $\vec{x}$ is a min–max fair vector on χ, then $-\vec{x}$ is max–min fair on $-\chi$ and vice versa. Thus, in the remainder of this chapter, we give theoretical results only for max–min fairness. The results for min–imax fairness can be extended with ease. Uniqueness of max–min or min–max fair vector if existed is assured by the following theorem, the proof of which is analog to the one in [139] and is omitted.

Theorem 5.2.1

If a max–min (or min–max) fair vector exists on a set χ, then it is unique. ■

Weighted min–max fairness is a classical variation, defined as follows.

Definition 5.2.3 [139] *Given some positive constants w_i (termed as the "weight"), a vector $\vec{x}$ is "weighted max–min fair" on set χ, if and only if increasing one component x_s must be at the expense of decreasing some other component x_t such that $x_t/w_t \le x_s/w_s$.*

This concept is generalized in [252], which introduces the concept of "util max–min fairness":

Definition 5.2.4 *Given N increasing functions ξ: $\Re \to \Re$ interpreted as utility functions, a vector $\vec{x}$ is util max–min fair" on set χ if and only if increasing one component x_s we are forced to decrease some other component x_t such that*

$$\xi_t(x_t) \le \xi_s(x_s).$$

Note that the above "util max–min fairness" is also called "weighted max–min fairness" in [255].

Consider the mapping ξ defined by

$$(x_1,\ldots,x_N) \to (\xi_1(x_1),\ldots,\xi_N(x_N)) \tag{5.3}$$

It follows immediately that a vector $\vec{x}$ is util max–min fair on set χ if and only if $\xi(\vec{x})$ is max–min fair on the set $\xi(\chi)$. The case of weighted max–min fairness is corresponding to $\xi_i(x_i) = x_i/w_i$. Thus, we now restrict our attention to unweighted max–min fairness.

5.2.2 MAX–MIN FAIRNESS AND LEXIMIN ORDERING

We will use leximin ordering later on, a concept from economy. Let us define the "order mapping" Γ: $\Re \to \Re$ in nondecreasing order. That is

$$\Gamma(x_1,\ldots,x_n) = (x_{(1)},\ldots,x_{(n)}),$$

with

$$x_{(1)} \le x_{(2)} \cdots \le x_{(n)},$$

and for all i, $x_{(i)}$ is one of the x_js. Let us also define the lexicographic ordering of vectors in χ by

$$\vec{x} >^{lex} \vec{z}$$

if and only if

$$(\exists i) x_i > z_i \text{ and } (\forall j < i) x_i = z_i.$$

We also say that

$$\vec{x} \ge^{lex} \vec{z}$$

if and only if

$$\vec{x} >^{lex} \vec{z}$$

or

$$\vec{x} = \vec{z}.$$

The latter relation is a total order on $\Re^N$. We now present the following definitions.

Definition 5.2.5 [253] *The vector* $\overrightarrow{x}$ *is leximin larger than or equal to* $\overrightarrow{z}$ *if* $\Gamma(\overrightarrow{x}) \geq^{lex} \Gamma(\overrightarrow{z})$.

Definition 5.2.6 [253] *The vector* $\overrightarrow{x} \in \chi$ *is leximin maximal on a set* χ *if for all* $\overrightarrow{z} \in \chi$ *we have* $\Gamma(\overrightarrow{x}) \geq^{lex} \Gamma(\overrightarrow{z})$.

Note that a leximin maximum is not necessarily unique.

Theorem 5.2.2

[256] Any compact subset of $\Re^n$ has a leximin maximal vector. ■

It has been observed in [252,254,257] that a max–min fair allocation is also leximin maximal, for the feasible sets defined in these papers. This statement has been generalized to an arbitrary feasible set in [256], which is described in the following theorem.

Theorem 5.2.3

If a max–min fair vector exists on a set χ, then it is the unique leximin maximal vector on χ. ■

Thus, the existence of a max–min fair vector implies the uniqueness of a leximin maximum. The reverse statement is, however, not true: see [248]. The paper [256] defines a weaker version of max–min fairness, "maximal fairness"; it corresponds to the notion of leximin maximal vector, hence it is not unique and exists on a larger class of feasible sets.

5.2.3 EXISTENCE AND MAX–MIN ACHIEVABLE SETS

As we discussed before, in many situations, a number of publications have established the existence of a max–min fair allocation by using various methods. We now present a generalized proof that applies to a larger class of continuous sets. Note that a max–min fair vector may not always exist on all feasible sets, even for compact and connected sets. Simple counterexamples are given in [248]. We herein give a sufficient condition for a max–min vector to exist.

Definition 5.2.7 *A set* χ *is max–min achievable if there exists a max–min fair vector on* χ.

Theorem 5.2.4

Consider a mapping ξ as defined in Equation 5.3. Assume that ξ_i is increasing and continuous for all i. If the set χ is convex and compact, then $\xi_{(}\chi)$ is max–min achievable. ■

Note that if $\overrightarrow{x}^*$ is max–min fair on χ and $\overrightarrow{z}^*$ is max–min fair on $\xi_{(}\chi)$, then generally we have

$$\overrightarrow{z}^* \neq \xi(\overrightarrow{x}^*).$$

As a special case, if letting $\xi_i(x) = x$, we observe that all convex and compact sets are max–min achievable. Taking $\xi_i(x) = x/w_i$, we also conclude that the weighted max–min fairness exists on all compact and convex sets. More generally, util max–min fairness exists on all compact and convex sets, if the utility functions are continuous (and increasing).

In [257], the utility functions ξ_i are assumed to be arbitrary continuous, increasing, and concave. With these assumptions, the set $\xi(\chi)$ is also convex and compact. Note that in general, though, the set $\xi(\chi)$ used in the above theorem is not necessarily convex. Counterexamples with nonconvex sets can be found in [252,255].

Given an initial allocation of bandwidth among a set of individual users, a change to a different allocation that makes at least one individual better off without making any other individual worse off is called a Pareto improvement. An allocation is defined as "Pareto efficient" or "Pareto optimal" when no further Pareto improvements can be made. If an economic allocation in any system is not Pareto efficient, there is a potential for a Pareto improvement—an increase in Pareto efficiency: through reallocation, improvements can be made to at least one participant's benefit without reducing any other participant's benefit.

It is important to note, however, that a change from an inefficient allocation to an efficient one is not necessarily a Pareto improvement. Thus, in practice, ensuring that nobody is disadvantaged by a change aimed at achieving Pareto efficiency may require compensation of one or more parties. For instance, if a change in economic policy eliminates a monopoly and that market subsequently becomes competitive and efficient, the monopolist will be made worse off. However, the loss to the monopolist will be more than offset by the gain in efficiency, in the sense that the monopolist could hypothetically be compensated for its loss while still leaving a net gain for others in the economy, a Pareto improvement.

From Theorem 5.2.3, we observe that the max–min fair vector, if it exists, is strictly Pareto optimal. It is also known that no continuous utility function exists that can represent leximin ordering [253]. Therefore, we cannot use a classical convex optimization approach to find a max–min fair vector. However, for the case of convex feasible sets, a max–min fair vector is a limiting case of utility fair allocations, which is discussed in [248] for the specific case of a single path network and in [248] for the general case.

5.3 MP ALGORITHM AND WF ALGORITHM

5.3.1 MP ALGORITHM

The paper [248] presents the MP algorithm, which finds the max–min fair vector on any feasible set, if it exists. It is outlined as following.

The MP Algorithm

10: let $\Psi^0 = \{1,\ldots,N\}$, $\chi^0 = \chi$ and $n = 0$

20: **do**

30: $n = n+1$

40: Problem MP^n: maximize T^n subject to:

$$\begin{cases} (\forall i \in \Psi^{n-1}) \quad x_i \geqslant T^n \\ \qquad\qquad \vec{x} \in \chi^{n-1} \end{cases} \tag{5.4}$$

50: let $\chi^n = \{\vec{x} \in \chi^{n-1} | (\forall i \in \Psi^{n-1})\, x_i \geqslant T^n,\ (\exists i \in \Psi^{n-1})\, x_i > T^n\}$
and $\Psi^n = \{i \in \{1,\ldots,N\} | (\forall \vec{x} \in \chi^n)\, x_i > T^n\}$

60: **until** $\Psi^n = \varnothing$;

70: return the only element in χ^n

Basically, the above algorithm maximizes in each step the minimal coordinate of the feasible vector, until all coordinates are processed. The nth step of the algorithm is a minimization problem called MP^n: χ^n represents the remaining search space and Ψ^n represents the direction of the searching process, in terms of coordinates that can be further increased. The algorithm will stop if the component χ is max–min achievable, and then χ^n is reduced to a single element, which is the desired max–min fair vector. We give the following theorem on that the outputted result of the above MP algorithm is exactly the max–min fair vector that are sought.

Theorem 5.3.5

The above MP algorithm terminates and finds the max–min fair vector on χ if it exists, in at most N steps. ∎

In [249], an algorithm is presented for calculating the leximax minimal allocation, which is seen to be a particular implementation of MP. Given the fact that, the feasible set considered therein is compact convex, it follows from Theorems 5.2.1 and 5.2.4 that the leximax minimal allocation obtained in [249] comes out to be a min–max fair allocation essentially.

5.3.2 WF ALGORITHM

We now compare MP with the WF approach used in the traditional settings [139]. We here present an extended version that includes the minimal rate guarantees, as discussed in [250].

The concept of free disposal is first needed for our purpose. In economy, it is defined as the right of each user to dispose of an arbitrary amount of owned commodities [253], or alternatively, to consume fewer resources than maximally allowed. Regarding the application of it to the context of resource allocation for communication network, one can slightly modify it as follows. Denote a unitary vector $\vec{e_i}$ in the way

$$(\vec{e_i})_j = \delta_{ij}.$$

Definition 5.3.1 *A set χ has the free disposal property if*

- *There exists $\vec{m}$ with $x_i \geqslant m_i$ for all $\vec{x} \in \chi$ and*
- *For all $i \in \{1,\ldots,N\}$ and for all α such that $\vec{x} - \alpha\vec{e_i} \geqslant \vec{m}$, we have $\vec{x} - \alpha\vec{e_i} \in \chi$.*

Essentially, free disposal applies to any set, in which each coordinate is independently lower bounded. In addition, one can always decrease a feasible vector, as long as the above lower bounds can be maintained. So far, we are now ready to describe the WF algorithm as follows.

The WF Algorithm
10. Assume χ is free disposal
20. let $\Psi^0 = \{1,\ldots,N\}$, $\chi^0 = \chi$ and $n = 0$
30. **do**
40. $n = n + 1$
50. Problem WF^n: maximize T^n subject to:

$$\begin{cases} (\forall i \in \Psi^{n-1}) & x_i = max(T^n, m_i) \\ & \vec{x} \in \chi^{n-1} \end{cases} \tag{5.5}$$

60. let $\chi^n = \{\vec{x} \in \chi^{n-1} | (\forall i \in \Psi^{n-1})\, x_i \geqslant T^n, (\exists i \in \Psi^{n-1})\, x_i > T^n\}$ and $\Psi^n = \{i \in \{1,\ldots,N\} | (\forall \vec{x} \in \chi^n)\, x_i > T^n\}$
70: **until** $\Psi^n = \varnothing$;
80: return the only element in χ^n

On free disposal sets, we have the following theorem to demonstrate the equivalence of MP and WF.

Theorem 5.3.6

Let χ be a max–min achievable set that satisfies the free disposal property. Then, at every step n, the solutions to problems WF^n and MP^n are the same. ■

In summary, under the conditions of the above theorem, WF and MP terminate and return the same result, i.e., the max–min fair vector if it exists. The above theorem is in essence stronger than Theorem 5.3.5, since the two algorithms lead to the same result at every step. However, if the free disposal property does not hold, then WF may not lead to the desired max–min fair allocation.

The examples provided for single path unicast routing [139], multicast util max–min fairness [251,252], and minimal rate guarantee [250,254] all have the free disposal property. Therefore, the WF algorithm can be used in all these situations. Contrarily, the load distribution situation discussed in [249] is not of free disposal and one is forced to use MP (for the discussion, one can refer to [249] for a specific example).

We outline the complexity of the algorithms in case of linear constraints into Table 5.1 (see [248]).

5.4 PROPORTIONAL FAIRNESS

Kelly [104] proposed the following concept of proportional fairness (PF).

Definition 5.4.1 *A vector of rates x^* is proportionally fair if it is feasible. It satisfies $x^* \geqslant 0$ and $A^T x^* \leqslant c$, and if for any other feasible vector x, the aggregate of proportional change is negative. That is*

$$\Sigma \frac{x_i - x_i^*}{x_i^*} \leqslant 0. \tag{5.6}$$

In [105], Kelly et al. suggested a simple algorithm that converges to the proportionally fair rate vector.

We now take a look at the PF from another perspective. Let us consider the following optimization problem (**P**):
maximize

$$g = \Sigma_i p_i f(x_i), \tag{5.7}$$

Table 5.1
Comparison of Complexity of MP and WF Algorithm

The Linear Constraints	χ: n-Dimensional Feasible Set with m Linear Inequalities	χ: Being Defined Implicity with an l-Dimensional Slack Variables	
		Implicit Set	Free Disposal
The complexity of MP algorithm	$O(nF(m))$	$O(nF(m))$	$O(nF(m))$
The complexity of WF algorithm	$O(nm)$	$O(nm)$	$O(nm) + F(m)$
Remarks	$F(m)$ is the complexity of linear programming	Each of the n steps of MP algorithm is a linear programming problem. There are n steps, each of complexity being $O(m)$ in WF algorithm	An additional linear programming problem for the explicit characterization is needed

subject to

$$A^T x \leqslant c, \tag{5.8}$$

over

$$x \geqslant 0, \tag{5.9}$$

where f is an increasing strictly concave function and the components p_i are positive numbers as the proportional factors.

Considering the optimization problem (**P**), given the fact that, the objective function (5.7) is strictly concave and the feasible region expressed in the constraints (5.8) and (5.9) is compact, we conclude that the optimal solution of (**P**) exists and is unique.

Let

$$L(x, \mu) = g(x) + \mu^T (c - A^T x).$$

From the Karush–Kuhn–Tucker conditions [144] in Theorem 3.1.1, we have

$$\nabla g^T - \mu^T A^T = 0, \tag{5.10}$$

$$\mu_j (c_j - A_j^T x^*) = 0, \quad for\ all\ j, \tag{5.11}$$

$$A^T x^* \leqslant c, \tag{5.12}$$

$$x^* \geqslant 0, \mu \geqslant 0, \tag{5.13}$$

where $\nabla g^T = (p_1 f'(x_1), \ldots, p_n f'(x_n))$. There is a very special situation, where there are only one link and N flows (connections) in the network, one finds that the optimal solution of (**P**) is $x_i = c/N$ for all i. This suggests that all the flows (connections) have an equal share of the bottleneck capacity, no matter whatever the increasing concave function f is. Particularly, Equation 5.10 implies $f'(x_i) = \mu$ for all i, so that $x_i = f'^{-1}(\mu)$ for all i. If x is a proportionally fair vector, then it solves (**P**) when we take the objective function (i.e., the utility function)

$$f(x) = \log(1 + x)$$

with $p_i = 1$ for all i. Therefore, one concludes that a proportionally fair vector is the set of bandwidth allocation that maximizes the sum of all the logarithmic utility functions [115].

To obtain the fair rate vector is not straightforward for the network with multiple bottlenecks. Let us consider the simple network (see Figure 5.1) with three bottleneck nodes and four flows (users). Under the assumption that the three links have the capacity c_1, c_2, and c_3, respectively, and they satisfy $c_1 = c_2 < c_3$. The max–min fair rate vector of this network is

$$(c_1/3,\ c_1/3,\ c_1/3,\ c_3 - (c_1/3)).$$

However, the proportionally fair rate vector is not the same as the max–min fair rate in this case because if we decrease the rate of User 1, the sum of the utility functions f will increase. Hence, the optimal vector x depends on the utility function f when there are more than two bottlenecks in the network.

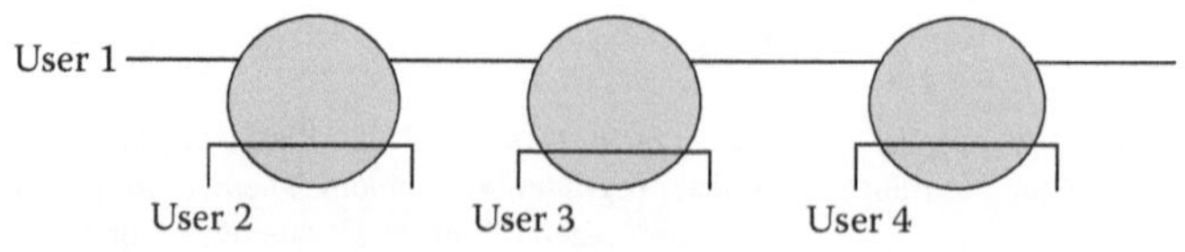

Figure 5.1 A simple network with three bottleneck nodes and four flows (users).

The concavity of the utility function f has impact on the fair sharing among flows (users). If f is a convex increasing function but not concave, then to maximize the objective function g of (**P**), the flow rate x_i that is larger should be further increased, since by increasing x_i one can increase $f(x_i)$. When f is linear, the rate of increase of f is the same for all x. When f is concave, the flow with a rate being smaller x_i will be favored, considering $f'(x) > f'(y)$ if $x < y$.

5.4.1 FAST TRANSMISSION CONTROL PROTOCOL: REALIZATION OF WEIGHTED PF IN BANDWIDTH ALLOCATION

Congestion control is needed for the aim of sharing network resources among competing users while guaranteeing quality of service (QoS) delivery. It is extremely important in the situations, where the resources are very limited and the competing users vary over time unpredictably, yet efficient sharing is desired. Feedback control is seen to be extremely important to be implemented in networks as the distributed algorithm to generate the proper actions to cope with the unpredictable supply and demand constraints. Under the feedback control, traffic sources are able to adapt their rates to congestion in their paths dynamically. Within the Internet, this is performed by the so-called transmission control protocol (TCP) in source and the proper mechanism implemented at destination computers involved in data transfers.

The current TCP congestion control algorithm is referred to as TCP Reno [168]. Its enhancements are described in [293–295]. TCP Reno has been working very well even though the Internet has been experiencing huge growth in terms of size, speed, load, and connectivity. However, as bandwidth-delay product continues to scale up, TCP Reno is found to be a bottleneck problem in performance. In networks with large bandwidth-delay products, TCP Reno has encountered the following four difficulties [172] with unsatisfactory performance:

- Linear increasing is too slow, but multiplicative decrease is too dramatic;
- To maintain large congestion window asks for small packet loss probability;
- Oscillation is a problem in dynamic;
- Packet loss probability is unpredictable, and flow dynamics is hard to be stabilized.

Motivated by the significant advantage of delay-based protocol (e.g., high-speed TCP (HSTCP) [170] and scalable TCP (STCP) [171]) over loss-based approach [296,297] at high data transmission speed, a new approach, FAST TCP, using delay as a congestion measure augmented with loss information is developed [172].

FAST TCP [172,290,334,352] is a new TCP congestion control algorithm for high-speed long-distance networks; it aims to rapidly stabilize networks into steady, efficient, and fair operating points. It uses queuing delay, in addition to packet loss, as a congestion signal. Queuing delay provides a finer measure of congestion and scales more naturally with network capacity than packet loss probability does [172]. Using the queuing delay as a congestion measure in its window-updating equation [334] allows FAST TCP to overcome difficulties [291] encountered by currently used algorithms (such as TCP Reno [292]) in networks with large bandwidth-delay products.

The paper [172] has presented experimental results on FAST TCP by the Linux prototype and compared its performance with TCP Reno, HSTCP, and STCP. Among them, FAST TCP outperforms the other three protocols in terms of throughput, fairness, stability, and responsiveness.

5.4.2 WEIGHTED PROPORTIONAL FAIR AND FAST TCP

A general network can be described as a set $L = \{1, \ldots, M\}$ of links, shared by a set $I = \{1, \ldots, N\}$ of flows. Each link $l \in L$ has a capacity c_l. Flow $i \in I$ traveled a router L_i consisting of a subset of links, i.e., $L_i = \{l \in L \,|\, i \text{ traverses } l\}$. A link l is shared by a subset I_l of flows where $I_l = \{i \in I \,|\, i \text{ traverses } l\}$. Let x_i be the rate of flow i and $x = \{x_i, i \in I\}$ be the rate vector. Let $A = (A_{li}, i \in$

$I, l \in L$) be the routing matrix, where $A_{li} = 1$ if flow i traverses link l and 0 otherwise. Throughout this section, the terms "flow," "sources," and "users" are used synonymously.

A rate vector $x \geq 0$ is called *feasible* if

$$\sum_{i \in I_l} x_i \leq c_l, \ \forall l \in L. \tag{5.14}$$

The notion of fairness characterizes how competing users should share the bottleneck resources subject to the above constraint. A feasible flow rate vector x is defined to be *max–min* fair if any rate x_i cannot be increased without decreasing some x_j which is smaller than or equal to x_i [139]. Kelly et al. [105] proposed the so-called PF. A rate vector x^* is α_i-weighted proportional fair if it is feasible, and if for any other feasible vector x_i, the aggregate of proportional change is nonpositive,

$$\sum_{i \in I} \alpha_i \frac{x_i - x_i^*}{x_i^*} \leq 0, \tag{5.15}$$

where α_i is positive numbers, $i = 1, 2, \ldots$.

Consider the following optimization problem (**P**):

$$\max_{x \geq 0} \sum_{i \in I} U_i(x_i), \tag{5.16}$$

subject to the constraint given by Equation 5.14, where U_i is the utility function of user i. We follow the standard approach [105] of taking the Lagrangian

$$L(x; p) = \sum_i (U_i(x_i) - x_i q_i) - \sum_l p_l c_l, \tag{5.17}$$

where p_l, called the *price* of link l, is the Lagrange multiplier of the constraint due to the capacity of link l. We assume that

$$q_i(t) = \sum_{l=1}^{M} A_{li} p_l(t - \tau_{li}^b) \tag{5.18}$$

is the aggregate price observed by source i in its path and link l observes the aggregate source rate

$$y_l(t) = \sum_{i=1}^{N} A_{li} x_i(t - \tau_{li}^f), \tag{5.19}$$

where τ_{li}^f is the forward feedback delay from source i to link l and τ_{li}^b is the backward feedback delay from link l to source i. For simplicity, we assume that the feedback delays τ_{li}^f and τ_{li}^b are constants.

For given link prices, each source i determines its optimal rate as

$$x_i(p) = \arg\max_{x_i} U_i(x_i) - x_i q_i = (U_i')^{-1}(q_i). \tag{5.20}$$

The primal optimization (**P**) can then be replaced by its dual (**D**) given by

$$\min_{p \geq 0} \sum_i (U_i(x_i(p)) - q_i x_i(p)) + \sum_l c_l p_l. \tag{5.21}$$

According to [105], α_i-weighted PF is achieved within a system of social welfare maximization, if all users have utility functions of the following form:

$$f_i(x_i) = \alpha_i \log x_i. \tag{5.22}$$

That is, a α_i-weighted proportional fair vector solves the above optimization problem (**P**) by maximizing the sum of all the logarithmic utility functions. In this case, Equation 5.20 becomes

$$x_i = \frac{\alpha_i}{q_i}. \tag{5.23}$$

For the existing version of FAST TCP, it is known [172] that the following source window-updating equation has a unique equilibrium point (x_i^*, q_i^*) satisfying Equation 5.23

$$w_i(t+1) = \gamma\left(\frac{d_i w_i(t)}{d_i + q_i(t)} + \alpha_i(w_i(t), q_i(t))\right) + (1-\gamma)w_i(t), \tag{5.24}$$

where

$$\alpha_i(w_i, q_i) = \begin{cases} \alpha_i w_i & \text{if } q_i = 0 \\ \alpha_i & \text{otherwise.} \end{cases}$$

Since this equilibrium point is known ([172], Theorem 1) to be the unique optimal solution of the above problem (**P**) with the specific utility functions given by Equation 5.22, FAST TCP maximizes the sum of logarithmic utility functions. This implies, in particular, that the current FAST TCP achieves α_i-weighted PF. Note that the fairness parameter α_i is also the number of flow i's packets that are buffered in the routers in its path at equilibrium. If there are N flows, the total number of packets buffered in the routers at equilibrium is $\sum_{i=1}^{N} \alpha_i$ (see [334]). A guideline on how to set this very important parameter for FAST TCP is provided in [298].

5.4.3 PROPORTIONAL FAIR RESOURCE ALLOCATION IN WIRELESS NETWORKS

In the third-generation wireless data network, the PF concept has been implemented as a scheduling policy [259] at the base station (BS) to allocate the downlink resources among different users. Proportional fair is used as a channel-state-based scheduling algorithm that utilizes user diversity. In a wireless network, there are a number of active users competing a wireless channel with the channel condition experienced by each user varying independently and individually. Better channel conditions generate higher data rate and poor channels lead to lower data rate. Each user continuously measures channel condition and then feedback it to its relay node or its BS with centralized PF scheduler.

Proportional fair scheduler is to schedule the user with the largest $maxR_i/A_i$ at every time slot. R_i is the rate achievable by user i and A_i is the average rate of the user i. A_i is computed over a time window. If user i is scheduled at the nth time slot, then at the $(n+1)$th time slot

$$A_i(n+1) = (1-\alpha)A_i(n) + \alpha R_i,$$

else

$$A_i(n+1) = (1-\alpha)A_i(n).$$

Let us consider the case of 10 user classes (refer to Figure 5.2), each user has disparate rates

$$R_1 = 2.4\,\text{Mbps},\ R_2 = 2.6\,\text{Mbps},\ R_3 = 2.8\,\text{Mbps},\ R_4 = 3\,\text{Mbps},\ R_5 = 3.2\,\text{Mbps},$$
$$R_6 = 3.4\,\text{Mbps},\ R_7 = 3.6\,\text{Mbps},\ R_8 = 3.8\,\text{Mbps},\ R_9 = 4\,\text{Mbps},\ R_{10} = 4.2\,\text{Mbps},$$

and has disparate average rate

$$A_1(0) = 2.0\,\text{Mbps},\ A_2(0) = 2.2\,\text{Mbps},\ A_3(0) = 2.4\,\text{Mbps},$$
$$A_4(0) = 2.6\,\text{Mbps},\ A_5(0) = 2.8\,\text{Mbps},\ A_6(0) = 3.0\,\text{Mbps},$$
$$A_7(0) = 3.2\,\text{Mbps},\ A_8(0) = 3.4\,\text{Mbps},\ A_9(0) = 3.6\,\text{Mbps},$$
$$A_{10}(0) = 3.8\,\text{Mbps}.$$

By taking, for example, $\alpha = 0.2$, when $n = 1$, we have

$$A_i(1) = A_i(0),\ \max\frac{R_i}{A_i} = \frac{R_1}{A_1},$$

thus the first time slot is distributed to the first user. Then, $A_i(2)$ is computed by

$$A_1(2) = (1-\alpha)A_1(1) + \alpha R_1,\ A_2(2) = (1-\alpha)R_2, \ldots, A_{10}(2) = (1-\alpha)A_{10},$$

the second time slot is, therefore, distributed to the second user. With continued computing, while $1 \leqslant n \leqslant 10$, the nth time slot is distributed to the nth user. While $n > 10$, the time slot is circulation distributed to the relevant user, the nth time slot is distributed to $(n/10)^{\text{th}}$ user. The sequence of proportional fair scheduler is related to $R_i/A_i(0)$: the larger of the $R_i/A_i(0)$, the sooner the users are scheduled. For the wireless communication system of Figure 5.2, we perform the above computing process of proportional fair time slots assignment for the case of $\alpha = 0.2$ and for the case of $\alpha = 0.8$. All computing results are tabulated into Tables 5.2 and 5.3, respectively. These results are then also reflected in Figure 5.2.

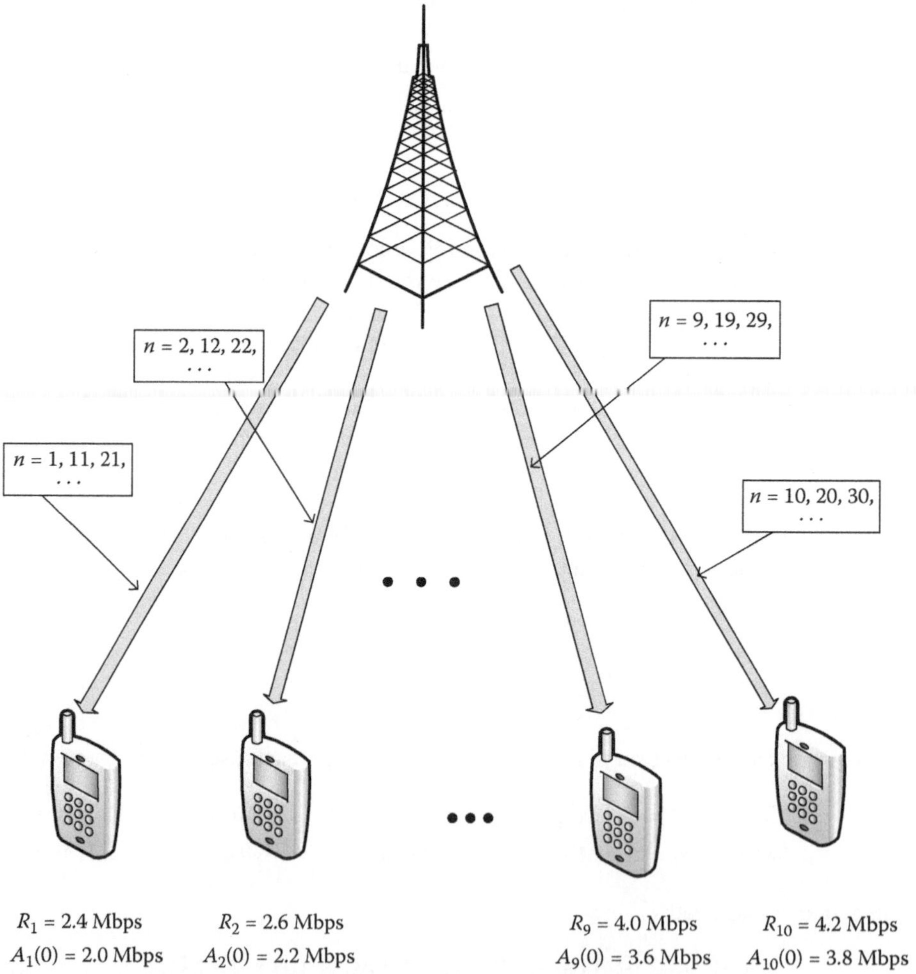

Figure 5.2 An example of proportional fair scheduling being employed at the BS to schedule downlink flows among different users, where the BS serves 10 users, R_i $(i = 1, 2, \ldots, 10)$ is the rate achievable by user i and $A_i(0)$ is the initial average rate of user i. According to the proportional fair policy, the BS serves with User 1 at time slots $n = 1, 11, 21, \ldots$, with User 2 at time slots $n = 2, 12, 22, \ldots$, and so on.

Since the proportional fair scheduler was suggested [259,299] to be used in bandwidth allocation in 3G network (the Evolution Data Optimized (EV-DO) wireless data system), there have been a lot of work dedicated to this area. These developments include the following:

- The proportional fair algorithm to maximize the sum of the logarithm of the throughput of the users served by each BS [300];
- The user-level performance of the proportional fair algorithm in a dynamic setting [301];
- Extensions to proportional fair algorithm to support QoS [302–305];

Table 5.2
Time Slot Assignment to 10 Users by the PF Criteria for the Wireless Communication System of Figure 5.2: Results for α =0.2

n	The rate achievable by user and the initial average rate of user: $A_1(0)=2;\ A_2(0)=2.2;\ A_3(0)=2.4;\ A_4(0)=2.6;\ A_5(0)=2.8$ $A_6(0)=3;\ A_7(0)=3.2;\ A_8(0)=3.4;\ A_9(0)=3.6;\ A_{10}(0)=3.8$	$\max \frac{R_i}{A_i}$
$n=1$	$A_1(1)=2;\ A_2(1)=2.2;\ A_3(1)=2.4;\ A_4(1)=2.6$ $A_5(1)=2.8;\ A_6(1)=3;\ A_7(1)=3.2;\ A_8(1)=3.4$ $A_9(1)=3.6;\ A_{10}(1)=3.8$	$\max \frac{R_i}{A_i}=\frac{R_1}{A_1}=1.2$
$n=2$	$A_1(2)=2.08;\ A_2(2)=1.76;\ A_3(2)=1.92;\ A_4(2)=2.08$ $A_5(2)=2.24;\ A_6(2)=2.4;\ A_7(2)=2.56;\ A_8(2)=2.72$ $A_9(2)=2.88;\ A_{10}(2)=3.04$	$\max \frac{R_i}{A_i}=\frac{R_2}{A_2}=1.47$
$n=3$	$A_1(3)=1.664;\ A_2(3)=1.928;\ A_3(3)=1.536;\ A_4(3)=1.664$ $A_5(3)=1.792;\ A_6(3)=1.92;\ A_7(3)=2.048;\ A_8(3)=2.176$ $A_9(3)=2.304;\ A_{10}(3)=2.432$	$\max \frac{R_i}{A_i}=\frac{R_3}{A_3}=1.82$
$n=4$	$A_1(4)=1.3312;\ A_2(4)=1.5424;\ A_3(4)=1.7888$ $A_4(4)=1.3312;\ A_5(4)=1.4336;\ A_6(4)=1.536$ $A_7(4)=1.6384;\ A_8(4)=1.74080;\ A_9(4)=1.8432$ $A_{10}(4)=1.9456$	$\max \frac{R_i}{A_i}=\frac{R_4}{A_4}=2.56$
$n=5$	$A_1(5)=1.06496;\ A_2(5)=1.23392;\ A_3(5)=1.43104$ $A_4(5)=1.66496;\ A_5(5)=1.14688;\ A_6(5)=1.2288$ $A_7(5)=1.31072;\ A_8(5)=1.39264;\ A_9(5)=1.47456$ $A_{10}(5)=1.556480$	$\max \frac{R_i}{A_i}=\frac{R_5}{A_5}=2.79$
$n=6$	$A_1(6)=0.851968;\ A_2(6)=0.987136;\ A_3(6)=1.144832$ $A_4(6)=1.331968;\ A_5(6)=1.557504;\ A_6(6)=0.98304$ $A_7(6)=1.048576;\ A_8(6)=1.114112;\ A_9(6)=1.179648$ $A_{10}(6)=1.2451844$	$\max \frac{R_i}{A_i}=\frac{R_6}{A_6}=3.46$
$n=7$	$A_1(7)=0.6815744;\ A_2(7)=0.7897088;\ A_3(7)=0.9158656$ $A_4(7)=1.0655744;\ A_5(7)=1.2460032;\ A_6(7)=1.466432$ $A_7(7)=0.8388608;\ A_8(7)=0.8912896;\ A_9(7)=0.9437184$ $A_{10}(7)=0.9961472$	$\max \frac{R_i}{A_i}=\frac{R_7}{A_7}=4.29$
$n=8$	$A_1(8)=0.54525952;\ A_2(8)=0.63176704;\ A_3(8)=0.73269248$ $A_4(8)=0.85245952;\ A_5(8)=0.99680256;\ A_6(8)=1.1731456$ $A_7(8)=1.39108864;\ A_8(8)=0.71303168;\ A_9(8)=0.75497472$ $A_{10}(8)=0.79691776$	$\max \frac{R_i}{A_i}=\frac{R_8}{A_8}=5.33$
$n=9$	$A_1(9)=0.436207616;\ A_2(9)=0.505413632;\ A_3(9)=0.586153984$ $A_4(9)=0.681967616;\ A_5(9)=0.797442048;\ A_6(9)=0.93851648$ $A_7(9)=1.112870912;\ A_8(9)=1.330425344;\ A_9(9)=0.603979776$ $A_{10}(9)=0.637534208$	$\max \frac{R_i}{A_i}=\frac{R_9}{A_9}=6.62$
$n=10$	$A_1(10)=0.3489660928;\ A_2(10)=0.4043309056;\ A_3(10)=0.4689231872$ $A_4(10)=0.5455740928;\ A_5(10)=0.6379536384;\ A_6(10)=0.750813184$ $A_7(10)=0.8902967296;\ A_8(10)=1.0643402752$ $A_9(10)=1.2831838208;\ A_{10}(10)=0.5100273664$	$\max \frac{R_i}{A_i}=\frac{R_{10}}{A_{10}}=8.23$

Table 5.3
Time Slot Assignment to 10 Users by the PF Criteria for the Wireless Communication System of Figure 5.2: Results for α = 0.8

n	The rate achievable by user and the initial average rate of user: $A_1(0) = 2$; $A_2(0) = 2.2$; $A_3(0) = 2.4$; $A_4(0) = 2.6$; $A_5(0) = 2.8$ $A_6(0) = 3$; $A_7(0) = 3.2$; $A_8(0) = 3.4$; $A_9(0) = 3.6$; $A_{10}(0) = 3.8$	$\max \frac{R_i}{A_i}$
$n = 1$	$A_1(1) = 2$; $A_2(1) = 2.2$; $A_3(1) = 2.4$; $A_4(1) = 2.6$ $A_5(1) = 2.8$; $A_6(1) = 3$; $A_7(1) = 3.2$; $A_8(1) = 3.4$ $A_9(1) = 3.6$; $A_{10}(1) = 3.8$	$\max \frac{R_i}{A_i} = \frac{R_1}{A_1} = 1.2$
$n = 2$	$A_1(2) = 2.32$; $A_2(2) = 0.44$; $A_3(2) = 0.48$ $A_4(2) = 0.52$; $A_5(2) = 0.56$; $A_6(2) = 0.6$; $A_7(2) = 0.64$ $A_8(2) = 0.68$; $A_9(2) = 0.72$; $A_{10}(2) = 0.76$	$\max \frac{R_i}{A_i} = \frac{R_2}{A_2} = 5.9$
$n = 3$	$A_1(3) = 0.464$; $A_2(3) = 2.168$; $A_3(3) = 0.096$ $A_4(3) = 0.104$; $A_5(3) = 0.112$; $A_6(3) = 0.12$ $A_7(3) = 0.128$; $A_8(3) = 0.136$; $A_9(3) = 0.144$ $A_{10}(3) = 0.152$	$\max \frac{R_i}{A_i} = \frac{R_3}{A_3} = 29.2$
$n = 4$	$A_1(4) = 0.0928$; $A_2(4) = 0.4336$; $A_3(4) = 2.2592$ $A_4(4) = 0.0208$; $A_5(4) = 0.0224$; $A_6(4) = 0.024$ $A_7(4) = 0.0256$; $A_8(4) = 0.0272$; $A_9(4) = 0.0288$ $A_{10}(4) = 0.0304$	$\max \frac{R_i}{A_i} = \frac{R_4}{A_4} = 144.2$
$n = 5$	$A_1(5) = 0.01856$; $A_2(5) = 0.08672$; $A_3(5) = 0.45184$ $A_4(5) = 2.40416$; $A_5(5) = 0.00448$; $A_6(5) = 0.0048$ $A_7(5) = 0.00512$; $A_8(5) = 0.00544$; $A_9(5) = 0.00576$ $A_{10}(5) = 0.00608$	$\max \frac{R_i}{A_i} = \frac{R_5}{A_5} = 714.3$
$n = 6$	$A_1(6) = 0.003712$; $A_2(6) = 0.017344$; $A_3(6) = 0.090368$ $A_4(6) = 0.480832$; $A_5(6) = 2.560896$; $A_6(6) = 0.00096$ $A_7(6) = 0.001024$; $A_8(6) = 0.001088$ $A_9(6) = 0.001152$; $A_{10}(6) = 0.001216$	$\max \frac{R_i}{A_i} = \frac{R_6}{A_6} = 3542$
$n = 7$	$A_1(7) = 0.0007424$; $A_2(7) = 0.0034688$; $A_3(7) = 0.0180736$ $A_4(7) = 0.0961664$; $A_5(7) = 0.5121792$; $A_6(7) = 2.720192$ $A_7(7) = 0.0002048$; $A_8(7) = 0.0002176$ $A_9(7) = 0.0002304$; $A_{10}(7) = 0.0002432$	$\max \frac{R_i}{A_i} = \frac{R_7}{A_7} = 17548$
$n = 8$	$A_1(8) = 0.00014848$; $A_2(8) = 0.00069376$; $A_3(8) = 0.00361472$ $A_4(8) = 0.01923328$; $A_5(8) = 0.10243584$; $A_6(8) = 0.5440384$ $A_7(8) = 2.88004096$; $A_8(8) = 0.00004352$ $A_9(8) = 0.00004608$; $A_{10}(8) = 0.00004864$	$\max \frac{R_i}{A_i} = \frac{R_8}{A_8} = 87316$
$n = 9$	$A_1(9) = 0.000029696$; $A_2(9) = 0.000138752$; $A_3(9) = 0.000722944$ $A_4(9) = 0.003846656$; $A_5(9) = 0.020487168$; $A_6(9) = 0.10880768$ $A_7(9) = 0.576008192$; $A_8(9) = 3.040008704$ $A_9(9) = 0.000009216$; $A_{10}(9) = 0.000009728$	$\max \frac{R_i}{A_i} = \frac{R_9}{A_9} = 434077$
$n = 10$	$A_1(10) = 0.0000059392$; $A_2(10) = 0.0000277504$; $A_3(10) = 0.0001445888$ $A_4(10) = 0.0007693312$; $A_5(10) = 0.0040974336$; $A_6(10) = 0.021761536$ $A_7(10) = 0.1152016384$; $A_8(10) = 0.6080017408$ $A_9(10) = 4.0000018432$; $A_{10}(10) = 0.0000019456$	$\max \frac{R_i}{A_i} = \frac{R_{10}}{A_{10}} = 2158717$

- Exponential rule performing best in gracefully trading-off delay versus throughput [303,304];
- The fairness aspect of the proportional fair algorithm [306,307];
- A monitoring-based feedback algorithm to correct the unfairness of the proportional fair algorithm under heterogeneous channel conditions [307];
- Unfairness using the proportional fair algorithm when users experiencing heterogeneous channel quality [301];

- A new definition of fairness that extends the absolute fairness bound measurement [306];
- The Generalized Processor Sharing discipline [308];
- An opportunistic proportional fair algorithm that is a weighted combination of the max rate scheduling [303];
- Load balancing approach [309,310];
- Load balancing and fairness (max–min fairness) being considered jointly in the wireless LAN context [311];
- Explicitly multihop transmission and the cases of Markovian traffic with a two-priority scheduler, as well as orthogonal modulation with power splitting [312].

We assume that, the channel measurement feedback speed is relatively faster than the channel rate variation, the scheduler then has a good enough estimate of all the users channel condition once it schedules a packet to be transmitted to the user. Since channel condition varies independently among different users, PF takes advantage of user diversity by selecting the user with the best condition to transmit during different time slots. Compared to a round-robin scheduler, this approach can improve overall throughput significantly. However, such a throughput maximizing scheme leads to the unfair disadvantage in the sense that users with relatively poorer channel conditions can be starved. Further, such an approach can result in non-Pareto optimal bandwidth allocation when considering the network as a whole. Therefore, it is important to consider PF in a network-wide context with user associations to BSs governed by optimizing a generalized proportional fairness (GPF) objective [260]. The paper [260] proposes efficient off-line optimal algorithms and heuristic-based greedy online algorithms to solve the above problems.

A rigorous formulation of the network-wide PF bandwidth allocation problem has been given by [260]. All notations and variables used in the GPF scheduling models and algorithms proposed by [260] are given by Table 5.4.

Let us consider a 3G wireless data network consisting of a set of BSs Γ and a set of users U. A user u's average data rate if associated with a BS a is denoted by $ra_{l\alpha}$. Let $S_l = \{\alpha | ra_{l\alpha} > 0, \forall \alpha \in \Gamma\}$. Let r_l be the actual bandwidth allocation to user l by the network. If we only consider

Table 5.4
Notations and Variables Used in the GPF Scheduling Models and Algorithms That Are Proposed by [260]

Variables and Notations	Denotations for
l	User
α	Base station (BS)
Γ	The set of BSs
$ra_{l\alpha}$	The average data rate of of l with a BS α
L	The set of users
r_l	The actual bandwidth allocation to user l by the network
$z_{l\alpha}$	The association variable, i.e., $z_{l\alpha} = 1$ if user l is associated with BS α; 0, otherwise
S_l	The set of α, for $\forall \alpha \in \Gamma, ra_{l\alpha} > 0$
Z_α	The sum of $z_{l\alpha}$, while $l \in L$
H	The mapping between l, α, and $ra_{l\alpha}$
θ_l	The scheduling priority of user l
w	The mapping between l and w_l
Φ_α	The user whose $z_{l\alpha} = 1$, while $\forall l \in L$
$g_a\{\Phi_\alpha, l, H, \theta_l\}$	The function defines actual bandwidth r_l by BS α
$M()$	Multiuser diversity gain

the situation, where 3G data users have the typical elastic traffic, then the following formulation describes the PF allocation of network resources

$$\max \sum_{l\in L} \log(r_l+1). \tag{5.25}$$

Note that the above formulation does not cover the situation where the 3G network has inelastic traffic, for example, the real-time multimedia traffic. For this case, we have to use the corresponding utility function of either a step function or a sigmoid function. A step utility function of the real-time multimedia traffic represents that the traffic is on hard requirement for bandwidth, while a sigmoid utility function represents that the traffic is flexible on requirement for bandwidth. In 3G network, there is an admissible number of users to the network. Regarding 3Glx-EVDO, it can accommodate a total of 60 Walsh codes for orthogonal transmission. Given this, one may want to achieve that limit of 60 active users in a BS and in a sector carrier by a proper scheduling policy. However, it should be noted that the maxima of 60 active user number is very difficult to achieve in practice because it is almost impossible to attain 600 kbps downlink channel by activating a total of 60 users simultaneously. If so, there is very little bandwidth to offer to each user.

Let $z_{l\alpha}$ be the association variable, i.e., $z_{l\alpha}=1$, if user l is associated with BS α; 0, otherwise. With adherence to the EV-DO standard [259], every user is only associated with one single BS, i.e., one BS only transmit, data through downlink to one single user. If all users must be admitted to the communications, we should have

$$\sum_{l\in L} z_{l\alpha}=1.$$

Let the bandwidth allocation for users associated with a given BS be proportional fair. Let the number of users associated with BS α be $Z_\alpha=\sum_{(l\in L)} z_{l\alpha}$. Let denote the set of users associated with a given BS α be Φ_α, i.e., $\Phi_\alpha=\{l|z_{l\alpha}=1,\forall l\in L\}$. In the general case, the multiuser diversity gain can depend on the set of users, not just the number of users. Let H be the mapping that, given l and α, returns $ra_{l\alpha}$. Users may have different scheduling priorities for service differentiation. We denote the scheduling priority of user l as θ_l. Let w be the mapping that given a user l returns w_l. If $z_{l\alpha}=1$ for a given user l and BS α, its actual bandwidth r_l allocation by BS α will be a general function of all the users associated with α. Denote this function by $g_\alpha(\Phi_\alpha,l,H,\theta)$. Of course, $g_\alpha(\Phi_\alpha,l,H,\theta)=0$, if $l\notin\Phi_\alpha$. We now derive the following problem formulation for GPF1 [260]:

$$\max \sum_{\alpha\in\Gamma}\sum_{l\in\Phi_\alpha} (\log(g_\alpha(\Phi_\alpha,l,H,\theta))) \tag{5.26}$$

subject to

$$\sum_{\alpha\in S_l} z_{l\alpha}=1,\forall l\in L \tag{5.27}$$

$$\Phi_\alpha=\{l|z_{l\alpha}=1,\forall l\in L\} \tag{5.28}$$

$$z_{l\alpha}=\{0,1\} \tag{5.29}$$

If the relative rate fluctuations are statistically identical, the multiuser diversity gain only depends on the number of users associated with a given BS, we obtain the following restricted version of the GPF1 problem, which is referred to as GPF2 [260].

$$\max \sum_{l\in L}\sum_{\alpha\in S_l} z_{l\alpha}\log\left(ra_{l\alpha}w_l\frac{\Phi(g_\alpha)}{\omega_\alpha}\right) \tag{5.30}$$

subject to

$$\sum_{\alpha\in S_l} z_{l\alpha}=1,\forall l\in L \tag{5.31}$$

$$\omega_\alpha = \sum_{l:\alpha \in S_l} z_{l\alpha} w_l, \forall \alpha \in \Gamma \tag{5.32}$$

$$Z_\alpha = \sum_{l:\alpha \in S_l} z_{l\alpha}, \forall \alpha \in \Gamma \tag{5.33}$$

$$z_{l\alpha} = \{0,1\} \tag{5.34}$$

When all the users have the same priority in GPF2, we have the following special case, GPF3 [260]:

$$\max \sum_{l \in L} \sum_{l \in S_l} z_{l\alpha} \log \left(ra_{l\alpha} \frac{M(Z_\alpha)}{Z_\alpha} \right) \tag{5.35}$$

subject to

$$\sum_{\alpha \in S_l} z_{l\alpha} = 1, \forall \alpha \in \Gamma \tag{5.36}$$

$$Z_\alpha = \sum_{l:\alpha \in S_l} z_{l\alpha}, \forall \alpha \in \Gamma \tag{5.37}$$

$$z_{l\alpha} = \{0,1\} \tag{5.38}$$

By studying the above three models GPF1, GPF2, and GPF3, we have the following results [260]:

- The GPF1 problem is NP-hard;
- The GPF2 problem is NP-hard;
- There does not exist a polynomial time algorithm that can approximate the solutions of GPF1 and GPF2;
- If we know Z_α in GPF3, then the problem can be solved optimally in polynomial time.

For the special case where multiuser diversity only depends on the number of users scheduled together, which is the model GPF3, efficient off-line optimal algorithms and heuristic-based greedy online algorithms are proposed in [260] to solve the optimization problem GPF3. The off-line optimal algorithm is the following Algorithm **OfflineOPT-*K* BS**, whereas the heuristic-based greedy online algorithm is the following Algorithm **KComponent**.

Algorithm **OfflineOPT-*K* BS**
Input:
Network H = (A,K),multiuser diversity
gain $M(), ra_{l\alpha}, \forall \alpha \in, l \in L$
1. for each $(Z_1, \ldots, Z_{|\Gamma|})$ such that $\sum_{i=1}^{|\Gamma|} = n$
2. if $Z_i > |\Phi_i|$, where $\Phi_i = \{l | ra_{li} > 0, \forall l \in L\}$,
3. then next iteration
4. if $ra_{li} = 1$ and $Z_i > 0$
5. $p_{li} = log \left(r_{li} \frac{G(Z_i)}{Z_i} \right)$
6. else $p_{li} = 0$
7. MatchingAlgo$(H, \{p_{li}\})$
end

Algorithm **KComponent**
Input:
Network H = (A,U),multiuser diversity
gain $M(), mapping(H) s.t. RA(l,\alpha) = ra_{l\alpha}, \forall \alpha \in, l \in L$
1. Run MinK-Cut algorithm to obtain
2. $H_1, \ldots, H_K$ connected components
3. for each $H_i, \forall i, \ldots, K$
4. OfflineOPT-K BS(H_i, G, RA)
5. for each user l whose edges cross components
6. Greedily assign l to BS α that improves
7. the objective function the most
end

5.5 (p,β)-PROPORTIONAL FAIRNESS

Considering the network depicted by Figure 5.1, controversy may arise on what is a fair rate allocation for it. One may argue that the max–min fair rate is mostly desirable on one hand. On the other hand, however, User 1 is using more resources than the other three flows under the max–min fair rate, for the situation where $c_3 > 2/3c_1$. In a general sense, the problem is reduced to how to compromise between the fairness among users and the utilization of resources. We will go back to this issue a bit later.

The max–min fairness enforces higher priority to the fairness while less consideration is given to resource utilization. The following definition of (p,β)-PF is a generalization of PF and max–min fairness. When $\beta = 1$, the (p,β)-PF is reduced to that of PF and as β approaches infinity, it converges to that of max–min fair.

Definition 5.5.1 *((p,β)-proportionally fair) [115]: Let $p = (p_1, \ldots, p_N)$ and β be positive numbers. A vector of rates x^* is (p,β)-proportionally fair if it is feasible and for any other feasible vector x*

$$\Sigma p_i \frac{x_i - x_i^*}{x_i^{*\beta}} \leqslant 0. \tag{5.39}$$

Note that Equation 5.39 reduces to Equation 5.6 when $p = (1, \ldots, 1)^T$ and $\beta = 1$.

The following theorem gives the relationship between the above definition and the optimization problem (**P**).

Theorem 5.5.7

[115] If we take the utility function f_β as

$$f_\beta(x) = \begin{cases} \log(x+1), & \text{if } \beta = 1 \\ (1-\beta)^{-1} x^{1-\beta}, & \text{otherwise.} \end{cases}$$

Then the rate vector x^* solves the problem (**P**) with $f = f_\beta$ iff x^* is (p,β)-proportionally fair. ■

Proof: If x^* is a solution of (**P**), then X^* is (p,β)-proportionally fair. Multiplying Equation 5.10 by $(x - x^*)$, we find

$$\nabla g^T (x - x^*) = \mu^T A^T (x - x^*).$$

Taking the summation of the identity Equation 5.11 over j, we obtain

$$\mu^T c = \mu^T A^T x^*.$$

From the constraint that the aggregate flow rate should not be over the link capacity, we immediately have

$$\mu^T A^T x \leqslant \mu^T c.$$

Combining the above relations, we see that

$$\mu^T A^T x \leqslant \mu^T A^T x^*.$$

Therefore,

$$\nabla g^T (x - x^*) = \mu^T A^T (x - x^*) \leqslant 0.$$

This contradicts to the fact that

$$\nabla g^T (x - x^*) = \Sigma_i (\frac{p_i(x - x^*)}{(x^{*\beta})}).$$

Hence, we have

$$\Sigma_i (\frac{p_i(x - x^*)}{(x^{*\beta})}) \leqslant 0.$$

So far, we have proved that the solution of (**P**) satisfies Equation 5.39.

Now we prove that if x^* is (p, β)-proportionally fair, then x^* is a solution of the optimization problem (**P**). Note that

$$\begin{aligned} g(x) &= g(x^*) + \nabla g(x^*)^T (x - x^*) + (1/2)(x - x^*)^T \nabla^2 g(x^*)(x - x^*) \\ &\quad + O(\| x - x^* \|^2). \end{aligned}$$

The second term

$$\nabla g(x^*)^T (x - x^*) = \Sigma_i (\frac{p_i(x - x^*)}{(x^{*\beta})}) \leqslant 0$$

for all feasible x and the third term is strictly negative by negative definiteness of $\nabla^2 g$. Hence, x^* is a local minimum. The vector x^* then solves the optimization problem (**P**) given that the problem (**P**) has a unique global solution.

We will now consider the case in which $p_i = 1$ for all users. If the component x^β is the optimal solution to the optimization problem (**P**) by taking the function f_β to be the following form

$$f_\beta = -(-h)^\beta,$$

and the constraint set is

$$\chi = \{x \mid A^T x \leqslant c, x \geqslant 0\}.$$

The above set χ is seen to be compact. If we take a sequence $\{x^\beta\}$ from this compact set χ, there must exist a subsequence denoted by $\{x^{\beta_k}, k \geqslant 1\}$ and some values of β_k, such that x^{β_k} converges to some

$$\bar{x} \in \chi, \quad k \to \infty.$$

We see that $\bar{x}$ is the max–min vector. Since that max–min vector is unique, one concludes that all the limit points of $\{x^\beta\}$ are the unique max–min vector. Therefore, the subsequence $\{x^\beta\}$ converges to the max–min vector.

On the contrary, if $\overline{x}$ is not a max–min vector, then there exists a user i whose rate $\overline{x_i}$ can be increased at the cost of decreasing the rates of other users $\overline{x_j}$, which are greater than $\overline{x_i}$. Let L_1 be the set of saturated (fully utilized) links used by i and L_2 be the set of the other links used by i. For each link $l \in L_1$, there exists a user, say $u(l)$, whose rate $\overline{x}_{u(l)}$ is greater than $\overline{x}_i$. That is, $\overline{x}_{u(l)} \geqslant x_i$ for $j \in L_1$. Define ζ by

$$\zeta = \frac{1}{N}\min\{\min_{l\in L_1}(\overline{x}_{u(l)} - \overline{x}_i), \min_{l\in L_2}(c_l - (A^T\overline{x})_l)\}, \tag{5.40}$$

where N is the number of users. For simplicity, we denote f_{β_k} and x^{β_k} by f_k and x^k, respectively. From the fact that x^k converges to $\overline{x}$, we can find x_0 such that for all $k \geqslant k_0$, for all j

$$\overline{x}_j - \frac{\zeta}{N+10} \leqslant x_j^k \leqslant \overline{x}_j + \frac{\zeta}{N}. \tag{5.41}$$

Now define the sequence of vectors y^k as follows:

$$y_j^k = \begin{cases} x_j^k + \zeta, & \text{if } j = i, \\ x_j^k - \zeta, & \text{if } j = u(l) \text{ for } l \in L_1, \\ x_j^k, & \text{otherwise.} \end{cases}$$

It can be shown $y^k \geqslant 0$ and $A^T y^T \leqslant c$ for $k \geqslant k_0$ without difficulty since we choose ζ small enough. We now establish a contradiction with the optimality of x^k. Consider the expression A_k defined by

$$A_k = \Sigma_j(f_k(y_j^k) - f_k(x_j^k)). \tag{5.42}$$

From the optimality of x^k, we have

$$A_k \leqslant 0. \tag{5.43}$$

Now

$$A_k = f_k(x_i^k + \zeta) - f_k(x_i^k) + \Sigma_{l\in L_1}(f_k(x_{u(l)}^k - \zeta) - f_k(x_{u(k)}^k)). \tag{5.44}$$

From the theorem of intermediate values, there exist numbers c_i^k such that

$$\begin{cases} x_i^k \leqslant c_i^k \leqslant x_i^k + \zeta, \\ f_k(x_i^k + \zeta) - f_k(x_i^k) = f_k'(c_i^k)\zeta. \end{cases}$$

Combining with Equation 5.41, we find

$$c_i^k \leqslant \overline{x}_i + (\zeta/N) + \zeta \leqslant \overline{x}_i + 2\zeta.$$

Similarly, one finds some numbers $c_{u(l)}^k$ such that

$$\begin{cases} x_{u(l)}^k - \zeta \leqslant c_{u(l)}^k \leqslant x_{u(l)}^k, \\ f_k(x_{u(l)}^k - \zeta) - f_k(x_{u(l)}^k) = -f_k'(c_{u(l)}^k)\zeta. \end{cases}$$

Taking into Equation 5.41 account, one finds that

$$c_{u(l)}^k \geqslant \overline{x}_i + 3\zeta.$$

Therefore, we have

$$\begin{aligned} A_k &= \zeta(f_k'(c_i^k) - \Sigma_{l\in L_1} f_k'(c_{u(l)}^k)), \\ &\geqslant \zeta(f_k'(\overline{x}_i + 2\zeta) - G f_k'(\overline{x}_i + 3\zeta)), \\ &= \zeta f_k'(\overline{x}_i + 2\zeta)(1 - G\frac{f_k'(\overline{x}_i + 2\zeta)}{f_k'(\overline{x}_i + 3\zeta)}), \end{aligned} \tag{5.45}$$

where G is the cardinality of L_1. The inequality follows from the concavity of f_k and the bounds on c_i^k and $c_{u(k)}^k$. Since the last term in the parenthesis tends to 1 as k increases and $f_k^{'} > 0$, for k large enough $A_k > 0$, which is a contradiction to Equation 5.43. Therefore, we have the following statement that outlines the relation between max–min fair rate and the parameter β.

Theorem 5.5.8

If $h(x)$ is a differentiable increasing concave negative function when $x \geqslant 0$, the solution of (**P**) with $f_\beta = -(-h)^\beta$ approaches the max–min fair rate vector as $\beta \to \infty$. ■

If we take

$$f_\beta(x) = (-1/(\beta - 1))(1/x)^{\beta-1},$$

which satisfies the conditions of the above theorem with

$$h = -(1/x),$$

we then have the following corollary.

Corollary 5.5.1 *The (p, β)-proportionally fair rate vector approaches the max–min fair rate vector as $\beta \to \infty$.*

5.6 UTILITY FAIRNESS INDEX: A NEW MEASURE OF FAIRNESS FOR UTILITY-AWARE BANDWIDTH ALLOCATION

Many resource allocation schemes aim to share the available capacity of a network in a fair way, but due to practical causes, this is not always achievable. Jain's fairness index (JFI) is used to evaluate quantitatively the fairness performance of a scheme. Because different users and different services may have different utility functions, they are affected in different ways by deviations from fair capacity apportionment. JFI does not consider users' utility functions, so we include them in our generalization of JFI termed utility fairness index (UFI) in this section. Numerical examples are given to demonstrate the benefit of UFI.

5.6.1 INTRODUCTION

Fairness measures are used in communication network to determine whether users or applications are receiving a fair share of network resources with respect to given fairness criteria. One well-known fairness criterion is *max–min fairness* [139] that for the case of a single resource (link), will share the resource equally among many greedy users. However, for the case of a network with multiple links, a max–min may lead to inefficiency and often provides lower average throughput than a *maximum throughput* policy [365] (see also [175,288] for discussions on fairness-efficiency trade-off). To meet the max–min fairness, the least expensive flows are assigned all capacity they can use, and no capacity may remain for the most expensive flows. On the contrary, a maximum throughput policy will starve the expensive flows, and may result fewer "happy customers."

Kelly [104] proposed PF whereby bandwidth is allocated among users so as to maximize the sum of the utilities of all flows in progress. In [366], a bandwidth allocation policy [106,107] called *minimum potential delay* is proposed. It aims to minimize the sum of the potential delays of the flows in progress where a delay of a flow is approximated by the reciprocal of its allocated rate. In [115], the following class of utility functions is proposed

$$U_i(x_i, \alpha) = \begin{cases} (1-\alpha)^{-1} x_i^{1-\alpha} & \text{if } \alpha \geq 0, \alpha \neq 1 \\ \log x_i & \text{if } \alpha = 1, \end{cases} \tag{5.46}$$

where x_i is the rate of source i, and α is a parameter representing fairness level. Bandwidth allocation using the above utility function is called *α-fair* [115]. This includes all the aforementioned allocation policies: maximum throughput ($\alpha = 0$), PF ($\alpha = 1$), minimum potential delay ($\alpha = 2$), and max–min fairness ($\alpha = \infty$). Moreover, it also characterizes the fairness situation in the major TCP congestion control protocols, for example, TCP Reno ($\alpha = 2$); HSTCP ($\alpha = 1.2$); and TCP Vegas, FAST TCP, STCP ($\alpha = 1$).

Considering the fact that users of real-time services have very different utilities than those of nonreal-time data services, Cao and Zegura [257] proposed the so-called *utility max–min fairness* that aims to maximize the minimum utility achieved by a user in the network. Another resource allocation criterion that caters for real-time services called *utility PF*, is proposed in [367]. The relationship between utility PF and the utility max–min fairness is analogous to the relationship between PF and max–min fairness.

Telecommunications protocols normally do not achieve exactly the fair allocation they are designed for. Therefore, there is a need to define a measure to quantify how fair is a protocol relative to a given fairness criterion. The most well-known measure is JFI [338]. For n flows, with flow i receiving a rate x_i, JFI for this allocation is defined as

$$f(x) = \frac{\left(\sum_{i=1}^{n} x_i\right)^2}{n \sum_{i=1}^{n} x_i^2}.$$

This metric ranges continuously in value from $1/n$ to 1, with 1 corresponding to the equal allocation for all users. This fairness index provides a convenient method to compare the performance of different congestion control algorithm. However, we find that this traditional index may not be appropriate for measuring the fairness among users in the utility-aware bandwidth allocation schemes. We, therefore, propose a generalization of it, termed as UFI, which offers the desirable features as JFI but takes into account various utility requirements from users.

5.6.2 UTILITY FAIRNESS INDEX

Although JFI has been frequently used to measure fairness among various resource allocation schemes, it may be unsuitable when comparing utility-aware fairness and traditional bandwidth fairness.

Let us look at the following example. Suppose a single link with bandwidth 20 units is shared by two sources. Source 1 attains a utility of $\log(x_1)$ and Source 2 attains a utility of $2\log(x_2)$. According to the max–min fairness criteria, each source gets equal bandwidth of 10 units, which results in JFI to be 1. However, in utility max–min fairness, the allocations for Source 1 and Source 2 are 16 units and 4 units, respectively. Its JFI is 0.735.

From this example, one may deduce that the max–min fairness is fairer than utility max–min fairness. In fact, considering the difference of sources' requirement, utility max–min fairness is better than max–min fairness. Because in utility max–min allocation, both sources receive the same high utility and hence both achieve good performance; in max–min allocation, Source 1 has poorer performance, whereas Source 2 has excellent performance. The key difference is that bandwidth is the resource to be allocated in max–min fairness while in utility max–min fairness, the more important factor for applications is the performance derived from a given bandwidth allocation. JFI is only concerned with bandwidth, so it is not suitable for using in measuring the fairness level among users in a utility max–min fair bandwidth allocation.

To overcome the above limitation and to facility measurement of fairness level in utility-aware bandwidth allocation schemes, we generalize JFI into the UFI to cope with utility-aware allocation policies.

Definition 5.6.1 *For n flows, with user i receiving a bandwidth x_i and attaining a utility function $U_i(x_i)$, the UFI of the allocation is defined as*

$$UFI(x) = \frac{\left(\sum_{i=1}^{n} U_i(x_i)\right)^2}{n \sum_{i=1}^{n} (U_i(x_i))^2}. \tag{5.47}$$

UFI has the desired features [338] to act as a fairness index:

- Independent of population size n;
- Bounded between $1/n$ and 1;
- Independent of amount of whole shared resource; and finally,
- Continuous so that any change in bandwidth allocation changes the fairness also.

The above UFI facilities measurement of fairness level under various utility-aware bandwidth allocation schemes. It is a natural extension to JFI, to see this one lets $U_i(x_i) = x_i$, for $i = 1, 2, \ldots, n$ in Equation 5.47. This UFI has actually provided us with a unified measurement of various traffic pattern, including elastic traffic and inelastic traffic (real-time applications), with different bandwidth and utility requirements.

5.6.3 NUMERICAL EXAMPLES

In this section, we use two examples to demonstrate the desired property of the proposed UFI.

Example 1:

Let us consider a network consisting of a single link of capacity 16 Mbps shared by two users. One user transmits Internet protocol television (IPTV) data with a nonconcave bandwidth utility function $U_1(x_1)$. The other user transmits data according to an elastic application with strictly increasing and concave bandwidth utility $U_2(x_2)$. Figure 5.3 shows how different bandwidth allocations affect the received utility.

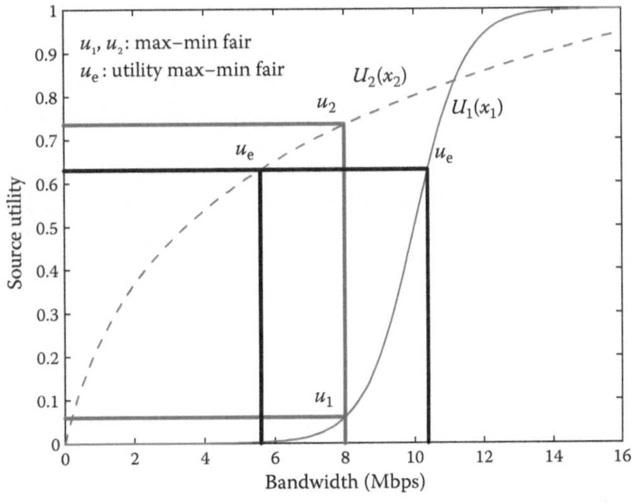

Figure 5.3 Max–min fairness and utility max–min fairness.

IPTV is a system through which television services are delivered using the Internet protocol suite over a packet-switched network such as a LAN or the Internet, instead of being delivered through traditional terrestrial, satellite signal, and cable television formats. Unlike downloaded media, IPTV offers the ability to stream the media in smaller batches, directly from the source. As a result, a client media player can begin playing the data (such as a movie) before the entire file has been transmitted. This is known as streaming media.

IPTV services may be classified into three main groups:

- Live television, with or without interactivity related to the current TV show;
- Time-shifted television: catch-up TV (replays a TV show that was broadcast hours or days ago), start-over TV (replays the current TV show from its beginning);
- Video on demand (VOD): browse a catalog of videos, not related to TV programming.

Depending on the network architecture of the service provider, there are two main types of video server architecture that can be considered for IPTV deployment: centralized and distributed.

The centralized architecture model is a relatively simple and easy to manage solution. For example, as all contents are stored in centralized servers, it does not require a comprehensive content distribution system. Centralized architecture is generally good for a network that provides relatively small VOD service deployment, has adequate core and edge bandwidth, and has an efficient content delivery network. Distributed architecture is just as scalable as the centralized model; however, it has bandwidth usage advantages and inherent system management features that are essential for managing a larger server network. Operators who plan to deploy a relatively large system should, therefore, consider implementing a distributed architecture model right from the start. Distributed architecture requires intelligent and sophisticated content distribution technologies to augment effective delivery of multimedia contents over service provider's network.

With the max–min fair resource allocation, both users would receive 8 Mpbs. This is not good enough for User 1, since its received utility is just 0.0594. But JFI is 1 for this allocation, which says this allocation is totally fair. In contrast to this, utility max–min fair bandwidth allocation gives 10.387 Mbps to User 1 and 5.613 Mbps to User 2, which allows the IPTV application having an acceptable performance. But JFI is just 0.9183, which is lower than that of max–min fairness. Obviously, JFI is not proper to gauge the fairness in this situation. We calculate the UFI: in the utility max–min fairness case, UFI = 1 and in the max–min fairness case, UFI = 0.5804. From this, we know that UFI is more appropriate for utility-aware bandwidth allocation policies.

Example 2:

Consider a linear network, as shown in Figure 5.4, which consists of two links L1 and L2 with a capacity of 10 Mbps and shared by three sources S1, S2, and S3. S1 traverses link L1 and L2, and S2 and S3 traverse L1 and L2, respectively.

We calculate five sets of bandwidth allocations for this model that meet PF, minimum potential delay fairness, max–min fairness, utility PF, and utility max–min fairness, respectively. To calculate the former three sets, we assume the sources have their utility functions given by Equation 5.46. We consider the following optimization problem of α-fair bandwidth allocation

$$\max \sum_{i=1}^{3} U_i(x_i, \alpha) \tag{5.48}$$

$$\text{Subject to } x_1 + x_2 \leq 10, \quad x_1 + x_3 \leq 10. \tag{5.49}$$

where α is a fairness degree parameter. Under $\alpha = 1$, $\alpha = 2$, and $\alpha = \infty$, the above problem leads to the solution of proportional fair, minimal potential fair, and max–min fair bandwidth allocation, respectively.

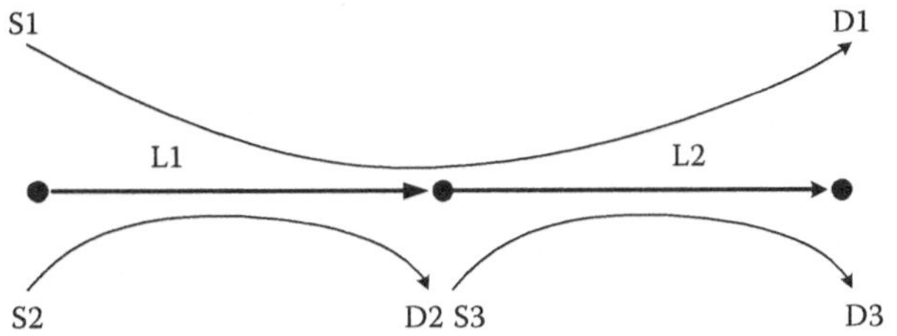

Figure 5.4 Network topology.

Solving the above problem, we obtain the rates $x_i(\alpha)$ as follows:

$$x_1(\alpha) = \frac{10}{1+2^{1/\alpha}},\ x_2(\alpha) = x_3(\alpha) = \frac{10\cdot 2^{1/\alpha}}{1+2^{1/\alpha}}. \quad (5.50)$$

By letting $\alpha = 1$, $\alpha = 2$, and $\alpha \to \infty$ in Equation 5.50, we obtain the following proportional fair bandwidth allocation, the minimal potential fair bandwidth allocation, and the max–min bandwidth allocation are:

$$[x_1(1), x_2(1), x_3(1)] = [10/3, 20/3, 20/3],$$

$$[x_1(2), x_2(2), x_3(2)] = \left[\frac{10}{1+\sqrt{2}}, \frac{10\sqrt{2}}{1+\sqrt{2}}, \frac{10\sqrt{2}}{1+\sqrt{2}}\right],$$

$$[x_1(\infty), x_2(\infty), x_3(\infty)] = [5,5,5].$$

Next, we assume that the source utility functions are

$$U_1(x_1) = 1/(1+e^{-2(x_1-5)}),\quad U_2(x_2) = 0.125x_2,$$
$$U_3(x_3) = \log(x_3+1)/\log 11. \quad (5.51)$$

Among them, Source 1 has the real-time application utility, which is delay sensitive and has real-time requirements. For Source 1, degradation in bandwidth may severely degrade its performance. Source 2 and 3 are the elastic traffic.

To obtain the utility max–min fair allocation [257], note that Source S1 and Source S2 share the same bottleneck Link L1, so their utilities must be the same and Link L1 should be saturated. That is, $U_1(x_1) = U_2(x_2)$ and $x_1 + x_2 = 10$. Simple calculations lead to the utility max–min fair bandwidth allocation: $x_1 = 5.2022$ Mbps, $x_2 = x_3 = 4.7978$ Mbps.

The utility PF bandwidth allocation can be computed using the optimization model suggested in [367], which is

$$\max \int \frac{1}{1/(1+e^{-2(x_1-5)})}dx_1 + \int \frac{1}{0.125x_2}dx_2 + \int \frac{1}{\log(x_3+1)/\log 11}dx_3 \quad (5.52)$$

$$\text{Subject to } x_1 + x_2 \le 10,\quad x_1 + x_3 \le 10. \quad (5.53)$$

Solving the above optimization problem, the utility PF allocations are $x_1 = 4.7024$ Mbps, $x_2 = 5.2976$ Mbps, and $x_3 = 5.2976$ Mbps.

Using the above results, we compute JFI and our UFI under these five sets of bandwidth allocation, which is shown in Table 5.5. We plot Table 5.5 into Figure 5.5. From Figure 5.5, we observe that max–min fairness, utility PF, and utility max–min fairness almost have the same value of JFI. In fact under these allocation policies, User 1 performs very differently. However, JFI has the difficulty in telling the difference of the real application performance because it does not consider

Table 5.5
Fairness Indexes under Various Policies

Bandwidth Allocation Policies	Jain's Fairness Index	Utility Fairness Index
PF ($\alpha = 1$)	0.9259	0.6936
Minimum potential delay fairness ($\alpha = 2$)	0.9771	0.7884
Utility PF	0.997	0.9206
Max–min fairness ($\alpha = \infty$)	1	0.9745
Utility max–min fairness	0.9985	0.9906

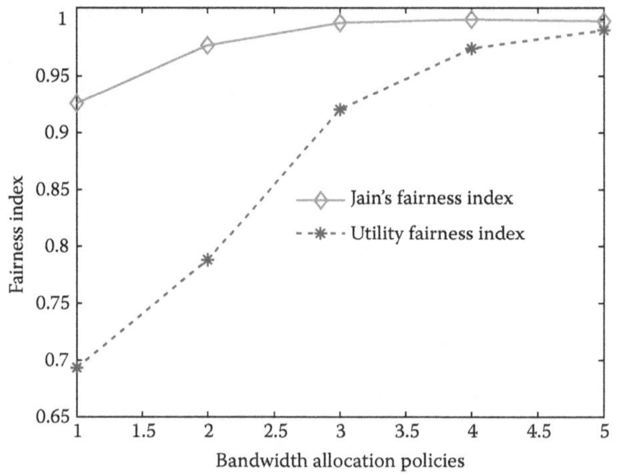

Figure 5.5 JFI and UFI.

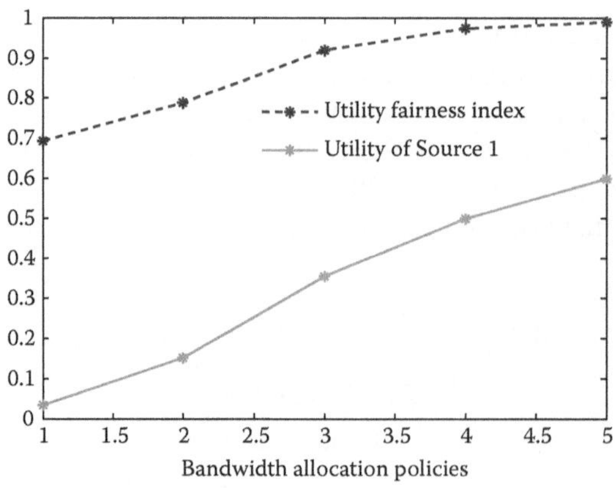

Figure 5.6 Utility of Source 1.

requirements of sources' utility. The UFI is more suitable for these situations. The performance of real-time application Source 1 mainly dominates the overall network performance. So we can use the utility of Source 1 to roughly represent the network performance. Figure 5.6 demonstrates the utility of Source 1 and UFI of the network under these bandwidth allocations. They have a similar curve, which suggests that the UFI can reflect the network performance with real-time application.

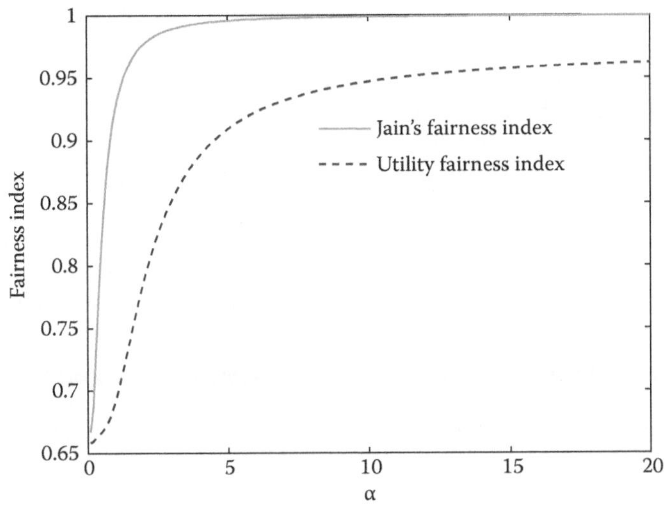

Figure 5.7 Fairness index of α-fair bandwidth allocation.

Figure 5.7 shows the two fairness indexes of α-fair bandwidth allocation. When $\alpha \geq 6$, JFI is very close to 1, which indicates that the network performs ideally well independent of α. But from the above analysis, we know that in utility max–min fairness allocation the network performs better than that in max–min fairness ($\alpha = \infty$). So the quantity of measurement of fairness for utility max–min fairness should be larger than max–min fairness. But this comparison is impossible because the JFI of max–min fairness is already 1. On the contrary, the UFI leaves much bigger room for measuring the fairness level in the utility-aware fairness allocation schemes such as utility max–min fairness and utility PF.

5.6.4 CONCLUSIONS

In this section, measurement of fairness level for bandwidth allocation schemes has been discussed. To cope with the recently proposed utility-aware fairness, we proposed a UFI, which has all the desired features of a fairness index. This new metric enables us to measure the users' fairness level under various utility-aware bandwidth allocation schemes.

5.7 FURTHER DISCUSSIONS ON FAIR SHARING RESOURCE POLICIES IN THE CONTEXT OF COMMUNICATION NETWORKS

Generally, policies for sharing resources that are characterized by low level of fairness provide high average throughput but low stability in the service quality, meaning that the achieved service quality is varying in time depending on the behavior of other users. If this instability is severe, it may result in unhappy users who will choose another more stable communication service.

Max–min fair resource sharing results in higher average throughput (or system spectral efficiency in wireless networks) and better utilization of the resources than a work-conserving equal sharing policy of the resources [283,284]. In equal sharing, some dataflows may not be able to utilize their "fair share" of the resources. A policy for equal sharing would prevent a dataflow from obtaining more resources than any other flow and from utilizing free resources in the network.

On the other hand, max–min fairness provides lower average throughput than maximum throughput resource management, where the least expensive flows are assigned all capacity they can use, and no capacity might remain for the most expensive flows. In a wireless network, an expensive user is typically a mobile station at far distance from the BS exposed to high signal attenuation. However, a maximum throughput policy would result in starvation of expensive flows and may result in fewer "happy customers."

A compromise between max–min fairness and maximum throughput scheduling is PF, where the resources are divided with the goal to achieve the same cost to each user or to minimize the maximum cost per unit that a dataflow reaches. Expensive dataflows achieve lower service quality than others in PF, but does not suffer from starvation. Max–min fairness results in more stable service quality and therefore perhaps "happier customers."

In packet radio wireless networks, the fairly shared spectrum efficiency (FSSE) can be used as a combined measure of fairness and system spectrum efficiency. The system spectral efficiency is the aggregate throughput in the network divided by the utilized radio bandwidth in hertz. The FSSE is the portion of the system spectral efficiency that is shared equally among all active users (with at least one backlogged data packet in queue or under transmission). In case of scheduling starvation, the FSSE would be zero during certain time intervals. In case of equally shared resources, the FSSE would be equal to the system spectrum efficiency. To achieve max–min fairness, the FSSE should be maximized.

FSSE is useful especially when analyzing advanced radio resource management schemes, for example, channel adaptive scheduling, for cellular networks with best-effort packet data service. In such systems, it may be tempting to optimize the spectrum efficiency (i.e., the throughput). However, this might result in scheduling starvation of "expensive" users at far distance from the access point, whenever another active user is closer to the same or an adjacent access point. Thus, the users would experience unstable service, perhaps resulting in a reduced number of happy customers. Optimizing the FSSE results in a compromise between fairness (especially avoiding scheduling starvation) and achieving high spectral efficiency.

If the cost of each user is known, in terms of consumed resources per transferred information bit, the FSSE measure may be redefined to reflect PF. In a proportional fair system, this "proportionally fair shared spectrum efficiency" (or "fairly shared radio resource cost") is maximized. This policy is less fair since "expensive" users are given lower throughput than others, but still scheduling starvation is avoided.

Multipath networks allow that each source–destination pair can have several different paths for data transmission, thus they improve the performance of increasingly bandwidth-hungry applications and well cater for traffic load balancing and bandwidth usage efficiency. The paper [258] studies fair resource allocation for users in multipath networks and formulates it as a multipath network utility maximization problem with several fairness concepts. By applying the Lagrangian method, subproblems for users and paths are derived from the resource allocation model and interpreted from an economic point of view. In order to solve the model, a novel rate-based flow control algorithm is proposed for achieving optimal resource allocation, which depends only on local information. In the presence of round-trip delays, sufficient conditions are obtained for local stability of the delayed algorithm. As for the end-to-end implementation in Internet, a window-based flow control mechanism is presented since it is more convenient to implement than rate-based flow control.

6 Fair Bandwidth Allocation Algorithms for Ring Metropolitan Area Networks

The Resilient Packet Ring (RPR) [391,527] IEEE 802.17 standard is a new technology developed for high-speed backbone metropolitan area networks (MANs). The key performance objectives of RPR are to simultaneously achieve fairness, high utilization, and spatial reuse for fairness eligible (FE) traffics over the ring, and as such designing efficient fair bandwidth allocation algorithms to achieve such goals is an important task. This chapter presents a survey of the existing fair bandwidth allocation approaches that are applied in RPR networks. Three representative approaches, namely aggressive mode (AM), conservative mode (CM), and distributed virtual-time scheduling in rings (DVSRs), are discussed in detail. The operating mechanism for them is analyzed, their algorithms are designed, and the performance comparisons among them are given on the basis of theoretical analyses and simulation demonstrations.

6.1 INTRODUCTION

The RPR [266,270,519] IEEE 802.17 standard is a new network structure and data transport technology for the ring MAN. The RPR network uses the dual-ring topology to optimize the packet transfer over the fiber media. Before the development of RPR, several other ring technologies such as SONET, Gigabit Ethernet, and FDDI have been used in metropolitan networks, but all these predecessors of RPR have their own respective shortcomings, which hinder their applications into such networks.

A SONET ring adopts the point-to-point circuit mechanism to ensure the bandwidth assignment between any pair of nodes on the ring, but this technology results in low utilization of the ring bandwidth. A Gigabit Ethernet ring achieves better utilization but suffers from unfairness, particularly, this unfairness is always encountered in a scenario where congestion-responsive TCP flows share bandwidth with nonresponsive UDP flows, which results in TCP flows obtaining throughput significantly below their fair share. In FDDI, the data transmission on the ring is depended on the rotating token; it thus fails to employ the spatial reuse and results in poor throughput or utilization.

RPR, however, overcomes the shortages of its predecessor technologies. It inherits the efficiency, robust, and flexibility attributes from those aforementioned ring technologies. Furthermore, RPR, which is designed to optimize the data transfer over the fiber media, improves the current ring technologies so that it meets the requirements of the high-speed metropolitan network more satisfactorily. On this stand, RPR is regarded as the promising technology for optimizing the data transfer and is likely to be the primary technology of the next metro IP networks.

Unlike the legacy ring technologies (such as FDDI) that only one node holding the token can transmit data packets at a time [266,520,521], RPR supports the destination release so that packets on the ring will be removed from the destination node instead of going back to the source node. Therefore, spatial reuse [267,343,393,522,523] can be achieved in RPR to better utilize the ring bandwidth. In this case, how to ensure the fairness among different RPR nodes competing for the ring bandwidth becomes a challenging task. In addition to fairness, achieving high utilization of the bandwidth is also the key performance requirement in RPR networks, including other telecommunication networks. Generally speaking, designing efficient bandwidth allocation (fairness) algorithms to simultaneously achieve the goals such as fairness, high utilization, and maximal spatial reuse becomes the key technical challenge of the RPR network.

6.2 BASIC CONCEPTS OF RPR FAIRNESS ALGORITHMS

In the RPR network, each node has two adjacent nodes, namely upstream node and downstream node. All the RPR nodes, which are connected by a fiber link pair, compose the entire RPR ring network. Each RPR node can act as a source node to send data, or as a destination node to receive (remove) data or as a switch node to transit the data traffic.

Generally, the RPR support three classes of service, namely classes A, B, and C, to meet the various performance requirements for different traffics over the RPR ring [266,267,520,523]. Class A traffic is divided into A_0 and A_1, Class B traffic is divided into B-CIR (committed information rate) and B-EIR (excess information rate). Class C traffic is best-effort transport traffic. Class A and the CIR portion of the Class B support the traffics with delay and rate guarantee, and the bandwidth needed for these traffics is preallocated. While Class C and the EIR portion of Class B only receive the low priority and can only use the available portion of the ring bandwidth less than the portion reserved for those high-priority traffics, and these traffics are called the FE traffic. That is, the available ring bandwidth is allocated to these traffics by some fairness algorithms. Therefore, different contending RPR stations (nodes) that have the FE traffic to send should get their share of the ring bandwidth through an RPR fairness algorithm.

We mainly focus on the FE traffics that are controlled by the RPR fairness algorithms, while the high-priority traffic classes (such as Class A and Class B-CIR) are considered as fairness invisible and are subject to the admission control. In RPR networks, the congestion control and fairness for FE traffics produced by different competing stations is a key design, which contributes to achieve fairness, high utilization, and spatial reuse over the entire RPR ring simultaneously.

A RPR station always deal with the upstream traffics transiting the station before it can add traffics into the ring, thus it is very easy for a downstream station to be starved by the upstream stations, which intend to be greedy. Generally, the solution to this starve problem is to force all the relevant stations to behave subject to a specialized RPR fairness algorithm. Specifically, the RPR fairness algorithm is to distribute the bandwidth among the contending stations that have the FE traffics to send over a congested link, so that each node gets no more than its fair share of the capacity of the link.

Recently, a number of RPR fairness algorithms have been proposed to meet such requirements. Among these algorithms, two representative operation modes can be clarified: one is the so-called AM [266,267,520,523,525] and the other is termed as CM [266,267,520,523–525]. Both modes operate with a similar mechanism in the sense that a congested downstream node conveys a control (fairness) message to its upstream nodes to throttle their traffic rates so as to eliminate the congestion downstream.

DVSR [267,393,523], inheriting the running mechanism of AM and CM in large, achieves better performance than the two modes by using new method for fairness message calculation and delivery. Despite the three mechanisms are different in the congestion detection and the fair message calculation, they have much similarity in the RPR node architecture that is constructed for the operation of the RPR bandwidth allocation. We illustrate the architecture of a generic RPR node in Figure 6.1.

The RPR node contains several modules that are associated with the RPR fairness algorithms. They are the data scheduler module (denoted as Scheduler), the traffic monitor module (denoted as Traffic Monitor), the bandwidth allocation module (denoted as Bandwidth Allocator), and the rate control module (denoted as Rate Controller). Most of these modules have the similar configuration for different fairness algorithms except for the rate control module. In AM and CM, only one rate controller is set in a RPR node to control/adjust its own sending rate; contrarily in DVSR, the rate controllers are at the per-destination granularity, i.e., each rate controller is in charge of a mini flow which is originated from this node while destined at another certain node.

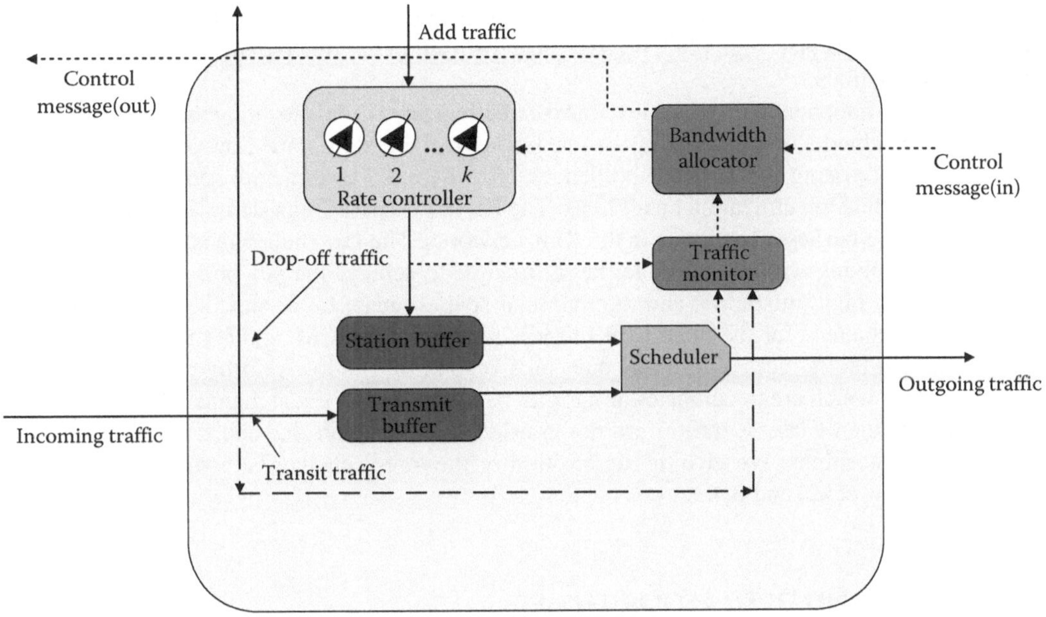

Figure 6.1 One general RPR node architecture.

Generally, an RPR station may have two kinds of traffics to deal with: one is called add traffic and the other is the transit traffic. The former is started from the station itself, while the latter is from the upstream nodes and need to be forwarded by the station. The Traffic Monitor is set to measure the add and transit traffics using the byte counters, and the measured results are used to inform the Bandwidth Allocator.

The RPR fairness algorithm is performed in Bandwidth Allocator according to the measured traffic rates and the control message from downstream, so as to determine the sending rate for this local node, on one hand, and also to produce the fairness control message which is delivered to the upstream nodes, on the other hand. Besides, two kinds of buffers, one is the Station Buffer and the other is the Transit Buffer, are also relevant to the RPR fair algorithms, which are used for the data traffic originating from the local node (termed Add Traffic) and the data traffic passing by this node (termed Transit Traffic). The Scheduler module determines how to serve these data traffics according to their buffer occupancy by some strategy.

As for the whole ring network, all the RPR nodes take the same architecture and the fairness algorithm can be performed at each RPR node; similarly, different algorithms will differ in how to produce the fairness control message at the congested node. The DVSR also differs from AM and CM in delivering the fairness messages.

Before the introduction of these algorithms, the specialized fairness principle termed ring ingress aggregated with spatial reuse (RIAS) [267,393–395,523] in RPR networks should be clarified at first. RIAS fairness has several components: the first component defines the basic unit for fairness determination at a link as an ingress-aggregated (or IA) flow, which is the aggregation of the individual flows originating from the same source node while ending at different destinations. Note that this fairness criterion is different from the flow-based max–min fairness (see, e.g., [139,401,405,526]), in which the traffic granularity for fairness determination at a certain link is the single flow.

The first component also implies that the traffics crossing a certain link are classified according to the distinguished stations (nodes), that is, the rate assigned to an IA flow crossing a certain link is determined by its relevant stations fair share of the ring bandwidth. Generally, all the stations (having the FE traffics to send) on the ring may get their fair shares in a weighted manner; hence the weighted fairness (see, e.g., [285,342,404]) can be used in the ring bandwidth allocation among different RPR stations.

The second component aims to achieve maximal efficiency while allocating the bandwidth fairly. Specifically, bandwidth on a certain link can be further reclaimed by some flows over this link when the other flows crossing this link are bottleneck elsewhere, so as to achieve the so-called spatial reuse to ensure the high utilization on all links. The RIAS fairness is regarded as a performance goal desired for all the fairness algorithms in the RPR networks. The key challenge is to design RPR fair algorithms complying with the RIAS fairness principle to achieve the key performance objectives such as fairness, high utilization, and the maximal spatial reuse. In the next section, we describe the running mechanism for the three RPR fairness schemes, AM, CM, and DVSR, respectively, in detail. Note that the fairness algorithms only take account of the low-priority data traffic streams (i.e., FE traffic), which are to compete for the ring bandwidth in an equal manner. The uncontrolled traffic streams (e.g., Class A traffic) are not considered as they are beyond the constraint of the RPR fairness algorithms. We give the description of the three schemes in a way as illustrated in Algorithms 6.3.1, 6.3.2, and 6.4.3.

6.3 AM, CM, AND DVSR ALGORITHMS

6.3.1 AGGRESSIVE MODE AND CONSERVATIVE MODE

AM and CM algorithms can represent most operations of the current RPR fairness algorithms.

There are two key parameters for both algorithms, one is the $forward\_rate$, which represents the sending rate of all transit traffics originating from other nodes passed by a RPR node; the other is the my_rate, which represents the sending rate of the nodes own add traffic. Both $forward\_rate$ and my_rate are measured by the Traffic Monitor using byte counts over a fixed time interval, say, T, and will be used for the RPR fairness algorithms. Despite their differences, both algorithms operate similarly in that when a downstream node detects the congestion, it will calculate a fairness message and convey it to the relevant upstream nodes. Upon receiving the fairness message, the upstream node may either adjust its own sending rate according to the fairness message or use it to compare with its own calculated fair rate according to its local congestion status.

Specifically, when an RPR node receives the fairness message within a control packet from downstream nodes, it will check the fairness message in the control packet. If the fairness message holds a null value, which indicates the noncongestion situation in downstream, then the RPR node may encounter two choices: one is to increase its own sending rate by a fixed value, if it is not congested, and further propagate a null value message to its upstream nodes; the other is to calculate a fair rate according to its congestion status, if the node is congested, and then propagate the message containing the fair rate value to its upstream nodes. If the arriving message contains a non-null fair rate value, which indicates the congestion has occurred downstream, the RPR node will act as follows: if the node is not congested, it then adjusts its sending rate to the advertised fair rate and go on delivering the fairness message upstream; otherwise, it will compare the fair rate calculated by itself with the advertised fair rate in the arrived message and take the minimum, which will then be used as the allowed sending rate of the node itself and its upstream nodes. The whole process described above can be specified by the flowchart as illustrated in Figure 6.2.

We take a general RPR Node n as the representative for all the RPR nodes in Figure 6.2, where $forward\_rate[n]$ and $my\_rate[n]$ are represented as its measured $forward\_rate$ and my_rate at Node n, respectively. The component $fair\_rate[n]$ designates the local fair rate value determined at Node n and will propagate fairness message to its upstream nodes. Generally speaking, the above process includes two steps: one is to get the allowed sending rate for Node n and the other is to produce and propagate the fairness message to the upstream nodes (that contribute the congestion at Node n).

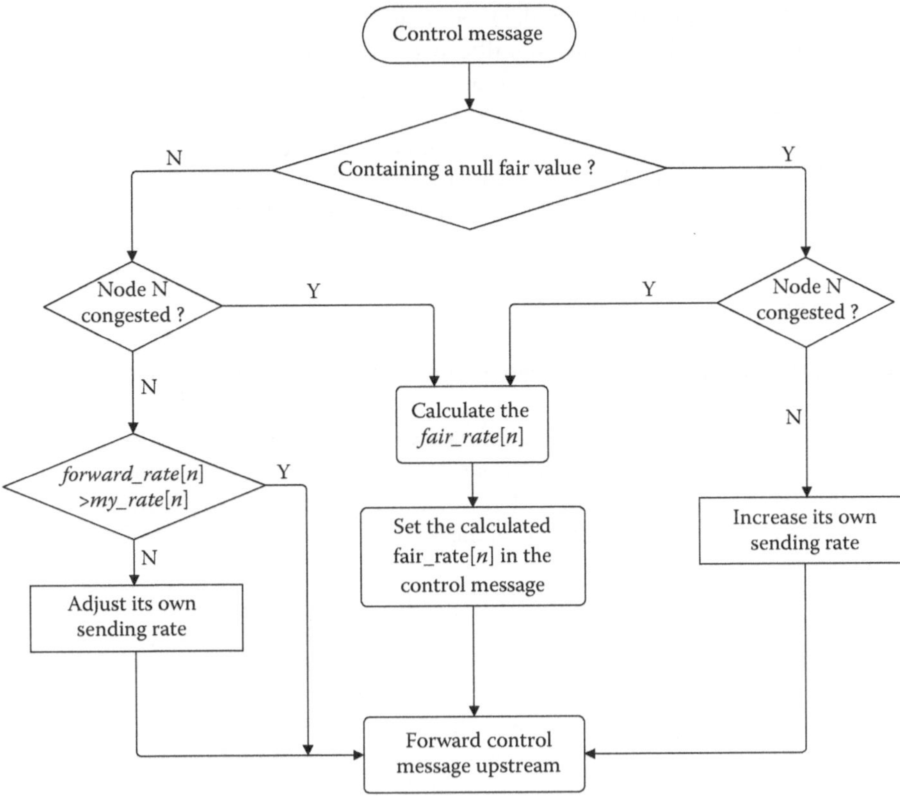

Figure 6.2 One general RPR node architecture.

Both AM and CM operate in the same mechanism as described above, their difference lies in the congestion detection and the calculation of fairness messages. Specifically, the congestion detection differs in the transit buffer (denoted as Transit Buffer) mode: in the single-queue mode, only a primary transit queue (PTQ) is used, and a node is said to be congested if its PTQ exceed a threshold (or its local add traffic has to be waited for a long time); while in the dual-queue mode, two transit buffers, one is the PTQ and the other is the second transit queue (STQ) are used, a node is said to be congested if the occupancy of the STQ excess its threshold.

Previously by default, the AM adopts the dual-queue mode and the CM adopts the single-queue mode; however, this distinction has become less critical and CM currently also adopts the dual-queue mode for its congestion control, thus the mechanism for congestion detection is similar between AM and CM and we regard that both the AM and CM adopt the dual-queue mode. There are two thresholds for the STQ of the dual-queue mode: one is the so-called *low_threshold* and the other is *high_threshold*. Relevant behaviors are defined when the occupancy of the STQ is bounded by *low_threshold* or *high_threshold*. There are two ways to detect the congestion at an RPR node for both modes, one is the rate-based mechanism and the other is the buffer-based mechanism.

For each mechanism, the essential reason for congestion at an RPR node is that the demand for bandwidth on its downstream link is greater than the supply, or the total transmit traffic (including the transit traffics and its own add traffic) is above a certain threshold. Therefore, the RPR fairness algorithm should fairly distribute the available bandwidth among all the contending stations that use this link bandwidth for their FE traffics. The rate-based mechanism determines the congestion at a RPR node according to whether the sum of its *forward_rate* and *add_rate* is beyond a threshold.

For AM, the threshold is typically set to the total unreserved rate (i.e., the congested link capacity minus the reserved rate for the guaranteed traffic, e.g., Class A_0); while for CM, the threshold is set to a fixed value less than the total unreserved rate.

The buffer-based mechanism determines the congestion at a given RPR node according to the STQ occupancy. For AM or CM, when the occupancy of the STQ at an RPR node goes beyond a certain threshold, the node is said to be congested.

Specifically, AM and CM differ significantly in producing the fairness control messages at congested node, one difference is the time interval to compute the local fair rate, the other is how to calculate the fair rate value that will be sent to upstream nodes that contributes the congestion. We will describe this for AM and CM, respectively.

6.3.2 AM ALGORITHM

A key issue for AM is how to calculate the fair rate when a node becomes congested, generally, the fair rate is equal to normalized measurement $my\_rate[n]$ at the end of a given time interval, say T. We denote $allow\_rate[n]$ as its own allowed sending rate of the congested Node n and $fair\_rate[n]$ as the fair rate value set in the fairness control message that is to be sent upstream. We take Node n as a representative in Algorithm 6.3.1.

Algorithm 6.3.1 describes the congested Node n how to calculate the local fair rate value and how to adjust its own sending rates under AM. Note that we consider a weight (e.g., wn) for each node in Algorithm 6.3.1, which makes each node get its weighted fair share of the capacity of the congested link.

We can further divide the state of Node n into three cases: the first is the uncongested state when its STQ occupancy (STQ$_depth[n]$) is below the $low\_threshold$, the second is congested state when its STQ occupancy is between the $low\_threshold$ and $high\_threshold$, and finally, the third is the severe congested state when its STQ occupancy exceeds the $high\_threshold$. Node n enters the congestion state, when

$$STQ\_depth[n] \geq low\_threshold.$$

Algorithm 6.3.1 AM Algorithm

Upon receiving the fairness control messages at Node n
if $fair\_rate[n-1] = null$ **then**
  if $Congested(Noden)$ **then**
    $fair\_rate[n] \leftarrow my\_rate[n]/\omega_n$
    $allow\_rate[n] \leftarrow (fair\_rate[n])\omega_n$
  else
    $fair\_rate[n] \leftarrow fair\_rate[n-1]$
    $allow\_rate[n] \leftarrow my\_rate[n] + \varepsilon$
  end if
else
  if $Congested(Noden)$ **then**
    $fair\_rate[n] \leftarrow min(fair\_rate[n-1], my\_rate[n]/\omega\_n)$
    $allow\_rate[n] \leftarrow (fair\_rate[n])\omega\_n$
  else
    $fair\_rate[n] \leftarrow fair\_rate[n-1]$
    $allow\_rate[n] \leftarrow (fair\_rate[n])\omega_n$
  end if
end if
Forward the fairness message containing $fair\_rate[n]$ upstream

Node n then calculates its local fair rate in the following way:

$$fair\_rate[n] \leftarrow my\_rate[n]/w_n \tag{6.1}$$

Note $fair\_rate[n]$ (in formula 6.1) is computed at a very short time interval (generally once 100 μs) and will be propagated upstream for the rate adjustment of those nodes, which contributes the congestion at Node n. Specifically, for an upstream Node i, which sends data traffic through Node n to a downstream Node j, get its allowed sending rate as follows:

$$allowed\_rate[i][j] \leftarrow \min_{i \leq n < j}(fair\_rate[n])w_i \tag{6.2}$$

When $STQ\_depth[n]$ exceeds the $high\_threshold$ (severely congested state), Node n stops adding traffic into the ring, so as to relieve its local severe congestion. When STQ$_depth[n]$ drops below the $low\_threshold$, Node n propagates a fairness message containing a null value upstream, and the upstream nodes receive such message will increase their sending rates periodically till the congestion occurs again at the downstream. One problem for AM algorithm is that the local fair rate value ($fair\_rate[n]$) determined at the congested Node n may not reflect the congested state at Node n and the true fair shares for relevant nodes, e.g., under the case of unbalanced traffic, a congested node may have a demand of bandwidth which is less than a common fair share, thus the fair rate advertised to upstream by this node may cause the severe oscillations of the upstream nodes and the throughput degradation. We will further discuss this through experiments in next sections.

6.3.3 CM ALGORITHM

CM algorithm differs significantly from the AM algorithm by the congested node to compute the local fair rate, as illustrated in Algorithm 6.3.2. Similarly, we can also distinguish the state of a RPR station according to its STQ occupancy as done in the AM algorithm. Under CM, when a Node n becomes congestion, it calculates its local fair rate as follows:

$$fair\_rate[n] \leftarrow Capacity/m \tag{6.3}$$

Note the Capacity is the (total or mainly) unreserved bandwidth of the congested link adjacent to Node n, which is to be used for all FE traffics passed by this link, m is the number of active stations that sends traffic through the link, which needs to be measured by Node n. However, the time interval for $fair\_rate[n]$ to be computed (adapted) by the congested Node n is different from that of AM; specifically, the congested Node n delivers a new fair rate until all the stations contributing the congestion at Node n have made the adjustment according to the last advertised fair rate. Hence, the congested node under CM adapts its local fair rate more slowly than under AM. Accordingly, a upstream Node i gets its allowed sending rate as follows:

$$allowed\_rate[i][j] \leftarrow \min_{i \leq n < j}(fair\_rate[n]) \tag{6.4}$$

As we can see from formula 6.3, when Node n becomes congested, the $fair\_rate[n]$ is calculated as the available capacity divided by the number of active stations that contributes the congestion at Node n, and this calculation keeps the same when STQ$_depth[n]$ is between the $low\_threshold$ and $high\_threshold$. When Node n enters the severe congestion state (STQ$_depth[n] > high_threshold$), it also stops adding the FE traffic into the ring as in the AM algorithm. A similar problem for CM is that the local fair rate ($fair\_rate[n]$) calculated by the congested node may be incorrect, thus this fair rate advertised to upstream by this node may also cause the severe oscillations of the upstream nodes and the throughput degradation (which will be revealed through experiments in next sections).

Algorithm 6.3.2 CM Algorithm

Upon receiving the fairness control messages at Node n
if $fair\_rate[n-1] = null$ **then**
  if $Congested(Noden)$ **then**
    $fair\_rate[n] \leftarrow Capacity/m$
    $allow\_rate[n] \leftarrow (fair\_rate[n])$
  else
    $allow\_rate[n] \leftarrow my\_rate[n] + \varepsilon$
    $fair\_rate[n] \leftarrow fair\_rate[n-1]$
  end if
else
  if $Congested(Noden)$ **then**
    $fair\_rate[n] \leftarrow min(fair\_rate[n-1], Capacity/m)$
    $allow\_rate[n] \leftarrow (fair\_rate[n])$
  else
    $fair\_rate[n] \leftarrow fair\_rate[n-1]$
    $allow\_rate[n] \leftarrow (fair\_rate[n-1])$
  end if
end if
Forward the control message containing $fair\_rate[n]$ upstream

6.4 DVSR ALGORITHM

DVSR is a new dynamic RPR fairness bandwidth allocation algorithm. Under the DVSR scheme, the RPR node architecture shares much similarity with that under AM and CM scheme, despite that the Rate Controller is constructed on the basis of the per-destination granularity. DVSR also differs obviously from both those schemes in the several aspects such as the congestion detection, fair rate value calculation, and the delivery of fairness control message. DVSR also adopts the byte counts mechanism to measure the sending rate of the transit traffic rate and its own add traffic rate over a fixed time interval T, which is similar to AM and CM. However, the fair rate value calculation at a congested RPR Node n is not the same as that obtained from AM or CM, which is embodied in Algorithm 6.4.3.

DVSR is performed at every RPR node in a distributive manner, and each node can calculate a local fair rate at the given time interval T. These fair values (which will be contained in a control packet) are used for rate adjustment at the relevant source nodes. Specifically, DVSR may adopt one common control packet to collect and deliver the control messages for all the RPR nodes, e.g., for a N-node ring, the control packet will be divided into n fields and the ith field is used for Node i to load its recently calculated fair rate value in the packet. As the control packet rotates around the ring, the sending rate(s) of each node can be adjusted to the desired fair value(s).

We suppose that DVSR is performed at a general RPR Node k. The DVSR measures the arriving bytes for each traffic (including the transit traffic and its add traffic) at Node k during the time interval T, which are denoted by $l_1, l_2, \ldots, l_n$, respectively. Moreover, we consider that all these byte counts are ordered such that $l_1, l_2, \ldots, l_n$, and then the local fair rate is computed at this node according to Algorithm 6.4.3. Though the DVSR in [267,393] is presented in an equal manner for each RPR node, however, we will give the weighted form for it in Algorithm 6.4.3.

In Algorithm 6.4.3, C is the available capacity on the outgoing (downstream) link of Node k for all the FE traffics passing through the link, li/T represents the measured traffic rate of the flow from ingress (source) Node i, while li/CT is regarded as the proportion of the link capacity C (which is designated as R_i). The parameter F_k is the local fair rate value calculated at Node k, note F_k, which is to be written into the kth field in the control packet is only a relative (referent) parameter. Node k and

Algorithm 6.4.3 DVSR Algorithm

At the given time interval T
$R_i \leftarrow l_i \,/\, CT$
$p \leftarrow \Sigma R_i$
if $p < 1$ **then**
  $F_k \leftarrow R_n/\omega_n + (1-p)/\omega_n$
else
  $i \leftarrow 1$;
  $Sum = \sum \omega_n$;
  $Rcapacity \leftarrow 1$;
  $F_k \leftarrow 1/Sum$;
  while $(R_i/\omega_i) and ((R_n/\omega_n) \geqslant F_k)$ **do**
    $Sum \leftarrow Sum - \omega_i$;
    $Rcapacity \leftarrow Rcapacity - R_i$;
    Rcapacity /Sum;
    $i \leftarrow i+1$;
  end while
end if
Set F_k in the relevant field of the arriving control packet

its upstream nodes may get their allowed sending rates according to this value (such as multiplying F_k by its weight and the adjacent link capacity).

Another issue is how to deliver the fairness messages and determine the sending rate at each node. Though the DVSR can also periodically produce the fairness messages at the congested node and deliver them upstream, just as done in AM and CM. However, more effectively, DVSR uses one common control packet containing the fairness message calculated by every RPR node (such as F_k) to run around the ring. When the control packet arrives some node, say Node i, Node i will adjust the sending rates for its add traffic according to the fairness message in the control packet and its own computed fair rate value F_i. F_i is also written into the ith field in the control packet for further propagation. The fairness message in the control packet is recently calculated (updated) by its downstream nodes so that Node i is aware of its fair share at each downstream link, e.g., according to formula 6.4, Node i therefore gets its fair share at a downstream Link n, which is given by

$$R_i^{\,n} = F_n \times W_i \tag{6.5}$$

Note Node i may have multiply individual flows (originating from Node i) with different destination nodes, thus it should further determine the sending rates for these flows by using its rate controllers. Specifically, if Node i has a flow ending at a downstream Node j, then this (individual) flow will get its fair share r_{ij} as determined by formula 6.6

$$r_{ij} = \min_{i \le n < j} \{(w_{ij} \times R_i^n)/w_i^n\} \tag{6.6}$$

where R_i^n is determined by formula 6.4 and w_i^n is the sum of all these flows' weights. Therefore, each flow originating from Node i will be assigned a minimum rate it can send along the path to its destination. As the control packet rotates around the ring, each flow on the RPR ring can eventually drag its sending rate to the desired fair share of the ring bandwidth. By only using one common control packet, the control messages on the ring for rate adjustment have been greatly reduced, which leads to faster convergence and higher efficiency than AM and CM algorithms.

6.4.1 PERFORMANCE COMPARISONS OF AM, CM, AND DVSR

In this section, we make a comparison about properties such as the fairness, utilization, and throughput among the three algorithms as discussed above. The comparisons are illustrated in Table 6.1. We also construct the following configuration as depicted by Figures 6.3 and 6.4 to compare the performance among the three algorithms by simulations.

Specifically, we suppose each node is assigned an equal weight to compete the ring bandwidth. For the first model in Figure 6.3, we set the capacity of each link to be 600 Mb/s and the propagation delay on each link to be 2 ms. We consider two flows with unbalanced input rates such that Flow (1, 3) has a demand for the full link capacity while Flow (2, 3) only has a demand of 100 Mb/s. Specifically, for DVSR algorithm, we suppose the global control packet is placed at Node 3 at the beginning of the session, which will run around the ring in a direction, which is opposite to that of the dataflow. The total simulation time is set to 100 ms. The simulation results about the sending rates of the two flows under AM, CM, and DVSR are presented in Figures 6.5 through 6.7, respectively.

Table 6.1
Performance Comparisons among AM, CM, and DVSR

Schemes Performance	Fairness	Utilization	Throughput	Oscillations	Converging Speed
AM	No	Low	Low	Severe	Slow
CM	Low	Low	Low	Severe	Slow
DVSR	Yes	High	High	Slight	Fast

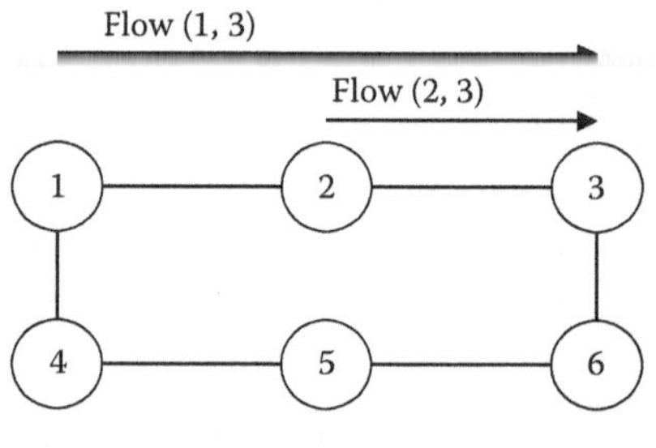

Figure 6.3 A six-node ring topology with the unbalanced traffic (Model 1).

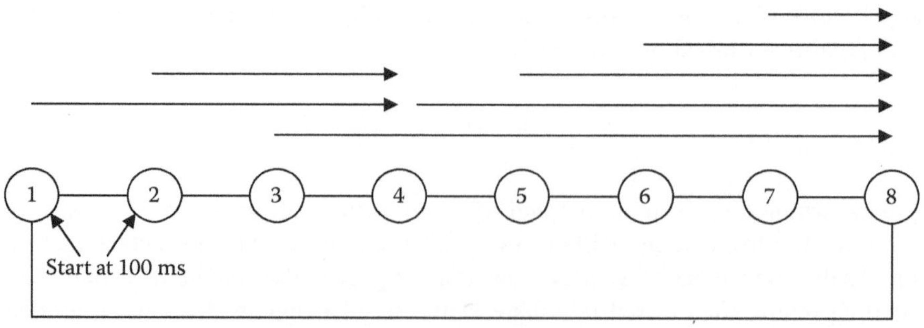

Figure 6.4 An eight-node topology with the balanced traffic (Model 2).

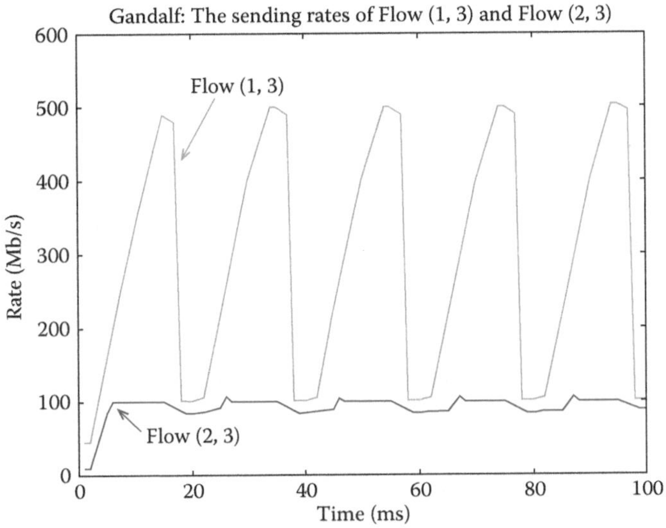

Figure 6.5 The sending rates under AM algorithm (Model 1).

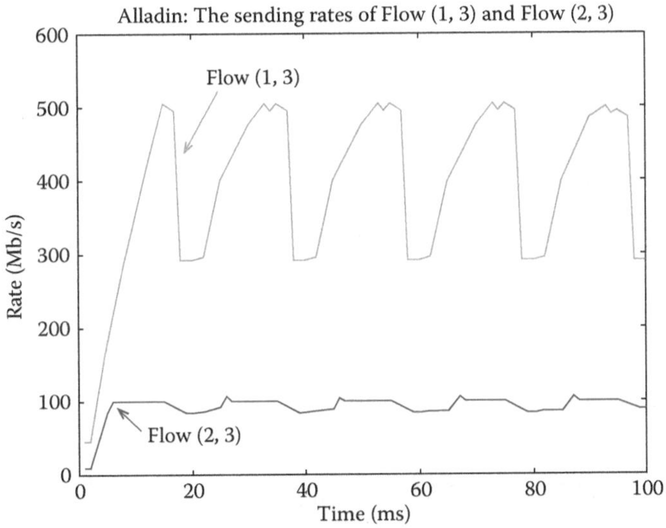

Figure 6.6 The sending rates under CM algorithm (Model 1).

For the second model in Figure 6.4, we set the capacity on each link to be 600 Mb/s and the propagation delay on each link to be 1 ms. All the dataflows on ring are greedy and have a balanced demand for the full link capacity. Note the five flows (3, 8), (4, 8), (5, 8), (6, 8), and (7, 8) are long-lived and active during the whole session, while the two flows (1, 4) and (2, 4) start at the middle of the simulation process. Specifically, for DVSR, the control packet starts from Node 8 to collect and propagate the fairness messages for each node around the ring. The total simulation time is set to 200 ms, which is further divided into two parts: the first part is from 0 to 100 ms, when only the five flows (3, 8), (4, 8), (5, 8), (6, 8), and (7, 8) are active while the two flows (1, 4) and (2, 4) are off; the second part is from 101 to 200 ms, when the total seven flows (including flows (1, 4) and (2, 4))

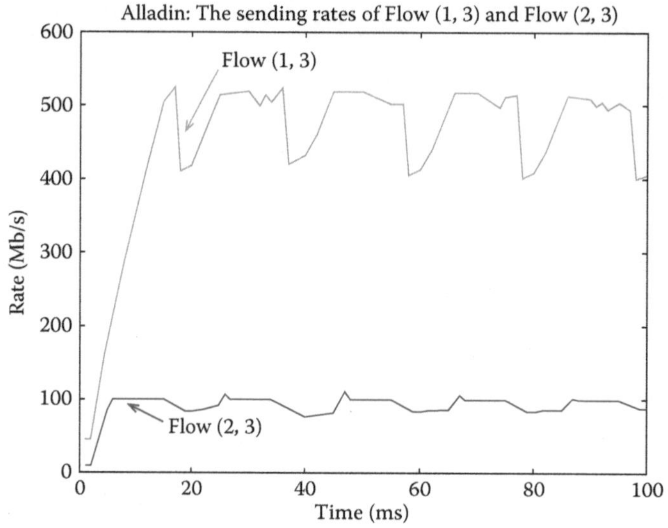

Figure 6.7 The sending rates under DVSR algorithm (Model 1).

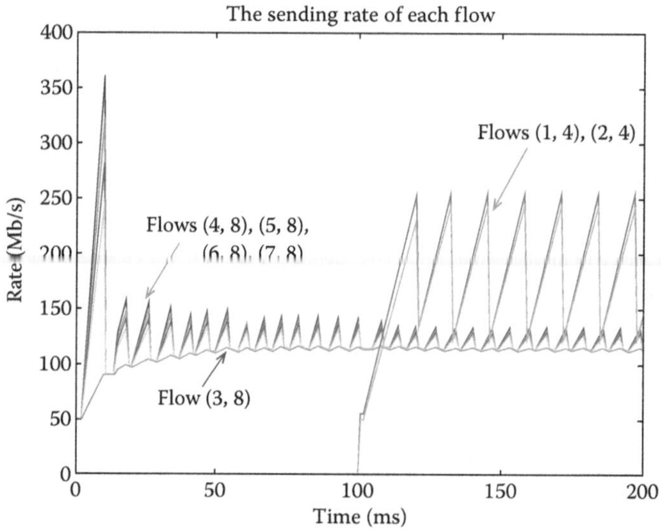

Figure 6.8 The sending rates under AM algorithm (Model 2).

are active. The simulation results about the sending rates of each flow under the three algorithms are presented in Figures 6.8 through 6.10, respectively.

As we can see from Figures 6.5 through 6.7, the unbalanced Flow (2, 3) keeps its demand traffic rate 100 Mb/s, which is lower than its fair share of the ring bandwidth. The traffic rate of Flow (1, 3) that has a full capacity demand is different under each of the three schemes.

Under AM, the sending rates of Flow (1, 3) oscillates between 100 and 500 Mb/s dramatically (see Figure 6.5), which results in the very low throughput for this flow; under CM, the sending rate of Flow (1, 3) oscillates between 300 and 500 Mb/s (see Figure 6.6), which also subsequently degrades its throughput; under DVSR (see Figure 6.7), the sending rates of Flow (1, 3) can be kept

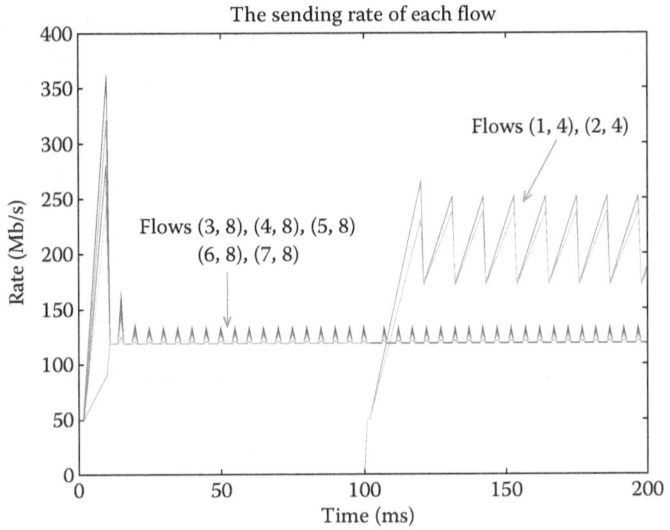

Figure 6.9 The sending rates under CM algorithm (Model 2).

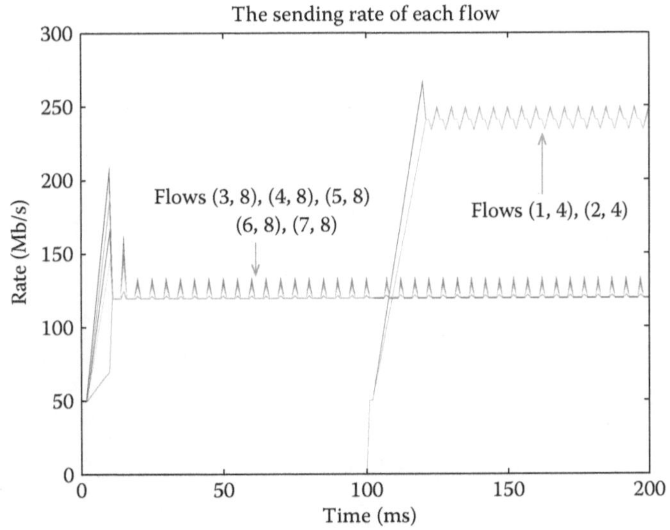

Figure 6.10 The sending rates under DVSR algorithm (Model 2).

near the desired value 500 Mb/s (see Figure 6.7), thus it achieves good throughput of Flow (1, 3), high link utilization and spatial reuse, which meets the key performance objectives of the RPR more satisfactorily.

As we can see from Figures 6.8, 6.9, and 6.10, during the first part of simulation (from 0 to 100 ms), the oscillations among the sending rates of the five flows (flows (3, 8), (4, 8), (5, 8), (6, 8), and (7, 8)) under each scheme are small (though the CM and DVSR show somehow slighter than AM), this is because all these five flows are under a balanced input scenario (each flow has a full capacity demand). However, during the second part of the simulation, the traffic rates for the new two add-in flows (Flow (1, 4) and Flow (2, 4)) under AM and CM schemes show severe oscillations. This is because the Flow (3, 8) is constrained at the downstream bottleneck link (between Nodes 7 and 8), which leads to the unbalanced scenario on the upstream bottleneck link (between Nodes 3 and 4); therefore, the oscillations for the two news flows, i.e., Flow (1, 4) and Flow (2, 4) are similar to that of the first simulation. However, under DVSR, the sending rates of the two flows can rapidly converge to the desired fair shares and few oscillations occur. This simulation further verifies the DVSR can achieve the RPR goals such as fairness and high efficiency much more satisfactorily than AM and CM.

7 Efficient and Fair Bandwidth Allocation for Wide and Metropolitan Area Networks

This chapter focuses on the designing issues with relation to efficiency, capability to meet quality of service (QoS) requirements, and fairness in wide and metropolitan area networks. The approaches presented herein are based on control theory, which successfully fulfill the aim of meeting efficiency, fairness, and spatial reuse requirements of a general network with ring topology.

7.1 INTRODUCTION

It has been established that metropolitan and wide-area network design should aim for efficiency, ability to meet QoS requirements, and fairness. This is clearly the case for the recently developed Resilient Packet Ring (RPR) technology [267,389,390,392,439,524,525] proposed in the IEEE 802.17 standard for metropolitan area networks. One requirement of this standard is to achieve the so-called ring ingress aggregated with spatial reuse (RIAS) fairness [267,393,394,525] in the RPR networks. The RIAS fairness is different from the well-known max–min fairness and proportional fairness criteria. According to the RIAS fairness reference model, the basic unit for fair allocation at a link is the aggregate of the flows originating from the same source node, and the maximum spatial reuse for the relevant links should then be ensured.

However, spatial reuse introduces a fairness challenge because upstream nodes must limit their traffic to a rate determined by congestion downstream. Consequently, the key performance objective of RPR is actually reduced to achieving high utilization and fairness simultaneously [267]. There have been several RPR fairness algorithms proposed [439,524,525]. However, they all suffer from severe and permanent oscillations [267,525]. Such oscillations decrease throughput and increase delay jitter and packet loss. Stability of rate allocation in terms of buffer occupancy at the destination node is desirable in improving network performance and in guaranteeing high efficiency. Normally, if the system is stable, then oscillation is reduced, thus higher link utilization and lower packet loss is achieved.

This chapter aims to achieve better efficiency, fairness, and stability than previously proposed algorithms. Using control theory, we reduce the negative effect of oscillations by designing a rate controller where the control parameters are selected to maintain stability. The so-called adaptive bandwidth allocation (ABA) algorithm we propose herein is a new traffic control scheme, although ABA is established on the basis of a model with a number of flows competing for a single-bottleneck link, we will show that it can also meet efficiency, fairness, and spatial reuse requirements of more general RPR networks.

The classical fairness definition is the so-called max–min fairness [139,142,395–401], which treats the heavy flows passing through a congested bottleneck equally. In addition, proportional fairness [105,366,402] where each source received allocation of bandwidth, during congestion, in proportion to its requirement, is considered of interest. The max–min fairness is modified by many schemes [115,285,342,403–406] to include a consideration for minimum rate (MR) for each flow. The authors of [405,406] have given a general definition of fairness called general weighted (GW) fairness, which implicitly provides MR guarantee and divides the excess bandwidth proportional to the predetermined weights. Different fairness criteria such as max–min fairness, and proportional

MR can be considered as special cases of this general fairness. Reference [406] also proposes a switch algorithm to achieve a GW fairness rate allocation scheme.

Under GW fairness, the MR is guaranteed to any source. After the MRs are set for flows, the next step is to allocate the excess bandwidth (which is the available bandwidth minus the sum of the MRs) fairly among the traffic flows in a weighted manner. Our ABA adopts this (GW) fairness idea and applies it to RPR networks to achieve the RIAS fairness. Considering both congestion and stability, we overcome a certain inefficiency drawback encountered by Algorithm A in [405] that tries to achieve GW fairness. The inefficiency there is caused by having an empty buffer at the bottleneck, while at the same time, there are packets queued for transmission at the source.

The remainder of this chapter is organized as follows: Section 7.2 discusses the bandwidth allocation approach among the flows. We define a new fair weight function for each flow in Section 7.3. In Section 7.4, we propose the ABA algorithm and the corresponding implementation method. In Section 7.5, we use a simple model to illustrate how the proposed ABA algorithm can be applied to RPR networks.

7.2 BANDWIDTH ALLOCATION ON A SINGLE LINK

If the link is shared by various flows and its bandwidth is less than the demand for it, the flows should be controlled to avoid congestion, while at the same time, the link utilization should be maximized and its bandwidth should be fairly apportioned for the different flows. That is, our bandwidth allocation algorithm should be both fair and efficient with the avoidance of congestion. Let us consider the model depicted in Figure 7.1 to shed more light on the relationships between fairness and efficiency performance.

As we can see, $Flow(1,k)$, $Flow(2,k)$, …, $Flow(N,k)$ will pass by Node P before they enter Link L, and $b(P,1)$, $b(P,2)$, … , $b(P,N)$ are the feedback signals to Node 1, 2, …, N from Node P, respectively. Node P should propagate the feedback information as soon as possible to Node i (from 1 to N) to control the traffic rate of $Flow(i,k)$ for achieving the high utilization of Link L as well as avoiding the congestion on it. On the other hand, the bandwidth on Link L should be fairly shared among these flows and the policy for fair allocation should be determined. Under the GW fairness

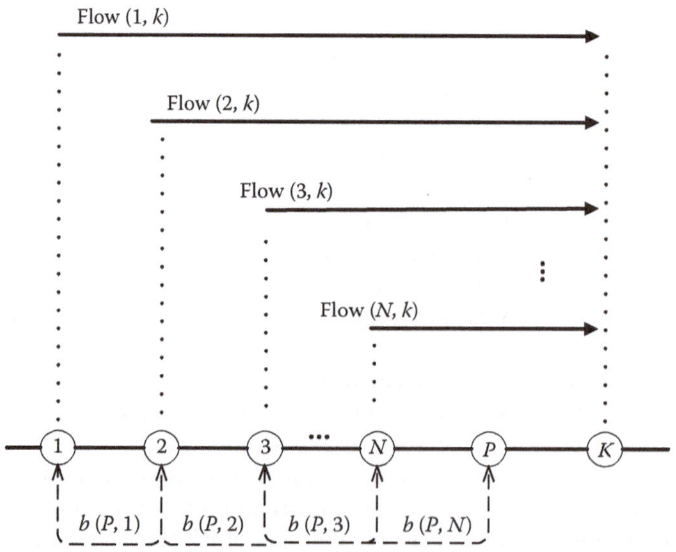

Figure 7.1 N flows competing for Link L capacity.

allocation scheme, each flow is given a minimum bandwidth, beyond that the excess bandwidth is allocated proportionally to the weight for each flow. This can be formulated as follows:

$$R(i) = \text{MR}(i) + \omega_i \left\langle B - \sum_{i=1}^{N} \text{MR} \right\rangle (i) / \sum_{i=1}^{N} \omega_j \quad (7.1)$$

where B is the total bandwidth for follows on the link L, MR(i) is the minimum rate of $Flow(i,k)$, and ω_i is the weight of $Flow(i,k)$, whereas $R(i)$ is the allowed traffic rate for $Flow(i,k)$. Therefore, an important question is how to determine the weight for each flow.

7.3 NEW WEIGHT FUNCTION FOR GW FAIRNESS

In this section, we define one weight function, which is based on the concept of price/charge mechanism [405,406]. As we choose to allocate the excess bandwidth proportionally to the weight of each flow in accordance with the weighted fairness allocation scheme, the primary problem is how to set the weights of the flows. Vandalore et al. [405,406] have introduced a certain weight function, but that function does not associate the weight with the relevant excess bandwidth directly. We first present the problem from the point of view of the user, the bandwidth allocated to a flow can be regarded as a quantity of service offered to its user. The more bandwidth for the flow, the more service quantity is offered to the user. Considering that cost is linear with the quantity of service, in our bandwidth allocation analogy, we assume that the charge is proportional to the service bandwidth requirements. Generally, the charge C is a function [405,406] of the number of bits W received by the user and average rate R for its flow:

$$C = f(W,R) = c + wW + rR \quad (7.2)$$

where c is the fixed cost per connection (flow) for each user, w is the cost per bit, and r is the cost per Mb/s. For the GW fairness allocation, where MR is given, the charge function can be rewritten as

$$C = f(W,R) = c + wW + rR, (M \leqq R) \quad (7.3)$$

where M is the MR, and m is dollars per Mb/s of MR. Since M is the reserved rate (bandwidth) allocated to each flow, the price of m/r is set for it. Suppose there are two users and their costs are C_1 and C_2, respectively. Suppose their allocated rates are R_1 and R_2 and, thus, they transmit W_1 and W_2 bits, respectively. Their costs are

$$C_1 = c + wW_1 + rR_1 + mM_1 C_2 = c + wW_2 + rR_2 + mM_2$$

For the same time period T, the charge C for which a user pays should be proportional to the number of bits W it receives from the network, that is

$$\frac{C_1}{C_2} = \frac{W_1}{W_2} \quad (7.4)$$

Then, we can obtain

$$\frac{c + wW_1 + rR_1 + mM_1}{W_1} = \frac{c + wW_2 + rR_2 + mM_2}{W_2} \quad (7.5)$$

Since $W = RT$, we have

$$\begin{aligned} \frac{c}{R_1 T} + w + \frac{rR_1}{R_1 T} + \frac{mM_1}{R_1 T} &= \frac{c}{R_2 T} + w + \frac{rR_2}{R_2 T} + \frac{mM_2}{R_2 T} \\ &\Rightarrow \frac{c + mM_1}{R_1 T} = \frac{c + mM_2}{R_2 T} \\ &\Rightarrow \frac{c + mM_1}{R_1} = \frac{c + mM_2}{R_2} \end{aligned} \quad (7.6)$$

As R includes the component M, we can assume that

$$\frac{R_1}{c+mM_1} = \frac{R_2}{c+mM_2} = k \tag{7.7}$$

where k can take any positive value. Then, we have

$$\frac{R_1 - M_1}{R_2 - M_2} = \frac{kc + (km-1)M_1}{kc + (km-1)M_2}, \tag{7.8}$$

where $(R-M)$ is the excess part of R over M, so we choose the $kc+(km-1)M$ or $kc/(km-1)+M$ as the weight for fair allocation of the excess bandwidth. Comparing to the weight function of [405,406], the above weight function associates the weight with the excess bandwidth allocated to each flow directly, so it may be more useful to achieve *GW* fairness in practice.

7.4 ABA ALGORITHM

7.4.1 TRAFFIC CONTROL

In this section, we describe our ABA algorithm. It is based on the proportional-derivative (PD) control approach and also provides GW fairness. Consider several flows sharing a common link and competing for its limited bandwidth. Then the aims of our scheme are to achieve the weighted fairness among the different flows, to maintain stable and controlled queuing delay, and to maximize links utilization. We first assign the MR for each flow; before allocating the total excess bandwidth, we should keep a reserved portion of it. This reserved portion is mainly used for congestion avoidance in the presence of bursty traffic and other emergency cases. After fairly allocating the targeted excess bandwidth to the flows competing for the common link in accordance with their weights, we further use the proposed PD approach to readjust the source rates of these flows. In this way, the remaining excess bandwidth is used to improve link utilization while at the same time maintaining a stable buffer occupancy within its target range. We propose the following PD-Controller to control the sending rate of a flow, which passes by a given link

$$R(t) = R_{\max} - a(Q(t-\tau_b) - Q_0) - b(Q(t-\tau_b) - Q(t-\tau_b-1)). \tag{7.9}$$

Here, $R_{\max}$ is the maximal sending rate at the source, which is determined by the *link's* capacity, a and b are tuning arguments, which will be determined based on the stability analysis in the next section. The component τ_b is the delay from the link back to the *flow's* source, $Q(t)$ is the buffer occupancy of the flow at the link and Q_0 is its target value. $R(t)$ is the allowed sending rate of the flow at its source.

7.4.2 DESCRIPTION OF THE ABA ALGORITHM

The above flow-controlling scheme can be realized by the ABA algorithm (which is based on the GW fairness scheme and on some ideas from Algorithm A in [406]). The ABA algorithm is composed of two processes: first, the congested link bandwidth is apportioned among the flows according to the GW fairness criterion using the iteration method. Next, we further stably increase the traffic rate for each flow based on a PD-Controller. The entire ABA process is illustrated in Figure 7.2. We now present a pseudocode for the ABA algorithm, as depicted in Figure 7.3.

We use the following variables and notations. The variable $Link_{\text{Cap}}$ is the total capacity of one common link shared by all competing flows, Cap is the total excess bandwidth of the $Link_{\text{Cap}}$ besides the MRs for these flows while Target is the available excess bandwidth after reserving a reserved portion. The component MR(i) is the minimum rate for Flow i, while $Fairshare(i)$ is the fair apportionment of the excess bandwidth (Target) for Flow i. The component $Total_{\text{Input}}$ is the sum of these

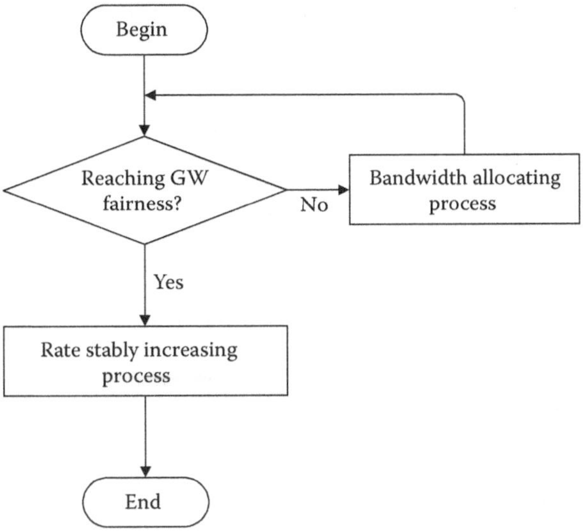

Figure 7.2 The flowchart for the ABA algorithm.

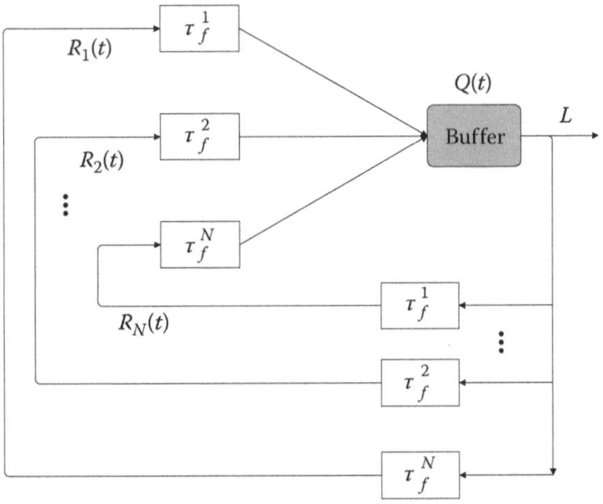

Figure 7.3 The traffic (rate) adjusting model.

flows input rates, and Input is a part of the input rates over MR $=$ MR(i). The notation EAL(i) indicates how much of $Fairshare(i)$ is actually being used by Flow i and this value can be used to compute the next $Fairshare(i)$ value. The argument $Flowshare(i)$ is used to compare with the $Fairshare(i)$ for Flow i at each iteration, the difference between them acts as one of the two necessary conditions to test whether the GW fairness is achieved (the other condition is whether the Input equals to the Target), when the GW fairness is achieved, the $Fairshare(i)$ converges to $Flowshare(i)$ and the Input equals to the Target. The variable $Source(i)$ is the traffic rate of Flow i measured at the destination or intermediate node and used by the GW fairness allocation procedure, while $Ri(t)$ is the traffic rate for Flow i through PD-control adjusting after the GW fairness allocation process. We suppose the receiver has the buffer with an enough large size to receive the data.

Algorithm ABA operates at each output port of the relevant intermediate nodes. ABA uses traffic measurements taken by the switch. We divide the time into consecutive equal-sized slots called averaging units. The measurements are taken at any consecutive averaging unit, during which the

Algorithm 7.4.4 Algorithm *ABA*

Start with:
$Cap \leftarrow Link_{Cap} - \sum MR(i);$
$Target \leftarrow Cap \times f;$
$Input \leftarrow Total_{Input} - \sum MR(i);$
$z \leftarrow Input/Target;$

For every $flow_i$ do
$Fairshare(i) \leftarrow (Target \times w_i)/\sum(w_j);$
end

L_1 :
$EAL(i) \leftarrow min(1, max(0, Source(i) - MR(i))/Fairshare(i);$
$Fairshare(i) \leftarrow (Target \times w_i)/\sum w_j \times EAL(j)$;
$Flowshare(i) \leftarrow max(0, Source(i) - MR(i))/z;$
$B(i) \leftarrow R(i) + max(Fairshare(i), Flowshare(i));$
$B(i) \leftarrow min(B(i), Target);$
$Source(i) \leftarrow B(i);$
If$(Flowshare(i) \neq Fairshare(i))\vee(Input \neq Target)$ Go to L_1
For each Flow i
$R_i(t) \leftarrow Source(i);$
L_2 :
$R(t) \leftarrow R_{max} - a(Q(t-t_b) - Q_0) - b(Q(t-t_b) - Q(t-1-t_b));$
$\Delta R(t) \leftarrow R(t) - \sum R_i(t);$
$\Delta R_i(t) \leftarrow (\Delta R(t) \times w_i)/\sum w_j;$
$R_i(t) \leftarrow R_i(t) + \Delta R_i(t);$
Go To L_2;

switch estimates the number of the active flows and calculates their traffic rates. Specifically, the Input, which is the difference of the sum of all these flows input rates minus the sum of their MRs is equal to the Target, when the $Flowshare(i)$ value converges to the value of $Flowshare(i)$ for each flow. The available bandwidth, i.e., $Link_{\text{Cap}} - Cap \times (1 - f)$, is shared among the flows according to the GW fairness criterion, thus the G fairness bandwidth allocation process is finished. After this bandwidth apportionment, the reserved portion of the *link's* bandwidth, namely $Cap \times (1 - f)$, remains unused and so we further use the PD-control approach to adjust the sending rate at the source for each flow. On one hand, the reserved portion of the *link's* bandwidth is still allocated for each flow in a weighted manner without violating the GW fairness policy; on the other hand, we increase the rate for each flow fairly and stably to maintain a desired queue size, and avoiding link congestion, unnecessary queuing delay, and severe oscillations. Subsequently, the ABA algorithm achieves high efficiency as well as fairness guarantee.

7.4.3 STABILITY ANALYSES TO THE ABA ALGORITHM

Oscillations appearing in rate allocation degrade the network performance [267,407]. Our algorithm alleviates this problem using control theory. The ABA algorithm uses the PD-control approach to adjust the *source's* send rate, so that buffer occupancy is stably near its target. In the PD-Controller given by Equation (7.9), the source-send rate is adjusted according to the feedback of the buffer occupancy at the destination node. If the control gains a and b are chose to guarantee the *system's* stability, the buffer occupancy at the destination node will converge to its target. This avoids buffer overflow or link underutilization. In this situation, on one hand, if the congestion situation worsens, the destination node can inform the source to reduce the send rate timely; on the other hand, if additional bandwidth in the output port is suddenly available, there is data in the buffer that can be transmitted to utilize the bandwidth. System stability is essential to ensure that the above manipulations be carried out smoothly (see, e.g., [337,407,408]) without oscillations.

Here, we use the *Routh–Hurwitz* [408] method to analyze the stability of the system and to find out the appropriate control parameters a and b of the PD-Controller, which stabilize the system. With reference to the model shown in Figure 7.1, we further consider the closed-loop system in Figure 7.3. As we can see from Figure 7.4, N traffic streams share the bottleneck Link L, the components τ_f^1, $\tau_f^2, \ldots, \tau_f^N$ are the forward time delays from the source node to Link L for $Flow(1,k)$, $Flow(2,k)$, and $Flow(N,k)$, whereas τ_b^1, τ_b^2 ,$\cdots$, τ_b^N are the relevant feedback delays, and $R_i(t)$ is the rate determined for source Node i (i = 1 to N) for Flow i. The buffer is in Node P, and $Q(t)$ is the buffer occupancy determined by the Link L's capacity and the *sources'* send rates. We can formulate $Q(t)$ for this model as follows:

$$Q(t) = \int_0^t \left[\sum_{i=1}^{N} R_i(t - \tau_f^i) - \mu \right] dt + Q(0). \tag{7.10}$$

In Equation 7.10, $Ri(t)$ is the send rate of source Node i, μ is the Link L's capacity, and $Q(0)$ is the initial value of buffer size. We adjust the total sending rate at the source according to the following PD-Controller.

$$R(t) = R_{\max} - a(Q(t - \tau_b) - Q_0) - b(Q(t - \tau_b) - Q(t - \tau_b - 1)). \tag{7.11}$$

Here, $\tau_b = \max\{\tau_b^i\}$. Suppose, *Source*($i$) is the existed rate of source i before the PD-control tuning and let $R_i(0) = Source(i)$, then we can obtain $\Delta R(t) = R(t) - \sum_{i=1}^{N} R_i(0)$, which can be regarded as the total rate increment for all the sources. Note that $\Delta R_i(t) = \omega_i \Delta R(t)$ is the rate increment for source i. So the new rate for each source after the PD-control tuning is given:

$$\begin{aligned} R_i(t) &= R_i(0) + \Delta R_i(t) \\ &= R_i(0) + \omega_i [R_{\max} - a(Q(t - \tau_b) - Q_0) \\ &\quad - b(Q(t - \tau_b) - Q(t - \tau_b - 1)) - \sum_{i=1}^{N} R_i(0)]. \end{aligned} \tag{7.12}$$

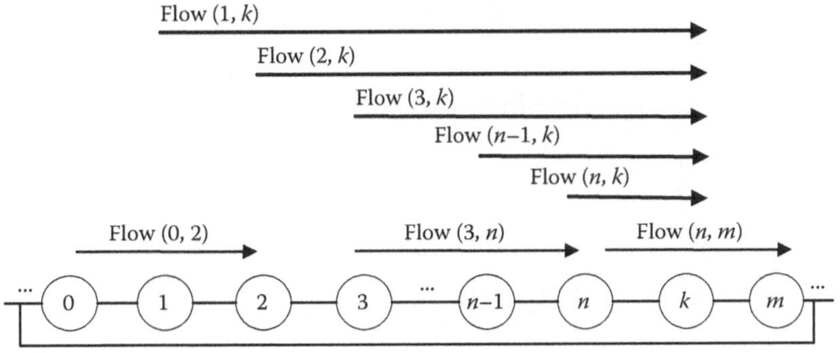

Figure 7.4 Illustration of ABA algorithm with application to RPR.

Combining Equation 7.10 with Equation 7.12, we obtain

$$\begin{aligned} Q_t(0) = \int_0^t \Bigg\{ & \sum_{i=1}^{N} [R_i(0) + \omega_i [R_{\max} - a(Q(t - \tau_f^i - \tau_b) - Q_0) \\ & - b(Q(t - \tau_f^i - \tau_b) - Q(t - 1 - \tau_f^i - \tau_b)) - \sum_{i=1}^{N} R_i(0)]] - \mu \Bigg\} \mathrm{d}t + Q(0). \end{aligned} \tag{7.13}$$

Then by differentiating both sides of Equation 7.13, we obtain the following differential equation:

$$\begin{aligned} Q(t) = \sum_{i=1}^{N} \Bigg[& R_i(0) + \omega_i [R_{\max} - a(Q(t - \tau_f^i - \tau_b) - Q_0) \\ & - b(Q(t - \tau_f^i - \tau_b) - Q(t - 1 - \tau_f^i - \tau_b)) - \sum_{i=1}^{N} R_i(0)] \Bigg] - \mu. \end{aligned} \tag{7.14}$$

Taking the *Laplace Transform*, one yields

$$\begin{aligned} sQ(s) = & \frac{R_{\max} + aQ_0 - \mu}{s} - aQ(s) \sum_{i=1}^{N} \omega_i e^{(-\tau_f^i - \tau_b)s} \\ & - bQ(s) \left[\sum_{i=1}^{N} \omega_i e^{(-\tau_f^i - \tau_b)s} - \sum_{i=1}^{N} \omega_i e^{(-1-\tau_f^i - \tau_b)s} \right]. \end{aligned} \tag{7.15}$$

Equation 7.15 can be further written as

$$\left(s + a \sum_{i=1}^{N} \omega_i e^{(-\tau_f^i - \tau_b)s} + b \sum_{i=1}^{N} \omega_i e^{(-\tau_f^i - \tau_b)s} - b \sum_{i=1}^{N} \omega_i e^{(-1-\tau_f^i - \tau_b)} \right) Q(s) = \frac{R_{\max} + aQ_0 - \mu}{s}. \tag{7.16}$$

So, we can obtain the *characteristic equation* (CE) of the closed-loop system:

$$\left(s + a \sum_{i=1}^{N} \omega_i e^{(-\tau_f^i - \tau_b)s} + b \sum_{i=1}^{N} \omega_i e^{(-\tau_f^i - \tau_b)s} - b \sum_{i=1}^{N} \omega_i e^{(-1-\tau_f^i - \tau_b)} \right) = 0. \tag{7.17}$$

By substitution of the *n-order Taylor series expansions* of $e^{(-\tau_f^i - \tau_b)s}$, $e^{(-1-\tau_f^i - \tau_b)s}$ into Equation 7.17, one yields

$$\begin{aligned} & \frac{(-1)^n}{n!} \sum_{i=1}^{N} \omega_i [(a+b)(\tau_f^i + \tau_b)^n - b(1 + \tau_f^i + \tau_b)^n] s^n \\ & + \frac{(-1)^{(n-1)}}{(n-1)!} \sum_{i=1}^{N} \omega_i [(a+b)(\tau_f^i + \tau_b)^{(n-1)} - b(1 + \tau_f^i + \tau_b)^{(n-1)}] s^{(n-1)} \\ & + \cdots + \frac{1}{2} \sum_{i=1}^{N} \omega_i [(a+b)(\tau_f^i + \tau_b)^2 - b(1 + \tau_f^i + \tau_b)] s^2 \\ & + [1 - \sum_{i=1}^{N} \omega_i [(a+b)(\tau_f^i + \tau_b) - b(1 + \tau_f^i + \tau_b)] s + a = 0. \end{aligned} \tag{7.18}$$

We now use *Routh–Hurwitz* approach to analyze Equation 7.18 to obtain the required stability range of the parameters a and b. To this end, we use the *Routh* table.

Table 7.1
The Routh Table

a_n	$a_n - 2$	$a_n - 4$	...
$a_n - 1$	$a_n - 3$	$a_n - 5$	...
$a_n - 2$	$a_n - 4$	$a_n - 6$	...
...	...	...	...

As seen from Table 7.1, the first two rows of the table are determined by the coefficients of [396] directly, starting from the leading coefficient a_n. The elements in the following rows are given by

$$
\begin{aligned}
b_{n-2} &= a_{n-2} - a_{n-3}\frac{a_n}{a_{n-1}} \\
&= (-1)^{(n-2)}\left[\frac{\sum_{i=1}^{N}\omega_i[(a+b)(\tau_f^i+\tau_b)^{(n-2)} - b(1+\tau_f^i+\tau_b)^{(n-2)}]}{(n-2)!}\right. \\
&\quad - \frac{(n-1)!\sum_{i=1}^{N}\omega_i[(a+b)(\tau_f^i+\tau_b)^{n} - b(1+\tau_f^i+\tau_b)^{n}]}{n!(n-3)!\sum_{i=1}^{N}\omega_i[(a+b)(\tau_f^i+\tau_b)^{(n-1)}]} \\
&\quad \left.\times\frac{\sum_{i=1}^{N}\omega_i[(a+b)(\tau_f^i+\tau_b)^{(n-3)} - b(1+\tau_f^i+\tau_b)^{(n-3)}]}{n!(n-3)!\sum_{i=1}^{N}\omega_i[(a+b)(\tau_f^i+\tau_b)^{(n-1)}]}\right]
\end{aligned}
\tag{7.19}
$$

$$
\begin{aligned}
b_{n-4} &= a_{n-4} - a_{n-5}\frac{a_n}{a_{n-1}} \\
&= (-1)^{(n-4)}\left[\frac{\sum_{i=1}^{N}\omega_i[(a+b)(\tau_f^i+\tau_b)^{(n-4)} - b(1+\tau_f^i+\tau_b)^{(n-4)}]}{(n-4)!}\right. \\
&\quad - \frac{(n-1)!\sum_{i=1}^{N}\omega_i[(a+b)(\tau_f^i+\tau_b)^{n} - b(1+\tau_f^i+\tau_b)^{n}]}{n!(n-5)!\sum_{i=1}^{N}\omega_i[(a+b)(\tau_f^i+\tau_b)^{(n-1)}]} \\
&\quad \left.\times\frac{\sum_{i=1}^{N}\omega_i[(a+b)(\tau_f^i+\tau_b)^{(n-5)} - b(1+\tau_f^i+\tau_b)^{(n-5)}]}{n!(n-5)!\sum_{i=1}^{N}\omega_i[(a+b)(\tau_f^i+\tau_b)^{(n-1)}]}\right]
\end{aligned}
\tag{7.20}
$$

$$
\begin{aligned}
b_{n-6} &= a_{n-6} - a_{n-7}\frac{a_n}{a_{n-1}} \\
&= (-1)^{(n-6)}\left[\frac{\sum_{i=1}^{N}\omega_i[(a+b)(\tau_f^i+\tau_b)^{(n-6)} - b(1+\tau_f^i+\tau_b)^{(n-6)}]}{(n-6)!}\right.
\end{aligned}
$$

$$-\frac{(n-1)!\sum_{i=1}^{N}\omega_i[(a+b)(\tau_f^i+\tau_b)^n-b(1+\tau_f^i+\tau_b)^n]}{n!(n-7)!\sum_{i=1}^{N}\omega_i[(a+b)(\tau_f^i+\tau_b)^{(n-1)}]} \times \frac{\sum_{i=1}^{N}\omega_i[(a+b)(\tau_f^i+\tau_b)^{(n-7)}-b(1+\tau_f^i+\tau_b)^{(n-7)}]}{n!(n-7)!\sum_{i=1}^{N}\omega_i[(a+b)(\tau_f^i+\tau_b)^{(n-1)}]}\Bigg] \tag{7.21}$$

The *Routh–Hurwitz* stability test states that the system is stable if and only if all the elements in the first column of the *Routh* table are strictly positive. Based on this test, we can determine the range of the control parameters a and b such that the system is stable in terms of the buffer occupancy of the destination node. If we take $n = 2$ as an example, Equation 7.18 can be written as

$$\frac{1}{2}\sum_{i=1}^{N}\omega_i[(a+b)(\tau_f^i+\tau_b)^2-b(1+\tau_f^i+\tau_b)^2]s^2 + \left[1-\sum_{i=1}^{N}\omega_i[(a+b)(\tau_f^i+\tau_b)-b(1+\tau_f^i+\tau_b)]\right]s+a=0. \tag{7.22}$$

And its relevant Routh table is

According to the Routh–Hurwitz stability principle, we have

$$\begin{cases} \frac{1}{2}\sum_{i=1}^{N}\omega_i[(a+b)(\tau_f^i+\tau_b)^2-b(1+\tau_f^i+\tau_b)^2]>0 \\ 1-\sum_{i=1}^{N}\omega_i[(a+b)(\tau_f^i+\tau_b)-b(1+\tau_f^i+\tau_b)]>0 \\ u \geq 0 \end{cases} \tag{7.23}$$

Let $a = b$, the range of a or b is given by

$$0<a<1/\sum_{i=1}^{N}\omega_i(\tau_f^i+\tau_b-1)$$

If the value for a or b satisfies this condition, all the roots of Equation 7.22 are in the open left-half plane (OLHP). In this case, the network is asymptotically stable in terms of the buffer occupancy. The above analyses provide a useful guideline on how to choose the control parameters to meet the system stability requirement.

Table 7.2
Routh Table for n = 2

$\frac{1}{2}\sum_{i=1}^{N}\omega_i[(a+b)(\tau_f^i+\tau_b)^2-b(1+\tau_f^i+\tau_b)^2]$	a
$1-\sum_{i=1}^{N}\omega_i[(a+b)(\tau_f^i+\tau_b)-b(1+\tau_f^i+\tau_b)]$	0
a	0

7.5 APPLICATION OF THE ABA ALGORITHM TO THE RING METROPOLITAN AREA NETWORKS

In this section, we discuss the ability of the ABA algorithm to achieve the key performance objectives of RPR networks, including RIAS fairness without oscillation and high utilization. RIAS fairness has two components. One defines the so-called ingress-aggregated (IA) flow [267], that is, the aggregate of all flows that are generated at the same source node while terminated at different destination nodes as one allocation unit for the bandwidth allocation algorithms.

With regard to the specific allocation unit, whose members share a given bottleneck link, the bandwidth allocation should be performed in a fair (equally or weighted) manner among those aggregated flows within the same IA flow. The other achieves maximal spatial reuse subject to the performed bandwidth allocation for the IA flows and the restriction of the link capacity. Specifically, bandwidth on some link can be further reclaimed by the IA flows over the link when it is unused due to the lack of demand or in cases of flows are bottlenecked elsewhere. Using the RPR model shown in Figure 7.4, we will show how the ABA algorithm can be applied to RPR networks to satisfy the RIAS fairness principle. As we can see from the model, $Flow(1,k)$, $Flow(2,k)$, $Flow(n,k)$, and $Flow(n,m)$ all pass through $Link(n,k)$; $Flow(0,2)$ and $Flow(1,k)$ share $Link(1,2)$; $Flow(3,n)$ shares $Link(3,4)$ with $Flow(1,k)$, $Flow(2,k)$, and $Flow(3,k)$. Suppose each link has the equal bandwidth C.

We perform the ABA algorithm (which is operated at Node n) for the $Link(n,k)$ and propagate the control messages to all the relevant upstream nodes in order to achieve the GW fairness among IA flows passing by $Link(n,k)$ and the full utilization on this link. As a result, each IA flow, which is generated from Node 1, Node 2, ..., Node n, respectively, obtains the fair rate R_{1k}, $R_{2k}, \ldots, R_{nk}$. Note $Flow(n,k)$ and $Flow(n,m)$ belong to the same IA flow and these two flows should be treated as a whole unit when executing the ABA algorithm at Node n. Likewise, the ABA algorithm is operated at Node 3 for the three aggregated flows pass on $Link(3,4)$, and $Flow(3,k)$ and $Flow(3,n)$ should also be regarded as an integrated (IA) flow. As the two flows $Flow(1,k)$ and $Flow(2,k)$ have already obtained their fair shares on the downstream $Link(n,k)$, which are determined by the ABA algorithm operated in Node n, so we keep these values for the two flows and we only need to consider $Flow(3,k)$ and $Flow(3,n)$ that form another integrated flow, when we execute the ABA algorithm at Node 3. In addition, we execute the ABA algorithm at Node 1 for the two flows passing through $Link(1,2)$. Consequently, we only need to consider $Flow(0,2)$ using the ABA algorithm not necessarily considering $Flow(1,k)$.

After we obtain the fair share for all the aggregated flows passing through these links, we further consider the inner bandwidth allocation of the IA flows that contain several single flows originating from the same source node according to the RIAS fairness principle, for this example; both Node 3 and Node n are relevant, so we should allocate Node $n's$ bandwidth (defined as the bandwidth apportioned to the IA flow originating from node n) for $Flow(n,k)$ and $Flow(n,m)$, and Node $3's$ bandwidth for $Flow(3,k)$ and $Flow(3,n)$ according to the same fairness policy. After all these processes, we are able to allocate fairly the bandwidth to each flow and to reuse the excess bandwidth on some links. Combining ABA bandwidth allocation with its feedback control mechanism described in Section 7.4.3, the sending rate of each node is stabilized and oscillations that exist in the previous proposals [267,439,524,525] are avoided. The ABA algorithm can thus realize the objectives of simultaneously achieving fairness, high utilization, and spatial reuse for RPR networks.

7.6 CONCLUSIONS

In this chapter, we have presented the ABA algorithm, which is based on the GW fairness bandwidth allocation scheme and adopts the PD control approach. The ABA scheme is able to achieve fairness as well as high utilization for flows competing for a certain common link at first. Then, it is further extended to fulfill the total bandwidth allocation task for the whole network with satisfactory performance in terms of fairness and efficiency. The stability of the PD-control approach has been analyzed to give guideline on how to choose the control parameters so that the scheme can avoid oscillations and congestion and further enhance the performance. We have explained how the ABA can be applied to the ring metropolitan area networks. The analyses show that the ABA algorithm can overcome the limitation of the existing bandwidth allocation schemes and achieve high utilization and fairness simultaneously.

8 Trade-Off Approach between the Efficiency and Fairness for Networks

Trade-off between efficiency (utilization, throughput, or revenue) and fairness in a general telecommunications network with relation to any fairness criterion, that is compromising between the fairness among users and the utilization of resources, is interesting and important within the context of resource allocation and performance optimization. This chapter is dedicated to this important topic, which largely follows [288,289].

8.1 GENERAL BACKGROUND ON FAIRNESS AND EFFICIENCY TRADE-OFF ISSUE

Efficiency–fairness trade-offs have been of interest to people from many walks of life [262–264, 268,269,278,287,288,472]. Different societies and countries make their choices on these trade-offs. This chapter focuses on such trade-offs in the context of telecommunications networks and provides a framework for evaluation and presentation of such trade-offs. It applies to any network, topology, and any fairness criterion. It can be applied to networks such as Resilient Packet Rings (RPR) [266,270] (the IEEE 802.17 standard for metropolitan area networks), local area networks (wireline and wireless), and wide area networks and problems associated with decisions on how much resources are required to serve remote communities, and decisions on what resources to provide to customers being cross-subsidized, such as those in remote communities. Fairness criteria compatible with the framework include ring ingress aggregated with spatial reuse (*RIAS*) *fairness* [267,393]; *max–min fairness* [139,265]; *proportional fairness* [104,105,278]; *general weighted* (GW) *fairness* [285,286]; and *minimum potential delay fairness* [273].

8.2 (α,β)-FAIRNESS: GENERAL CONCEPT FOR BALANCING THE FAIRNESS AND EFFICIENCY

We introduce the concept of (α,β)-*fairness*. In particular, we define capacity assignment to be (α,β)-*fair* if the rate allocated to a flow is neither less than α times its fair allocation for $0 \leq \alpha \leq 1$ nor higher than β times its fair allocation for $1 \leq \beta$. In this chapter, we provide a framework for maximizing the *efficiency*, under a constraint of, say, (90%, 150%)-fairness. By efficiency, we mean the value of a general utility function of the flow rate allocation. Examples of such utility functions are profit, revenue, utilization, and throughput. In contrast to the approach used in [115,328], we do not seek to choose a particular utility function as a trade-off between efficiency and fairness. In our framework, the utility function that defines efficiency is unrelated to the fairness criterion. Our framework addresses the question: How much can the efficiency be improved by compromising on fairness to a certain extent? Because decreasing α and increasing β increases the feasible set, the efficiency–fairness function is monotonically nonincreasing with α for a fixed β and nondecreasing with β for a fixed α.

In the real world, markets are not efficient. This is also the case in the telecommunications industry. There are many effects that distort market efficiency and lead to revenue functions that are nonsmooth, noncontinuous, nonconcave, and even nonmonotonic. Clearly, if a service provider has two customers one of which pays for the services and the other does not, the service provider will try to allocate more resources to serve the paying customer. However, if other considerations (social, regulations, etc.) force the service provider to serve the nonpaying customer, it will try to do it in a way that will maximize its revenues subject to certain "fairness" constraints. We provide here a framework that achieves this. In particular, we employ recently developed nonlinear programming (NLP) methods that can accurately solve the particular global optimization problems associated with maximizing efficiency subject to fairness constraints.

In this chapter, we generalize our earlier work reported in [288] in two ways. First, we consider here a large class of nonlinear (including nonconcave, nonsmooth, noncontinuous, and nonmonotonic) utility functions, while [288] considered only linear utility functions. The use of nonconcave, nonsmooth, and noncontinuous utility functions is motivated by the fact that markets are often inefficient. Another difference of [288] is that there we only considered the lower bound fairness parameter α, and here we also consider the upper bound parameter β. Such an upper bound is motivated as it avoids situations whereby a user is allocated significantly less bandwidth than his/her neighbor. In such a case, it will not be much consolation that the bandwidth the user allocated is not significantly lower than a certain overall "fair" value (the α constraint).

There have been many publications on fairness and other resource allocation problems in telecommunications networks (see, for example, [104,105,112,115,128,278,325,328,472], and references therein). Mathematically, the general approach is to maximize aggregate utility subject to linear capacity constraints. For tractability and to allow distributed flow control algorithms, the utility functions considered have usually been concave. Some formulations, such as [112], impose the additional constraint that each flow has a minimum and maximum transmission rate. The parameters α and β in our formulation provide a specific interpretation for these minimum and maximum rates.

The focus here is the problem of a network operator allocating virtual private links to users, rather than flow control. Instead of seeking to maximize the aggregate benefit to the users, this chapter seeks to maximize the benefit to the service provider. More significantly, this problem allows centralized algorithms to be used, which allows a wider class of problems to be studied. By using global optimization algorithms, flows with nonconcave, nonsmooth, and noncontinuous utility functions can be considered.

Network designers must also choose the degree of unfairness allowed, by setting α and β. This can be done in terms of an efficiency–fairness function, which quantifies the trade-off as follows. First, the "fair" rates are chosen, in terms of a fairness criterion such as max–min fairness or proportional fairness. Then, for a range of α and β values, the operator's utility is numerically optimized given the (α, β)-fairness constraint, namely that each flow obtains between α and β times its fair allocation. Unlike the earlier papers, we solve many optimization problems not one. First, we compute the rates according to a given fairness criteria. Then, we add the so-called fairness constraints to the capacity constraints to make sure that no flow is either less than α times its fair allocation or it is higher than β times its fair allocation. And we maximize the efficiency subject to these sets of constraints for range of α and β values. This gives us the efficiency as a function of α and β, which we call the efficiency–fairness function. Our efficiency function is an aggregate utility each of which may be nonconcave, nonsmooth, and noncontinuous.

In this chapter, we consider two global optimization algorithms. The first algorithm is the so-called Lipschitz Global Optimization (LGO). It is a well-known algorithm that uses the branch-and-bound global search method [280]. This algorithm is one of the best algorithms in solving optimization problems with constraints [279].

The second algorithm is the so-called Algorithm for Global Optimization Problems (AGOP)—a recently developed algorithm presented in [276]. This algorithm is designed for solving continuous optimization problems with box constraints, that is, problems where the feasible region is the Cartesian product of intervals. It is, therefore, relevant to the type of problems we consider in this chapter to maximize efficiency subject to capacity and fairness constraints. The efficiency of the algorithm has been demonstrated in solving many difficult practical and test problems (see, for example, [275–277]).

Throughout this chapter, we use the notation $\langle u, v\rangle$ for the link that connects nodes u and v and $[u, v]$ for the dataflow from node u to node v.

As an illustration of the (α, β)-fairness concept, consider the two-node single-link network presented in Figure 8.1 with the link capacity equal to 1. There are two flows from node u to node v, designated as Flow 1 and Flow 2. Both flows aim to transmit at unlimited rate. Assume that Flow 1 pays 1 [$/unit capcacity] and Flow 2 pays 1.5 [$/unit capacity]. Assume that we choose max–min as our fairness criterion, and total revenue [$] as our utility. Accordingly, if each of the flows is assigned a rate of $1/2$, this will yield a utility of 1.25. However, if we relax the fairness constraint to (α, β)-fairness, say for all $\{\alpha, \beta\}$ pairs that obey the relation $\beta = 1 - \ln \alpha$ for $1 \geq \alpha > 0$ (with $\beta = \infty$ for $\alpha = 0$), in which case the lower bound (set by α) is the tighter constraint, then Flow 1 will be assigned the rate of $\alpha/2$ and Flow 2 will be assigned the rate of $1 - \alpha/2$. This will give utility of $1.5 - \alpha/4$ plotted in Figure 8.2. Maximum utility is achieved if fairness is completely ignored ($\alpha = 0$ and $\beta = \infty$) and Flow 2 is assigned the full bus capacity. Clearly, the slope of the efficiency–fairness function, for the previous example, can be made significantly steeper if we further increase the charge of Flow 2 (and/or decrease the charge of Flow 1).

Notice that utilization can be viewed as a special case of revenue by setting the cost of the total capacity of each link to 1 and the cost of each flow per link is equal to the proportion of the link capacity it uses. Figure 8.3 illustrates our framework. The inputs are network topology, a set of α, β values, efficiency utility function and fairness criterion, and the output is the efficiency fairness function for the set of the (α, β) values. Given the general setting of our framework, we can also answer questions of fairness associated with serving individuals or communities in remote locations. It is an important political and socioeconomical problem in many countries how much society and telecommunications providers should spend in serving remote communities. One extreme view is that people in remote communities should have "equal access." That is, they have the same access to telecommunications services, and at the same cost, people in major cities have. This corresponds in our framework to $\alpha = \beta = 1$. Another extreme view is that services to remote communities should be left to market forces $\alpha = 0$ and $\beta = \infty$. Of course, there are many views in between these two extreme views. We will now use a three-node example to demonstrate how our framework can apply to the efficiency–fairness trade-off related to the question of servicing remote communities.

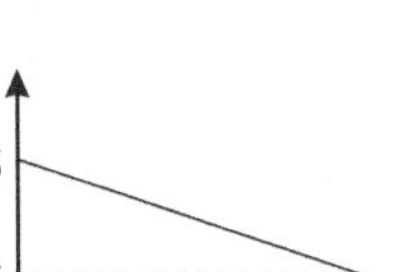

Figure 8.1 Topology and flows of the two-node single-link example.

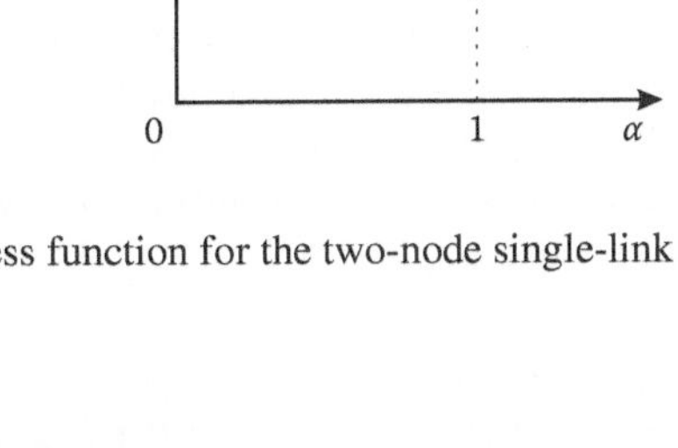

Figure 8.2 The efficiency–fairness function for the two-node single-link example.

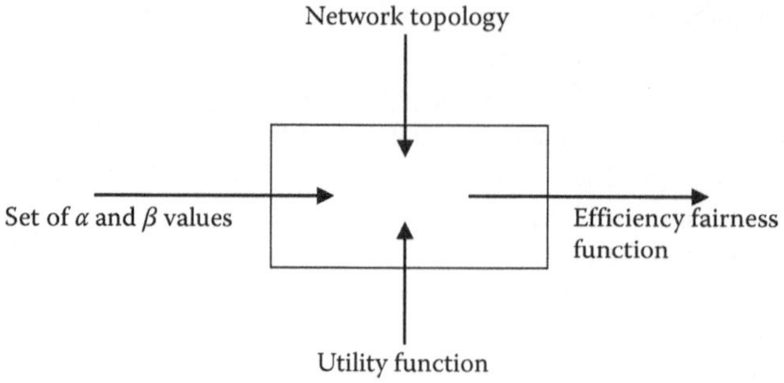

Figure 8.3 The framework.

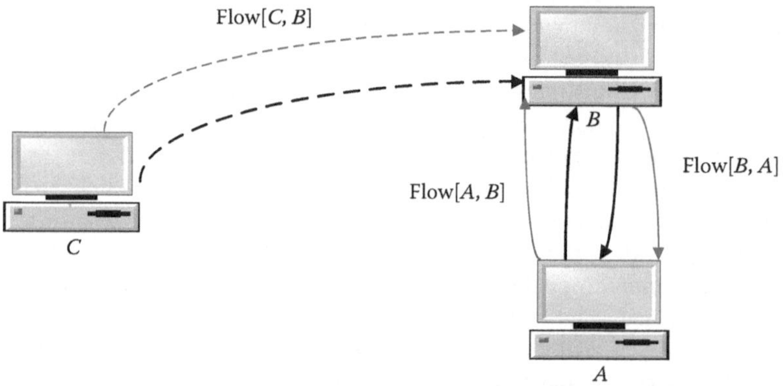

Figure 8.4 Topology and flows of the remote node example.

Consider the three-node network presented in Figure 8.4. Nodes A and B represent major cities. Node C is a remote community. Flow $[A,B]$ represents the flow from A to B on the directed link $\langle A,B\rangle$, and Flow $[B,A]$ represents the flow from B to A on a different directed link $\langle B,A\rangle$. The two links between Nodes A and B have capacity 1. The link $\langle C,B\rangle$ does not yet exist in practice. There is a need to make a major investment in infrastructure to make this link a reality. To apply our framework to this problem, we include the link $\langle C,B\rangle$ with capacity bounded above by 1. The actual capacity of this link is to be determined, and this infrastructure investment will be included in the utility function associated with flow $[C,B]$; the cost of installing capacity is assumed to be twice the revenue raised over the life of the infrastructure. In other words, we assume for simplicity that allocating rate x_{CB} to flow $[C,B]$ will contribute to the aggregate utility,

$$-2x_{CB} + x_{CB} = -x_{CB}.$$

This assumption of linear relationship between the rate provided and its infrastructure cost is made here for simplicity.

Normally, a more appropriate model for this relationship is a step function, which we consider in Section 8.4. The fairness criterion we consider is equal rate fairness; that is, each flow is allocated rate 1. The aggregate utility of all three flows includes the actual values of the flows $[A,B]$ and $[B,A]$ each of which obtains its maximum values at 1 and the utility of flow $[C,B]$, which is *minus* the flow on $[C,B]$. Altogether, the utility is equal to

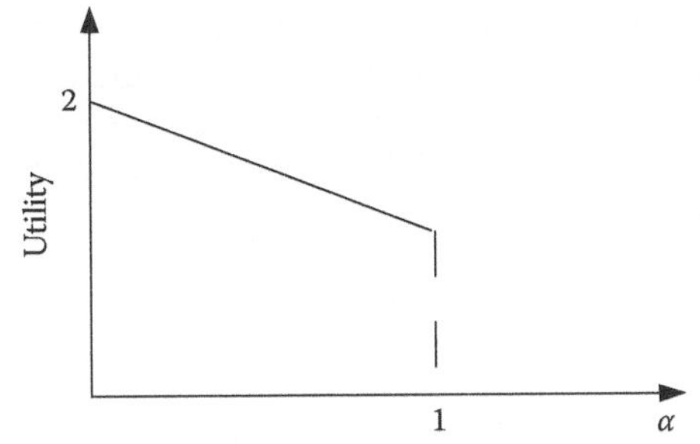

Figure 8.5 The efficiency–fairness function for the remote node example.

$$1+1-x_{CB}=2-x_{CB}.$$

Notice that the fair allocation is 1 to each of the three flows and that the capacity of each link is also equal to 1, the efficiency fairness curve does not change with β (no flow can exceed its fair allocation), so the allocation is independent of β. The rate obtained by flow $[C,B]$ will be α, and the efficiency–fairness function is $2-\alpha$, as shown in Figure 8.5. We acknowledge that having capacity on a link from a remote location equal to capacity on a link between major cities is unrealistic. It is also unrealistic to expect that such two links will be utilized in the same way. It is, therefore, more reasonable to assign capacity of say 0.01 for the $[C,B]$ in which case the value of the utility function for $0<\alpha\leq 1$ will be 0.01.

Of course, the utility functions we considered above are simplistic. For example, serving customers at remote locations involve a much more complex cost function than the one we have considered. Nevertheless, the utility functions we consider in this chapter have characteristics of very wide generality that can be applicable to realistic cases.

Having introduced the concepts of (α,β)-fairness, the efficiency–fairness function and having demonstrated their applicability, we are ready to formalize these notions. In Section 8.3, we formulate an NLP that leads to the efficiency–fairness function, and in Section 8.4, we further discuss this NLP and the methods used to solve it. In Section 8.5, we provide several network examples to demonstrate how to formulate an NLP that finds the optimal bandwidth allocation for a general network under the fairness and capacity constraints and how to obtain the efficiency–fairness function.

8.3 NONLINEAR PROGRAM FORMULATION IN TERMS OF (α,β)-FAIRNESS

We consider an M node network. The nodes are designated $1,2,3,\ldots,M$. All sources are assumed to be greedy. Each source may transmit one or more flows. Let R_{ij} be the rate assigned to flow $[i,j]$. The aim is to set the R_{ij} values to maximize the utility subject to fairness and capacity constraints.

Let F be the set of all flows. Consider the case that there is a utility function $U_{ij}(R_{ij})$ for flow $[i,j]$, and the total utility is the sum of the utility per flow,

$$U=\sum_{[i,j]\in F}U_{ij}(R_{ij}). \tag{8.1}$$

It makes sense in many applications. Users are charged according to usage, and the total revenue is the total sum of revenues collected from the individual users. The functions $U_{ij}(R_{ij})$ may also not be linear. They may not even be concave or smooth or continuous. We, therefore, can allow the functions $U_{ij}(R_{ij})$ to be nonconcave, nonsmooth, and noncontinuous.

Let $f(i,j)$ be the fair allocation for flow $[i,j]$ according to our chosen fairness criterion. Let us assume that we require that the allocation is (α,β)-fair. Then, R_{ij} will be bounded below by $\alpha f(i,j)$ and above by $\beta f(i,j)$. This leads to the following NLP problem.

$$\text{Maximize } U = \sum_{[i,j]\in F} U_{ij}(R_{ij}) \tag{8.2}$$

$$\text{subject to: } \alpha f(i,j) \le R_{ij} \le \beta f(i,j), \quad \forall\, [i,j]; \tag{8.3}$$

$$\sum_{[i,j]\in F\langle m,n\rangle} R_{ij} \le C\langle m,n\rangle, \quad \forall\, \langle m,n\rangle; \tag{8.4}$$

where $C\langle m,n\rangle$ denotes the capacity of link $\langle m,n\rangle$ (this may be the current capacity or the maximum allowed value after a capacity upgrade) and $F\langle m,n\rangle$ is the set of all flows that use link $\langle m,n\rangle$.

Concerning the above proposed optimization model, according to the well-known Karush–Kuhn–Tucker (KKT) conditions [162,163], we have the following theorem to obtain its solution.

Theorem 8.3.1 *For the NLP problems 8.2 through 8.4, there exist the extremal point R^*_{ij}, and the KKT multipliers μ_{ij}, λ_{ij}, and ω_{mn}, such that*

$$\left.\frac{\partial U_{ij}(R_{ij})}{\partial R_{ij}}\right|_{R^*_{ij}} + \mu_{ij} - \lambda_{ij} - \omega_{mn} = 0, \tag{8.5}$$

$$\mu_{ij}, \lambda_{ij} \ge 0, \quad \textit{for all } i,j \tag{8.6}$$

$$\omega_{mn} \ge 0, \quad \textit{for all } m,n \tag{8.7}$$

$$\mu_{ij}\left(\alpha f(i,j) - R^*_{ij}\right) = 0, \quad \textit{for all } i,j \tag{8.8}$$

$$\lambda_{ij}\left(R^*_{ij} - \beta f(i,j)\right) = 0, \quad \textit{for all } i,j \tag{8.9}$$

$$\omega_{mn}\left(\sum_{[i,j]\in F\langle m,n\rangle} R^*_{ij} - C\langle m,n\rangle\right) = 0, \quad \textit{for all } m,n \tag{8.10}$$

$$\alpha f(i,j) - R^*_{ij} \le 0, \quad \textit{for all } i,j \tag{8.11}$$

$$R^*_{ij} - \beta f(i,j) \le 0, \quad \textit{for all } i,j \tag{8.12}$$

$$\sum_{[i,j]\in F\langle m,n\rangle} R^*_{ij} - C\langle m,n\rangle \le 0, \quad \textit{for all } m,n. \tag{8.13}$$

8.4 ANALYSES AND SOLUTION METHODOLOGIES

For the sake of simplicity, we will use the following notations. We denote the number of flows in the set $F = \{[i,j]\}$ by n. Let the variables x_k, $k = 1,2,\ldots,n$, stand for the rates R_{ij} assigned to flows $[i,j] \in F$. We also denote by m the number of all links in the network, and let C_l, $l \in \{1,2,\ldots,m\}$ be the capacity of link l. Let the usage of link l be specified as $r_l = [r_{l1}, r_{l2}, \ldots, r_{ln}]$, where

$$r_{lk} = \begin{cases} 1 & \text{if flow } k \text{ uses link } l, \\ 0 & \text{otherwise.} \end{cases}$$

Then, given fairness parameters α and β, the problems 8.2 through 8.4 can be formulated as follows:

$$\text{maximize} \quad U(x), \tag{8.14}$$

$$\text{s.t.} \quad x \in X \subset RR^n_+; \tag{8.15}$$

where

$$X = \{x \geq 0 : \ x \in B(\alpha,\beta) \text{ and } r_l x \leq C_l, \ \ l = 1,\ldots,m\}, \tag{8.16}$$

and RR^n_+ is the set of points in RR^n with nonnegative coordinates. Throughout this chapter, inequalities between vectors are taken component-wise.

The notation $r_l x$ stands for the scalar product of vectors r_l and x. The set $B(\alpha,\beta) \subset RR^n_+$ is a box defined by

$$B(\alpha,\beta) = \{x = (x_1,\ldots,x_n) : \ \alpha f_k \leq x_k \leq \beta f_k, \ k = 1,\ldots,n\},$$

where f_k is the fair allocation for flow k.

As mentioned above, we consider $\alpha \in [0,1]$ and $\beta \geq 1$. It is clear that the feasible set X is a convex set (polytope). In this model, the fairness parameters α and β are taken into account in box constraints.

In (8.14), $U(x)$ is a utility function of x representing what we call the efficiency.

In this chapter, we consider the class of efficiency (utility) functions defined by

$$U(x) = \sum_{k=1}^{n} \left(\max_{q=1,\ldots,Q} a_k^q \ln(d_k x_k + b_k^q) - c_k S(x_k) \right), \tag{8.17}$$

where $a_k^q > 0$, $b_k^q > 0$, $c_k > 0$, and $d_k > 0$, for all $q = 1,\ldots,Q$ and $k = 1,\ldots,n$. Larger values a_k^q indicate the importance of flow k. The terms $c_k S(x_k)$ in Equation 8.17 represent the costs of infrastructure upgrades. The function $S(x_k)$ may be discontinuous, such as a nondecreasing step function. Typically, $S(0) = 0$. The framework does not require that these costs have the form of a sum over k, but that suffices for the network examples studied here.

Therefore, we consider the following Problem (**P1**):

$$\text{maximize} \ \ U(x) = \sum_{k=1}^{n} \left(\max_{q=1,\ldots,Q} a_k^q \ln(d_k x_k + b_k^q) - c_k S(x_k) \right),$$

$$\text{s.t.} \ \ x \in X \subset RR^n_+;$$

where X is defined by (8.16).

We say that x^* is a locally optimal solution (maximum) in Problem **P1** if there is a neighborhood $x \in \mathscr{V}$ such that $U(y) \leq U(x^*)$ for all $y \in \mathscr{V} \cap X$.

Let us consider certain special cases with $c_k = 0$, $k = 1,\ldots,n$. If $Q = 1$ and $a_k^1 = b_k^1 = 1$ for all k, then we have the following objective function:

$$U_1(x) = \sum_{k=1}^{n} \ln(d_k x_k + 1). \tag{8.18}$$

This is the simplest version of Equation 8.17. In the calculation below, we also consider the case when $Q > 1$. For example, if $Q = 2$ and $a_k^1 = 1, a_k^2 = 4$, $b_k^1 = 1, b_k^2 = 0.9$, and $d_k = 1$, then we have the following objective function:

$$U_2(x) = \sum_{k=1}^{n} \max\{\ln(x_k + 1), \ 4\ln(x_k + 0.9)\}. \tag{8.19}$$

Note that in the latter, $\ln(x_k + 1)$ dominates for $0 \leq x \leq 0.124$ and $4\ln(x_k + 0.9)$ dominates for $0.124 < x$.

The followings are a few comments about the utility function (8.17).

Unlike the purely logarithmic utility function of [104,105,112,174], the utility function is well-behaved for all nonnegative x because $b_k > 0$. In [104,105,112,174], the utility tends to $-\infty$ as a user's rate tends to zero, to ensure fairness. In our approach, a degree of fairness is imposed explicitly by requiring (α,β)-fairness, and so there is no need for an unbounded utility function. Bounded utility functions will generally better reflect the operator's true profit from allocating given rates.

The particular form of Equation 8.17 is used here as an example of a utility function that is nonsmooth and nonconcave. In an inefficient market, which is what we have in the real world, an operator may be able to charge users a higher price when the bandwidth becomes sufficient to provide a new service.

We explain the meaning of objective function 8.17. The coefficients a_k^q define different levels of importance of flows depending on rates. For instance, function 8.19 defines two intervals for each flow: [0, 0.13] and [0.13, +∞]. In the first interval, the efficiency of flow defined by "ln," meanwhile in the second interval it is four times "ln," which results in giving more importance to flows exceeding 0.13.

There is another advantage of using (8.19). If $B(\alpha,\beta) = RR_+^n$, then the optimal solution in terms of objective function 8.18 may contain zero coordinates; that is zero flow rates. It is quite possible that some of these flows are important. In this case, applying Equation 8.19, we may have nonzero flow rates for these important flows. This may happen as a result of assigning a larger amount of flows, which indicates the efficiency (necessity) of shifting to the next level of importance. We note that this is the case we have for Examples 2 and 3 in Section 8.5.

Now we note the following properties of Problem **P1**.

Proposition 8.4.1 *If $Q = 1$ and $c_k = 0$ for all k, then Problem* **P1** *has a unique locally optimal solution.*

The uniqueness of local maxima follows from the fact that the feasible set X is convex and the objective function is strictly concave because $a_k^1 > 0$ and $d_k^1 > 0$. Moreover, the condition $b_k^1 > 0$ guarantees that the objective function is well defined on the feasible set.

Note that, if $f_k = 0$ were allowed, the box constraints could collapse to $x_k = 0$, in which case the objective function would be undefined for $b_k^1 = 0$. However, since $f_k > 0$ for all k, the objective function is well behaved on the interior of RR_+^n for all b_k^1.

Proposition 8.4.2 *If $c_k = 0$, then all locally optimal solutions to Problem* **P1** *belong to the boundary of the feasible set X.*

This proposition follows from the fact that the feasible set X is convex and the objective function is monotonic because $c_k = 0$; that is, $U(x) \leq U(y)$ if $x \leq y$.

From Proposition 8.4.1, it follows that Problem **P1** with Equation 8.18 has a unique locally optimal solution. However, the situation is completely changed when $Q > 1$. In this case, objective function $U(x)$ may be nonconcave and nonsmooth. Therefore, it may have many local maxima, which are on the boundary of the feasible set. As an example, consider a simple case of two variables ($n = 2$) for objective function (8.19) with constraints $X = \{(x_1, x_2) \geq 0 : x_1 + x_2 \leq 1\}$. In this case, we have three locally optimal solutions: (0, 1), (0.5, 0.5), and (1, 0).

The number of locally optimal solutions may drastically increase as the number of variables n increases. In this case, the reasonable goal could be to find solutions that are close to the global solution.

The existence of many locally optimal solutions belonging to the boundary of the feasible set is the main difficulty that complicates the finding of a global solution to Problem **P1**. This difficulty is similar to those in the concave minimization problem, which is NP-hard and is one of the challenging optimization problems (see, for example, [281]). Another difficulty is that the objective function (8.17) is nonsmooth.

Finally, if $c_k \neq 0$, the objective function may be discontinuous. In most applications, S (and hence U) will be piecewise continuous. If S has continuous pieces, then the box $B(\alpha,\beta)$ can be broken into a^n continuous sub-boxes. However, each sub-box may still contain multiple local minima.

These issues require the use of efficient (global) optimization techniques for solving Problem **P1**. However, existing algorithms cannot in general guarantee to find a global solution. In many examples, we know only "the best known solution." The efficiency of an algorithm can be determined by comparison with other algorithms.

As mentioned above, we consider here the two global optimization algorithms LGO and AGOP. While LGO is well known, AGOP has only recently been developed. AGOP [276] is especially designed for continuous optimization problems with box constraints. It uses a line search mechanism where the descent direction is obtained via a dynamical systems approach. It is applicable to a wide range of optimization problems requiring only function evaluations to work. In particular, it does not require gradient information and can be used to find minima of nonsmooth functions. The efficiency of the algorithm has been demonstrated in solving many difficult practical and test problems (see, for example, [275–277]).

The AGOP algorithm will now be described in terms of minimizing a cost function $g : RR^n \to RR$. AGOP must first be given a set of initial points, say $\Omega = \{x^1,\ldots,x^q\} \subset RR^n$, $q \geq 2$. Generally, a suitable choice for an initial set of points is the set of some vertices of a given box. Let $x^\star \in \Omega$ be the point in Ω with the smallest cost, that is, $g(x^\star) \leq g(x)$ for all $x \in \Omega$. The set Ω and the values of g at each of the points in Ω allow us to determine a vector v to be used as a possible descent direction from point $x^\star$, as outlined below. An inexact line search along the direction of v provides a new point $\hat{x}^{q+1} \neq x^\star$. A local search about $\hat{x}^{q+1}$ is then carried out using a method called *local variation*. This is an efficient local optimization technique that does not explicitly use derivatives and can be applied to nonsmooth functions. A good survey of direct search methods can be found in [272]. Letting x^{q+1} denote the optimal solution of this local search, the set Ω is augmented to include x^{q+1}. Starting with this updated Ω, the whole process can be repeated. The process is terminated when v is approximately 0 (or a prescribed bound on the number of iterations is reached). The solution returned is the current $x^\star$, that is, the point in Ω with the smallest cost.

The success of global optimization algorithms mainly depends on their ability to escape the best local minimum found so far in order to find "deeper" local minima. In our case, this is the determination of a possible descent direction v from currently found local minimum x^*. We present here the formula used in the calculations below. For more details and motivations behind it, see [20].

Let $I = \{1,\ldots,q\}$ and let $\Omega = \{x^m : m \in I\}$ be a set of initial points, and $g^m = g(x^m)$. Let $g^* = g(x^*)$, with $x^* = (x_1^*,\ldots,x_n^*)$, be the smallest cost of the points in Ω. For each coordinate $i \in \{1,\ldots,n\}$, define the following sets: $X_i^+ = \{m \in I :\ x_i^m > x_i^*\}$, $X_i^- = \{m \in I :\ x_i^m < x_i^*\}$, $G^+ = \{m \in I :\ g^m > g^*\}$, $G_i^{++} = G^+ \cap X_i^+$, $G_i^{+-} = G^+ \cap X_i^-$. Let $|A|$ denote the cardinality of a set A.

Define $v = (v_1,\ldots,v_n)$ as follows:

$$v_i = \sum_{m \in G_i^{+-}} \frac{1}{|G^+|} \frac{-\Delta x_i^m}{\Delta g^m} \alpha_i^m \cdot \sum_{m \in G_i^{+-}} \frac{1}{|X_i^-|} \frac{\Delta g^m}{\|\Delta x^m\|} \alpha_i^m$$
$$- \sum_{m \in G_i^{++}} \frac{1}{|G^+|} \frac{\Delta x_i^m}{\Delta g^m} \alpha_i^m \cdot \sum_{m \in G_i^{++}} \frac{1}{|X_i^+|} \frac{\Delta g^m}{\|\Delta x^m\|} \alpha_i^m. \tag{8.20}$$

Here, $\Delta x_i^m = x_i^m - x_i^*$, $\Delta g^m = g^m - g^*$, $\alpha_i^m = |\Delta x_i^m|/\|\Delta x^m\|$, and $\|\Delta x^m\|$ is the Euclidean norm. We use the conventions that an empty sum is 0 and that $0/0 = 0$.

Formula 8.20 represents a deterministic approach to calculate v, in contrast to randomized global search algorithms. It tries to take into account the contribution of each coordinate i on the increase in function values. The final value v_i is, in some sense, the average value over the all set Ω.

We consider v as a possible "global" descent direction because it uses information obtained from points that may be quite far from each other. If this direction fails to provide a better point x^{q+1}, that is $g(x^{q+1}) > g(x^*)$, then we add this point to set Ω, which supplies a new direction v calculated by formula 8.20 with this updated set. We note that the role of a direction v is to find a basin of a new local minimum, that is, to find $\hat{x}^{q+1}$ rather than x^{q+1}. In other words, a search over direction v is successful if it finds a basin of a deeper local minima.

In global optimization, usually, we cannot guarantee that we will find a successful descent direction even if x^* is not a global minima. This is the main difficulty that global optimization algorithms encounter. The success of an algorithm can be checked only on numerical experiments. The results obtained in this chapter once more emphasize the efficacy of the proposed approach. Formula 8.20 can be considered as an alternative (deterministic) method to those methods that are based on random search for finding a possible descent direction.

If we have just one initial point x^1, to run this algorithm starting from this particular point, first we apply a local search about this point to get a new point (local minimum) x^2. Then, the above procedure is performed using the set $\Omega = \{x^1, x^2\}$.

To solve optimization problems with constraints, AGOP uses the following scheme, which is demonstrated on Problem **P1**.

First, we transform this problem to a minimization problem with box constraints, applying penalty functions to the linear constraints (see [282] and references therein). Given a penalty coefficient $\gamma > 0$, set

$$g_\gamma(x) = -U(x) + \gamma \sum_{l=1}^{m} (\max\{r_l x - C_l, 0\})^2. \tag{8.21}$$

Consider the Problem (**P2**):

$$\text{Minimize } g_\gamma(x), \quad \text{s.t. } x \in B(\alpha, \beta).$$

Then, we perform the following steps to solve Problem **P1**.

The algorithm to solve Problem **P1**.

10. Take any penalty coefficient $\gamma > 0$ and any number $\lambda > 1$.
20. Apply AGOP to Problem **P2**, which has only box constraints and denote the solution by x^0.
30. Set $p = 1$ and $x^p = x^0$.
40. Set $\gamma = \gamma\lambda$. Apply AGOP to **P2** starting from initial point x^p. Let the solution found be x^{p+1}.
50. If $|g_\gamma(x^{p+1}) - g_\gamma(x^p)| < \varepsilon$, then stop. Otherwise, set $p = p + 1$ and go to 40.

The convergence of the algorithm to a local minimum can be proved for smooth functions, in particular, for Problem **P1** with Equation 8.18. AGOP, like any other algorithms using direct search methods [272], does not guarantee to get a local minimum for nonsmooth objective functions. However, in practice, these methods (having global search character) often perform well, when dealing with nonsmooth functions (see, for example, [275]) and also even noncontinuous functions. The results obtained for Examples 2 and 3 in Section 8.5 also confirm this fact (see Tables 8.3 and 8.5 for a comparison of LGO and AGOP on nonsmooth objective function U_2).

8.5 NUMERICAL STUDIES

The flexibility of the form of the objective function U in **P1** means that powerful global optimization techniques are required to perform the design. This section uses two such techniques, LGO and AGOP, to determine the efficiency–fairness curve for a range of topologies, with objective functions of the form (8.17).

The (α, β)-fairness concept specifies how far a rate allocation can deviate from the "fair" rate but does not specify what rate is to be used as the reference. For these results, we take the max–min fair rates [139] to be "fair." Given α and β, this yields box constraints on the rates of the form

$$B(\alpha, \beta) = \{x = (x_1, \ldots, x_n) : \alpha f_k \le x_k \le \beta f_k, \ k = 1, \ldots, n\},$$

where f_k is the fair allocation for flow k. Given box $B(\alpha,\beta)$, we will denote by $\xi(\alpha,\beta)$ the optimal (maximal) value of objective function in Problem **P1**:

$$\xi(\alpha,\beta) = \max_{x\in X} U(x).$$

In each of the examples below, the set X represents the set of feasible points for the example under consideration.

The efficiency–fairness trade-off is governed by the values of α and β. In order to make the efficiency–fairness function easier to plot, we express both α and β in terms of a single parameter, s. In particular, we set

$$\alpha_s = (10-s)/10, \ \ \beta_s = (\sqrt{2})^s, \quad s = 0,1,\ldots,10.$$

It is clear that

$$B(\alpha_0,\beta_0) \subset B(\alpha_1,\beta_1) \subset \cdots \subset B(\alpha_{10},\beta_{10});$$

and therefore

$$\xi(\alpha_0,\beta_0) \leq \xi(\alpha_1,\beta_1) \leq \cdots \leq \xi(\alpha_{10},\beta_{10}).$$

In all examples below, we will consider objective functions given by Equation 8.17 with $Q = 1$ or $Q = 2$.

8.5.1 EXAMPLE 1: LINEAR NETWORK WITH UNIFORM CAPACITY

Consider the network shown in Figure 8.6 of [328]. There are n flows and $(n-1)$ concatenated links in the network. We consider the case that each link has bandwidth of 1 unit. Flow x_n travels through all the links, flow x_1 travels through link 1, flow x_2 travels through link 2, and so on, and flow x_{n-1} travels through link $n-1$, as shown in Figure 8.6.

The max–min fair allocation for this network is to assign the same rate to all the flows:

$$x_1 = x_2 = x_3 = \ldots x_n = \frac{1}{2}$$

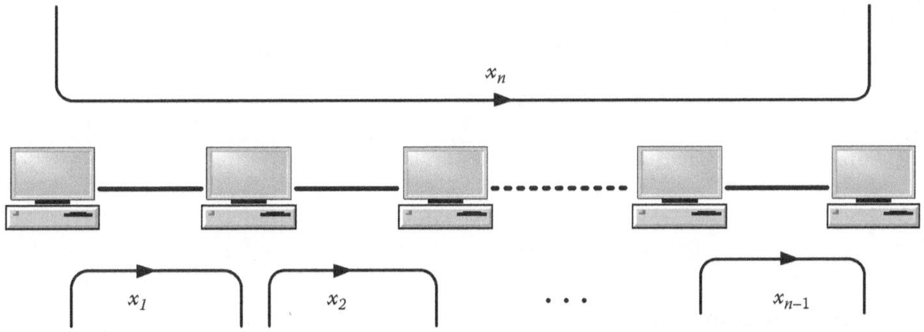

Figure 8.6 The classical network model used in Example 1 in Section 8.5.1.

where x_i is the rate allocation to flow i. We describe the optimizing problem of efficiency–fairness trade-off as follows:

$$\text{Maximize} \quad U(x)$$

subject to:

$$\begin{aligned} \tfrac{1}{2}\beta \geq x_i \geq \tfrac{1}{2}\alpha & \quad \text{for } i = 1,2,\ldots,n \\ x_i + x_n \leq 1 & \quad \text{for } i = 1,2,\ldots,n-1 \\ x_i \geq 0 & \quad \text{for } i = 1,2,\ldots,n \end{aligned}$$

By taking, for example, $n = 10$ and

$$U(x) = \sum_{i=1}^{10} \log(x_i + 1),$$

we solve the above linear programming problem for the following values of α and β

$$\begin{aligned} (\alpha,\beta) &= (0.1, 1), \\ (\alpha,\beta) &= (0.2, 1.1), \\ (\alpha,\beta) &= (0.3, 1.2), \\ (\alpha,\beta) &= (0.4, 1.3), \\ (\alpha,\beta) &= (0.5, 1.4), \\ (\alpha,\beta) &= (0.6, 1.5), \\ (\alpha,\beta) &= (0.7, 1.6), \\ (\alpha,\beta) &= (0.8, 1.7), \\ (\alpha,\beta) &= (0.9, 1.8), \\ (\alpha,\beta) &= (1.0, 1.9), \end{aligned}$$

to obtain the optimal solution of the flow rate vector $(x_1, x_2, x_3, x_4, x_5, x_6, x_7, x_8, x_9, x_{10})$, the optimal total utility, and further calculate the fairness index (**FI**) [141] by the following formula:

$$FI = \frac{(x_1 + x_2 + \ldots + x_{10})^2}{10(x_1^2 + x_2^2 + \ldots + x_{10}^2)}.$$

We tabulate all the results into Table 8.1. The relation of the calculated FI values and the chosen values of α and β are plotted into a three-dimensional figure: Figure 8.7 and the relation of the calculated maximum utility values and the chosen values of α and β are also plotted into a three-dimensional figure: Figure 8.8.

For this example, we next consider the following two objective functions:

$$U_{11}(x) = \sum_{i=1}^{n} \ln(x_i + 1),$$

and

$$U_{12}(x) = \sum_{i=1}^{n} \max\{\ln(x_i + 1),\ 4\ln(x_i + 0.9)\}.$$

Given α and β, the maximal values of objective functions will be denoted as

$$\xi_{11}(\alpha,\beta) = \max U_{11}(x),$$

Table 8.1

Optimal Flow Rate Solution, Optimal Total Utility, and Fairness Index Values by Solving the Efficiency–Fairness Trade-Off Problem for the Bus Network of Figure 8.6 ($n = 10$)

(α, β)	Solution: $(x_1, x_2, x_3, x_4, x_5, x_6, x_7, x_8, x_9, x_{10})$	Total Utility	Fairness Index
(0.1, 1)	(0.50, 0.50, 0.50, 0.50, 0.50, 0.50, 0.50, 0.50, 0.50, 0.50)	4.0546	1.0000
(0.2, 1.1)	(0.55, 0.55, 0.55, 0.55, 0.55, 0.55, 0.55, 0.55, 0.55, 0.45)	4.3159	0.9969
(0.3, 1.2)	(0.60, 0.60, 0.60, 0.60, 0.60, 0.60, 0.60, 0.60, 0.60, 0.40)	4.5665	0.9894
(0.4, 1.3)	(0.65, 0.65, 0.65, 0.65, 0.65, 0.65, 0.65, 0.65, 0.65, 0.35)	4.8071	0.9794
(0.5, 1.4)	(0.70, 0.70, 0.70, 0.70, 0.70, 0.70, 0.70, 0.70, 0.70, 0.30)	5.0380	0.9680
(0.6, 1.5)	(0.70, 0.70, 0.70, 0.70, 0.70, 0.70, 0.70, 0.70, 0.70, 0.30)	5.0380	0.9680
(0.7, 1.6)	(0.65, 0.65, 0.65, 0.65, 0.65, 0.65, 0.65, 0.65, 0.65, 0.35)	4.8071	0.9794
(0.8, 1.7)	(0.60, 0.60, 0.60, 0.60, 0.60, 0.60, 0.60, 0.60, 0.60, 0.40)	4.5665	0.9894
(0.9, 1.8)	(0.55, 0.55, 0.55, 0.55, 0.55, 0.55, 0.55, 0.55, 0.55, 0.45)	4.3159	0.9969
(1, 1.9)	(0.50, 0.50, 0.50, 0.50, 0.50, 0.50, 0.50, 0.50, 0.50, 0.50)	4.0547	1.0000
(1, 1)	(0.50, 0.50, 0.50, 0.50, 0.50, 0.50, 0.50, 0.50, 0.50, 0.50)	4.0547	1.0000

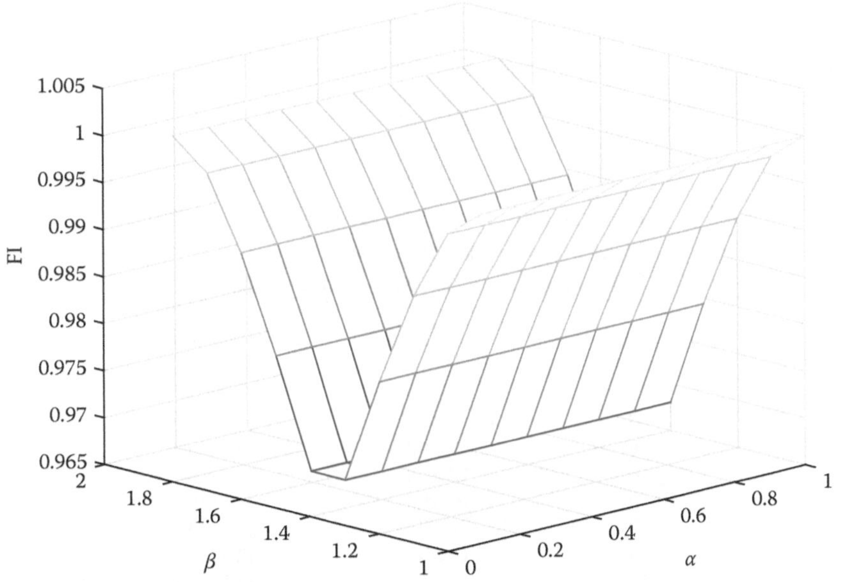

Figure 8.7 A three-dimensional plot of the relation between the calculated FI values and the chosen values of α and β for the bus network of Figure 8.6.

and

$$\xi_{12}(\alpha, \beta) = \max U_{12}(x),$$

respectively.

It is not difficult to show that objective functions U_{11} and U_{12} have a unique optimal solution for this simple example. Both algorithms, LGO and AGOP, could easily find the optimal solution for all boxes $B_s = B(\alpha_s, \beta_s)$, $s = 0, \ldots, 10$. The results are presented in Table 8.2. The efficiency–fairness functions for Example 1 for the objective functions $U_{11}(x)$ and $U_{12}(x)$ are presented in Figure 8.9.

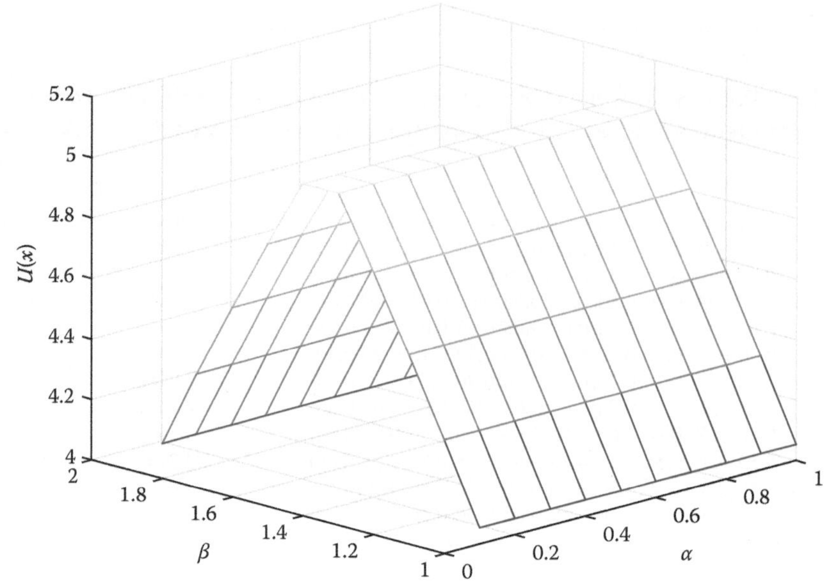

Figure 8.8 A three-dimensional plot of the relation between the calculated maximal utility values and the chosen values of α and β for the bus network of Figure 8.6.

Table 8.2
Results Obtained by Algorithms LGO and AGOP for Objective Functions U_{11} and U_{12} with $n = 20$

Box	ξ_{11}	ξ_{12}
B_0	8.1093	26.9178
B_1	8.6984	29.4393
B_2	9.2665	31.8648
B_3	9.8148	34.1999
B_4	10.3443	36.4496
B_5	10.8558	38.6180
B_6	11.3503	40.7090
B_7	11.8283	42.7260
B_8	12.2905	44.7671
B_9	12.7376	46.8029
B_{10}	13.1698	48.7809

Note: $B_s = B(\alpha_s, \beta_s)$ $(s = 0, 1, \ldots, 10)$ and ξ_{11}, ξ_{12} are the optimal values of objective functions. Note that the results are the same for both algorithms.

8.5.2 EXAMPLE 2: LINEAR NETWORK WITH TWO LONG FLOWS

Consider now a somewhat more complicated network—the linear network with two long flows [328], as shown in Figure 8.10. In this model, we choose the capacity $C = (500, 400, 300, 200, 500)^T$ and calculate the max–min fairness rate allocation. It is given by $x_1 = 400$, $x_2 = 300$, $x_3 = 100$, $x_4 = 100$, $x_5 = 400$, $x_6 = 100$, and $x_7 = 100$ (Figure 8.11).

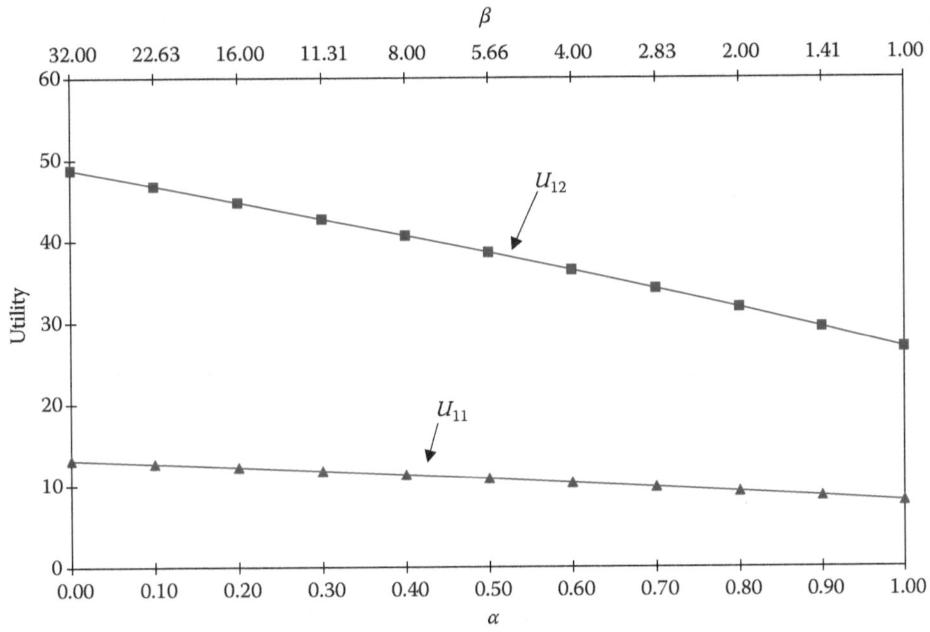

Figure 8.9 Efficiency–fairness functions for Example 1.

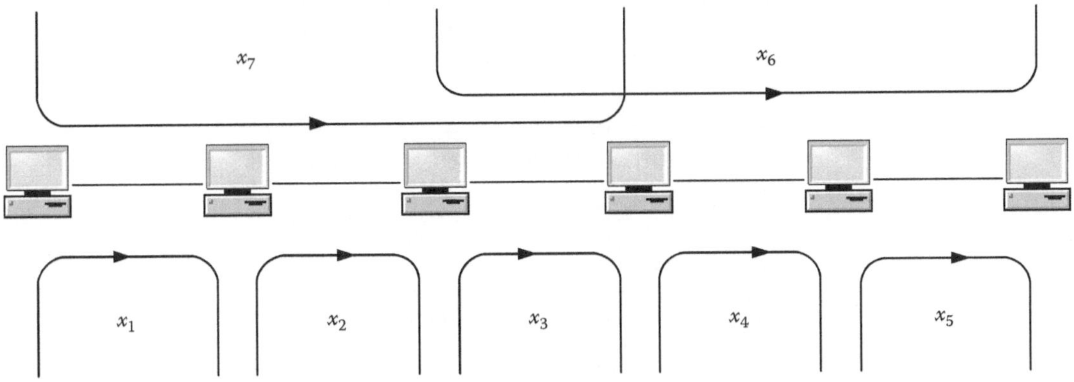

Figure 8.10 The linear network with two long flows used in Example 2.

The optimization problem is as follows:

$$\text{Maximize} \quad U(x)$$

subject to:

$$\begin{gathered}400\beta \geq x_1 \geq 400\alpha,\ 300\beta \geq x_2 \geq 300\alpha,\\ 100\beta \geq x_3 \geq 100\alpha,\ 100\beta \geq x_4 \geq 100\alpha,\\ 400\beta \geq x_5 \geq 400\alpha,\ 100\beta \geq x_6 \geq 100\alpha,\\ 100\beta \geq x_7 \geq 100\alpha,\ x_1 + x_7 \leq 500,\\ x_2 + x_7 \leq 400,\ x_3 + x_6 + x_7 \leq 300,\\ x_4 + x_6 \leq 200,\ x_5 + x_6 \leq 500,\\ x_i \geq 0 \text{ for } i = 1, 2, \ldots, 7.\end{gathered}$$

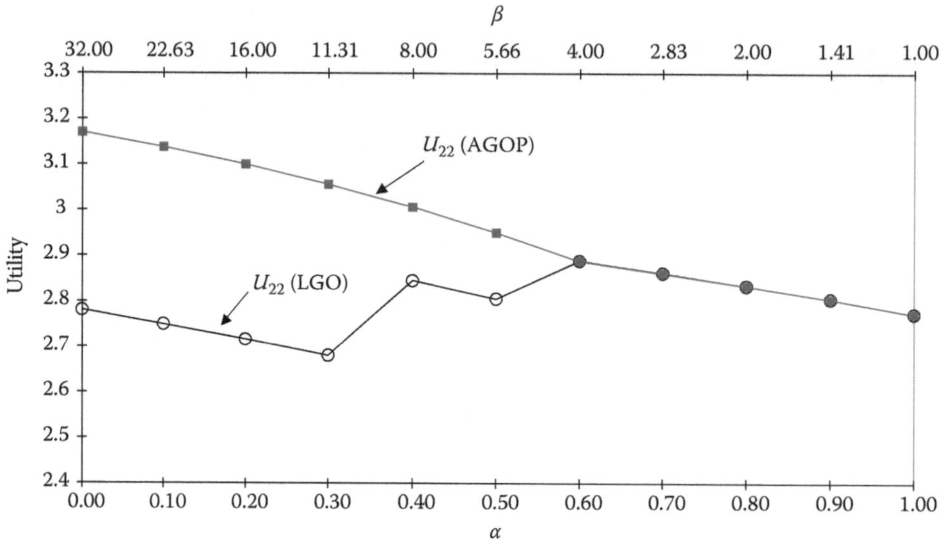

Figure 8.11 Efficiency–fairness function for Example 2 with objective function $U_{22}(x)$.

Table 8.3
Results Obtained by Algorithms LGO and AGOP for Objective Functions U_{21} and U_{22}

Box	ξ_{21}	ξ_{22}	
	LGO and AGOP	LGO	AGOP
B_0	2.3749	2.7727	2.7727
B_1	2.4251	2.8052	2.8052
B_2	2.4730	2.8340	2.8340
B_3	2.5188	2.8618	2.8617
B_4	2.5623	2.8886	2.8883
B_5	2.6038	2.8053	2.9508
B_6	2.6432	2.8451	3.0067
B_7	2.6805	2.6805	3.0566
B_8	2.7158	2.7158	3.1004
B_9	2.7492	2.7492	3.1383
B_{10}	2.7806	2.7806	3.1703

Note: $B_s = B(\alpha_s, \beta_s)$ $(s = 0, 1, \ldots, 10)$ and ξ_{21}, ξ_{22} are the optimal values of objective functions. Note that the results for U_{21} are the same for both algorithms.

For this example, we consider the following two utility functions:

$$U_{21}(x) = \sum_{i=1}^{7} \ln(x_i/500 + 1)$$

and

$$U_{22}(x) = \sum_{i=1}^{5} \ln(x_i/500 + 1) + \sum_{i=6}^{7} \max\{\ln(x_i/500 + 1),\ 4\ln(x_i/500 + 0.9)\}.$$

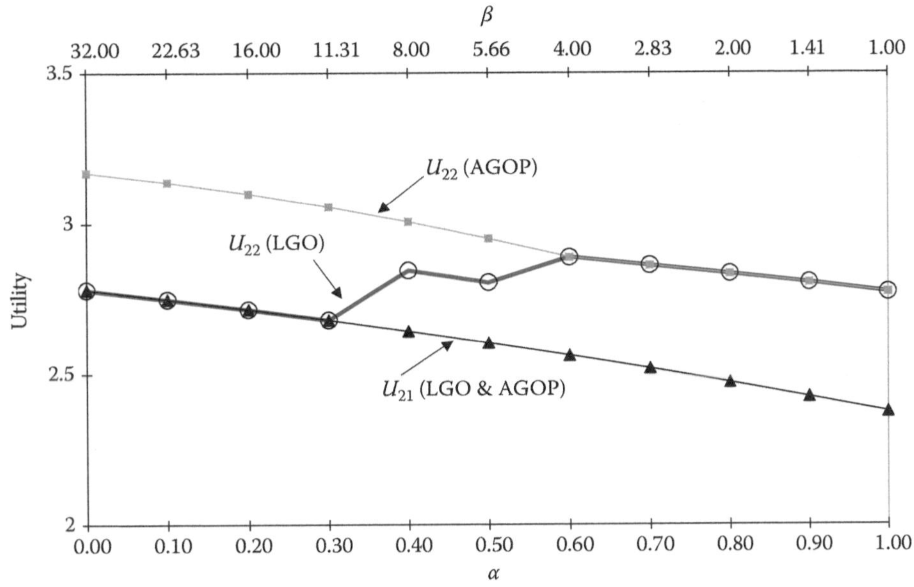

Figure 8.12 Efficiency–fairness functions for Example 2 with objective functions $U_{21}(x)$ and $U_{22}(x)$.

These two utility functions will lead to the efficiency–fairness functions:

$$\xi_{21}(\alpha,\beta) = \max_{x \in X} U_{21}(x)$$

and

$$\xi_{22}(\alpha,\beta) = \max_{x \in X} U_{22}(x),$$

respectively.

The results are presented in Table 8.3 and Figure 8.12. As mentioned above, objective function U_{21} has a unique optimal solution. Both algorithms, LGO and AGOP, could easily find this solution for all boxes $B(\alpha_s, \beta_s)$, $s = 0, \ldots, 10$. However, objective function U_{22} has many locally optimal solutions. As can be seen from Figure 8.12, the LGO algorithm has a difficulty in finding deep locally optimal solutions when the box $B(\alpha_s, \beta_s)$ becomes larger (for larger s values). The estimate from LGO can be improved slightly by enforcing the monotonicity constraint, increasing the estimates for certain (α, β) pairs as necessary; this is not necessary using AGOP.

To study the effect of the α and β parameters of the efficiency–fairness function, we used only AGOP to produce the results presented in Figures 8.14 and 8.15.

The results show that an appreciable gain in utility can be achieved by relaxing the fairness constraints.

Note that in Figures 8.14 and 8.15, the curves are flat for a wide range of α and β values. The operating point, and hence utility, becomes independent of β, when the β constraint ceases to be tight, that is, when all rates are constrained from decreasing by the α constraints or constrained from increasing by the capacity constraints. Clearly, the further α is from 1, the larger the utility is at the point where this happens.

An analogous effect is observed in Figure 8.15. Note, however, that the curves for different β values are not entirely coincident until they reach the β-dependent ceiling as α decreases. Rather, there is a gradual divergence before the limit is reached. That occurs when some of the flows' rates are constrained by α and some by β.

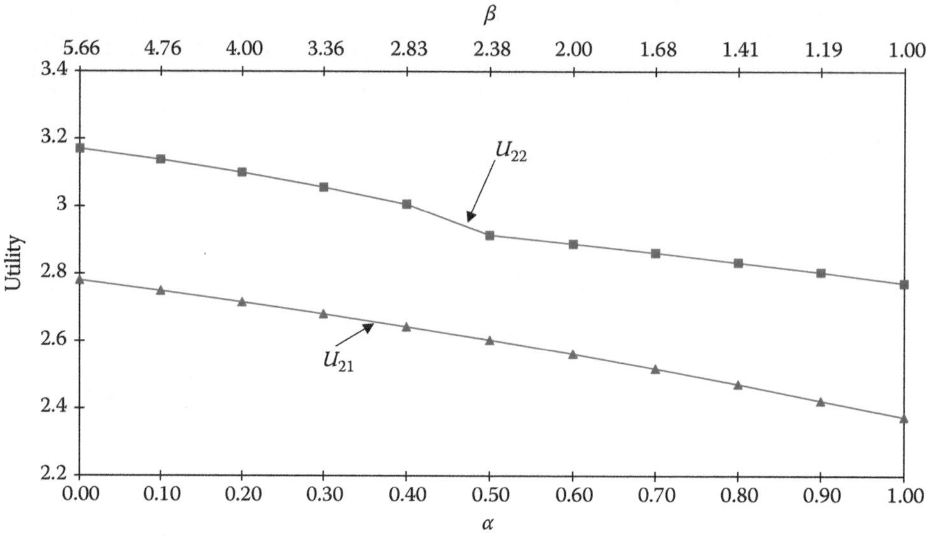

Figure 8.13 Efficiency–fairness functions for Example 2 for a small range of β.

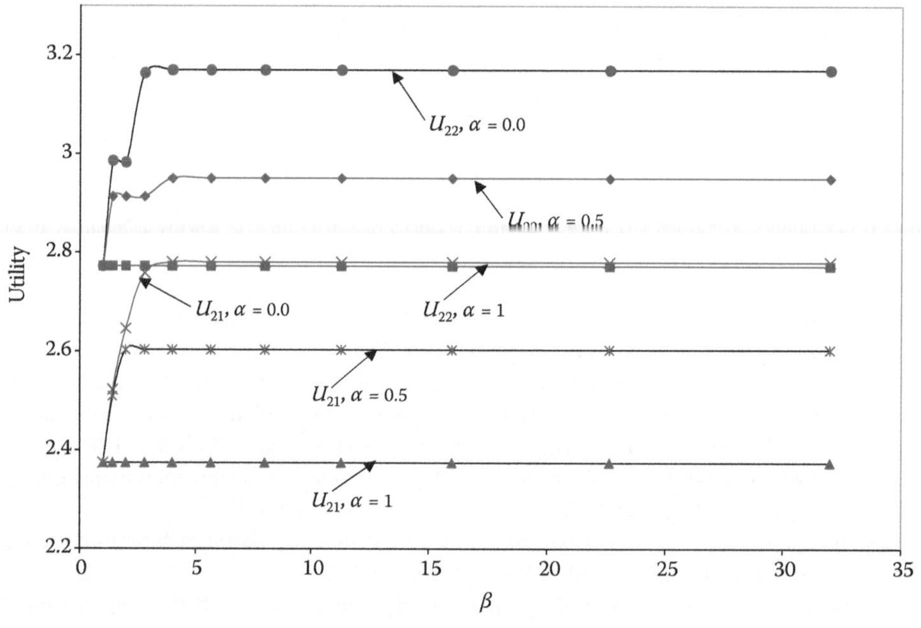

Figure 8.14 Efficiency–fairness functions for Example 2, where α is fixed and β changes.

8.5.3 EXAMPLE 3: 12-NODE NETWORK

The topology of this network is presented in Figure 8.16. Assume that the capacity of each link is set at $C = 2$ and that there are 35 flows in this model. The flows' route information and the max–min rate allocations are given in Table 8.4.

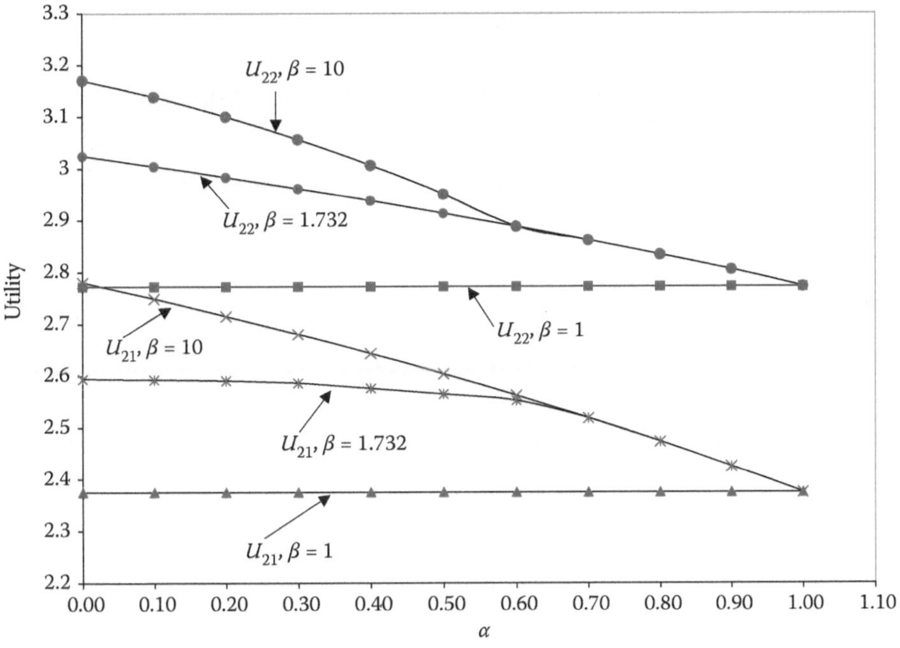

Figure 8.15 Efficiency–fairness functions for Example 2, where β is fixed and α changes.

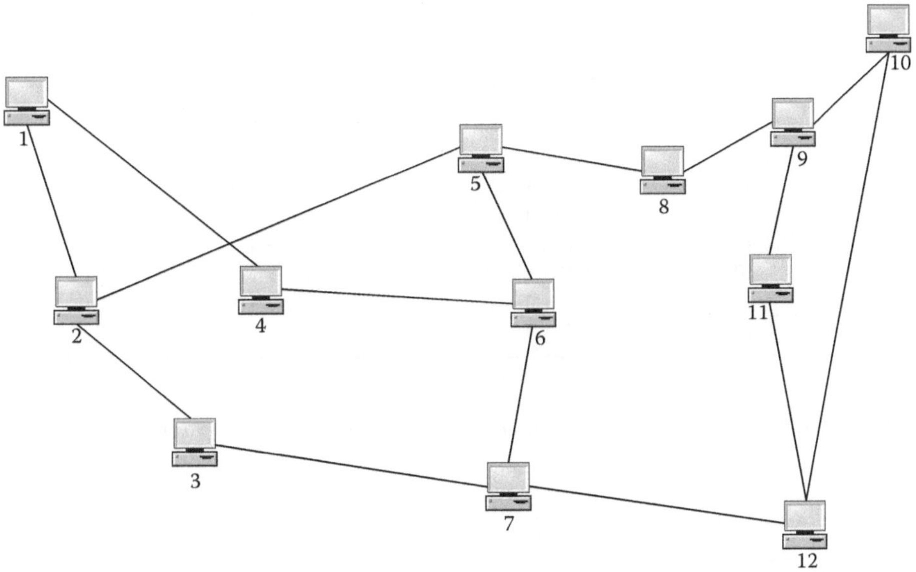

Figure 8.16 The 12-node network used in Example 3.

Table 8.4
Routes and Calculated Max–Min Rate Allocation of the 12-Node Network

Flows (O–D)	Route	Max–Min Rate Allocation (c)
1–2	1–2	$x_1^\dagger = 2/15$
1–3	1–2–3	$x_2^\dagger = 2/15$
1–4	1–4	$x_3^\dagger = 61/240$
1–5	1–2–5	$x_4^\dagger = 2/15$
1–8	1–2–5–8	$x_5^\dagger = 1/10$
1–6	1–4–6	$x_6^\dagger = 61/240$
1–7	1–4–6–7	$x_7^\dagger = 9/80$
2–3	2–3	$x_8^\dagger = 1/6$
2–4	2–1–4	$x_9^\dagger = 2/15$
2–5	2–5	$x_{10}^\dagger = 1/5$
2–6	2–5–6	$x_{11}^\dagger = 1/5$
2–7	2–3–7	$x_{12}^\dagger = 1/6$
3–4	3–2–1–4	$x_{13}^\dagger = 2/15$
3–5	3–2–5	$x_{14}^\dagger = 1/6$
3–6	3–7–6	$x_{15}^\dagger = 9/80$
3–7	3–7	$x_{16}^\dagger = 2/15$
3–8	3–2–5–8	$x_{17}^\dagger = 1/10$
5–6	5–6	$x_{18}^\dagger = 11/40$
5–7	5–6–7	$x_{19}^\dagger = 9/80$
5–8	5–8	$x_{20}^\dagger = 1/10$
5–9	5–8–9	$x_{21}^\dagger = 1/10$
5–10	5–8–9–10	$x_{22}^\dagger = 1/10$
5–11	5–8–9–11	$x_{23}^\dagger = 1/10$
5–12	5–6–7–12	$x_{24}^\dagger = 9/80$
6–7	6–7	$x_{25}^\dagger = 9/80$
6–8	6–5–8	$x_{26}^\dagger = 1/10$
6–9	6–5–8–9	$x_{27}^\dagger = 1/10$
6–10	6–7–12–10	$x_{28}^\dagger = 9/80$
6–12	6–7–12	$x_{29}^\dagger = 9/80$
7–8	7–6–5–8	$x_{30}^\dagger = 1/10$
7–11	7–12–11	$x_{31}^\dagger = 5/24$
7–12	5–12	$x_{32}^\dagger = 5/24$
1–10	1-2-5-8-9-10	$x_{33}^\dagger = 1/10$
1–12	1–4–6–7–12	$x_{34}^\dagger = 9/80$
1–11	1–2–3–7–12-11	$x_{35}^\dagger = 1/10$

Considering max–min fairness and a general utility function, the relevant NLP is thus formulated as follows:

$$
\begin{aligned}
&\text{Maximize } U(x);\\
&\text{subject to:}\\
&(2/15)\beta \geq x_1 \geq (2/15)\alpha, && (2/15)\beta \geq x_2 \geq (2/15)\alpha,\\
&(61/240)\beta \geq x_3 \geq (61/240)\alpha, && (2/15)\beta \geq x_4 \geq (2/15)\alpha,\\
&(1/10)\beta \geq x_5 \geq (1/10)\alpha, && (61/240)\beta \geq x_6 \geq (61/240)\alpha,\\
&(9/80)\beta \geq x_7 \geq (9/80)\alpha, && (1/6)\beta \geq x_8 \geq (1/6)\alpha,\\
&(2/15)\beta \geq x_9 \geq (2/15)\alpha, && (1/5)\beta \geq x_{10} \geq (1/5)\alpha,
\end{aligned}
$$

$$
\begin{aligned}
&(1/5)\beta \geq x_{11} \geq (1/5)\alpha, && (1/6)\beta \geq x_{12} \geq (1/6)\alpha,\\
&(2/15)\beta \geq x_{13} \geq (2/15)\alpha, && (1/6)\beta \geq x_{14} \geq (1/6)\alpha,\\
&(9/80)\beta \geq x_{15} \geq (9/80)\alpha, && (2/15)\beta \geq x_{16} \geq (2/15)\alpha,\\
&(1/10)\beta \geq x_{17} \geq (1/10)\alpha, && (11/40)\beta \geq x_{18} \geq (11/40)\alpha,\\
&(9/80)\beta \geq x_{19} \geq (9/80)\alpha, && (1/10)\beta \geq x_{20} \geq (1/10)\alpha,\\
&(1/10)\beta \geq x_{21} \geq (1/10)\alpha, && (1/10)\beta \geq x_{22} \geq (1/10)\alpha,\\
&(1/10)\beta \geq x_{23} \geq (1/10)\alpha, && (9/80)\beta \geq x_{24} \geq (9/80)\alpha,\\
&(9/80)\beta \geq x_{25} \geq (9/80)\alpha, && (1/10)\beta \geq x_{26} \geq (1/10)\alpha,\\
&(1/10)\beta \geq x_{27} \geq (1/10)\alpha, && (9/80)\beta \geq x_{28} \geq (9/80)\alpha,\\
&(9/80)\beta \geq x_{29} \geq (9/80)\alpha, && (1/10)\beta \geq x_{30} \geq (1/10)\alpha,\\
&(5/24)\beta \geq x_{31} \geq (5/24)\alpha, && (5/24)\beta \geq x_{32} \geq (5/24)\alpha,\\
&(1/10)\beta \geq x_{33} \geq (1/10)\alpha, && (9/80)\beta \geq x_{34} \geq (9/80)\alpha,\\
&(1/10)\beta \geq x_{35} \geq (1/10)\alpha,\\
&x_1 + x_2 + x_4 + x_5 + x_9 + x_{13} + x_{33} + x_{35} \leq C,\\
&x_3 + x_6 + x_7 + x_9 + x_{13} + x_{34} \leq C,\\
&x_2 + x_8 + x_{12} + x_{13} + x_{14} + x_{17} + x_{35} \leq C,\\
&x_4 + x_5 + x_{10} + x_{11} + x_{14} + x_{17} + x_{33} \leq C,\\
&x_{12} + x_{15} + x_{16} + x_{35} \leq C, \;\; x_6 + x_7 + x_{34} \leq C,\\
&x_{11} + x_{18} + x_{19} + x_{24} + x_{26} + x_{27} + x_{30} \leq C,\\
&x_5 + x_{17} + x_{20} + x_{21} + x_{22} + x_{23} + x_{26} + x_{27} + x_{30} + x_{33} \leq C,\\
&x_7 + x_{15} + x_{19} + x_{24} + x_{25} + x_{28} + x_{29} + x_{30} + x_{34} \leq C,\\
&x_{24} + x_{28} + x_{29} + x_{31} + x_{32} + x_{34} + x_{35} \leq C, \;\; x_{23} \leq C,\\
&x_{21} + x_{22} + x_{23} + x_{27} + x_{33} \leq C, \;\; x_{22} + x_{33} \leq C,\\
&x_{28} \leq C, \;\; x_{31} + x_{35} \leq C.
\end{aligned}
$$

For this example, we consider the following two utility functions:

$$U_{31}(x) = \sum_{i=1}^{35} \ln(x_i + 1);$$

and

$$U_{32}(x) = \sum_{i \notin I} \ln(x_i + 1) + \sum_{i \in I} \max\{\ln(x_i + 1),\ 4\ln(x_i + 0.9)\},$$

where the set

$$I = \{5, 13, 17, 24, 26, 27, 30, 33, 34, 35\}$$

represents routes on which there is demand for a "premium" service, if the rate is sufficient.

These two utility functions will lead to the efficiency–fairness functions:

$$\xi_{31}(\alpha, \beta) = \max_{x \in X} U_{31}(x)$$

and

$$\xi_{32}(\alpha, \beta) = \max_{x \in X} U_{32}(x),$$

respectively.

Table 8.5
Results Obtained by Algorithms LGO and AGOP for Objective Functions U_{31} and U_{32}

Box	ξ_{31}	ξ_{32}	
	LGO and AGOP	LGO	AGOP
B_0	4.5537	4.5597	4.5597
B_1	6.2648	6.8061	6.8061
B_2	8.5403	10.8493	10.8493
B_3	9.5563	13.0041	13.0041
B_4	10.2515	14.2254	14.1813
B_5	10.6974	14.9410	14.8482
B_6	10.9704	15.0653	15.0728
B_7	11.1267	15.0614	15.2953
B_8	11.1747	14.4944	15.5612
B_9	11.2196	13.8754	15.7534
B_{10}	11.2624	13.8102	15.8245

Note: $B_s = B(\alpha_s, \beta_s)$ $(s = 0, 1, \ldots, 10)$ and ξ_{31}, ξ_{32} are the optimal values of objective functions. Note that the results for U_{31} are the same for both algorithms.

The results are presented in Table 8.5. The performance of algorithms LGO and AGOP is similar to those in Example 2. The objective function U_{31} has a unique optimal solution and both algorithms find this solution for all boxes $B_s = B(\alpha_s, \beta_s)$, $s = 0, \ldots, 10$. However, objective function U_{32} has many locally optimal solutions.

LGO provides slightly better solutions for boxes B_4 and B_5; however, AGOP performs better on larger boxes and provides a monotonic estimate of the efficiency–fairness curve.

LGO was run in the General Algebraic Modeling System and AGOP in Lahey Fortran. The aggregate elapsed times for all 11 cases $(s = 0, \ldots, 10)$ were around 30% more for AGOP than LGO (Figure 8.17).

8.5.4 EXAMPLE 4: CASE WITH A REMOTE NODE

We now consider an example with the four nodes network shown in Figure 8.18, where $c_l = 1$ for all links. In this example, nodes 2, 3, and 4 are three cities with existing telecommunications infrastructure and node 1 represents a remote location. The link from node 1 to node 2 does not exist and a major cost is required to connect node 1 to the rest. We consider the following four flows: $x_1 = [1,4], x_2 = [2,3], x_3 = [3,4]$, and $x_4 = [4,2]$. Each of these flows is an aggregate of many individual flows. Let us assume that the three flows x_2, x_3, x_4 are equal in size, that is, each of them has the same number of individual flows. However, flow x_1, that is associated with traffic from the remote node 1, is much smaller than the other three. In this example, we consider the case that the number of individual flows that it carries is 19 times smaller than that of each of the other three flows. In this case, we will use the weighted max–min fairness criterion to reflect the difference in the number of individual flows carried by x_2, x_3, x_4 versus x_1 in a way that individual users will obtain "equal access." Accordingly, the weighted max–min rates are as follows:

$$x_1 = \frac{1}{20}, \quad x_2 = x_3 = \frac{19}{20}, \quad x_4 = 1.$$

Considering this fairness criterion, relevant NLP is formulated as follows:

$$\text{Maximize} \quad U(x)$$

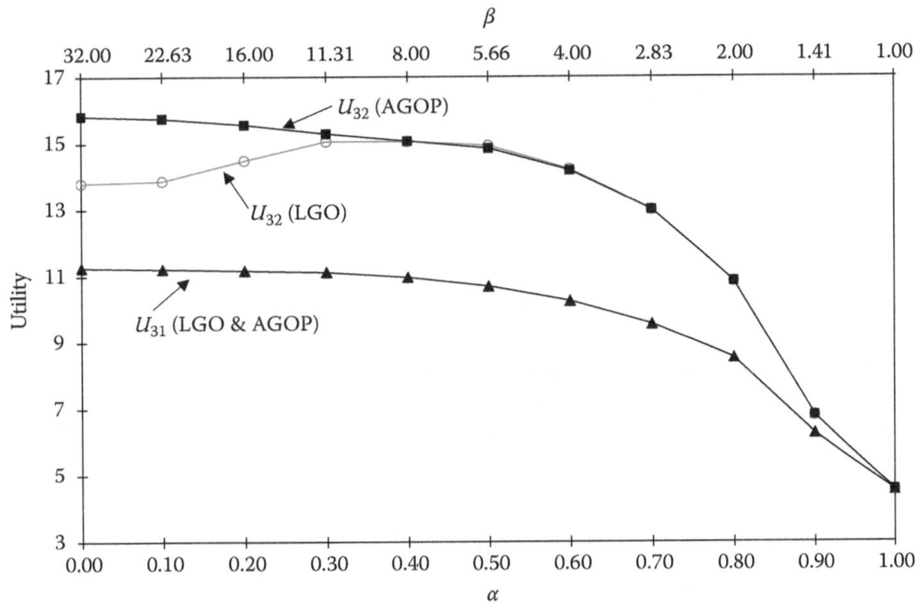

Figure 8.17 Efficiency–fairness functions for the 12-node network of Example 3.

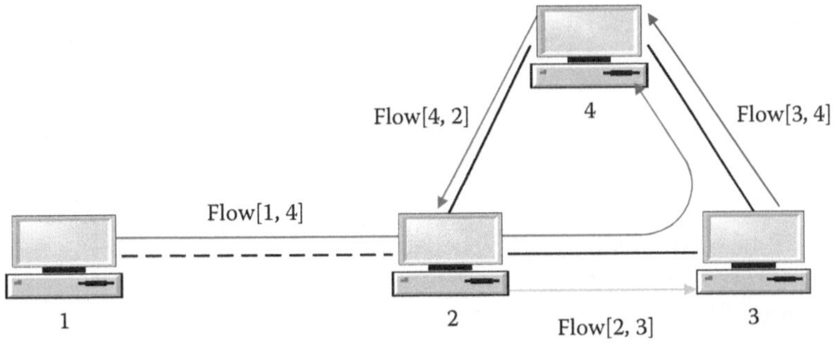

Figure 8.18 Topology and routes for the case with a remote node used in Example 4.

subject to:

$$
\begin{aligned}
\tfrac{1}{20}\beta \geq x_1 \geq \tfrac{1}{20}\alpha&,\\
\beta \geq x_4 \geq \alpha&\\
\tfrac{19}{20}\beta \geq x_i \geq \tfrac{19}{20}\alpha& \quad \text{for } i = 2,3,\\
x_1 + x_2 \leq 1&\\
x_1 + x_3 \leq 1&\\
x_4 \leq 1&\\
x_i \geq 0& \quad \text{for } i = 1,2,3,4.
\end{aligned}
$$

For this example, we consider the objection functions U_{40}, U_{41}, and U_{42}, where

$$U_{4k}(x) = -S(x_1) + 2^k \ln(x_1 + 1) + \sum_{k=2}^{4} \ln(x_k + 1),$$

Table 8.6
Results Obtained by Algorithm AGOP for Objective Functions U_{40}, U_{41}, and U_{42}

Box	ξ_{40}	ξ_{41}	ξ_{42}
B_0	1.0776	1.1264	1.2240
B_1	1.0779	1.1441	1.2807
B_2	1.0783	1.1675	1.3581
B_3	1.0785	1.1973	1.4619
B_4	1.0788	1.2334	1.5980
B_5	1.0790	1.2726	1.7708
B_6	1.0791	1.2902	1.8655
B_7	1.0793	1.2902	1.8655
B_8	1.0794	1.2902	1.8655
B_9	1.0794	1.2902	1.8655
B_{10}	2.0794	2.0794	2.0794

Note: $B_s = B(\alpha_s, \beta_s)$ $(s = 0, 1, \ldots, 10)$ and ξ_{40}, ξ_{41}, ξ_{42} are the optimal values of objective functions.

$S(x) = 0$ if $x = 0$, $S(x) = 1$ if $x \in (0, 1/3]$, $S(x) = 1.5$ if $x \in (1/3, 2/3]$, and $S(x) = 2$ if $x > 2/3$. These utility functions lead to the efficiency–fairness functions:

$$\xi_{40}(\alpha,\beta) = \max_{x \in X(\alpha,\beta)} U_{40}(x),\ \xi_{41}(\alpha,\beta) = \max_{x \in X(\alpha,\beta)} U_{41}(x),$$

and

$$\xi_{42}(\alpha,\beta) = \max_{x \in X(\alpha,\beta)} U_{42}(x),$$

respectively.

We used AGOP to solve this problem. The results are presented in Table 8.6. The efficiency–fairness functions, $\xi_{4i}(\alpha,\beta)$, $i = 0, 1, 2$, are presented in Figure 8.19.

Note that when $\alpha = 0$, it is permissible to set $x_1 = 0$ and not to build the new link at all. For the three objective functions considered, this is the optimal strategy in that case, giving $x^* = (0, 1, 1, 1)$. When $x_1 = 0$ is not allowed, there is little to be gained by relaxing the fairness requirements if the utility function is U_{40} or U_{41}, but significantly higher revenue can be obtained by sacrificing fairness if the utility has the form of U_{42}.

We note the following on the example of objective function U_{41}. The similar situation is valid for U_{42}. In particular, we make the following remarks:

- The optimal solution $x^* = (0.3333, 0.6667, 0.6667, 1)$ is the same for the boxes $B_9, \ldots, B_6$, having, for the first variable, the intervals [0.005, 1.13], [0.01, 0.8], [0.015, 0.566], and [0.02, 0.4], respectively. In other words, letting x_1 to be in the interval [0.005, 1.13], the optimal solution would be 0.3333. This means that if we established a new link (with cost $S(x_1) = 1$) connecting node 1 to the others, then the optimal efficiency of the whole network is achieved if we use this link in a full capacity.
- For boxes $B_5, \ldots, B_0$, the optimal efficiency of the whole network is achieved at the maximal value (the upper bound $\beta/20$) for the first variable x_1, that is, 0.2828, 0.2, 0.1414, 0.1, 0.0707, and 0.05, respectively. This means that we have to use this new link in a maximal level allowed by β-fairness.
- In U_{40}, all flows have the same price. The optimal solution for this function is always achieved at the minimal value (the lower bound $\alpha \cdot 1/20$) for the first variable x_1, that is, 0, 0.005, 0.01,

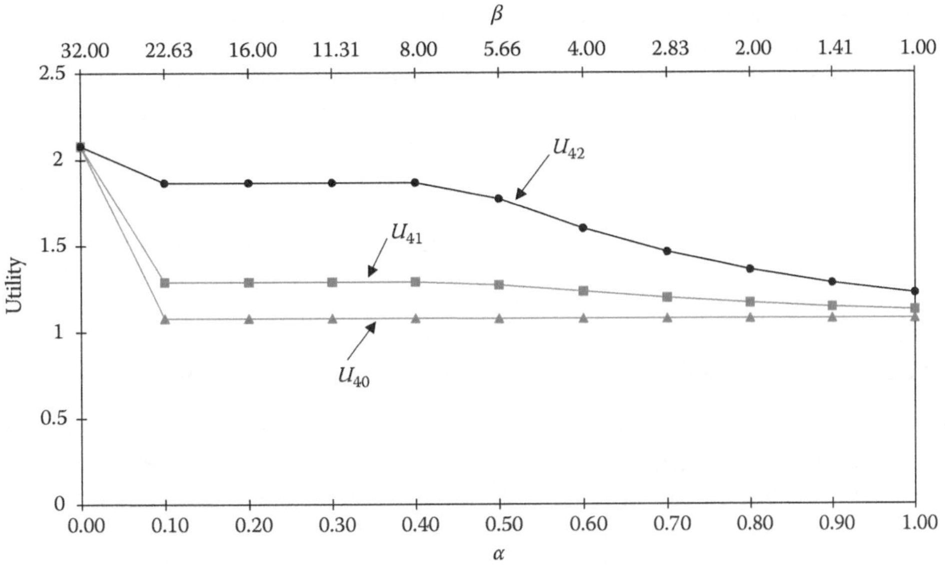

Figure 8.19 Efficiency–fairness functions for Example 4.

0.015, 0.02, 0.025, 0.03, 0.035, 0.04, 0.045, and 0.05. This means that for the efficiency of whole network, we have to use this new link in a minimal level defined by α-fairness.

The first two situations are due the higher rate (price) for the first flow ($2\ln(x_1+1)$ and $4\ln(x_1+1)$).

8.6 RESOURCE ALLOCATION IN TERMS OF α-FAIRNESS

In the previous sections, we have discussed resource allocation by using the concept of (α,β)-*fairness*. That is, we define source sending rate assignments to be (α,β)-*fair* if the rate allocated to a flow is neither less than α times its fair allocation for $0 \leq \alpha \leq 1$ nor higher than β times its fair allocation for $1 \leq \beta$. It is interesting to take a look at the particular situation, where the source sending rate allocation is restricted only by the condition that the rate is not less than α times its fair allocation for $0 \leq \alpha \leq 1$, while the upper bound of the fair allocation is kept open. This particular approach of resource allocation for networks is proposed by [288], where the concept of α-fairness is presented. Thereby, we are trying to trade-off between the fairness and efficiency (utilization, throughput, or revenue). We formulate in [288] a linear program that finds the optimal bandwidth allocation by maximizing efficiency subject to α-fairness constraints. This leads to what we have found on the relation between the efficiency and fairness, which discloses the benefit in efficiency as a function of the extent to which fairness is compromised.

We focus herein on the trade-offs in the context of telecommunication networks and provide a framework for evaluation and presentation of such trade-offs. Our approach applies to any network, topology, and any fairness criterion.

The paper [288] introduces the following concept of α-*fairness*.

Definition 8.6.1 [288] *In a network, a set of bandwidth allocation is termed to be α-fair if no flow receives less than α times its fair allocation for* $0 \leq \alpha \leq 1$.

We address the interesting issue, that is, how to maximize *efficiency*. The efficiency can be defined as revenue that the network provider can obtain under a specific pricing scheme, the total utilization, throughput, etc., under constraint of, for example, 95% fairness. By studying the above issue, another interesting question that we will be able to answer is how much efficiency level can be achieved by compromising on fairness to a certain degree.

The discussions presented herein apply to either wireless or wireline networks. We use the notation $\langle z,y \rangle$ for the link (connection) that connects nodes z and y and $[z, y]$ for the dataflow (user) from node z to node y. Let us consider the three-node bus network presented in Figure 8.20. The capacity on each link is assumed to be a unity. Node a aims to transmit at unlimited rate to node c (denoted as flow $[a,c]$), and node b also tries to transmit at unlimited rate to node c (denoted as flow $[b,c]$). If we choose max–min as the fairness criterion. Subsequently, each of the flows $[a,c]$ and $[b,c]$ will be assigned a rate of 1/2. Assuming all links are equal, this will mean that the utilization is 3/4 as link $\langle b,c \rangle$ is fully utilized and link $\langle a,b \rangle$ is only half utilized. However, if we relax the fairness constraint to α-fairness, then flow $[b,c]$ will be assigned the rate of $\alpha/2$ and flow $[a,c]$ will be allocated the rate of $1-\alpha/2$. This will give utilization of $1-\alpha/4$ plotted in Figure 8.21. Full utilization is achieved if fairness is completely ignored ($\alpha = 0$) and flow $[a,c]$ is assigned the full capacity.

Notice that utilization can be viewed as a special case of revenue by setting the cost of the total capacity of each link to unity and the cost of each flow per link is equal to the proportion of the link capacity it uses. Obviously, the slope of the fairness and efficiency relation curve, for the previous example, can be adjusted. For example, if efficiency is measured by revenue and flow $[a,c]$ pays much more than flow $[b,c]$ per bit/second transmitted per link. We will consider only revenue as a measure of efficiency.

8.6.1 α-FAIRNESS AND AN OPTIMIZATION MODEL FOR TRADE-OFF BETWEEN TOTAL REVENUE AND FAIRNESS

We now formulate a linear optimization model, from which one can find the optimal bandwidth allocation for a general network under the constraints of α-fairness. We will show the solution procedures, the use of the model, and its importance in balancing the total revenue and fairness.

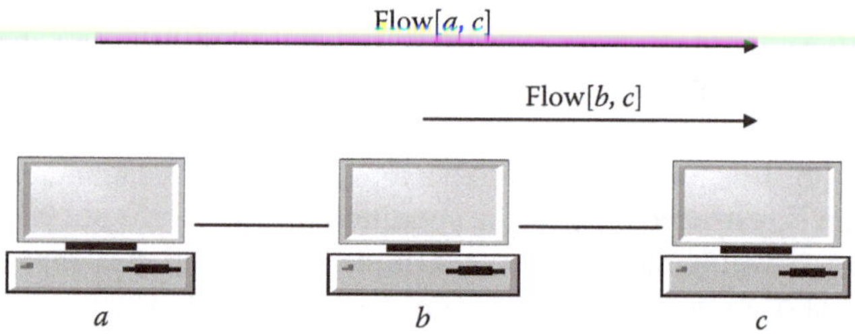

Figure 8.20 A three-node bus network.

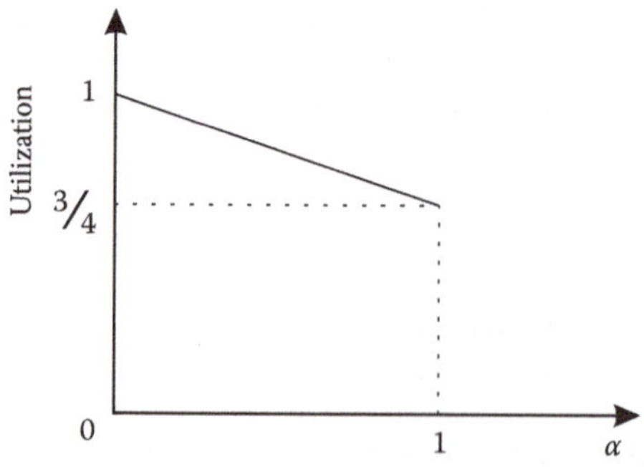

Figure 8.21 The fairness–efficiency relation curve for the three-node bus network.

An N node network is considered. The nodes are designated by a number $1, 2, 3, \ldots, N$. All sources (flows) are assumed to be greedy. Let R_{ij} be the rate assigned to flow $[i, j]$. We look at the issue how to set the R_{ij} values such that an objective function such as revenue or utilization is maximized subject to fairness and capacity constraints.

Let $P_{ij}\langle m,n\rangle$ be the price per bit/second transmitted on link $\langle m,n\rangle$ by flow $[i, j]$. Assume a unique route between any two endpoints, by which we exclude the multicast transmission scenario) and let L_{ij} be the set of links in the route from i to j. Let F be the set of all flows (users). Normally, the revenue obtained from flow $[i, j]$ grows linearly with its rate, the total revenue obtained by the network per second for transmission of flow $[i, j]$ is equal to

$$R_{ij} \times \sum_{\langle m,n\rangle \in L_{ij}} P_{ij}\langle m,n\rangle.$$

Therefore, the total revenue per second obtained by the network is given by

$$\sum_{[i,j] \in F} R_{ij} \sum_{\langle m,n\rangle \in L_{ij}} P_{ij}\langle m,n\rangle.$$

Defining the prices per link and per flow as above covers a wide range of pricing schemes. For example, if a pricing scheme is based on sources paying according to throughput, only the first link in a route may incur a positive cost. We acknowledge that pricing scheme may not be linear in which case other algorithms will be required to evaluate the efficiency–fairness curve.

Let $f(i, j)$ be the fair allocation for flow $[i, j]$ according to our chosen fairness criterion. Let us assume that we require that the allocation is α-fair. Then, R_{ij} will be bounded below by $\alpha f(i, j)$. This leads to the following linear programming formulation.

$$\text{Max } E = \sum_{[i,j] \in F} R_{ij} \sum_{\langle m,n\rangle \in L_{ij}} P_{ij}\langle m,n\rangle \tag{8.22}$$

subject to

$$R_{ij} \geq \alpha f(i, j) \tag{8.23}$$

$$\sum_{[i,j] \in F\langle m,n\rangle} R_{ij} \leq C\langle m,n\rangle \quad \text{for each link } \langle m,n\rangle \tag{8.24}$$

where $C\langle m,n\rangle$ denotes the capacity of link $\langle m,n\rangle$ and $F\langle m,n\rangle$ is the set of all flows that use link $\langle m,n\rangle$. This optimization problem gives rise to the function $E(\alpha)$, which is our efficiency–fairness curve. This function can be obtained by using well-known linear programming techniques [261]. All that is required is one run of the simplex algorithm for, say, $\alpha = 1$ and one simplex pivot per each segment of $E(\alpha)$. It is known from Chapter 6 of [261] that the function $E(\alpha)$ is piecewise linear and concave. The endpoints of each of its segments correspond to the values of α at which alternative optimal solutions exist for E.

Concerning the above proposed optimization model, according to the well-known KKT conditions [162,163], we have the following theorem to obtain its solution.

Theorem 8.6.1 *For the NLP problems 8.22 through 8.24, there exist the extremal point R_{ij}^*, and the KKT multipliers μ_{ij}, and ω_{mn}, such that*

$$\sum_{\langle m,n\rangle \in L_{ij}} P_{ij}\langle m,n\rangle + \mu_{ij} - \omega_{mn} = 0, \tag{8.25}$$

$$\mu_{ij} \geq 0, \quad \textit{for all } i, j \tag{8.26}$$

$$\omega_{mn} \geq 0, \quad \textit{for all } m, n \tag{8.27}$$

$$\mu_{ij}\left(\alpha f(i, j) - R_{ij}^*\right) = 0, \quad \textit{for all } i, j \tag{8.28}$$

$$\omega_{mn}\left(\sum_{[i,j]\in F\langle m,n\rangle} R^*_{ij} - C\langle m,n\rangle\right) = 0, \quad \textit{for all } m,n \tag{8.29}$$

$$\alpha f(i,j) - R^*_{ij} \leq 0, \quad \textit{for all } i,j \tag{8.30}$$

$$\sum_{[i,j]\in F\langle m,n\rangle} R^*_{ij} - C\langle m,n\rangle \leq 0, \quad \textit{for all } m,n \tag{8.31}$$

8.6.2 TWO ILLUSTRATING EXAMPLES

In this subsection, we demonstrate the application of our linear programming formulation to the two examples illustrated in Figures 8.22 and 8.24. RIAS fairness [267,393] is assumed to be used in both examples. For the three-node ring of Figure 8.22, we assume that there are three flows: $[1,3]$, $[2,1]$, and $[2,3]$. These three flows are competing for the bandwidth on link $\langle 2,3\rangle$.

Assume that the capacities on links $\langle 1,2\rangle$, $\langle 2,3\rangle$, and $\langle 3,1\rangle$ are 1, 2, and 1.5 Mb/s, respectively. Let $P_{13}\langle 1,2\rangle = 1$ cents/Mb/s, $P_{13}\langle 2,3\rangle = 2$ cents/Mb/s, $P_{23}\langle 2,3\rangle = 3$ cents/Mb/s, $P_{21}\langle 2,3\rangle = 3$ cents/Mb/s, $P_{21}\langle 3,1\rangle = 2$ cents/Mb/s. According to the RIAS fairness [267,393], $f(1,3) = 1$ Mb/s, $f(2,3) = 0.5$ Mb/s, and $f(2,1) = 0.5$ Mb/s. In this example, the efficiency/revenue function can be written as

$$E = 3R_{13} + 3R_{23} + 5R_{21}.$$

The efficiency–fairness trade-off problem is thus obtained by the following linear programming formulation.

$$\begin{aligned}
&\text{Max} \quad E = 3R_{13} + 3R_{23} + 5R_{21},\\
&\text{subject to}\\
&\quad R_{13} \geq \alpha f(1,3) = \alpha\\
&\quad R_{23} \geq \alpha f(2,3) = 0.5\alpha\\
&\quad R_{21} \geq \alpha f(2,1) = 0.5\alpha\\
&\quad R_{13} + R_{23} + R_{21} \leq 2,\\
&\quad R_{13} \leq 1,\\
&\quad R_{21} \leq 1.5\\
&\quad x_i \geq 0 \text{ for } i = 1,2,3,
\end{aligned}$$

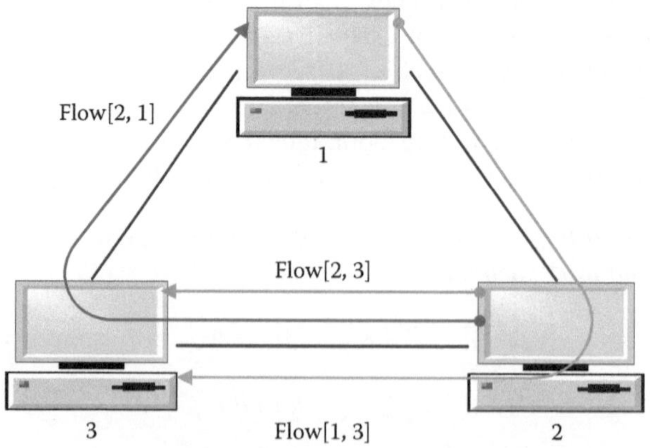

Figure 8.22 Topology and flows for the three-node ring example.

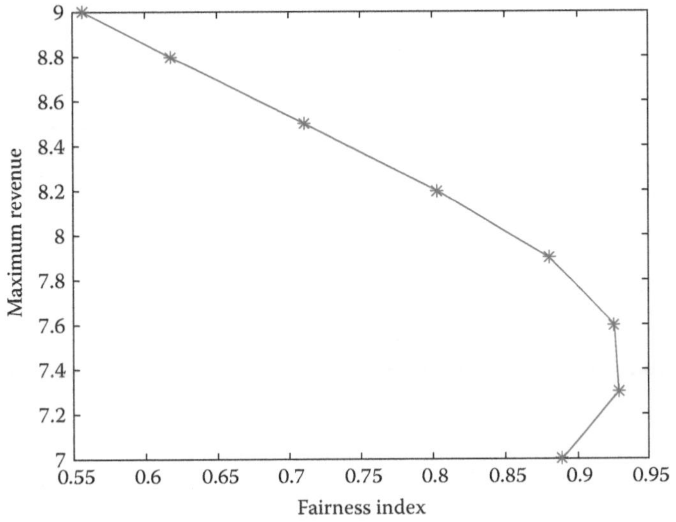

Figure 8.23 A plot of the relation between the values of the maximized total revenue and the fairness index values for the three-node ring example of Figure 8.22.

where the component α is the above-defined parameter related to our α-fairness definition.

We solve the above linear programming problem for $\alpha = 0.1, 0.2, \ldots, 0.9, 1.0$ to obtain the optimal solution of (R_{13}, R_{23}, R_{21}), the optimal total revenue, and further calculate the **FI** by the following formula [141]:

$$FI = \frac{(R_{13}+R_{23}+R_{21})^2}{3(R_{13}^2+R_{23}^2+R_{21}^2)}.$$

We tabulate all the results into Table 8.7.

Further, we plot the relation between the values of the maximized total revenue and the fairness index values into Figure 8.23.

For the model of Figure 8.24, the total cost (end-to-end) of each of the flows excluding [3,7] is assumed to be 1 cent/Mb/s. The total cost (end-to-end) of flow [3,7] is 48 cent/Mb/s. Therefore, the

Table 8.7
Optimal Solution, Optimal Total Revenue, and Fairness Index Values by Solving the Efficiency–Fairness Trade-Off Problem for the Three-Node Ring Example of Figure 8.22

α	Solution: (R_{13}, R_{23}, R_{21})	Total Revenue	Fairness Index
0.1	(0.3324, 0.1676, 1.5000)	9.0000	0.5582
0.2	(0.3466, 0.1534, 1.5000)	9.0000	0.5570
0.3	(0.3431, 0.1569, 1.5000)	9.0000	0.5573
0.4	(0.4000, 0.2000, 1.4000)	8.8000	0.6173
0.5	(0.5000, 0.2500, 1.2500)	8.5000	0.7111
0.6	(0.6000, 0.3000, 1.1000)	8.2000	0.8032
0.7	(0.7000, 0.3500, 0.9500)	7.9000	0.8801
0.8	(0.8000, 0.4000, 0.8000)	7.6000	0.9259
0.9	(0.9000, 0.4500, 0.6500)	7.3000	0.9292
1.0	(1.0000, 0.5000, 0.5000)	7.0000	0.8889

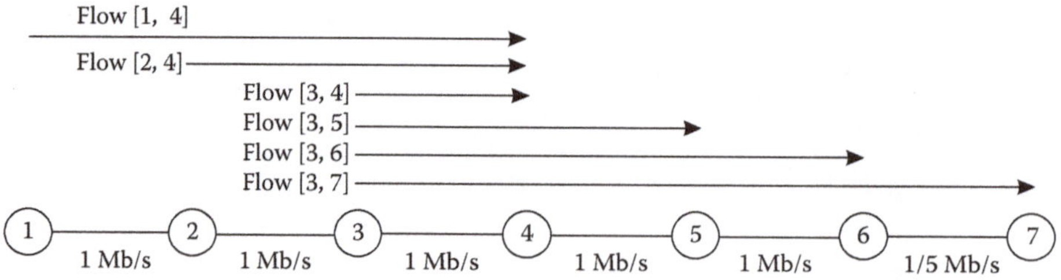

Figure 8.24 Topology and flows for the seven-node bus example.

efficiency/revenue function can be written as

$$E = R_{13} + R_{24} + R_{34} + R_{35} + R_{36} + 48R_{37}.$$

The efficiency–fairness trade-off problem is thus formulated into the following:

$$\text{Max} \quad E = R_{13} + R_{24} + R_{34} + R_{35} + R_{36} + 48R_{37},$$

subject to

$$\begin{aligned}
&R_{13} \geq \alpha f(1,4) = \alpha/3 \\
&R_{24} \geq \alpha f(2,4) = \alpha/3 \\
&R_{34} \geq \alpha f(3,4) = \alpha/12 \\
&R_{35} \geq \alpha f(3,5) = \alpha/12 \\
&R_{36} \geq \alpha f(3,6) = \alpha/12 \\
&R_{37} \geq \alpha f(3,7) = \alpha/12 \\
&R_{13} \leq 1 \\
&R_{13} + R_{24} \leq 1 \\
&R_{13} + R_{24} + R_{34} + R_{35} + R_{36} + R_{37} \leq 1 \\
&R_{35} + R_{36} + R_{37} \leq 1 \\
&R_{36} + R_{37} \leq 1 \\
&R_{37} \leq \frac{1}{5} \\
&x_i \geq 0 \text{ for } i = 1,2,3,4,5,6,
\end{aligned}$$

where α is the above-defined parameter related to our α-fairness definition.

We solve the above linear programming problem for $\alpha = 0.1, 0.2, \ldots, 0.9, 1.0$ to obtain the optimal solution of $(R_{13}, R_{24}, R_{34}, R_{35}, R_{36}, R_{37})$, the optimal total revenue, and further calculate the **FI** [141] by the following formula:

$$FI = \frac{(R_{13} + R_{24} + R_{34} + R_{35} + R_{36} + R_{37})^2}{7(R_{13}^2 + R_{24}^2 + R_{34}^2 + R_{35}^2 + R_{36}^2 + R_{37})^2}.$$

We tabulate all the results into Table 8.8.

Further, we plot the relation between the values of the maximized total revenue and the FI values into Figure 8.25.

The above two examples produce efficiency–fairness curves that are different in the slope of the right-hand segment. The seven-node bus example clearly produces a much steeper slope for that right-hand segment, meaning that giving up a small percentage of fairness can lead to a significant increase in revenue. Moreover, the seven-node bus example demonstrates that there are cases

Table 8.8
Optimal Solution, Optimal Total Revenue, and Fairness Index Values by Solving the Efficiency–Fairness Trade-Off Problem for Seven-Node Bus Example of Figure 8.24

α	Solution (R_{13},R_{24},R_{34},R_{35},R_{36},R_{37})	Total Revenue E	Fairness Index
0.1	(0.1574, 0.0975, 0.0839, 0.0614, 0.3996, 0.2000)	10.4000	0.6807
0.2	(0.1633, 0.1127, 0.0777, 0.0634, 0.3830, 0.2000)	10.4000	0.7060
0.3	(0.1678, 0.1331, 0.0747, 0.0654, 0.3590, 0.2000)	10.4000	0.7421
0.4	(0.1709, 0.1615, 0.0795, 0.0674, 0.3207, 0.2000)	10.4000	0.7974
0.5	(0.1691, 0.2066, 0.1149, 0.0748, 0.2347, 0.2000)	10.4000	0.9003
0.6	(0.2016, 0.2422, 0.1199, 0.0822, 0.1540, 0.2000)	10.4000	0.9049
0.7	(0.2353, 0.2580, 0.0992, 0.0782, 0.1293, 0.2000)	10.4000	0.8564
0.8	(0.2690, 0.2808, 0.0799, 0.0757, 0.0946, 0.2000)	10.4000	0.7852
0.9	(0.3000, 0.3000, 0.0750, 0.0750, 0.0750, 0.1750)	9.2250	0.7326
1	(0.3333, 0.3333, 0.0833, 0.0833, 0.0833, 0.0833)	4.9167	0.6667

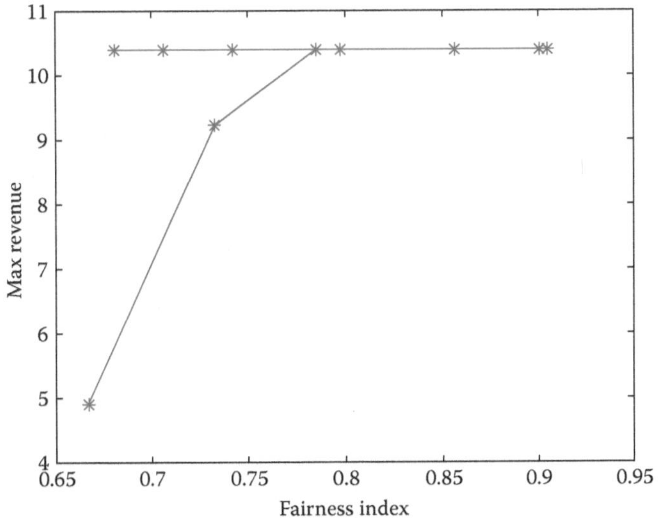

Figure 8.25 A plot of the relation between the values of the maximized total revenue and the fairness index values for seven-node bus example of Figure 8.24.

whereby after trading significant increase in revenue for a small percentage of fairness, no further revenue increase is possible even if further compromise on fairness is made. Such considerations can be made by designers using the efficiency–fairness curve.

8.7 CONCLUSIONS

This chapter has presented the (α, β)-fairness concept and the α-fairness, which specify how far a resource allocation is allowed to be from an "ideally fair" allocation, according to any fairness criterion. This decouples the efficiency criterion from the fairness criterion and quantifies the familiar concept of the trade-off between efficiency–fairness.

We have described a framework to produce "efficiency–fairness functions" that allow network operators to first set fairness constraints and then to optimize their efficiency. We have formulated an NLP problem, which finds the optimal rate allocation for a general network and any ideally fair rate allocation, under the (α, β)-fairness constraints and the α-fairness constraints. This leads to what we call the efficiency–fairness function, which shows the increase in efficiency as a function of the extent to which fairness is compromised.

This framework applies both when the network is fixed and only the rates can be chosen and also when new capacity is to be added to the network. In the latter case, the "efficiency" reflects both the revenue and the infrastructure cost associated with the rate allocation.

We have applied two global optimization algorithms, LGO and AGOP, to solving the NLP for a variety of networks. For the examples tested, AGOP seems to be a particularly promising algorithm.

We have presented the theoretical analyses on the proposed models and the numerically solving procedures for the problems. The approach presented here is important in network design when one faces the issue of balancing the fairness (between users) and the running efficiency of the network provider.

9 Fairness Comparisons among the Leading High-Speed Protocols

In networking, it is often required to quantify by how much one protocol is fairer than another and how certain parameter setting and/or protocol enhancements improve fairness. This chapter provides a framework to evaluate the fairness of various protocols in a general telecommunications network. Within this framework, there are two key components:

- A benchmark
- A single-dimension metric.

We suggest to use the max–min fairness bandwidth allocation as the benchmark and the Euclidean distance between any bandwidth allocation under any protocol and the max–min bandwidth allocation as the metric. Explicitly, we provide a method to compare the fairness of two sets of bandwidth allocation under two different protocols for a given network by using this metric. On the basis of this new framework, we evaluate the fairness of FAST TCP and TCP Reno relative to the max–min fairness criteria. The distance between the max–min fair allocation and allocations based on each of the two protocols is measured using the Euclidian norm. We derive explicit expressions for these distances for a general network and compare the fairness of these two protocols by using their corresponding utility functions. Finally, we numerically demonstrate how this method can be applied to compare the fairness of FAST TCP and TCP Reno for a "Parking Lot" linear network and for the NSFNET Backbone network. In addition to merely a comparison between protocols, such numerical results can also provide guidelines for better choice of parameters to make a protocol fairer in a given scenario.

We also compare the equilibrium properties of FAST TCP and TCP Vegas. Although the two have the same equilibrium point when all sources know their true propagation delays, FAST is fairer when there are estimation errors. The performance of Vegas approaches that of FAST when the queuing delay is very much less than the propagation delay.

9.1 FAIRNESS COMPARISON BETWEEN FAST TCP AND TCP RENO

In the past several decades, fairness has been considered an important criterion in design of computer and telecommunications networks. Its importance has given rise to many discussions in the networking literature on how to define and provide fairness [115,128,139,142,269,278,287,288,328, 338–347]. Given the importance of fairness in networking, it is often required to quantify by how much one protocol is fairer than another and by how much certain parameter setting and/or protocol enhancements improve fairness.

The current TCP congestion control algorithm, TCP Reno, and its variants have experienced scalability problems [172,314]. FAST TCP [172] has recently gained popularity as a viable TCP alternative and is considered promising in terms of system stability, throughput, responsiveness, and fairness [330,334,348]. The question of fairness comparing FAST TCP and TCP Reno has therefore been an important and popular research topic [325,349–354].

In [349], the authors investigate goodput, fairness, and friendliness properties presented in Westwood+, New Reno, and Vegas TCP by using both ns2 simulations and live Internet measurements. The work in [350] studies the equilibrium and fairness properties of networks, where both TCP Reno and FAST TCP are used. In particular, they show that the fairness property of FAST TCP and TCP Reno is preserved "locally" within each protocol when TCP Reno flows and FAST TCP flows share the same network bandwidth. In [351], the authors compare the fairness of FAST TCP and TCP Vegas for a single-bottleneck link model in the persistent congestion situation by using a queue equilibrium formulation proposed in [325,352]. For a single-bottleneck link, the earlier works in [353] analyzes the performance of TCP Vegas in comparison with TCP Reno in terms of fairness using a simple closed fluid model. In [354], the authors investigate the fairness issue in the situation where TCP Reno and TCP Vegas flows share a single-bottleneck link. However, their results are only related to the fairness of networks where heterogeneous protocols coexist.

This section provides a framework to evaluate the fairness of various protocols in a general telecommunications network. We promote the use of a "measure" of fairness because it is important to be able to quantify by how much one protocol is fairer than another and how certain parameter setting and/or protocol enhancements improve fairness. Such a measure requires (1) a benchmark and (2) a single-dimension metric. We suggest to use the max–min fairness bandwidth allocation as the benchmark and the Euclidean distance between any bandwidth allocation under any protocol and the max–min bandwidth allocation as the metric. Explicitly, we provide explanations on how to compare the fairness of two sets of bandwidth allocation under two different protocols for a given network by using this metric. We illustrate the implication and intuition of this metric by using simple examples. These examples also show when the bandwidth allocation satisfies the max–min fairness, the Euclidean distance is minimized and equal to zero indicating the point of maximum fairness. We apply these new concepts of benchmark and measure to compare between FAST TCP and TCP Reno both for a "Parking Lot" network and the NSFNET backbone network. We also demonstrate the importance of parameter setting by showing that one protocol can be made closer to max–min than the other by choosing better parameter settings.

We acknowledge that our particular choices of measure and metric may be controversial. We therefore first clarify that the choice of max–min fairness as benchmark is without loss of generality, and the results of this section are readily applied to any other fairness benchmark. The choice of the Euclidian norm may remain controversial as the measure of fairness depends on users' utility functions. Nevertheless, the Euclidean distance function measures the "as-the-crow-flies" distance, which means the shortest distance between two points. It has been used in [347,355] to evaluate the closeness of two rate allocations; it is consistent with the fairness index proposed by Jain et al. [338] for the special case of a single-bottleneck link, and it extends the fairness index of [338] to a general network with multiple-bottleneck links.

9.1.1 THE METRIC

A huge amount of recent works have considered the bandwidth allocation problem as an optimization problem aiming to find an allocation policy that maximizes the sum of all sources' utilities subject to link capacity constraints.

Consider a network with l links shared by n flows. Link j ($j = 1,2,\ldots,l$) has capacity c_j, and x_i is the rate of flow i ($i = 1,2,\ldots,n$). We define the routing matrix $R = (R_{ij}, i = 1,\ldots,n, j = 1,\ldots,l)$, where $R_{ij} = 1$ if flow i uses the link j, and $R_{ij} = 0$, otherwise. Consider the following utility optimization problem:

Problem **P**

$$\begin{cases} \max_{x\geq 0} \sum_i U_i(x_i) \\ \text{subject to } Rx \leq c, \end{cases} \tag{9.1}$$

where U_i is the utility function of flow i, and c is the capacity vector.

In [115], the following class of utility functions is proposed:

$$U(x,\alpha) = \begin{cases} (1-\alpha)x_i^{1-\alpha} & \text{if } \alpha \neq 1 \\ \log x_i & \text{if } \alpha = 1, \end{cases}$$

where $\alpha \geq 0$. This includes four special cases: maximum throughput ($\alpha = 0$), proportional fairness ($\alpha = 1$), minimum potential delay ($\alpha = 2$), and max–min fairness ($\alpha = \infty$) [129,278,328]. The papers [129,328] point out that one allocation policy with larger α is fairer. However, an exact measure of fairness is not considered there.

Jain et al. [338] have proposed a fairness index applicable to any system where all users are sharing the same resource. In other words, it can be used for a quantitative evaluation of the fairness of an allocation policy in a single-bottleneck link network. In this section, we apply the fairness index of [338] to a general network by comparing the fairness of any two protocols with concave utility functions by considering their deviations from a pre-chosen rate allocation benchmark.

In this section, the rate allocation benchmark we choose is max–min fairness [139]. The choice is without loss of generality as the method we use can apply to any other rate allocation benchmark. The choice of max–min fairness is based on its popularity and importance in computer networks. Note that for a general network, the max–min fair rate allocation can be easily obtained using the well-known progressive filling algorithm [347,355].

Definition 9.1.1 *Given a general network, for two protocols (λ and μ) with different utility functions, their rate allocations are solutions to the optimization problem P, which are denoted by*

$$X^\lambda = \{x_i^\lambda, i = 1, \ldots, n\}$$

and

$$X^\mu = \{x_i^\mu, i = 1, \ldots, n\},$$

respectively. The max–min rate allocation for this network is $X = \{x_i, i = 1, \ldots, n\}$. If

$$||X^\lambda - X||_2 < ||X^\mu - X||_2,$$

we say that the protocol λ is closer to max–min than protocol μ. The above norm is the standard Euclidean norm defined by

$$||X^\mu||_2 = \sqrt{\sum_{i=1}^{n} (x_i^\mu)^2}. \tag{9.2}$$

Here, we use Euclidean norm to measure the deviation from the rate allocation benchmark because Euclidean norm reflects the distance between any two vectors. It has been used to evaluate the closeness of two rate allocations in [347,355].

To illustrate the above concept, let us consider the following two simple networks. The first network is shown in Figure 9.1. It comprises only a single link with two flows passing through it. The capacity of this link is assumed to be one unit. Clearly, the max–min fair bandwidth allocation for this network is $(1/2, 1/2)$. For any bandwidth allocation of this network under any protocol, denoted by (x_1, x_2), we must have $x_1 + x_2 = 1$. The standard Euclidean norm that measures the distance of the bandwidth allocation (x_1, x_2) from the max–min bandwidth allocation is

$$Eu = \sqrt{(x_1 - 1/2)^2 + (x_2 - 1/2)^2} = \sqrt{2(x_1 - 1/2)^2} = \sqrt{2}\,|\,x_1 - 1/2\,|\,.$$

In Figure 9.2, we plot this standard Euclidean norm as a function of x_1. The figure illustrates that the Euclidean distance, in consistent with our notion of fairness, obtains higher values as x_1 is further from its max–min fair value at $x_1 = 1/2$, and that this distance is minimized when $x_1 = 1/2$, i.e., when the bandwidth allocation is of max–min fair.

The second network is described in Figure 9.3. In this linear network, there are three flows passing three links. The capacity on each of these three links is assumed to be one unit. One easily finds that the max–min fair bandwidth allocation for this network is $(1/3, 1/3, 1/3)$. Let us for simplicity consider a set of rate allocations, denoted by (x_1, x_2, x_3), that obey the following two

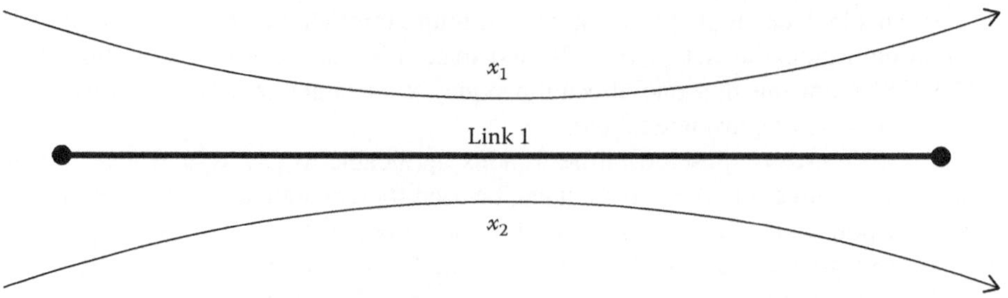

Figure 9.1 The single link two-flow scenario.

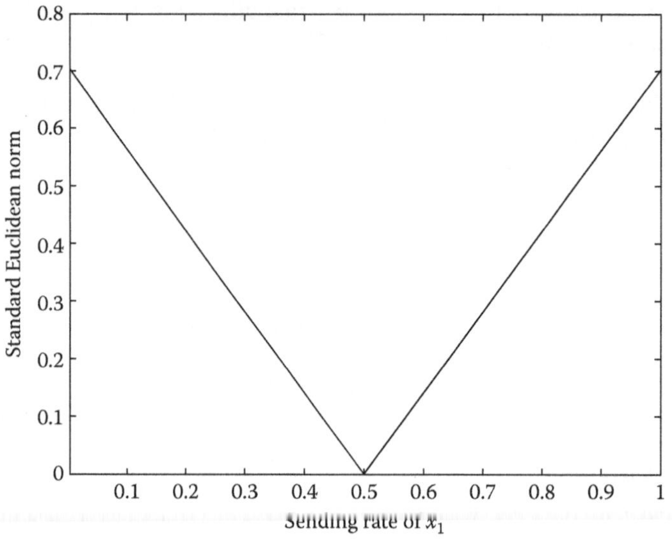

Figure 9.2 The standard Euclidean norm Eu as a function of x_1.

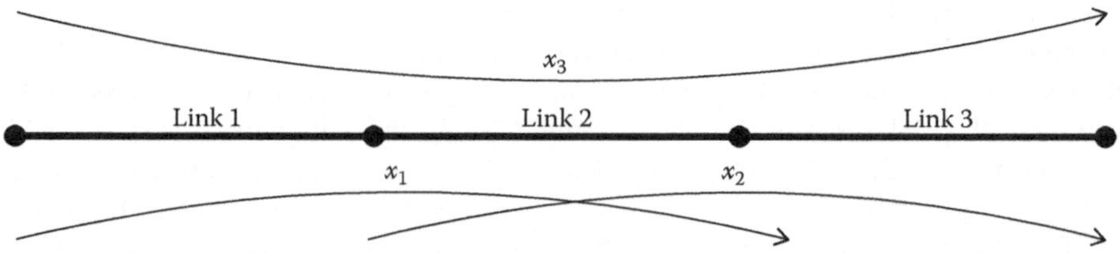

Figure 9.3 A case of three links and three flows.

conditions: (1) $x_1 = x_2$ and (2) Link 2 is fully utilized. From the second assumption, we have $x_1 + x_2 + x_3 = 1$. Subsequently, the standard Euclidean norm that measures the distance of the bandwidth allocation (x_1, x_2, x_3) from the max–min bandwidth allocation should be

$$\begin{aligned} Eu &= \sqrt{(x_1 - 1/3)^2 + (x_2 - 1/3)^2 + (x_3 - 1/3)^2} \\ &= \sqrt{2(x_1 - 1/3)^2 + (1 - 2x_1 - 1/3)^2} \\ &= \sqrt{6}\,|\,x_1 - 1/3\,|\,. \end{aligned}$$

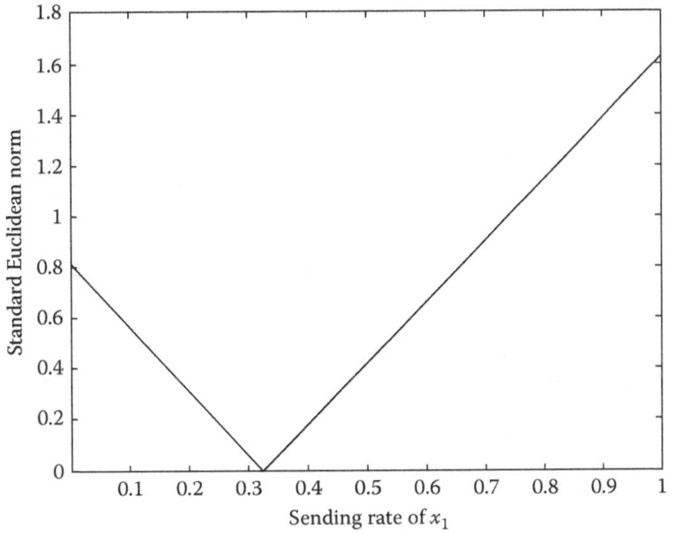

Figure 9.4 The standard Euclidean norm Eu as a function of x_1.

We plot the above standard Euclidean norm as a function of x_1 in Figure 9.4. Again, the figure shows that the standard Euclidean distance increases as x_1 is further away from its max–min fair value at $x_1 = 1/3$, and that this distance is minimized when $x_1 = 1/3$, i.e., when the bandwidth allocation is max–min fair.

When there is only a single-bottleneck link, the fairness comparison result of two rate allocations by using our Euclidean distance metric is the same as the one obtained using the fairness index of [338]. We clarify this issue in the following. Assume that there is one single-bottleneck link; the capacity of bottleneck link is C. We do two separate runs for two different protocols μ and λ. In the first run n λ flows pass through the bottleneck link, and in the second n μ flows pass through it. Let the sending rate of each λ flow be denoted by x_i^λ, where $i = 1, 2, \ldots, n$, and the sending rate of each μ flow is x_i^μ, where $i = 1, 2, \ldots, n$.

The max–min rate allocation for this network is x_i, where $i = 1, 2, \ldots, n$. If the protocol λ is fairer than μ, using the fairness index, then

$$\frac{\left(\sum_{i=1}^{n} x_i^\lambda\right)^2}{n\sum_{i=1}^{n} \left(x_i^\lambda\right)^2} > \frac{\left(\sum_{i=1}^{n} x_i^\mu\right)^2}{n\sum_{i=1}^{n} \left(x_i^\mu\right)^2}. \tag{9.3}$$

For the bottleneck link, we have

$$\sum_{i=1}^{n} x_i^\lambda = \sum_{i=1}^{n} x_i^\mu = C. \tag{9.4}$$

Inequality 9.3 is equivalent to

$$\sum_{i=1}^{n} \left(x_i^\lambda\right)^2 < \sum_{i=1}^{n} \left(x_i^\mu\right)^2. \tag{9.5}$$

Using the new metric proposed by Definition 9.1.1, that is,

$$\sqrt{\sum_{i=1}^{n} \left(x_i^\lambda - x_i\right)^2} < \sqrt{\sum_{i=1}^{n} \left(x_i^\mu - x_i\right)^2}, \tag{9.6}$$

we obtain

$$\sum_{i=1}^{n}\left(x_i^{\lambda}\right)^2-2\sum_{i=1}^{n}x_i^{\lambda}\cdot x_i+x_i^2<\sum_{i=1}^{n}(x_i^{\mu})^2-2\sum_{i=1}^{n}x_i^{\mu}\cdot x_i+x_i^2. \tag{9.7}$$

From Equations 9.7 and 9.4, we obtain Equation 9.5. This shows that using these two metrics to compare fairness gives the same result.

The above fact also indicates that our new metric is an extension of the fairness index [338] to a general network with multiple-bottleneck links. We will further illustrate this extension in the following sections using two popular TCP protocols, TCP Reno (a widely used protocol) and FAST TCP (a leading future option), in a general network's settings.

9.1.2 FAIRNESS COMPARISON OF FAST TCP VERSUS TCP RENO FOR A GENERAL NETWORK

We now compare the fairness of FAST TCP and TCP Reno using the proposed metric. We assume that for a FAST TCP flow i, if it passes multiple links, the summation of the total queue length in equilibrium at these links is α_i. Consider utility optimization problem (9.1), for FAST TCP, we use the following utility function as suggested by [172,350,351]:

$$U_i^f(x_i^f)=\alpha_i\log x_i^f, \tag{9.8}$$

where x_i^f is the rate of FAST TCP flow i in equilibrium. We take the utility function of TCP Reno [111]:

$$U_i^r(x_i^r)=\frac{1.5}{D_i^2}\cdot\left(-\frac{1}{x_i^r}\right), \tag{9.9}$$

where D_i is the equilibrium round-trip time (RTT), i.e., propagation plus queuing delay, that TCP Reno flow i experienced, and x_i^r is the rate of TCP Reno flow i. The choice of this utility function is without loss of generality, and it is made for the sake of simplicity; a similar analysis can also be performed choosing an alternative utility function.

Solving Problem **P** may require coordination among possibly all flows; sometimes, we solve it by solving its dual problem. By solving its dual problem, the rate of the flow i is obtained by (see [112])

$$x_i=\left(U_i'(p)\right)^{-1}, \tag{9.10}$$

where $(U_i'(p))^{-1}$ is the inverse of $U_i'(p)$, and p is the path bandwidth price for flow i. A procedure has been given in [112] on how to calculate the bandwidth allocation for a certain protocol that optimizes the aggregate network utility. We hereby follow that procedure. Using Equations 9.8 through 9.10, we are able to obtain the bandwidth allocation for FAST TCP and TCP Reno that optimizes the aggregate utility. They are

$$x_i^f=\frac{\alpha_i}{p_i^f}, \tag{9.11}$$

$$x_i^r=\sqrt{\frac{1.5}{p_i^r}}\Big/D_i, \tag{9.12}$$

respectively, where p_i^f and p_i^r are the path bandwidth price of FAST TCP flow i and TCP Reno flow i, respectively. From Definition 9.1.1, if

$$\sqrt{\sum_{i=1}^{n}\left(x_i^f-x_i\right)^2}\le\sqrt{\sum_{i=1}^{n}(x_i^r-x_i)^2}, \tag{9.13}$$

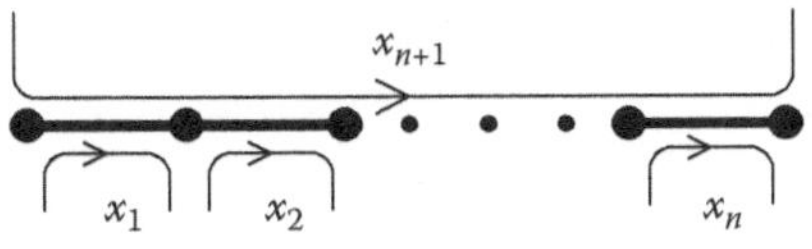

Figure 9.5 The "Parking Lot" example.

FAST TCP is closer to max–min than TCP Reno. Equation 9.13 requires a solution of Problem **P** by global optimization. However, combining Equations 9.11 through 9.13 leads to the following condition:

$$\sqrt{\sum_{i=1}^{n}\left(\frac{\alpha_i}{p_i^f}-x_i\right)^2} \le \sqrt{\sum_{i=1}^{n}\left(\sqrt{\frac{1.5}{p_i^r}}\Big/D_i-x_i\right)^2}, \tag{9.14}$$

which can be solved in a distributed manner by having each flow compute its own price as suggested in [112] and provides condition for fairness comparison between FAST TCP and TCP Reno.

9.1.3 TWO NUMERICAL EXAMPLES

In this section, we present two examples to demonstrate how to compare FAST TCP and TCP Reno relative to max–min.

Example 1: The "Parking Lot" network

Consider the "Parking Lot" network shown in Figure 9.5. Assume that there are n links each of which has unit capacity, flow i only passes through the ith link, for $i = 1,2,\ldots,n$, and flow $n+1$ passes through all the links. Let the sending rate of each FAST TCP flow be denoted by x_i^f, where $i = 1,2,\ldots,n+1$, when all the flows are FAST TCP flows, and the sending rate of each TCP Reno flow is x_i^r, where $i = 1,2,\ldots,n+1$, when all the flows are TCP Reno flows.

Let the FAST TCP and TCP Reno Problem **P** utility functions be according to Equations 9.8 and 9.9, respectively; then, the rates of FAST TCP and TCP Reno flows are obtained as follows. To calculate the FAST TCP flows' rate, let

$$\gamma = \frac{1}{1+\frac{\sum_{i=1}^{n}\alpha_i}{\alpha_{n+1}}},$$

by using Equations 9.8 and 9.10, we have the FAST TCP bandwidth allocation:

$$x_i^f = 1-\gamma,\ (i = 1,2,\ldots,n),\ x_{n+1}^f = \gamma. \tag{9.15}$$

To calculate the TCP Reno flows' rate, let

$$\omega = \frac{1}{1+D_{n+1}\cdot\sqrt{\sum_{i=1}^{n}\frac{1}{D_i^2}}},$$

by using Equations 9.9 and 9.10, we have the TCP Reno bandwidth allocation:

$$x_i^r = 1-\omega,\ (i = 1,2,\ldots,n),\ x_{n+1}^r = \omega. \tag{9.16}$$

The max–min fairness rates are computed to be

$$x_i = 1/2\ (i = 1,2,\ldots,n+1). \tag{9.17}$$

Here, we use Equation 9.13 for fairness comparison. Let

$$\Gamma = \sqrt{\sum_{i=1}^{n}(1-\gamma-1/2)^2+(\gamma-1/2)^2} = \sqrt{n+1}\cdot|1/2-\gamma|, \tag{9.18}$$

$$\Omega = \sqrt{\sum_{i=1}^{n}(1-\omega-1/2)^2+(\omega-1/2)^2} = \sqrt{n+1}\cdot|1/2-\omega|. \tag{9.19}$$

For $D_{n+1} > D_i$ $(i = 1,2,\ldots,n)$, then

$$D_{n+1}\cdot\sqrt{\sum_{i=1}^{n}\frac{1}{D_i^2}} = \sum_{i=1}^{n}\left(\frac{D_{n+1}}{D_i}\right)^2 > 1,$$

so $\omega < 1/2$. Then, Equation 9.19 is reduced to

$$\Omega = \sqrt{n+1}\cdot(1/2-\omega). \tag{9.20}$$

If FAST TCP is closer to max–min than TCP Reno, i.e., $\Gamma < \Omega$, which is equivalent to

$$\sqrt{n+1}\cdot|1/2-\gamma| < \sqrt{n+1}\cdot(1/2-\omega), \tag{9.21}$$

that is,

$$\omega < \gamma < 1-\omega, \tag{9.22}$$

from this, we further obtain

$$\frac{1}{D_{n+1}\cdot\sqrt{\sum_{i=1}^{n}\frac{1}{D_i^2}}} < \frac{\sum_{i=1}^{n}\alpha_i}{\alpha_{n+1}} < D_{n+1}\cdot\sqrt{\sum_{i=1}^{n}\frac{1}{D_i^2}}. \tag{9.23}$$

By using Equation 9.23, we can set the FAST TCP parameter α to make FAST TCP closer to max–min fairness than TCP Reno.

Example 2: The NSFNET backbone network example

Let us consider the NSFNET Backbone network shown in Figure 9.6. Again, we set all link capacities to unity, and consider 10 flows. The routing information is given in Table 9.1. We run the progressive filling algorithm to compute the max–min fair allocation.

Considering a general utility function, the relevant optimization problem is thus formulated as follows:

$$
\begin{aligned}
&\text{Maximize } U(x)\\
&\text{subject to:}\\
&x_1+x_2+x_5 \le 1,\\
&x_1+x_2+x_3+x_4 \le 1,\\
&x_1+x_3+x_4 \le 1,\\
&x_1+x_4 \le 1,\\
&x_6+x_7+x_8+x_9 \le 1,\\
&x_7+x_8+x_9+x_{10} \le 1,\\
&x_8+x_9+x_{10} \le 1,\\
&x_9+x_{10} \le 1.
\end{aligned}
$$

Assume that all α are equal for all FAST TCP flows. We use the tools provided by AMP Raw Data Query [356] to measure the RTTs between the source and destination pairs of the 10 flows. We set the source and destination nodes as the university machines in those cities. We use FAST TCP utility

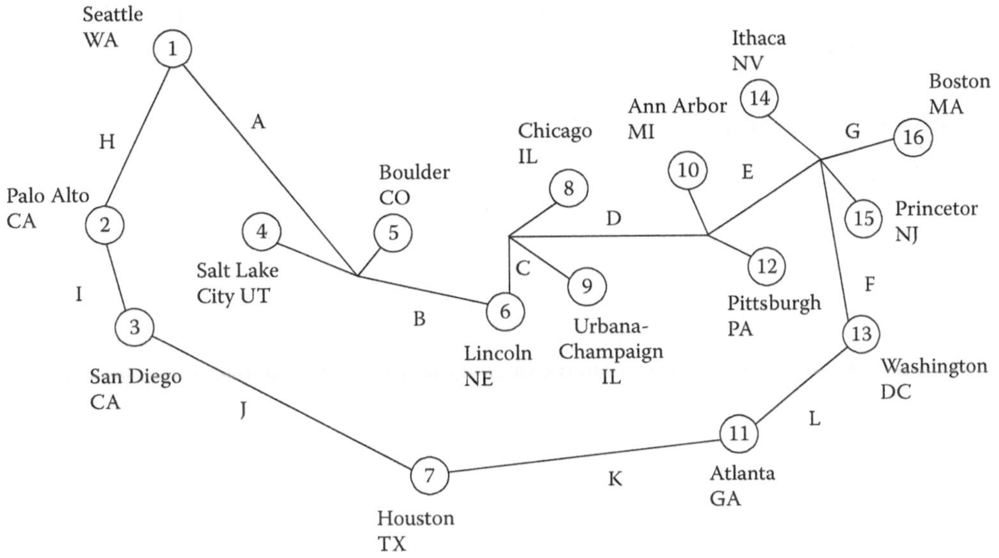

Figure 9.6 The NSFNET backbone network.

Table 9.1
Bandwidth Allocation Results for the NSFNET Example

Flows	Route	RTT (ms)	Max–Min Rate	FAST TCP Rate	TCP Reno Rate
1:1–16	A–B–C–D–E–G	75	0.2500	0.2	0.2712
2:1–8	A–B–C	50	0.2500	0.2	0.1808
3:5–15	B–C–D–E	72.80	0.2500	0.3	0.2708
4:5–16	B–C–D–E–G	74.51	0.2500	0.3	0.2772
5:1–4	A	36.18	0.5000	0.6	0.5480
6:1–2	H	34.42	0.2500	0.4	0.2961
7:1–3	H–I	27.66	0.2500	0.2	0.1783
8:1–7	H–I–J	60.06	0.2500	0.2	0.2590
9:1–11	H–I–J–K	57.03	0.2500	0.2	0.2666
10:2–11	I–J–K	61.11	0.2500	0.4	0.2961

functions given by Equation 9.8 and TCP Reno utility functions given by Equation 9.9. These two utility functions will lead to two rate allocation results. We list the flows' route information, max–min rate allocations, the FAST TCP rate allocation, and the TCP Reno rate allocations in Table 9.1.

We calculate $\left(\sum_{i=1}^{10}(x_i^f - x_i)^2\right)^{1/2} = 0.185$ and $\left(\sum_{i=1}^{10}(x_i^r - x_i)^2\right)^{1/2} = 0.1050$, so TCP Reno is fairer than FAST TCP in this case. However, when parameters change, different outcomes may arise.

9.1.4 CONCLUDING REMARKS

We have provided further insights into the issue of how to quantify by how much one TCP protocol is fairer than another and how parameter setting improves the fairness of one protocol over another. Using max–min fairness as a benchmark and the Euclidean norm as a measure of fairness, we have compared FAST TCP and TCP Reno for general networks. Two examples for the use of this measure have been provided.

9.2 FAIRNESS COMPARISON BETWEEN FAST TCP AND TCP VEGAS

Given the importance of controlling congestion and stability of the Internet, there have been many proposals aiming to improve the well-known TCP Reno protocol. Two such popular proposals that have received significant attention in recent years are TCP Vegas [169] and FAST TCP [172]. This section compares the fairness between FAST TCP and TCP Vegas at their equilibrium states. Although the two protocols have the same equilibrium point when all sources know their true propagation delays, we find that FAST TCP is fairer when there are estimation errors. The performance of TCP Vegas approaches that of FAST when the queuing delay is very much less than the propagation delay.

9.2.1 GENERAL BACKGROUND, NOTATIONS, AND MODELS OF TCP VEGAS AND FAST TCP

TCP Vegas and FAST TCP aim to improve throughput and fairness over their predecessors, the most popular of which is TCP Reno, by using queuing delay as a congestion signal because queuing delay provides a finer measure of congestion and scales more naturally with network capacity than packet loss probability does [172].

Because both FAST TCP and TCP Vegas use queuing delay as congestion signal, their window-updating algorithm relies on the propagation delay, which is estimated by a measure called *baseRTT*. This measure is defined as the minimum RTT observed so far. Because sometimes the routers' queues never become empty, the actual propagation delay may be inaccurately estimated by *baseRTT*, which results in unfairness [172,175] and excessive variations of routers' queues.

Considering steady-state (equilibrium) conditions, we aim to analyze and compare the fairness of FAST TCP versus TCP Vegas. However, it is important to clarify at the outset a certain confusion that one may encounter considering the definition of TCP Vegas. A certain ambiguity exists in this definition, and there are two interpretations of TCP Vegas, one of which makes TCP Vegas very similar to FAST TCP. In particular, TCP Vegas uses queuing delay as the congestion signal which is different from TCP Reno. It is parameterized by a parameter α. As pointed out in [174], the original description of Vegas [314] was ambiguous about whether α measures the rate per second, or per RTT. In this section of exposition herein, the term "Vegas" is used for the form analyzed in [174] (specifically, Section 4.2 of [174]). This matches the prose description in [314], in which α is the rate per second. Where ambiguity may arise, it will be called "the prose version of Vegas." The term "FAST" is used for the form adopted by FAST TCP [172], which matches the implementation of Vegas, in which α is the rate per RTT. As we consider equilibrium conditions, we ignore the slow start phase for both Vegas and FAST TCP in what follows.

Let $L(i)$ denote the set of links used by flow i. Let d_i [seconds] be the true propagation delay of flow i, let

$$\hat{d}_i = d_i + \delta_i$$

be the estimated propagation delay, and let $D_i(t) = d_i + q_i(t)$ be the RTT of flow i, including queuing delay of $q_i(t)$. Let $w_i(t)$ [packets] and $x_i(t)$ [packets/s] be the window size and rate for flow i, which are related by

$$w_i(t) = x_i(t)D_i(t). \tag{9.24}$$

Let c_l [packets/s] be the capacity of link l and $b_l(t)$ [packets] be the backlog at link l. Quantities without explicit time dependence are either constants or equilibrium values; for example, b_l is the equilibrium backlog at link l.

Both FAST and Vegas use a parameter called α, although the meaning of each is subtly different, as alluded to in the introduction. Flows using FAST and Vegas aim to keep a fixed number of packets in queues throughout the network. Under FAST, flow i aims to keep α_i packets, while under Vegas it aims to keep $\alpha_i d_i$. To avoid confusion, the alpha values for Vegas will be denoted by α^+. Our comparisons detailed here will use the following scenario, called *Persistent Congestion* in [174]. All flows share a single-bottleneck link of capacity c [seconds], have equal α [packets] (or α^+ [packets/s]), and have equal propagation delays d [seconds]. Flows arrive consecutively, spaced far enough apart for the system to reach equilibrium between arrivals, and keep transmitting greedily and persistently. When the ith flow arrives, it causes the queue size at the bottleneck link to increase by $B(i)$ [packets]. If d were known exactly, then $B(i)$ would be α under FAST or $\alpha^+ d$ under Vegas. However, the estimate $\hat{d}$ will be assumed to be the RTT seen when the flow first arrives, given by

$$d(i) = d + p(i-1) \text{ [seconds]},$$

and $B(i)$ will consequently be larger. Here,

$$p(i) = \sum_{j=1}^{i} B(j)/c \text{ [seconds]}$$

is the total queuing delay after the arrival of flow i.

To distinguish between the equilibrium of FAST and of Vegas, quantities pertaining to Vegas will have a superscript $+$.

9.2.2 EQUILIBRIUM CONDITIONS, UTILITY FUNCTIONS, AND PERSISTENT CONGESTION

As we ignore the slow start phase, for Vegas parameters of α_i^+ and β_i^+ for flow i, the update rule for Vegas [314] can be expressed as

$$w_i(t+1) = \begin{cases} w_i(t) + \frac{1}{D_i(t)}, & \text{if } \frac{w_i(t)}{d_i} - \frac{w_i(t)}{D_i(t)} < \alpha_i^+, \\ w_i(t) - \frac{1}{D_i(t)}, & \text{if } \frac{w_i(t)}{d_i} - \frac{w_i(t)}{D_i(t)} > \beta_i^+, \\ w_i(t), & \text{otherwise.} \end{cases} \tag{9.25}$$

Recall the equilibrium results for Vegas, derived in [174], when the ith flow estimates its propagation delay by $\hat{d}_i$, and $\beta_i^+ = \alpha_i^+$. A Vegas flow is in equilibrium if

$$\alpha^+ \hat{d}_i = w_i \left(1 - \frac{\hat{d}_i}{D_i}\right). \tag{9.26}$$

The equilibrium state of Vegas maximizes the sum of the flows' utilities, where the utility of flow i is

$$U_i(x_i) = \alpha_i^+ (d_i + \delta_i) \log x_i + \delta_i x_i. \tag{9.27}$$

Under Persistent Congestion, the increments in queue occupancy, $p^+(i)$, satisfy

$$\sum_{j=1}^{i} \frac{d + p^+(j-1)}{p^+(i) - p^+(j-1)} = \frac{c}{\alpha^+} \tag{9.28}$$

with

$$p^+(1) = \frac{\alpha^+ d}{c}, \tag{9.29}$$

and the rate of flow j between the arrival of flow i and flow $i+1$ is

$$x_j^+(i) = \frac{\alpha^+(d+p^+(j-1))}{p^+(i)-p^+(j-1)}. \tag{9.30}$$

Analogous results will now be derived for FAST.

The update rule for FAST can be written as [172]

$$w_i(t+1) = \gamma\left(\frac{d_i w_i(t)}{D_i(t)} + \alpha\right) + (1-\gamma)w_i(t) \tag{9.31}$$

for some constant $\gamma \in (0,1]$, giving the equilibrium condition

$$\alpha = w_i\left(1 - \frac{\hat{d}_i}{D_i}\right). \tag{9.32}$$

(Note that this rule also applies if Vegas adapts its α^+ to its estimate of d_i in an attempt to achieve proportional fairness. In that case, $\alpha^+ := \alpha/\hat{d}_i$.)

By arguments analogous to [174], it can be shown that the FAST equilibrium maximizes the sum of flows' utilities, where the utilities are now given by

$$U_i(x_i) = \alpha_i \log x_i + \delta_i x_i. \tag{9.33}$$

The core argument is to show that the derivative of the utility of flow i is the sum of suitable Lagrange multipliers, p_l, corresponding to the price of each link, evaluated at the equilibrium. That is,

$$U_i'(x_i) \equiv \frac{\alpha_i}{x_i} + \delta_i = \sum_{l\in L(i)} p_l. \tag{9.34}$$

That can be seen as follows. The number of packets queued by flow i at link l is $b_l x_i/c_l$. The total number of packets queued by flow i in equilibrium is

$$\sum_{l\in L(i)} \frac{b_l x_i}{c_l}.$$

Since the total number of packets from flow i in flight, w_i, is equal to the sum of those in propagation, $x_i d_i$, and those queued, in equilibrium

$$w_i - x_i d_i = \sum_{l\in L(i)} \frac{b_l x_i}{c_l}. \tag{9.35}$$

Combining Equations 9.24, 9.32, and 9.35 gives

$$\alpha_i = w_i\left(1 - \frac{\hat{d}_i}{D_i}\right) = w_i - x_i(d_i + \delta_i) = \sum_{l\in L(i)} \frac{b_l x_i}{c_l} - \delta_i x_i.$$

Rearranging and setting $p_l = b_l/c_l$ yield Equation 9.34, as required.

On the surface, it seems that Equations 9.27 and 9.33 are equivalent under the substitution $\alpha_i = \alpha_i^+ \hat{d}_i$. However, it turns out to be very significant that the number of packets that Vegas attempts to maintain in the queue, $\alpha_i^+ \hat{d}_i$, depends on the error in the estimate of the propagation delay. To see that, consider the equations for persistent congestion analogous to Equations 9.28 and 9.30, which will now be derived, following [175].

The equilibrium rates satisfy

$$x_1 + x_2 + \cdots + x_i = c \tag{9.36}$$

and

$$x_j = \alpha / q(j), \tag{9.37}$$

where $q(j)$ is the queuing delay as observed by flow j. When there is only one flow in the link, this yields $x_1 = c$ and $q(1) = B(1)/c$. Using yields 9.36 and 9.37, we have $B(1) = \alpha$ and $p(1) = \alpha/c$. When the second flow enters the link, it estimates $\hat{d}(2) = d + B(1)/c$ and perceives a queuing delay of $q(2) = B(2)/C$, while the first flow sees the true queuing delay of $q(1) = (B(2)+B(1))/c$. Again using Equations 9.36 and 9.37 gives

$$1/(B(1)+B(2)) + 1/B(2) = 1/\alpha$$

and

$$1/p(2) + 1/(p(2)-p(1)) = c/\alpha.$$

By induction,

$$\sum_{j=1}^{i} \frac{1}{p(i)-p(j-1)} = \frac{c}{\alpha}, \tag{9.38}$$

and the rate of flow j between the arrival of flow i and flow $i+1$ is

$$x_j(i) = \frac{\alpha}{p(i)-p(j-1)}. \tag{9.39}$$

Note that, unlike Equations 9.28 and 9.30, these expressions are independent of d. If $d \gg p^+(j)$ for all j, then the queuing delay and rates of Vegas reduce to those of FAST (with the substitution of $\alpha^+ d = \alpha$). However, if $p^+(j)$ is not negligible, the results will be different.

9.2.3 COMPARISON OF THE FAIRNESS OF FAST TCP AND TCP VEGAS

Since the equilibria are different, the following question arises: among the protocols of FAST TCP and TCP Vegas, which one is fairer?

When $i = 2$, Equation 9.28 for Vegas becomes

$$\frac{d}{p^+(2)} + \frac{d+p^+(1)}{p^+(2)-p^+(1)} = \frac{c}{\alpha^+} = \frac{d}{p^+(1)}$$

giving

$$\begin{aligned}\frac{p^+(2)}{p^+(1)} &= \frac{3d+p^+(1) \pm \sqrt{(3d+p^+(1))^2-4d^2}}{2d} \\ &= \frac{3d+p^+(1)+\sqrt{(d+p^+(1))(5d+p^+(1))}}{2d},\end{aligned} \tag{9.40}$$

where the $\pm$ becomes $+$ since otherwise $p^+(2) < p^+(1)$, which is not possible. The rate of the worse-off flow for Vegas is thus

$$x_1^+(2) = \frac{\alpha^+ d}{p^+(2)} = \frac{2c}{3+\alpha^+/c+\sqrt{(1+\alpha^+/c)(5+\alpha^+/c)}}, \tag{9.41}$$

which is clearly less than

$$x_1(2) = \frac{2c}{3+\sqrt{5}} \tag{9.42}$$

yielded by FAST.

Note that this trend continues, and FAST is fairer than Vegas for any number of sources. More specifically, if $\alpha^+ d = \alpha$, then $p^+(i) > p(i)$ for all $i > 1$. Furthermore, $x_1^+(i) < x_1(i)$ for all $i > 1$ and all α, α^+. For $i = 2$, this follows from Equations 9.41 and 9.42. The result can then be shown by induction on the hypothesis $p^+(j) > p(j)$ for all $1 < j < i$. To see that this implies $p^+(i) > p(i)$, assume instead that $p^+(i) \le p(i)$. Then, $p^+(i) - p^+(j-1) < p(i) - p(j-1)$, whence

$$\frac{c}{\alpha^+} = \frac{d}{p^+(i)} + \sum_{j=2}^{i} \frac{d + p^+(j-1)}{p^+(i) - p^+(j-1)} > \frac{d}{p^+(1)} + \sum_{j=2}^{i} \frac{d + p^+(j-1)}{p(i) - p(j-1)}$$
$$> \frac{d}{p(1)} + \sum_{j=2}^{i} \frac{d}{p(i) - p(j-1)} = \frac{c}{\alpha/d},$$

where the second inequality uses $p(1) = p^+(1)$. This contradiction establishes $p^+(i) > p(i)$. Since $p(0) = 0$, Equations 9.30 and 9.39 show that, for this α, $x_1^+(i) < x_1(i)$. However, $x_1(i)$ is independent of α by Equation 9.39, since $p(i) \propto \alpha$. Thus, $x_1^+(i) < x_1(i)$ for all $i > 1$ in general.

This finding has two implications. On the one hand, it shows that the prose version of Vegas is less fair than FAST and the implemented form of Vegas. On the other hand, it shows that persistent congestion is less of a problem than is predicted by the analysis of [174].

The authors of Vegas suggested using small values of α. However, the optimal value is difficult to set in practice. If it is too small, then the queuing it induces in a high-speed link may be small compared to the jitter of RTTs, making measurement inaccurate. Hence, a large value must often be used, as is proposed for FAST [290]. However, if such an α is used over a low-capacity link, it is possible for α^+/c to be large. As α^+/c becomes large, Vegas becomes arbitrarily unfair. In the case of two sources,

$$x_1^+(2) < \frac{c}{2 + \alpha^+/c} \to 0.$$

Consider also the inductive hypothesis (in i) that $p^+(j) \gg p^+(k) \gg d$ for all $k < j \le i$. This is true for $i = 1$ by Equation 9.29, when $\alpha^+/c \gg 1$. Then, Equation 9.28 becomes

$$\frac{p^+(i-1)}{p^+(i)} \approx \frac{c}{\alpha^+} \ll 1,$$

showing that $p^+(i) \approx d(\alpha^+/c)^i$ for all i. Thus, by Equation 9.30, the most recently arriving flow obtains almost all of the capacity. Although this case is pathological, it is in principle possible under Vegas. However, it cannot occur under FAST, since $p(i)$ is independent of d.

This scaling of Vegas is in contrast to that of FAST. Although $B(k) = (p(k) - p(k-1))c$ are known to diverge [175], they diverge slower than any power of k. To see that, assume instead that there exist $\eta > 0$ and $\gamma > 0$ such that $B(k) > \eta k^\gamma$ for all k. Approximating sums by integrals in (9.38) gives

$$\frac{1}{\alpha} = \sum_{j=1}^{i} \frac{1}{\sum_{k=j}^{i} \eta k^\gamma}$$
$$\approx \frac{\gamma+1}{\eta} \sum_{j=1}^{i} \frac{1}{i^{\gamma+1} - (j-1)^{\gamma+1}}$$

$$\approx \frac{\gamma+1}{\eta}\left(\frac{1}{i^{\gamma+1}}+\sum_{j=2}^{i}\frac{1}{i^{\gamma}(i-j+1)(\gamma+1)}\right)$$
$$= O\left(\frac{\log(i)}{i^{\gamma}}\right) \to 0.$$

However, α is a constant, which is a contraction and shows that there is no positive power of k that $B(k)$ consistently grows faster.

Empirically, it appears that

$$\frac{B(i)}{B(1)} = \log(i) + o(1).$$

If this is indeed the case, then the lowest throughput of any source is

$$\frac{B(1)}{(B(1)+\cdots+B(i))} = O(1/i\log(i)).$$

9.2.4 SIMULATION RESULTS

To verify the above analysis, we consider a persistent congestion scenario with 10 flows. The bottleneck link bandwidth is 100 Mbps (= 12,500 packet/s), and the round trip propagation delay is 40 ms (20 ms each way). Simulations were performed using ns2 [321] disabling the slow start phase. The Vegas module in ns2 interprets α as the rate per propagation delay, as FAST does. The results presented here for "Vegas" were obtained by modifying the code to implement the prose version of Vegas. The FAST implementation was [325]. We measure the queue size and every source's stable rate for three flow control rules: FAST ($\alpha = 200$ packet), Vegas with $\alpha^+ = 250$ packet/s, and Vegas with $\alpha^+ = 2475$ packet/s. For Vegas, $\beta = \alpha + 25$ packet/s; we chose $\alpha^+ < \beta^+$ to prevent oscillation [314]. These induce additional queuing delays of approximately $\alpha^+ d/c = 0.4$ and 4 ms, respectively.

To validate Equations 9.28 and 9.38, we compare the actual simulation rates with theoretical analysis for FAST (in Figure 9.7), Vegas with $\alpha^+ = 250$ packet/s (in Figure 9.8), and Vegas with $\alpha^+ = 2475$ packet/s (in Figure 9.9) for the second and sixth sources. The results match almost perfectly for FAST and Vegas with $\alpha^+ = 2475$ packet/s. The slight difference for Vegas with $\alpha^+ = 250$ packet/s is because the difference $\beta - \alpha$, needed for stability, is a non-negligible fraction of α.

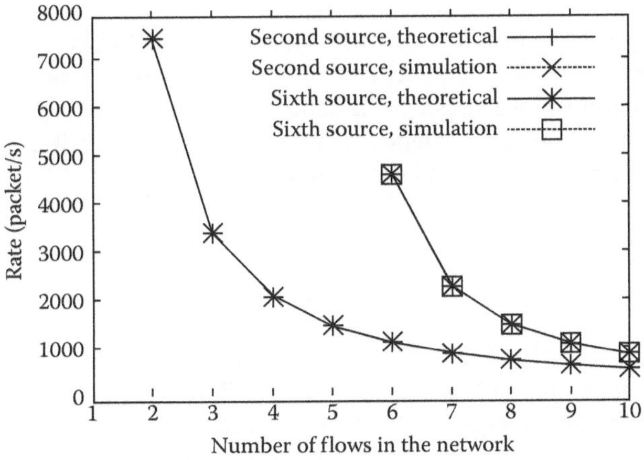

Figure 9.7 Second and sixth sources rate for FAST.

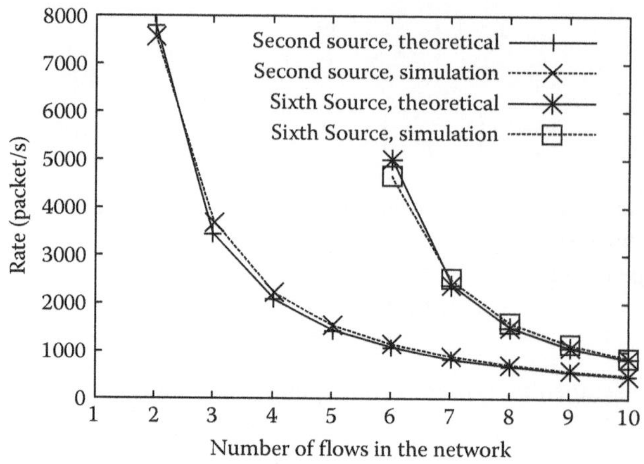

Figure 9.8 Second and sixth sources rate for Vegas ($\alpha^+ = 250$ packet/s and $\beta^+ = 275$ packet/s).

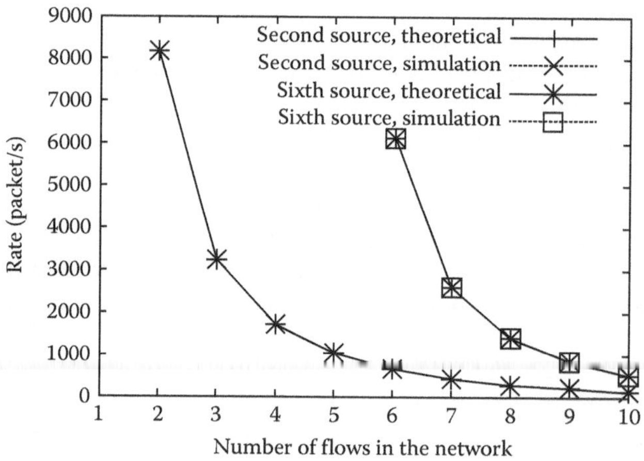

Figure 9.9 Second and sixth sources rate for Vegas ($\alpha^+ = 250$ packet/s and $\beta^+ = 275$ packet/s).

To compare the fairness of FAST, Vegas with small α^+, and Vegas with large α^+, we use two different criteria. The first criterion considers the most disadvantaged node (as in [124]). In our case, it is the first node. Therefore, we consider the ratio of its source rate, $x_1(n)$, to its fair rate which is c/n for the three cases. Figure 9.10 depicts the values for the first criterion for each of the three schemes as new flows arrive. The second criterion is defined by

$$F = \frac{(\sum_{i=1}^{n} x_i)^2}{n\sum_{i=1}^{n} x_i^2}. \tag{9.43}$$

By Equation 12.26, the case $F = 1$ reflects the situation where all flows are equal, while the smaller the value of F is, the larger the differences between the flows are. The three curves resulting from the second criteria are plotted in Figure 9.11. A very consistent message emerges from Figures 9.10 and 9.11. Fairness is adversely affected by the increase in the number of sources. Also, the larger α^+ is, the more unfairly the first Vegas flow is treated, which is consistent with our analysis in the previous section. This is illustrated for a wider range of α^+ in Figure 9.12.

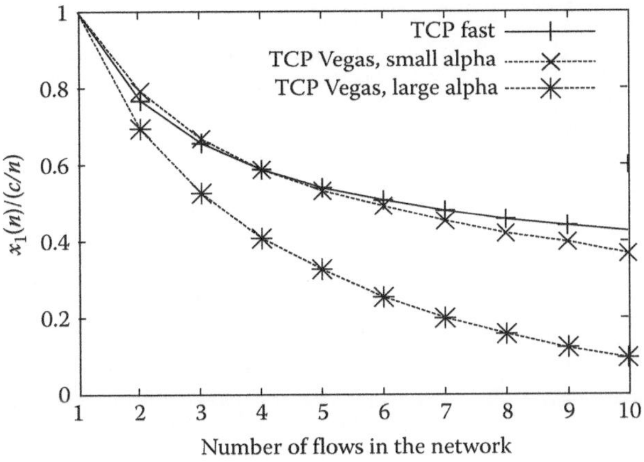

Figure 9.10 Fairness to the first flow.

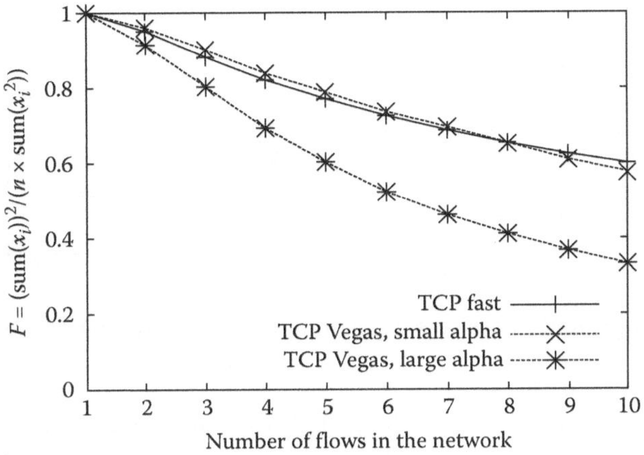

Figure 9.11 Overall fairness comparison.

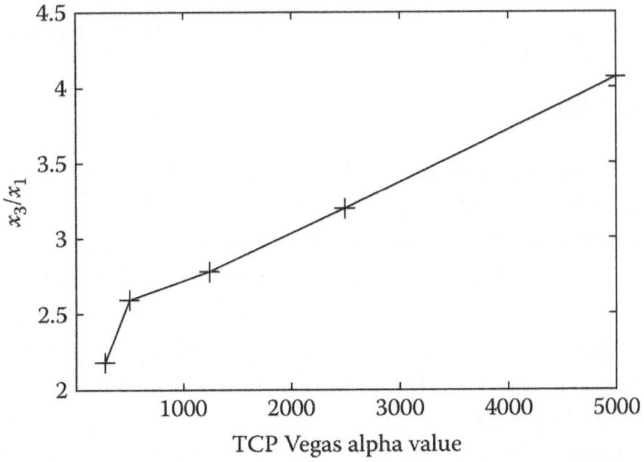

Figure 9.12 Comparison of third and first flow rates for Vegas at different α^+.

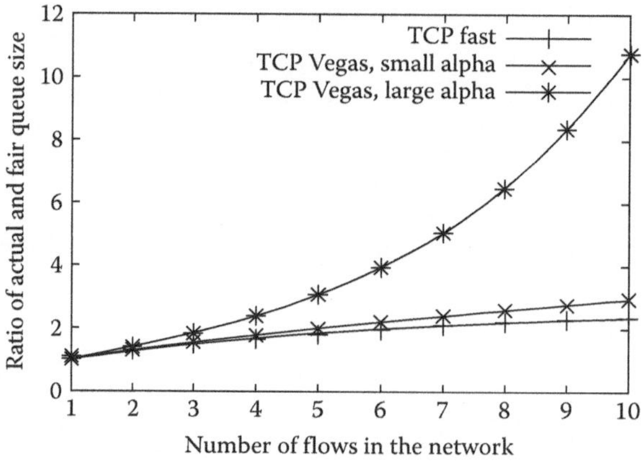

Figure 9.13 Buffer size needed at different flow numbers for Vegas (α^+ = 275 packet/s), Vegas (α^+ = 2500 packet/s), and FAST TCP.

Table 9.2
Queuing Delay, $p(i)$ [seconds], for $\alpha^+ \gg c$

Flow Number	**1**	**2**	**3**	**4**	**5**	**6**	**7**	**8**	**9**	**10**
$d(\alpha^+/c)^i$	0.8	16	320	6.4e3	1.3e5	2.5e6	5.1e7	1.0e9	2.0e10	4.1e11
Equation 9.28	0.8	18.4	403.3	8.8e3	1.9e5	4.2e6	9.6e7	2.1e9	4.5e10	9.6e11
Simulation	0.976	21.68	426.19	–	–	–	–	–	–	–

Another disadvantage of Vegas is that it generally requires more buffer space than FAST. Moreover, the rate of increase of the required buffer size will be greater for larger α^+, as shown in Figure 9.13.

For completeness, we also simulated the pathological case of $\alpha^+ \gg c$. We consider 10 flows with $\alpha^+ = 2500$ packet/s sharing a 1 Mbps bandwidth link with round trip propagation delay 40 ms. This gives $\alpha^+ = 20c$. The analysis shows that $p(i)$ should be roughly equal to $d(\alpha^+/c)^i$. Table 9.2 compares this approximation with both the simulated and the theoretical $p(i)$. Simulation results were only obtained for $i \leq 3$, due to the very large queue sizes involved. The 20% discrepancy between the theoretical and simulation results for the first source is because the measured *baseRTT* includes the 8 ms packetization delay, which is 20% of the 40 ms propagation delay.

9.2.5 CONCLUDING REMARKS

Both FAST TCP and TCP Vegas adjust their rates based on the estimated propagation delay and the measured RTT. If they both know their true propagation delays, then they have identical equilibrium rates and buffer occupancies. However, in the presence of estimation error, differences arise. In particular, the equilibrium of the prose form of Vegas differs from that of FAST and the form of Vegas commonly implemented. We have found expressions for the rates of different FAST and Vegas flows. In the particular case of persistent congestion, we have quantified the amount by which FAST is fairer than Vegas. In particular, the previous analysis of Vegas under persistent congestion [174] does not apply exactly to FAST or the implemented version of Vegas, and the unfairness due to persistent congestion is actually less severe than it predicts. In the analysis of [174], Vegas's fairness is influenced by its parameter α^+, whereas FAST's fairness is independent of α. The larger α^+ is, the more unfairly Vegas will treat the old flows, which have an accurate estimate of their propagation delay. When α^+ is small compared to the bottleneck link capacity, Vegas's equilibrium is almost the same as that of FAST. However, in the extreme case of $\alpha^+ \gg c$, Vegas becomes highly unfair, and the buffer occupancy increases geometrically. In contrast, for FAST and Vegas with small α^+, we demonstrated numerically that the increase in queue size is of order $\log(i)$ as the ith flow arrives.

9.3 EXPERIMENTAL FAIRNESS COMPARISON AMONG FAST TCP, HSTCP, STCP, AND TCP RENO

In the paper [172], the authors fully describe FAST TCP from design to implementation. Four difficulties corresponding to throughput, fairness, stability, and responsiveness are discussed for the current TCP protocols at large windows. Equilibrium and stabilities are well explored for FAST TCP. It is demonstrated how FAST TCP overtakes the current TCP protocols on the above four aspects in the algorithms, which are implemented in the designed prototype.

In [172], experiments on dummynet test-bed are conducted with the aim of comparing new TCP algorithms as well as the Linux TCP implementations. Simulation results on ns2 are presented as well. The authors conducted experiments for two scenarios: (1) Scenario 1: there are three TCP flows, with propagation delays of 100, 150, and 200 ms, that started and terminated at different times and (2) Scenario 2: Pretty much similar to Scenario 1, but there are eight flows being active according to deferent joining and departing schedule.

In both scenarios, FAST TCP converges to its new equilibrium rate allocation rapidly and stably as new flows join and old flows left. Similar smooth dynamics of TCP Reno can be seen from its throughput trajectory. However, HSTCP, STCP, and BIC-TCP exhibit significant fluctuation, though they can also respond very quickly to the competing situation at the link.

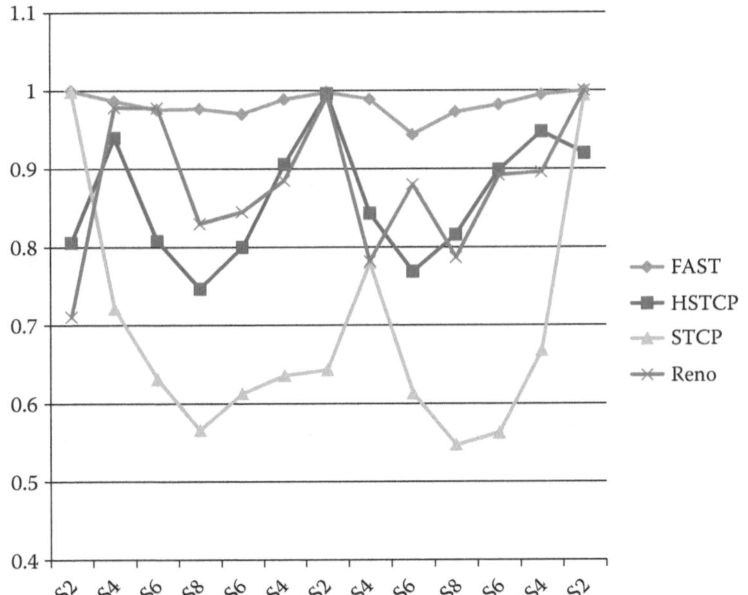

Figure 9.14 Fairness comparison among FAST TCP, HSTCP, STCP, and TCP Reno: an experimental case study presented in Table IV of [172] where TP1, TP2, ..., TP13 denote the time periods [0, 1800], [1800, 3600], ..., [21600, 23400], respectively. S2, S4, S6, and S8 denote the number (2, 4, 6, and 8) of flows.

Based on the throughput trajectories of each protocol, Jains fairness indexes for the rate allocation corresponding to the above two scenarios are calculated (see Tables 3 and 4 in [172]). We plot those results of Table 4 in [172] into Figure 9.14. The results for the above two scenarios show that FAST TCP obtains the best intraprotocol fairness, very close to 1, followed by HSTCP, TCP Reno, BIC-TCP, and then STCP.

10 Bidirectional Transmission and an Extension of Network Utility Maximization Model

This chapter focuses on the bidirectional data transmissions of a connection, in which TCP and ACK packets are flowing together but in opposite directions. We discuss the performance of FAST TCP in a bidirectional transmission scenario. We further take the first step to investigate the network utility maximization (NUM) theory in the two-way flow scenario. The modeling and solution lead to some straightforward discoveries of a critical problem with network bandwidth allocation and performance optimization in the NUM framework to account for the impact of ACK packet flow.

10.1 INTRODUCTION

In a communication system, the data are transmitted between devices or parties. In a point-to-point system composed of two connected devices, called Devices A and B, we often refer to the data transmission direction. The directions may be from Device A to Device B, from Device B to Device A, or both. The terms *simplex* and *duplex* are used to describe the data transmission direction. In a simplex system, the data can only be transmitted from one device to another device over a path. In a duplex system, two clearly defined data transmissions are defined, and the data carrying information can be transmitted in both directions, Device A to Device B over one path and Device B to Device A over the other. According to the simultaneity of transmissions, the types of the communication system include full-duplex systems and half-duplex systems.

In a full-duplex system, both devices can transmit data between each other simultaneously. Many communications are full-duplex systems, and one typical example is the telephone invented by Alexander Graham Bell in 1875. As we all know, the device is composed of an earphone and a microphone. The microphone transmits the speech signal of one end, and the reproduced voice from the other end can be heard by the earphone simultaneously. To achieve this, a bidirectional communication channel should be built between two telephones, or more strictly speaking, two communication paths should be built between the telephones.

In a half-duplex system, both devices can transmit data between each other, but not simultaneously. There are still two clearly defined paths for bidirectional transmissions, i.e., from Device A to Device B and from Device B to Device A. Different from a full-duplex system, devices in a half-duplex system can either transmit or receive data at a time. A typical example of a half-duplex device is a handheld transceiver having bidirectional radio. The device, also called walkie-talkie, has a "push-to-talk" switch that starts transmission. When the local person wants to speak to the remote one, he pushes the switch. The device turns on the transmitter but turns off the receiver, so the person cannot hear the remote voice. To listen to the remote voice, the switch should be released, which turns on the receiver but turns off the transmitter.

In the network, i.e., the OSI model, the term "duplex" is actually determined at physical layer (layer 1) and data link layer (layer 2). TCP works in transport layer (layer 4), which is used for making connections between nodes on a network. TCP is bidirectional, not duplex. When workstation A transmits a TCP packet to workstation B, workstation B sends an acknowledgment (ACK) of sorts, but the communications are not simultaneous as in the case of duplex, but rather one after another.

The transmission of a connection is bidirectional, because the directions of the connection's TCP packets and ACKs are opposite to each other. We call this bidirectional transmission type as *simple bidirectional*. In another pattern, data are transmitted through two or more connections in opposite directions over the same network path. We call this bidirectional transmission type as *composed bidirectional*. In the composed bidirectional pattern, TCP packets transmitted through the connections in one direction share the same network path capacity with ACKs of connections in the opposite direction. The resulting situation may be that a large number of data buffers on the path are occupied by the ACKs, which leads to the performance degradation.

To better understand the performance of the bidirectional TCP flow, it is essential to describe the symmetric link and the asymmetric link. The "asymmetric" means that the bandwidth (and bit rate) of a link is greater in one direction (i.e., downstream) than that in the reverse direction (i.e., upstream). A typical technology is asymmetric digital subscriber line (ADSL). When ADSL is used to access Internet, it is able to provide the higher speed for the download from Internet services, but it does not provide high speed in the other direction. Actually, "asymmetric" is very popular in current communication links or network paths, such as wireless network, mobile communication network, and satellite communication.

10.2 PERFORMANCE OF BIDIRECTIONAL FLOWS

In popular TCP protocols, ACK is designed to return a signal to denote the correct transmission. The result shows that all transmission protocols can be equivalent or approximately equivalent to those unidirectional protocols. This expectation is realized for most Internet services in early years because people always download a large amount of data from Internet server, but seldom upload data to the server.

In most flow control models [109], the ACK packet flow is also simplified to feedback signals. However, an ACK is a packet with several bytes, and the network must handle ACKs. Of course, the above models are reasonable and may not change the throughput performance when the size of the ACK packet is small enough. Thus, the time to handle ACKs is zero, and ACKs do not affect the data packet rate. For example, in popular Internet Client/Server (C/S) services on symmetric links, the direction of the TCP data flows is unidirectional, i.e., from the server to the client. ACK flows are also transmitted in one way direction, i.e., from the client to the server. In this case, ACK flows and data packet flows are in opposite directions, and they do not disturb each other. Furthermore, the bandwidth of the ACK packet flow is very small when compared with the uplink capacity. The newly developed technology such as P2P enables any users to be a server and a client at the same time. In these scenarios, the performance of TCP flows should be rechecked.

Concerns about the performance of bidirectional TCP flows were raised in a very early year. In 1991, two phenomena, ACK compression and an out-of-phase queue synchronization mode, are reported [358]. An ACK is a signal transmitted between communicating processes or computers to signify acknowledgment, or receipt of response, as part of a communications protocol. An ideal ACK is just a signal without bandwidth requirement. Actually, an ACK is composed of several bytes and needs bandwidth when it is transmitted. Furthermore, an ACK is synchronized with a TCP data packet. When TCP data packets in one direction share the same path with ACKs in the opposite direction, ACKs share a common buffer with TCP data packets in the end system. The result is that bunched ACKs of a connection arrive in the sender, the so-called ACK compression. ACK compression can result in unfair throughput received with competing connections, and TCP packets are led to an out-of-phase queue synchronization mode. Furthermore, the throughput of TCP connection with ACK compression is less compared to what could be expected without ACK compression.

The behavior of bidirectional TCP flows in a rate-controlled network is different from that of unidirectional TCP flows. In [357], the dynamics of bidirectional TCP flow in a rate-controlled network is analyzed by developing the system models. If the TCP data packets and ACKs are mixed in a common shared queue buffer at the end systems, ACK compression exists and results in persistent throughput degradation in such networks. The common queue shared by TCP packets and ACKs may be either at the data link layer or at the IP layer. The throughput degradation due to bidirectional flows is significant because of periodic burst behavior of the source IP queue buffer.

In later years, Balakrishnan and Padmanabhan [359] describe that the bidirectional TCP throughput degradation performance in links with asymmetric bandwidth comes from adverse interaction between TCP data packet flows in one direction and ACK flows in the opposite direction. Lakshman et al. [362] determine the TCP throughput as a function of buffering round-trip times (RTTs) and a normalized asymmetric factor, and show how TCP throughput degradation can be alleviated through individual connection buffer and individual bandwidth allocation in the reverse direction.

The bidirectional TCP flows also appear in wireless networks, and the performance is investigated. In wireless ad hoc networks, an upper bound and an approximation for the bidirectional transmission capacity are provided in [360]. In wireless LANs, bidirectional transmission capacity in Asymmetric Access Points is listed in [361]. Almost all TCP variants must face the bidirectional flows, e.g., TCP Vegas and FAST TCP. The effort for improving these bidirectional TCP throughput is also made in [363].

To completely investigate the effect of bidirectional flows in the network is difficult compared to unidirectional flows. In the following subsections, to be well understood, we simplify the analysis of bidirectional flow performance by just considering a single link. However, two bidirectional flow patterns are included.

10.2.1 SIMPLE BIDIRECTIONAL MODEL IN A SINGLE ASYMMETRIC BOTTLENECK LINK

In a single asymmetric bottleneck link, TCP flows in simple bidirectional pattern show performance degradation. The scenario is shown in Figure 10.1, where each source sends a TCP packet flow and receives an ACK flow. The subscript or superscript f and b denote the forward and backward transmission directions of sources, respectively. The link is asymmetric. The forward capacity is c_f [B/s], and the forward propagation delay is d_f. The backward capacity is $c_b < c_f$ [B/s], and the backward propagation delay is d_b. The link has infinite forward and backward buffering storages.

The asymmetry capacity ratio (ACR) and the asymmetry packet ratio (APR) are used to denote the asymmetry. The two factors are defined as follows:

$$k_c = \frac{c_f}{c_b}$$

and

$$k_P = \frac{P_b}{P_f},$$

where P_f is the packet size [bytes], and P_b is the ACK packet size [bytes].

Overall, asymmetry factor (AF) is the product of k_c, k_P

$$k_f = \frac{c_f \cdot P_b}{c_b \cdot P_f}. \tag{10.1}$$

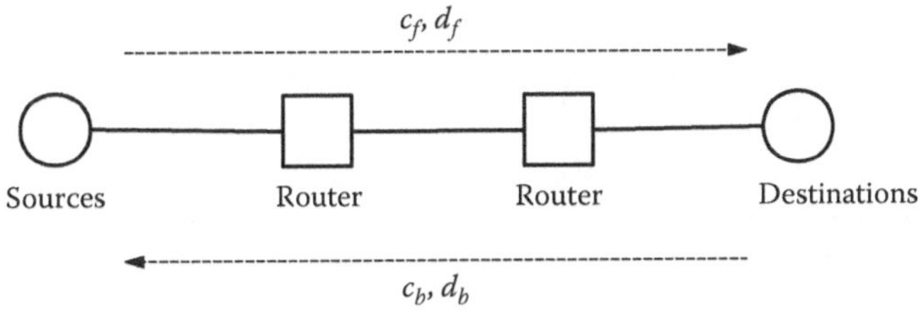

Figure 10.1 Simple bidirectional flows in the single asymmetric link.

Let $w_i(t)$ be the congestion window of source i; then, the throughput [B/s] of source i at time t is

$$x_i^f(t) = \frac{w_i(t) \cdot P_i^f}{d + q_i^f(t)}. \tag{10.2}$$

The ACKs in the backward link are response signal with fixed size, and the source i ACK rate in the backward direction is proportional to its data flow throughput but delayed by a time. The proportion is P_i^b / P_i^f thus

$$x_i^b(t) = \frac{P_i^b}{P_i^f} x_i^f(t - \tau_i^f).$$

Substituting the APR, it is

$$x_i^b(t) = k_{Pi} x_i^f(t - \tau_i^f). \tag{10.3}$$

The aggregate rates, or the bandwidth occupied by flows, in the forward and backward links are

$$y^f(t) = \sum_i x_i^f(t - \tau_i^f), \tag{10.4}$$

$$y^b(t) = \sum_i x_i^b(t - \tau_i^b). \tag{10.5}$$

Considering a continuous-time TCP model, the queuing delay of packets in the forward link buffer has been traditionally evaluated by an average value:

$$\dot{p}^f(t) = \frac{1}{c_f}(y^f(t) - c_f). \tag{10.6}$$

In the same way, the queuing delay of packets in the backward link buffer is averagely obtained by

$$\dot{p}^b(t) = \frac{1}{c_b}(y^b(t) - c_b). \tag{10.7}$$

Then, the source i end-to-end queuing delay, $q_i(t)$, can be obtained by the sum of the two queue delays. It is

$$q_i(t) = p^f(t - \tau_i^b) + p^b(t). \tag{10.8}$$

Here, we take a recently developed high-speed TCP variant called FAST TCP [172] as an instance, to look into the throughput of sources. A FAST TCP has a special algorithm to update its congestion window. The algorithm is based on the source average RTT and a queuing delay the source evaluated. The dynamics of FAST TCP congestion window is

$$\dot{w}(t) = \gamma\left(\frac{dw(t)}{d + q(t)} + \alpha - w(t)\right), \tag{10.9}$$

where $\gamma \in (0, 1]$, and α is a constant.

From the earlier equations, the dynamic behavior of FAST TCP sources in the scenario can be obtained. Here, we suppose that the sources achieve a stable status. The time t can be omitted in the earlier equations. Thus, it is easy to investigate the throughput of TCP flows.

For TCP flows in stable status, the TCP data packet rate of a source is equal to its ACK rate. Here, introduce a rate of source i in packets/s. $x'_f(i)$ denotes the TCP data packet velocity, and $x'_b(i)$ is the ACK velocity:

$$x'_f(i) = \frac{x_f(i)}{P_f},$$

$$x_b'(i) = \frac{x_b(i)}{P_b}.$$

The two velocities are equal due to self-clock of TCP, and thus

$$x_f'(i) = x_b'(i).$$

For all TCP sources, the sum of their bandwidth should be equal or less than the link capacity. It is obtained that

$$\sum_i x_f(i) \leq c_f$$

and

$$\sum_i x_b(i) \leq c_b.$$

The above can be changed to

$$\sum_i x_f'(i) \leq \frac{c_f}{P_f}$$

and

$$\sum_i x_b'(i) \leq \frac{c_b}{P_b}.$$

The throughput of TCP should be constrained by these two inequations. However, in different cases, the results are different. The first case is when AF is small, i.e., $k_f \leq 1$. Therefore,

$$\sum_i x_f'(i) \leq \frac{c_f}{P_f} \leq \frac{c_b}{P_b}.$$

The achievable bandwidth for all TCP sources is c_f/P_f:

$$\sum_i x_f'(i) = \frac{c_f}{P_f}.$$

For α proportional fair FAST TCP, it is

$$x_f'(i) = \frac{\alpha_i}{\sum \alpha_i} \sum x_f'(f).$$

Thus, the throughput of each flow

$$x_f'(i) = \frac{\alpha_i}{\sum \alpha_i} \frac{c_f}{P_f}$$

Another case is when $k_f > 1$, and thus

$$\frac{c_f}{P_f} > \frac{c_b}{P_b}$$

and

$$\sum_i x_b'(i) \leq \frac{c_b}{P_b}.$$

The achievable bandwidth for all TCP sources is

$$\sum_i x_b'(i) = \frac{c_b}{P_b}.$$

Fast TCP sources achieve α proportional fair, and then

$$x_b'(i) = \frac{\alpha_i}{\sum \alpha_i} \sum x_b'(i) = \frac{\alpha_i}{\sum \alpha_i} \frac{c_b}{P_b}.$$

The TCP flow throughput $x_f(i)$ does not include the ACK, and it is only related to TCP data packet. Thus, the TCP throughput in bytes/s is the product of packet size P_f and the rate in packets/s:

$$x(i) = \begin{cases} x'_f(i)P_f & k_f \geq 1, \\ \frac{1}{k_f}x'_b(i)P_b & k_f < 1. \end{cases}$$

The following results can be obtained:

$$x(i) = \begin{cases} \frac{\alpha_i}{\sum \alpha_i} c_f, & k_f \geq 1, \\ \frac{1}{k_f}\frac{\alpha_i}{\sum \alpha_i} c_f, & k_f < 1. \end{cases} \tag{10.10}$$

The meaning of 10.10 is easily understood. If c_f/P_f is large enough, the TCP throughput is determined by c_b/P_b, while if c_b/P_b is large enough, the TCP throughput is determined by c_f/P_f. In a simple scenario where the packet size is fixed, the results are that the throughput is determined by c_b if c_f is large enough, or vice versa.

In fact, we always expect that TCP flows can utilize the capacity of the forward link.

10.2.2 COMPLEX BIDIRECTIONAL IN A SINGLE-BOTTLENECK LINK

Consider a single asymmetric link network with complex bidirectional flows, as shown in Figure 10.2. Different from the scenario shown in Figure 10.1, in this network, two TCP flows have completely different transmission directions. The data packet of Flow 1 is from Source 1 to Destination 1 over the forward link. The data packet of Flow 2 is from Source 2 to Destination 2 in the backward link. Of course, ACKs of Source 1 are queued in the buffer with data packets of Source 2. The capacity and propagation delay of the forward link are c_f and τ_f, respectively. The capacity and propagation delay of the backward link are c_r and τ_r, respectively.

We also take FAST TCP as an instance to look into the throughput of two sources. Especially, we focus on the performance in stable state, but omit the transient state.

According to the congestion window of FAST TCP [172], each FAST TCP at the source periodically adjusts the congestion window by the network parameters, which includes the average measured RTT and the estimated queuing delay. A discrete algorithm to adjust the congestion window is

$$w \leftarrow \min\left\{2w, (1-\gamma)w + \gamma\left(\frac{\text{base}RTT}{\text{avg}RTT}w + \alpha\right)\right\}, \tag{10.11}$$

where the parameter $\gamma \in (0,1]$. The base*RTT* is the minimum RTT observed so far, which is regarded as the propagation time of data. The parameter α is always a positive integer value that implies the total number of packets buffered in end systems in stable state over all links on the path. The avg*RTT* is the averaged value of RTT$_i$, and FAST TCP uses the moving average method. The average value, but not real-time RTT, actually indicates every packet's RTT to adjust congestion window.

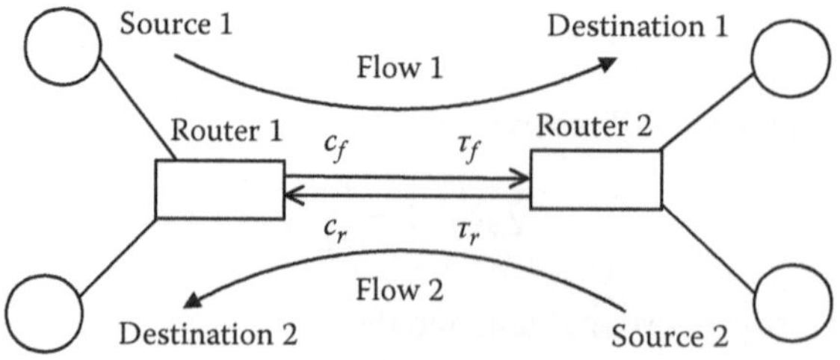

Figure 10.2 Complex bidirectional flows in single asymmetric link.

The data packet flow throughput of FAST TCP Sources 1 and 2, is by a simple formula,

$$x_i = \frac{w_i}{\text{avg}RTT}. \tag{10.12}$$

If the system reaches its equilibrium, the unique equilibrium solution (x^*, q^*) of the system is

$$x_i^* = \frac{\alpha_i}{q_i^*}. \tag{10.13}$$

The queuing delay is obtained by

$$q_i^* = RTT_i^* - \text{base}RTT_i. \tag{10.14}$$

In the scenario discussed in this subsection, the network only has one duplex link. For each source, the queue delay is the same, i.e.,

$$q^* = q_1^* = q_2^*.$$

Each RTT of packets is obtained by getting the interval between the packet sending time and the arrival time of its ACK. Ideally, we do not expect the ACK be buffered in the system and brings extra time. It is the truth when only considering unidirectional flows. However, in the scenario shown in Figure 10.2, the RTT includes at least four elements: queue delay time of data packets q_{df} in the forward path, fixed forward delay time d_{pf} including forward propagation delay and data packet processing time, ACK queue delay time q_{dr} in the backward path, and fixed backward delay time d_{pr} including backward propagation delay and ACK packet processing time.

Queuing delay q does not include fixed delay time, but the sum of q_{df} and q_{dr}. Of course, $q_{dr} = 0$ if ACKs are not queued in the buffer. The change of either q_{df} or q_{dr} leads to the change of queue delay q. In an asymmetric link, backward queue delay q_{dr} may be primary part due to the less backward capacity or the greater backward ACK flows. Furthermore, the data packet of the backward flow can occupy the ACK bandwidth of the forward flow, which results in throughput degradation.

Consider an equilibrium point where q_{df}^* and q_{dr}^* are fixed, and then (10.13) is changed to

$$x_i^* = \frac{\alpha_i}{q_{df_i}^* + q_{dr_i}^*}. \tag{10.15}$$

In complex bidirectional TCP scenario, we cannot expect $q_{dr_i}^*$ to be zero because TCP flows are aggressive. Thus, the throughput degradation is the doomed result. We define a throughput degradation ratio to describe this

$$\eta = \frac{X^*}{x^*}, \tag{10.16}$$

where X^* is the data throughput of a source in a single source, a single link without opposite flows; x^* is the data throughput of a source when in a single source, a single link with opposite flows.

The value of η, if α is identical, is determined by queuing delays

$$\eta = \frac{q_{df_i}^* + q_{dr_i}^*}{q_{df_i}^*}. \tag{10.17}$$

The throughput degradation, from (10.17), is seen to come from increased delay. The analysis does not provide the value of source throughput. If we return to read the last subsection, the throughput of the two sources in two cases can be obtained by similar deduction.

10.3 EXTENDED NUM MODEL FROM ONE-WAY FLOWS TO BIDIRECTIONAL FLOWS

A communication network usually has data packets and acknowledge (ACK) packets being transmitted in opposite directions. ACK packet flows may affect the performance of data packet flows, which is unfortunately not considered in the usual NUM model. The paper by Ge and Tan [364] presents a NUM model in networks with two-way flows (NUMtw) by adding a routing matrix to cover ACK packet flows. The source rates are obtained by solving the dual model, and the relation to the routing matrix of ACK packet flows is disclosed. Furthermore, the source rates in networks with one-way flows by the usual NUM model are compared to those in networks with two-way flows by the NUMtw model. This section gives a brief introduction to the NUM model (theory) proposed by Ge and Tan [364] in the following.

The NUM model of bidirectional TCP flows in a network is different from that of unidirectional ones. In fact, a network is composed of a set of duplex links in most cases. Consider a network with K duplex links. The data can be transmitted in the two opposite directions of one link. Its data flow and its corresponding ACK flow are carried by each connection. The data flow is from its source node and directed to its destination node, while the ACK flow is from the destination node to the source node. To clearly describe the link, one duplex link is denoted as two simplex links. Denote one transmission direction by an integer, e.g., k, while the other by an integer $K+k$. Suppose that F connections are in the network. Number a data flow as k, and its ACK flow is denoted as $K+k$. The corresponding capacities of the links are c_k, c_{K+k}. The flow rates, $x_f, x_f > 0, x_{F+f}, x_{F+f} \geq 0$, are the data flow rate and the ACK flow rate. x_{F+f} is, in most cases, in proportion to x_f, i.e., $x_{F+f} = \theta_f x_f$, where θ_f is the proportional factor. The users always focus on the data flow, but not on the ACK flow. The NUM model, when considering bidirectional flows, is

$$\max \sum_{f=1}^{F} U_f(x_f). \tag{10.18}$$

The utility function $U_f(x_f)$ is only related to the data flow. The function is, as usual, assumed to be increasing and continuously differentiable function. The constraints are

$$M \begin{bmatrix} X_1 \\ X_2 \end{bmatrix} \leq \begin{bmatrix} C_1 \\ C_2 \end{bmatrix}, \tag{10.19}$$

where M is the routing matrix, the dimensions of which is $2K \times 2F$,

$$M = \begin{bmatrix} N_{11} & N_{12} \\ N_{21} & N_{22} \end{bmatrix}.$$

The routing matrix is divided in to four submatrices: A_{11},A_{12},A_{21}, and A_{22}. Each is with $K \times F$ dimensions. The capacity vector of one direction is $C_1 = \{c_k\}$, and that of the other direction is $C_2 = \{c_{K+k}\}$. Accordingly, $X_1 = \{x_f\}$, $X_2 = \{x_{F+f}\}$ are the vectors denoting the flow directions. Another constraint is

$$x_{F+f} = \theta_f x_f \quad 1 \leq f \leq F. \tag{10.20}$$

The routing information $N_{11} = \{m_{kf}\}$ denotes the relation of flow f and link k, $m_{kf} = 1$ if flow f through link k; otherwise, $m_{kf} = 0$. Similarly, $N_{12} = \{m_{k(F+f)}\}$, $N_{21} = \{m_{(K+k)f}\}$, and $N_{22} = \{m_{(K+k)(F+f)}\}$ denote the corresponding routing information.

By substituting Equation 10.20 into 10.19, the constraint in the NUM model becomes

$$\begin{bmatrix} N_{11} & N_{12} \\ N_{21} & N_{22} \end{bmatrix} \begin{bmatrix} X_1 \\ \theta X_1 \end{bmatrix} \leq \begin{bmatrix} C_1 \\ C_2 \end{bmatrix}, \tag{10.21}$$

where θ is the diagonal matrix.

The rates of all data flows can be obtained by solving the above model. Let $\Phi_1 = \{\lambda_k, \lambda_k \geq 0\}$, $\Phi_2 = \{\lambda_{K+k}, \lambda_{K+k} \geq 0\}$, and then the Lagrangian form is defined as

$$L(X_1, \Phi_1, \Phi_2) = \sum_{f=1}^{F} U_f(x_f) + \sum_{k=1}^{2K} \lambda_l (c_k - \sum_{f=1}^{F} x_f(b_{kf} + b_{l(F+f)}\theta_f)).$$

The dual model is

$$\min D(\Phi_1, \Phi_2). \quad (10.22)$$

The objective function is $\max L(X_1, \Phi_1, \Phi_2)$:

$$D(\Phi_1, \Phi_2) = \sum_{f=1}^{F} Y_f(p_f) + \sum_{k=1}^{2K} \lambda_k c_k$$

$$Y_f(p_f) = \max(U_f(x_f) - x_f p_f), \quad (10.23)$$

where p_f is the entry of the vector P:

$$P = \left[\; N_{11}^T + \theta N_{12}^T, \, N_{21}^T + \theta N_{22}^T \; \right] \begin{bmatrix} \Phi_1 \\ \Phi_2 \end{bmatrix}. \quad (10.24)$$

The Lagrangian multiplier, $\lambda_k, 1 \leq k \leq 2K$, means the price per unit bandwidth on link k, and p_f is the sum of link prices observed by the connection.

The solution to bidirectional NUM model is the solution to maximization (10.23). Each p_f, a unique maximizer, determines $x_f(p_f)$. Because $U_f(x_f)$ is a function having increasing, strictly concave characteristics in its argument and it is continuously differentiable, $U_f^{'}$ is decreasing. Thus,

$$p_f = U_f^{'}(x_f). \quad (10.25)$$

The source rate can be obtained by the inverse of $U_s^{'}$, and

$$x_f(\lambda) = \left[U_f^{'-1} \left(\sum_{l}^{2K} \lambda_k (b_{kf} + b_{k(F+f)}\theta_f) \right) \right]^+. \quad (10.26)$$

Thus, if given λ_k and the routing information $[N_{11}^T + \theta N_{12}^T, \, N_{21}^T + \theta N_{22}^T]$, an individual flow can obtain its flow rate.

10.4 TWO EXAMPLES OF THE NUM MODEL OF BIDIRECTIONAL FLOWS

This section gives two examples of the NUM model of bidirectional flows to demonstrate the application of the extensions to the NUM model from one-way flows to two-way flows.

Example 1:

Consider a network with three duplex links, as in Figure 10.3.

The capacities of the six links are, in order, $100c_2$, c_2, $100c_2$, $100c_5$, c_5, and $100c_5$. The utility functions are $U_1(x_1) = \log x_1$ and $U_2(x_2) = 9 \log x_2$.

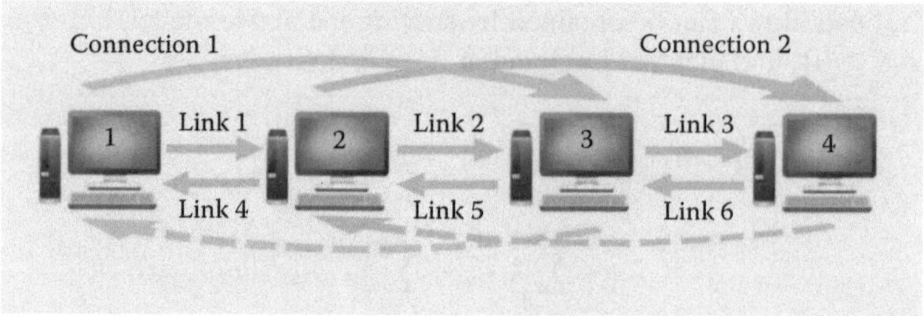

Figure 10.3 A network with two connections: the thick solid lines represent the data packet flows, and the thin and dashed lines represent the ACK packet flows.

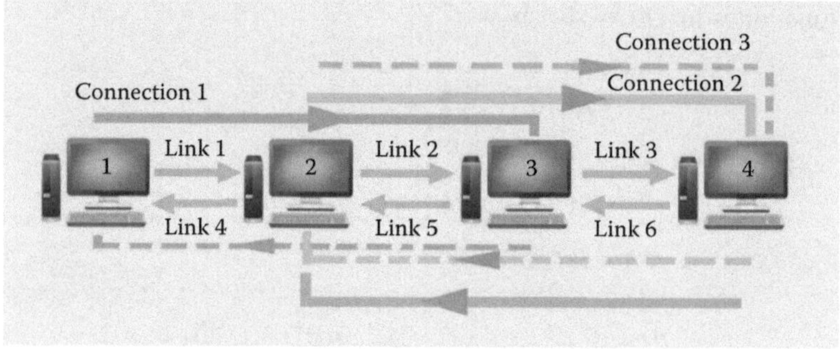

Figure 10.4 A network with three connections, where the transmission directions of Connection 3 are opposite to those of Connection 2.

The data packet flows of Connections 1 and 2 are from Node 1 to Node 3 and from Node 2 to Node 4, respectively. The constraints of the NUM model in S1A are as follows: $x_1 \leq 100c_2$, $x_1 + x_2 \leq c_2, x_2 \leq 100c_2$. By solving the optimization problem, which is formulated by Equations 10.18 and 10.19, the flow rates in S1A are

$$x_1 = 0.1c_2, \quad x_2 = 0.9c_2. \tag{10.27}$$

The ACK packet flows of Connections 1 and 2 are from Node 3 to Node 1 and from Node 4 to Node 2, respectively. Another three constraints are added to the NUMtw model in S1B compared to the NUM model in S1A, which are $\theta_1 x_1 \leq 100c_5, \theta_1 x_1 + \theta_2 x_2 \leq c_5, \theta_2 x_2 \leq 100c_5$. There are two possible solutions: one is

$$x_1 = c_5/(10\theta_1) \quad x_2 = 9c_5/(10\theta_2)$$

under the Condition $1/\theta_1 + 9/\theta_2 < 10c_2/c_5$; the other is as Equation 10.27 under $\theta_1 + 9\theta_2 < 10c_5/c_2$ or $\theta_1 = \theta_2, c_5 = \theta_1 c_2$.

In this example, it can be verified by the accurate flow rates; if $c_5/c_2 > \theta_{MAX}$, the flow rates in S1B are identical to those in S1A, while if $c_5/c_2 < \theta_{MIN}$, the flow rates in S1B are less than those in S1A.

Specially, setting the data packet size to 1000 bytes and the ACK packet size to 40 bytes for all connections, and $c_2 = 500$ Mb/s, $c_5 = 10$ Mb/s in Example 1, the flow rates x_1 and x_2 in S1A are 50 and 450 Mb/s, while in S1B, they are 25 and 225 Mb/s, respectively.

Example 2:

Add Connection 3 to the network in Example 1, and each connection has the same utility function. The data packet flow and the ACK packet flow of the connection are from Node 4 to Node 2 and from Node 2 to Node 4, respectively, as shown in Figure 10.4.

The solution to the NUM model in S2A can be solved by $U_1'(x_1) = U_2'(x_2), x_1 + x_2 = c_2, x_3 = c_5$. The extended routing matrix G in S2B is

$$G^T = \begin{pmatrix} 1 & 1 & 0 & \theta_1 & \theta_1 & 0 \\ 0 & 1 & 1 & 0 & \theta_2 & \theta_2 \\ 0 & \theta_3 & \theta_3 & 0 & 1 & 1 \end{pmatrix}.$$

There are four possible solutions in S2B: one possible solution satisfying

$$\theta_1 U_1'^{-1}(\theta_1 U_3'(x_3)) + \theta_2 U_2'^{-1}(\theta_2 U_3'(x_3)) + x_3 = c_5,$$
$$x_1 = U_1'^{-1}(\theta_1 U_3'(x_3)),$$
$$x_2 = U_2'^{-1}(\theta_2 U_3'(x_3)),$$
$$x_1 + x_2 + \theta_3 x_3 < c_2 \quad \mu_5 > 0,$$

the other possible solution satisfying

$$U_1'^{-1}(U_3'(x_3)/\theta_3) + U_2'^{-1}(U_3'(x_3)/\theta_3) + \theta_3 x_3 = c_2,$$
$$x_1 = U_1'^{-1}(U_3'(x_3)/\theta_3),$$
$$x_2 = U_2'^{-1}(U_3'(x_3)/\theta_3),$$
$$\theta_1 x_1 + \theta_2 x_2 + x_3 < c_5 \quad \mu_2 > 0,$$

the third possible solution satisfying

$$c_2 - \theta_3 c_5 - (1 - \theta_1\theta_3)x_1 = (1 - \theta_2\theta_3)U_2'^{-1}(U_1'(x_1)),$$
$$x_1 = U_1'^{-1}(U_2'(x_2)),$$
$$x_2 = U_2'^{-1}(U_1'(x_1)),$$
$$\mu_2 = \frac{U_1'(x_1) - \theta_1 U_3'(x_3)}{1 - \theta_1\theta_3} > 0,$$
$$\mu_5 = \frac{U_3'(x_3) - \theta_3 U_1'(x_1)}{1 - \theta_1\theta_3} > 0,$$

and the fourth possible solution satisfying

$$U_3'\left(\frac{c_5 - \theta_2 c_2 - (\theta_1 - \theta_2)x_1}{1 - \theta_2\theta_3}\right) =$$
$$\frac{(\theta_2\theta_3 - 1)U_1'(x_1) - (\theta_1\theta_3 - 1)U_2'\left(\frac{c_2 - \theta_3 c_5 - (1-\theta_1\theta_3)x_1}{1-\theta_2\theta_3}\right)}{\theta_2 - \theta_1},$$
$$x_2 = \frac{c_2 - \theta_3 c_5 - (1 - \theta_1\theta_3)x_1}{1 - \theta_2\theta_3},$$
$$x_3 = c_5 - \theta_1 x_1 - \theta_2 x_2,$$
$$\mu_5 = \frac{U_1'(x_1) - U_2'(x_2)}{\theta_1 - \theta_2} > 0,$$
$$\mu_2 = \frac{\theta_2 U_1'(x_1) - \theta_1 U_2'(x_2)}{\theta_2 - \theta_1} > 0,$$

In this example, if $U_s(x_s) = \log x_s$, $\theta_s = \theta_0$, the flow rates in S2A are

$$x_1 = x_2 = c_2/2 \quad x_3 = c_5,$$

while when the settings satisfy

$$\theta_0 < \frac{3\theta_0}{1+2\theta_0^2} < \frac{c_2}{c_5} < \frac{2+\theta_0^2}{3\theta_0} < \frac{1}{\theta_0},$$

the flow rates in S2B are

$$x_1 = x_2 = \frac{c_2 - \theta_0 c_5}{2(1-\theta_0^2)}, \quad x_3 = \frac{c_5 - \theta_0 c_2}{1-\theta_0^2},$$

which are less than those in S2A.

Specially, setting the data packet size to 500 bytes and the ACK packet size to 40 bytes for all connections, and $c_2 = c_5 = 100$ Mb/s in Example 2, the flow rates x_1, x_2, and x_3 in S1A are 50, 50, and 100 Mb/s, while in S1B, they are 46.3, 46.3, and 92.6 Mb/s, respectively.

The paper by Ge and Tan [364] is the first attempt to investigate the NUM theory in the two-way flow scenario. The modeling and solution lead to some straightforward discoveries of a critical problem with network bandwidth allocation and performance optimization in the NUM framework to account for the impact of ACK packet flow.

Flow control approaches consist of two components: a source algorithm that dynamically adjusts rate (or window size) in the response to congestion in its path, and a link algorithm that updates, implicitly or explicitly, a congestion measure and sends it back to sources that uses that link. These algorithms, well known as the primal-dual algorithm, are all resulted from the NUM model. However, these algorithms unfortunately failed to address the impacts of ACK packet flow. With an extension of the usual NUM model being proposed, it opens the possibility of developing the new primal-dual algorithms for the two-way scenario by following the NUMtw model to take into account the ACK packet flow. Guided by the NUMtw model and theory, new congestion control approaches and protocols are then possibly extended for a general network with data packet flow and ACK packet flow.

11 Traffic Matrix Estimation

It is very useful to estimate traffic matrix (TM) from link measurements and routing information, especially for the tasks of capacity planning, traffic engineering, and network reliability analysis. However, this problem is ill-posed as it involves more unknowns than data and the challenge lies in this problem is subsequently its ill-posed nature. This chapter proposes a new method to estimate IP TM. By using a sophisticated matrix analyzing theory: singular value decomposition (SVD) and optimization technique, we are able to overcome the challenge of ill-posed nature of the problem. The inference of TM is described into an optimization problem and is then solved by calculating the SVD of the routing matrix and by applying the Lagrange multipliers method. We propose the mathematical model and solution procedure. As the methodology is based on SVD and Lagrange multiplier, we termed this proposal as SVDLM. Further, in order to capture the time-varying characteristic of traffic volume, we introduce the covariance matrix to estimate the TM for network and present a modified adaptive algorithm. We term this method as SVDLM-I. Theoretical analyses and numerical results are provided to demonstrate the accuracy and efficiency of these two methods.

11.1 INTRODUCTION

With the exponential increase in the size of the IP network, the TM estimation is an important index of network design and measuring. A TM is a representation of the volume of traffic that flows between origin–destination (OD) node pairs in a communication network. It will provide data support and determination basis for network topology structure design, link measurements, and routing information [368–371]. In recent years, IP network TM estimation problem has attracted more and more research efforts. The direct measurement approaches [372] for TM are generally unavailable because of the technique restriction, now the main approaches are inference approaches to estimate the TM. Generally, the relationship between the TM, the routing matrix, and the link counts can be described by a linear equation $Y = AX$. Therefore, the research efforts [125,126,373–380] in the area of the TM estimation have focused on the statistical inference problem about the traffic volume between OD pairs from Y and A.

In IP network, the amount of the OD pairs is usually much more than the link counts and it will lead to uncertainty of solution for equation $Y = AX$. Therefore, the challenge lies in the TM estimation problem is its high ill-posed nature.

Most of the reported efforts aforementioned have been focused on modeling method, which relies on statistical inference techniques. They may not be always efficient and workable in real networks due to lack of a mathematical tool to face the ill-posed challenge. We suggest a novel method to overcome this difficulty by using SVD and matrix transformations on the routing matrix, then to obtain the extreme value of the optimization problem as the solution for the TM estimation. Numerical results are provided to demonstrate the accuracy of this new method.

The rest of this chapter is organized as follows: Section 11.2 summarizes the related works. Then, Section 11.3 presents our methodology and main results and analyzes the computational complexity for the SVDLM algorithm. In Section 11.4, we introduce the covariance matrix to capture the time-varying nature of network and present the algorithms. This algorithm is based on SVDLM. We termed it as SVDLM-I. In Section 11.5, we present the simulation results that demonstrate the limitations of the conventional method and the advantages of our methods. Finally, conclusion is provided in Section 11.6.

11.2 RELATED WORK

The state-of-the-art research efforts in TM estimation have been focused on the modeling method, which relies on statistical inference techniques [373].

We herein summarize a variety of methods that have been proposed so far to address the subject of TM estimation. In [374], the authors propose changing the IGP link weights in order to obtain more information to reduce the uncertainty in the estimates. Although this technique is powerful in collapsing errors, it requires carriers to alter their routing in order to obtain a TM. It is not clear that carriers are willing to do this. The authors of [375] have introduced the optimization approach; they pose the problem as a linear program (LP) and have attempted to compute X directly. In [375], a reduction of the feasible solutions space is achieved by imposing linear constraints on X.

A time-varying statistical approach has been proposed in [376] to estimate an evolving TM over time by using link counts at router interfaces. It is referred to as the EM methods in [377] since the core of the approach is based on an expectation maximization algorithm to compute the maximum likelihood estimates. The Bayesian method [378] and the EM method all make assumptions about the distribution of X_j, i.e. the modeling assumptions about OD flows. These methods use $E(X/Y)$ as their estimations for X. The LP method uses the link counts Y as hard constraints, while the inference methods use Y to compute conditional distributions. In [377], the authors evaluate the previous three methods, and outline their corresponding advantages and disadvantages.

The authors of [375] apply the gravity models in telephone networks to estimate Internet OD volume. Gravity models are based on the assumption that the Origin and the Destination of a traffic flow are independent. In this model, the volume of traffic flow with Origin O and Destination D is proportional to the total volume of traffic with Source O, and to the total volume of traffic with Destination D. Another method based on this OD independence assumption and depended on information theoretic has been proposed in [379]. This method can be seen as a generalization of gravity models [375]. Furthermore, another information-theoretic approach has been introduced by the authors of [380] to estimate point-to-point and point-to-multipoint traffic matrices. Tan et al. [125,126] have proposed methods to estimate TM which fits the monitored link loads and is nearest to the initial TM obtained by assuming the Gaussian distribution.

Our approach, in contrast, is aimed at discarding the disadvantage of the existing methods in generating the prior, improving the Gaussian distribution to generate the prior. Our approach is to solve the TM estimation problem by SVD and optimization theory. It will overcome the ill-posed restrictions by SVD, and matrix transformations for the routing matrix and then to obtain the extreme value of the optimization problem leading to the solution of the TM estimation.

11.3 METHODOLOGY AND MAIN RESULTS

In this section, we first state the problem of IP traffic estimation in detail and then discuss how to generate the prior. Second, we present our new method and the main results, and finally we valuate the performance of our approach.

11.3.1 PROBLEM STATEMENT

Let a network be described by a graph that has p nodes. There will be $n = p \times (p-1)$ OD pairs. Based on this fact, the TM X is introduced to describe the volume of traffic that flows between OD pairs, $X_{i,j}$ being the element of this matrix. It is more convenient to transfer this TM into a vector representation. So, we number the OD pairs from 1 to n and let $X_j (1 \leq j \leq n)$ denote the amount of data transmitted by OD pair j. Let $Y = (Y_1, \ldots, Y_m)^T$ be the vector of link counts, where Y_l represents the link count for link l, and m denotes the total number of links in the network. Through an $m \times n$ routing matrix $A = [a_{ij}]_{m \times n}$, the vectors X and Y can be related. Element $a_{i,j} = 1$ if link i belong to the path associated with OD pair j, otherwise $a_{i,j} = 0$. The relation between link counts, routing information, and TM can be represented as the linear matrix equation

$$Y = AX \tag{11.1}$$

Given the ill-posed nature of the TM estimation problem, to determine the solution of Equation 11.1, it is required to make use of additional information, i.e., the prior X'. X' can be an initial TM. So, the "best" solution to the TM estimation problem will be the one that minimizes the deviation between X' and the solution X under the constraint imposed by Equation 11.1.

We suggest that, the initial TM X' can be derived from the addition information such as link measurements and routing information, it will be described later, and the Euclidean distance can serve as this metric–Euclidean distance function measures the "as-the-crow-flies" distance, which is the nearest distance between two points.

The Euclidean squared distance does not take the square root; nevertheless, in essence, there is no difference between Euclidean distance and Euclidean squared distance when finding the solution that minimizes the distance between two points. Therefore, we choose the Euclidean squared distance as the metric for the sake of simplicity.

For a network with the known link counts $Y \in R^m$ and the routing matrix $A \in R^{m\times n}$, the TM estimation problem can be described into the following optimization problem:

$$OP_1 : \min \left(X - X'\right)^T \left(X - X'\right) \tag{11.2}$$

$$\text{Subject to: } AX = Y,$$

where X' is the initial TM, which is termed as prior. In the following section, we will give our method to generate the prior.

11.3.2 PRIOR GENERATING

By the Bayesian approach, it is assumed that the prior follows a Poisson distribution with the mean λ_i, and the random values are drawn for each $X_i' \sim Poisson(\lambda_i)$. In the EM method, the authors assume that the prior follows a Gaussian distribution, they generate μ_i to uniformly distribute in an interval and assign the same variance for all OD pairs, i.e., σ_i is a constant. The random values are drawn for each $X_i' \sim N(\mu_i, \sigma_i)$. In [125,126], the author uses the link measurements and routing information to get the prior. In this method, μ_i is the mean link count of every link and each $X_i' \sim N(\mu_i, \sigma_i)$. There, σ_i is the 40 which is the same as [377] for the convenience of comparing the results with it.

We propose a new method to generate prior X'. Assume that the prior used in our setup follows a Gaussian distribution because the Gaussian distribution is effective than the Poisson distribution when generating the prior [377]. Let $sum\_A_i (1 \le i \le m)$ denotes the sum of row i in the matrix A and $sum\_A_j (1 \le j \le n)$ denotes the sum of column j in the matrix A. The following steps are listed to determine the prior mean rates:

Step 1: For each $j(1 \le j \le n)$, get $i(1 \le i \le m)$, if $a_{i,j} = 1$ in matrix A, then $p = i(1 \le p \le m)$.
Step 2: Let $sum\ \mu_j' = \sum_p \left(Y_p / sum\_A_p\right)$.
Step 3: $\mu_j = sum\ \mu_j' / sum\ A_j$.

We also assign the same variance for all OD pairs, $\sigma_j = 40$, as assigned in [126,377]. Therefore, the prior can be generated by the following Gaussian distribution:

$$X_j' \sim N\left(\mu_j, \sigma_j\right) \tag{11.3}$$

Compared with [126], our method takes account of the difference for the traffic volume between the OD pairs. In [126], μ_j is same for each OD pair, so our method is efficient to generate the prior, and then get the solution for the estimation problem.

11.3.3 METHODOLOGY AND MAIN RESULTS FOR SVDLM

Let $X = [X_1, X_2, \ldots, X_n]^T$, in which element is $X_j (1 \leq j \leq n)$ and denote the traffic volume between OD pairs. Let the prior $X' = [X_1', X_2', \ldots, X_n']^T$, in which element is $X_j' (1 \leq j \leq n)$. So, the optimizing problem of OP_1 can be formulated as

$$OP_2 : \min\left[(X_1 - X_1')^2 + (X_2 - X_2')^2 + \cdots + (X_n - X_n')^2\right] \tag{11.4}$$

$$\text{Subject to: } AX = Y.$$

Get SVD for the routing matrix A and let k be the rank of the routing matrix $A(k \leq m)$, so A can be expressed as

$$A = U\begin{bmatrix} \Sigma & 0 \\ 0 & 0 \end{bmatrix} V^T, \tag{11.5}$$

where U is an $m \times m$ unitary matrix and V is an $n \times n$ unitary matrix. $\Sigma = \text{diag}(\sigma_1, \sigma_2, \ldots, \sigma_k)$, $\sigma_1 \geq \sigma_2 \geq \cdots \geq \sigma_k > 0$ is the positive singular value of matrix A. The matrices U and V are orthogonal matrices and satisfy

$$\begin{aligned} U^T U &= U^{-1} U = I_m \\ V^T V &= V^{-1} V = I_n, \end{aligned} \tag{11.6}$$

where I_m is an $m \times m$ identity matrix and I_n is an $n \times n$ identity matrix. Let $Z = V^T X$, so $X = VZ$ and $X' = VZ'$. We then have

$$\begin{aligned} (X - X')^T (X - X') &= (VZ - VZ')^T (VZ - VZ') \\ &= (Z - Z')^T V^T V (Z - Z') \\ &= (Z - Z')^T (Z - Z'), \end{aligned} \tag{11.7}$$

$$\begin{aligned} AX = Y \Rightarrow U\begin{bmatrix} \Sigma & 0 \\ 0 & 0 \end{bmatrix} V^T X &= Y \\ \Rightarrow U\begin{bmatrix} \Sigma & 0 \\ 0 & 0 \end{bmatrix} Z &= Y \\ \Rightarrow \begin{bmatrix} \Sigma & 0 \\ 0 & 0 \end{bmatrix} Z &= U^T Y. \end{aligned} \tag{11.8}$$

Therefore, through the proposed matrix transformations, the original TM estimation problem OP_1 is reduced to the following optimization problem OP_3:

$$OP_3 : \min (Z - Z')^T (Z - Z') \tag{11.9}$$

$$\text{Subject to: } \begin{bmatrix} \Sigma & 0 \\ 0 & 0 \end{bmatrix} Z = U^T Y.$$

Because the rank of matrix A is k, so we partition Z and Z' in the form of

$$\begin{aligned} Z &= [Z_1^T, Z_2^T]^T \\ Z' &= [(Z_1')^T, (Z_2')^T]^T, \end{aligned} \tag{11.10}$$

respectively, where Z_1 and Z_1' are the vectors of k dimension, and Z_2 and Z_2' are the vectors of $n - k$ dimension. Now, the main result of the optimization problem OP_3 can be stated as the following:

Theorem 11.3.1

The optimization problem OP_3 has the following optimized solution:

$$Z = [Z_1^T, Z_2^T]^T,$$

where $Z_1 = \Sigma^{-1}\left[U^T Y\right]_k$, and $Z_2 = Z_2'$; $\left[U^T Y\right]_k$ is the first k-dimension elements of $U^T Y$.

Proof: Apply the method of Lagrange multipliers based on optimization technique, we can obtain

$$L(Z_1, Z_2, \lambda) = (Z_1 - Z_1')^2 + (Z_2 - Z_2')^2 + \lambda^T \left[\begin{bmatrix} \Sigma Z_1 \\ 0 \end{bmatrix} - U^T Y\right].$$

Differentiating L with respect to each element of the vectors Z_1 and Z_2 and setting these derivatives equal to zero give

$$\frac{\partial L}{\partial Z_1} = 2(Z_1 - Z_1') + \begin{bmatrix} \Sigma \\ 0 \end{bmatrix}^T \lambda = 0 \tag{11.11}$$

$$\frac{\partial L}{\partial Z_2} = 2(Z_2 - Z_2') = 0 \tag{11.12}$$

$$\begin{bmatrix} \Sigma Z_1 \\ 0 \end{bmatrix} - U^T Y = 0. \tag{11.13}$$

From Equation 11.12, we obtain

$$Z_2 = Z_2' \tag{11.14}$$

From Equation 11.13, $\begin{bmatrix} \Sigma Z_1 \\ 0 \end{bmatrix} = U^T Y \Rightarrow Z_1 = \Sigma^{-1}\left[U^T Y\right]_k$, where $\left[U^T Y\right]_k$ is the first k-dimension elements of $U^T Y$ and the next $n-k$ dimension elements of $U^T Y = 0$.

So

$$Z_1 = \Sigma^{-1}\left[U^T Y\right]_k. \tag{11.15}$$

This finishes the proof.

The proof procedure provides the theory basis for the SVDLM algorithm. U, V, and Σ can be obtained by the matrix A by the SVD method. The optimization solution Z of Equation 11.9 can be calculated by Theorem 11.3.1. We finally obtain the required TM by

$$X = VZ. \tag{11.16}$$

■

11.3.4 SVDLM ALGORITHM DESCRIPTION

The following steps provide an overview of the SVDLM algorithm:

Step 1: Obtain U, V, and Σ by the SVD method to the matrix A.
Step 2: Get X' by Equation 11.3.
Step 3: Calculate $Z' = V^T X'$, and partition Z into Z_1' and Z_2', where Z_1' are the vectors of first k dimension and Z_2' are the vectors of next $n-k$ dimension.
Step 4: Calculate Z_1 and Z_2 by Equations 11.14 and 11.15.
Step 5: Construct Z with Z_1 and Z_2.
Step 6: Get the estimation value X by Equation 11.16.

11.3.5 COMPUTATIONAL COMPLEXITY

To evaluate the performance of the SVDLM method, we analyze the computational complexity of it in this section.

Theorem 11.3.2

The computational complexity of MPLM is $O(m^3)$, where m is the number of link counts.

Proof: From Equations 11.14 and 11.15, we can see that the formula is made up of multiplication, inversion, and transposition of the matrices or vectors. The computational complexity of the multiplication operation for matrix $A_{m\times n}$ and $A_{n\times k}$ is $O(m\times n\times k)$, the computational complexity of the inversion operation for matrix $A_{n\times n}$ is $O(n^3)$, and the computational complexity of the transposition operation for matrix $A_{m\times n}$ is $O(m\times n)$. Based on these, we easily proved that to compute Z_1, one needs $O(2m^2+k^2+k^3)$ computations. Z_2 can be directly assigned a value so that the computational complexity is omitted. The total computation is $O(2m^2+k^2+k^3)$. Recall that m is the number of link counts and k is the rank of matrix A, k is less than or equal to m, so the computational complexity of our algorithm is $O(k^3)$.

In [125], the author proposes the MPLM method to estimate the TM, its computational complexity is $\left((k+1)^2 n+k^3+3mk^2+3mk+2m\right)$. Compared with this method, our method's computational complexity is less than it distinctly.

The computational complexity of the EM method [377] is at least proportional to N_e^5, where N_e is the number of edge nodes. To the best of our knowledge, the best computational complexity, among that of the known TM estimation algorithms reported in the literature [377] so far is $O\left(N_e^2\right)$ and the computational complexity of MPLM is proportional to $O\left(N_e^2\right)$ [125]. So, the computational complexity of SVDLM clearly demonstrates the attraction of the algorithm in real network implementation. ■

11.4 IMPROVED ALGORITHM FOR TIME-VARYING NETWORK

Using SVDLM algorithm, the OD matrix can be estimated from the link values as a solution of the linear system $Y=AX$ and this can be treated as stationary estimation. In practice, on a backbone network, the OD demands are time varying and periodicity. There are, for example, some day/night effects, as well as some variations of the traffic demand over the successive days (works/nonworking day for example).

In order to capture the time-varying and periodicity essence, we first introduce the concept, the covariance matrix, and then estimate the TM based on SVDLM algorithm.

11.4.1 COVARIANCE MATRIX

Based on the network measurements and studies, in an IP network, especially IP backbone network, the TM has approximate for every day, every week, every month, or even every year. For example, the traffic volume which passed through a certain router on 8:00 am in Monday morning of a week may be very approximate to that passed through the same router at the same time in the previous week. It suggests that the TM has an obvious time correlations and has a period varying. Using the history data for the additional information, we can attain the more accurate estimation value for the OD traffic.

Next, we will analyze the covariance matrix for the TM.

If $X(t), t=1,\ldots,T$, is a series solutions of the linear system $Y=AX$, where t refers to a given period of time of typical duration one hour or one day. The SNMP measurements routinely provide link volume at a high rate. Based on SVDLM algorithm, we can obtain the series corresponding $X(t)$, so we can suppose the TMs $X(1), X(2),\ldots,X(t)$ are known. The covariance matrix for the TM can be expressed by the samples of the TM, $X(1), X(2),\ldots,X(t)$.

$$C_M = \frac{1}{T}\sum_{i=1}^{T}\left[X(i) - \overrightarrow{X}\right]\left[X(i) - \overrightarrow{X}\right]^T, \tag{11.17}$$

where $\overrightarrow{X} = \frac{1}{T}[X(1)+X(2)+\cdots+X(T)]$, and

$$C_M = C_M^T. \tag{11.18}$$

Examining the covariance matrix CM of flow matrix, the covariance matrix, elements reflect the correlation properties of network flow between each OD at different times, and diagonal elements manifest the time correlation of flow matrix, the nondiagonal elements manifest the spatial correlation of the flow matrix, so using sample covariance matrix, C_M can effectively capture spatial–temporal correlation of flow matrix.

11.4.2 SVDLM-I ALGORITHM DESCRIPTION FOR TIME-VARYING NETWORK

The following steps provide an overview of the SVDLM-I algorithm:

Step 1: SNMP measurements $Y(t)$ are stored, $t = 0, 1, 2, \ldots, T, T+1, \ldots$.
Step 2: Calculate the estimation of TM for $X(0)$ using SVDLM algorithm.
Step 3: Suppose there are T samples in the TM to calculate, we obtain the covariance matrix.
Step 4: Get the $X(T+1)$, the solution for $X(T+1)$ can be obtained by OP_3:

$$OP_3 : \min\left(X - C_M X(t)\right)^T\left(X - C_M X(t)\right) \tag{11.19}$$

Subject to: $AX = Y$.

Step 5: From Equation 11.16, we can obtain $X(T+1)$

$$\begin{cases} X(t+1) = VZ \\ Z_1 = \Sigma^{-1}\left[U^T Y\right]_k \\ Z_2 = Z_2' \\ Z = [Z_1^T, Z_2^T]^T \\ Z' = [(Z_1')^T, (Z_2')^T]^T \\ Z' = V^{-1} C_M X(t). \end{cases} \tag{11.20}$$

From the procedure of the solution for OP_3, we find that $X(T+1)$ obtains a more accurate TM estimation by using the covariance matrix and the component $X(T)$.

11.5 NUMERICAL RESULTS

To illustrate the performance of the proposed SVDLM method, we use the same four-node topology network as [377], which is depicted in Figure 11.1. We will generate the numerical results by using SVDLM method and compare them with those that are reported in this literature. As shown in Figure 11.1, we use the same link counts, i.e.,

$$Y = [318, 601, 559, 903, 882, 1154, 851]^T.$$

We deduce the matrix A from the topology.

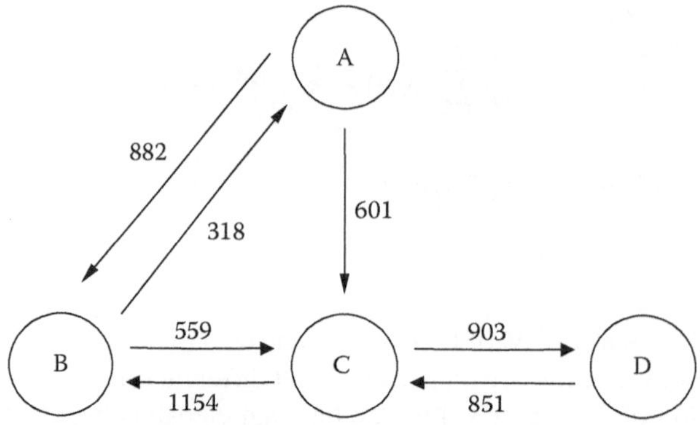

Figure 11.1 A network with four-node topology.

$$
A = \begin{array}{c}
\begin{array}{cccccccccccc} AB & AC & AD & BA & BC & BD & CA & CB & CD & DA & DB & DC \end{array} \\
\left[\begin{array}{cccccccccccc}
1 & 0 & 0 & 0 & 0 & 0 & 0 & 0 & 0 & 0 & 0 & 0 \\
0 & 1 & 1 & 0 & 0 & 0 & 0 & 0 & 0 & 0 & 0 & 0 \\
0 & 0 & 0 & 0 & 1 & 1 & 0 & 0 & 0 & 0 & 0 & 0 \\
0 & 0 & 1 & 0 & 0 & 1 & 0 & 0 & 1 & 0 & 0 & 0 \\
0 & 0 & 0 & 1 & 0 & 0 & 1 & 0 & 0 & 1 & 0 & 0 \\
0 & 0 & 0 & 0 & 0 & 0 & 1 & 1 & 0 & 1 & 1 & 0 \\
0 & 0 & 0 & 0 & 0 & 0 & 0 & 0 & 0 & 1 & 1 & 1
\end{array}\right]
\end{array}
\begin{array}{l} \\ AB \\ AC \\ BC \\ CD \\ BA \\ CB \\ DC. \end{array}
$$

After doing SVD to matrix A, we can obtain U, V, and Σ.

$$
U = \begin{bmatrix}
0 & 0 & 0 & 0 & 0 & 0 & 1 \\
0 & -0.4082 & 0.7071 & 0 & 0 & -0.5774 & 0 \\
0 & -0.4082 & -0.7071 & 0 & 0 & -0.5774 & 0 \\
0 & -0.8165 & 0 & 0 & 0 & 0.5774 & 0 \\
-0.5 & 0 & 0 & 0.7071 & 0.5 & 0 & 0 \\
-0.7071 & 0 & 0 & 0 & -0.7071 & 0 & 0 \\
-0.5 & 0 & 0 & -0.7071 & 0.5 & 0 & 0
\end{bmatrix}
$$

$$
V = \begin{bmatrix}
0 & 0 & 0 & 0 & 0 & 0 & 1 & 0 & 0 & 0 & 0 & 0 \\
0 & -0.20 & 0.5 & 0 & 0 & -0.57 & 0 & -0.23 & 0.22 & -0.33 & -0.37 & -0.13 \\
0 & -0.61 & 0.5 & 0 & 0 & 0 & 0 & 0.23 & -0.22 & 0.33 & 0.37 & 0.13 \\
-0.19 & 0 & 0 & 0.5 & 0.46 & 0 & 0 & 0.35 & -0.25 & -0.49 & 0.14 & -0.20 \\
0 & -0.20 & -0.5 & 0 & 0 & -0.57 & 0 & 0.29 & 0.42 & 0.07 & 0.32 & 0.03 \\
0 & -0.61 & -0.5 & 0 & 0 & 0 & 0 & -0.29 & -0.42 & -0.07 & -0.32 & -0.03 \\
-0.46 & 0 & 0 & 0.5 & -0.19 & 0 & 0 & -0.39 & 0.12 & -0.02 & 0.14 & 0.54 \\
-0.27 & 0 & 0 & 0 & -0.65 & 0 & 0 & 0.58 & -0.12 & -0.15 & -0.33 & 0.08 \\
0 & -0.40 & 0 & 0 & 0 & 0.57 & 0 & 0.05 & 0.64 & -0.25 & -0.05 & -0.10 \\
-0.65 & 0 & 0 & 0 & 0.27 & 0 & 0 & 0.04 & 0.12 & 0.52 & -0.29 & -0.34 \\
-0.46 & 0 & 0 & -0.5 & -0.19 & 0 & 0 & -0.23 & -0.12 & -0.34 & 0.48 & -0.28 \\
-0.19 & 0 & 0 & -0.5 & 0.46 & 0 & 0 & 0.18 & 0 & -0.18 & -0.18 & 0.63
\end{bmatrix}
$$

$$
\Sigma = \begin{bmatrix}
2.6131 & 0 & 0 & 0 & 0 & 0 & 0 \\
0 & 2 & 0 & 0 & 0 & 0 & 0 \\
0 & 0 & 1.4142 & 0 & 0 & 0 & 0 \\
0 & 0 & 0 & 1.4142 & 0 & 0 & 0 \\
0 & 0 & 0 & 0 & 1.0824 & 0 & 0 \\
0 & 0 & 0 & 0 & 0 & 1 & 0 \\
0 & 0 & 0 & 0 & 0 & 0 & 1
\end{bmatrix}.
$$

According to the SVDLM-I algorithm, we obtain our results after generating the prior and replacing the variables in Equations 11.14 through 11.16 with the actual values of the parameters.

In [377], the authors performed the simulations for two groups of data. One assumes that the original TMs follow a Poisson distribution. The authors use Bayesian method to obtain the estimations of TMs, and the results are shown in the Column 2 of Table 11.1. The other assumes that the original TMs follow a Gaussian distribution and the authors use EM method to obtain the TMs, results of which are shown in the Column 2 of Table 11.2. Herein, we use our proposed SVDLM method to obtain the estimations of TMs for the above two scenarios, and our results are listed in the third column of Tables 11.1 and 11.2, respectively.

From Tables 11.1 and 11.2, the SVDLM method shows a substantial improvement over the Bayesian method and the EM method. By the Bayesian method, 3 of the 12 OD pairs have the error over 20%, and 4 of them have the error between 10% and 20%. However, there are all OD pairs

Table 11.1
Estimation of TMs Using Poisson Original TMs

Prior	Bayesian		SVDLM	
(Poisson)	Estimated TM	Error (%)	Estimated TM	Error (%)
AB: 318	318	0	318.0000	0
AC: 289	342	18	281.5000	2.60
AD: 312	259	17	319.5000	2.40
BA: 294	334	14	288.2500	1.96
BC: 292	310	6	268.5000	8.05
BD: 267	249	7	290.5000	8.80
CA: 305	291	5	318.7750	4.52
CB: 289	361	25	284.5250	1.55
CD: 324	395	22	293.0000	9.57
DA: 283	257	9	274.9750	2.84
DB: 277	245	12	275.7250	0.46
DC: 291	349	20	300.3000	3.20

Table 11.2
Estimation of TMs Using Gaussian Original TMs

Prior	EM		SVDLM	
Gaussian	Estimated TM	Error (%)	Estimated TM	Error (%)
AB: 318.65	318.65	0	318.6500	0
AC: 329.48	286.98	13	310.2802	5.83
AD: 277.18	318.36	15	300.7198	8.49
BA: 298.14	298.14	0	275.2642	7.67
BC: 354.81	360.97	1.6	342.2435	3.54
BD: 355.39	347.94	2	359.7565	1.23
CA: 327.20	317.34	3	307.7951	5.93
CB: 330.04	373.65	13	353.8231	7.21
CD: 253.01	217.32	14	227.5237	10.07
DA; 320.50	329.07	3	306.9407	4.23
DB: 291.52	246.60	15	276.4411	5.17
DC: 310.40	344.82	11	287.6182	7.34

below 10% under the SVDLM method. The average error was 13% for the Bayesian method, while 3.83% for the SVDLM method. By the EM method, six OD pairs have the error over 10%, while there are only one OD pair having the error between 10% and 15% and all the others below 10% under the SVDLM method. The average error was 7.6% for the EM method, and 5.56% for ours.

We further plot those numerical results of Tables 11.1 and 11.2 in Figures 11.2 and 11.3, respectively. In these two figures, the vertical axis represents the traffic matrices and the horizontal axis denotes the 12 OD pairs orderly, each OD pair having 5 units space from each other. These two figures clearly demonstrate the advantage of our method: the curve of SVDLM follows that of the original TMs well, while the Bayesian and EM curves have great deviations from the original TMs.

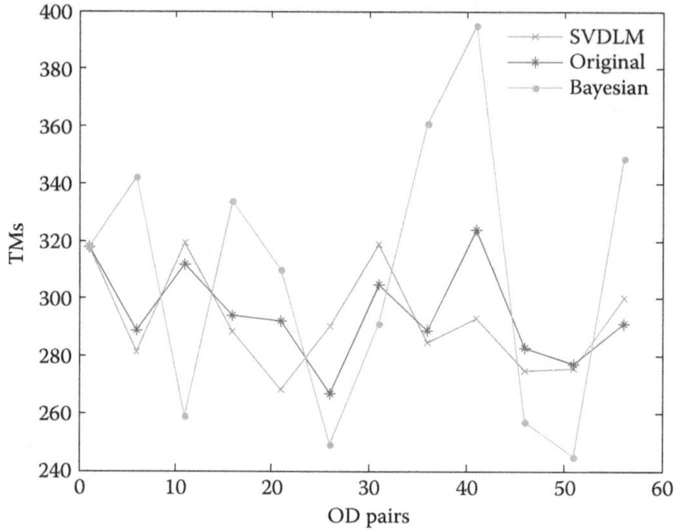

Figure 11.2 Estimated TMs with original TMs of Poisson distribution.

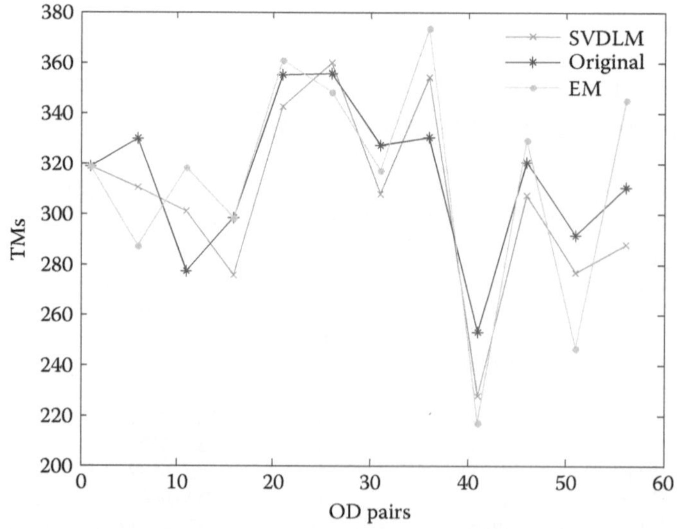

Figure 11.3 Estimated TMs with original TMs of Gaussian distribution.

In [125], the authors use the link measurements and routing information to get the prior and μ_i are the mean link counts of every link and each $X_i^{\prime} \sim N(\mu_i, \sigma_i)$. Our method takes account of the difference for the traffic volume between the OD pairs.

Table 11.3 gives the numerical results to illustrate that our method for generating prior is more reasonable than MPLM.

Figure 11.4 clearly demonstrates the advantage of our method: the error wave of SVDLM is less than those of MPLM.

To illustrate the advantage of SVDLM method further, we use these two methods to generate the prior in several times and get the average error. Table 11.4 gives the numerical results with

Table 11.3
Comparison of SVDLM and MPLM

Prior MPLM	MPLM		Prior SVDLM	SVDLM	
	Estimated TM	Error (%)		Estimated TM	Error (%)
230.7209	318.0000	37.83	292.9162	318.0000	8.56
367.1864	310.1252	15.54	318.1060	346.9156	9.06
298.0211	290.8748	2.40	240.6444	254.0844	5.59
230.8199	274.4813	18.92	285.6706	286.7990	0.39
309.9913	261.1491	15.76	219.2950	215.5125	1.72
296.7782	297.8509	0.36	362.6396	343.4875	5.28
269.8462	305.2146	13.11	286.5711	285.9554	0.21
312.3825	304.0896	2.65	337.5484	335.8043	0.52
264.3595	314.2743	18.88	320.7977	305.4281	4.79
238.739	302.3041	26.63	325.0038	309.2456	4.85
222.4879	242.3917	8.95	239.8812	222.9947	7.04
278.1076	306.3042	10.14	333.9022	318.7597	4.54

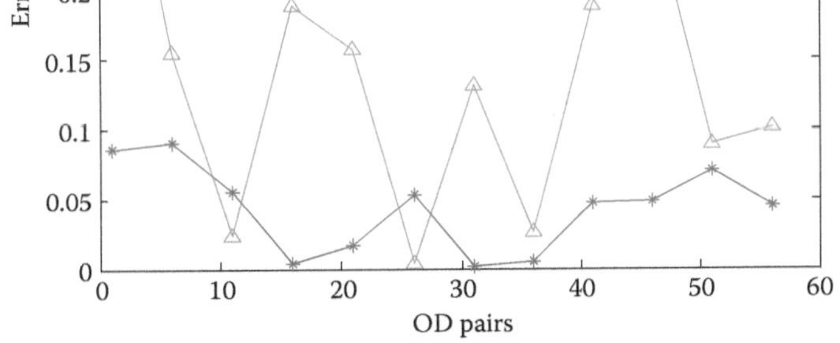

Figure 11.4 Comparison of the error wave between SVDLM and MPLM.

two methods in 300 times. Figures 11.5 and 11.6 give the average error for these two methods, respectively.

Table 11.4 shows that the number of the average error less than 5% for SVDLM is 9%, by contract, MPLM is 6.33%. It tells us that the SVDLM method can generate more times prior with smaller average error. Figures 11.5 and 11.6 illustrate this point more clearly, the range of average error wave for MPLM is bigger, and it is from 0 to 30%. Whereas the range of average error wave for SVDLM is from 0% to 15%.

In the following, we will show the performance of our two methods to estimate the time-varying TMs.

As for the original Gaussian distributed TMs, six OD pairs have the error over 10% for the stationary estimation with random prior, and five OD pairs have the error over 10% for the time-varying estimation with $X(T)$ and covariance matrix as prior. The average error is 12.56% for the former one and 8.92% for the latter one.

To gain intuition, we plot the numerical results in Figure 11.7. From Figure 11.7, we can conclude that it is better to use $X(T)$ and covariance matrix as prior when estimating $X(T+1)$. Given the fact that the OD demand values have time-varying and periodicity property at different timescales, when estimating $X(T+1)$, it is reasonable to choose $X(T)$ and covariance matrix as prior.

From the numerical results, it indicates that our method of generating the prior and calculating the estimated results is more efficient.

Table 11.4
Average Error Comparison between SVDLM and MPLM

Average error (%)	<5	<7	<9	<11	<13
MPLM (%)	6.33	20	44	72.33	88.33
SVDLM (%)	9	26	58.33	80,67	92

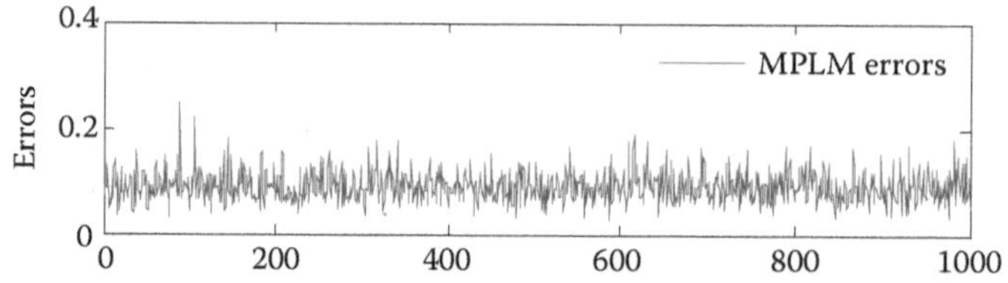

Figure 11.5 The average error of MPLM.

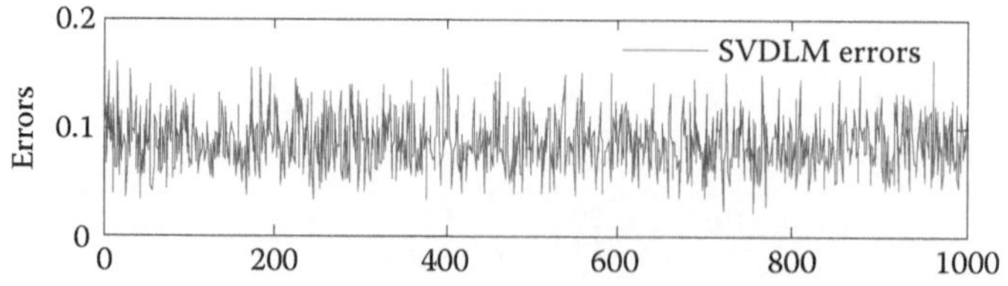

Figure 11.6 The average error of SVDLM.

Table 11.5
Comparison of SVDLM-I with SVDLM

Prior	SVDLM (with Random Prior)		SVDLM-I ($X(T)$ as Prior)	
	Estimated TM	Error (%)	Estimated TM	Error (%)
AB: 331.0076	347.0000	4.83	347.0000	4.83
AC: 310.3029	284.8203	8.21	309.1989	0.36
AD: 285.5619	342.4797	19.93	318.1011	11.39
BA: 295.5393	226.9325	23.21	289.9704	1.88
BC: 275.4103	303.7049	10.27	259.8392	5.65
BD: 300.5921	320.6951	6.69	364.5608	21.28
CA: 274.0183	294.9580	7.64	295.4183	7.81
CB: 301.0143	306.8230	1.93	338.0663	12.31
CD: 299.4048	343.8252	14.84	324.3382	8.33
DA: 287.4579	351.9096	22.42	288.4112	0.33
DB: 269.4548	294.2095	9.19	326.0041	20.99
DC: 328.7083	257.7809	21.58	289.4846	11.93

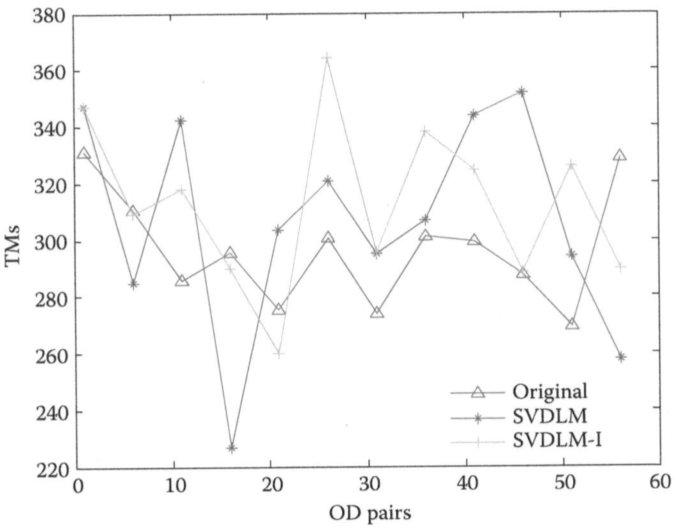

Figure 11.7 Comparison of SVDLM-I with SVDLM.

11.6 CONCLUSIONS

In this chapter, we have proposed a new approach to estimate the TM. We apply the optimization technique to deduce the TM, with the key mathematical theory being the matrix SVD method. Through both theoretical analysis and numerical results, it is demonstrated that our SVDLM and SVDLM-I approaches, calculate the TM more accurately and more efficiently.

12 Utility-Optimized Aggregate Flow Level Bandwidth Allocation

The usual utility-aware network bandwidth allocation problems and solutions are aimed to maximize the aggregate utility subject to link capacity constraints. They are formulated and solved by using the individual flow rate vector. Due to the architecture of the current networks like the Internet, the individual flow rates are generally not measurable directly at the routers for the network service provider. However, the aggregate flow rates are more convenient to obtain and to adjust in some situations.

In this chapter, we study the utility maximization problems for communication networks but from a router-level (aggregate flow level) bandwidth allocation standpoint. It should be noted that when we are talking about the router-level bandwidth allocation, we also mean the aggregate flow-level bandwidth allocation for wire-lined networks, including the Internet, as in this case flows are aggregated at the routers. However, once wireless networks are concerned, it is more precise to use the term "the aggregate flow level bandwidth allocation." By using the generalized matrix inverse, we propose a general model of utility-optimized router-level bandwidth allocation and its solution, where the objective function and the constraints are formulated in terms of the aggregate flow rate vector rather than from the individual flow rate vector as in the usual optimization problem. With the same aim to maximize the total utility of a network, we find that, the new proposed models are equivalent to the usual optimization problem in the sense that they lead to the same optimum and their solutions satisfy the given routing scheme. We also discuss the special cases where the routing matrix is of full row rank and where there is one single-hop flow in every link in the network. We suggest a direct application to IP-based virtual private network (IP-VPN) of the latter case. We present the mathematical models and solution procedures that lead to the utility-optimized aggregate flow rate vector and further illustrate them by numerical examples. We believe our bandwidth allocation approach is more appropriate for deployment in communication networks than the usual solutions. We discuss the direct application of the utility-optimized aggregate flow-level bandwidth allocation model in resource allocation for the optical burst switching networks in the next chapter.

12.1 INTRODUCTION

Following Kelly's pioneering works [104], there have been numerous researches (see, e.g., [100,105,112,115,119,124,128,132,190,199]) to provide utility-aware network bandwidth allocation schemes based on optimization theory. The objective of these schemes is to maximize the source aggregate utility subject to link capacity constraints. Specifically, a bandwidth allocation policy can be expressed in terms of a utility function $U_i(x_i)$ in the sense that the desired bandwidth allocation $x^* = (x_i^*$, all sources $i)$ solves the utility maximization problem:

$$\max_x \sum_i U_i(x_i) \text{ subject to link capacity constraints.} \tag{12.1}$$

It is seen that this traditional solution was to allocate bandwidth directly among the individual flows in a network. By "flow," we also refer to source–destination (SD) pair, or original–destination (OD) traffic in the sequels.

Recently, the authors in [116] have introduced the theory of *mechanism design* to the study of autonomous networks, which is subsequently used by [103,123] to differentiate QoS in bandwidth sharing problems. Related works also include studies [131] on bandwidth allocation in selfish overlay networks. In [131], the authors propose to use a market-driven approach to regulate the behaviors of selfish nodes. However, the main aim in these papers is the provision of the autonomous nodes but not to maximize the total utility of the network.

In a network, routers and switches direct traffic by forwarding data packets between nodes according to a routing scheme. Flows are subsequently aggregated at the relevant links. From this standpoint, all the known utility-optimized bandwidth allocation methods are of an individual flow level basis.

There are several reasons to study the aggregate flow-level bandwidth allocation:

- One advantage of aggregate flow-level bandwidth allocation over that of flow level may be that it facilities implementation in real networks. For a typical network, the set of traffic between all OD pairs known as traffic matrix is not readily available to regional registries or Internet Service Providers, but the aggregated link traffic measurements are. The aggregate flow rates at links are thus easily adjustable due to the fact that, current link counts are readily available through the Simple Network Management Protocol (SNMP) which is provided by nearly all commercial routers and OS; the traffic matrix or OD counts, however, are not collected directly by most local area networks (LANs). To network administrators, measurements of individual flow rate, in this case typically require specialized router software and hardware dedicated to data collection.
- In current network design practices, communication networks should be designed to cope with the frequent variations of load so as to provide a level of reliability [113]. In order to guarantee a level of reliability, networks are expected to hold alternate paths and sufficient additional capacity to carry the traffic even if a link or a node fails. In the case of unpredicted traffic shifts, dynamic link capacity adjustment is especially important. Aggregate flow-level bandwidth allocation is seen to be important in attaining the above design objectives.
- An ideal bandwidth allocation scheme needs to be both flexible enough to accommodate a wide variety of traffic sources, and enforceable so that a well-behaved source can retain its allocation in spite of the actions of malicious or greedy sources. While the traditional flow-level bandwidth allocation has been successful and popular to a large extent, it is not robust against the behavior of greedy or malicious users. When compared to individual flow-level bandwidth allocation, aggregate flow level bandwidth allocation can be more easily incorporated with the ability of isolating user behavior [102].
- With the presence of very high-bandwidth traffic sources such as video traffic, the network must be prepared for increase in load fluctuations. In the one hand, a single traffic of this kind will be able to pump data into a network at a high rate to consume a large fraction of the peak bandwidth of a network link. On the other hand, the traffic source can complete its data transmission in a short time because of the high transmitting rate. Once the transmission is complete, the network load will suddenly drop. In addition, for high-speed networks, there is a problem of increased mismatches in bandwidth. Under these situations, aggregate flow-level bandwidth allocation is imperative [108] to provide rapid feedback to allow a network to adapt to load changes and maximizes its performance.
- The complete Internet consists of a large number of interconnected autonomous systems (ASs) each of which constitutes a distinct routing domain. Such AS are usually run by a single organization such as a company or university. Within an AS, routers communicate with each other using one of the several possible intradomain routing protocols also known as interior gateway protocols. ASs are connected via gateways, these exchange information using interdomain routing protocol also known as exterior gateway protocols. In light of the

above feature of Internet, the utility-optimized aggregate flow-level bandwidth allocation is believed to have direct usefulness of charging and rate control in Internet.

This chapter studies the fundamental problem of bandwidth allocation at an aggregate flow level. We propose a general model of utility-optimized aggregate flow-level bandwidth allocation and its solution. We also discuss the special cases where the routing matrix is of full row rank and where there is one single-hop flow in every link in the network. We suggest a direct application to IP-VPN of the latter case. We present the mathematical models and solution procedures and further illustrate them by numerical examples.

12.2 AGGREGATE FLOW-LEVEL BANDWIDTH ALLOCATION: GENERAL MODEL AND GENERAL SOLUTION

A general network can be described as a set $L = \{1,\ldots,n\}$ of links, shared by a set $I = \{1,\ldots,m\}$ of flows. One flow is corresponding to one source, one OD pair or one SD pair. Each link $l \in L$ has capacity c_l. Flow $i \in I$ travels a route L_i consisting of a subset of links, i.e., $L_i = \{l \in L | i \text{ traverses } l\}$. A link l is shared by a subset I_l of flows where $I_l = \{i \in I | i \text{ traverses } l\}$. We assume there are n links and m flows. Let x_i be the rate of flow i and let $X = \{x_i, i \in I\} = [x_1, x_2, \ldots, x_m]^T$ be the *individual* flow rate vector, where $(\bullet)^T$ denotes the transpose. Let $A = (A_{li}, i \in I, l \in L) = [a_{il}] \in R^{n\times m}$ be the routing matrix, where $A_{li} = 1$ if flow i traverses link l, and 0 otherwise. The *aggregate* flow rate at link l is denoted by y_l, and the aggregate flow rate vector is $Y = \{y_l, l \in L\} = [y_1, y_2, \ldots, y_n]^T$.

The individual flow rate vector, the aggregate flow rate vector, and the routing matrix are associated by

$$AX = Y.$$

Under a certain routing scheme, the above routing association tells us that the individual flow rate vector can determine the aggregate flow rate vector. In the case of the routing matrix A which is invertible, the individual flow rate vector and the aggregate flow rate vector are of one-to-one correspondence. That is, for a certain routing scheme, the aggregate flow rate vector can also determine the individual flow rate vector uniquely. In relation to this issue, the inference of OD byte counts from link byte counts measured at router interfaces under a fixed routing schemes has been discussed in the so-called *network tomography* [130], where the authors use the statistics method.

For most networks, typically, the number of columns is much greater than the number of rows in the routing matrix A because there are usually much more OD pairs than links. This is one of the reasons that often cause the routing matrix A not to be invertible. Therefore, this chapter studies a general case of a general network with the routing matrix A not necessarily being invertible, or even not necessarily being square by using the *generalized matrix inverse*. We propose a general framework that yields the aggregate flow-level bandwidth allocation for a general network with a general routing topology under the utility-optimizing scheme. It is established that the optimization of total utility of a network also leads an optimum of the aggregate flow rate vector, and thus can further determine the individual flow rate vector.

An individual flow rate vector $X \geq 0$ and its corresponding aggregate flow rate vector $Y \geq 0$ is called *feasible* if

$$Y = AX \leq C, \tag{12.2}$$

where $C = [c_1, c_2, \ldots, c_n]^T$ is the link capacity vector. The inequality (12.2) is just the link capacity constraints stated in (12.1). If for some link l, $y_l = c_l$, we say this link is *saturated* or *bottlenecked*.

Let $U(\bullet)$ be an increasing, strictly concave, and continuously differentiable function. This function is the so-called utility function, which indicates the user's want-satisfaction degree. The following nonlinear flow-level optimization problem (we call it **FP**):

$$Q_F^* = \max \sum_{i=1}^{m} U_i(x_i), \tag{12.3}$$

$$\text{subject to } AX \le C, \tag{12.4}$$

$$\text{over } X \ge 0, \tag{12.5}$$

is the well-known flow-level bandwidth allocation problem, which solution of which is well established by Lagrangian method in the literatures (see, e.g., [104,105]). It should be noted that the above optimization problem has to be decomposed [104,105] into two simpler optimization problems, one for each user and the other for the network due to the difficulty that it involves utilities U that are unlikely known by the network.

From [120], for any $n \times m$ routing matrix A, there is an $m \times n$ matrix A^- (not necessarily unique) such that

$$AA^-A = A. \tag{12.6}$$

Here, A^- is called a $\{1\}$-*INVERSE* of A. As a particular $\{1\}$-*INVERSE* of A, its *Moore–Penrose inverse* $A^+ \in m \times n$ is uniquely determined by the following four equalities:

$$AA^+A = A, \tag{12.7}$$

$$A^+AA^+ = A^+, \tag{12.8}$$

$$(AA^+)^T = AA^+, \tag{12.9}$$

$$(A^+A)^T = A^+A. \tag{12.10}$$

The general form of any $\{1\}$-*INVERSE* of A is given by

$$A^- = A^+ + K - A^+AKAA^+, \tag{12.11}$$

where $K = [K_{ij}]$ is an arbitrary $m \times n$ matrix. Equation 12.11 suggests that the set of $\{1\}$-*INVERSE* can be parameterized by a matrix K.

The equation $AX = Y$ is *consistent* if and only if

$$(I - AA^-)Y = 0, \tag{12.12}$$

for some $\{1\}$-*INVERSE* of A. The general solution [120] is

$$X = A^-Y + (I - A^-A)H, \tag{12.13}$$

where $H = [h_1, h_2, \ldots, h_m]^T$ is an arbitrary $m \times 1$ vector.

We now still consider the optimization of aggregate utility of a network but at an aggregate flow-level (router-level) basis. We define the objective function by using the aggregate flow rate vector and describe this aggregate flow-level optimization problem (we call it **RP**) as follows: to find an optimum aggregate flow rate vector Y such that

$$\begin{aligned} Q_R^* &= \max_{Y,H,K} Q(Y,H,K) \\ &= \max \sum_{i=1}^{m} U_i(F_i(Y,H,K)), \end{aligned} \tag{12.14}$$

where

$$F(Y,H,K) := \begin{bmatrix} F_1(Y,H,K) \\ \vdots \\ F_m(Y,H,K) \end{bmatrix} := A^- Y + (I - A^- A)H, \tag{12.15}$$

and the $\{1\}$-*INVERSE* of A is given by

$$A^- = A^+ + K - A^+ AKAA^+, \tag{12.16}$$

subject to

a. The *consistency constraint*

$$(I - A^- A)Y = 0, \tag{12.17}$$

b. The *positive constraint* that the vector $F(Y,H,K)$ should be positive

$$\begin{bmatrix} F_1(Y,H,K) \\ \vdots \\ F_m(Y,H,K) \end{bmatrix} \geq 0, \tag{12.18}$$

c. The capacity constraints

$$Y \leq C. \tag{12.19}$$

With regard to the optimization problem **RP** described by Equations 12.14 through 12.19, we have the following statement:

Theorem 12.2.1

The two optimization problems **FP** and **RP** have the same optimized value, i.e., $Q_R^* = Q_F^*$, and further their solutions (optimums) Y and X satisfy the routing association, i.e., $Y = AX$.

Proof: To verify the above statement, consider the Lagrangian form in Equations 12.14 and 12.19

$$\begin{aligned} &L(F(Y,H,K),\lambda,Z) \\ &= \sum_{i=1}^{m} U_i(F_i(Y,H,K)) - \lambda^T (C - Y - Z), \end{aligned} \tag{12.20}$$

where $\lambda \in R^n$ is a vector of Lagrange multipliers and $Z \in R^n$ is a vector of slack variables. If Y, H, and K satisfy Equations 12.15 through 12.18, we then have

$$\begin{aligned} AF(Y,H,K) &= A(A^- Y + (I - A^- A)H) \\ &= AA^- Y \\ &= Y. \end{aligned}$$

By letting $X = F(Y,H,K)$ and substituting it into the Lagrangian form (12.20), one can see that, this Lagrangian form is exactly

$$\begin{aligned} &L(X,\lambda,Z) \\ &= \sum_{i=1}^{m} U_i(x_i) - \lambda^T (C - AX - Z), \end{aligned} \tag{12.21}$$

which is exactly the Lagrangian form for the problem of FP described by Equations 12.3 through 12.5. From the general theory of constrained convex optimization [114], it follows the statement. ■

So far, we have actually proposed a nonlinear optimization model for aggregate flow-level bandwidth allocation, which maximizes the total utility of a network. The following steps provide an overview of the *solution strategy*:

Step 1 For any network with its routing matrix A, calculate its Moore–Penrose inverse A^+;
Step 2 Using Equation 12.11, calculate the general form of any $\{1\}$-*INVERSE* of A by incorporating an arbitrary $m \times n$ matrix K as parameter;
Step 3 Calculate $F(Y,H,K)$ using Equation 12.15;
Step 4 Formulate the nonlinear optimization problem RP, where the objective function is given by Equation 12.14, the constrains are Equations 12.17 through 12.19;
Step 5 Solve the nonlinear optimization problem RP to obtain the aggregate flow rate vector Y.

The following example is developed to demonstrate the equivalence of our new optimization problem **RP** to the **FP** and the solution strategy of **RP**.

Example 1: A linear network

Consider the linear network, topology, and flow pattern of which are shown in Figure 12.1. We assume that all links have ten unit capacity and the utility function takes a logarithm form, i.e.,

$$U(x) = \log x.$$

For this network, construct the routing matrix A as follows:

$$A = \begin{bmatrix} 1 & 1 & 0 & 0 & 0 & 1 \\ 0 & 1 & 1 & 0 & 0 & 1 \\ 0 & 0 & 0 & 0 & 1 & 1 \\ 0 & 0 & 0 & 1 & 1 & 1 \\ 0 & 0 & 0 & 1 & 0 & 0 \end{bmatrix}.$$

The *Moore–Penrose* inverse of A is calculated to be

$$A^+ = \begin{bmatrix} 0.6250 & -0.3750 & -0.0833 & -0.0417 & 0.0417 \\ 0.2500 & 0.2500 & -0.1667 & -0.0833 & 0.0833 \\ -0.3750 & 0.6250 & -0.0833 & -0.0417 & 0.0417 \\ 0.0000 & 0.0000 & -0.3333 & 0.3333 & 0.6667 \\ -0.1250 & -0.1250 & 0.4167 & 0.2083 & -0.2083 \\ 0.1250 & 0.1250 & 0.2500 & 0.1250 & -0.1250 \end{bmatrix}.$$

According to Equation 12.11, the set of $\{1\}$-*INVERSE* of A is parameterized by a matrix K as

$$A^- = A^+ + K - A^+AKAA^+,$$

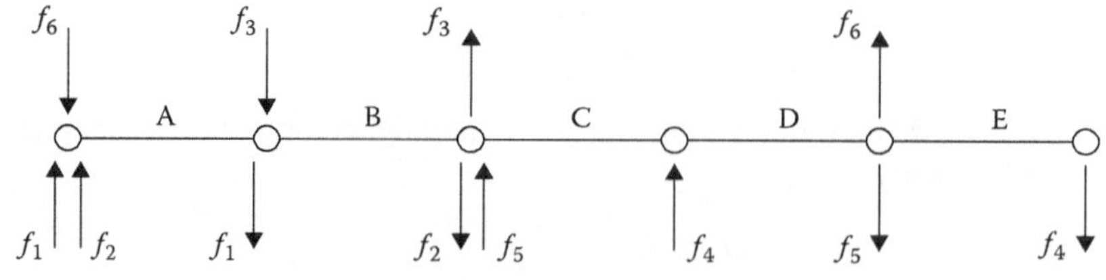

Figure 12.1 A linear network with five links and six flows.

where $K = [K_{ij}]_{6\times 5}$.

According to Equation 12.15,

$$F(Y,H,K) = A^{-}Y + (I - A^{-}A)H,$$

where Y is a 5×1 vector and H is a 6×1 vector. The objective function of the optimization problem **RP** is expressed as

$$Q_R^* = \max \sum_{i=1}^{6} \log(e_i F(Y,H,K) + 1),$$

where e_i is the ith identity vector with the size 6×1. The objective function contains Y, H, and K, it has $5+6+6\times 5 = 41$ variables. Solving the optimization problem **RP** with the constrain conditions 12.17 through 12.19, one obtains the following optimum and the aggregate flow rate vector and the other associated parameters

$$Q_R^* = 9.3904,$$
$$Y = [10, 10, 5.6450, 10, 4.3550]^T,$$
$$H = [2.5494, 1.2643, 2.5085, 1.3891, 1.0205, 5.2038]^T,$$

$$K = \begin{bmatrix} 0.9654 & 1.1704 & 0.1555 & 0.5943 & 0.6778 \\ 0.2716 & 0.2745 & 0.3575 & 0.7957 & 0.6404 \\ 0.9654 & 0.9672 & 0.1392 & 0.3537 & 0.6669 \\ 0.5000 & 0.4684 & 0.6309 & 0.6438 & 0.6309 \\ 0.7373 & 0.7420 & 0.1752 & 0.4894 & 1.0191 \\ 0.2627 & 0.2580 & 0.9463 & 0.4363 & -0.0433 \end{bmatrix}.$$

According to Equation 12.16, the corresponding $\{1\}$-*INVERSE* of A is calculated as follows:

$$A^{-} = A^{+} + K - A^{+}AKAA^{+} = \begin{bmatrix} 1.3405 & 0.4185 & -0.1585 & 0.1214 & 0.5188 \\ 0.0218 & -0.0262 & -0.1829 & 0.0193 & 0.4044 \\ 0.3405 & 1.4185 & -0.0873 & 0.0503 & 0.5899 \\ 0.0000 & 0.0000 & -0.1274 & 0.1274 & 0.8726 \\ 0.3622 & 0.3923 & 0.4123 & 0.3861 & 0.6778 \\ -0.3622 & -0.3923 & 0.6440 & -0.4433 & -0.6206 \end{bmatrix}.$$

To verify that the optimization problem **RP** and the original optimization problem **FP** have the same optimum and their solutions X and Y satisfy the routing scheme, i.e., $AX = Y$. To this purpose, we formulate the original optimization problem **FP** as follows:

$$Q_F^* = \max \sum_{i=1}^{6} U(x_i) = \sum_{i=1}^{6} \log(x_i + 1),$$

subject to

$$AX \leq 1,\ X \geq 0.$$

Solving this problem yields the optimum and the individual flow rate vector, they are

$$Q_F^* = 9.3905,$$
$$X = [6.2222, 2.6111, 6.2222, 4.4167, 4.4167, 1.1667]^T.$$

One thus verifies that $Y \approx AX$. With reference to **RP** problem, using the above already obtained values of A^{-} and H, we can further calculate the individual flow rate vector X by Equation 12.13

$$X = [6.2149, 2.6103, 6.2149, 4.3550, 4.4702, 1.1748]^T.$$

Considering the errors introduced in the calculations of the optimization problems, it is thus seen that the optimums and their corresponding individual flow rate vector in the two optimization problems match each other quite well. This then validates our model of the proposed optimization problem RP and the method to calculate the utility-optimized bandwidth allocation in router level for a general network.

For the nonlinear optimization problem RP, the objective function contains a total of $(n+m+n\times m)$ variables as it contains the variable vector $Y\in R^n$, the intermediate variable vector $H\in R^m$, and the intermediate variable matrix $K\in R^{m\times n}$; the feasible region [the constraints (a), (b), and (c)] is confined by a total of $(n+2m)$ inequalities. By constructing the Lagrangian function and by using the Karush–Kuhn–Tucker condition [117], the nonlinear optimization problem RP can be solved if the optimal solution exists. When compared to RP, in the original optimization problem FP, the objective function contains m variables; the feasible region is confined by a total $n+m$ inequalities. Solving the problem RP is thus seen to be much more complex than solving the problem FP, as the size (the number of links and the number of flows) of the network scales up. However, if the routing matrix A is of full row rank, the complexity of solving the optimization problem RP can be reduced significantly. In the following, we focus on this special case.

12.3 CASE OF THE ROUTING MATRIX BEING FULL-ROW RANK

This section studies the proposed nonlinear optimization problem RP for aggregate flow-level bandwidth allocation in detail for the special case of the routing matrix being full row rank.

For a general model of network, there are usually much more flows than links, rendering the routing matrix to be rectangular. However, in real networks, the routing matrix can frequently be of full row rank [129]. For example, for the topology of a network, if there is one single-hop flow in every link, such routing matrix is seen to be of full row rank. To verify this, we can rearrange by interchanging the relevant columns of the routing matrix $A\in R^{m\times n}$ $(m>n)$ to express A as $A=[I_{n\times n},A_1]$, where $I_{n\times n}$ is the $n\times n$ identity matrix and A_1 is a $n\times(m-n)$ matrix.

The above models are realistic that appear in practice frequently: most peer-to-peer applications involve one-link flows between a data source and a receiver (downloader). For other types of overlay communication sessions such as overlay multicast, each edge in the corresponding topology (single tree, multiple trees, or mesh) corresponds to a single-hop flow [131].

Let $A\in R^{n\times m}$ $(m>n)$ be the routing matrix of a network with full row rank. The *singular value decomposition* (**SVD**) [110] of A is the factorization

$$A=U\left[\sum\nolimits_{n\times n},\,\mathbf{0}_{n\times(m-n)}\right]V^T, \tag{12.22}$$

where

$$U=[u_1,u_2,\ldots,u_n]\in R^{n\times n}, \tag{12.23}$$

$$V^T=[v_1^T,v_2^T,\ldots,v_n^T,v_{n+1}^T,\ldots,v_m^T]^T\in R^{m\times m}, \tag{12.24}$$

are orthogonal, and $\sum_{n\times n}=$ diag $(\sigma_1,\sigma_2,\ldots,\sigma_n)$ with $\sigma_1\geq\sigma_2\geq\cdots\geq\sigma_n>0$.

For a full row rank matrix $A\in R^{n\times m}$ $(m>n)$, it is noted that AA^T is invertible. In this case, the Moore–Penrose inverse of A is turned out to be its "right pseudoinverse," denoted by A_R^+. We further have

$$A_R^+=A^T(AA^T)^{-1}. \tag{12.25}$$

For the network with a full row rank routing matrix A, we define the objective function by using the aggregate flow rate vector Y and the right pseudoinverse of A and describe this aggregate flow-level optimization problem **RP** in the following way. Let v_i $(i=1,\ldots,m)$ be the vectors in the matrix

V shown in Equation 12.24, $\alpha = [\alpha_{n+1}, \alpha_{n+2}, \ldots, \alpha_m]^T$ is any scalar vector with the size $m-n$ (just for convenience, we order the elements from $n+1$ there), by denoting

$$F(Y,\alpha) := \begin{bmatrix} F_1(Y,\alpha) \\ \vdots \\ F_m(Y,\alpha) \end{bmatrix} := A_R^+ Y + \sum_{i=n+1}^{m} \alpha_i v_i, \tag{12.26}$$

the **RP** problem is to find an optimum aggregate flow rate vector Y such that

$$Q_R^* = \max_{Y,\alpha} Q(Y,\alpha) = \max \sum_{i=1}^{m} U_i(F_i(Y,\alpha)), \tag{12.27}$$

subject to the constraints

$$\begin{aligned} F(Y,\alpha) &\geq 0 \\ Y &\leq C. \end{aligned} \tag{12.28}$$

The optimization problem **RP** described by Equations 12.27 and 12.28 has a unique optimum for the vector Y, since the objective function is strictly concave function of Y and the feasible region is compact. Furthermore, we have the following result:

Theorem 12.3.2

For a network where the routing matrix is of full row rank, the two optimization problems **FP** and **RP** have the same optimized value, i.e., $Q_R^* = Q_F^*$, and further their solutions (optimums) Y and X satisfy the routing scheme, i.e., $Y = AX$.

Proof: For the full row rank routing matrix A, one has its SVD

$$\begin{aligned} A &= U\left[\Sigma_{n\times n}, \mathbf{0}_{n\times(m-n)}\right]V^T \\ &= [u_1, u_2, \ldots, u_n] \times \left[\Sigma_{n\times n}, \mathbf{0}_{n\times(m-n)}\right] \\ &\quad \times [v_1^T, v_2^T, \ldots, v_m^T]^T. \end{aligned}$$

One thus has

$$Av_i = 0, \; i = n+1, \ldots, m.$$

Based on this, we can write an explicit expression for the complete space of solutions to $AX = Y$:

$$X = A_R^+ Y + \sum_{i=n+1}^{m} \alpha_i v_i, \text{ for any } \alpha_i. \tag{12.29}$$

Now consider the Lagrangian form in Equations 12.27 and 12.28

$$L(F(Y,\alpha),\lambda,Z) = \sum_{i=1}^{m} U_i(F_i(Y,\alpha)) - \lambda^T(C - Y - Z), \tag{12.30}$$

where $\lambda \in R^n$ is a vector of Lagrange multipliers and $Z \in R^n$ is a vector of slack variables. By letting $F(Y,\alpha) = X$ in Equation 12.30, we have $AF(Y,\alpha) = AX$, we thus see that the Lagrange form for the RP problem is exactly that of the original FP problem described in Equations 12.3 through 12.5. From the general theory of constrained convex optimization [114], it follows the statement. ■

Observing the objective function and the constraints of the optimization problem RP described in Equations 12.27 and 12.28, one notes that there are a total of $(2m-n)$ variables in the objective function and a total of $m+n$ inequalities to confine the feasible region. This computing complexity is thus comparable to the original optimization problem FP.

The following steps provide an overview of the *solution strategy* for the optimization problem RP in the case of the routing matrix being full row rank:

Step 1. For any network with a full row rank routing matrix A, calculate its right pseudoinverse, A_R^+ using Equation 12.25;
Step 2. Calculate its SVD and obtain v_i $(i = 1, \ldots, m)$;
Step 3. Calculate $F(Y, \alpha)$ using Equation 12.26;
Step 4. Formulate the nonlinear optimization problem RP, where the objective function is given by (12.27), the constrains are given by Equation 12.28;
Step 5. Solve the nonlinear optimization problem RP to obtain the aggregate flow rate vector Y.

Example 2: NSFNET backbone network

Let us consider the NSFNET Backbone network shown in Figure 12.2. The routing information and link capacity are assumed to be given by Table 12.1. For this network, we assume that there are 10 links and 12 flows.

We construct the routing matrix:

$$A_R^+ = \begin{bmatrix} 1 & -1 & 1 & -1 & 0 & 0 & 0 & 0 & 0 & 0 \\ 0 & 1 & -1 & 1 & 0 & 0 & 0 & 0 & 0 & 0 \\ 0 & 0 & 1 & -1 & 0 & 0 & 0 & 0 & 0 & 0 \\ 0 & 0 & 0 & 0.7857 & -0.5714 & 0.3571 & -0.3571 & 0.2413 & -0.2413 & 0.2857 \\ 0 & 0 & 0 & 0.2143 & 0.3714 & -0.3571 & 0.3571 & -0.2413 & 0.2413 & -0.2857 \\ 0 & 0 & 0 & -0.2143 & 0.4286 & 0.3571 & -0.3571 & 0.2413 & -0.2413 & 0.2857 \\ 0 & 0 & 0 & 0 & 0 & 0 & 1 & -1 & 1 & -1 \\ 0 & 0 & 0 & -0.1429 & 0.2857 & -0.4286 & 0.4286 & 0.1429 & -0.1429 & -0.1429 \\ 0 & 0 & 0 & 0 & 0 & 0 & 0 & 0 & 1 & -1 \\ 0 & 0 & 0 & 0.0714 & -0.1429 & 0.2143 & -0.2143 & -0.0714 & 0.0714 & 0.5714 \\ 0 & 0 & 0 & 0.2143 & -0.4286 & 0.6429 & -0.6429 & 0.7857 & -0.7857 & 0.7143 \\ 0 & 0 & 0 & -0.0714 & 0.1429 & -0.2143 & 0.2143 & 0.0714 & -0.0714 & 0.4286 \end{bmatrix}.$$

The above routing matrix is seen to be of full row rank. We calculate its right pseudoinverse A_R^+ by using Equation 12.25.

$$A = \begin{bmatrix} 1 & 1 & 0 & 0 & 0 & 0 & 0 & 0 & 0 & 0 & 0 & 0 \\ 0 & 1 & 1 & 0 & 0 & 0 & 0 & 0 & 0 & 0 & 0 & 0 \\ 0 & 0 & 1 & 1 & 1 & 0 & 0 & 0 & 0 & 0 & 0 & 0 \\ 0 & 0 & 0 & 1 & 1 & 0 & 0 & 0 & 0 & 0 & 0 & 0 \\ 0 & 0 & 0 & 0 & 1 & 1 & 0 & 0 & 0 & 0 & 0 & 0 \\ 0 & 0 & 0 & 0 & 0 & 1 & 1 & 0 & 0 & 0 & 1 & 0 \\ 0 & 0 & 0 & 0 & 0 & 0 & 1 & 1 & 0 & 0 & 1 & 1 \\ 0 & 0 & 0 & 0 & 0 & 0 & 0 & 1 & 1 & 0 & 1 & 1 \\ 0 & 0 & 0 & 0 & 0 & 0 & 0 & 0 & 1 & 1 & 0 & 1 \\ 0 & 0 & 0 & 0 & 0 & 0 & 0 & 0 & 0 & 1 & 0 & 1 \end{bmatrix}.$$

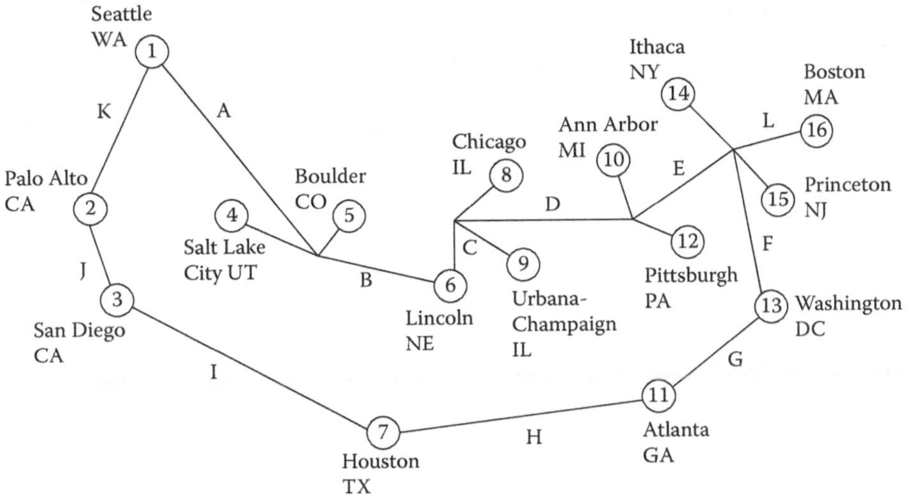

Figure 12.2 NSFNET backbone network.

Table 12.1
Routing Information and Link Capacity

Flows	Rout	Link	Capacity (Gbps)
1: 1–5	A	A	20
2: 1–6	A–B	B	15
3: 4–9	B–C	C	9
4: 6–10	C–D	D	12
5: 6–14	C–D–E	E	18
6: 12–13	E–F	F	8
7: 14–11	F–G	G	20
8: 13–7	G–H	H	16
9: 11–3	H–I	I	17
10: 7–2	I–J	J	30
11: 15–7	F–G–H		
12: 13–2	G–H–I–J		

We further calculate the SVD of A and obtain v_i $(i = 11, 12)$ as follows:

$$[v_{11}, v_{12}] = \begin{bmatrix} 0.0000 & 0.0000 \\ 0.0000 & 0.0000 \\ 0.0000 & 0.0000 \\ -0.4608 & -0.0447 \\ 0.4608 & 0.0447 \\ -0.4608 & -0.0447 \\ 0.0000 & 0.0000 \\ -0.2515 & -0.6044 \\ 0.0000 & 0.0000 \\ 0.2093 & -0.5598 \\ 0.4608 & 0.0447 \\ -0.2093 & 0.5598 \end{bmatrix}.$$

The utility function used here is

$$U(x) = \log(x+1).$$

Now we are ready to formulate the optimization problem RP according to Equations 12.26 through 12.28. The solutions to the RP are

$$\begin{aligned} Q_R^* &= 20.3178, \\ Y &= [20.0000,\ 12.9870,\ 9.0000,\ 6.0061,\ 6.1393, \\ &\quad 8.0000,\ 15.5194,\ 16.0000,\ 17.0000,\ 13.3864]^T, \\ \alpha &= [-1.1217,\ -7.7910]^T. \end{aligned}$$

If we substitute the above Y and α into Equation 12.26, one further obtains the individual flow rate vector

$$\begin{aligned} X &= F(Y, \alpha) \\ &= [10.0069,\ 9.9931,\ 2.9939,\ 3.0006, \\ &\quad 3.0055,\ 3.1338,\ 3.1330,\ 7.0546, \\ &\quad 3.6136,\ 9.7879,\ 1.7333,\ 3.5985]^T. \end{aligned}$$

One can also formulate the original FP and solve it to obtain its optimum and the bandwidth allocation. They are

$$\begin{aligned} Q_F^* &= 20.3178, \\ X &= [10.0000,\ 10.0000,\ 3.0000,\ 3.0000, \\ &\quad 3.0000,\ 3.1343,\ 3.1343,\ 7.0505, \\ &\quad 3.6090,\ 9.7819,\ 1.7315,\ 3.6090]^T. \end{aligned}$$

One observes that the two optimums outputted from the two optimization problems **FP** and **RP** are equal, i.e., $Q_F^* = Q_R^*$, and their corresponding individual rate vectors are very close.

12.4 UTILITY FUNCTION OF THE AGGREGATE FLOW

When optimizing the aggregate utility of a network, we choose the function $U(\bullet)$ to be a utility function of the individual flow rate; it is an increasing, strictly concave, and continuously differentiable function with regard to the individual flow rate. In the optimization problem RP, we have, however, formulated the total utility of a network in terms of the aggregate flow rate.

This section is to study whether or not the utility function originally chosen for the individual rate is still an increasing and concave function and thus is a utility function for the aggregate flow rate. In the case of the utility function for the individual flow is also a one for the aggregate flow, when optimizing the total utility of a network; the network is able to specify its utilities here with refer to the guideline given below.

For the general case as stated in the optimization problem **RP** described by Equations 12.14 through 12.19, for the functions $U_i(F_i(Y,H,K))$ ($i = 1$ to m) to be increasing and strictly concave, and thus still to be the utility functions of the aggregate rate vector, the following two conditions must be satisfied:

1. The first-order differentiation of Y should be positive, i.e.,

$$\begin{bmatrix} \frac{\partial U}{\partial y_1} \\ \vdots \\ \frac{\partial U}{\partial y_n} \end{bmatrix} = (A^-)^T \begin{bmatrix} \frac{\partial U}{\partial F_1} \\ \vdots \\ \frac{\partial U}{\partial F_m} \end{bmatrix} > 0,$$

2. The Hessian matrix of Y should be negative definite, i.e.,

$$\begin{bmatrix} \frac{\partial^2 U}{\partial y_1^2} & \frac{\partial^2 U}{\partial y_1 \partial y_2} & \cdots & \frac{\partial^2 U}{\partial y_1 \partial y_n} \\ \vdots & \vdots & \ddots & \vdots \\ \frac{\partial^2 U}{\partial y_n \partial y_1} & \frac{\partial^2 U}{\partial y_n \partial y_2} & \cdots & \frac{\partial^2 U}{\partial y_n^2} \end{bmatrix}$$

$$= (A^-)^T \mathrm{diag}\left\{\frac{\partial^2 U}{\partial F_1^2}, \cdots, \frac{\partial^2 U}{\partial F_m^2}\right\}(A^-) < 0,$$

where F_i denotes $F_i(Y, H, K)$ $(i = 1, \ldots, m)$.

For the case of the routing matrix being full row rank, in the optimization problem RP described by Equations 12.26 through 12.28, for the functions $U_i(F_i(Y, \alpha))$ $(i = 1, \ldots, m)$ to be increasing and strictly concave, and thus still to be the utility functions of the aggregate rate vector Y, the following two conditions must be satisfied:

1. The first-order differentiation of Y should be positive, i.e.,

$$\begin{bmatrix} \frac{\partial U}{\partial y_1} \\ \vdots \\ \frac{\partial U}{\partial y_n} \end{bmatrix} = (AA^T)^{-1}A \begin{bmatrix} \frac{\partial U}{\partial F_1} \\ \vdots \\ \frac{\partial U}{\partial F_m} \end{bmatrix} > 0, \quad (12.31)$$

2. The Hessian matrix of Y should be negative definite, i.e.,

$$\begin{bmatrix} \frac{\partial^2 U}{\partial y_1^2} & \frac{\partial^2 U}{\partial y_1 \partial y_2} & \cdots & \frac{\partial^2 U}{\partial y_1 \partial y_n} \\ \vdots & \vdots & \ddots & \vdots \\ \frac{\partial^2 U}{\partial y_n \partial y_1} & \frac{\partial^2 U}{\partial y_n \partial y_2} & \cdots & \frac{\partial^2 U}{\partial y_n^2} \end{bmatrix} \quad (12.32)$$

$$= \quad (AA^T)^{-1}A\mathrm{diag}\left\{\frac{\partial^2 U}{\partial F_1^2}, \cdots, \frac{\partial^2 U}{\partial F_m^2}\right\}A^T(AA^T)^{-1} < 0,$$

where F_i denotes $F_i(Y, \alpha)$ $(i = 1, \ldots, m)$.

From Equation 12.32, it is observed that as soon as

$$\mathrm{diag}\left\{\frac{\partial^2 U}{\partial F_1^2}, \ldots, \frac{\partial^2 U}{\partial F_m^2}\right\} < 0, \quad (12.33)$$

the above Hessian matrix of Y is negative definite considering the matrix A is of full row rank. The following theorem establishes this observation.

Theorem 12.4.3

Given that the routing matrix is of full row rank, the aggregate utility of the RP problem is a strictly concave function with regard to Y.

Proof: Since the aggregate utility of the FP problem is a concave function with regard to X, its Hessian matrix

$$\begin{aligned} &\operatorname{diag}\left\{\frac{\partial^2 U}{\partial x_1^2}, \frac{\partial^2 U}{\partial x_2^2}, \ldots, \frac{\partial^2 U}{\partial x_m^2}\right\} \\ &= \operatorname{diag}\left\{\frac{\partial^2 U}{\partial F_1^2}, \frac{\partial^2 U}{\partial F_2^2}, \ldots, \frac{\partial^2 U}{\partial F_m^2}\right\} < 0, \end{aligned}$$

where "<0" means that the matrix is negative definite. For any m dimension vector $X \neq 0$, one has

$$X^T \operatorname{diag}\left\{\frac{\partial^2 U}{\partial F_1^2}, \frac{\partial^2 U}{\partial F_2^2}, \ldots, \frac{\partial^2 U}{\partial F_m^2}\right\} X < 0.$$

Since the routing matrix A ($n \times m$) is of full row rank. It is easily concluded that $(AA^T)^{-1}A$ and $A^T(AA^T)^{-1}$ are of full row rank and full column rank, respectively. Denote $A^T(AA^T)^{-1}$ as C, one obtains that

$$CX \neq 0, \tag{12.34}$$

for any m dimension vector $X \neq 0$. Since otherwise it follows that

$$C^T CX = 0. \tag{12.35}$$

Because C is full column rank, $C^T C$ is an $m \times m$ invertible matrix. According to Equation 12.35, one concludes $X = 0$, which contradicts the previous statements. Thus, we verify Equation 12.34. Since

$$\operatorname{diag}\left\{\frac{\partial^2 U}{\partial F_1^2}, \frac{\partial^2 U}{\partial F_2^2}, \ldots, \frac{\partial^2 U}{\partial F_m^2}\right\}$$

is negative definite, one has

$$\begin{aligned} &X^T\{(AA^T)^{-1} A \operatorname{diag}\left\{\frac{\partial^2 U}{\partial F_1^2}, \ldots, \frac{\partial^2 U}{\partial F_m^2}\right\} A^T (AA^T)^{-1}\} X \\ &= X^T\{C^T \operatorname{diag}\left\{\frac{\partial^2 U}{\partial F_1^2}, \ldots, \frac{\partial^2 U}{\partial F_m^2}\right\} C\} X \\ &= (CX)^T \operatorname{diag}\left\{\frac{\partial^2 U}{\partial F_1^2}, \ldots, \frac{\partial^2 U}{\partial F_m^2}\right\} (CX) < 0, \end{aligned}$$

for any m dimension vector $X \neq 0$. So

$$(AA^T)^{-1} A \operatorname{diag}\left\{\frac{\partial^2 U}{\partial F_1^2}, \ldots, \frac{\partial^2 U}{\partial F_m^2}\right\} A^T (AA^T)^{-1}$$

is negative definite, which indicates that the aggregate utility of the RP problem is a concave function with regard to Y. ■

The interesting implication of the above theorem is that the utility function original chosen for the individual flow is also a utility function for the aggregate flow once condition (12.31) is satisfied. We further illustrate this observation through the following two examples.

Example 3:

Assume that every link is of one unit capacity. The utility takes the form

$$U(x) = \log(x + 1).$$

The routing matrix here is

$$A = \begin{bmatrix} 1 & 0 & 0 & 1 & 1 \\ 0 & 1 & 0 & 1 & 1 \\ 0 & 0 & 1 & 1 & 0 \end{bmatrix}.$$

So, Equation 12.31 becomes

$$\left[\frac{\partial U}{\partial y_1}, \frac{\partial U}{\partial y_2}, \frac{\partial U}{\partial y_3}\right]^T = \frac{1}{8}\begin{bmatrix} \frac{5}{1+x_1} - \frac{3}{1+x_2} - \frac{1}{1+x_3} + \frac{1}{1+x_4} + \frac{2}{1+x_5} \\ -\frac{3}{1+x_1} + \frac{5}{1+x_2} - \frac{1}{1+x_3} + \frac{1}{1+x_4} + \frac{2}{1+x_5} \\ -\frac{1}{1+x_1} - \frac{1}{1+x_2} + \frac{5}{1+x_3} + \frac{3}{1+x_4} - \frac{2}{1+x_5} \end{bmatrix}.$$

It should be noted that the variables in the above equation are actually the source rates, so they must satisfy the following *feasible* constraints

$$X \geq 0,$$
$$AX \leq C,$$

where $X = [x_1, x_2, x_3, x_4, x_5]^T$. Under these constraints, the minimal value that satisfies Equation 12.31 is

$$[0.18019,\ 0.18019,\ 0.156082]^T > 0,$$

which means that $U(F(Y,\alpha))$ is a strict increasing function of Y. Due to the limitation of the space, we do not present the Hessian matrix of Equation 12.32 here. But it can be seen that it is a negative definite matrix, which indicates that the aggregate utility is concave with regard to Y.

Let us consider another example.

Example 4:

The utility function is the same as previous example. The routing matrix is

$$A = \begin{bmatrix} 1 & 0 & 1 \\ 0 & 0 & 1 \\ 0 & 1 & 1 \end{bmatrix}.$$

Then, Equation 12.31 becomes

$$\begin{bmatrix} \frac{\partial U}{\partial y_1} \\ \frac{\partial U}{\partial y_2} \\ \frac{\partial U}{\partial y_3} \end{bmatrix} = \begin{bmatrix} \frac{1}{1+x_1} \\ -\frac{1}{1+x_1} - \frac{1}{1+x_2} + \frac{1}{1+x_3} \\ \frac{1}{1+x_2} \end{bmatrix}.$$

Under the similar constraints as discussed in the previous example, the minimum value of $\partial U/\partial y_2$ is -1.5, the corresponding source rates are $X = [0,0,1]^T$. It means that increasing the aggregate flow rate at link 2 does not necessarily increase the aggregate utility. On the contrary, it may decrease the aggregate utility under certain conditions. The Hessian matrix (12.32) here is

$$\begin{bmatrix} \frac{\partial^2 U}{\partial y_1^2} & \frac{\partial^2 U}{\partial y_1 y_2} & \frac{\partial^2 U}{\partial y_1 y_3} \\ \frac{\partial^2 U}{\partial y_2 y_1} & \frac{\partial^2 U}{\partial y_2^2} & \frac{\partial^2 U}{\partial y_2 y_3} \\ \frac{\partial^2 U}{\partial y_3 y_1} & \frac{\partial^2 U}{\partial y_3 y_2} & \frac{\partial^2 U}{\partial y_3^2} \end{bmatrix} =$$

$$\begin{bmatrix} -\frac{1}{(1+x_1)^2} & \frac{1}{(1+x_1)^2} & 0 \\ \frac{1}{(1+x_1)^2} & -\sum_{i=1}^{3} \frac{1}{(1+x_i)^2} & \frac{1}{(1+x_2)^2} \\ 0 & \frac{1}{(1+x_2)^2} & -\frac{1}{(1+x_2)^2} \end{bmatrix}.$$

Also under the feasible constraints, the Hessian matrix above is negative definite, indicating that the aggregate utility is concave with regard to Y.

The above two examples demonstrate that unlike increasing property, the concave property of utility function with regard to Y always holds as long as the routing matrix is of full row rank.

12.5 CASE OF THE NETWORK WITH EVERY LINK HAVING SINGLE-HOP FLOW

For a network, if every link has one single-hop flow, with regard to the aggregate flow-level bandwidth allocation modeled by RP, we have the following interesting observation:

Theorem 12.5.4

For the network with one single-link flow at each link, the solution to the optimization problem RP is the capacity vector, i.e., the utility-optimized bandwidth allocation is the one that makes every link be fully saturated.

Proof: Due to the equivalence of the RP and FP problems, we only demonstrate that Theorem 12.5.4 is true for the FP problem, so it automatically holds for the RP problem.

Suppose there exists an optimal solution $X^* = [x_1^*, x_2^*, \ldots, x_m^*]^T$ that solves the problem FP, thus the aggregate utility $U(X^*) = \sum_{i=1}^{m} U_i(x_i^*)$ is the maximum. Also, assume that there exists one link, which is not saturated. Without lose of generality, we denote this link as l and its corresponding one link flow rate as x_l^*. Since Link l is not saturated, there must be some residual bandwidth Δ which is not used. Thus, one can allocate this residual bandwidth to the one link flow, which yields another feasible solution

$$X' = [x_1^*, x_2^*, \ldots, x_l^* + \Delta, \ldots, x_m^*]^T.$$

Due to the fact that the utility is a strict increasing function, the aggregate utility

$$U(X') = \sum_{i=1}^{l-1} U_i(x_i^*) + U_l(x_l^* + \Delta) + \sum_{i=l+1}^{m} U_i(x_i^*)$$

is larger than

$$U(X^*) = \sum_{i=1}^{m} U_i(x_i^*),$$

which contradicts the assumption that

$$X^* = [x_1^*, x_2^*, \ldots, x_m^*]^T$$

is the optimal solution. Thus, it is proved by contradiction that every link is saturated and the solution to the optimization problem RP is the capacity vector. ■

In light of Theorem 12.5.4, for the network with one single-hop flow at each link, the computation of the utility-optimized individual flow rate vector X becomes easier than the original optimization problem FP. In this case, we follow the following procedures:

- To solve the RP problem described by

$$Q_R^* = \max_{\alpha} Q(\alpha) = \max \sum_{i=1}^{m} U_i(F_i(\alpha)), \tag{12.36}$$

subject to the constraints

$$F(\alpha) \geq 0, \tag{12.37}$$

where

$$F(\alpha) := \begin{bmatrix} F_1(\alpha) \\ \vdots \\ F_m(\alpha) \end{bmatrix} := A_R^+ C + \sum_{i=1}^{m} \alpha_i v_i. \tag{12.38}$$

- By substituting the optimum α^* of the RP problems 12.36 and 12.37 into Equation 12.38, one thus obtains the utility-optimized individual flow rate vector $X = F(\alpha^*)$.

With compared to the original optimization problem FP in obtaining the individual flow rate optimum, where the objection function contains m variables and the feasible region is confined by a total $n+m$ inequalities, in the above simplified RP problem, its objection function has only $m-n$ variables, and the feasible region is confined by only a total m inequalities.

12.6 APPLICATION TO BANDWIDTH PROVISION IN IP-VPN NETWORKS

A VPN extends a private network across a public network, such as the Internet. It enables users to send and receive data across shared or public networks as if their hboxcomputing devices were directly connected to the private network. Thereby one is thus benefitted from the functionality, security, and management policies of the private network. A VPN is created by establishing a virtual point-to-point connection through the use of dedicated connections, virtual tunnelling protocols, or traffic encryption.

The way that a VPN spains the Internet is just similar to that a wide area network (WAN) does. From a user perspective, the extended network resources are accessed in the same way as resources available within the private network. Traditional VPNs are characterized by a point-to-point topology, and they do not tend to support or connect broadcast domains. Therefore, communication, software, and networking, which are based on OSI layer 2 and broadcast packets, such as NetBIOS used in Windows networking, may not be fully supported or work exactly as they work on a LAN. VPN variants, such as Virtual Private LAN Service (VPLS), and layer 2 tunneling protocols, are designed to overcome this limitation.

VPNs can allow employees to securely access a corporate intranet while travelling outside the office. Similarly, VPNs can securely connect geographically separated offices of an organization, creating one cohesive network. VPN technology is also used by individual Internet users to secure their wireless transactions, to circumvent geographical restrictions and censorship, and to connect to proxy servers for the purpose of protecting personal identity and location.

The traditional dial-up network service on the Internet is for registered IP addresses only. A new class of virtual dial-up application which allows multiple protocols and unregistered IP addresses is also desired on the Internet. Examples of this class of network application are support for privately addressed IP, IPX, and AppleTalk dial-up via SLIP/PPP across existing Internet infrastructure.

The support of these multi-protocol virtual dial-up applications is of significant benefit to end users and Internet Service providers as it allows the sharing of very large investments in access and core infrastructure and allows local calls to be used. It also allows existing investments in non-IP protocol applications to be supported in a secure manner while still leveraging the access infrastructure of the Internet.

IP-VPNs [101] can be constructed on an IP network at a lower cost than with conventional dedicated lines. In an IP-VPN network, the provision of fair IP-VPN services is particularly important for IP-VPN service providers due to its best-effort feature. It is proposed in [108,118] that the bandwidth provision can be realized between routers by using an AIMD window flow control. These approaches achieve fair bandwidth allocation among IP-VPNs by aggregating multiple flows accommodated in a VPN into a single flow and by performing an AIMD window flow control for those aggregated flows. Bandwidth can be further distributed to users accommodated in the same VPN by manipulations in the customer edge (CE) routers. This kind of aggregate flow-level rate control is easily deployed into an existing IP network, because only modifications to IP-VPN service provider's edge routers (i.e., PE routers) are necessary. Viewing this characteristic of IP-VPN services, our proposed aggregate flow-level bandwidth allocation method is seen to be particularly attractive for deploying in IP-VPN networks.

The following example validates Theorem 12.5.4 and displays the direct usefulness of the proposed aggregate flow-level bandwidth allocation approach to IP-VPN networks.

Now consider multiple VPNs reside in a generic network with multiple-bottleneck links. The network topology of this IP-VPN is shown in Figure 12.3. The capacities of bottleneck link 1, 2, and 3 are 10, 20, and 30 Mbps, respectively. The round-trip time of the flows are constants, which are 3, 1, 1, and 1 μs. The routing matrix is

$$A = \begin{bmatrix} 1 & 1 & 0 & 0 \\ 1 & 0 & 1 & 0 \\ 1 & 0 & 0 & 1 \end{bmatrix}.$$

As every VPN has the TCP service at its link, we suppose that the utility of the aggregate flow of every VPN takes the following form of TCP Reno-2 [111]:

$$U(x_s, D_s) = \frac{1}{D_s} \times \log \frac{x_s D_s}{2x_s D_s + 3},$$

where x_s and D_s are the source rate and the round-trip time, respectively. According to Equation 12.38, the rate vector can be described as

$$\begin{aligned} X &= A_R^+ C + \Sigma_{i=n+1}^{m} \alpha_i v_i \\ &= A_R^+ C + \alpha_4 v_4 \\ &= [15 - 0.5\alpha_4, -5 + 0.5\alpha_4, 5 + 0.5\alpha_4, 15 + 0.5\alpha_4]^T. \end{aligned}$$

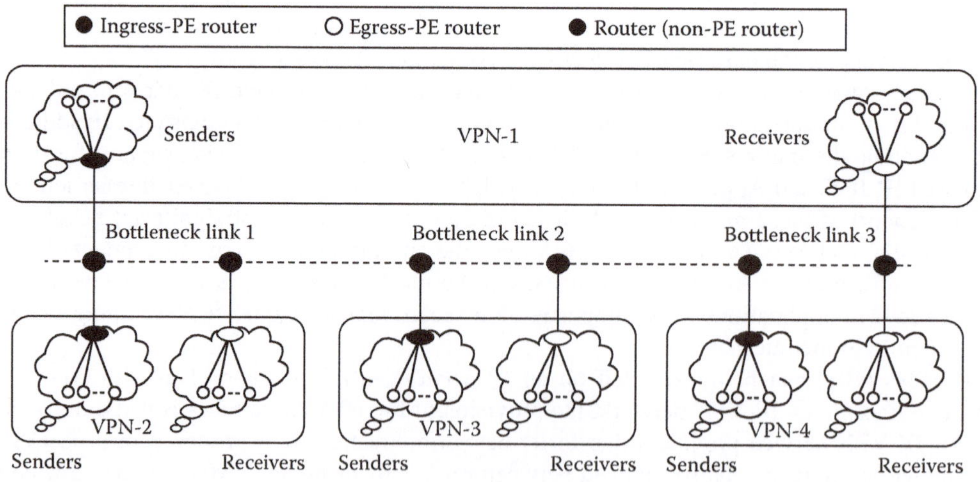

Figure 12.3 Network topology of an IP-VPN network.

So, the optimization problem becomes

$$\max \sum_{s=1}^{4} U(x_s, D_s)$$
$$= \max\left(3\times 10^6 \times \log \frac{(15-0.5\alpha_4)\times 3\times 10^{-6}}{2\times(15-0.5\alpha_4))\times 3\times 10^{-6}+3}\right.$$
$$+10^6 \times \log \frac{(-5+0.5\alpha_4)\times 10^{-6}}{2\times(-5+0.5\alpha_4))\times 10^{-6}+3}$$
$$+10^6 \times \log \frac{(5+0.5\alpha_4)\times 10^{-6}}{2\times(5+0.5\alpha_4))\times 10^{-6}+3}$$
$$\left.+10^6 \times \log \frac{(15+0.5\alpha_4)\times 10^{-6}}{2\times(15+0.5\alpha_4))\times 10^{-6}+3}\right),$$

subject to

$$X \geq 0.$$

Solving it yields

$$\alpha_4 = 26.8037.$$

So, the allocation is

$$X = [1.5982, 8.4018, 18.4018, 20.4018]^T.$$

It means that the aggregate rates of VPN1, VPN2, VPN3, and VPN4 are 1.5982, 8.4018, 18.4018, and 20.4018 Mbps, respectively.

For the feasibility of our scheme, special treatment should be done in ingress PE routers, which carry out some flow rate adjustment dictated by formula 12.38, to regulate the aggregate rate of certain VPN, thus the aggregate utility is maximized. The following steps provide a sketch of the procedures done in ingress PE routers.

Step 1: For any routing matrix A with each link having a one-link flow; calculate its right pseudoinverse A_R^+ by using Equation 12.25;
Step 2: Calculate its SVD and obtain v_i $(i = n+1, \ldots, m)$;
Step 3: Calculate $F(\alpha)$ using Equation 12.38;
Step 4: Solve the optimization problem formulated by Equation 12.36 and 12.37 to obtain α;
Step 5: Use Equation 12.38 to obtain the aggregate flow rate for certain VPN and adjust the rate according to this value.

12.7 CONCLUSION

This chapter has addressed a fundamental problem of maximizing the total utility of a network subject to the bandwidth constraints. With deviation to the usual formulation and solution, where the models and solutions are given in terms of the individual flow rates, we propose a general model and its corresponding solution procedure on how to obtain the utility-optimized aggregate flow rate. Further, we have shown the method of utility-optimized aggregate flow-level bandwidth allocation for the special case where the routing matrix is of full row rank and where there is one single-hop flow in every link of a network. We have discussed the relationship between the utility function of the individual flows and the aggregate flows with specific to a routing scheme. The obtained solutions are believed to be directly deployable to communication networks. To the best of our knowledge, this chapter is the first attempt to investigate the issue of bandwidth allocation in a aggregate flow level to optimize the total utility of a network. Our modeling and solution methodology leads to a straightforward discovery of a critical problem with network bandwidth allocation. The uniqueness of our approach comes from the use of matrix generalized inverse.

Flow control can be designed to allocate bandwidth resource fairly among competing users with the aggregate utility of a network being maximized. The current flow control approaches consist of two components: a source algorithm that dynamically adjusts rate (or window size) in the response to congestion in its path, and a link algorithm that updates, implicitly or explicitly, a congestion measure and sends it back to sources that uses that link. These algorithms are all resulted from the conventional flow-level optimization. Our future research would study the design of congestion control scheme along the aggregate flow-level bandwidth allocation scheme.

13 Bandwidth Allocation of OBS Networks Using the Aggregate Flow-Level Network Utility Maximization Approach

To meet the increasing bandwidth demands and to reduce costs, optical burst switching (OBS) network [635–637] is a promising candidate for the next-generation optical Internet. This chapter presents the novel methods of bandwidth allocation in OBS networks, using utility-maximized aggregate flow approach.

13.1 INTRODUCTION

With the recent advances in wavelength division multiplexing (WDM) technology, the amount of raw bandwidth available in fiber links has increased by many orders of magnitude. Meanwhile, the rapid growth of Internet traffic requires high transmission rates, which are far beyond a conventional router's capability. To meet the increasing bandwidth demands and to reduce costs, several approaches have been proposed to take advantage of optical communications via some particular optical switching technologies. Of all these paradigms, optical circuit switching (OCS) is relatively easy to implement but lacks flexibility to cope with the fluctuating traffic and the dynamically changing link states. Optical packet switching (OPS) is conceptually ideal, but the required optical technologies, such as optical buffer and optical logic, are too immature for it to come into reality. OBS [384,385,387] is a new approach that inherits the advantages of both OCS and OPS. It is a promising candidate for the next-generation optical internet.

On studying the network utility maximization (NUM) problem, Kelly et al. have proposed a fundamental approach for bandwidth allocation and congestion control for computer networks, including the Internet. After that, numerous research results have been published to provide utility-aware and utility-maximized network bandwidth allocation schemes on the basis of the optimization theory. Mathematically, in all the optimization problem, the objective is to maximize the source aggregate utility subject to link capacity constraints. Among all these schemes, a notable bandwidth allocation policy in the NUM model can be expressed in terms of a utility function $U_i(x_i)$, in the sense that the desired bandwidth allocation $x^* = (x_i^*, all\ source\ i)$ solves the utility maximization problem

$$\max \sum_i U_i(x_i)$$

subject to the link capacity constraints. Traditional solutions are solved for the individual flows, which can also be referred to as the source–destination pair or the origin–destination (OD) traffic in a network to reach the desired bandwidth allocation. In a network, routers and switches direct traffic by forwarding data packets between nodes according to the specified routing scheme. Flows are subsequently aggregated at the relevant links. Considering the above fact, resource allocation approach on the basis of the aggregate flow will be desired and natural for being implemented in real networks in some situations. Unfortunately, all the known utility-optimized bandwidth allocation methods are on an individual flow-level basis. Until recently, Tan et al. [164] have proposed a novel approach that solves the fundamental NUM problem of bandwidth allocation at an aggregate flow level (i.e., the router level) and proved that the network utility reaches the same maxima as Kelly's NUM model. According to this approach, we can further calculate the optimal individual flow rates on the basis of the optimal aggregate flow rates.

When the utility of network achieves the maximum, the aggregate flow rate vector must saturate the link capacity to ensure that there is no bandwidth waste. But in traditional Transmission Control Protocol (TCP) network, the requirement that the links operating in the full-load state all the time is difficult to realize. Although an ideal bandwidth allocation scheme needs to be both flexible enough to accommodate a wide variety of traffic sources and enforceable so that a well-behaved source can retain its allocation despite the actions of malicious or greedy sources. Nevertheless, the link capacity is already fixed in each network. So, the price to achieve the optimum network utility is to abandon the robustness of network, which is clearly disadvantageous.

Let us take a look at the assembly process of packets in OBS networks [381–383]. Each burst consists of two parts: a control packet and a data burst. The control packet contains information of the data burst, including the burst length information. It is sent out first to secure and allocate the data channel, and then the channel will set aside a certain amount of bandwidth for the incoming data burst in accordance with the information of control packet in order to make full use of the channel bandwidth. According to this characteristics of OBS network [386], the utility-maximized bandwidth allocation approach at an aggregate flow level is seen to be directly applicable to the OBS networks. This chapter explores the application of this approach to the OBS networks.

The remainder of this chapter is organized as follows: In Section 13.2, we introduce the architecture and the data transmission mechanism of an OBS network. In Section 13.3, we describe the models of router-level bandwidth allocation including a general model, a full row rank routing model, and a model of networks with single-hop in every link. We then discuss the solutions to these models. Section 13.4 establishes an OBS network NUM model, which is formulated and solved in terms of the aggregate flow rate by numerical examples. We present a novel algorithm for bandwidth allocation of OBS networks using the utility maximization approach in aggregate flow level in Section 13.6. The conclusions are drawn in Section 13.7.

13.2 OPTICAL BURST SWITCHING NETWORK

13.2.1 EDGE NODES OF OBS NETWORK

An OBS network consists of edge nodes (the ingress nodes and the egress nodes) and core nodes. Various types of client data are aggregated at the ingress node and then transmitted as bursts as demonstrated in Figure 13.1, the bursts are afterward routed over a buffer-less core network and will be disassembled at the egress node as shown in Figure 13.2. A burst consists of two separate parts, a control packet and a data burst. Before sending the data burst, the control packet (burst control packet, BCP) is sent separately over an out-of-band channel. Figure 13.3 describes the transmission

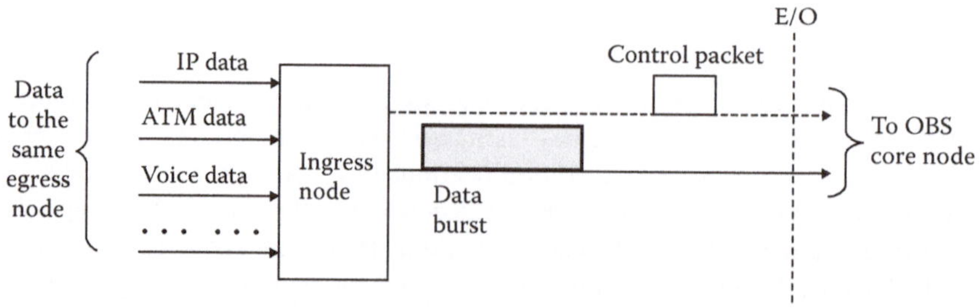

Figure 13.1 Process of burst assembly.

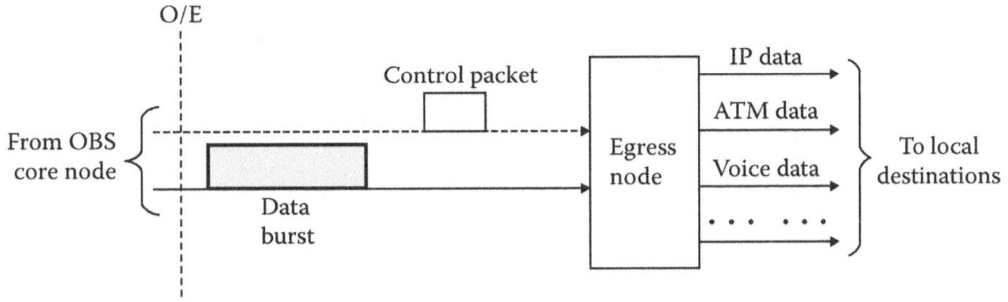

Figure 13.2 Process of burst disassembly.

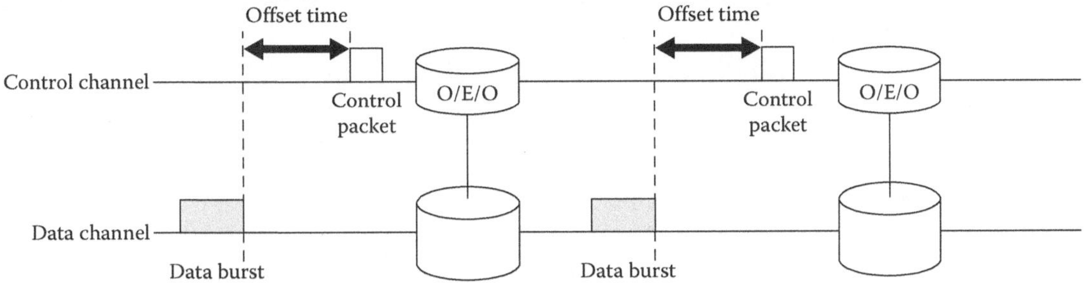

Figure 13.3 Transmission of data burst and control packet.

mechanism of data burst and control packet within the core of an OBS network. For each data burst, a control packet contains its "header" information, including the burst length information, which will be transmitted to multiple link data channels. A control packet goes through an O/E/O conversion at each intermediate OBS node and then is processed electronically to configure the underlying switching fabric. There is an offset time between a control packet and the corresponding data burst to compensate for the processing/configuration delay. If the offset time is large enough, the data burst will be switched *all-optically* and in a "cut-through" manner, i.e., without being delayed at any intermediate core node. In this way, no optical fiber delay lines (FDLs) are necessary at any intermediate node.

13.2.2 CORE NODES OF OBS NETWORK

In an OBS network, the available sets of wavelengths on the links are classified into two groups: switching control units and optical switch fabrics. The switching control units are used to transmit the burst control packets while the data bursts are sent in the optical switch fabrics. First, the edge node collects IP packets, which are destined to a common egress edge node and are belonged to the same service class, up to the sum of their packet sizes reaching the predefined maximum burst size or the burst timeout fires. Then, the control packet is generated and scheduled to send in order to set up the path and make the reservation for the following data burst.

A burst is generated either when the sum of the collected packet sizes reaches the maximum threshold or the burst assembling time reaches the timeout limit. The burst scheduler is in charge of creating bursts and their corresponding control packets, adjusting the offset time for each burst, scheduling bursts on each output link, and forwarding the bursts and their control packets to the OBS core network. This process is displayed by Figure 13.4.

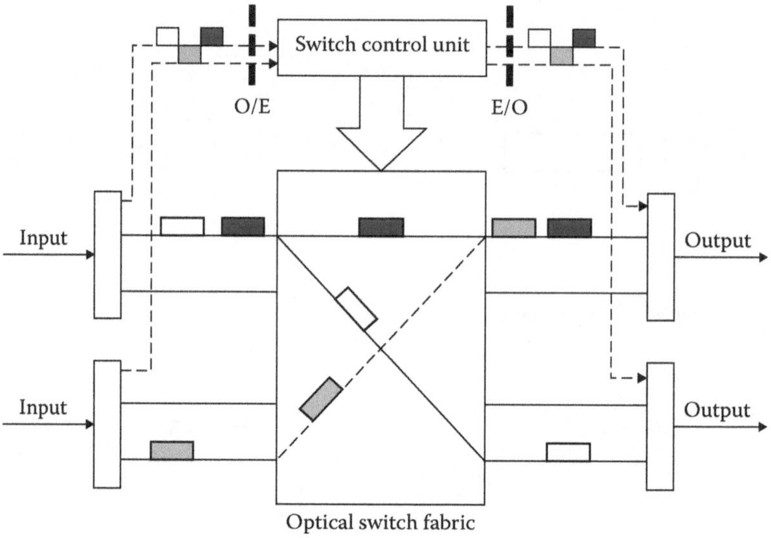

Figure 13.4 The working mechanism of OBS core node.

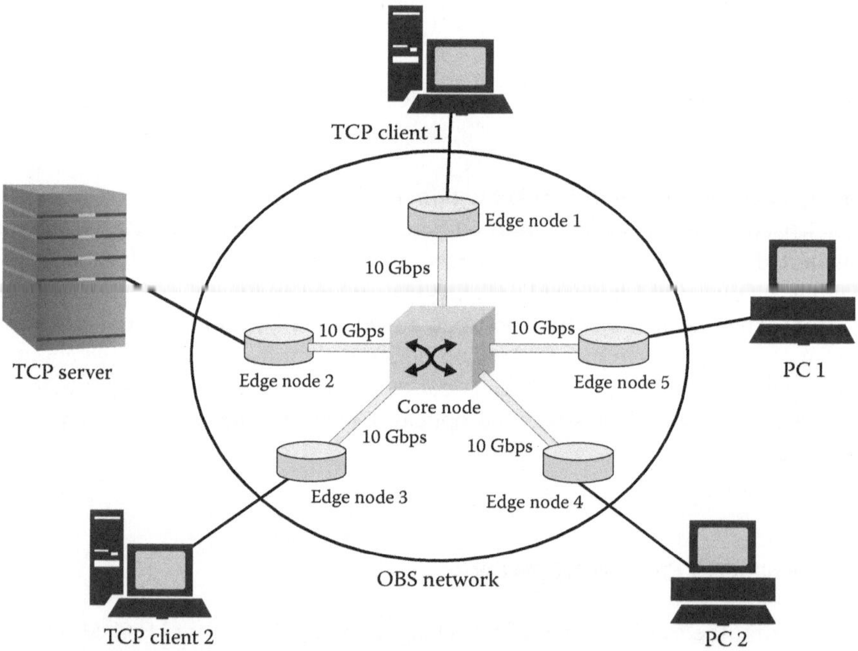

Figure 13.5 An OBS network.

13.2.3 TCP OVER OBS

Given that TCP is the prevailing transport protocol and likely to be adopted in future optical networks, so the evaluation of TCP performance over OBS networks is an important issue. To realize TCP/Internet Protocol (IP) over OBS network in reality, TCP sender and receiver are connected to an OBS network through local IP access networks, as presented in Figure 13.5.

We assume that the control packet channel has enough bandwidth for control packets and all the control packets can set up their maximum delay time for buffering their corresponding bursts

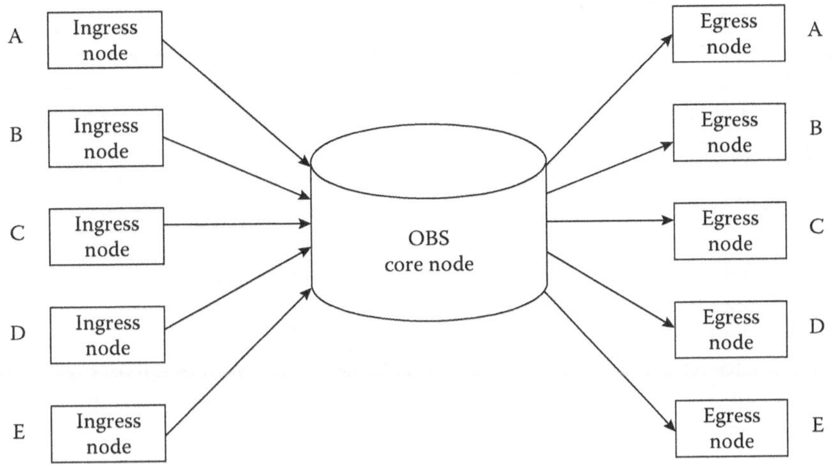

Figure 13.6 Network topology of Figure 13.5.

at the edge switch. Therefore, there is no control packet contention at control channels and there is plenty of time for packets to complete the assembly. To reduce the burst blocking probability at the ingress edge nodes, the electronic buffers are employed to postpone the bursts before the outgoing wavelengths become available. Considering this fact, the network topology of Figure 13.5 can be simplified into Figure 13.6.

13.3 ROUTER-LEVEL BANDWIDTH ALLOCATION APPROACH

A general network can be described as a set $L = 1, \ldots, n$ of links, shared by a set $I = 1, \ldots, m$ of flows. One flow is corresponding to one source, one OD pair or one source–destination pair. Each link $l \in L$ has a capacity c_l. Flow $i \in I$ travels a route L_i consisting of a subset of links that can be written as $L_i = \{l \in L \mid \text{flow } i \text{ traverses } l\}$. Then, we assume that there are n links and m flows. Let x_i be the rate of flow i and let $X = \{x_i, i \in I\} = [x_1, x_2, \ldots, x_m]^T$ be the *individual* flow rate vector, where $(\bullet)^T$ denotes the transpose. Let

$$A = (A_{li},\ i \in I,\ l \in L) = [a_{il}] \in R^{(n \times m)}$$

be the routing matrix, where $A_{li} = 1$ if flow i traverses link l and otherwise $A_{li} = 0$. Then, we define the *aggregate* flow rate at link l as y_l, so the aggregate flow rate vector is

$$Y = \{Y_l, l \in L\} = [y_1, y_2, \ldots, y_n]^T.$$

According to these definitions, the relationship of the individual flow rate vector, the aggregate flow rate vector, and the routing matrix can be described as

$$AX = Y. \tag{13.1}$$

The aforementioned routing association tells us that the aggregate flow rate vector is determined by the individual flow rate vector if the routing matrix is known. But the individual flow rate vector cannot be uniquely determined by a certain aggregate flow rate vector.

Without loss of generality, the routing is assumed to be fixed during the time interval of interest. For the case where the routing is dynamically changeable due to link roaming or link failure, our approach can still be applicable by updating the time interval and the corresponding arguments. In the case of the routing matrix A is invertible, the individual flow rate vector and the aggregate flow rate vector can become one-to-one correspondence. To put it in another way, under this routing scheme, not only the individual flow rate vector can determine the aggregate flow rate vector but the aggregate flow rate vector can also determine the individual flow rate vector solely. In relation to this issue, the inference of OD byte counts from link byte counts measured at router interfaces under a fixed routing schemes has been discussed in the so-called *network tomograph* approach, which uses the statistics method.

With the routing matrix A not necessarily being invertible or even not necessarily being square, the approach applied the *generalized matrix inverse* to the general case of a general network. A general framework that yields the router-level bandwidth allocation for a general network with a general routing topology under the utility-optimizing scheme has been proposed. When the network achieves the optimization of the total utility, the aggregate flow rate vector is the optimal solution and it can further determine the individual flow rate vector.

A flow rate vector is always positive and constrained. The individual flow rate vector $X \geq 0$ and the corresponding aggregate flow rate vector $Y \geq 0$. We describe the link capacity constrain condition as follows:

$$Y = AX \leq C. \tag{13.2}$$

where $C = [c_1, c_2, \ldots, c_n]^T$ is the link capacity vector. If $y_l = c_l$ in some link, we say this link is *saturated* or *bottlenecked*. For any $n \times m$ routing matrix A, there is a $m \times n$ matrix A^- (not necessarily unique) such that

$$AA^-A = A. \tag{13.3}$$

A^- is called an $\{1\}$-*INVERSE* of A. As a particular $\{1\}$-*INVERSE* of A, its *Moore–Penrose inverse* $A^+ \in m \times n$ is uniquely determined by the following four equalities:

$$AA^+A = A, \tag{13.4}$$

$$A^+AA^+ = A^+, \tag{13.5}$$

$$(AA^+)^T = AA^+, \tag{13.6}$$

$$(A^+A)^T = A^+A. \tag{13.7}$$

The general form of any $\{1\}$-*INVERSE* of A is given by

$$A^- = A^+ + K - A^+AKAA^+, \tag{13.8}$$

where $K = [K_{ij}]$ is an arbitrary $m \times n$ matrix. Equation 13.8 suggests that the set of $\{1\}$-*INVERSE* can be parameterized by a matrix K.

For some $\{1\}$-*INVERSE* of A, Equation 13.1 is consistent if

$$(I - AA^-)Y = 0. \tag{13.9}$$

The general solution of Equation 13.1 is

$$X = A^-Y + (I - A^-A)H, \tag{13.10}$$

where $H = [h_1, h_2, \ldots, h_m]^T$ is an arbitrary $m \times 1$ vector.

We now consider to formulate the optimization of aggregate utility of a network in terms of the router-level aggregate flow rate i.e., the aggregate flow rate). The router-level optimization problem is described as follows: to find an optimal aggregate flow rate vector Y such that

$$\begin{aligned} Q_R^* &= \underset{Y,H,K}{\text{Max}} Q(Y,H,K) \\ &= \text{Max} \sum_{i=1}^{m} U_i(F_i(Y,H,K)). \end{aligned} \tag{13.11}$$

In the above, $U(x)$ is the so-called *utility function*, which indicates the user's want-satisfaction degree, it is an increasing, strictly concave, and continuously differentiable function for the elastic traffic. The component Q is the objective function formulated by using the aggregate flow rate vector. Further

$$F_i(Y,H,K) := \begin{bmatrix} F_1(Y,H,K) \\ \vdots \\ F_m(Y,H,K) \end{bmatrix}$$
$$:= A^{-}Y + (I - A^{-}A)H. \tag{13.12}$$

The constraint conditions are given as follows:

a. The *consistency* constraint

$$(I - A^{-}A)Y = 0. \tag{13.13}$$

b. The *positive constraint* that the vector $F(Y,H,K)$ should be positive

$$\begin{bmatrix} F_1(Y,H,K) \\ \vdots \\ F_m(Y,H,K) \end{bmatrix} \geq 0. \tag{13.14}$$

c. The capacity constraints

$$Y \leq C. \tag{13.15}$$

According to the constrain conditions that are presented by Equations 13.13 through 13.15, we can work out the optimum value of the aggregate flow rate vector Y and the individual flow rate vector X, such that they satisfy the routing association $Y = AX$ and maximize the objective function as well. That is, they both maximize the total utility of the network.

13.3.1 NETWORK MODEL WITH A FULL ROW RANK ROUTING MATRIX

For most routing matrices, the number of columns is much greater than the number of rows typically because there are usually much more OD pairs than links in the architecture of networks, which means the routing matrix A cannot be invertible all the time. However, in the real network, the routing matrix can frequently be of full row rank. For example, for the topology of a network, if there is one single-hop flow in every link, such a routing matrix is seen to be of full row rank.

For a full row rank matrix $A \in R^{(n \times m)} (m > n)$, it is noted that AA^{+} is invertible. In this case, the *Moore–Penrose inverse* of A is turned out to be its *right pseudoinverse*, which is denoted by A_R^{+}. In this case, we have

$$A_R^{+} = A^T (AA^T)^{-1}. \tag{13.16}$$

The *singular value* decomposition (SVD) of A is the factorization

$$A = U[\Sigma_{n \times n}, \mathbf{0}_{n \times (m-n)}]V^T, \tag{13.17}$$

where

$$U = [u_1, u_2, \ldots, u_n] \in R^{n \times n}, \tag{13.18}$$

$$V^T = [v_1^T, v_2^T, \ldots, v_n^T, v_{n+1}^T, \ldots, v_m^T]^T \in R^{m \times m}. \tag{13.19}$$

In Equation 13.19, $\alpha = [\alpha_{n+1}, \alpha_{n+2}, \ldots, \alpha_m]^T$ is any scalar vector with the size $m-n$ (just for convenience, we order the elements from $n+1$ there), then one can describe the constrain conditions as the following:

$$\begin{aligned} F(Y,\alpha) &:= \begin{bmatrix} F_1(Y,\alpha) \\ \vdots \\ F_m(Y,\alpha) \end{bmatrix} \\ &:= A_R^+ Y + \sum_{i=n+1}^{m} \alpha_i v_i, \end{aligned} \tag{13.20}$$

$$\begin{aligned} &F(Y,\alpha) \geq 0, \\ &Y \leq C. \end{aligned} \tag{13.21}$$

The problem of router-level utility maximization is to find an optimal aggregate flow rate vector Y such that

$$\begin{aligned} Q_R^* &= \max_{Y,\alpha} \mathrm{Q(Y,}\alpha) \\ &= \max \sum_{i=1}^{m} U_i(F_i(Y,\alpha)). \end{aligned} \tag{13.22}$$

Considering the optimization problem 13.11 together with the constrain conditions as formulated by Equations 13.20 and 13.21, there is an unique optimum: the aggregate flow rate vector Y that maximizes the total utility of the network with a full row rank routing matrix. One can further find the individual flow rate vector X satisfying the routing scheme $Y = AX$.

13.3.2 NETWORK MODEL WITH EVERY LINK HAVING A SINGLE-HOP FLOW

For the network with one single-link flow at each link, the solution to the optimization problem (13.11) is the capacity vector. That is, the utility-optimized bandwidth allocation is the one that makes every link be fully saturated.

For the case, where there is a single-hop flow in every link in the considered network, the solution to the router-level utility optimization problem can be described by

$$Q_R^* = \max \alpha \sum_{i=1}^{m} U_i(F_i(\alpha)), \tag{13.23}$$

subject to the constraints

$$F(\alpha) \geq 0, \tag{13.24}$$

where

$$F(\alpha) := \begin{bmatrix} F_1(\alpha) \\ \vdots \\ F_m(\alpha) \end{bmatrix} := A_R^+ C + \sum_{i=n+1}^{m} \alpha_i v_i. \tag{13.25}$$

By substituting the optimum α^* of the optimization problem 13.23 and 13.24 into 13.25, one thus obtains the utility-optimized individual flow rate rector $X = F(\alpha^*)$.

13.4 APPLICATIONS TO OBS NETWORKS: DEMONSTRATING EXAMPLES

In this section, we provide two examples to demonstrate how one allocates bandwidth of OBS networks by using the router-level (aggregate flow-level) NUM approach.

13.4.1 OBS NETWORK MODEL WITH THREE TO SEVEN NODES

Let us consider an OBS network with a simple architecture: there are six flows traversing from three ingress nodes to their corresponding egress nodes. For example, Flow f_1 and Flow f_2 both are first traversing from the ingress node A. They are aggregated at the ingress link to form the aggregate flow f_{AI}. Next they are switching at the core node. Finally, they are redirected to their corresponding destination egress nodes B and C. The basic structure of aggregation and switching of the OBS network is shown in Figure 13.7. The routing information of this three to seven nodes network is given in Table 13.1. According to the origin router and destination router, we can obtain the capacity constraint of every flow, which is shown by Table 13.2.

For the OBS network shown in Figure 13.7, we partition it into two parts: the ingress network and the egress network as shown in Figure 13.8. According to the aggregate flow-level utility maximization theory, which was proposed by Tan et al. [164], the NUM problem can be formulated and solved in terms of the aggregate flow rate vector. The resulted aggregate flow rate vector that is the optimum of the problem, and the individual flow rate vector that is solved from the normal NUM problem formulated in terms of the individual flow rate vector, are associated with each other through the routing matrix.

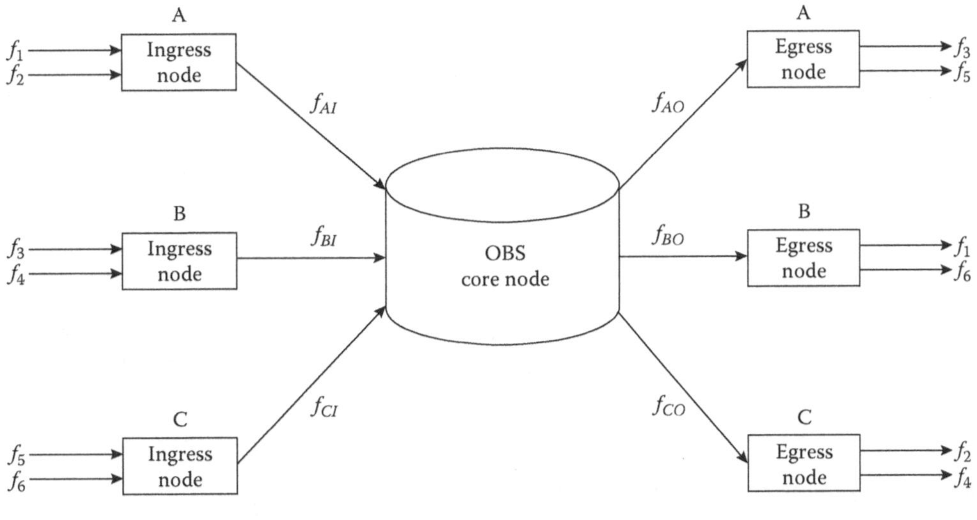

Figure 13.7 An OBS network with three to seven nodes.

Table 13.1
Routing Architecture and Link Capacity of the OBS Network of Figure 13.7

Aggregate Flows	Route	Capacity (Gbps)
f_{AI}	The ingress node A	10
f_{AO}	The egress node A	10
f_{BI}	The ingress node B	15
f_{BO}	The egress node B	15
f_{CI}	The ingress node C	20
f_{CO}	The egress node C	20

Table 13.2
Routing Information and Link Capacity of the OBS Network of Figure 13.8

Flows	Source	End	Capacity(Gbps)
f_1	The ingress node A	The egress node B	$\min\{10, 15\} = 10$
f_2	The ingress node A	The egress node C	$\min\{10, 20\} = 10$
f_3	The ingress node B	The egress node A	$\min\{15, 10\} = 10$
f_4	The ingress node B	The egress node C	$\min\{15, 20\} = 15$
f_5	The ingress node C	The egress node A	$\min\{20, 10\} = 10$
f_6	The ingress node C	The egress node B	$\min\{20, 15\} = 15$

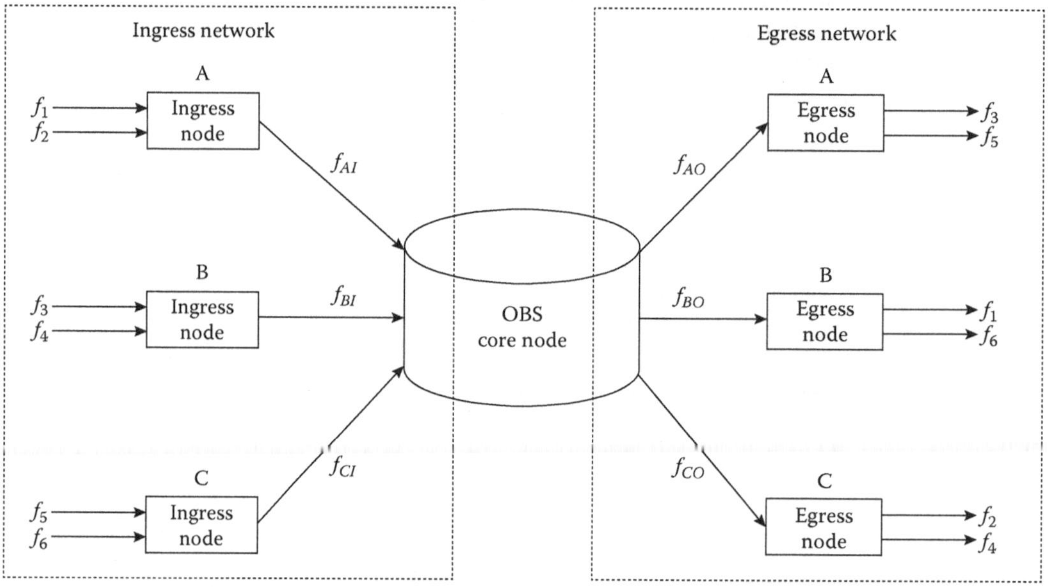

Figure 13.8 Ingress network and egress network.

In what follows, we show how one is able to use the aggregate flow-level utility maximization theory [164] to calculate the aggregate flow rate (bandwidth) and then further to obtain the individual flow rate (bandwidth) for this OBS network.

For these two networks, the ingress network and the egress network, we construct the routing matrices A and D

$$A = \begin{bmatrix} 1 & 1 & 0 & 0 & 0 & 0 \\ 0 & 0 & 1 & 1 & 0 & 0 \\ 0 & 0 & 0 & 0 & 1 & 1 \end{bmatrix}$$

$$D = \begin{bmatrix} 0 & 0 & 1 & 0 & 1 & 0 \\ 1 & 0 & 0 & 0 & 0 & 1 \\ 0 & 1 & 0 & 1 & 0 & 0 \end{bmatrix}$$

and calculate the aggregate flow Y_I in the ingress network and the aggregate flow Y_E in the egress network as follows:

$$Y_I = AX_{IA} = \begin{bmatrix} 1 & 1 & 0 & 0 & 0 & 0 \\ 0 & 0 & 1 & 1 & 0 & 0 \\ 0 & 0 & 0 & 0 & 1 & 1 \end{bmatrix} \begin{bmatrix} f_1 \\ f_2 \\ f_3 \\ f_4 \\ f_5 \\ f_6 \end{bmatrix}$$

$$Y_E = DX_{ID} = \begin{bmatrix} 0 & 0 & 1 & 0 & 1 & 0 \\ 1 & 0 & 0 & 0 & 0 & 1 \\ 0 & 1 & 0 & 1 & 0 & 0 \end{bmatrix} \begin{bmatrix} f_1 \\ f_2 \\ f_3 \\ f_4 \\ f_5 \\ f_6 \end{bmatrix}$$

For the ingress network, the routing matrix A is seen to be of full row rank. From the last section, the *Moore–Penrose inverse* of A is turned out to be its *right pseudoinverse*, which is denoted by A_R^+, and is given by

$$A_R^+ = A^T(AA^T)^{-1}.$$

It is calculated to be

$$A_R^+ = \begin{bmatrix} 0.5 & 0 & 0 \\ 0.5 & 0 & 0 \\ 0 & 0.5 & 0 \\ 0 & 0.5 & 0 \\ 0 & 0 & 0.5 \\ 0 & 0 & 0.5 \end{bmatrix}$$

We further use Equation 13.17 to calculate the SVD of A and to obtain v_i $(i = 4, 5, 6)$ as follows:

$$[v_4, v_5, v_6] = \begin{bmatrix} 0.5 & -0.3536 & -0.3536 \\ -0.5 & 0.3536 & 0.3536 \\ -0.5 & -0.3536 & -0.3536 \\ 0.5 & 0.3536 & 0.3536 \\ 0 & 0.5 & -0.5 \\ 0 & -0.5 & 0.5 \end{bmatrix}.$$

According to Equation 13.20, one has

$$F(Y_I, \alpha) = A_R^+ Y_I + \sum_{i=4}^{6} \alpha_i v_i = \begin{bmatrix} 0.5f_1 + 0.5f_2 + 0.5\alpha_4 - 0.3536\alpha_5 - 0.3536\alpha_6 \\ 0.5f_1 + 0.5f_2 - 0.5\alpha_4 + 0.3536\alpha_5 + 0.3536\alpha_6 \\ 0.5f_3 + 0.5f_4 - 0.5\alpha_4 - 0.3536\alpha_5 - 0.3536\alpha_6 \\ 0.5f_3 + 0.5f_4 + 0.5\alpha_4 + 0.3536\alpha_5 + 0.3536\alpha_6 \\ 0.5f_5 + 0.5f_6 + 0.5\alpha_5 - 0.5\alpha_6 \\ 0.5f_5 + 0.5f_6 - 0.5\alpha_5 + 0.5\alpha_6 \end{bmatrix}.$$

According to Equations 13.21 and 13.22, we can formulate the constraint conditions and the objective function of the NUM model in terms of the aggregate flow rate vector for the ingress network as follows:

$$Q_R = \max \alpha \sum_{i=1}^{6} U_i(F(Y_I, \alpha)),$$
$$Y_I \leq C,$$
$$F(Y_I, \alpha) \geq 0.$$

If we take the utility function as the following:

$$U_i(x_i) = \log(x_i + 1), \ i = 1, 2, \ldots, 6,$$

the solutions to the above optimization problem are obtained as follows:

$Q_R^* = 11.1519$: the maximized total utility of the ingress network,
$Y_I = [10, 15, 20]^T$: the optimum of the aggregate flow rate of the ingress network,
$\alpha \doteq [\alpha_4, \alpha_5, \alpha_6]^T = [7.5, 5.3176, 5.3176]^T$: the parameters under which the optimization of the total utility is obtained for the ingress network.

If we substitute the obtained Y_I, α, and v_i $(i = 4, 5, 6)$ into Equation 13.20, we further obtain the desired optimum of the individual flow rate vector for the ingress network, which is given by

$$X_{IA} = F(Y_I, \alpha) = [5, 5, 0, 15, 10, 10]^T.$$

Next, we work on the egress network using the same method to obtain the optimal aggregate bandwidth allocation and then further to obtain the desired individual flow bandwidth allocation.

The routing matrix D in the egress network is seen to be of full row rank. Consider in the equation

$$Y_E = DX_{ID}$$

the *Moore–Penrose Inverse* of D is turned out to be its *right pseudoinverse*, which is denoted by D_R^+, $D_R^+ = D^T(DD^T)^{-1}$. So, the *right pseudoinverse* of D is calculated to be

$$D_R^+ = \begin{bmatrix} 0 & 0.5 & 0 \\ 0 & 0 & 0.5 \\ 0.5 & 0 & 0 \\ 0 & 0 & 0.5 \\ 0.5 & 0 & 0 \\ 0 & 0.5 & 0 \end{bmatrix}$$

We further use Equation 13.17 to calculate the SVD of D and obtain the singular vectors v_i' $(i = 4, 5, 6)$ as follows:

$$[v_4', v_5', v_6'] = \begin{bmatrix} 0.1847 & -0.2612 & -0.6306 \\ -0.6306 & 0.1847 & -0.2612 \\ -0.2612 & -0.6306 & 0.1847 \\ 0.6306 & -0.1847 & 0.2612 \\ 0.2612 & 0.6306 & -0.1847 \\ -0.1847 & 0.2612 & 0.6306 \end{bmatrix}.$$

According to Equation 13.20, we have the following equation to formulate the aggregate flow vector in the egress network:

$$F(Y_E,\beta) = D_R^+ Y_E + \sum_{i=4}^{6} \beta_i v_i'$$
$$= \begin{bmatrix} 0.5f_1+0.5f_6+0.1847\beta_4-0.2612\beta_5-0.6306\beta_6 \\ 0.5f_2+0.5f_4-0.6306\beta_4+0.1847\beta_5-0.2612\beta_6 \\ 0.5f_3+0.5f_5-0.2612\beta_4-0.6306\beta_5+0.1847\beta_6 \\ 0.5f_2+0.5f_4+0.6306\beta_4-0.1847\beta_5+0.2612\beta_6 \\ 0.5f_3+0.5f_5+0.2612\beta_4+0.6306\beta_5-0.1847\beta_6 \\ 0.5f_1+0.5f_6-0.1847\beta_4+0.2612\beta_5+0.6306\beta_6 \end{bmatrix},$$

where Y_E is a 3×1 vector and $\beta_i,\ i = 4,5,6$ are the parameters. According to Equations 13.21 and 13.22, we can formulate the constraint conditions and the objective function of the NUM model in terms of the aggregate flow rate vector for the egress network as follows:

$$Q_R = \max \beta \sum_{i=1}^{6} U_i(F(Y_E,\beta)),$$
$$Y_E \leq C,$$
$$F(Y_E,\beta) \geq 0.$$

If we take the utility function as the following:

$$U_i(x_i) = \log(x_i+1), i = 1,2,\ldots,6,$$

the solutions to the above optimization problem are

$Q_R^* = 11.1519$: the maximized total utility of the egress network,
$Y_E = [10,\ 15,\ 20]^T$: the optimum of the aggregate flow rate of the egress network,
$\beta \doteq [\beta_4,\ \beta_5,\ \beta_6]^T = [-2.7705,\ 3.9180,\ 9.4590]^T$: the parameters. under which the optimization of the total utility is achieved for the egress network.

By substituting the above Y_E, $[v_4', v_5', v_6']$ and β into Equation 13.20, one further obtains the individual flow rate vector in the egress network as the following:

$$X_{ID} = F(Y_E,\beta)$$
$$= [0,\ 10,\ 5,\ 10,\ 5,\ 15]^T.$$

Because α and β are the arbitrary scalar vectors, X_{IA} and X_{ID} will change with α and β. However, from the perspective of router level, we can see from Table 13.3 that the ingress network and the egress network both reach their maximum utility if they are both fully occupied in terms of their link capacities.

Through the above methods, we are able to calculate the proportions of bandwidth allocation both for ingress network and the egress network, which drive the total utility of the two networks to their maxima at the router level. We believe that, as long as the bandwidth allocation for the entire network satisfies the following condition, from the router level, the network can achieve the maximum utility:

$$X = \lambda X_{IA} + (1-\lambda) X_{ID};\ (0 \leq \lambda \leq 1).$$

Note that, the components X_{IA} and X_{ID} are the individual flow rate vectors in the ingress network and the egress network, respectively.

Table 13.3
Total Utility of Router

$f_{AI} = 10$	$f_{BI} = 15$	$f_{CI} = 20$	$U(Y_I) = \log(f_{AI}+1)+\log(f_{BI}+1)+\log(f_{CI}+1) = 8.2150$
$f_{AO} = 10$	$f_{BO} = 15$	$f_{CO} = 20$	$U(Y_E) = \log(f_{AO}+1)+\log(f_{BO}+1)+\log(f_{CO}+1) = 8.2150$

Table 13.4
Bandwidth Allocation and Network Utility with Various λ

λ	x_1	x_2	x_3	x_4	x_5	x_6	$Q(X) = \sum_{i=1}^{6} \log(x_i + 1)$	$Q(Y) = \sum U(Y)$
0	0	10	5	10	5	15	11.152	8.215
0.05	0.25	9.75	4.75	10.25	5.25	14.75	11.357	8.215
0.1	0.5	9.5	4.5	10.5	5.5	14.5	11.517	8.215
0.15	0.75	9.25	4.25	10.75	5.75	14.25	11.643	8.215
0.2	1	9	4	11	6	14	11.744	8.215
0.25	1.25	8.75	3.75	11.25	6.25	13.75	11.824	8.215
0.3	1.5	8.5	3.5	11.5	6.5	13.5	11.886	8.215
0.35	1.75	8.25	3.25	11.75	6.75	13.25	11.933	8.215
0.4	2	8	3	12	7	13	11.966	8.215
0.45	2.25	7.75	2.75	12.25	7.25	12.75	11.985	8.215
0.5	2.5	7.5	2.5	12.5	7.5	12.5	11.991	8.215
0.55	2.75	7.25	2.25	12.75	7.75	12.25	11.985	8.215
0.6	3	7	2	13	8	12	11.966	8.215
0.65	3.25	6.75	1.75	13.25	8.25	11.75	11.933	8.215
0.7	3.5	6.5	1.5	13.5	8.5	11.5	11.886	8.215
0.75	3.75	6.25	1.25	13.75	8.75	11.25	11.824	8.215
0.8	4	6	1	14	9	11	11.744	8.215
0.85	4.25	5.75	0.75	14.25	9.25	10.75	11.643	8.215
0.9	4.5	5.5	0.5	14.5	9.5	10.5	11.517	8.215
0.95	4.75	5.25	0.25	14.75	9.75	10.25	11.357	8.215
1	5	5	0	15	10	10	11.152	8.215

Recall that we have obtained the following optimal individual flow rate vectors for the ingress network and the egress network that achieve the maximized utility for both networks:

$$X_{IA} = [5,\ 5,\ 0,\ 15,\ 10,\ 10]^T,\ X_{ID} = [0,\ 10,\ 5,\ 10,\ 5,\ 15]^T.$$

Using the above two vectors and letting λ take the value from 0 to 1 increased by the step size 0.05, the rate of individual flows and the total network utility on both in the flow level and router level corresponding to λ are calculated and they are then outlined by Table 13.4. In the table, we denote $X = [x_1, x_2, \ldots, x_6]^T$, and we calculate the total utility of routers in the following way:

$$\begin{aligned} Q(Y) = \sum U(Y) = U(Y_I) &= \log(f_{AI}+1)+\log(f_{BI}+1)+\log(f_{CI}+1) \\ = U(Y_E) &= \log(f_{AO}+1)+\log(f_{BO}+1)+\log(f_{CO}+1). \end{aligned}$$

From Table 13.4, we find that the router-level network utility is fixed at the value 8.215 no matter what value of λ is because the aggregate flow rate vector is always fixed at $Y = [10, 15, 20]^T$, under which every router is saturated. However, the flow-level network utility increases at first and then decreases, the peak value is 11.991 ($\lambda = 0.5$). Therefore, the bandwidth allocation

$$X = [2.5, 7.5, 2.5, 12.5, 7.5, 12.5]^T$$

is the mostly desirable and the mostly ideal. That is, if we allocate the bandwidth for the three to seven nodes OBS network as this, the whole network utility will reach its maximum both in the router level and flow level.

13.4.2 OBS NETWORK MODEL WITH FOUR TO NINE NODES

In order to further verify the method that we proposed in this chapter, let us consider another OBS network with more complex architecture: there are 12 flows traversing from four ingress nodes to their corresponding egress nodes. For example, Flow f_1, Flow f_2, and Flow f_3 are all first traversing from the ingress node A. They are all aggregated at the ingress link to form the aggregate flow f_{AI}. Next, they are switched at the core node. Finally, they are redirected to their corresponding destination egress nodes B, C, and D. The basic structure of aggregation and switching of the OBS network is shown in Figure 13.9.

The routing information of this four to nine node OBS network and the link capacity for the ingress nodes and the egress nodes are given in Table 13.5. According to the origin router and destination router, we can obtain the capacity constraints of every flow, which are given by Table 13.6.

We partition this OBS network into two parts: the ingress network and the egress network, which is shown in Figure 13.10.

For these two networks, we construct their routing matrices A and D and then formulate the relations between the aggregate flow rate vector and the individual flow rate vector through the routing structure. The routing matrix for the ingress network is

$$A = \begin{bmatrix} 1 & 1 & 1 & 0 & 0 & 0 & 0 & 0 & 0 & 0 & 0 & 0 \\ 0 & 0 & 0 & 1 & 1 & 1 & 0 & 0 & 0 & 0 & 0 & 0 \\ 0 & 0 & 0 & 0 & 0 & 0 & 1 & 1 & 1 & 0 & 0 & 0 \\ 0 & 0 & 0 & 0 & 0 & 0 & 0 & 0 & 0 & 1 & 1 & 1 \end{bmatrix}$$

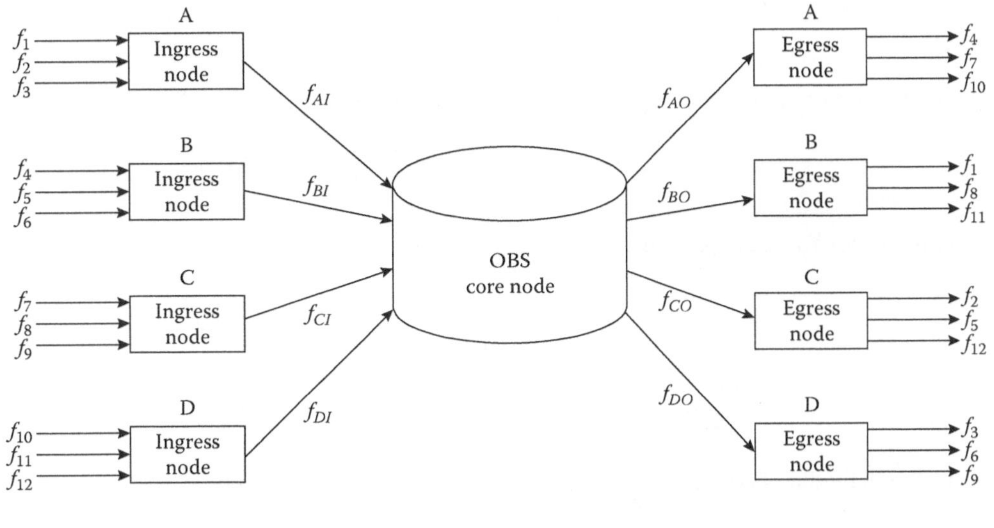

Figure 13.9 An OBS network with four to nine nodes.

Table 13.5
Routing Information and Link Capacity of Every Router

Flows	Route	Capacity (Gbps)
f_{AI}	The ingress node A	15
f_{AO}	The egress node A	15
f_{BI}	The ingress node B	15
f_{BO}	The egress node B	15
f_{CI}	The ingress node C	20
f_{CO}	The egress node C	20
f_{DI}	The ingress node D	30
f_{DO}	The egress node D	30

Table 13.6
Routing Information of Every Flow and the Capacity Constraints

Flows	Source	End	Capacity (Gbps)
f_1	The ingress node A	The egress node B	$\min\{15, 15\} = 15$
f_2	The ingress node A	The egress node C	$\min\{15, 20\} = 15$
f_3	The ingress node A	The egress node D	$\min\{15, 30\} = 15$
f_4	The ingress node B	The egress node A	$\min\{15, 15\} = 15$
f_5	The ingress node B	The egress node C	$\min\{15, 20\} = 15$
f_6	The ingress node B	The egress node D	$\min\{15, 30\} = 15$
f_7	The ingress node C	The egress node A	$\min\{20, 15\} = 15$
f_8	The ingress node C	The egress node B	$\min\{20, 15\} = 15$
f_9	The ingress node C	The egress node D	$\min\{20, 30\} = 20$
f_{10}	The ingress node D	The egress node A	$\min\{30, 15\} = 15$
f_{11}	The ingress node D	The egress node B	$\min\{30, 15\} = 15$
f_{12}	The ingress node D	The egress node C	$\min\{30, 20\} = 20$

The individual flow vector in the ingress network is denoted by

$$X_{IA} = [f_1, f_2, f_3, f_4, f_5, f_6, f_7, f_8, f_9, f_{10}, f_{11}, f_{12}]^T.$$

The routing matrix for the egress network is

$$D = \begin{bmatrix} 0 & 0 & 0 & 1 & 0 & 0 & 1 & 0 & 0 & 1 & 0 & 0 \\ 1 & 0 & 0 & 0 & 0 & 0 & 0 & 1 & 0 & 0 & 1 & 0 \\ 0 & 1 & 0 & 0 & 1 & 0 & 0 & 0 & 0 & 0 & 0 & 1 \\ 0 & 0 & 1 & 0 & 0 & 1 & 0 & 0 & 1 & 0 & 0 & 0 \end{bmatrix}$$

Note that the individual flow vector in the ingress network is still denoted by

$$X_{ID} = [f_1, f_2, f_3, f_4, f_5, f_6, f_7, f_8, f_9, f_{10}, f_{11}, f_{12}]^T.$$

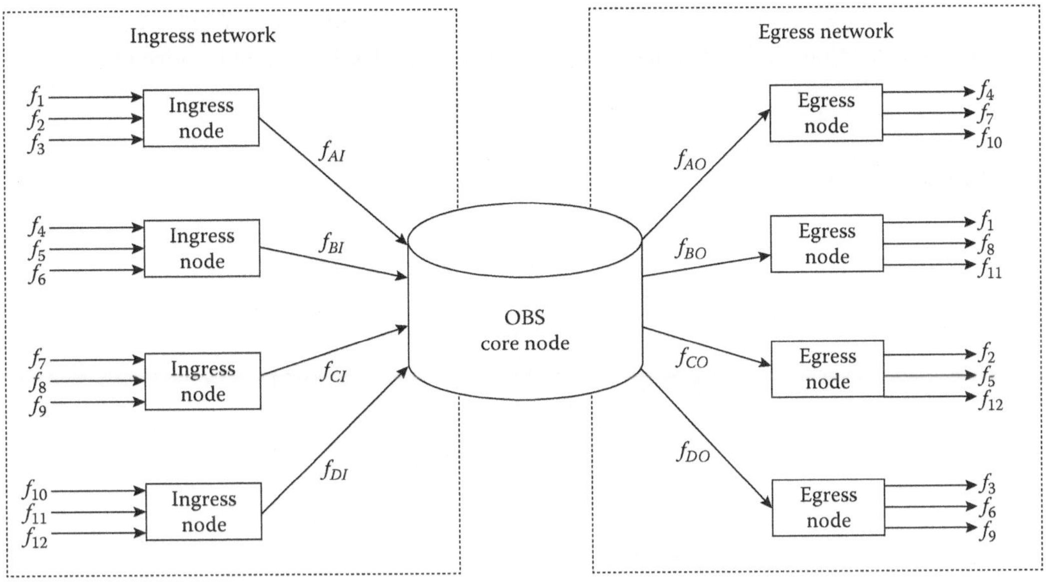

Figure 13.10 Ingress network and egress network.

Therefore, we have the aggregate flow rate vectors in the ingress network and the egress network: Y_I and Y_E, which are calculated in the following manner:

$$Y_I = AX_{IA}$$

$$= \begin{bmatrix} 1 & 1 & 1 & 0 & 0 & 0 & 0 & 0 & 0 & 0 & 0 & 0 \\ 0 & 0 & 0 & 1 & 1 & 1 & 0 & 0 & 0 & 0 & 0 & 0 \\ 0 & 0 & 0 & 0 & 0 & 0 & 1 & 1 & 1 & 0 & 0 & 0 \\ 0 & 0 & 0 & 0 & 0 & 0 & 0 & 0 & 0 & 1 & 1 & 1 \end{bmatrix} \begin{bmatrix} f_1 \\ f_2 \\ f_3 \\ f_4 \\ f_5 \\ f_6 \\ f_7 \\ f_8 \\ f_9 \\ f_{10} \\ f_{11} \\ f_{12} \end{bmatrix}$$

$$= DX_{ID} = \begin{bmatrix} 0 & 0 & 0 & 1 & 0 & 0 & 1 & 0 & 0 & 1 & 0 & 0 \\ 1 & 0 & 0 & 0 & 0 & 0 & 0 & 1 & 0 & 0 & 1 & 0 \\ 0 & 1 & 0 & 0 & 1 & 0 & 0 & 0 & 0 & 0 & 0 & 1 \\ 0 & 0 & 1 & 0 & 0 & 1 & 0 & 0 & 1 & 0 & 0 & 0 \end{bmatrix} \begin{bmatrix} f_1 \\ f_2 \\ f_3 \\ f_4 \\ f_5 \\ f_6 \\ f_7 \\ f_8 \\ f_9 \\ f_{10} \\ f_{11} \\ f_{12} \end{bmatrix}$$

For the ingress network, the routing matrix A is seen to be of full row rank. The *Moore–Penrose inverse* of A is therefore its *right pseudoinverse*, which is denoted by A_R^+, and is given by

$$A_R^+ = A^T(AA^T)^{-1}.$$

In particular, one has

$$A_R^+ = \begin{bmatrix} 0.3333 & 0 & 0 & 0 \\ 0.3333 & 0 & 0 & 0 \\ 0.3333 & 0 & 0 & 0 \\ 0 & 0.3333 & 0 & 0 \\ 0 & 0.3333 & 0 & 0 \\ 0 & 0.3333 & 0 & 0 \\ 0 & 0 & 0.3333 & 0 \\ 0 & 0 & 0.3333 & 0 \\ 0 & 0 & 0.3333 & 0 \\ 0 & 0 & 0 & 0.3333 \\ 0 & 0 & 0 & 0.3333 \\ 0 & 0 & 0 & 0.3333 \end{bmatrix}$$

One further uses Equations 13.17 through 13.19 to calculate the SVD of A and to obtain the vector $V_{5-12} \triangleq [v_5, v_6, v_7, v_8, v_9, v_{10}, v_{11}, v_{12}]^T$, which are given by

$$V_{5-12} = \begin{bmatrix} 0.3333 & 0.3333 & 0.3333 & 0.3333 & 0.3333 & -0.1925 & -0.1925 & -0.1925 \\ -0.4553 & -0.4552 & 0.1220 & 0.1220 & 0.1220 & 0.2629 & 0.2629 & 0.2629 \\ 0.1220 & 0.1220 & -0.4553 & -0.4553 & -0.4553 & -0.0704 & -0.0704 & -0.0704 \\ -0.3333 & -0.3333 & 0 & 0 & 0 & -0.3849 & -0.3849 & -0.3849 \\ 0.6667 & -0.3333 & 0 & 0 & 0 & 0.1925 & 0.1925 & 0.1925 \\ -0.3333 & 0.6667 & 0 & 0 & 0 & 0.1925 & 0.1925 & 0.1925 \\ 0 & 0 & 0.6667 & -0.3333 & -0.3333 & 0 & 0 & 0 \\ 0 & 0 & -0.3333 & 0.6667 & -0.3333 & 0 & 0 & 0 \\ 0 & 0 & -0.3333 & -0.3333 & 0.6667 & 0 & 0 & 0 \\ 0 & 0 & 0 & 0 & 0 & 0.6667 & -0.3333 & -0.3333 \\ 0 & 0 & 0 & 0 & 0 & -0.3333 & 0.6667 & -0.3333 \\ 0 & 0 & 0 & 0 & 0 & -0.3333 & -0.3333 & 0.6667 \end{bmatrix}$$

According to Equation 13.20, we have the following formula to calculate the aggregate flow rate vector in the ingress network:

$$F(Y_I, \alpha) = A_R^+ Y_I + \sum_{i=5}^{12} \alpha_i v_i,$$

where Y_I is a 4×1 vector, and α denotes the vector that consists of eight parameters α_i, $i = 5, 6, \ldots, 12$. According to Equations 13.21 and 13.22, we formulate the NUM problem in terms of the aggregate flow rate vector Y_I for the ingress network as follows:

$$Q_R^* = \max \alpha \sum_{i=1}^{12} U_i(F(Y_I, \alpha)),$$

$$\text{subject to } F(Y_I, \alpha) \geq 0,$$

$$Y_I \leq C,$$

where C stands for the link capacity at the ingress links, which is assumed to be $[15, 15, 20, 30]^T$. By taking the following utility functions:

$$U_i(x_i) = \log(x_i + 1), i = 1, 2, \ldots, 12,$$

we obtain the solutions to the above optimization problem: Y_I and α and the maximized total utility value, Q_R^*. They are

$$\begin{aligned} &Q_R^* = 22.2186, \\ &\alpha = [4.7472, 5.7522, 0.5211, 0.5211, 0.5211, 6.4935, 6.4935, 6.4935]^T, \\ &Y_I = [15,\ 15,\ 20,\ 30]^T. \end{aligned}$$

If we substitute the above Y_I and α into Equation 13.20, one further obtains the individual flow rate vector

$$\begin{aligned} X_{IA} &= F(Y_I, \alpha) \\ &= [4, 5, 6, 4, 5, 6, 1, 1, 18, 10, 10, 10]^T. \end{aligned}$$

Next we work on the egress network in the same manner. The routing matrix D of the egress network is seen to be of full row rank as well. According to the equations

$$Y_E = DX_{ID}$$

the *Moore–Penrose inverse* of D is turned out to be its *right pseudoinverse*, which is denoted by D_R^+, $D_R^+ = D^T(DD^T)^{-1}$. So, the *right pseudoinverse* of D is calculated to be

$$D_R^+ = \begin{bmatrix} 0 & 0.3333 & 0 & 0.0000 \\ 0 & 0 & 0.3333 & 0.0000 \\ 0 & 0 & 0 & 0.3333 \\ 0.3333 & 0.0000 & 0 & -0.0000 \\ 0 & 0 & 0.3333 & 0.0000 \\ 0 & 0 & 0 & 0.3333 \\ 0.3333 & 0.0000 & 0 & 0.0000 \\ 0 & 0.3333 & 0 & -0.0000 \\ 0 & 0 & 0 & 0.3333 \\ 0.3333 & 0.0000 & 0 & 0.0000 \\ 0 & 0.3333 & 0 & -0.0000 \\ 0 & 0 & 0.3333 & 0.0000 \end{bmatrix}$$

We further use Equations 13.17 and 13.19 to calculate the SVD of D and obtain the vector $V_{5-12} \triangleq [v_5, v_6, v_7, v_8, v_9, v_{10}, v_{11}, v_{12}]^T$ as follows:

$$V_{5-12} = \begin{bmatrix} 0.1155 & -0.2 & -0.3464 & -0.4 & -0.2 & -0.3464 & -0.4 & 0.1155 \\ -0.4 & 0.1155 & 0.2 & -0.3464 & 0.1155 & 0.2 & -0.3464 & -0.4 \\ -0.3464 & -0.4 & -0.1155 & 0.2 & -0.4 & -0.1155 & 0.2 & -0.3464 \\ -0.2 & 0.3464 & -0.4 & 0.1155 & 0.3464 & -0.4 & 0.1155 & -0.2 \\ 0.7 & -0.0577 & -0.1 & 0.1732 & -0.0577 & -0.1 & 0.1732 & -0.3 \\ 0.1732 & 0.7 & 0.0577 & -0.1 & -0.3 & 0.0577 & -0.1 & 0.1732 \\ 0.1 & -0.1732 & 0.7 & -0.0577 & -0.1732 & -0.3 & -0.0577 & 0.1 \\ -0.0577 & 0.1 & 0.1732 & 0.7 & 0.1 & 0.1732 & -0.3 & -0.0577 \\ 0.1732 & -0.3 & 0.0577 & -0.1 & 0.7 & 0.0577 & -0.1 & 0.1732 \\ 0.1 & -0.1732 & -0.3 & -0.0577 & -0.1732 & 0.7 & -0.0577 & 0.1 \\ -0.0577 & 0.1 & 0.1732 & -0.3 & 0.1 & 0.1732 & 0.7 & -0.0577 \\ -0.3 & -0.0577 & -0.1 & 0.1732 & -0.0577 & -0.1 & 0.1732 & 0.7 \end{bmatrix}$$

According to Equation 13.20, we have

$$F(Y_E,\beta) = D_R^+ Y_E + \sum_{i=5}^{12} \beta_i v_i,$$

where Y_E is a 12×1 vector and $\beta = [\beta_5, \beta_6, \ldots, \beta_{12}]^T$ is a scalar parameter vector. According to Equations 13.21 and 13.22, we formulate the NUM problem in terms of the aggregate flow rate vector Y_E for the egress network as follows:

$$Q_R^* = \max_\beta \sum_{i=1}^{12} U_i(F(Y_E,\beta)),$$
$$\text{subject to } F(Y_E,\beta) \geq 0,\ Y_E \leq C,$$

where the capacity vector for the egress network is assumed to be $C = [15, 15, 20, 30]^T$. By taking the following utility function:

$$U_i(x_i) = \log(x_i + 1), \quad i = 1, 2, \ldots, 12,$$

we then obtain the solutions to the above optimization problem as follows:

$$\beta = [-7.5154, -0.088, 2.3427, -1.5576, -0.088, 4.8427, 0.9424, 4.9846]^T$$
$$Y_E = [15,\ 15,\ 20,\ 30]^T.$$

Under the above optimal solutions, the total utility for the egress network reaches its maximum:

$$Q_R^* = 22.8410.$$

If one substitutes the above Y_E and β into Equation 13.20, one further obtains the desired individual flow rate vector:

$$\begin{aligned} X_{ID} &= F(Y_E,\beta) \\ &= [2.5,\ 2.5,\ 10,\ 2.5,\ 2.5,\ 10,\ 5,\ 5,\ 10,\ 7.5,\ 7.5,\ 15]^T. \end{aligned}$$

Looking from the routing level, as previously assumed, if we fix all the link capacities for the ingress routes and the egress routes as the following:

$$f_{AI} = 15 \text{ Gbps}, f_{BI} = 15 \text{ Gbps}, f_{CI} = 20 \text{ Gbps}, f_{DI} = 30 \text{ Gbps},$$
$$f_{AO} = 15 \text{ Gbps}, f_{BO} = 15 \text{ Gbps}, f_{CO} = 20 \text{ Gbps}, f_{DO} = 30 \text{ Gbps},$$

the total utility for the ingress network and the egress network is calculated to be

$$U(Y_I) = \log(f_{AI}+1) + \log(f_{BI}+1) + \log(f_{CI}+1) + \log(f_{DI}+1) = 12.0237,$$
$$U(Y_E) = \log(f_{AO}+1) + \log(f_{BO}+1) + \log(f_{CO}+1) + \log(f_{DO}+1) = 12.0237.$$

Note that the value 12.0237 is the maximal total utility both for the ingress network and the egress network under the situation, where the bandwidth allocation for the aggregate flows reaches the full capacities of the ingress routes and the egress routes. That is, only in the situation where both the ingress routes and the egress routes are fully saturated, their aggregate flow rates achieve their maximal total utility for the ingress network and the egress network.

Recall that we have obtained the following optimal individual flow rate vectors for the ingress network and the egress network that achieve the maximized utility for both networks:

$$X_{IA} = [4, 5, 6, 4, 5, 6, 1, 1, 18, 10, 10, 10]^T,$$
$$X_{ID} = [2.5,\ 2.5,\ 10,\ 2.5,\ 2.5,\ 10,\ 5,\ 5,\ 10,\ 7.5,\ 7.5,\ 15]^T.$$

Using the above two vectors and letting λ take the value from 0 to 1 and be increased by the step size 0.05, we propose the following convex combination formula:

$$X = \lambda X_{IA} + (1-\lambda)X_{ID};\ (0 \leq \lambda \leq 1)$$

to calculate the rate of individual flows for the whole OBS network. The rate of individual flows for the whole network and the total network utility both in the flow level and router level

Table 13.7
Bandwidth Allocation and Network Utility with Different λ

λ	x_1	x_2	x_3	x_4	x_5	x_6	x_7
0	2.5	2.5	10	2.5	2.5	10	5
0.05	2.575	2.625	9.8	2.575	2.625	9.8	4.8
0.1	2.65	2.75	9.6	2.65	2.75	9.6	4.6
0.15	2.725	2.875	9.4	2.725	2.875	9.4	4.4
0.2	2.8	3	9.2	2.8	3	9.2	4.2
0.25	2.875	3.125	9	2.875	3.125	9	4
0.3	2.95	3.25	8.8	2.95	3.25	8.8	3.8
0.35	3.025	3.375	8.6	3.025	3.375	8.6	3.6
0.4	3.1	3.5	8.4	3.1	3.5	8.4	3.4
0.45	3.175	3.625	8.2	3.175	3.625	8.2	3.2
0.5	3.25	3.75	8	3.25	3.75	8	3
0.55	3.325	3.875	7.8	3.325	3.875	7.8	2.8
0.6	3.4	4	7.6	3.4	4	7.6	2.6
0.65	3.475	4.125	7.4	3.475	4.125	7.4	2.4
0.7	3.55	4.25	7.2	3.55	4.25	7.2	2.2
0.75	3.625	4.375	7	3.625	4.375	7	2
0.8	3.7	4.5	6.8	3.7	4.5	6.8	1.8
0.85	3.775	4.625	6.6	3.775	4.625	6.6	1.6
0.9	3.85	4.75	6.4	3.85	4.75	6.4	1.4
0.95	3.925	4.875	6.2	3.925	4.875	6.2	1.2
1	4	5	6	4	5	6	1

λ	x_8	x_9	x_{10}	x_{11}	x_{12}	$Q(X)=\sum_{i=1}^{12}\log(x_i+1)$	$Q(Y)$
0	5	10	7.5	7.5	15	22.841	12.024
0.05	4.8	10.4	7.625	7.625	14.75	22.898	12.024
0.1	4.6	10.8	7.75	7.75	14.5	22.947	12.024
0.15	4.4	11.2	7.875	7.875	14.25	22.988	12.024
0.2	4.2	11.6	8	8	14	23.021	12.024
0.25	4	12	8.125	8.125	13.75	23.046	12.024
0.3	3.8	12.4	8.25	8.25	13.5	23.062	12.024
0.35	3.6	12.8	8.375	8.375	13.25	23.070	12.024
0.4	3.4	13.2	8.5	8.5	13	23.069	12.024
0.45	3.2	13.6	8.625	8.625	12.75	23.061	12.024
0.5	3	14	8.75	8.75	12.5	23.042	12.024
0.55	2.8	14.4	8.875	8.875	12.25	23.015	12.024
0.6	2.6	14.8	9	9	12	22.978	12.024
0.65	2.4	15.2	9.125	9.125	11.75	22.930	12.024
0.7	2.2	15.6	9.25	9.25	11.5	22.871	12.024
0.75	2	16	9.375	9.375	11.25	22.800	12.024
0.8	1.8	16.4	9.5	9.5	11	22.716	12.024
0.85	1.6	16.8	9.625	9.625	10.75	22.618	12.024
0.9	1.4	17.2	9.75	9.75	10.5	22.504	12.024
0.95	1.2	17.6	9.875	9.875	10.25	22.372	12.024
1	1	18	10	10	10	22.219	12.024

corresponding to λ are calculated and they are then outlined by Table 13.7. In the table, we denote $X = [x_1, x_2, \ldots, x_{12}]^T$, and we calculate the total utility of routers in the following way:

$$\begin{aligned} Q(Y) &= \sum U(Y) \\ &= U(Y_I) = \log(f_{AI}+1) + \log(f_{BI}+1) + \log(f_{CI}+1) + \log(f_{DI}+1) \\ &= U(Y_E) = \log(f_{AO}+1) + \log(f_{BO}+1) + \log(f_{CO}+1) + \log(f_{DO}+1). \end{aligned}$$

Observing in Table 13.7, we find that, the router-level network utility is stabilized at 12.024 no matter what the value of λ is. This is because the aggregate flow rate vector is always fixed at $Y = [15, 15, 20, 30]^T$, which drives every router to be fully saturated to yield the maximal total utility. However, a different story arises from the flow level. At the flow level, the network utility is increased at first and then decreased, the peak value is 23.070 (where $\lambda = 0.35$). Therefore, when we allocate this four to nine nodes network's bandwidth as

$$X = [3.025, 3.375, 8.6, 3.025, 3.375, 8.6, 3.6, 3.6, 12.8, 8.375, 8.375, 13.25]^T,$$

the whole network utility will reach its maxima both in the router level and flow level.

13.5 NUMERICAL PLOTS AND ANALYSES

This section first plots some numerical results that are presented in the previous section and then provides discussions to these plots to yield some insights on how to allocate the bandwidth for the OBS networks efficiently by using the NUM theory from the perspective of the aggregate flow level.

Using the convex combination

$$X = \lambda X_{IA} + (1-\lambda) X_{ID},$$

the relationship between flow rate X and λ is depicted in Figure 13.11, where the x-axis presents the individual flow number, the y-axis presents the value of λ, and the z-axis presents the flow rate. When $\lambda = 0$, the pillars of all sorts of color match the individual flows' rates, which leads to the optimized utility in the ingress network; while when $\lambda = 1$, the pillars match the individual flows' rate, which leads to the optimized utility in the egress network. For the three to seven nodes network model, the relation between the network utility and the convex combination parameter λ is depicted by Figure 13.12. The bar chart in Figure 13.12 describes the relationship between the network utility on flow level and λ. The x-axis presents the value of λ and the y-axis presents the network utility on flow level. As the value of λ increases from 0 to 1, we can clearly see that the utility increases first, peaks at $\lambda = 0.5$, and then starts to fall. According to the two figures, Figures 13.11 and 13.12, we find that why X_{IA} ($\lambda = 1$) or X_{ID} ($\lambda = 0$) is not the optimal individual flow rate vector. Even though they lead to the maximization of network utility on router level, due to some individual flow rate is 0, which means there exists idle channel, they cannot lead to the maximization of network utility on flow level. Therefore, we draw the conclusion that to achieve the maximization of network utility on flow level, each channel must be used without any channel being idle. We describe the following bandwidth allocation scheme:

$$X = \lambda X_{IA} + (1-\lambda) X_{ID}, \quad 0 \le \lambda \le 1, \tag{13.26}$$

$$Q_R^*(X) = \max \sum_{i=1}^{m} X_i, \tag{13.27}$$

$$Q_R^*(Y) = \max \sum_{i=1}^{n} Y_i, \tag{13.28}$$

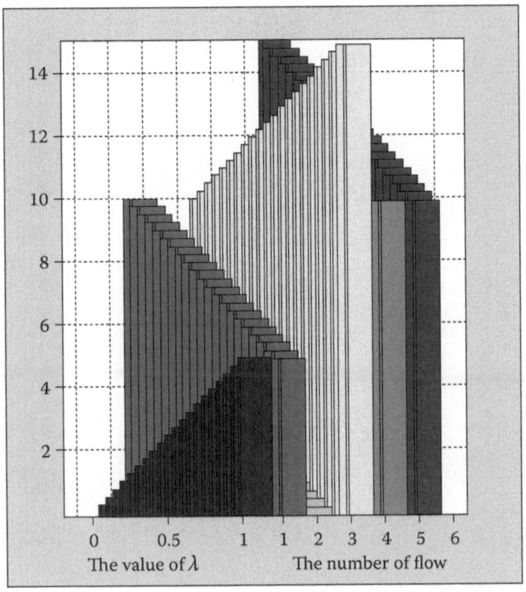

Figure 13.11 Rates of flow and λ in the three to seven nodes network.

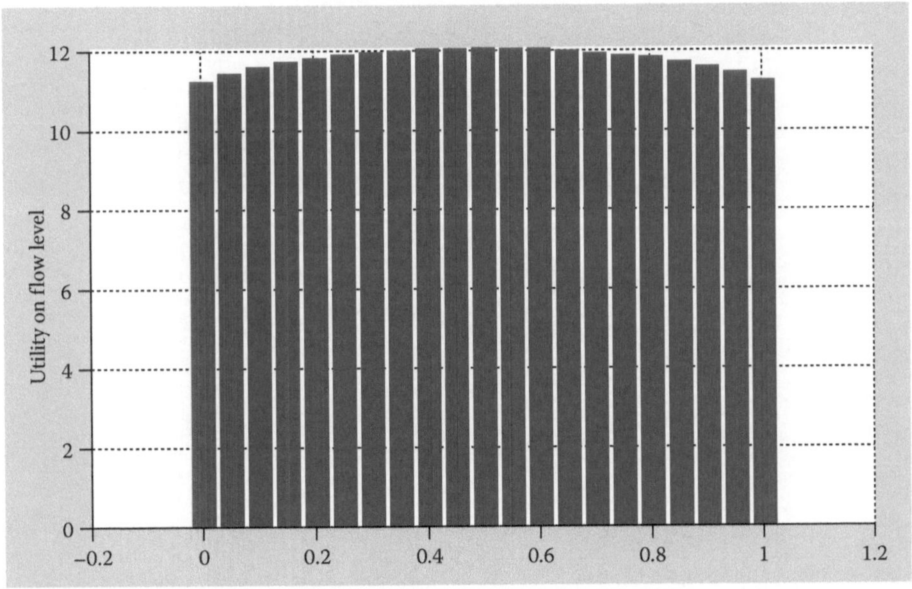

Figure 13.12 Utility on flow level and λ in the three to seven nodes network.

which is then displayed in the 3D coordinate system in Figure 13.13. In this figure, the x-axis presents the value of λ, the y-axis presents the utility on router level, which is $Q_R^*(Y)$, and the z-axis presents the utility on flow level, which is $Q_R^*(X)$. Observing from Figure 13.13, we can see that no matter how λ changes, $Q_R^*(Y)$ is always equal to 8.215, and the aggregate flow rate vector is fixed at $Y = [10, 15, 20]^T$ during the parameter λ changes, which means all aggregate links are fully saturated. In this situation, the routers will maintain the full-load state and the utility of every edge router always reaches its maximization. But the utility on flow level increases first and then

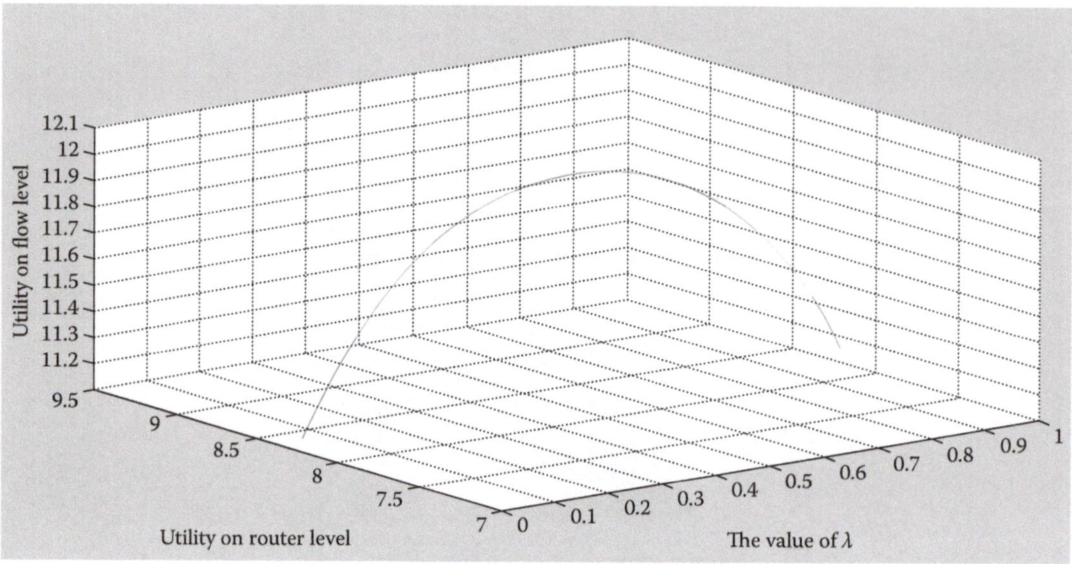

Figure 13.13 The curve of relation of utility and λ in the three to seven nodes network.

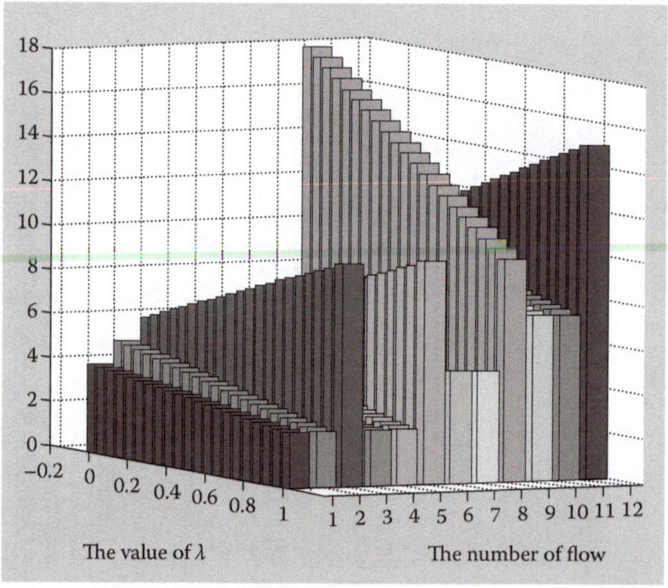

Figure 13.14 Rates of flow and λ in the four to nine nodes network.

decreases, arrives its peak value $Q_R^*(X) = 11.9910$ when $\lambda = 0.5$, under which the individual flow rate vector reaches

$$X = [2.5, 7.5, 2.5, 12.5, 7.5, 12.5]^T.$$

Under this situation, the three to seven nodes network reaches the maximization of total network utility both in the flow level and router level.

For the four to nine node network model, we use the similar approach to analyze. According to the equation $X = \lambda X_{IA} + (1 - \lambda)X_{ID}$, we can plot the relationship between the flow rate X and λ in Figure 13.14, where the x-axis presents the individual flow number, the y-axis presents the value of λ, and the z-axis presents the flow rate. When $\lambda = 0$, the pillars of all sorts of color match the individual flows' rate, which leads to the optimized utility in the ingress network. On the contrary, when $\lambda = 1$, the pillars match the individual flows' rate, which leads to the optimum of utility in the egress network.

The bar chart in Figure 13.15 describes the relationship between the network utility on flow level and the parameter λ. The x-axis presents the value of λ and the y-axis presents the network utility on flow level. As the value of λ increases from 0 to 1, we can clearly see that the utility increases first, peaks at $\lambda = 0.35$, and then starts to decline. According to Figures 13.14 and 13.15, we find that why either X_{IA} ($\lambda = 1$) or X_{ID} ($\lambda = 0$) is not the optimal individual flow rate vector. Even though they lead to the maximization network utility on router level, where some individual flow rate is 0, which means there exists a certain idle channel, they cannot lead to the maximization of network utility on flow level. Subsequently, we can come to a conclusion that, to achieve the maximization of network utility on flow level, each channel must be used to avoid any channel being idle.

As before, we describe the bandwidth allocation schemes 13.26 through 13.28 which are then displayed into the 3D coordinate system in Figure 13.16. In this figure, the x-axis presents the value of λ, the y-axis presents the utility on router level, which is $Q_R^*(Y)$, and the z-axis presents the utility on flow level, which is $Q_R^*(X)$. Observing from Figure 13.16, we can see that no matter how λ changes, $Q_R^*(Y)$ is always equal to 12.024, and the aggregate flow rate vector is fixed at $Y = [10, 15, 20, 30]^T$ during the parameter λ changes, which means all aggregate links are fully saturated. In this situation, the routers will maintain the full-load state and the utility of every edge router always reaches its maximization. But the utility on flow level increases first and then decreases, arrives at its peak value $Q_R^*(X) = 23.070$ when $\lambda = 0.35$, under which the individual flow rate vector reaches

$$X = [3.025, 3.375, 8.6, 3.025, 3.375, 8.6, 3.6, 3.6, 12.8, 8.375, 8.37, 13.25]^T.$$

Under this situation, the four to nine nodes network reaches the maximization of total network utility both in the flow level and router level.

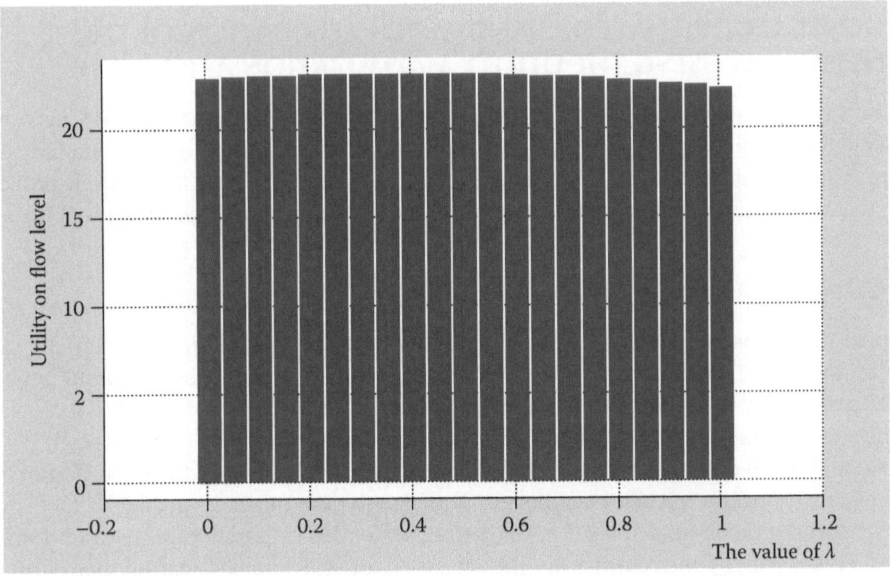

Figure 13.15 Utility on flow level and λ in four to nine nodes network.

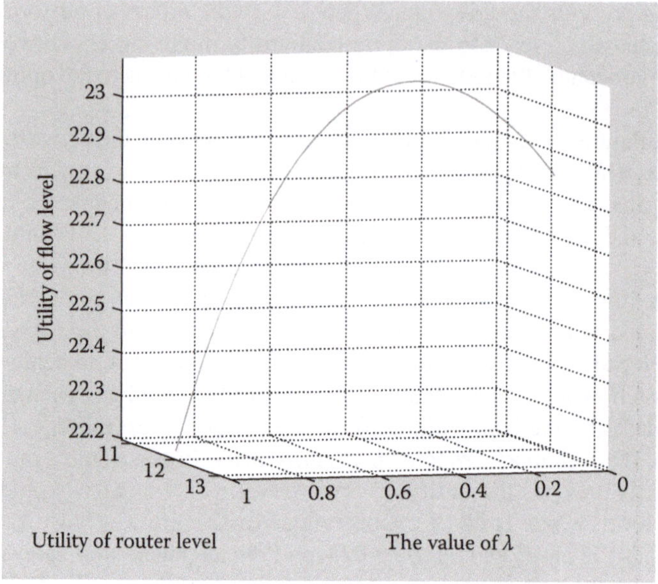

Figure 13.16 The curve of relation of utility and λ in four to nine nodes network.

13.5.1 SUMMARY

Through the two examples given in the previous sections, we summarize some general rules in dealing with the bandwidth allocation issue for OBS networks. The issue can be tackled by utilizing the NUM theory. First, the NUM problem needs to be solved both from the perspective of flow level and router level (i.e., the aggregate flow level). To achieve the maximum of utility on router level is the foundation of utility maximization on flow level. The second rule is that no free channel/link is allowed to exist for the network to achieve the maximal utility.

13.6 NOVEL ALGORITHM FOR BANDWIDTH ALLOCATION OF OBS NETWORKS USING THE UTILITY MAXIMIZATION APPROACH

This section is dedicated to outline an algorithm for bandwidth allocation of OBS networks using the utility maximization approach. This algorithm, namely Algorithm 13.6.1, enables us to find the optimal bandwidth allocation for a general OBS network following the explicit procedures by utilizing the utility maximization approach both in the router level and the individual flow level.

13.7 CONCLUSION

The purpose of OBS is to dynamically provision subwavelength granularity by optimally combining electronics and optics. OBS considers sets of packets with similar properties called bursts. Therefore, OBS granularity is finer than OCS. OBS provides more bandwidth flexibility than wavelength routing but requires faster switching and control technology. OBS can be used for realizing dynamic end-to-end all optical communications. OBS network is a special case with single-hop in every link and the capacity of every link can be adjusted. This nature determines its attraction that each link in any OBS network can be fully loaded in operation, which is not possible in the traditional TCP/IP networks. This is exactly the reason why the recent novel utility maximization theory in aggregate flow level (i.e., the route level) [164] can be successfully used in bandwidth allocation for OBS networks.

Algorithm 13.6.1 Algorithm for bandwidth allocation of OBS networks using the aggregate flow-level utility maximization approach

1: The whole network is partitioned into two networks: the ingress network and the egress network
2: Identifying the routing matrix $A \in R^{n\times m}$ for the ingress network, such that the aggregate flow rate and the individual flow rate satisfy $Y_I = AX_{IA}$. There are n aggregate links (flows) and m individual flows.
3: Calculate the pseudoinverse of A using $A_R^+ = A^T(AA^T)^{-1}$, calculate its SVD and obtain the singular vector $[v_n, v_{n+1}, \ldots, v_m]^T$
4: Calculate $F(Y_I, \alpha)$ by $F(Y_I, \alpha) = A_R^+ Y_I + \Sigma_{i=n}^m \alpha_i v_i$
5: Solve the following optimization problem:

$$Q_R = max_\alpha \sum_{i=1}^{m} log(F(Y_I, \alpha) + 1),$$

$$Y_I \le C,$$

$$F(Y_I, \alpha) \ge 0,$$

to obtain Q_R^*, Y_I, and α
6: Calculate the individual flow rate vector X_{IA} using $X_{IA} = F(Y_I, \alpha) = A_R^+ Y_I + \Sigma_{i=n}^m \alpha_i v_i$
7: Identifying the routing matrix $D \in R^{n\times m}$ for the egress network, such that the aggregate flow rate and the individual flow rate satisfy $Y_E = DX_{ID}$. There are n aggregate links (flows) and m individual flows.
8: Calculate the pseudoinverse of D using $D_R^+ = D^T(DD^T)^{-1}$, calculate its SVD and obtain the singular vector $[v_n, v_{n+1}, \ldots, v_m]^T$
9: Calculate $F(Y_E, \beta)$ by $F(Y_E, \beta) = D_R^+ Y_E + \Sigma_{i=n}^m \beta_i v_i$
10: Solve the following optimization problem:

$$Q_R = max_\beta \sum_{i=1}^{m} log(F(Y_E, \beta) + 1),$$

$$Y_E \le C,$$

$$F(Y_E, \beta) \ge 0,$$

to obtain Q_R^*, Y_E, and β
11: Calculate the individual flow rate vector X_{ID} using $X_{ID} = F(Y_E, \beta) = D_R^+ Y_E + \Sigma_{i=n}^m \beta_i v_i$
12: To obtain the desired optimal bandwidth allocation by the convex combination

$$X = \lambda X_{IA} + (1 - \lambda) X_{ID}$$

for the whole network

In this chapter, we have proposed the NUM bandwidth allocation problem for the next-generation optical network by studying the basic structure of OBS network and provided the effective solutions to the NUM models. We propose a new and general algorithm based on Tan et al. NUM solution on router level and its corresponding solution procedure on how to obtain the utility-optimized individual flow rate for the OBS network. Further, we have discussed two general OBS network models and disclosed the relationship between utility and the bandwidth allocation of individual flows. To the best of our knowledge, the approach presented in this chapter is the first attempt to apply the utility maximization model into the next-generation optical networks. The uniqueness of this approach comes from the use of matrix generalized inverse. We believe the method and the algorithm proposed herein are directly applicable to the OBS networks in designing the architecture and in the optimizing performance.

In future research, it would be interesting to take a close look at the issues of how to realize the utility-maximized bandwidth allocation by aggregating packets into bursts, and how to reduce the guardband impact on data channel throughput and other related issues in OBS networks. For example, the issue of how to provide QoS in optical burst-switched WDM networks with limited FDLs under the maximization utility scheme would be interesting. How the NUM model and its solution applied for buffer-based QoS schemes [381] and the offset-time-based QoS scheme [388] deserve further study as well would also be interesting.

14 Power Adjusting Algorithm on Mobility Control for Mobile Ad Hoc Networks

Power saving is one of the key issues in mobile ad hoc networks (MANETs), which can be realized both in medium access control or media access control (MAC) layer and network layer. However, previous researches in MAC layer are mostly focused on improving the channel utilization by adopting variable-range transmission power control. In this chapter, we focus on the power savings in the network layer and propose a power adjusting algorithm (PAA). In the presence of mobile host (MHs) mobility, PAA is designed to conserve energy by adjusting the transmission power to maintain the routes connectivity and restarting the route discovery periodically to find the new better route dynamically. After analyzing the operations of PAA, we find that the length of route discovery restarted period is a critical argument, which will affect power saving; so we propose an optimizing model that finds the optimal value of this argument by analyzing the energy consumption of this algorithm. PAA can handle the mobility of MANET by adjusting the transmission power and in the meantime save energy by restarting route discovery periodically to balance the energy consumption on route discovery and packets delivering. It is suggested that PAA can be implemented in the dynamic source routing (DSR) protocol. Simulation results are provided, which demonstrate that DSR embedded with PAA saves nearly 40% of the energy compared to DSR without PAA in the presence of high mobility.

14.1 INTRODUCTION

MANET is a kind of wireless network with MHs; it was deployed without any fixed routers and all nodes are capable of movement and can be connected dynamically in an arbitrary manner. Two MHs can communicate with each other either directly or indirectly. MHs in MANET operate not only as hosts but also as routers, such network is highly self-configured, that is, nodes in the network automatically establish and maintain connectivity with other nodes. Representative environments of the MANET applications are fleets in oceans, natural disasters, and battle fields. Some or all of the nodes in an MANET may rely on batteries or other exhaustible energy resources. As such, one of the most important system design requirements in MANETs is power saving [409].

Power-saving operations in MAC protocol is initially discussed in [410,411]. By introducing the appropriate distributed active-sleep schedule for each node, the authors of [410] propose an efficient power-saving MAC protocol for multihop mobile ad hoc networks called p-MANET, which avoids power consumption by activating mobile node during one beacon interval for every n interval, where n is the size of a superframe. The authors of [411] propose an on-demand power management framework for ad hoc networks. In this framework, nodes that are not involved into delivering may go to sleep to adapt to the traffic load, which will save energy in the ad hoc network. These two algorithms are energy-saving schemes, which are adaptive to traffic load. This chapter, however, studies the power-saving issue in MANETs from a different perspective. Rather than adaptive to traffic load, our algorithm is adaptive to MHs mobility, which can be implemented in the existing reactive routing protocols in MANETs to achieve energy efficiency and maintain network connectivity simultaneously. Recent power control developments for MANETs also include the transmission power control mechanisms (e.g., [412–414]) for increasing channel utilization and the distributed protocol [415] for interplaying between the MAC and network layers. Motivated by these works, we introduce a novel power control mechanism into the MAC layer to control the mobility, which can conserve energy and maintain network connectivity in MANETs. With the aid of this power control method, we are able to design an algorithm that is shown to be promising in conserving energy in both MAC layer and network layer.

The recently reactive routing protocols designed for wireless ad hoc network include the notable DSR protocol [416] and Ad hoc On-Demand Distance Vector Routing (AODV) protocol [417]. DSR is a routing protocol for wireless ad hoc networks. This protocol uses source routing, in which all the routing information is maintained and dynamically updated at mobile nodes. It has two major phases, namely, route discovery and route maintenance. These two functions work together to enable any host in the ad hoc network to dynamically discover and maintain a route to any other host in the network. The source broadcasts a Route Request (RREQ) message to find a route, Route Reply (RREP) is then generated if the message has reached the intended destination node. AODV builds routes using a route request and route reply query cycle. When a source node desires a route to a destination for which it does not already have a route, it broadcasts an RREQ packet across the network. A node receiving the RREQ may unicast an RREP back to its source or rebroadcasts the RREQ depending on if it is the destination or not. Nodes keep track of the RREQ source IP address and broadcast ID. As the RREP propagates back to the source, nodes set up forward pointers to the destination. Once the source node receives the RREP, it may begin to forward data packets to the destination.

These routing protocols are usually sensitive to the mobility of MHs. The MHs mobility frequently causes the route to be broken because the receiver frequently moves out of the transmission range of the sender. In DSR, if some active link is broken, the downstream neighbor is currently unreachable. Then, the node broadcasts a Route Error (RERR) packet back to the source, indicating that the route topology has changed, the source node must start route discovery by broadcasting the RREQ packet to find a new path in this case. In AODV, some periodic hello messages can be used to ensure symmetric links, as well as to detect link failures [418]. Once the next hop becomes unreachable, the node upstream of the break first repairs the failed link locally; if this fails then the node propagates an unsolicited RREP to all active upstream nodes. Upon receiving notification of a broken link, source nodes can restart the discovery process if they still require a route to the destination. Even though there is a certain mechanism of route maintenance, implemented in these protocols to maintaining network connectivity, it is still a challenge how to schedule the two schemes, namely route discovery and route maintenance, so as to reach a state of global power saving. Therefore, this chapter is aimed at designing an algorithm to control mobility by maintaining routes and also save energy by balancing the energy consumption on route discovery and packets delivering.

Power control is applied in design of routing protocols in MANET for long-lived flow, for example, the TCP flow in MANET. We focus on power saving in both MAC layer and network layer and propose a PAA. The main techniques in PAA are

- Conserving power in the routing protocol, this includes the power consumed on both route discovery and data transmission.
- Adjusting the transmission power in the presence of MHs mobility to keep the routes connectivity when the packets are transmitting on this route.
- Restarting the route discovery after an appropriate period to find a new better route for data transmission.

Therefore, PAA is essentially a period schedule that introduces the adjustable transmission power control mechanism, periodically restarts the route discovery to balance the energy consumption on route discovery and packets delivering. This algorithm can be implemented both within DSR and AODV. We analyze this algorithm theoretically and detail its parameter settings. Its merits are displayed by simulations. In the simulation, DSR is selected to be implemented with PAA for its simple schedule on route maintenance, and the simulation result shows that the DSR with PAA saved nearly 40% of the energy compared to the DSR without PAA.

The remainder of this chapter is organized as follows: Research background on energy-saving routing protocols is discussed in Section 14.2. In Section 14.3, the propagation model, mobility model, and network assumptions are introduced. Section 14.4 proposes and describes the PAA. Section 14.5 discusses the parameters of the PAA, where we propose an energy model to obtain the desired parameter for PAA in order to minimize the energy consumption. In Section 14.6, we implement the PAA and perform simulations to analyze and compare the DSR with PAA and DSR without PAA on the power saving performance. Finally, the conclusions are drawn in Section 14.7.

14.2 MAIN BACKGROUND

With the proliferation of portable mobile devices such as cell phones, laptops, and handheld digital devices, MANET has attracted significant attention in recent years as a method to provide data communications among these devices without any infrastructure. Fundamental characteristics of MANET include

Multihop routing. No default router available, every node acts as a router and forward each others packets to enable information sharing between MHs.

Dynamically changing network topologies. In MANET, because nodes can move arbitrarily, the network topology, which is typically multihop, can change frequently and unpredictably, resulting in route changes, frequent network partitions, and possibly packet losses.

Energy-constrained operation. Because batteries carried by each mobile node have limited power supply, processing power is limited, which in turn limits services and applications that can be supported by each node. This becomes a bigger issue in MANETs because as each node is acting as both an end system and a router at the same time, additional energy is required to forward packets from other nodes.

The specific MANET issues and constraints described above pose significant challenges in routing protocol design. A large body of research has been accumulated to address these specific issues and constraints [419–424].

Previous works on multihop routing can be classified into two routing methodologies. One approach is proactive routing protocol, which is also called as table-driven protocols. Examples of reactive routing protocols are Optimized Link State Routing protocol [425] and Dynamic destination-Sequenced Distance-Vector protocol [426]. In the proactive routing protocol, nodes continuously exchange route information with all reachable nodes and attempt to maintain consistent, up-to-date routing information. Therefore, the source node can get a routing path immediately if it needs one. A different approach from the proactive routing is the reactive routing, it is also called as on-demand routing protocol. Examples of reactive routing protocols are DSR [416] protocol and AODV protocol [417]. This type of routing protocols creates routing only when desired by the source node. When a node requires a routing to a destination, it initiates a routing discovery process to find a routing path. However, routes may be disconnected due to node mobility; therefore, route maintenance is an important operation of reactive routing protocols. Compared with proactive routing protocols, less control overhead and better scalability are the main strengths in reactive routing protocols.

Researches on changing network topologies are mainly focused on mobility model of hosts [427, 428] and route recover [416,417]. Mobility model poses theoretical analysis on the mobility of nodes and network connectivity, while route recover aims to recover the broken routes. DSR protocol recovers the broken route by restarting routing discovery process, However, AODV first tries to repair this broken route, only when this attempting fails AODV will restart the routing discovery process.

To conserve energy, energy-efficient routing protocols can be divided into two categories: one is minimum energy routing protocols [410–414,429–431] and the an other one is maximum-lifetime routing [410–414,432,433]. Minimum energy routing protocols search a route to minimize the energy consumed to forward a packet from the source to the destination, while maximum network lifetime, for the general case, selects routes and adjusts the corresponding power levels achieving a close to the optimal lifetime. Since minimum energy routing scheme is also an important part in most recent maximum network lifetime routing protocols such as Conditional Max–Min Battery Capacity Routing [432] and Conditional Maximum Residual Packet Capacity routing [433], we are focused on developing more efficient minimum energy routing protocols in this chapter.

14.3 PROPAGATION MODEL, MOBILITY MODEL, AND NETWORK ASSUMPTIONS

We first introduce the propagation model and mobility model used in this section, and then some network assumptions and notations are listed, which will help us to propagate simple problems and models.

14.3.1 PROPAGATION MODEL

Here, we use free space propagation model to forecast the power level of sender and receiver within line of sight. Let P_t and P_r be the power level when the packet is transmitted at the sender and received at the receiver, respectively. Let the distance between the sender and receiver be d. Then, the power level of the receiver is given by Friis formula [434]

$$P_r = \frac{P_t G_t G_r \lambda^2}{(4\pi)^2 d^2 L}, \tag{14.1}$$

where λ is the carrier wavelength, L is the system wastage factor, and G_t and G_r are the antenna gains at the sender and receiver, respectively. Note that λ, L, G_t, and G_r are constants in this formula.

In the free space, we let the path loss exponent be 2 and then the power consumed on the transmitter side by sending one unit of data is

$$P_t = \varepsilon_{11} + \varepsilon_2 d^2 \tag{14.2}$$

And the power consumed on the receiver side by receiving one unit of data is

$$P_r = \varepsilon_{12} \tag{14.3}$$

Here, ε_{11} is the power to run the transmitter circuitry, ε_{12} is the power to run the receiver circuitry, and ε_2 is the power for the transmit amplifier to achieve an acceptable signal-to-noise ratio.

Then, the power consumed by the network to forward one unit of data can be calculated as follows:

$$P_f = P_t + P_r = \varepsilon_1 + \varepsilon_2 d^2 \tag{14.4}$$

Here, we let $\varepsilon_1 = \varepsilon_{11} + \varepsilon_{12}$.

As the power consumption on computation is less than the energy for radio transmission by order of magnitude, we ignore the power consumption on computation in this section.

14.3.2 MOBILITY MODEL

Many mobility models (e.g., random walk model, pursue mobility model, ant mobility model) [427] are proposed in recent years; we use the random walk model to simplify our simulation work here.

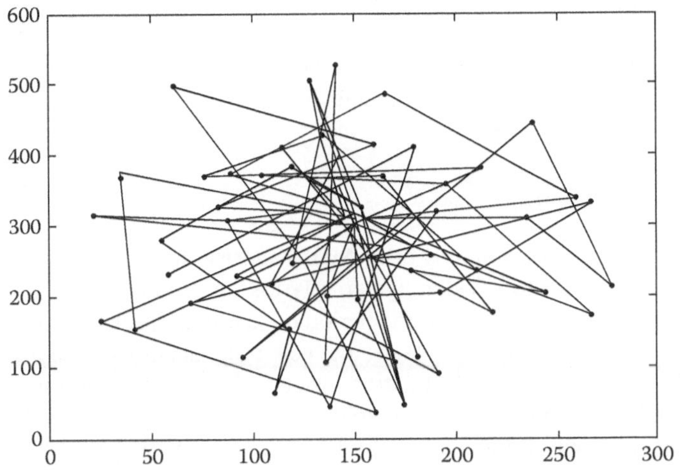

Figure 14.1 Traveling pattern of an MH using 2D random walk mobility model.

In the random walk model, an MH moves from its current location to a new location by randomly choosing a direction and speed in which to travel. The new speed and direction are both chosen from predefined ranges, [0, *maxspeed*] and [0, 2π], respectively. Each movement in the random walk model occurs in a constant time interval t, at the end of which a new direction and speed are calculated. If an MH, which moves according to this model, reaches a simulation boundary, it bounces off the simulation border with an angle determined by the incoming direction. The MH then continues along this new path. Figure 14.1 is cited from [427], which shows an example of the movement observed from this 2D model. The MH begins its movement in the center of the 300m × 600 m simulation area or position (150, 300). At each point, the MH randomly chooses a direction between 0 and 2 and a speed between 0 and 10 m/s. The MH is allowed to travel for 60 s before changing direction and speed.

14.3.3 NETWORK ASSUMPTIONS

Without sacrificing the generality, we make the following assumptions:

- There are N MHs uniformly displayed in an area with radius R.
- We adapt the random walk mobility model as our MHs mobility model.
- The optimal transmission radius of the MH, which minimized the power consumption in the multihop ad hoc network, is the characteristic distance [435], denoted as r_{char}, where $r_{\text{char}} = \sqrt{\frac{\varepsilon_1}{\varepsilon_2}}$.
- We also make sure that r_{char} and n satisfy

$$\pi r_{\text{char}}^2 \geq \frac{\log n + c(n)}{n},$$

this condition will keep the network being connected [428].
- The MHs max speed is set to be v_m m/s in the random walk model.
- The data rate of the flow through the network is r_f packets/s.
- The length of the message RREQ, RREP, and DATA is L_{RREQ}, L_{RREP}, and L_{DATA}, respectively,
- The main cause to the break of route is the mobility of the mobile host, which means that when the receiver moves out of the senders transmission range, a link break will happen, and this will cause the route break. We ignore the result of host failure, unreliable channel, and network congestion in this section.

14.4 PAA DESIGN

In this section, we present the PAA, which can be embedded in the existing reactive routing protocols [416,417] such as DSR for power conserving. We give the full description of PAA and discuss its implementations in real networks.

14.4.1 DESCRIPTION OF PAA

For a flow with data rate r_f originating from MH s and routing to MH d, as PAA constructs a periodic schedule on packet transmission, we simply let the time of sending every k packet to be one period T, where $T = k/r_f$; then after one period of sending k packets, the old route may not be energy efficient due to the mobility of hosts. So the new route needs to be found, so route discovery is restarted to find a new route to the destination instead of the old one. Figure 14.2(a) has shown a route from MH 2 to MH 6 constructed by route discovery. After a period of T, this route becomes less energy efficient as shown in Figure 14.2(b). Finally, as shown Figure 14.2(c), we have found a new and more energy-efficient route (2, 8, 1, 5, 6) by restarting route discovery.

We first describe the algorithm of PAA as follows:

Step 1: The source of this flow will start route discovery to find a route $(n_0 n_1, \ldots, n_l)$ to the destination, where $n_0 = s$ and $n_l = d$. In order to save power, we use the characteristic distance r_{char} as the transmission radius of the RREQ and RREP messages.

Step 2: By analyzing the power level of RREQ and RREP messages received by the host, every host on the route path will compute the distance to its adjacent host. We denote the distance between adjacent MHs n_i and n_j on the route path to be d_{ij}^0.

Step 3: Every $1/r_f$ second, one packet will be transmitted along this route, let us say that it is the first $(1 \leq t \leq l)$ packet on the fly; we forward it with the transmission radius of $(d_{ij}^{t-1} + 2v_m/r_f)$ from n_i to n_j, where d_{ij}^{t-1} is the distance that was estimated when the $(t-1)$th packet passed through the adjacent hosts n_i and n_j on the route path.

Step 4: If there are no more packets to send, the flow should be canceled from the network.

Step 5: For every period of $T = k/r_f$ seconds, after sending every k packet, we restart the route discovery to find a new route $(n_0 n_1, \ldots, n_l)$ to the destination host and then go to Step 2 to continue.

14.4.2 ANALYSIS OF PAA

In Step 1, the source broadcast the RREQ messages to find the destination. After receiving the RREQ messages, the destination will reply to the source by sending an RREP message to construct a route $(n_0 n_1, \ldots, n_l)$ from the source to the destination. By setting the transmission power to be

$$P_t = \varepsilon_1 + \varepsilon_2 r_{\text{char}}^2,$$

the transmission radius of RREQ and RREP message will be r_{char}, then a route $(n_0 n_1, \ldots, n_l)$ from the source to the destination is constructed by route discovery.

In Step 2, every host i on the route path needs to calculate the distance d_{ui}^0 from the upstream host u (except the source) and the distance d_{iv}^0 to the downstream host v (except the destination) by analyzing the power level p_t of the RREQ and RREP message received by the host. The calculation on how the distance can be computed from transmitted power level p_r and received power level will be discussed in the following section.

In Step 3, k data packets can be delivered to the destination within a period $T = k/r_f$ seconds. In every $1/r_f$ seconds, one packet will pass through the route. When the t-th packet passes through this route, we can estimate the distance d_{ij}^t between hosts n_i and n_j from the distance d_{ij}^{t-1}, which is

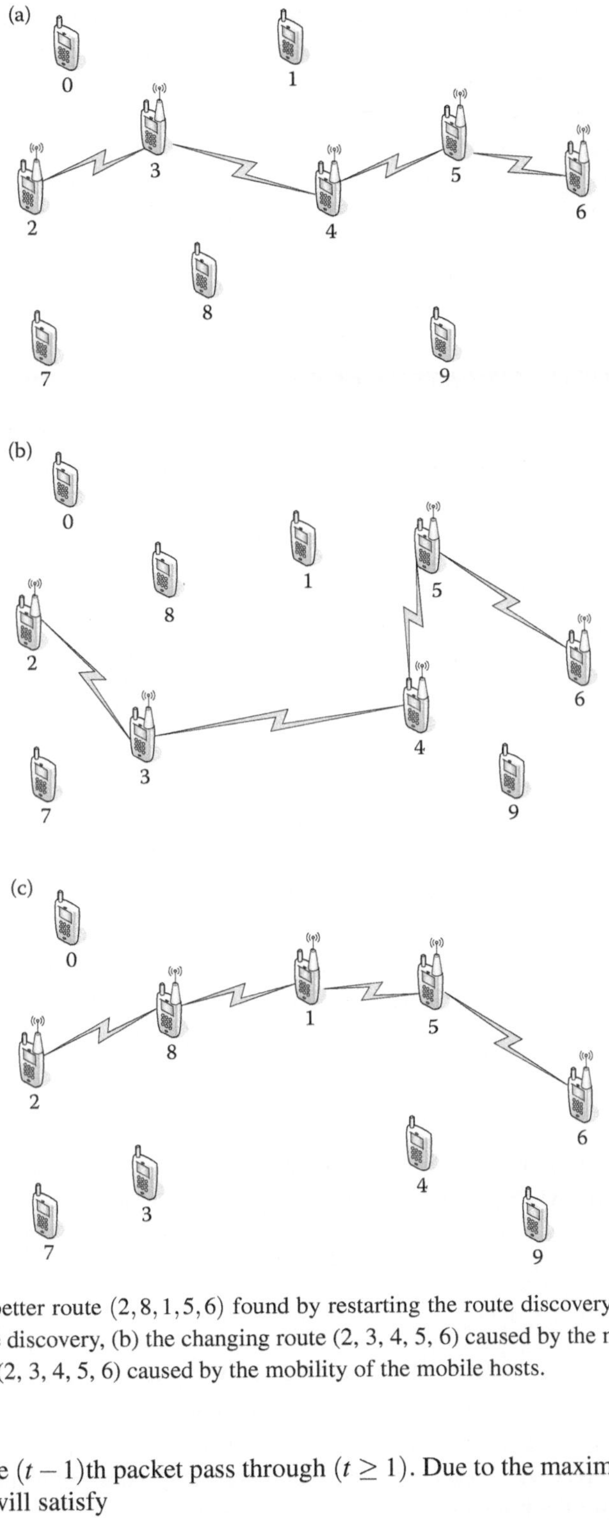

Figure 14.2 A new better route $(2,8,1,5,6)$ found by restarting the route discovery. (a) A route (2, 3, 4, 5, 6) constructed by route discovery, (b) the changing route (2, 3, 4, 5, 6) caused by the mobility of the MHs and (c) the changing route (2, 3, 4, 5, 6) caused by the mobility of the mobile hosts.

the distance when the $(t-1)$th packet pass through $(t \geq 1)$. Due to the maximal speed of the MH is v_m, the distance d_{ij}^t will satisfy

$$d_{ij}^{t-1} - 2\frac{v_m}{r_f} \leq d_{ij}^t \leq d_{ij}^{t-1} + 2\frac{v_m}{r_f} \tag{14.5}$$

In order to make sure that Packet t can be successfully delivered to the host n_j from host n_i, we just let the host $n_i^{\prime}s$ radio frequency (RF) power level cover a transmission distance of $d_{ij}^{t-1}+2v_m/r_f$. Then, after the transmission of Packet t, we can estimate the distance d_{ij}^t from the transmitted power level P_t and the received power level. Subsequently, this updated distance d_{ij}^t can be used to estimate the transmission range of the packet $t+1$.

In Step 4, when the flow has no packet to send, the network needs to cancel this flow.

In Step 5, when the final packet (Packet k) was delivered along this route, the source will restart route discovery to find new route, which will find a new energy efficient one, compared to the current one, according to the current position of the MH in the network. Next, we further discuss the parameter setting on k in a period in Section 14.5.

14.4.3 METHOD OF DISTANCE ESTIMATION

1. Estimate the initial distance d_{ij}^0 for the adjacent MH n_i and n_j on the route path $(n_0 n_1,\ldots,n_l)$.

As in Step 1 of PAA, the power level P_{ij}^i of the RREQ or RREP packet transmitted by the sender n_i is

$$P_{ij}^i = \varepsilon_1 + \varepsilon_2 r_{\text{char}}^2 \tag{14.6}$$

The power level P_{ij}^j of the RREQ or RREP packet received by the receiver n_j can be got from the receiver side. As the transmitted power level P_{ij}^i and the received power level P_{ij}^j satisfy Equation 14.1, which is our propagation model, then the distance d_{ij}^0 between the adjacent host sender n_i and receiver n_j on the route path $(n_0 n_1,\ldots,n_l)$ can be calculated from Equation 14.1 as follows:

$$d_{ij}^0 = \frac{\lambda}{4\pi}\sqrt[n]{\frac{P_{ij}^i G_t G_r}{P_{ij}^j L}} \tag{14.7}$$

2. Estimate d_{ij}^t from the preceding distance d_{ij}^{t-1} of the adjacent sender n_i and receiver n_j on the route path $(n_0 n_1,\ldots,n_l)$, where $1 \le t \le k$.

As in Step 3 of PAA, the power level P_{ij}^i at the sender side, which is used to transmit packet t, will cover a transmission radius of $d_{ij}^{t-1}+2v_m/r_f$. Then, it can be calculated in the following manner:

$$P_{ij}^i = \varepsilon_1 + \varepsilon_2\left(d_{ij}^{t-1} + 2\frac{v_m}{r_f}\right)^2. \tag{14.8}$$

The power level P_{ij}^j that the receiver receives the packet t will be obtained from the receiver side, and then the distance d_{ij}^t when packet t passes through sender n_i and receiver n_j can be calculated as follows:

$$d_{ij}^t = \frac{\lambda}{4\pi}\sqrt{\frac{P_{ij}^i G_t G_r}{P_{ij}^j L}} \tag{14.9}$$

When a packet comes to the sender n_i, the sender will forward the data packet with the transmission radius $(d_{ij}^{t-1}+2v_m/r_f)$ to ensure the reception of the data packet on the receiver n_j, the receiver will reply with a ack packet with the same transmission radius $(d_{ij}^{t-1}+2v_m/r_f)$ as the sender, and finally the sender and the receiver will estimate the current distance d_{ij}^t between them. This new estimated distance will be used for the transmission of the next packet.

Finally, the algorithm of PAA together with distance estimation method will ensure the successful packet transmission under our assumption. The periodic schedule of route recovery can be used to adjust the route energy state for power saving.

14.5 PARAMETERS SETTING OF PAA

The length of the period is vital in this algorithm. On the one hand, if the period is too short, route discovery is restarted too frequently, which will waste more energy on route discovery. On the other hand, if the period is too long, the link state along the route [e.g., the route (b) in Figure 14.2] will become worse and worse as the MH moves around in the network area. This will also waste much energy on delivering the packet along this energy inefficient route. This motivates us to find an appropriate length of period for route discovery. We will find this parameter based on the consideration that, if the energy wasted on the current route becomes unacceptable, the restarted route discovery will find a new better route to save as much energy as possible on delivering the packets.

Before finding the optimal length of the period, we need to first analyze the average distance of the MHs displayed in the network, then the energy consumption of the route discovery, and finally the distance variety between adjacent hosts on the route path in the presence of MHs mobility.

14.5.1 AVERAGE DISTANCE OF ANY TWO MHs

Concerning the average distance of any two MHs, we now present the following result.

Theorem 14.5.1

If two vectors $v_1(r_1,\theta_1)$, $v_2(r_2,\theta_2)$ where $0 \le r_1, r_2 \le R$, $0 \le \theta_1, \theta_2 \le 2\pi$, are uniformly distributed in a circle with radius R. By denoting the sum of them as $v(r,\theta) = v_1 + v_2$, the probability of v locating at $v(r,\theta)$ (where $0 \le r \le R$, $0 \le \theta \le 2\pi$) is

$$P(r,\theta,R) = \frac{1}{\pi^2 R^2}\left(2\arccos\frac{r}{2R} - \frac{r}{R}\sqrt{1-\left(\frac{r}{2R}\right)^2}\right) \tag{14.10}$$

Proof: Figure 14.3 displays the two vectors $v_1(r_1,\theta_1)$, $v_2(r_2,\theta_2)$, and their sum $v(r,\theta)$, where the vector $v(r,\theta)$ is located in a circle with radius $2R$. Two small circles with radius R are centered at the start point and the end point of the vector $v(r,\theta)$. All the possibility position of vector v_1 should be

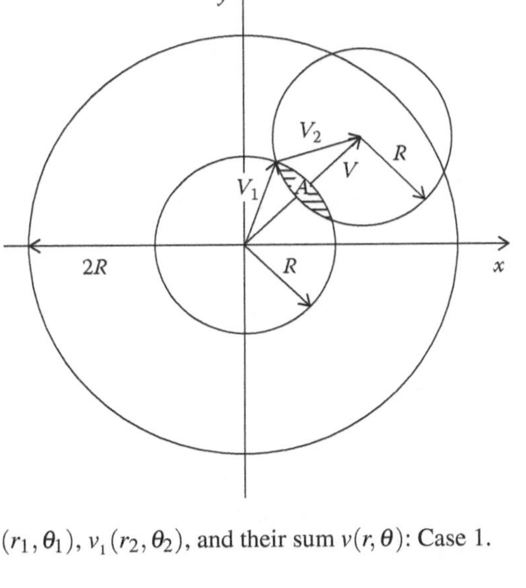

Figure 14.3 Two vectors $v_1(r_1,\theta_1)$, $v_1(r_2,\theta_2)$, and their sum $v(r,\theta)$: Case 1.

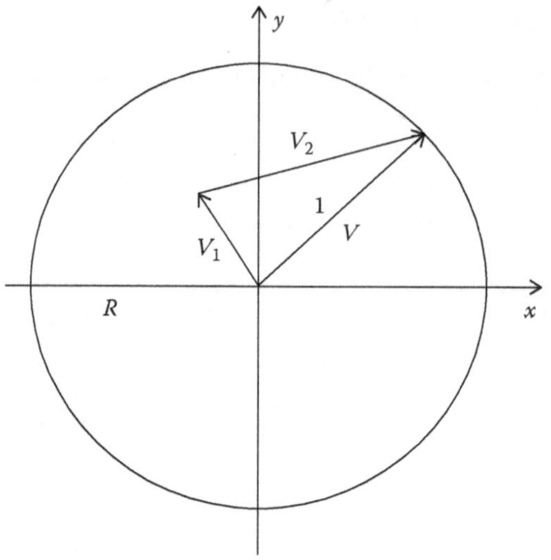

Figure 14.4 Two vectors $v_1(r_1, \theta_1)$, $v_2(r_2, \theta_2)$, and their sum $v(r, \theta)$: Case 2.

located in the overlapping regions of the two small circles. Then, the area of the overlapping regions of the two small circles is

$$A = R^2(\theta - \sin\theta) = R^2\left(2\arccos\frac{r}{2R} - \frac{r}{R}\sqrt{1-\left(\frac{r}{2R}\right)^2}\right) \tag{14.11}$$

The variable area of the vector $v_1(r_1, \theta_1)$ or $v_2(r_2, \theta_2)$ is πR^2, so the probability of the sum of two vectors v to locate at (r, θ) is

$$P(r,\theta,R) = \frac{A}{\pi R^2 \pi R^2} = \frac{1}{\pi^2 R^2}\left(2\arccos\frac{r}{2R} - \frac{r}{R}\sqrt{1-\left(\frac{r}{2R}\right)^2}\right) \tag{14.12}$$

This concludes the proof. ■

Now for any two hosts uniformly displayed in the network with radius R, the location of the two host is $v_1(r_1, \theta_1)$, $v_2(r_2, \theta_2)$, respectively, where $0 \le r_1$, $r_2 \le R$, $0 \le \theta_1$, and $\theta_2 \le 2\pi$. This can be seen in Figure 14.4. Then, the average distance l of this two MHs will be

$$E(l) = \int_0^{2R}\int_0^{2\pi} P(r,\theta,R)r \times r\mathrm{d}\theta\mathrm{d}r = \frac{128}{45\pi} \tag{14.13}$$

Therefore, for any two hosts, which are uniformly located in a network area with radius R, their average distance is $128/45\pi R$.

14.5.2 ENERGY CONSUMPTION ON ROUTE DISCOVERY

In a network with N MHs located, the energy consumption of route discovery is mainly by broadcasting the RREQ messages and unicasting the RREP messages, this means

$$E_{\text{restart}} = E_{\text{RREQ}} + E_{\text{RREP}} \tag{14.14}$$

The source host broadcast an RREQ message and this message will flood the whole network, so nearly every host will receive an RREQ message and they will rebroadcast the RREQ message to their neighbors. As the transmission range of the broadcasting is r_{char}, which is suggested in PAA, then the energy consumption on broadcasting the RREQ message can be calculated approximately as follows:

$$E_{\text{RREQ}} = NL_{\text{RREQ}}(\varepsilon_1 + \varepsilon_2 r_{\text{char}}^2) \tag{14.15}$$

The destination host will reply an RREP message after receiving an RREQ message; this RREP message will be unicasted back to the source. All the hosts who relay the RREP message will form a route $(n_0 n_1, \ldots, n_l)$ from the source to the destination. As the routing protocol selects the route with minimum number of hops to be the final route, so the average number of hops of the active route $(n_0 n_1, \ldots, n_l)$ will be

$$L = \left\lceil \frac{E_{(l)}}{r_{\text{char}}} \right\rceil = \left\lceil \frac{128R}{45\pi r_{\text{char}}} \right\rceil \tag{14.16}$$

Then, the energy consumption of unicasting the RREP message can be calculated approximately as follows:

$$E_{\text{RREP}} = LL_{\text{RREP}}(\varepsilon_1 + \varepsilon_2 r_{\text{char}}^2) = \left\lceil \frac{128R}{45\pi r_{\text{char}}} \right\rceil L_{\text{RREP}}(\varepsilon_1 + \varepsilon_2 r_{\text{char}}^2) \tag{14.17}$$

Finally, the energy consumption of the route discovery is:

$$E_{\text{restart}} = \left(NL_{\text{RREQ}} + \left\lceil \frac{128R}{45\pi r_{\text{char}}} \right\rceil L_{\text{RREP}}\right)(\varepsilon_1 + \varepsilon_2 r_{\text{char}}^2) \tag{14.18}$$

14.5.3 DISTANCE VARIETY BETWEEN ADJACENT MHs ON THE ROUTING PATH

For any two objects randomly moving in the space, the distance between them will become larger and larger, this phenomenon is called diffusion in physics. Here, we will estimate the distance variety of two adjacent MHs on the routing path, and this result will be used to build the energy consumption model of PAA in the following section, and then the optimal length of the period can be calculated from this model.

The average distance of any two adjacent hosts n_i and n_j along the route path $(n_0 n_1, \ldots, n_l)$ when packet $t-1$ and packet t comes are set to be d_{t-1} and d_t, respectively. We define the distance variety to be $\Delta d = d_t^2 - d_{t-1}^2$. As the routing protocol selects the route with minimum number of hops to be the final route, the preliminary distance between any two adjacent hosts along the route path just after the route discovery restarted is nearly r_{char}, we just set $d_0 = r_{\text{char}}$ for simplicity.

As we have assumed that the flow rate is r_f, so the time between any two consecutive packets is $1/r_f$, and the MHs max speed in their random walk mobility model is set to be v_m, so the maximum distance that the MHs can move during the interim of two continuous packets is $r_m = v_m/r_f$.

The displacement of the two MHs n_i and n_j during the interim of two packets is denoted as $s_1(r_1, \theta_1)$ and $s_2(r_2, \theta_2)$, respectively, where $0 \le s_1$, $s_2 \le r_m$, $0 \le \theta_1$, and $\theta_2 \le 2\pi$. The movement of the two hosts n_i and n_j is displayed in Figure 14.5(a), and from the viewpoint of host n_i, the movement of host n_j can be seen in Figure 14.5(b). The length of O_1O_2 is d_{t-1}, the length of O_1M is d_t. Then, we can estimate the distance variety $\Delta d a$ follows:

$$\begin{aligned}\Delta d &= d_t^2 - d_{t-1}^2 \\ &= \int_0^{2r_m} \int_0^{2\pi} P(s, \theta, r_m) \times (d_{t-1}^2 + s^2 + 2d_{t-1} s \cos\theta - d_{t-1}^2) s \mathrm{d}s \mathrm{d}\theta \end{aligned} \tag{14.19}$$

where $P(s, \theta, r_m)$ is the probability of the final position of the MH n_i located at from the viewpoint of MH, where $s = (-s_1) + s_2$, and $\sqrt{d_{t-1}^2 + s^2 + 2d_{t-1} s \cos\theta}$ is the length of vector O_1M, which is the final distance of the two hosts.

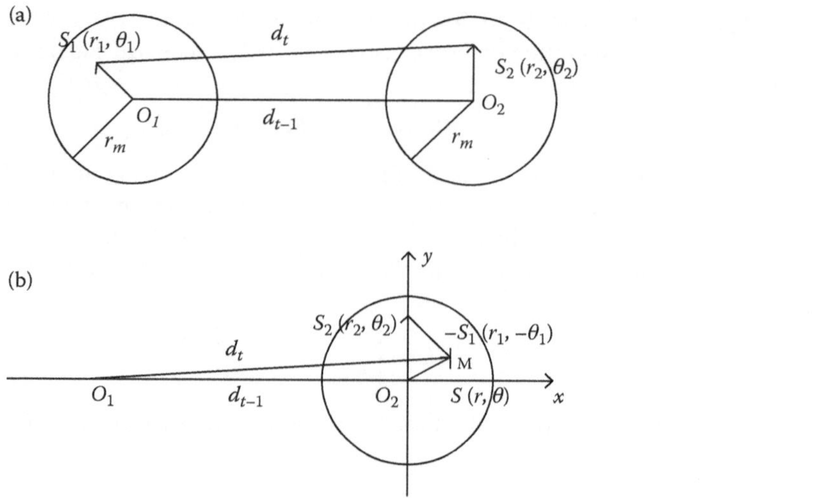

Figure 14.5 The movement of two adjacent MHs n_i and n_j. (a) The movement of host n_i and host n_j and (b) the movement of host n_j from the view of host n_i.

By some mathematical manipulations, Equation 14.19 is reduced to the following equation:

$$\Delta d = d_t^2 - d_{t-1}^2 = r_m^2 \tag{14.20}$$

This equation means that the distance variety Δd of two adjacent MHs n_i and n_j is the square of the maximum distance that the MH can move during the interval of two continuous packets.

14.5.4 FINDING THE OPTIMAL LENGTH OF THE PERIOD

In a period T, we let that k packets be transmitted along the route, then the problem of finding optimal period T is transformed to find the optimal number of packets that can be transmitted in a period. The data rate r_f of the flow is assumed to be a constant here.

After the route discovery, a route path $(n_0 n_1, \ldots, n_l)$ from the source to the destination is constructed. As in Step 3 of PAA, the energy of delivering t-th packet on this route path can be calculated in terms of

$$E(t) = \sum_{i=1}^{l} L_{\text{DATA}}(\varepsilon_1 + \varepsilon_2 d_t^2) = L L_{\text{DATA}}(\varepsilon_1 + \varepsilon_2 d_t^2), 1 \le t \le k \tag{14.21}$$

Then, the energy of delivering k packet in a period on an active route $(n_0 n_1, \ldots, n_l)$ should be

$$E_p = \sum_{t=1}^{k} E(t) = L L_{\text{DATA}}(k\varepsilon_1 + \varepsilon_2 \sum_{t=1}^{k} d_t^2) \tag{14.22}$$

Now, we need to minimize the average energy consumption on delivering one packet in a period, so this problem can be modeled as follows:

Minimize

$$\frac{1}{k}(E_p + E_{\text{restart}}) \tag{14.23}$$

Subjected to

$$d_0 = r_{\text{char}} \tag{14.24}$$

$$d_1 = d_0 + 2v_m \tag{14.25}$$

$$d_t^2 = d_{t-1}^2 + v_m^2, 2 \leq t \leq k \tag{14.26}$$

where E_{restart} is the energy cost on route discovery, Equation 14.23 means that we need to minimize the average energy consumption on delivering one packet. Condition 14.25 is derived from Step 3 of PAA and condition 14.26 is derived from Equation 14.20. Now by substituting condition 14.26 into the object function 14.23, we will get the function as follows

$$\begin{aligned} E(k) = & \frac{LL_{\text{DATA}}\varepsilon_2 r_m^2}{2}k + (\frac{LL_{\text{DATA}}\varepsilon_2 r_m^2}{2} + E_{\text{restart}})\frac{1}{k} \\ & + LL_{\text{DATA}}(\varepsilon_1 + \varepsilon_2(d_0 + 2v_m)^2 - \frac{3}{2}\varepsilon_2 r_m^2 \end{aligned} \tag{14.27}$$

In order to obtain the optimal length of period, we need to minimize the energy consumption (14.27) to make the energy on delivering packet in the period minimized. So, we can obtain the first-order derivative of function 14.25 and then obtain the optimal number of packets delivering in a period

$$k_{opt} = \sqrt{1 + \frac{2E_{\text{restart}}}{LL_{\text{DATA}}\varepsilon_2 r_m^2}} \tag{14.28}$$

When all the parameters are determined, combining with Equations 14.16 and 14.18, we obtain the optimal number of packets, then the length of the period in this network to restart the route discovery schedule will be $T = k_{opt}/r_f$.

14.6 SIMULATION RESULTS

In this section, we have implemented the routing protocol DSR with PAA embedded and DSR without PAA embedded. The parameter setting of these simulations is listed in Table 14.1. In the simulation, 100 flows are randomly selected in an MANET and every flow needs to deliver 1000 packets through the network. First, we verify whether the optimal number of packets can be sent in a period, then we compare the number of route discovery, the average energy consumption on delivering one packet, and the total energy consumption between the routing protocol DSR with PAA and DSR without PAA by changing the MHs maximum speed in the random walk mobility model. These results show that the algorithm of PAA balances the energy consumed on route discovery and data delivering; this enables the routing protocol DSR with PAA conserve energy in the presence of mobility.

In Figure 14.6, as the number of packets k changes from 1 to 200, the average energy consumption on delivering one packet is plotted in this figure; here, the max speed of the MH is set to be 10 m/s. From this picture, one observes that when the number of packet k is in the range of [20,60], more energy will be conserved on packet delivering. The optimal value that is theoretically

Table 14.1
Parameter Settings in the Simulation

N	1000
R	1000 m
ε_1	18,0000 pJ/bit/m^2
ε_2	10 pJ/bit/m^2
L_{RREQ}	16 byte
L_{RREP}	16 byte
L_{DATA}	512 byte
r_f	1 packet/s

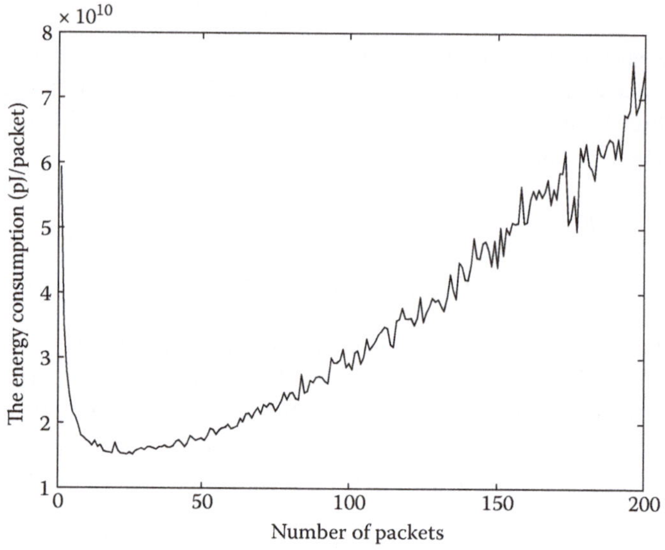

Figure 14.6 Verification of the optimal length of period.

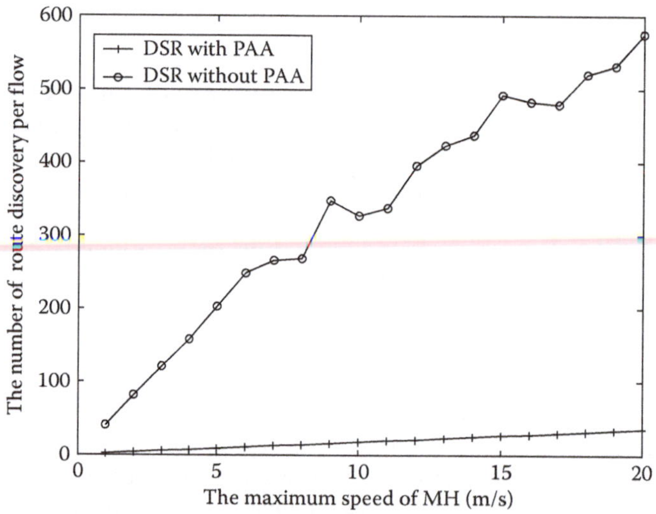

Figure 14.7 The number of route discovery per flow on delivering 1000 packets.

computed from our model is 57. Therefore, we have verified that this theoretical value agrees with the simulation results in the algorithm PAA.

By increasing the maximum speed of MHs from 1 to 20 m/s in the mobility model, the number of restarted route discovery per flow and the average energy consumption on delivering one packet are plotted in Figures 14.7 and 14.8, respectively. In Figure 14.7, the number of route discovery restarted is increased greatly in DSR without PAA compared to DSR with PAA. That is because the increasing max speed will enlarge the possibility of link break along the route path caused by MHs mobility in DSR without PAA, whereas this number increases slowly in DSR with PAA. In DSR with PAA, the number of route discovery restarted is significantly decreased by adjusting the transmission power dynamically to adapt to the mobility of MHs.

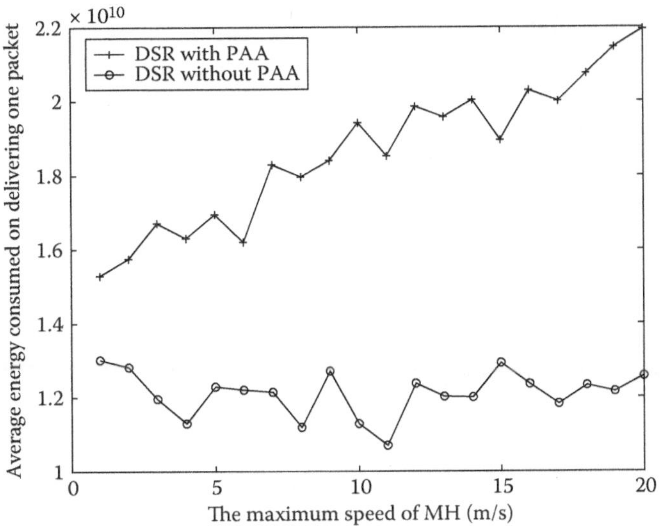

Figure 14.8 Average consumption on delivering one packet.

Figure 14.8 demonstrates the route energy state on delivering packets. In this figure, the route energy state of DSR without PAA is always stable at 1.25×10^{10}, while the route energy state of DSR with PAA increases, as the MHs max speed increases. So, the route energy state of DSR without PAA is maintained in good state, and the route energy state of DSR with PAA is becoming worse as the max speed increases. As it has been shown in Figure 14.7, this good state maintained by DSR without PAA is at the cost of much larger number of restarted route discovery, while PAA is aimed to reach a global balance not only in energy consumption on restarting route discovery but also on maintaining route energy consumption state on packet delivering. Clearly, less number of route discovery leads to more energy consumption on packet delivering; here, the number of route discovery should be selected carefully in PAA to minimize the total energy consumption.

The simulation results about the total energy consumption of the two protocols are plotted in Figure 14.9. From the figure, one can see that 40% of the energy consumed in the DSR without PAA was saved in the protocol DSR with PAA when the max speed is becoming high. In the presence of mobility of MH, the protocol DSR without PAA always restart the route discovery when a link break happens, it is suffering the mobility of MHs. Whereas in the protocol DSR with PAA, the mobility of MH was under the control of PAA. Furthermore, with the appropriate setting of the length of restarted period, the route discovery is restarted to adjust the route path, which always balances, the energy consumed on route discovery and packet delivering to reach a global power saving effect. In this simulation, 40% of the energy was saved by using PAA to control the mobility of MH.

14.7 CONCLUSIONS

An ad hoc network is a collection of wireless MHs forming a temporary network without the aid of any established infrastructure or centralized administration. In this section, we have proposed a PAA, which can be embedded in the routing protocol (e.g., DSR and AODV) to save power in the network layer for wireless ad hoc networks. The algorithm of PAA introduced the adjustable transmission power control to control the mobility of MHs. By properly setting of the length of period to restart the route discovery, PAA has the capability of balancing the energy consumption

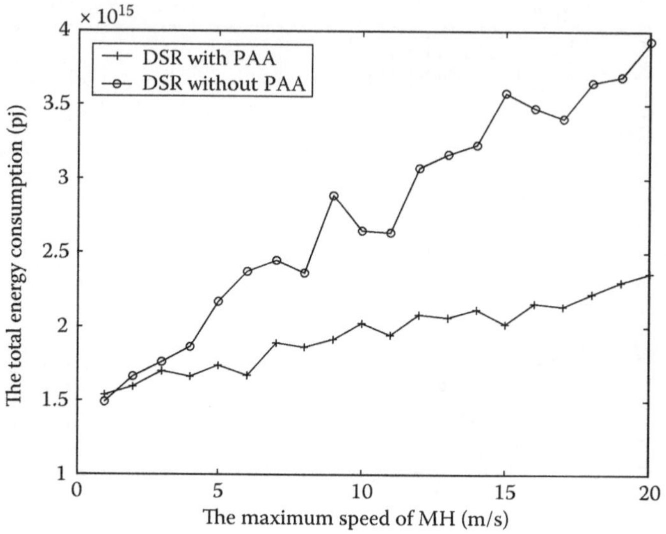

Figure 14.9 The total energy consumption on delivering 1000 packets.

on route discovery and packets delivering to save energy in network layer. By theoretical analysis and simulations, we first establish how the optimal length of period can be calculated from our model, and then by investigating the number of restarted route discovery and the route energy state, we also verified that PAA has indeed balanced the energy consumption on route discovery and packets delivering. Finally, the simulation results of total energy consumption show that the routing protocol DSR with PAA saves nearly 40% of the energy consumption under high-speed mobility model.

15 PCA-Guided Routing Algorithm for Wireless Sensor Networks

An important performance concern for wireless sensor networks (WSNs) [638–641] is the total energy dissipated by all the nodes in the network over the course of network lifetime. In this chapter, we propose a routing algorithm termed as principal component analysis (PCA)-guided routing algorithm (PCA-RA) by exploring the PCA approach. Our algorithm remarkably reduces energy consumption and prolongs network lifetime by realizing the objective of minimizing the sum of distances between the nodes and the cluster centers in a WSN. It is demonstrated that the PCA-RA can be efficiently implemented in WSNs by forming a nearly optimal K-means-like clustering structure. In addition, it can decrease the network load while maintaining the accuracy of the sensor measurements during data aggregating process. We validate the efficacy and efficiency of the proposed algorithm by simulations. Both theoretical analyses and simulation results demonstrate that this algorithm can perform significantly with less energy consumption and thus prolong the system lifetime for the networks.

15.1 INTRODUCTION

WSNs [436,638–641] consist of battery-powered nodes, which inherit sensing, computation, and wireless communication capabilities. Although there have been significant improvements in processor design and computing issues, limitations in battery provision still exist, bringing energy resource considerations as the fundamental challenge in WSNs. Consequently, there have been active research efforts devoted to lifting the performance limitations of WSNs. These performance limitations include network throughput, energy consumption, and network lifetime. Network throughput typically refers to the maximum amount of packets that can be successfully collected by the cluster heads (CHs) in the network, energy consumption refers to the minimize energy dissipation that nodes in the network consume, and network lifetime refers to the maximum time limit that nodes in the network remain alive until one or more nodes drain up their energy.

The routing algorithms have been specifically designed for WSNs because the energy optimization is an essential design issue. A good routing scheme is helpful in improving these performance limits such as reducing the energy consumption, prolonging the network lifetime, increasing the network throughput. Network researchers have studied a great variety of routing protocols in WSNs differing based on the application and network architecture. As demonstrated in [437,438], it can be classified into four categories: flit, hierarchical clustering, location-based routing, and quality of service (QoS)-based routing. The current routing protocols have design trade-offs between energy and communication overhead savings, as well as the advantages and disadvantages of each routing technique.

As the representative hierarchical clustering protocol, Low Energy Adaptive Clustering Hierarchy (LEACH) [439] has simplicity, flexibility, and scalability because its manipulations rely on randomized rotation of the CHs, but its features of unregulated distribution, unbalanced clustering structure, uniform initial energy, etc., hinder its performance. Based on LEACH, there are many variants, such as [440–443]. LEACH-E [440] more likely selects the nodes with higher energy as the CHs. LEACH-C [440] analytically determines the optimum number of CHs by taking into account the energy spent by all clusters. Power efficient gathering in sensor information system (PEGASIS) [441], threshold-sensitive energy-efficient sensor network (TEEN) [442], and adaptive threshold-sensitive energy-efficient sensor network (ATEEN) [443] improve the energy consumption by optimizing the data transmission pattern. Hybrid energy-efficient distributed clustering (HEED) [444] is a complete distributed routing protocol, which has different clustering formations and CHs selecting measures. These protocols have many restrictive assumptions, applicable limitations, so it has great improvement space and extensibility. The rapid development of wireless communications technology, and the miniaturization and low cost of sensing devices, has accelerated the development of WSNs [445,446]. As in [447], Zytoune et al. proposed a uniform balancing energy routing protocol. The balanced parallel K-means-based clustering protocol (BPK)-means [448] and balanced serial K-means-based clustering protocol (BSK)-means [449] can improve the structure of clusters and perform better load balance and less energy consumptions. Hybrid multihop routing algorithm (HMP-RA) [450] proposes a solution to address this issue through a hybrid approach that combines two routing strategies, flat multihop routing and hierarchical multihop routing. Exponential and sine cost function-based route (ESCFR) and double cost function-based route (DCFR) can map small changes in nodal remaining energy to large changes in the function value and consider the end-to-end energy consumption, nodal remaining energy [451]. Biologically inspired intelligent algorithms build a hierarchical structure on the network for different kinds of traffic, thus maximizing network utilization, while improving its performance [452]. In the case where sensor nodes are mobile, as in [453,454], the nodes can adjust their position to help balance energy consumption in areas that have high transmission load and/or mitigate network partition.

In this chapter, we consider an overarching algorithm that encompasses both performance metrics. It desires to minimize the sum of distances in the clusters. We show that the PCA [455], a useful statistical technique that has found application in fields such as face recognition and image compression, and a common technique for finding patterns in data of high dimension, can form a near-optimal K-means-like clustering structure, in which the distance between the non-CH nodes and CHs is nearly minimized.

Moreover, the data aggregating issue associated with the measurements accuracy calls for a careful consideration in scheme about data collecting and fusing. In this chapter, we investigate the PCA technology in a high relative measurements context for WSNs. Our objective is to obtain a good approximation to sensor measurements by relying on a few principal components while decreasing the network load.

The remainder of this chapter is organized as follows: In Section 15.2, we describe the network and energy model. Section 15.3 presents the PCA-RA model and gives numerical results to demonstrate the working mechanism of PCA-RA. Section 15.4 discusses the PCA-RA algorithm solution strategies. In Section 15.5, we simulate the PCA-RA and compare it with LEACH and LEACH-E. Finally, Section 15.6 concludes this chapter.

15.2 SYSTEM MODEL

15.2.1 NETWORK MODEL

Let us consider a two-tier architecture for WSNs. Figure 15.1 shows the physical network topology for such a network. There are three types of nodes in the networks, namely, a base station (BS), the cluster-head nodes (CHNs), and the ordinary sensor nodes (OSNs).

For each cluster of sensor nodes, there is one CHN, which is different from an OSN in terms of functions. The primary functions of a CHN are data aggregation for data measurements from the local clusters of OSN and relaying the aggregated information to the BS. For data fusion, a CHN analyzes the content of each measurement it receives and exploits the correlation among the data measurements. A CHN has a limited lifetime, so we need to consider rotating the CHN to balancing the energy consumption.

The third component is the BS. We assume that there is sufficient energy resource available at the BS and thus there is no energy constraint at the BS.

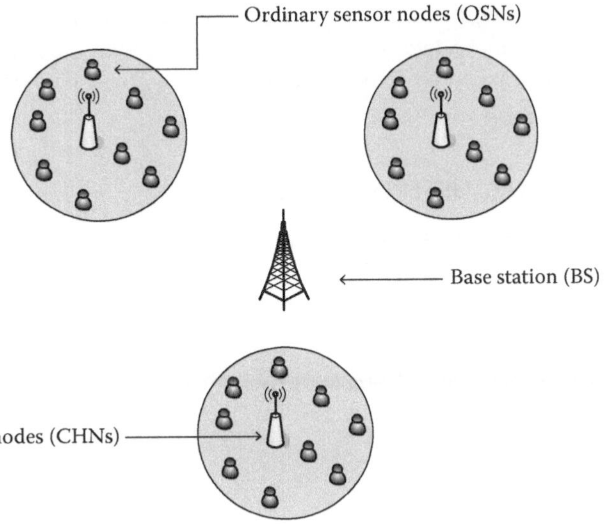

Figure 15.1 Physical topology for two-tier WSNs.

15.2.2 ENERGY CONSUMPTION MODEL

We compute the energy consumption using the first-order radio model [440]. The equations, which are used to calculate transmission costs and receiving costs for an L-bit message to cross a distance d, are shown as follows:

$$E_{TX}(L,d) = \begin{cases} L(E_{\text{elec}} + \varepsilon_{fs}d^2), & d < d_0 \\ L(E_{\text{elec}} + \varepsilon_{mp}d^4), & d \geq d_0 \end{cases} \tag{15.1}$$

$$E_{RX}(L) \doteq LE_{\text{elec}}$$

In Equation 15.1, the electronics energy, E_{elec}, depends on factors such as the digital coding, modulation, filtering, and spreading of the signal, whereas the amplifier energy, $\varepsilon_{fs}d^2$ or $\varepsilon_{mp}d^4$, depends on the distance to the receiver and the acceptable bit-error rate. d_0 is the distance threshold.

15.3 PCA-GUIDED ROUTING ALGORITHM MODEL

PCA is a classic technique in statistical data analysis, data compression, and image processing. PCA transforms a number of correlated variables into a number of uncorrelated variables called principal components. The objective of PCA is to reduce the dimensionality of the data set but retain most of the original variability in the data. The first principal component accounts for as much of the variability in the data as possible. Mathematically, how to pick up the dimensions with the largest variances is equivalent to finding the best low-rank approximation of the data via the singular value decomposition (SVD) [456].

The design of routing algorithm is important in WSNs. Although plenty of interests are drawn on it, there is still a challenge to face on the aspect of efficiency and energy consumption. In this section, we describe the notations about PCA-RA model first and then propose the PCA-guided clustering model, and finally we present the PCA-guided data aggregating model.

15.3.1 NOTATIONS

Let $X \doteq \{x_1, x_2, \ldots, x_n\}$ represents the location coordinate matrix of a set of n sensors; $Y \doteq \{y_1, y_2, \ldots, y_n\}$ represents the centered data matrix, where $y_i = x_i - \bar{x}$, which defines the centered distance vector columnwise, and $\bar{x} = \sum_i X_i/n$ is the mean vector columnwise of X matrix.

Let $M \doteq \{m_1, m_2, \ldots, m_p\}$ be a group of measurements collecting from the sampling period. Each sensor generates a stream of data. Let $D_{N\times p}$ be a matrix with elements D_{ij}, $1 \le i \le n, 1 \le j \le p$, being the measurement taken by sensor i at point j. Let $D_{N\times p}$ be a centered matrix with elements $q_{ij} = d_{ij} - d_i/n, d_i = \sum_j d_{ij}$.

15.3.2 PCA-GUIDED CLUSTERING MODEL

We define the equation for the SVD of matrix Y [22]

$$Y \doteq \sum_k \lambda_k \mu_k v_k$$

The covariance matrix (ignoring the factor $1/n$) is

$$\sum_i (x_i - \overline{x})^T (x_i - \overline{x}) \doteq Y^T Y$$

The principal components V_k are eigenvectors satisfying

$$Y^T Y v_k \doteq (\lambda_k)^2 V_k, V_k \doteq Y^T u_k / \lambda_k^2 \tag{15.2}$$

15.3.2.1 K-Means Clustering Model

According to [457,458], we find the PCA dimension reduction automatically by performing data clustering according to the K-means objective function [459,460]. Using K-means algorithm, it can form a better cluster structure by minimizing the sum of squared errors. We define the squared distance between sensor nodes and cluster centers as

$$J_K \doteq \sum_{k=1}^{K} \sum_{i \varepsilon C_k} (x_i - m_k)^2 \tag{15.3}$$

where $m_k \doteq \sum_{x \varepsilon C_k} x_i / n_k$ is the center of cluster C_k and n_k is the number of sensor nodes in C_k. Given the fact that by minimizing the distance between sensor nodes and cluster centers, the energy consumption can be effectively reduced. Our clustering algorithm is thus designed to be capable of minimizing the above metric J_K.

For the sake of convenience, let us start with the case of $K = 2$. To obtain the explicit expression for J_K, let

$$d(C_P, C_l) \doteq \sum_{i \varepsilon C_P} \sum_{j \varepsilon C_l} (X_i - X_j)^2 \tag{15.4}$$

be the sum of squared distances between two clusters C_P and C_l; after some algebra one obtains

$$J_2 = \frac{d(C_1, C_1)}{2n_1} + \frac{d(C_2, C_2)}{2n_2} \tag{15.5}$$

where n_1 and n_2 are the number of sensor nodes in C_1 and C_2, respectively, and n is the total number of sensor nodes; therefore, we obtain $n = n_1 + n_2$. If denoting

$$\overline{y^2} = \sum_i y_i^T y_i / n = \frac{d(C_1, C_1)}{2n^2} + \frac{d(C_2, C_2)}{2n^2} + \frac{d(C_1, C_2)}{n^2} \tag{15.6}$$

$$J_D = \frac{n_1 n_2}{n} \left[2\frac{d(C_1, C_2)}{n_1 n_2} - \frac{d(C_1, C_1)}{n_1^2} - \frac{d(C_2, C_2)}{n_2^2} \right] \tag{15.7}$$

we thus have

$$
\begin{aligned}
& n\overline{y^2} - \frac{1}{2}J_D \\
&= n\left(\frac{d(C_1,C_1)}{2n^2} + \frac{d(C_2,C_2)}{2n^2} + \frac{d(C_1,C_2)}{n^2}\right) - \frac{1}{2}\left(\frac{n_1 n_2}{n}\left[2\frac{d(C_1,C_2)}{n_1 n_2} - \frac{d(C_1,C_1)}{n_1^2} - \frac{d(C_2,C_2)}{n_2^2}\right]\right) \\
&= \frac{d(C_1,C_1)}{2n} + \frac{d(C_2,C_2)}{2n} + \frac{d(C_1,C_2)}{n} - \frac{2n_1 n_2 d(C_1,C_2)}{2nn_1 n_2} + \frac{n_1 n_2 d(C_1,C_1)}{2nn_1^2} + \frac{n_1 n_2 d(C_2,C_2)}{2nn_2^2} \\
&= \left(\frac{d(C_1,C_1)}{2n} + \frac{n_1 n_2 d(C_1,C_1)}{2nn_1^2}\right) + \left(\frac{d(C_2,C_2)}{2n} + \frac{n_1 n_2 d(C_2,C_2)}{2nn_2^2}\right) \\
&\quad + \left(\frac{d(C_1,C_2)}{n} - \frac{2n_1 n_2 d(C_1,C_2)}{2nn_1 n_2}\right) \\
&= \left(\frac{n_1 d(C_1,C_1)}{2nn_1} + \frac{n_2 d(C_1,C_1)}{2nn_1}\right) + \left(\frac{n_2 d(C_2,C_2)}{2nn_2} + \frac{n_1 d(C_2,C_2)}{2nn_2}\right) \\
&= \frac{(n_1+n_2)d(C_1,C_1)}{2nn_1} + \frac{(n_2+n_1)d(C_2,C_2)}{2nn_2} \\
&= \frac{d(C_1,C_1)}{2n_1} + \frac{d(C_2,C_2)}{2n_2} \\
&= J_2
\end{aligned}
$$

That is

$$J_2 = n\overline{y^2} - \frac{1}{2}J_D \tag{15.8}$$

where $\overline{y^2}$ is a constant and denotes the distance between the sensor nodes and the center for all nodes; thus, min J_K is equivalent to max J_D and because the two resulting clusters are as separated and compact as possible. Because the averaged intracluster distance is greater than the sum of the averaged intercluster distances, that is

$$\frac{d(C_1,C_2)}{n_1 n_2} - \frac{d(C_1,C_1)}{n_1^2} - \frac{d(C_2,C_2)}{n_2^2} > 0 \tag{15.9}$$

From Equation 15.7, it is seen that J_D is always positive. This is to say, evidenced from Equation 15.8, for $K = 2$ minimization of cluster objective function, J_K is equivalent to maximization of the distance objective J_D, which is always positive.

When $K > 2$, we can do a hierarchical divisive clustering where each step uses the $K = 2$ clustering procedure. This procedure can get an approximated K-means clustering structure.

15.3.2.2 PCA-Guided Relaxation Model

In [458], it proves that the relaxation solution of J_D can obtain via the principal component. It sets the cluster indicator vector be

$$q(i) = \begin{cases} \sqrt{n_2/nn_1}, & \text{if } i\varepsilon C_1 \\ -\sqrt{n_1/nn_2}, & \text{if } i\varepsilon C_2 \end{cases} \tag{15.10}$$

The indicator vector satisfies the sum-to-zero and normalization conditions. Consider the squared distance matrix $H = (h_{ij})$, where $h_{ij} = \|x_i - x_j\|^2 \cdot q^T Hq = -J_D$ is easily observed.

1. The first relaxation solution

Let q take any value in $[-1, 1]$, the solution of minimization of $J(q) = q^T Hq/q^T q$ is given by the eigenvector corresponding to the lowest eigenvalue of the equation $Hz = \lambda z$.

2. The second relaxation solution

Let $\widehat{H} = (\widehat{h_{ij}})$, where the element is given by

$$\widehat{h_{ij}} = h_{ij} - h_{i\bullet}/n - h_{\bullet j}/n + h_{\bullet\bullet}/n^2 \tag{15.11}$$

in which $h_{i\bullet} = \sum_j h_{ij}$, $h_{\bullet j} = \sum_i h_{ij}$, and $h_{\bullet\bullet} = \sum_{ij} h_{ij}$.

After computing, we have $q^T \widehat{H} q = q^T Hq = -J_D$, then relaxing the restriction that q, the desired cluster indicator vector is the eigenvector corresponding to the lowest eigenvalue of $\widehat{H}z = \lambda z$.

3. The third relaxation solution

With some algebra, we can obtain $\widehat{H} = -2Y^T Y$. Therefore, the continuous solution for cluster indicator vector is the eigenvector corresponding to the largest eigenvalue of the covariance matrix $Y^T Y$, which by definition, is precisely the principal component v_1.

15.3.2.3 PCA-Guided Clustering Model

For K-means clustering where $K = 2$, the continuous solution of the cluster indicator vector is the principal component v_1, i.e., clusters C_1 and C_2 are given by

$$C_1 = \{i|v_1(i) \le 0\}, C_2 = \{i|v_1(i) > 0\} \tag{15.12}$$

We can consider using PCA technology to clustering sensor nodes for WSNs. It near minimizes the sum of the distances between the sensor nodes and cluster centers.

Example 1:

Let us assume in WSN that there are 20 sensors distributed in the network. X represents the 2D coordinate matrix.

$$X = \begin{bmatrix} 29.471 & 4.9162 & 69.318 & 65.011 & 98.299 \\ 55.267 & 40.007 & 19.879 & 62.52 & 73.336 \\ 37.589 & 0.98765 & 41.986 & 75.367 & 79.387 \\ 91.996 & 84.472 & 36.775 & 62.08 & 73.128 \\ 19.389 & 90.481 & 56.921 & 63.179 & 23.441 \\ 54.878 & 93.158 & 33.52 & 65.553 & 39.19 \\ 62.731 & 69.908 & 39.718 & 41.363 & 65.521 \\ 83.759 & 37.161 & 42.525 & 59.466 & 56.574 \end{bmatrix}$$

Compute the eigenvector of the matrix, $Y^T Y$, i.e., the principal component v_1.

$$v_1^T = \begin{bmatrix} -0.11623 & -0.47282 & 0.10624 & 0.058301 \\ 0.40896 & 0.0013808 & -0.20539 & -0.22552 \\ 0.033439 & 0.17823 & -0.15446 & -0.45626 \\ -0.067391 & 0.18908 & 0.16501 & 0.22147 \\ 0.2697 & -0.11445 & 0.04397 & 0.13673 \end{bmatrix}$$

If $v_1(i) \le 0$, sensor i belongs to C_1, otherwise it belongs to C_2. We depicted the above clustering results in Figure 15.2.

When $K > 2$, we do a hierarchical divisive clustering where each step using the $K = 2$ clustering procedure. In summary, a PCA-guided clustering model can be used to form a nearly optimal K-means-like clustering structure.

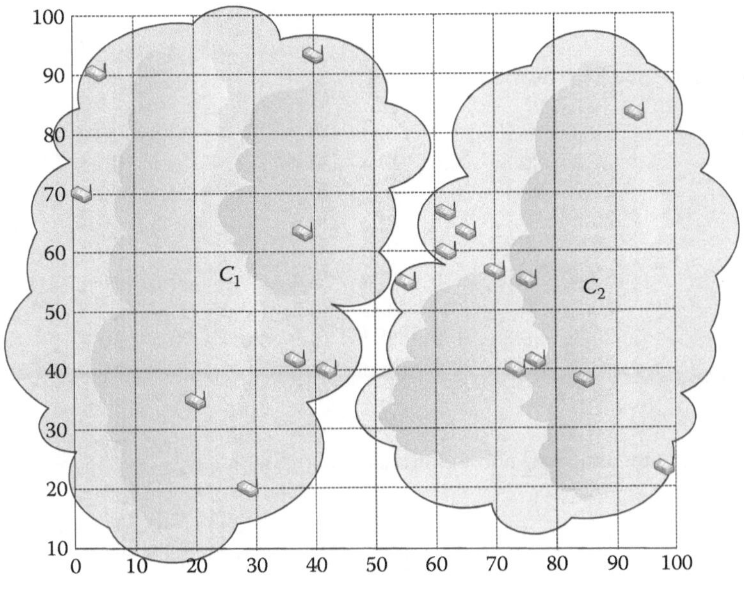

Figure 15.2 The clustering structure for 20 sensor nodes.

15.3.3 PCA-GUIDED DATA AGGREGATING MODEL

As mentioned earlier, $D_{N\times P}$ represents the data measurement matrix collecting from the sampling period by the CHs. Q is a centered matrix about D.

The vector $\{w_k\}_{1\leq k\leq N}$ is the principal components satisfying

$$QQ^T w_k = \lambda_k w_k \tag{15.13}$$

Because in most cases, there exist high correlations between sensor measurements, good approximations to sensor measurements can be obtained by relying on few principal components. The first principal component accounts for as much of the variability in the data as possible, so we can find the first principal component vector, such that their projection data can effectively express the original measurements. Approximations $\widehat{Q}$ to Q are obtained by

$$\widehat{Q} = w_1 w_1^T Q = w_1 Z \tag{15.14}$$

where

$$Z = w_1^T Q \tag{15.15}$$

The CHs only send three packets about Z, w_k, and the mean vector columnwise G of D matrix to the BS:

$$G = (g_i = d_{i\bullet}/p), d_{i\bullet} = \sum_j d_{ij} \tag{15.16}$$

Because the mean vector columnwise G is subtracted by the CHs prior to the aggregation of its value, the BS can add back after the computation of the approximation.

Example 2:

If the CH collects the matrix D as follows:

$$D = \begin{bmatrix} 20.3 & 20.2 & 20.8 & 20.3 & 20.3 & 20.4 & 20.5 & 20.4 \\ 20.4 & 20.3 & 20.6 & 20.4 & 20.4 & 20.5 & 20.6 & 20.5 \\ 20.2 & 20.1 & 20.4 & 20.2 & 20.2 & 20.3 & 20.4 & 20.3 \\ 20.3 & 20.2 & 20.8 & 20.3 & 20.3 & 20.4 & 20.5 & 20.4 \\ 20.3 & 20.2 & 20.8 & 20.3 & 20.3 & 20.4 & 20.5 & 20.4 \\ 20.4 & 20.3 & 20.6 & 20.4 & 20.4 & 20.5 & 20.6 & 20.5 \\ 20.2 & 20.1 & 20.4 & 20.2 & 20.2 & 20.3 & 20.4 & 20.3 \\ 20.3 & 20.2 & 20.8 & 20.3 & 20.3 & 20.4 & 20.5 & 20.4 \\ 20.3 & 20.2 & 20.8 & 20.3 & 20.3 & 20.4 & 20.5 & 20.4 \\ 20.2 & 20.1 & 20.4 & 20.2 & 20.2 & 20.3 & 20.4 & 20.3 \end{bmatrix}$$

The CH can compute the matrix Q and the principal components w_1,

$$\mathbf{Z} = \mathbf{w}_1^{\mathbf{T}}\mathbf{Q} = \begin{bmatrix} -0.26306 & -0.56555 & 0.93394 & -0.26306 \\ -0.26306 & 0.039433 & 0.34193 & 0.039433 \end{bmatrix}$$

$$\mathbf{w}_1 = \begin{bmatrix} 0.39468 & 0.21031 & 0.21031 & 0.39468 & 0.39468 \\ 0.21031 & 0.21031 & 0.39468 & 0.39468 & 0.21031 \end{bmatrix}$$

$$\mathbf{G} = \begin{bmatrix} 20.4 & 20.462 & 20.262 & 20.4 & 20.4 \\ 20.462 & 20.262 & 20.4 & 20.4 & 20.262 \end{bmatrix}$$

then the packet about Z, w_1, and G is delivered to the BS. The BS will compute the approximation $\widehat{Q}$. After adding back the subtracted mean value G, we can obtain

$$\widehat{\mathbf{D}} = \begin{bmatrix} 20.296 & 20.177 & 20.769 & 20.296 & 20.296 & 20.414 & 20.535 & 20.416 \\ 20.407 & 20.344 & 20.659 & 20.407 & 20.407 & 20.471 & 20.534 & 20.471 \\ 20.207 & 20.144 & 20.459 & 20.207 & 20.207 & 20.271 & 20.334 & 20.271 \\ 20.296 & 20.177 & 20.769 & 20.296 & 20.296 & 20.414 & 20.535 & 20.416 \\ 20.296 & 20.177 & 20.769 & 20.296 & 20.296 & 20.414 & 20.535 & 20.416 \\ 20.407 & 20.344 & 20.659 & 20.407 & 20.407 & 20.471 & 20.534 & 20.471 \\ 20.207 & 20.144 & 20.459 & 20.207 & 20.207 & 20.271 & 20.334 & 20.271 \\ 20.296 & 20.177 & 20.769 & 20.296 & 20.296 & 20.416 & 20.535 & 20.416 \\ 20.296 & 20.177 & 20.769 & 20.296 & 20.296 & 20.416 & 20.535 & 20.416 \\ 20.207 & 20.144 & 20.459 & 20.207 & 20.207 & 20.271 & 20.334 & 20.271 \end{bmatrix}$$

15.4 PCA-GUIDED ROUTING ALGORITHM SOLUTION STRATEGIES

In Section 15.3, we have actually proposed a PCA-guided clustering and data aggregating model for routing optimization problem in WSNs by theoretical analyses and numerical examples. The following steps provide an overview of the solution strategy.

15.4.1 INITIALIZATION STAGE

In the first stage, we assume that a set of N location coordinates are gathered at the BS. The BS computes the first principal component v_1. The two clusters C_1 and C_2 are determined via v_1 according to (15.12) by the BS.

Table 15.1
Parameter Description

Parameters	Parameter's Description
κ	The number of clusters
N	The total number of sensor nodes
E_{elec}	The energy consumption per bit when sending and receiving
l	The sending data bit
$\varepsilon_{mp}/\varepsilon_{fs}$	The energy consumption about the amplifier
a	The data aggregating rate. It is application specific, where we assume $a = 3$ as mentioned in Section 15.2.
d_{toBS}	The average distance between the CHs and the BS, we assume it = 75 m
d_{toCH}	The average distance between the CHs and the non-CH nodes. From [440], we obtain $d_{toCH} = \sqrt{\frac{1}{2\pi}\frac{M^2}{k}}$ in which, M denotes the area of this field.

15.4.2 CLUSTERS SPLITTING STAGE

When the number of sensor nodes is huge, two clusters are not enough and can induce the energy consume rapidly. Considering splitting these clusters whose memberships are more than the CH can support.

If there are K clusters, there are on average N/K nodes per cluster (one CH node and non-CH nodes). We define that

$$\text{Ave} = N/K \tag{15.17}$$

We can estimate the average energy dissipated per round to get the most energy efficient number of the clusters:

$$\begin{aligned} E_{\text{round}} &= k(E_{\text{CH}} + E_{\text{non-CH}}) \\ &= k\left[lE_{\text{elec}}\left(\frac{N}{k}-1\right) + alE_{\text{elec}} + al\varepsilon_{mp}d_{toBS}^4 + \left(\frac{N}{k}-1\right)\left(lE_{\text{elec}} + l\varepsilon_{fs}d_{toCH}^2\right)\right] \end{aligned} \tag{15.18}$$

The notation and definition of the parameters in Equation 15.18 are described in Table 15.1.

We can obtain

$$E_{\text{round}} = l\left[(2N+k)E_{\text{elec}} + 3k\varepsilon_{mp}d_{toBS}^4 + \varepsilon_{fs}\frac{M^2}{2\pi}\frac{N}{k} - \varepsilon_{fs}\frac{M^2}{2\pi}\right] \tag{15.19}$$

According to the average energy dissipated per round, the scope of the clusters number can be estimated when it is the most energy efficient. In [439,448,449], the authors use the average energy dissipated per round to obtain the optimal cluster number. Based on this methodology, here in this chapter, we study the appropriate upper limit of the cluster nodes to perform the clusters splitting and thus extend this methodology.

Example 3:

If we assume that the number of sensor nodes is 100, average energy dissipated per round as the number of clusters is varied between 1 and 20. Figure 15.3 shows that it is most energy efficient

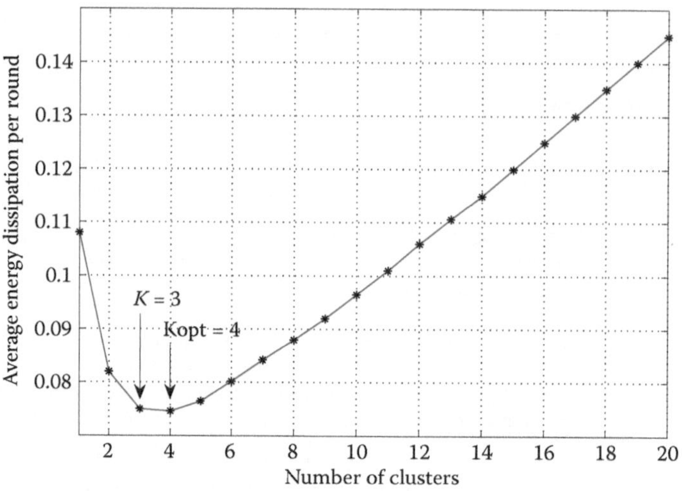

Figure 15.3 Average energy dissipation per round with varying the number of clusters.

when there are between three and five clusters in the 100-node network. We define the most appropriate number of clusters is varied from 3 to 5 because

$$\begin{aligned}&(E_{\text{round}}(3) - E_{\text{round}}(min))/E_{\text{round}}(min) < 0.03\\&(E_{\text{round}}(5) - E_{\text{round}}(min))/E_{\text{round}}(min) < 0.03\end{aligned} \tag{15.20}$$

We obtain the appropriate upper limit of the cluster nodes as $100/3 = 34$. If the number of cluster nodes is more than 34, PCA-guided clustering algorithm will be implemented to split it.

15.4.3 CLUSTER BALANCING STAGE

We use the BS running the PCA-guided clustering algorithm to divide sensor nodes based on the geographical information. We get K clusters from N nodes in the field rapidly and form the better clusters by dispersing the CHs throughout the network.

Now let us introduce the basic idea of the cluster balancing stage. Above, we obtain the number of clusters K if there are N nodes. We define the average node number per cluster as Ave. The cluster-balanced step is added in each iteration process. If $|C_j| > Ave, 1 \leq j \leq K$, the BS computes $dist(s_i, \overline{u_q})$, where $s_i \varepsilon C_j, j \neq q$ and $|C_q| < Ave$. This means to compute the distances between the nodes and each cluster center whose clusters node number is less than Ave. The BS gets s_i if its value is minimum and adjusts the node into the corresponding cluster computed. After implementing the cluster-balanced step, the BS limits the node number in each cluster and changes the clusters' unequal distribution in the space of nodes originally.

Example 4:

Figure 15.4 gives an example to illustrate how the cluster-balanced step works. In this example, if we only consider the geographical information of sensor nodes when using PCA-guided splitting algorithm, the sensor nodes s_1–s_6 should belong to the C_1. However, the nodes in C_1 are more than $Ave(= 5)$. Thus, this scenario motivates the cluster-balanced step. To have a balance among all the clusters, in our method, we suggest that the sensor s_6 should be grouped into C_2 because its distance is nearest to C_2. We obtain the final balanced cluster structure by a serial standard PCA-guided splitting stage and the cluster balancing stage.

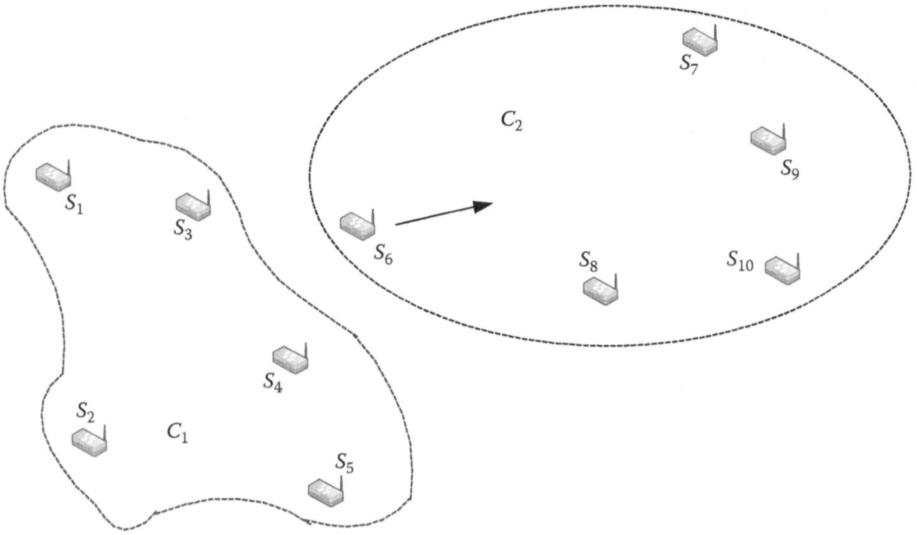

Figure 15.4 An example of the cluster balancing stage.

15.4.4 CHs SELECTING STAGE

After the BS divides the appropriate clusters using PCA technique, it needs to select the optimal CHs in these clusters. Assume that the initial energy is the same, the BS can rank matrix Y for each cluster, and section the sensor nodes, which is nearest to the cluster centers.

15.4.5 DATA AGGREGATING STAGE

The sensor nodes begin to transfer the data to the CHs after completing the cluster formation. The CHs collect the measurements from the sensor nodes and then compute the first principal component w_1 [as Equation 15.13], the mean vector G [as Equation 15.16] and Z [as Equation 15.15]. They can be delivered to the BS with a constant packet size for each CH. Finally, the BS will compute the approximate measurements by these packets.

15.4.6 DESCRIPTION FOR PCA-GUIDED ROUTING ALGORITHM

The procedure taken by the BS is outlined in Algorithm 15.4.1.

Algorithm 15.4.1 The Description for PCA-Guided Routing Algorithm

Start with:
Step 1: Compute the first principal component v_1
Step 2: According to (16.12) the two clusters C_1, C_2 are determined via v_1
Step 3: Compute E_{round} and get the appropriate upper limit of the cluster nodes
While the cluster node number are more than the appropriate upper limit.
Repeat Steps 2 and 3
end
IF needed, implement the cluster balancing stage
Select the cluster heads;
Compute w_1 [according to Equations 15.13, 15.15, and 15.16] for the approximate measurements.

15.5 SIMULATION RESULTS

To evaluate the performance of PCA-RA, we simulate it, LEACH and LEACH-E using a random 100-node network. The BS is located at (50, 150) in a 100 m × 100 m field.

Figures 15.5 and 15.6 show the clustering structure for using LEACH and PCA-RA, respectively. Comparing Figures 15.5 and 15.6, one finds that each cluster is as compact as possible and the CHs locate more closely to the cluster centers by using PCA-RA. This gives us an intuition that it is more efficient to balance the load of network and to even distribute the nodes among clusters by using PCA-RA.

The benefits of using PCA-RA are further demonstrated in Figures 15.7 and 15.8, where we compare the network performance of network lifetime and throughput under the PCA-RA with that under LEACH and LEACH-E. Figure 15.7 shows the total number of nodes that remain alive over the simulation time. While the first dead node remains alive for a longer time in PCA-RA, this is

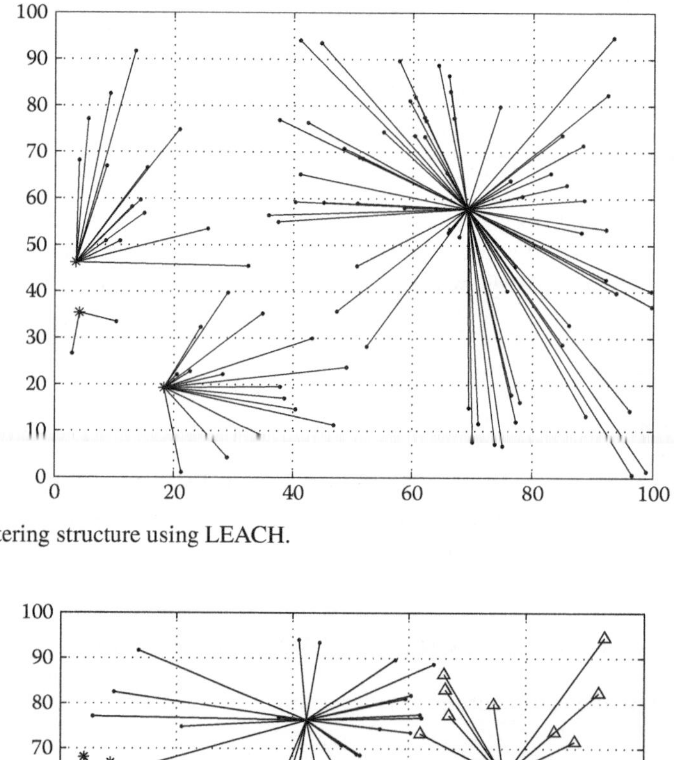

Figure 15.5 Clustering structure using LEACH.

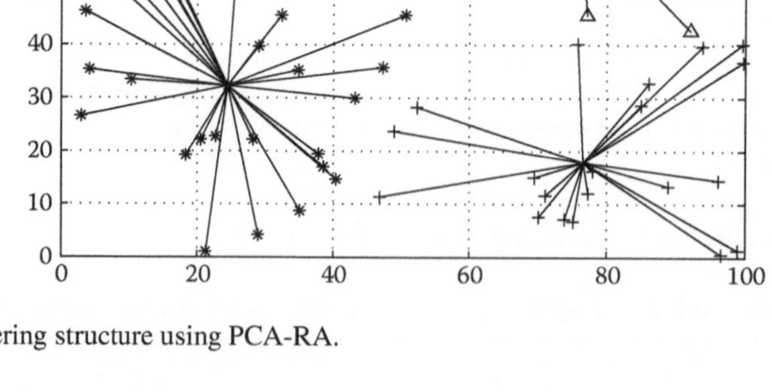

Figure 15.6 Clustering structure using PCA-RA.

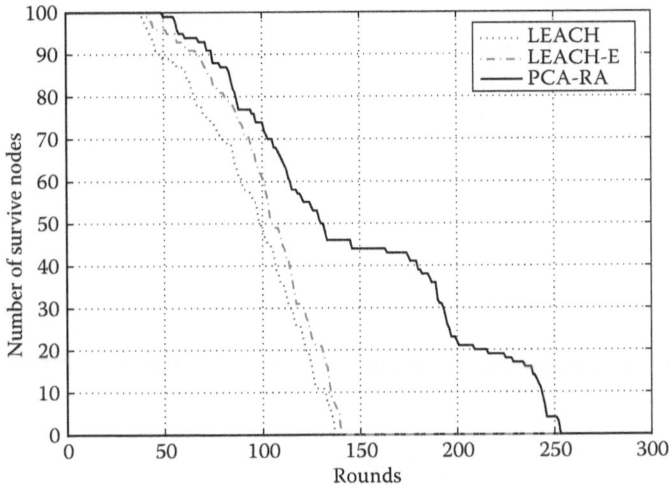

Figure 15.7 System lifetime using LEACH, LEACH-E, and PCA-RA.

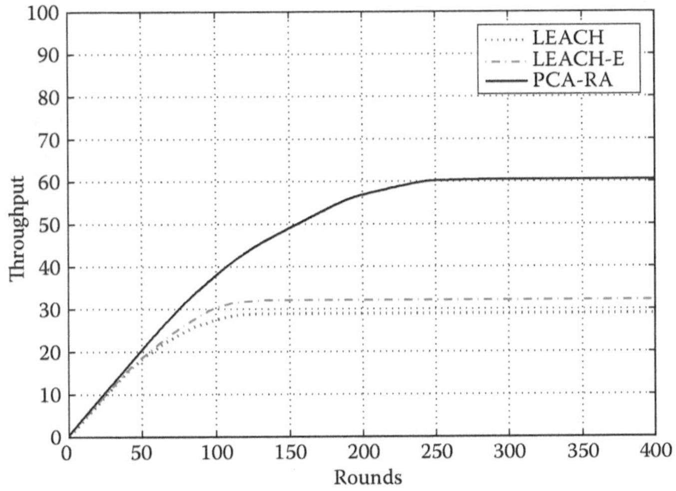

Figure 15.8 Throughput using LEACH, LEACH-E, and PCA-RA.

because PCA-RA takes into account the structure of clusters and the location of the CHs. Figure 15.8 shows that PCA-RA sends much more data in the simulation time than LEACH and LEACH-E.

Note that the survey nodes in each round are NS_i, each node can send D_i data. Then, the maximum throughput can be expressed as follows:

$$Throughput_{PCA\text{-}RA} = \sum_{i=1}^{T_{max}} NS_i \cdot D_i \tag{15.21}$$

From Figure 15.7, the $T_{\max}$ for CA-RA is more than the $T_{\max}$ for LEACH and LEACH-E. Then

$$Throughput_{PCA\text{-}RA} > Throughput_{LEACH}$$

$$Throughput_{PCA\text{-}RA} > Throughput_{LEACH\text{-}E}$$

For example, under the given test data, there are 60.456×10^3 bits data sent in whole network lifetime with PCA-RA. And there are 33.394×10^3 bits and 28.235×10^3 bits by using LEACH-E and LEACH, respectively. The mathematics demonstrates that PCA-RA has 80.01% increase about throughput compared with LEACH-E and 114.12% increase about throughput compared with LEACH.

PCA-RA is a centralized algorithm, and the complexity and communication cost of PCA-RA mostly happen in BS. About the balanced structure stage, we think that the effect on time complexity is small and consider that the time complexity is comparable to LEACH-C.

Table 15.2 displays the network lifetime (in terms of time that the first node becoming dead) and the resulting square error function of the senor node structure under K-means algorithms and PCA-RA.

From Table 15.2, we can find that K-means algorithm can get a minimum square error. Because of the cluster-balanced step, PCA-RA can get a bigger square error but the sensor nodes can survive a longer time. This implies that one can reach a certain trade-off between the total spatial distance of sensor structure and the network lifetime. The suboptimal solution in PCA-RA can achieve such trade-off.

In Figure 15.9, we simulate the sensor nodes collecting the measurements about temperature in some regions. Assume that the BS receives that the packets from the cluster head in a 2-min interval. Figure 15.9 demonstrates the approximations obtained by the BS about a certain sensor node in some intervals.

15.6 CONCLUSIONS

In this chapter, we propose the PCA-RA for WSNs. By disclosing the connection between PCA and K-means, we design a clustering algorithm by utilizing PCA technique, which efficiently develops

Table 15.2
Numerical Example

Parameters	*K*-means	PCA-RA
Square error	34,392.980	38,405.656
The first node dead time	88 rounds	99 rounds

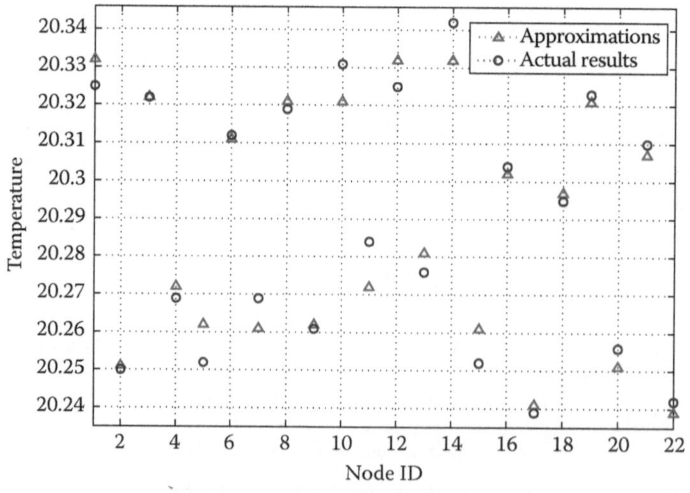

Figure 15.9 Approximation obtained for the sensor nodes.

a clustering structure in WSNs. Moreover, as a compression method, we demonstrate that the PCA technique can be used in data aggregation for WSNs as well. We establish that the explicit procedure of PCA-RA for WSNs by incorporating PCA technique into both the data aggregating and routing process. The advantages of the proposed algorithm are demonstrated through both theoretical analyses and simulation results. The simulation results show that the PCA-guided routing algorithm significantly reduces the energy consumption, prolongs the lifetime of network, and improves network throughput when compared with LEACH and LEACH-E. Further, it keeps the accuracy about the measurements while reducing the network load.

Future research will focus on the distributed strategies of PCA-guided data aggregation and will investigate the performance of PCA-RA with different values of parameter K.

16 Wireless Sensor Networks: Optimally Configuring and Clustering Approaches

This chapter presents the following approaches of optimally configuring and clustering a wireless sensor network (WSN):

To optimally configure a WSN by time matrix: Any practical clustering protocol in a WSN should maintain long connectivity, low and balanced energy consumption. In the existing clustering algorithms, the distance between regular node and cluster head (CH) is normally used as a unique metric for regular nodes to choose CH to join. However, they ignore the impact of the location of the base station (BS) on network lifetime. With regard to the energy consumption, in this section, we study not only the impact of distance between regular node and CH but also the impact of distance between the CHs and the BS. A novel matrix *Time* is introduced and is used as a new criterion for optimally assigning the regular nodes into clusters. By using this metric and solving a *min/min* problem in the static scenario and a *max* problem in the dynamic scenario, we are able to evaluate the network lifetime for each scenario. For the two scenarios, we formulate the calculation of network lifetime into an optimizing problem; we then give procedures on how to achieve its solution. Based on this, we present a distributed approach for network configuration in WSNs with the optimal energy efficiency, which is termed as DOCE. Simulation results demonstrate that DOCE significantly outperforms Low-Energy Adaptive Clustering Hierarchy (LEACH) and ad hoc network design algorithm (ANDA).

To optimally cluster a WSN into a multitier architecture: We propose a comprehensive energy consumption model for multitier clustered sensor networks, in which all the energy consumptions, not only in the phase of data transmissions but also in the phase of CH rotations (or cluster setup), are taken into account. By using this new model, we are able to obtain the solutions of optimal tier number and the resulting optimal clustering scheme, which defines how to group all the sensors into tiers, by the suggested numerical method. This then enables us to propose an energy-efficiency optimized distributed multitier clustering algorithm for WSNs. This algorithm is theoretically analyzed in terms of time complexity.

16.1 GENERAL BACKGROUND

WSN consisting of a large number of sensor nodes can be easily deployed in a variety of environments to offer reliable monitoring of a variety of environment for many applications in surveillance, machine failure diagnosis, and biological detection. Sensors in a WSN are equipped with sensing, data processing, and radio transmission units. The energy resource of the deployed sensor nodes is highly limited, which restricts the network lifetime. Therefore, any protocol for a WSN should consider the energy-efficiency issue with the aim to prolong the network lifetime.

Clustering is an efficient and scalable way to configure WSNs, which helps energy saving. In clustered environments, there are two kinds of nodes: CHs and regular nodes. A CH is responsible for conveying the gathered data from the regular nodes in its cluster and compresses the data before transmitting it to the BS. This responsibility of CHs leads to a higher level of energy exhaustion. The lifetime of clustered WSN is strongly related to CHs failure. The energy consumption of CH increases with the number of its members and the distances between CH and its BS. In order to maximize the network lifetime, it is important to distribute the energy consumption of sensors evenly across the network. Therefore, it is necessary to optimize the energy consumption of CHs by efficiently grouping regular nodes into clusters.

A number of clustering schemes [461–473,478,481] have been proposed to prolong the lifetime of sensor networks. The LEACH approach is first proposed [461] as a distributed, single-hop clustering method, where the sensor nodes elect themselves as CHs with some probability and then broadcast their decisions. Although LEACH periodically rotates the CH in a cluster, it still leads to unbalanced energy consumption as the rotation is only based on a random probability and it does not take into account the energy consumption of CH itself. LEACH-Centralized (LEACH-C) [462] is a variant of LEACH. In LEACH-C, the cluster formation is performed at the beginning of each round using a centralized algorithm by the sink. LEACH-C uses simulated annealing to search for near-optimal clusters. LEACH-C outperforms LEACH in several performance aspects, but it consumes much more energy. PEGASIS [463] improves the performance of LEACH and prolongs the network lifetime greatly with a chain topology. However, the delay is significant despite the energy is saved. A further development is LEACH-F (LEACH-C with Fixed clusters) [464]. LEACH-F is based on clusters that are formed once and then fixed. The CH position rotates among the nodes within the cluster. It consumes less energy due to the fact that once the clusters are formed, there is no setup overhead at the beginning of each round. But the scalability is limited by the fixed clusters. In all aforementioned clustering algorithms [461–464], regular nodes choose the CH that requires minimum communication according to the received signal strength. In [474], an energy-efficient method for nodes assignment in cluster-based network (ANDA) is proposed. Based on the minimum transmission power criterion, the proposed method maximizes the network lifetime by determining the optimal assignment of nodes to the CHs. It considers a network scenario where CHs are chosen as a priori and CHs dynamically adjust the size of the clusters through power control. Although ANDA can extend network lifetime to some extent by considering the failure of CH, such improvement can be very limited due to the fact that it ignores the impact of the location of the BS on network lifetime.

In the most existing clustering algorithms [461–473,478,481], the distance between regular node and CH is normally used as a unique metric for regular nodes to choose CH to join. However, they ignore the impact of the location of the BS on network lifetime. Merely accounting for the minimum energy consumption of regular nodes may lead to the preexhaustion of CH. In this section, we define the lifetime of a WSN as the minimal lifetime among CHs. That is, the network lifetime is the time period from the instant at which the first CH runs out of energy. During the data transmission phase, CHs transmit the aggregate data to BS over a long distance, which is a large energy-spending process. Therefore, in order to maximize the network lifetime, it is important to optimize the energy consumption of CHs by efficiently grouping nodes into clusters and minimizing the energy consumption of communications between CHs and BS.

With regard to the energy consumption, we study not only the impact of distance between regular node and CH but also the impact of distance between the CHs and the BS. A novel matrix *Time* is introduced and is used as a new criterion for optimally assigning the regular nodes into clusters. By using this metric and solving a *min/min* problem in the static scenario and a *max* problem in the dynamic scenario, we are able to evaluate the network lifetime for each scenario. For the two scenarios, we formulate the calculation of network lifetime into an optimizing problem; we then give procedures on how to achieve the solution. Based on this, we present a distributed algorithm for network configuration in WSNs with the optimal energy efficiency, which is termed as DOCE. Simulation results demonstrate that DOCE significantly outperforms LEACH and ANDA.

16.2 NOVEL METRIC FOR OPTIMAL CLUSTERING

Let us take the cluster formation shown in Figure 16.1 as an example. Regular node j may receive invitation message from CHs $1, 2, \ldots, 7$ to join their clusters. However, which CH it chooses can

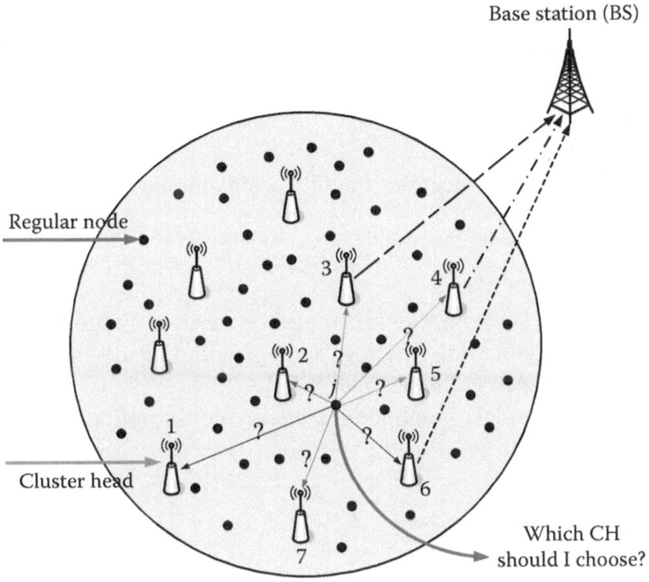

Figure 16.1 Cluster formation.

make difference on network lifetime. In the most existing clustering algorithms, node j chooses its CH according to the length or intensity of invitation message. The message length is strongly related to the distance between the CH and the regular node. In the existing clustering algorithms, node j will choose CH 2 to join because it is the nearest one. But this selection may be not the best one because it does not consider the distance between the CH and the BS. If we let node j join the CHed by CH 3, CH 4, or CH 5, the resulted cluster may have longer lifetime as CHs 3, 4, and 5 are nearer the BS than CH 2.

In this section, we analyze the energy consumption of CH, and we present a novel metric *Time* for optimal clustering configuration. We further give a numerical example to demonstrate how this new metric can be used in assigning regular nodes to clusters.

16.2.1 NETWORK ASSUMPTIONS

We consider a WSN under the following assumptions:

1. The BS without constrained energy resource is fixed and located far away from the sensors.
2. All nodes in the network are homogeneous and energy constrained.
3. All sensors transmit at the same power level and hence have the same transmission range r.
4. The CHs compress the data from the sensors in their cluster before forwarding it further.
5. The environment is contention free and error free, and thus no retransmission.
6. The initial energy of sensor node is uniformly distributed.

16.2.2 RADIO MODEL AND ENERGY CONSUMPTION OF A CH

In a WSN, each node has a transceiver for communication. Any energy-efficient protocol should be based on the specific network topology and radio system. We adopt here the radio model suggested in [461,462] to build the energy model for WSN. There are three radio parts, including sending circuitry, sending amplifier, and receiving circuitry in each sensor node. As in [461,462], the radio dissipates the energy units of E_{elec} to run the sending circuitry and receiving circuitry, and ε_{amp} is used to run the sending amplifier. We also adopt the free space (d^2 power loss) as in [462]. The energy, which is consumed in transmitting a single k-bit message over a distance d, can be calculated by

$$E_t(k,d) = E_{\text{elec}} \times k + \varepsilon_{\text{amp}} \times k \times d^2. \tag{16.1}$$

To receive this message, the radio consumes the following amount of energy:

$$E_r(k,d) = E_{\text{elec}} \times k, \tag{16.2}$$

where the radio dissipates E_{elec} to run the transmitter or receiver circuitry that depends on factors such as the digital coding, modulation, filtering, and spreading of the signal. The amplifier energy $\varepsilon_{\text{amp}} \times d^2$ depends on the distance to the receiver [462].

Based on the above radio model, we obtain the energy consumption for a CH with size n transmitting a k-bit message over d distance:

$$E(CH) = (n-1) \times k \times E_{\text{elec}} + n \times k \times E_{DA} + k \times (E_{\text{elec}} + \varepsilon_{\text{amp}} \times d^2), \tag{16.3}$$

where E_{DA} denotes the energy consumption by a CH in the operations of gathering the data from one regular node.

As observed from Equation 16.3, the energy consumption of a CH increases with the number of transmitted, received, gathered packets, and the device output transmission distance. We partition the energy consumption of a CH into two parts: intracluster communication consumption and output transmission consumption.

16.2.3 NOVEL METRIC: THE TIME MATRIX

A clustered WSN consists of two types of nodes: CHs and regular nodes. Let $S_C = \{1,\ldots,C\}$ be the set of CHs and $S_N = \{1,\ldots,N\}$ be the set of regular nodes. The duration that CH i is engaged in intracluster transmission during cluster formation phase can be calculated as follows:

$$t_1(i) = \frac{E_i}{n_i \times E_{DA} + \delta \times p_i + f(n_i)}, \tag{16.4}$$

where E_i is the initial energy available at CH i, $n_i \times E_{DA}$ is the energy consumption of data aggregation of CH i, p_i is the transmit power level of CH i, δ is a constant weighting factor, n_i is the number of nodes under the control of CH i, and $f(n_i)$ is the energy consumption of the CH in transmitting and receiving activity, which can be modeled as a function of n_i. For the sake of simplicity, we assume a linear relation between the power consumption of the CH transceiver in transmitting and receiving mode and the number of covered nodes [474]. That is, $f(n_i) = \gamma \times |n_i|$, where γ is a constant weighting factor.

In order to study the impact of intracluster transmission consumption on the lifetime of CHs, in this section, we define the following matrix:

$$t_1 = [t_1(i,j)]_{C\times N},$$

whose elements are given by

$$t_1(i,j) = \frac{E_i}{|n_{ij}| \times E_{DA} + \delta \times p_{ij} + \gamma \times |n_{ij}|}, \tag{16.5}$$

where p_{ij} is the power level needed at CH i to reach node j and $n_{ij} = \{k \in S_N | p_{ik} \leq p_{ij}\}$. The above element $t_1(i, j)$ represents the lifetime of CH i when communication happens between CH i and those regular nodes j which are controlled by CH i.

We define the duration of CH i in transmitting the aggregate data to the BS during data transmission phase as follows:

$$t_2(i) = \frac{E_i}{\varepsilon_{\text{amp}} \times \text{dis}(CH_i, BS)^2}, \tag{16.6}$$

where $\text{dis}(CH_i, BS)$ represents the distance between CH i and the BS, the parameter ε_{amp} is usually set as 100 pJ = bit = m^2 for a transmit amplifier.

As what can be observed from Equation 16.6, the value of $t_2(i)$ depends on the distance between CH and BS. That is, the location of BS strongly affects the lifetime of CH.

The element $t_1(i, j)$ guarantees that regular nodes choose the closest CH in order to minimize energy consumption of the cluster members, while $t_2(i)$ makes the node join the cluster to alleviate the load of the CH. To study how these two aspects of energy consumption of CH affects the lifetime of the CH, a weighted factor w is used to trade-off between t_1 and t_2 so as to construct the optimal network configuration. We define a novel weighted element $Time(i, j)$ for regular nodes to choose the CH to join as follows:

$$Time(i, j) = w \times t_1(i, j) + (1 - w) \times t_2(i). \tag{16.7}$$

By substituting Equations 16.5 and 16.6 into Equation 16.7, we have the following description of *Time* matrix:

$$Time = [Time(i, j)]_{C \times N},$$

where the element of *Time* matrix is given by

$$\begin{aligned} Time(i, j) = {} & w \times \frac{E_i}{\delta \times p_{ij} + \gamma \times |n_{ij}| + |n_{ij}| \times E_{DA}} \\ & + (1 - w) \times \frac{E_i}{\varepsilon_{\text{amp}} \times \text{dis}(CH_i, BS)^2}. \end{aligned} \tag{16.8}$$

The elements $Time(i, j)$ will be used as a new metric for regular nodes to make a decision to choose a best suitable CH to join during the cluster formation phase. The procedure of cluster configuration algorithm consists of two phases: CHs election and assignment of nodes to clusters. First, the CHs election procedure is done according to some mechanism which we will present in the following section. Second, the generic CHs collect the values p_{ij} of nodes that are within its maximum transmission range. Then, the CHs send its reachable nodes the following information:

- The distance between CH and BS,
- The CH energy level,
- The value of transmit power required to communicate with the node,
- The number of nodes a CH covers.

Based on the above information, every node can calculate the value of $t_1(i, j)$ and $t_2(i)$ in a distributed manner. Meanwhile, sensor nodes in the network can calculate the Matrix *Time* based on the received information mentioned above.

The procedure is executed in a distributed manner, once the power level of output transmission p_{ij} for CHs is known, the corresponding entry of matrix $Time(i, j)$ can be worked out sequentially. In our proposed optimal configuration approach, regular node j ($j \in S_N$) will choose the CH i ($i \in S_C$) rather than k if

$$Time(i, j) > Time(k, j), \quad \forall\, k \in S_C.$$

Furthermore, node j choose the maximum value of $Time(i,j)$ in its corresponding column j in Matrix *Time*. Namely, node j will be covered by CH i if

$$Time(i,j) = \max_{k \in S_C} Time(k,j).$$

16.2.4 NUMERICAL EXAMPLE

Let us consider a network scenario where the CHs are chosen as a priori, and all CHs in the network are fixed with initial energy $E_i = 5\ J$. We assume there are 5 CHs and 12 regular nodes being uniformly distributed over a rectangular network area. Under this example, we consider the scenario of DOCE executing at the point of 25th Round. We set the weight factor $w = 0.4$, other parameters are set according to Table 16.3. We calculate the *Time* matrix using Equation 16.8 to obtain

$$Time = [Time(i,j)]_{5\times 12} =$$

$$\begin{bmatrix} 8.8543 & 10.371 & 17.663 & 27.947 & 17.117 & 10.903 & \underline{61.156} & 21.247 & 19.135 & 27.726 & \underline{25.465} & 11.665 \\ 10.229 & 9.5948 & 21.392 & \underline{32.526} & 8.5245 & 8.315 & 22.226 & 22.737 & 18.737 & 36.641 & 14.39 & \underline{27.75} \\ 9.7231 & 9.7489 & 18.597 & 30.059 & \underline{24.869} & \underline{22.52} & 26.485 & 24.724 & \underline{19.606} & 14.993 & 15.319 & 11.18 \\ \underline{14.112} & \underline{23.639} & 23.878 & 20.553 & 9.0327 & 11.474 & 48.721 & \underline{32.159} & 18.965 & 25.382 & 14.251 & 11.21 \\ 11.749 & 6.3524 & \underline{36.543} & 19.705 & 8.2911 & 10.479 & 17.68 & 21.454 & 18.608 & \underline{37.783} & 17.611 & 13.199 \end{bmatrix}.$$

In the above matrix, the rows represent the CHs and the columns represent the regular nodes. The matrix element $Time(i,j)$ represents how long the CH i can function well when the regular node j choose the CH i. In our algorithm DOCE, each regular node chooses the CH as its best suitable CH based on the maximal value in each column in this *Time* matrix. For example, the element 14.112 is the maximal value in the first column, therefore, in terms of DOCE, the first regular node will choose the fourth CH to join so as to obtain the optimal cluster configuration.

Correspondingly, we define the following binary matrix to signify the corresponding cluster formation process:

$$Choose = [Choose(i,j)]_{5\times 12} =$$

$$\begin{bmatrix} 0 & 0 & 0 & 0 & 0 & 0 & 1 & 0 & 0 & 0 & 1 & 0 \\ 0 & 0 & 0 & 1 & 0 & 0 & 0 & 0 & 0 & 0 & 0 & 1 \\ 0 & 0 & 0 & 0 & 1 & 1 & 0 & 0 & 1 & 0 & 0 & 0 \\ 1 & 1 & 0 & 0 & 0 & 0 & 0 & 1 & 0 & 0 & 0 & 0 \\ 0 & 0 & 1 & 0 & 0 & 0 & 0 & 0 & 0 & 1 & 0 & 0 \end{bmatrix}.$$

The matrix element $Choose(i,j)$ represents whether the regular node j choose the CH i. If $Choose(i,j) = 0$, the regular node j does not choose the CH i; otherwise, it chooses the CH i.

The previous proposed cluster configuration method ANDA only considers the intracluster communication consumption but does not consider the load's balance of CHs. By using Equation 16.5, we can obtain a similar matrix t_1 as *Time*. ANDA takes the matrix t_1 as the criterion for regular nodes to choose their corresponding CHs. In ANDA, one has the following time matrix:

$$t_1 = [t_1(i,j)]_{5\times 12} =$$
$$\left[\begin{array}{cccccc} 7.0154 & \underline{9.846} & 2.9238 & 6.224 & 4.4415 & \underline{16.374} \\ 3.9408 & 8.761 & \underline{16.634} & 4.2313 & 6.0654 & 5.9131 \\ 5.9806 & 7.9344 & 2.5911 & 5.9167 & 5.3437 & 10.978 \\ 5.6277 & 8.7985 & 4.0047 & 9.3034 & \underline{12.385} & 15.85 \\ \underline{10.271} & 4.1009 & 7.3445 & \underline{10.428} & 4.2634 & 11.118 \end{array}\right.$$
$$\left.\begin{array}{cccccc} 13.804 & 4.9031 & 5.1413 & 5.1489 & 7.547 & 4.4068 \\ \underline{15.927} & 7.082 & 10.398 & 4.1774 & 5.8997 & \underline{12.136} \\ 12.051 & \underline{14.660} & 7.4075 & \underline{10.097} & \underline{9.5085} & 5.1557 \\ 5.1188 & 3.322 & \underline{14.368} & 7.6477 & 7.6454 & 3.9608 \\ 6.307 & 5.1043 & 11.013 & 4.7719 & 7.0812 & 9.6359 \end{array}\right].$$

Similarly, we obtain the following binary matrix to signify the corresponding cluster formation process in ANDA:

$$choose = [choose(i,j)]_{5\times 12} =$$
$$\begin{bmatrix} 0 & 1 & 0 & 0 & 0 & 1 & 0 & 0 & 0 & 0 & 0 & 0 \\ 0 & 0 & 1 & 0 & 0 & 0 & 1 & 0 & 0 & 0 & 0 & 1 \\ 0 & 0 & 0 & 0 & 0 & 0 & 0 & 1 & 0 & 1 & 1 & 0 \\ 0 & 0 & 0 & 0 & 1 & 0 & 0 & 0 & 1 & 0 & 0 & 0 \\ 1 & 0 & 0 & 1 & 0 & 0 & 0 & 0 & 0 & 0 & 0 & 0 \end{bmatrix}.$$

By comparing the matrix *Time* and t_1, we observe that the value of all elements in the matrix *Time* is larger than the elements in the matrix t_1. It means that DOCE achieves a longer network lifetime than ANDA. It is because that DOCE balances the load of CHs by considering a trade-off between cluster size and the distance between CH and BS, but ANDA only considers the intracluster communication consumption and ignores the impact of the location of the BS on network lifetime. The matrix *Choose* and *choose* store the corresponding cluster formation state. With the comparison of matrix *Choose* and *choose*, we observe that in the two different cluster configuration approaches, the size of cluster can be different as the regular nodes choose different clusters to join for a given WSN and its fixed CHs.

As observed from Equation 16.7, the value of $Time(i, j)$ depends on the weighted factor w. Therefore, let us take a look at how the component w influences the network lifetime. When other network parameters are set, the network lifetime is a function of w when DOCE is executed. We plot this function in Figure 16.2. The parameter w varies between 0.1 and 1 with a step 0.1. As observed from Figure 16.2, there is an optimal value of w ($w = 0.4$), under which the network lifetime achieves the maximum. When the value of w is set as 1, only the energy consumption of communication between regular nodes and CH is considered, yet the impact of the location of the BS on network lifetime is ignored. We consider these two factors together by using the parameter w. In other words, the parameter w cannot be set to 1 and the optimal value of w is 0.4 in this case. This idea will be incorporated later in our proposed algorithm: DOCE.

16.3 DOCE

In this section, we present a distributed optimal network configuration approach with energy-efficiency by considering two different scenarios (*static and dynamic scenario*) in a cluster-based WSN. For the static scenario and dynamic scenario, using the matrix *Time*, we formulate the calculation of network lifetime into an optimizing problem; we then give the procedures on how to achieve its solution.

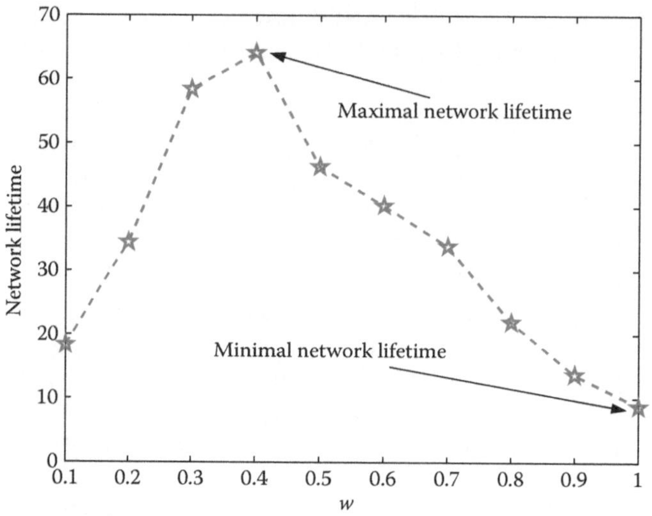

Figure 16.2 Network lifetime while varying w.

The clustering algorithm consists three parts: CH election, optimal assignments of nodes to clusters (cluster configuration), and data transmission. Our major work focuses on the cluster configuration phase. The election of CHs is probability based and the probability is based on the ratio of current residual energy and the initial energy of each node in the network. After the CHs are chosen, regular nodes choose their best suitable CHs, which maximize the network lifetime and energy efficiency according to our proposed novel metric.

16.3.1 STATIC SCENARIO OF DOCE

In this scenario, the CHs are chosen as a prior and the role of CH is fixed. The assignment of nodes to CHs is performed through the procedure *assignments of node to clusters*, which associates each node with the corresponding CH that leads to the longest functioning time. The generic CHs collect the values p_{ij} of all nodes that are within its maximum transmission range. Then, the CHs send its reachable nodes the following information: (1) the distance between CH and BS, (2) the CH energy level, (3) the value of transmit power required to communicate with the node, and (4) the number of nodes that the CH covers. The procedure is executed in a distributed manner, once the power level of output transmission for CHs is known, the corresponding entry of matrix $Time\ (i, j)$ can be worked out sequentially. In our proposed optimal configuration method, regular node j chooses the CH i if $Time(i, j) > Time_{k \in S_C}(k, j)$. Thus, node j ($\forall\ j \in S_N$) will be covered by CH i if

$$Time(i, j) = \max_{k \in S_C} Time(k, j).$$

Therefore, the resulting network configuration guarantees that energy consumption is minimized.

16.3.1.1 Evaluate the Network Lifetime by Solving a Min/Min Problem

As aforementioned, the matrix *Time* is used as a metric for regular node to choose an optimal cluster to join. During the cluster formation phase, each regular node chooses its CH in terms of the maximum value in the corresponding column; at the end, we can obtain a maximum value set, whose element number is equal to C. As we have known, the network lifetime is strongly related to the CHs failure. Hence, the lifetime can be defined as the minimal lifetime of CHs. We, thus, can evaluate the lifetime of the network theoretically based on *Time* matrix and the solution of a *min/min* problem. The *Time* matrix can be abbreviated as follows:

$$Time = [Time(i,j)]_{C\times N} = [Time(1)^T_{1\times N}, Time(2)^T_{1\times N}, \dots, Time(C)^T_{1\times N}]^T,$$

where

$$Time(i)_{1\times N} = [Time(i,1), Time(i,2), \dots, Time(i,N)],\ \forall\, i \in S_C.$$

Then, the network lifetime can be obtained by finding the minimal lifetime of CHs. That is

$$NETWORK_{\text{lifetime}} = L = \min_{i\in S_C}\left\{ \min_{j\in S_N}(Time(1,j)_{1\times N}), \min_{j\in S_N}(Time(2,j)_{1\times N}), \dots, \min_{j\in S_N}(Time(C,j)_{1\times N}) \right\}. \quad (16.9)$$

Once the matrix *Time* is computed by sensor nodes, the theoretical network lifetime can be found. First, we need to find the lifetime of each CH i ($i \in S_C$), which is the minimal value of the ith column of the *Time* matrix that has been computed already. Next, when the set of all CHs get their lifetime, the network lifetime is equal to the minimal value among those lifetimes of CHs. And the optimal assignment of nodes to CHs is executed based on the matrix *Time*. The assignments are recorded by matrix *Choose*, whose element binary variable $Choose(i,j)$ is set as 1 if CH i covers node j; otherwise, set as 0.

16.3.2 DYNAMIC SCENARIO OF DOCE

In the dynamic scenario, the network's topology is no longer static; the cluster configuration is updated periodically. The cluster configuration contains two procedures: *CH Election* and *Assignments of Regular Nodes*.

16.3.2.1 CH Election

The CHs election of our proposed cluster configuration method is probability based. In each round, when cluster are being created, each node become candidate CH based on the probability P and the ratio of current residual energy and initial energy of each node. Take sensor node s_i as an example. Sensor node s_i chooses a random number between 0 and 1. If the random number is less than our threshold $T(n)_{\text{new}}$, node s_i becomes a candidate CH during the current round. Otherwise, s_i gives up the right to be a CH at this round and waits for the next round. Then, the candidate nodes broadcast the *CANDE_HEAD* within radio range R_{compete} to advertise their decision. Each candidate node checks whether there is a candidate node with more residual energy within the diameter R_{compete}. Once the candidate node finds one with more current residual energy, it will give up the competition. Otherwise, it will be elected as CH at this round in the end. The threshold is set as [475]

$$T(n)_{\text{new}} = \frac{P}{1-P[r \bmod (1/P)]} \times \left[\frac{E_{n\_\text{current}}}{E_{n\_\max}} + r_s \text{div} \frac{1}{P}\left(1 - \frac{E_{n\_\text{current}}}{E_{n\_\max}}\right)\right], \quad (16.10)$$

where E_{n_current} represents the current residual energy of sensor node and $E_{n\_\max}$ is the initial energy of each node, and r_s is the number of consecutive rounds in which a node has not been the CH. When r_s reaches the value $1/P$, the threshold $T(n)_{\text{new}}$ is reset to the value it had before the inclusion of the remaining energy into the threshold equation.

16.3.2.2 Assignments of Regular Nodes

In previous clustering algorithms for WSN [462,467,468], the distance between regular node and CH is normally used as a unique metric for regular nodes to choose CH to join. The nodes choose the

CH such that the communication cost is minimal. And usually the communication cost is denoted by the received signal strength. However, merely pursuing efficient energy consumption of the regular nodes only may lead CHs being exhausted quickly during the data transmission phase. That is, although regular nodes choose the CH minimize its energy consumption, the CH which has more important impact on the network lifetime may be exhausted quickly due to long distance data transmission to the BS and immoderate cluster size. In this phase, we use metric $Time(i, j)$ proposed before which takes consideration both on the distance between regular node and its corresponding CH ($\text{dis}(R_i, CH)$) and the distance between CH and BS ($\text{dis}(CH_i, BS)$). Dimension of $Time(i, j)$ is $C \times N$, its generic element $Time(i, j)$ is used as a metric for regular nodes to make an optimal decision. Since the CHs have been elected, each CH broadcasts the message *HEAD_AD_MSG* to its reachable nodes. The regular nodes receive several *HEAD_AD_MSGs* and decide which cluster to join is the major work of our optimal network configuration. Take node s_i for example, s_i will choose the maximum value of $Time(i, j)$ in its corresponding column in matrix $Time(i, j)$ presented in Section 16.3.1.

16.3.2.3 Evaluate Network Lifetime by Solving a Max Problem

We propose here an optimization (*Max*) problem, in which solution leads to the desired network lifetime. First, let us define the following matrices:

$$\begin{aligned} p_{\text{Round}} &= [p(i, j)_{\text{Round}}]_{C \times N}, \\ Choose_{\text{Round}} &= [Choose(i, j)_{\text{Round}}]_{C \times N}, \\ E &= [E(i)]_{C \times 1} = [e_1, e_2, \ldots, e_C]^T, \end{aligned}$$

where the element $p(i, j)_{\text{Round}}$ in the energy matrix p_{Round} represents the energy consumption at the current Round when communication happens between CH i and the regular node j. The element $Choose(i, j)_{\text{Round}}$ in the binary matrix $Choose_{\text{Round}}$ represents whether the regular node j choose the CH i in the current Round: if yes, it is equal to 1; otherwise, it is equal to 0. The component e_i ($i \in S_C$) is the initial energy of each CH, and the vector E represents the initial energy of all CHs.

Let K be a constant representing that how long one round can last, and we assume all rounds have the same time duration. It is thus seen that, the Network Lifetime L can be obtained by solving the following *max* problem:

$$\text{Maxmize } L = K \times Round.$$

subject to the following constrains:

$$\sum_{\text{Round}} \text{diag}\{p_{\text{Round}} \times Choose^T_{\text{Round}}\} \leq E, \tag{16.11}$$

$$\sum_{i \in S_C} \sum_{j \in S_N} Choose_{\text{Round}}(i, j) = N, \tag{16.12}$$

$$\sum_{i \in S_C} \sum_{j \in S_N} Choose_{\text{Round}}(i, j) > C. \tag{16.13}$$

The first constraint (16.11) requires that each CHs total energy consumption cannot exceed its initial energy; the second constraint (16.12) says that each regular node choose only one CH to join; and the final constraint (16.13) requires each cluster covers at least one regular nodes. The procedure of solving the above *max* problem is given in Table 16.1.

16.3.3 DESCRIPTION OF DOCE

Now, we present the configuration procedure of WSN by using our proposed optimal novel metric. The cluster configuration algorithm composes two procedures: *election of CHs* and *assignments of*

Table 16.1
Pseudo-code of Solution to the Optimization Problem

1. Initial phase
// initializing the correlative variants and matrices in this phase.

```
Round=1        // Initialize Round and the total number of Round is 700.
K=20           // K is a constant number representing
                  how long one round can last
L = K × Round  // L is the network lifetime
```

$A_{\text{Round}} = [A_{\text{Round}}(i)]_{C\times 1} = \sum_{r=1}^{Round} diag[p_r \times Choose_r^T], \forall i \in S_C$

$p_{\text{Round}} = [p_{\text{Round}}(i,j)]_{C\times N}$ //energy consumption set of CHs to its regular nodes

$E = [E(i)]_{C\times 1} = [e_1, e_2, e_3, \ldots, e_C]^T$ // Residual energy of each CH at current Round

2. Iterative phase
// iterative process for evaluating network lifetime in this phase.

while($A_{\text{Round}}(i) \leq E_{\text{Round}}(i), \forall i \in S_C$ and $Round < 700$)
  Compute $Time(i,j)$ and $Choose_{\text{Round}}$;
  if ($\sum_i \sum_j Choose_{\text{Round}}(i,j) \neq N$ or $\sum_i \sum_j Choose_{\text{Round}}(i,j) < C$)
    break
  else
    $L = K \times Round$
    $Round = Round + 1$
  end if
end while
return L // Current L is the network lifetime calculated by DOCE

regular nodes optimally. The pseudocode of the distributed optimal network configuration is given in Table 16.2. The following procedure is executed in a distributed and periodic manner at the beginning of each round. Once the procedure is done at the beginning of each round, the configuration of sensor network is updated at the same time. In each round, the data transmission phase begins to function at the end of the following procedure.

The DOCE algorithm is implemented in a distributed manner by sensors. First, the CH i needs to collect the values c_{ij} related to the nodes that are within its maximum transmission range. Then, the CHs send its reachable nodes the following information: (1) the distance between CH and BS, (2) the CH energy level, (3) the value of transmit power required to communicate with the node, and (4) the number of nodes a CH covers. Based on this information, every node can calculate the value of l_{ij} and $\text{dis}(CH_i, BS)$.

16.3.4 COMPLEXITY ANALYSIS

The clustering algorithm composes three procedures: election of CHs, assignments of regular nodes, and data transmission. The CHs election is obtained by circularly checking, the eligibility of sensor nodes in the network. Hence, the complexity of CHs election procedure is $O(N)$, in which N is the number of sensors in the network. The assignment of nodes to clusters is obtained by determining for every node's minimum value $Time(i,j)$. Therefore, the complexity of the cluster formation algorithm is $O(C \times N)$. Based on the above analysis, the complexity of our DOCE algorithm is $O(N)$.

Table 16.2
Optimal Cluster Configuration Procedure of DOCE

//The following procedure is executed at the beginning of each round in a distributed manner

1. CHs Election phase

for (every sensor nodes i in WSN)
    $R(i) = rand$ // node i generates a random number between 0 and 1
    $$T(i) = \frac{P}{1-P[r\text{mod}(1/P)]} \times \left[\frac{E_{i\_current}}{E_i} + r_s\text{div}\frac{1}{P}\left(1-\frac{E_{i\_current}}{E_i}\right)\right]$$
    if($R(i) < T(i)$)
        $CH(i) = 1$ // node i to be a CH
        CH i sends HEAD_MSGs to its neighbor nodes
    else
        $CH(i) = 0$ //node i is still a regular node
    end if
end for

2. Assignments of regular nodes to clusters

each node collects the following information:
$$\text{dis}(CH_i, BS), E_i, p_{ij}, n_{ij}, \forall i \in S_C, \forall j \in S_N$$
calculate the matrix $Time$
$$Time = [Time(i,j)]_{C\times N}$$
$$= w \times \frac{E_i}{\delta \times p_{ij} + \gamma \times |n_{ij}| + n_i \times E_{DA}} + (1-w) \times \frac{E_i}{\varepsilon_{\text{amp}} \times \text{dis}(CH_i, BS)^2}$$
for(every regular node j in WSN)
    set $max = 0$
    for (every CH i in WSN)
        if ($Time(i,j) \geq max$)
            set $max = Time(i,j)$
            set $sel = i$ // j choose i as its CH
        end if
    end for
end for
for (every sensors i in WSN)
    if (CH i = 1)
        CH i creates TDMA schedule to its cluster members
end for

16.4 SIMULATION RESULTS OF THE OPTIMAL NETWORK CONFIGURATION ALGORITHM: DOCE

In this section, we evaluate the performance of our proposed optimal network configuration algorithm through simulation implemented with MATLAB®. Without loss of generality, we assume the BS is far away from the sensing region, and there is an unlimited amount of data to send to the sink node. Simulation time for each round is 20 s. We adopt the same MAC protocols as in [476]. The used parameters are listed in Table 16.3.

Lifetime and energy consumption are the two primary criterions in our approach for evaluating the performance of sensor networks that are characterized by the limited availability of energy within network nodes. In the simulation, network lifetime is defined as the time period from the instant when the network starts functioning to the instant when the first CH node runs out of energy.

Table 16.3
Parameters for Simulation

Rectangular region $R_0 \times R_0$	300×300
Number of nodes N	100
Data rate	2000 bits/s
Initial energy per node	[5, 10] J/battery
E_{elec}	50 nJ/bit
ε_{amp}	100 PJ/bit/m^2
E_{DA}	5 nJ/bit
δ	200
γ	4

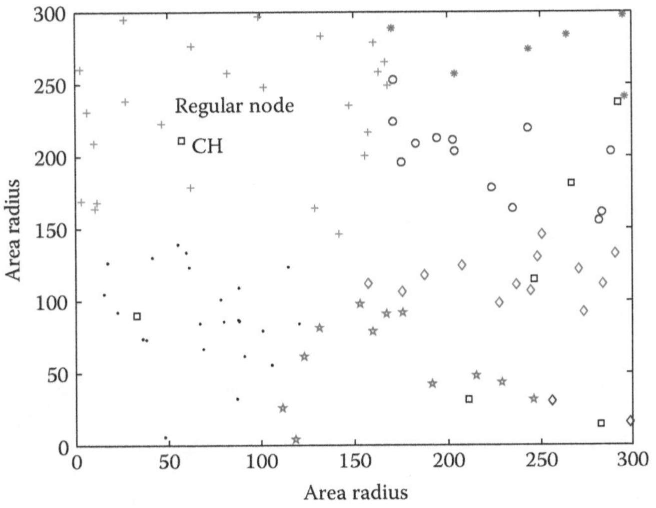

Figure 16.3 Cluster formation during Round = 20 ($BS.x = -100$ and $BS.y = 500$).

16.4.1 DYNAMIC SCENARIO OF DOCE

In the dynamic networking scenario, configuration procedure is periodically updated to adapt to the evolution of the CHs energy status. We derive the performance of DOCE in terms of lifetime, amount data received by BS, lifetime, alive node, energy efficiency, amount data, distribution of nodes, and average residual energy per node.

Figures 16.3 and 16.4 show the simulation results of our proposed algorithm with parameters area diameter $R_0 = 300$ and $w = 0.4$ on a network of 100 sensors distributed uniformly in a square area. The position of BS is set as $BS.x = -100$, $BS.y = 500$ and $BS.x = 150$, $BS.y = 600$, respectively. The different shapes of spots in the figures represent that regular nodes enter into different clusters; the nodes with the same shape belong to the same cluster. As expected, the closer to the BS, the larger the size of the cluster is. It is because that the CHs that are closer to BS cover more regular node while they deplete less energy when communicating with BS during transmission phase. The load of CHs is balanced by optimally adjusting the size of cluster.

We compare the lifetime and amount of data received by BS of DOCE with LEACH, in the situation where the assignments of nodes to clusters are based on the minimum transmission power criterion (denoted by the length of received message). Figure 16.5 shows the variance of alive node in the sensor network and the amount data received by BS as functions Time Rounds when DOCE and LEACH are executed. We set $w = 0.4$ and $P_{ch} = 0.03$ in DOCE and the data transmission rate

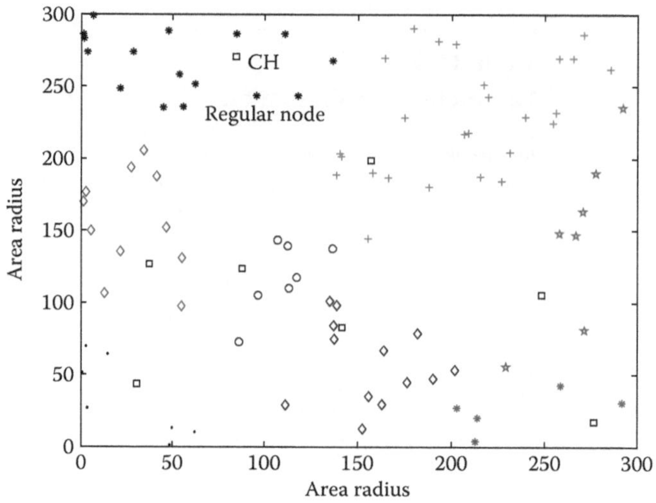

Figure 16.4 Cluster formation during Round = 20 ($BS.x = 150$ and $BS.y = 600$).

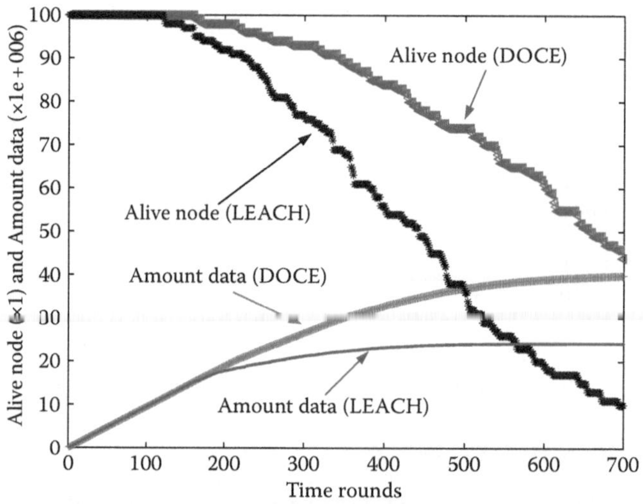

Figure 16.5 Alive node and amount data of LEACH and DOCE.

is 4000 bytes/s. As shown in the figure, DOCE outperforms LEACH not only in terms of the alive node but also in terms of the amount data. DOCE has a longer stable period than LEACH, and the amount of data received by BS is much more than LEACH. DOCE provides up to 30% in alive node and 20% in amount data improvements over LEACH.

In order to observe the impact of number of CHs on the performance of cluster configuration, we vary P_{ch} that represents the desired percentage of CHs next. When DOCE is executed, Figure 16.6 shows the variance of alive node in the sensor network and the amount data received by BS as functions of P_{ch} with the same time elapse (700 Rounds). In this scenario, the number of nodes in the network $N = 100$ and $w = 0.4$ and we consider the cases that the probability to be CH $P_{ch} = 0.02$, 0.04, and 0.05. As shown in Figure 16.6, with the increase of the P_{ch}, or in other words, the number of CH, the amount data received by the BS increases, while the lifetime of DOCE is not always in the way. Therefore, there may be an optimal P_{ch} that exists between 0.04 and 0.05 during the simulation under the concrete parameters setting.

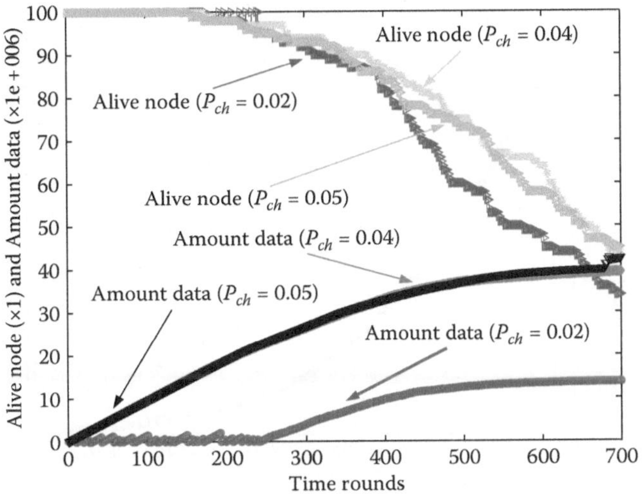

Figure 16.6 Alive node and amount data of DOCE while varying P_{ch} (the desired percentage of CHs).

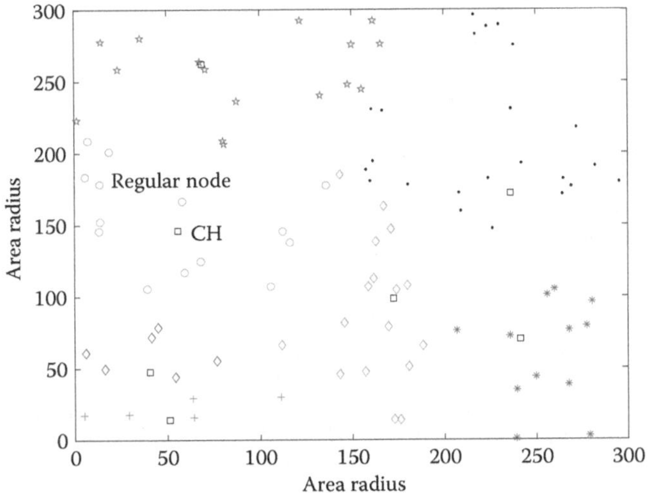

Figure 16.7 C luster formation of ANDA ($BS.x = -100$ and $BS.y = -300$).

Figure 16.7 shows the distribution of sensor nodes of ANDA. We assume that all the CHs are chosen and what need to do is assigning regular nodes to their CHs. In this scene, we set region diameter is 300, the total number of nodes is 100 and $w = 0.4$, $BS.x = -100$, and $BS.y = -300$. The result shows that the cluster formation of ANDA is contrary to our DOCE. Unlike in DOCE, in ANDA, the longer the distance between CH i and BS, the more regular nodes are grouped into the ith cluster. So, these CHs would die more quickly as a larger amount of energy is used to reach the BS which is far away. In this way, the disadvantage of ANDA is very obvious.

Figure 16.8 shows the variance of average residual energy per node as functions of the Time Rounds when DOCE is executed. We set $N = 100$, $R_0 = 300$, and $w = 0.4$. As expected, the variance of DOCE is smaller than LEACH and DOCE has a much more average residual energy. That is, the load of CH in DOCE is fairly balanced. It is because that DOCE assigns the regular nodes to cluster based on a novel metric which fully considers the load of CHs.

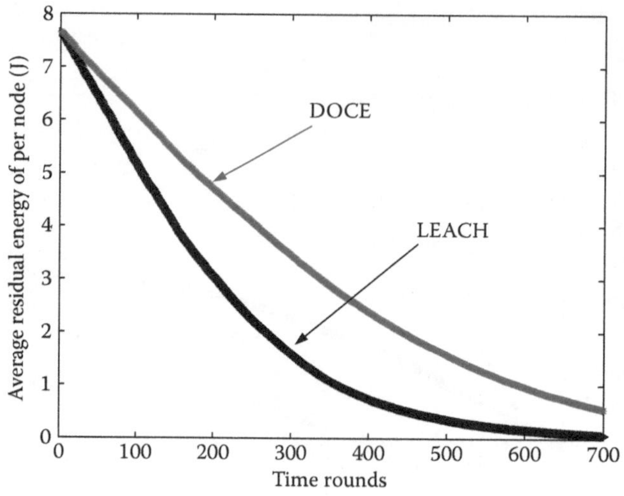

Figure 16.8 Comparison of average residual energy per node.

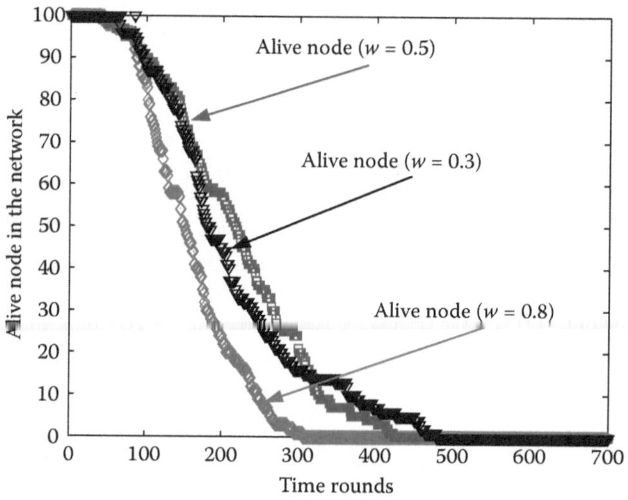

Figure 16.9 Alive node versus time rounds.

Figure 16.9 presents the alive node in the network of DOCE with the time elapsed while varying parameter w. We set the bits of a node transmitting a packet every time $Kbit = 6000$ byte/s. As shown in the figure, when $w = 0.8$ the number of alive node is less than when $w = 0.5$. It demonstrates that the energy consumption of communication between CH and BS has heavier impact on the lifetime of network than intracluster communication consumption. However, when w is decreased to 0.3, the number of alive node decreases and is less than that of $w = 0.5$. It is found that the alive node not always increases with the decreasing of w, so there is an optimal w existing between 0.3 and 0.5 during the simulation under the specific parameter setting.

Figure 16.10 shows the mean residual energy per node as functions of the time rounds while varying the parameter w (the trade-off weight). We assume that the initial energy of sensor node is uniformly distributed in the range $[5, 10]$ and we set $Kbit = 3000$ byte/s. As what can be observed in the figure, when we set the weighted factor $w = 0.3$, 0.5, and 0.8, the mean residual energy increases with the decrease of w regularly. However, when $w = 0.15$, coming forth a reverse phenomena,

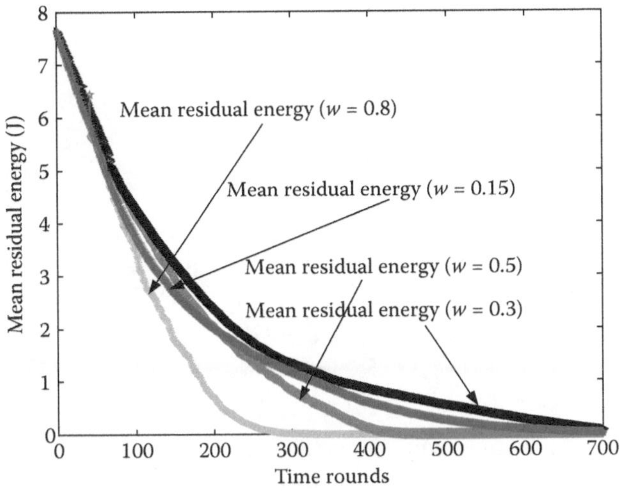

Figure 16.10 Mean residual energy versus time rounds.

that is, the mean residual energy per node of $w = 0.15$ is less than $w = 0.3$ rather than more than $w = 0.3$. So, there is an optimal w exists between 0.15 and 0.3 during the simulation under the specific parameters setting.

16.4.2 STATIC SCENARIO OF DOCE

Next, we consider the static scenario, where network configuration is executed only once. That is, the CHs are chosen as a prior and the role of CH is fixed, what DOCE need to do is to assign regular nodes to those fixed CHs. In this networking scenario, we set the number of CH $C = 100$, the number of regular node $N = 1000$. We compare the performance of DOCE with the results obtained by using a simple network configuration algorithm criterion based on the consideration of only one aspect of energy consumption (ANDA). With contrary to DOCE, in ANDA, the situation is to assign each node to the CH only considering the communication consumption between CH and regular nodes, yet the energy consumption between CH and BS is ignored absolutely. Results are plotted as functions of the ratio of the output transmit power to the power consumption due to the transmitting and receiving activity, denoted by K. Looking at the denominator in (16.5), similar as [474], we derive K as a function of δ and γ

$$K = \frac{(\delta \times p_{ij})}{(\gamma \times n_i)}. \tag{16.14}$$

Figure 16.11 shows the network lifetime as a function of the number of CHs, C. Curves are obtained for $N = 1000$ regular nodes and the varying values of the ratio of the output transmitting power to the energy consumption due to the transmitting and receiving activity, K. As expected, the lifetime increases with the number of CHs, the value of K increases with the increase of the number of CHs obviously. From the comparison with the performance of the ANDA scheme, we observe that the improvement achieved through DOCE is equal to 25% for $K = 0.1$, while it becomes negligible for $K = 10$ when the output transmit power contribution dominates. For both the DOCE scheme and ANDA, for a number of CHs being greater than 10, the performance is improved as K increases. This is because, when there are few CHs with respect to the number of nodes, the distribution of the nodes among the clusters appears to be fairly even. Thus, if energy consumption mainly depends on the number of nodes per cluster, that is, the sensor network with a larger cluster size, a less

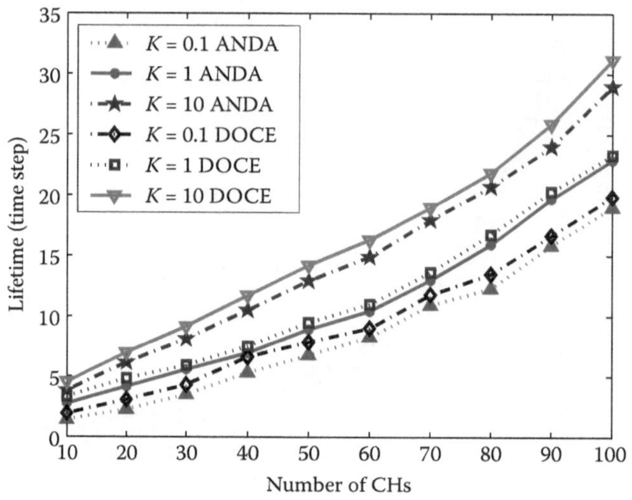

Figure 16.11 Comparison of lifetime between DOCE and ANDA.

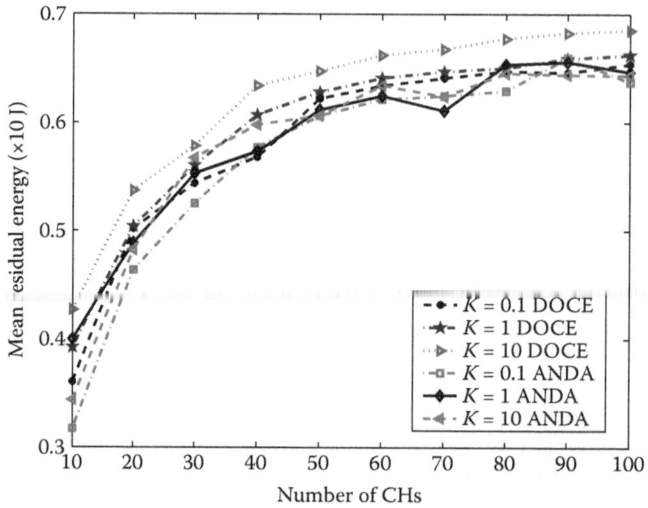

Figure 16.12 Comparison of mean residual energy at the CHs between DOCE and ANDA.

number of cluster, and a smaller value of K (i.e., $K = 0.1$), DOCE obtains a longer network lifetime compared to ANDA. As more CHs are available, the number of nodes per cluster decreases and the differences in the number of nodes assigned to the CHs become significant. If K gets a small value, that is, the output transmit power consumption is considered much more significant than the intracluster consumption. On the contrary, if K gets a large value, the intracluster consumption becomes the leading consumption. Since both the DOCE and the ANDA scheme obtain a larger lifetime when the number of CHs is increased. And we can get that the value of K increases with the increase of the number of CHs easily. Therefore, a longer lifetime is obtained for $K = 10$ both in DOCE and ANDA. However, for any value of K in the figure, our DOCE still outperforms the ANDA scheme in terms of lifetime.

Figure 16.12 shows the mean residual energy at the CHs as a function of the number of CHs, C. The number of regular nodes in the network is set as $N = 1000$. Curves are obtained for the specific parameters setting by varying values of K. Results obtained through DOCE and the ANDA schemes are compared. As expected, in Figure 16.12, the mean value of the residual energy increases with the number of CHs. We note that, for any value of K, DOCE outperforms the ANDA scheme in terms of mean residual energy.

16.5 MULTITIER CLUSTERED NETWORK TOPOLOGY ANALYSIS

In each round of multitier clustering scheme, we construct a k-tier clustering scheme (where $0 \leq k \leq K_{\max}$) to collect the data packets in the region. The parameter of tier-i is (N_i, D_i), where N_i is the number of tier-i CHs and D_i is the radius of tier-i cluster. We set the sink node to be the tier-0 CH with the radius of the cluster being D. That is to say, $N_0 = 1$ and $D_0 = D$. An example of three-tier clustered network topology structure is depicted by Figure 16.13. We also define the number of the common sensors in a k-tier clustering network to be N_{k+1}, and all the number of sensors at each tier N_i, $i = 0, 1, \ldots, k+1$ satisfy the following equation:

$$\sum_{i=0}^{k+1} N_i = N + 1, \tag{16.15}$$

where $N_i \in [1, N]$. Note that for our purpose of optimizing the energy consumption and to obtain the solution of optimal tier, we have allowed N_i, $i = 0, 1, \ldots, k+1$ to be real (float) numbers, though in real networks they are logically needed to be integer numbers.

If each sensor chooses the closest aggregate as its CH at each tier, the sensors essentially form a Voronoi diagram of the network region where each cluster corresponds to a Voronoi cell [482]. For each tier, a typical cluster can be approximated as a circle with the CH at the center, according to [483], we have the following relationship:

$$D_i = \frac{D}{\sqrt{N_i}}, \quad i = 0, 1, \ldots, k+1. \tag{16.16}$$

Equations 16.15 and 16.16 consist of the topology constraints to our energy model which will be used later on in deducing the energy consumption model.

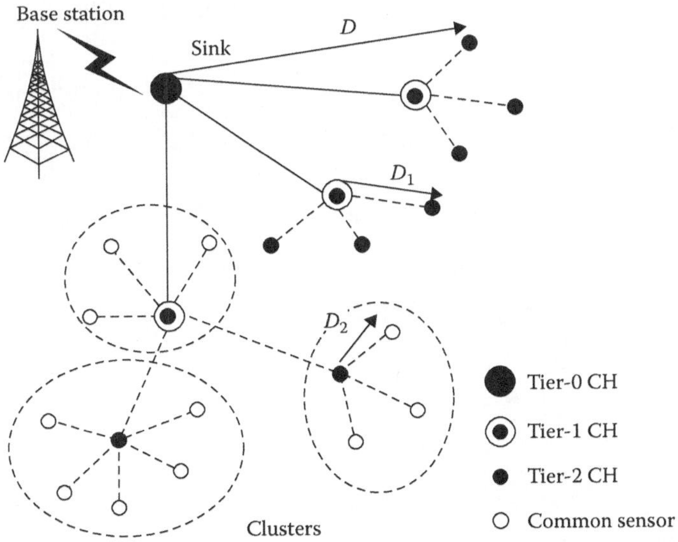

Figure 16.13 An example of three-tier clustered network topology structure.

16.5.1 TRANSMISSION ENERGY MODEL

In our assumptions, the sensors are distributed according to a homogeneous spatial Poisson process and, hence, N sensors are distributed uniformly in a cluster with radius D. For the sake of simplicity, as in [473], we also make the assumption that the CH is at the center of the circle and all the clusters in the same tier have the same size. Then, let d be a random variable that denotes the total length of the segments from the common sensors to the CH (see Figure 16.14). Therefore

$$E(D) = \int_A x \frac{dA}{\pi D^2} = \int_0^{2\pi} \int_0^D \frac{x^2\, dx\, d\theta}{\pi D^2} = \frac{2D}{3}. \tag{16.17}$$

Then, we define $E_t(D)$ to be the energy used by a sensor in the cluster with radius D to communicate one unit of data to the CH. We have

$$E_t(D) = 2\frac{E(D)}{r} = \frac{4D}{3r}. \tag{16.18}$$

16.5.2 ENERGY CONSUMPTION OF THE TIER-*I*

According to the topology setting of multitier clustering scheme in Section 16.5, there are N_i clusters with radius D_i in tier-i (see Figure 16.15 for tier-i topology structure in the multitier clustered network). The main source of energy consumption in the tier-i is the energy used in the setup phase and the energy used in the data communication phase at each round.

We define $E_i^{(s)}$ to be the energy used in the setup phase for CH Rotation of tier-i. It is given by

$$E_i^{(s)} = N_i m + 2m\Big(N + 1 - \sum_{j=0}^{i} N_j\Big), \quad i = 0, 1, \ldots, k. \tag{16.19}$$

The first part of Equation 16.19 is the energy consumed by the tier-i CHs broadcasting its control information to all the sensors under the tier-i, and the second part of Equation 16.19 is the energy consumed by all the sensors under the tier-i to receive control information and send out the control information to disperse it. The tier-i routing table can be built at the same time, so that every sensor node under tier-i knows how to deliver data packet to reach the tier-i CHs.

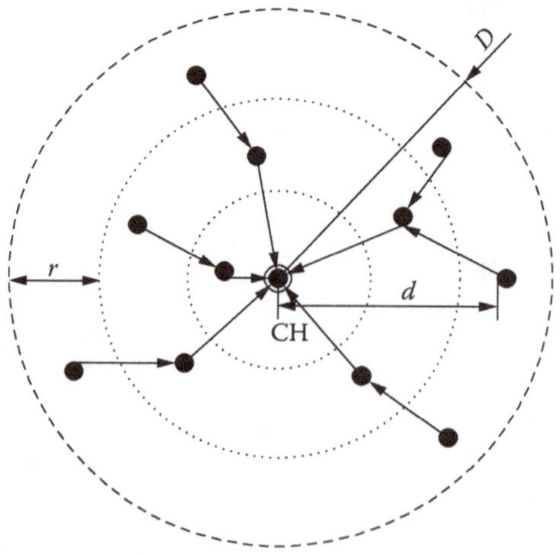

Figure 16.14 A typical cluster in a multihop clustered network.

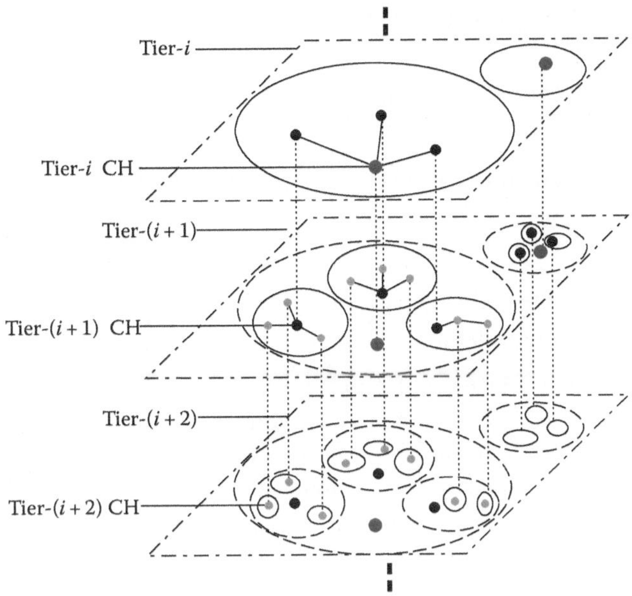

Figure 16.15 The tier-i network topology structure in multihop clustered network.

We define $E_i^{(d)}$ to be the energy used in the data communication phase of tier-i. It is given by

$$E_i^{(d)} = N_i \frac{N_{i+1}}{N_i} E_t(D_i) h = \frac{4 D_i N_{i+1} h}{3r}, \quad i = 0, 1, \ldots, k, \tag{16.20}$$

where h is the number of units of data (transmitted to the corresponding CH per sensor).

In tier-i, there are N_i clusters and every cluster has N_{i+1}/N_i sub-CHs at tier-$(i+1)$ or common sensors which will send the data it received or sensed to its CH, so a value of N_i is multiplied. The energy used by N_{i+1}/N_i sensors deployed uniformly in a circle with radius D_i each transmitting h units of data to its CH should be $(N_{i+1}/N_i)E_t(D_i)h$.

Combining Equations 16.19 and 16.20, we can obtain the energy consumption E_i of tier-i:

$$\begin{aligned} E_i &= E_i^{(s)} + E_i^{(d)} \\ &= N_i m + 2m\left(N + 1 - \sum_{j=0}^{i} N_j\right) + \frac{4 D_i N_{i+1} h}{3r}, \end{aligned} \tag{16.21}$$

where $i = 0, 1, \ldots, k$. We will use this energy model to calculate the energy consumption of each tier in the multitier clustering scheme.

16.5.3 ENERGY MODEL OF MULTITIER CLUSTERING SCHEME

According to the topology construction of k-tier clustering scheme, we obtain the energy consumption $E_k^{(t)}$ of a k-tier clustering scheme as follows:

$$E_k^{(t)} = \sum_{i=0}^{k} E_i. \tag{16.22}$$

where $k = 0, 1, \ldots, K_{\max}$.

By substituting Equations 16.16 and 16.21 into Equation 16.22, we can obtain:

$$E_k^{(t)} = 2km(N+1) - m\sum_{i=0}^{k}(2k+1-2i)N_i + \frac{4Dh}{3r}\sum_{i=0}^{k}\frac{N_{i+1}}{\sqrt{N_i}}, \tag{16.23}$$

where $k = 0, 1, \ldots, K_{\max}$. Equation 16.23 is the energy model of the k-tier clustering scheme.

16.5.4 ENERGY MODEL ANALYSIS

Our purpose here is to find the optimal number of tiers, k_{opt}, and the corresponding solution $N^{k_{\text{opt}}} = \{N_0, N_1, \ldots, N_{k_{\text{opt}}}, N_{k_{\text{opt}}+1}\}$ to reach the smallest energy consumption $\min(E_{k_{\text{opt}}}^{(t)})$ among all the minimized energy consumptions $\min(E_k^{(t)})$ of the k-tier clustering scheme, where $k = 0, 1, \ldots, K_{\max}$.

The method we use here to obtain the optimal number of tiers is to find out all the minimized energy consumptions $\min(E_k^{(t)})$ of the k-tier clustering schemes first, and then we select the tier number that reaches the smallest energy consumption among all the minimized energy consumption $\min(E_k^{(t)})$ of the k-tier clustering scheme as the optimal tier.

Now the only problem left is that for a k-tier clustering scheme whether there exists a optimal solution that minimizes the energy consumption of each k-tier clustering scheme and if it does exist, how to obtain the corresponding optimal solution.

We first analyze the energy model (16.23) of k-tier clustering scheme (where $k = 0, 1, \ldots, K_{\max}$) to get its first-order derivative and second-order derivative of Equation 16.23 as follows:

$$\frac{\partial E_k^{(t)}}{\partial N_i} = -m(2k+1-2i) + \frac{4Dh}{3r}\left(\frac{1}{\sqrt{N_{i-1}}} - \frac{N_{i+1}}{2N_i^{3/2}}\right), \tag{16.24}$$

and

$$\frac{\partial^2 E_k^{(t)}}{\partial N_i^2} = \frac{DhN_{i+1}N_i^{-5/2}}{r}, \tag{16.25}$$

where $i = 1, 2, \ldots, k$.

As the energy consumption $E_k^{(t)}$ is continuous, and the second-order derivative (16.25) is larger than 0, so there exists a solution $N^k = \{N_0, N_1, \ldots, N_k, N_{k+1}\}$ which reaches the minimum energy consumption to this k-tier clustering scheme.

Now we set all the first-order derivatives to be 0, to yield the following equations:

$$-m(2k+1-2i) + \frac{4Dh}{3r}\left(\frac{1}{\sqrt{N_{i-1}}} - \frac{N_{i+1}}{2N_i^{3/2}}\right) = 0, \qquad i = 1, 2, \ldots, k. \tag{16.26}$$

Therefore, we have k conditions here, and together with Equation 16.15 and the setting $N_0 = 1$, making $k+2$ equations from which we are able to obtain the optimal solution theoretically. We herein give a recursive method to obtain the required optimal solutions.

Note that Equation 16.26 is reduced to the following equations by some mathematical manipulations:

$$N_{i+1} = 2N_i^{3/2}\left(\frac{1}{\sqrt{N_{i-1}}} - \frac{3mr(2k+1-2i)}{4Dh}\right), \qquad i = 1, 2, \ldots, k. \tag{16.27}$$

Given we already know that $N_0 = 1$, and if we also know the number of tier-1 CHs N_1, we can get all the numbers for the other tier's CH (i.e., $N_2, N_3, \ldots, N_{k+1}$) by Equation 16.27, subject to Equation 16.15 also being satisfied. Therefore, it is sufficient to perform a binary search to find a value for N_1 satisfying Equation 16.15 and, thereby, we can get the optimal solution for k-tier clustering scheme.

From the above discussions, one observes that the complexity of computing the optimal solution for minimizing the energy consumption of a k-tier clustering scheme is $O(k\log(N))$ where N is the number of sensors. Hence, the complexity of obtaining the optimal tier of multitier clustering scheme is $O(K_{\max}^2 \log(N))$. Here, $K_{\max}$ will not exceed 10 even for a network consisting of million of sensors deployed in a vast area according to our experiences from simulations.

16.6 EXAMPLE OF FINDING OPTIMAL TIERS OF MULTITIER CLUSTERING SCHEME

In this section, an example is given. We first describe the energy consumption of multitier clustering scheme by the proposed energy model and then we calculate, by the proposed numerical method, the corresponding optimal solution $N^k = \{N_0, N_1, \ldots, N_{k+1}\}$, where $k = 0, 1, \ldots, K_{\max}$.

The parameters of this example are set according to Table 16.4. Substituting these parameters into Equation 16.23, the energy consumption model of this network is

$$E_k^{(t)} = 2002k - \sum_{i=0}^{k}(2k-2i+1)N_i + \frac{200}{3}\sum_{i=0}^{k}\frac{N_{i+1}}{\sqrt{N_i}}, \tag{16.28}$$

where $k = 0, 1, \ldots, K_{\max}$.

In order to get the optimal solution to minimize the energy consumption of k-tier clustering scheme, the $k+2$ conditions are worked out as follows:

$$\begin{aligned} N_0 &= 1 \\ \sum_{i=0}^{k+1} N_i &= 1001 \\ N_2 &= 2N_1^{3/2}\left(\frac{1}{\sqrt{N_0}} - \frac{3(2k-1)}{200}\right) \\ N_3 &= 2N_2^{3/2}\left(\frac{1}{\sqrt{N_1}} - \frac{3(2k-3)}{200}\right) \\ &\vdots \\ N_{k+1} &= 2N_k^{\frac{3}{2}}\left(\frac{1}{\sqrt{N_{k-1}}} - \frac{3}{200}\right) \end{aligned} \tag{16.29}$$

where $k = 0, 1, \ldots, K_{\max}$.

Now, for every k-tier clustering scheme, by solving these $k+2$ equations, we will get the minimum energy consumption and the corresponding optimal solution. This procedure is broken down as follows.

Table 16.4
Parameters for the Example

N	1000
D	10
r	1
h	5
m	1
$K_{\max}$	5

When $k = 0$, there are no CHs in the network, this is our energy model in the special case of DD [479], and the optimal solution calculated by 16.29 is $N^0 = \{1, 1000\}$, then the energy consumption of this scheme can be calculated directly by Equation (16.28) to be $\min(E_0^{(t)}) = 66664.67$.

When $k = 1$, there is only one tier of CHs in this network, this is our energy model in the special case for LEACH [477]. Certain numerical calculations by Equation 16.29 yield the optimal solution: $N^1 = \{1, 60.4, 939.6\}$, then the optimal energy consumption of one-tier clustering scheme obtained from (16.28) is $\min(E_1^{(t)}) = 14024.83$.

When $k \geq 2$, this is the energy model of multitier clustering scheme, and the optimal energy consumption of those clustering schemes together with the corresponding solution $N^k = \{N_0, N_1, \ldots, N_k, N_{k+1}\}$ can also be readily computed. The optimal energy consumption and the corresponding solution of all k-tier clustering scheme are given in Table 16.5.

Now we have obtained the minimum energy consumptions when $k = 0, 1, 2, 3, 4, 5$. Then, we can determine that $E_2^{(t)}$ consumed the minimum energy, so the optimum number of tiers is $k_{\text{opt}} = 2$, and the optimal solution is $N^2 = \{1.0, 17.4, 145.0, 837.6\}$. Subsequently, the energy consumption will be minimized by constructing the network to be a two-tier clustering scheme with one tier-0 CHs (or sink node precisely), 17.4 tier-1 CHs, 145.0 tier-2 CHs, and 837.6 tier-3 CHs (or common sensors precisely).

16.7 DISTRIBUTED MULTITIER CLUSTER ALGORITHM

Now, we will present a distributed multitier cluster algorithm to generate a k-tier clustering scheme for the multihop sensor network. By this we mean that there are k tiers of CHs, with tier-0 being the highest and tier-k being the lowest tier. In this clustered environment, the common sensors deliver the gathered data to the tier-k CHs. The CHs in tier-k aggregate this data and deliver the aggregated data to the tier-$(k-1)$ CHs and so forth. Finally, the CHs in tier-1 deliver the aggregated data to the sink node which is the tier-0 CH.

16.7.1 ALGORITHM DESCRIPTION

The algorithm works in an top-bottom fashion, that is, it first elects the tier-1 CHs, then tier-2 CHs, and so on. The tier-1 CHs are chosen as follows. Each sensor decides to become a tier-1 CH with probability N_1/N and broadcast a message to advertise itself as a tier-1 CH to the sensors nearby. This advertisement is forwarded to all the sensors without hop limitation which is dissimilar to the algorithm proposed in [473,480]. Each sensor that receives an advertisement will join the cluster by registering the route to the CH into their routing table and relays the advertisement to its neighbors.

Table 16.5
Optimal Energy Consumption and the Corresponding Solution of the Clustering Scheme in the Example

k	N_0	N_1	N_2	N_3	N_4	N_5	N_6	$\min(E_k^{(t)})$
0	1	1000						66,664.67
1	1	60.4	939.6					14,024.83
2	1	17.4	145.0	837.6				11,916.44
3	1	7.2	38.8	179.8	774.2			12,884.36
4	1	3.4	12.5	48.1	188.7	747.2		14,605.14
5	1	1.7	4.3	14.0	50.2	190.0	739.8	16,486.53

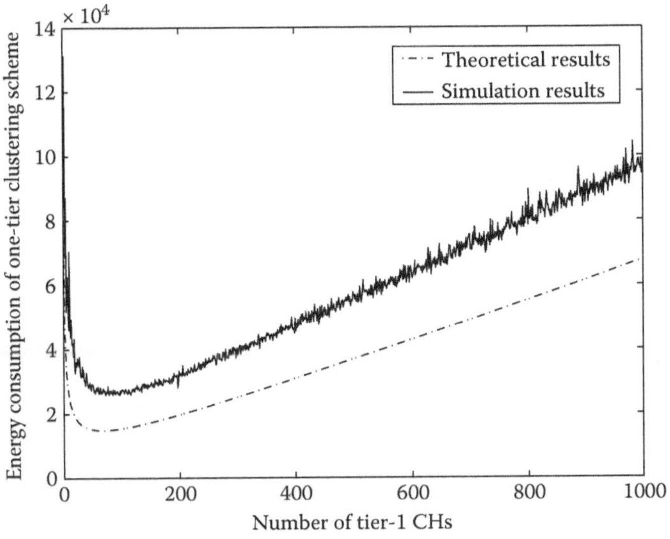

Figure 16.16 Energy consumption versus the number of CHs in the one-tier clustering scheme.

The rest of the sensors then elect N_2 sensors as tier-2 CHs with a probability $N_2/(N-N_1)$ and broadcast their decision of becoming a tier-2 CHs. Each sensor that receives an advertisement will join the tier-2 cluster by registering the route to the tier-2 CH into their routing table and relays the advertisement to its neighbors. All the tier-2 CH will route data according to the tier-1 routing table to reach the tier-1 CHs. CHs at the tier-i are chosen in the similar fashion with probability $N_i/(N-\sum_{j=1}^{i-1} N_j)$, where $i = 3,4,\ldots,k$. This algorithm generates the k-tier clustering scheme.

The pseudocode of the algorithm is given as follows:

```
/* Initialize */
Calculate the optimal solution N^k = {N_0, N_1, ..., N_k, N_{k+1}} for the k-tier clustering scheme
/* Set up routing table */
If (Sensor has not received a tier-i message and has not been elected as a CH at this round)
      Listen for message
      If (a message (k, i, id, hop) is received)
            /* Register route to the CHs tier-i routing table */
            Routing_table[i] = id
            Broadcast message (k, i, my_id, hop + 1)
/* Elect tier-(i + 1) CH */
If (i < k and the sensor has not been elected as a CH)
      rand = generate a random number between 0 and 1
      if (rand < N_{i+1}/(N − Σ_{j=1}^{i} N_j))
            ch_tier = i + 1
            routing_table[i + 1] = −1
            Calculate t = time taken by one message to pass through one hop
            /* A period of 2 × ⌈D/r⌉ × t time will be used to build each tier */
            Waiting for (2 × ⌈D/r⌉ − hop) × t time
            Broadcast message (k, i + 1, my_id, 0)
```

In this algorithm, we have not limited the diffuse hop of the advertisement message broadcasted by the CH as is done in [473,480,481], as the limitation of diffuse hop will lead to many forced CH at each tier. Such forced CH will disturb our energy model, and in our simulation, the forced CH will waste more energy by becoming a CH, instead of joining a cluster.

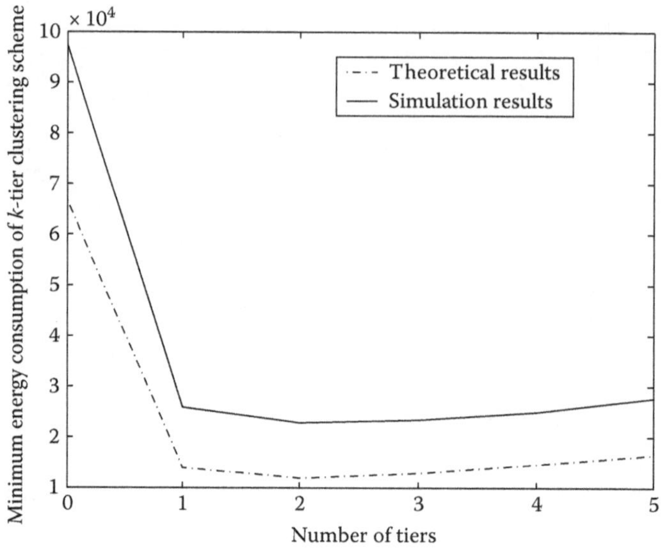

Figure 16.17 Energy consumption versus the number of tiers in the multitier clustering scheme.

At the beginning of each round, the sink node will broadcast a message (k, 0, *sink_id*, 0) to start the algorithm to generate a k-tier clustering scheme. If the sink node broadcasts a message (k_{opt}, 0, *sink_id*, 0), then the algorithm will generate a k_{opt}-tier clustering scheme which will minimize the energy consumption among all multitier clustering networks.

Continuing to the example given in Section 16.6, by using the network scenario and the parameter settings, we run the above proposed distributed multi-tier cluster algorithm. The theoretical results are compared with the simulation results in Figures 16.1 and 16.17 with regard to the energy consumption. In particular, Figure 16.1 plots the energy consumption versus to the number of CH's in the one-tier clustering scheme while Figure 16.17 plots Energy consumption versus to the number of tiers in the multi-tier clustering scheme, both presenting the theoretical results with comparison to the simulation results.

16.7.2 TIME COMPLEXITY OF THE CLUSTERING ALGORITHM

The algorithm can take advantage of opportunities for concurrency in communication, the run-time of the algorithm is the time to transmit a message from all CHs at all levels in the cluster hierarchy to the members of their clusters. The cluster members in each tier are at most $2 \times \lceil D/r \rceil$ hops away from the CH at the worst case. So, a period of $2 \times \lceil D/r \rceil$ hops delivery time can be used to build each tier. The time complexity of the proposed algorithm to generate a multitier clustering scheme with k tier is $O(kD/r)$ in a contention-free environment, where D is the radius of the sensor area and r is the radius of sensor's transmission range. This will make the proposed algorithm more suitable for large networks.

16.8 CONCLUSION

We propose a distributed optimal cluster configuration approach with energy efficiency (termed as DOCE) in WSN to maximize the network lifetime and energy efficiency. We present a novel *Time* matrix that constitutes a criterion for assigning a regular node into the most suitable cluster with the purpose of balancing the energy consumption and extending the network lifetime. Based on the *Time* matrix, we demonstrate that the network lifetime can be derived by solving a *min/min* problem in the static scenario and a *max* problem in the dynamic scenario. Numerical results are given to illustrate the optimal configuration method. By simulation, we compare DOCE with ANDA in terms of network lifetime and energy efficiency. We also compare DOCE with the classical clustering algorithm LEACH in terms of lifetime, alive node, energy efficiency, amount data, distribution of nodes, average residual energy per node, and so on. Simulation results show that our proposed DOCE has significant advantages over the existing algorithms in performance.

We have also focused on finding the optimal tiers in the multitier clustered scheme to minimize the energy consumption for data collection in WSNs. In particular, we first propose an energy consumption model for the multitier clustered network. Distinct from previously reported energy models, in our model, we have accounted for the energy consumption of the CH rotations. We then propose an approach to find out the optimal number of tiers to minimize the energy consumption by numerical methods.

We present a distributed multitier clustering algorithm to generate the energy-efficiency optimized multitier clustering scheme for WSNs. Simulation results corroborate the proposed energy model. The optimal solution of tier number can be worked out in a computationally efficient way by our energy model as demonstrated by an example.

We have assumed that the sensors are distributed according to a homogeneous Poisson process, but the local difference of the sensor's density also affects the energy consumption. Therefore, extending the multitier clustering algorithm to cover the heterogeneous clustering situation according to the local density will bring the algorithm to be more suitable for large-scale networks. This deserves further studies.

17 Big Data Collection in Wireless Sensor Networks: Methods and Algorithms

One of the most widespread and important applications in wireless sensor networks (WSNs) is the continuous data collection, such as monitoring the variety of ambient temperature and humidity. Due to the sensor nodes with a limited energy supply, the reduction of energy consumed in the continuous observation of physical phenomenon plays a significant role in extending the lifetime of WSNs. In a large-scale WSN, the data generated by the sensors may experience an exponential growth as time goes by. Subsequently, we may gather hundreds of thousands of data that have to be processed efficiently in order to make the right decisions. Although there are some support for the (limited) analysis of the data acquired by WSN in the conventional information technologies for data processing, storage, and reporting, such as servers, relational databases, and data warehouses/data marts, these technologies are not efficient enough to deal with very big amounts of data. Most importantly, they cannot cope with the processing needs that can be required for real-time processes where WSNs have been deployed, like fire detection, natural disasters, and traffic control. Furthermore, the high redundancy of sensing data leads to great waste of energy as a result of over-deployed sensor nodes.

We develop herein a structure fidelity data collection (SFDC) framework leveraging the spatial correlations between nodes to reduce the number of the active sensor nodes while maintaining the low structural distortion of the collected data. A structural distortion based on the image quality assessment approach is used to perform the nodes work/sleep scheduling, such that the number of the working nodes is reduced while the remainder of nodes can be put into the low-power sleep mode during the sampling period. The main contribution of SFDC is to provide a unique perspective on how to maintain the data fidelity in terms of structural SIMilarity (SSIM) in the continuous sensing applications for WSNs. The simulation results based on synthetic and real-world data sets verify the effectiveness of SFDC framework both on energy saving and data fidelity.

17.1 INTRODUCTION

WSN has been well suited for use with a variety of applications, including environmental monitoring, biological detection, smart spaces, and battlefield surveillance. These applications can be divided into two broad categories in terms of the way of data collection based on the observer interest: event-based detection and continuous data sampling. In event-based detection, the sensors report the data expected to collect to the sink only when an event of interest occurs. It means that the significant characteristic of the event-based detection in WSNs is delay intolerant and error sensitive. In other words, the data delivered to the sink need to be reliable once an emergency occurred within the sensing range of sensors is detected in real time. In this context, the detection success rate of interested events is crucial to the efficiency of the application.

Coverage-based scheduling methods can achieve full coverage of the target area by finding the disjoint cover sets of sensor nodes, and thus greatly improve the event detection rate. The network lifetime can be increased if only one cover set is active and the nodes in this set send their readings, while other nodes go to sleep and wait for the next time to be roused. To avoid the area uncovered, it is necessary to guarantee that the number of sensor nodes in each set is enough to cover the whole network.

In the continuous data sampling applications, the local measures in the sensor node's sensing area are regularly sampled and reported, such as the ambient temperature and humidity. After receiving all samples, the sink generates a data snapshot of the area at one point in time. Some applications may be more tolerant to discrepancies in the sensed values, not demanding the receiving data without any error. In a general case, the observers are more focused on the evolving process or spatial structure evolution of certain physical phenomenon in the monitored area. Such frequent sampling behavior expends more energy than the event detection. Thus, reducing energy waste is even more important to the continuous data sampling applications. Unlike the event-based detection, the continuous data sampling applications are delay and error tolerant. Differences between them lead to that the coverage-based scheduling methods are unusable for the applications characterized by continuous sampling.

WSNs have great potential that are exploited for many applications in real-world scenarios. However, it has been concluded that the problems of energy constraint and data redundancy emerge inevitably in the course of designing an application of WSNs, especially for high-density large-scale network.

First, due to the relatively less energy consumption on computation and sensing, wireless communications resulting from data delivery consume significant amount of power of sensor nodes only equipped with limited battery. In order to achieve longer lifetime, one of the design challenges for WSNs is to reduce the amount of communication as much as possible without sacrificing measurement integrity.

Second, to satisfy full network coverage requirements [484] and compensate for the impact of nodes failure [485,486], sensor nodes are deployed with much greater density than is needed. Therefore, for the general case in which sensor nodes are highly over-deployed, it is common that nodes redundancy results in data redundancy, whereas we can infer the degree of node redundancy through analyzing the sensed data set. Further, the redundant sensor nodes can be scheduled to sleep mode to save energy when the predefined observation fidelity is satisfied. The problem behind this approach is how to determine the observation fidelity. Much of the existing work in this area is still based on the error sensitivity approach, where the mean square error (MSE) is considered as the sole type of distortion. The inherent disadvantage of this method, however, is that MSE does not provide a good approximation of perceived structural information in the continuous data sampling applications. One distinct example is presented in research for image quality assessment [487], where larger MSE does not show lower structure level on the test images. As mentioned before, the over-deployed sensor nodes are highly correlated or structured, and thus meaning that the sampling data collected by spatial correlated nodes have strong dependencies carrying important structure information. We attempt to exploit the dependencies to reduce the number of nodes required to work for sampling and data transmission. Such reduction is bound to save energy and prolong the network lifetime of WSNs.

In this section, we propose a novel approach to implement sensor scheduling by maintaining the continuous sampling data fidelity that is defined by the SSIM successfully applied to image quality assessment methods. The goal of image quality assessment research is to provide a direct approach that quantifies the similarity or difference between the test and reference images. The simplest implementation to achieve the goal is the error sensitivity approach, which determines a better quality image based on lower MSE. Some limitations on the traditional approach have been highlighted by many previous researchers from image processing area [488–490], e.g., quality definition problem, decorrelation problem, and cognitive interaction problem. These problems also exist in the quality assessment of the continuous sampling applications for WSNs. A new philosophy for the assessment of image quality is the SSIM proposed in [487,491,492], which calculates the SSIM index of the degraded images by incorporating known human visual system (HVS) properties. With this method, the structural information of image signals consisting of abundant pixels exhibiting high spatial correlation can be extracted. Despite a big MSE relative to the reference image, the tested image still obtains a very high-fidelity value as long as the most of the structural information of the reference image can be preserved. Using the SSIM to process large-scale data in WSNs, especially when data are continuously sampled, it is unquestionable that we can get benefits in energy saving. The following are the two strong motivations for this:

- The extracted structural information is just the main concern in the continuous data sampling applications for WSNs. The application domain that needs to capture the data structural changes is highly adapted with the SSIM approach, which is inapplicable in event detection because of the character of error sensitivity.
- The limitations from the error sensitivity approaches are avoided when the SSIM index is used to be a data fidelity assessment criterion. In this way, we refer the data collected by sensor nodes at any time as a tested image to judge the node's contribution to the whole data collection. So that, only a part of sensor nodes need to work and sample while the senor nodes with a negligible contribution will enter a sleep mode.

Inspired by the above ideas, we propose an SFDC framework leveraging the spatial correlations between nodes to reduce the number of the active sensor nodes while maintaining the low structural distortion of the collected data. The ultimate goal of this section is to save energy in the continuous data sampling applications for WSNs by designing a novel node scheduling method, which considers the data fidelity in terms of spatial structure instead of the traditional MSE. To get the most out of spatial correlation, our framework is based on the spatial clustering algorithm and splitted into two phases: the learning phase and data collection phase. During the learning phase, all sensor nodes report their values to the sink. The sink assigns the sensor nodes with close distance and similar trend of the historical data to the same cluster. Within each cluster, the one with the least mean deviation on the readings is selected as the cluster head (CH). In order to apply SSIM index to better in sensor scheduling, we develop a model to calculate a node contribution to the cluster in terms of data SSIM. When given a predefined structure threshold, the CH can make the scheduling decisions on cluster members' work/sleep during the subsequent data collection phase. Therefore, our SFDC framework is well aware of the spatial correlation between nodes and can achieve significant energy saving without too much structure loss.

The rest of this chapter is organized as follows: In Section 17.2, we discuss the related works on data collection techniques for various applications in WSNs. In Section 17.3, we introduce the image quality assessment approach based on the SSIM. Section 17.4 presents our SFDC framework, where a clustering algorithm maximizing the utility of the structure fidelity approach and the nodes scheduling scheme guaranteeing the level of structural fidelity are proposed. Moreover, in order to avoid the structure index instability resulting from the change of the internode relationships, we propose an adaptive data collection method to adjust the data collection period against the cluster structure deterioration. Simulation results are provided in Section 17.5 to verify the correctness of SFDC framework both on energy saving and data fidelity and show the effectiveness of adaptive data collection. Finally, Section 17.6 presents the conclusions of this chapter and future works.

17.2 RELATED WORK

Many data collection approaches aimed to reduce the energy consumption have been proposed for WSNs recently. LEACH [439] is the most famous clustering-based data collection protocol that utilizes randomized rotation of local CHs to evenly distribute the energy load among the sensors in the network. Since LEACH uses localized coordination to enable scalability and robustness for dynamic networks, and incorporates data fusion into the routing protocol, it is not surprising that the amount of information that must be transmitted to the base station would be greatly reduced. Thereafter, a large number of data collection approaches based on clustering emerged as required. A centralized version of LEACH is LEACH-C proposed in Ref. [493]. Unlike LEACH, where nodes self-configure themselves into clusters, LEACH-C utilizes the base station for cluster formation. By using a central control algorithm, the base station selects the nodes with residual energy above the average node energy throughout the network as the CH nodes. A modification to LEACH is presented in Ref. [494], in which a two-level hierarchy clustering protocol (TL-LEACH) is designed for sensor networks where end user wants to remotely monitor the environment. TL-LEACH permits to better distribute the energy load among the sensors in the network, especially when the density of network is higher. The impact of heterogeneity of nodes in terms of their energy in WSNs that are hierarchically clustered is studied in Ref. [495]. Adapting this approach, an energy-efficient heterogeneous clustered scheme for WSNs based on weighted election probabilities of each node to become a CH according to the residual energy in each node is proposed.

Clustering techniques have an added benefit that provides an architectural framework for exploring data correlation in sensor networks. Some recent research on spatial clustering investigates the correlation measurement on sensor data in a highly correlated region. By using the spatial distance of sensors as the judgment of correlation, an approach presented in Ref. [496] defines a weight for each sensor's data that depend on the distance from the sample position to the target position. The Clustered AGgregation (CAG) proposed in [497] provides a mechanism that reduces the number of transmissions by calculating the deviation between different sensor readings to measure the spatial correlation. The cluster construction behind this concept is that a node is included in the cluster only when the deviation between the readings of it and the CH is less than the a user-provided error threshold, otherwise this node becomes a new CH. In the response phase, CAG transmits a single value per cluster, where only the CHs contribute to the data aggregation. By focusing on a few representative values rather than a large number of redundant data, CAG achieves a significant energy saving at the expense of a negligible quality. Another judgment method proposed in Ref. [498] is that a probability distribution function and correlation coefficients of nodes readings are used as a means of showing the representation of spatial correlation in the sensor network data model. A common feature in the approaches described above is that the CH nodes are selected stochastically. This raises the obvious problem that the selected CH is not the best one for performing the high-quality data collection in a specified application. In our SFDC framework, the CH is the one that achieve the maximization of structure fidelity in the cluster data model, thus overcoming the shortcoming of the random selection of CHs. Both the spatial distance and the correlation coefficient are considered in the design of our cluster construction such that the clustering algorithm provides a better foundation for obtaining the high structure fidelity.

The CH selection approach based on the dominating set theory is proposed in Ref. [499], where the authors define a weight to calculate the spatial correlation between sensor data. When the clusters are constructed, the data sampled in each CH have very high correlation with the data sampled in its cluster members. Consequently, only the CHs that can represent the data features of its members need to do the data sampling work, which means the data transmitted in the sensor network are reduced remarkably without any extra data aggregation algorithm. However, the biggest problem for large-scale WSNs is a massive amount of valuable information from member nodes that cannot be found by the end users. In Ref. [500], an adaptive sampling approach (ASAP) is developed for energy-efficient periodic data collection in WSNs. In ASAP, the sampler is composed of a dynamically changing subset of the nodes in which the readings are directly collected, whereas the readings of nonsamplers are predicted through the use of probabilistic models that are locally and periodically constructed. ASAP can be effectively used to increase the network lifetime while avoiding the data loss from nonsampler nodes such that keeping the high quality of the collected data.

Another approach of energy-efficient data collection in WSNs is data prediction technique leveraging the spatial or temporal correlation between sensor data. The data prediction applying in WSNs has a trade-off between reducing the communication cost and limiting the prediction cost. Based on the analysis for the performance of the trade-off in Refs. [501,502], an energy-efficient framework for clustering-based data collection is proposed in Ref. [503], where the benefit of adaptive scheme to enable/disable prediction operations is exploited. To improve prediction accuracy, the authors of Refs. [504,505] perform the prediction of data based on the multivariate correlation and multiple linear regression methods, respectively.

Our purpose of organizing sensor nodes into clusters is to realize the energy-efficient wake-up/sleep scheduling scheme by which some active nodes provide sensing services, while the others are inactive to conserve their energy. For nodes scheduling problem, many attempts have mainly focused on getting complete coverage of the monitored area, where a target of interest is within the sensing range of at least one sensor. A survey on target coverage problem in WSNs is presented in Refs. [484,506]. In Ref. [507], the authors propose a target coverage scheduling scheme based on a genetic algorithm that can find the optimal cover sets to extend the network lifetime. In each cover set, a sufficient number of sensor nodes cover the specified targets by using the evolutionary global search technique. The problem of proper scheduling and putting unnecessary sensor nodes into sleep mode was also explored in Ref. [508]. The proposed scheme is a kind of an adaptive/periodic on–off scheduling scheme in which sensor nodes use only local information to make scheduling decisions. A sensing topology management strategy proposed in Ref. [509] has similar characteristics with our approach in terms of exploiting redundancy for continuous data sampling applications and no a priori statistical assumptions on the underlying phenomenon need to be made. By introducing the finite-dimensional Hilbert space framework of sensors as random variables, sensor locations map onto vectors in this Hilbert space, and inner products between them are defined by the correlation structure of the sensed physical process. As a consequence of adopting the methodology, the number of disjoint sets of sensors can be maximized (or equivalently, the average number of sensors in each set is minimized), while ensuring that each one can provide no more than the user-specified distortion in the data sampling period. However, the proposed algorithms do not consider the cluster-based networking architecture that proved to have more advantages than other network model such as energy saving, scalability, ease of data fusion, and robustness. Moreover, the method of randomly selecting the candidate nodes into the active set is not optimal for achieving the minimum expected distortion. Instead, our active nodes selection algorithm maintains the biggest increase in the structure fidelity for each addition.

17.3 SSIM TO IMAGE QUALITY ASSESSMENT

In this section, we introduce the image fidelity metric that quantifies the distortion of a test image by comparing it with a reference image assumed to have perfect quality. How to assess the quality of a distorted image is the key issue in image processing applications. A number of metrics from different aspects have been proposed, e.g., visible errors, human perception, and the performance of image compression algorithm. Since the HVS is highly sensitive to the structural information, a new philosophy to compare the structures of the reference and the distorted images is proposed in Refs. [487,491,492].

Let $x = \{x_i | i = 1,2,\ldots,N\}$ and $y = \{y_i | i = 1,2,\ldots,N\}$ be the reference and distorted image signals, respectively, where i is the sample index and N is the number of signal samples. The SSIM index between signals x and y can be defined by

$$\mathrm{SSIM}(x,y) = l(x,y)c(x,y)s(x,y), \tag{17.1}$$

where the three functions $l(x,y)$, $c(x,y)$, and $s(x,y)$ represent the luminance, contrast, and structure comparison measures between x and y, respectively.

Let μ_x, μ_y, σ_x^2, σ_y^2, and σ_{xy} be the mean of x, the mean of y, the variance of x, the variance of y, and the covariance of x and y, respectively. Here, the mean and the variance of a signal are treated as estimates of the luminance and the contrast of the signal. The covariance can be considered as a measurement of how much one signal is changed nonlinearly to the other signal being compared. The luminance, contrast, and structure comparison measures are then given as follows:

$$l(x,y) = \frac{2\mu_x\mu_y}{\mu_x^2+\mu_y^2},\ c(x,y) = \frac{2\sigma_x\sigma_y}{\sigma_x^2+\sigma_y^2},\ s(x,y) = \frac{\sigma_{xy}}{\sigma_x\sigma_y}, \tag{17.2}$$

where

$$\mu_x = \bar{x} = \frac{1}{N}\sum_{i=1}^{N} x_i,\ \mu_y = \bar{y} = \frac{1}{N}\sum_{i=1}^{N} y_i \tag{17.3}$$

$$\sigma_x^2 = \frac{1}{N-1}\sum_{i=1}^{N}(x_i-\bar{x})^2,\ \sigma_y^2 = \frac{1}{N-1}\sum_{i=1}^{N}(y_i-\bar{y})^2 \tag{17.4}$$

$$\sigma_{xy} = \frac{1}{N-1}\sum_{i=1}^{N}(x_i-\bar{x})(y_i-\bar{y}). \tag{17.5}$$

Equations 17.1 and 17.2 are combined and the SSIM index can be rewritten as

$$\text{SSIM}(x,y) = \frac{4\mu_x\mu_y\sigma_{xy}}{(\mu_x^2+\mu_y^2)(\sigma_x^2+\sigma_y^2)}. \tag{17.6}$$

Based on Equation 17.6, the SSIM index has the following three properties:

1. Symmetry: $\text{SSIM}(x,y) = \text{SSIM}(y,x)$.
2. Boundedness: $\text{SSIM}(x,y) \in [-1,1]$ since the three components of Equation 17.1 range from $[0,1]$, $[0,1]$ and $[-1,1]$, respectively.
3. Unique maximum: $\text{SSIM}(x,y) = 1$ if and only if $x = y$ (in discrete representations, $x_i = y_i$ for all $i = 1,\ldots,N$.

In practice, we usually want to evaluate the fidelity of a distorted video composed of some continuous images over time. Using the SSIM index, an approach to measure the statistical properties locally is more appropriate for video quality assessment. Suppose the total number of images in a video is M, a video data matrix is denoted by X as

$$X = \begin{bmatrix} X_1 \\ X_2 \\ \vdots \\ X_M \end{bmatrix} = \begin{bmatrix} x_{11} & x_{12} & \cdots & x_{1N} \\ x_{21} & x_{22} & \cdots & x_{2N} \\ \vdots & \vdots & \ddots & \vdots \\ x_{M1} & x_{M2} & \cdots & x_{MN} \end{bmatrix}. \tag{17.7}$$

Each row of X corresponds to an image signal with N elements. With the use of SSIM index, the video fidelity index between the distorted video X and the reference video Y can be obtained as follows:

$$\text{SSIM}(X,Y) = \frac{1}{M}\sum_{j=1}^{M}\text{SSIM}(X_j,Y_j), \tag{17.8}$$

where $X_j = [x_{j1}, x_{j2}, \ldots, x_{jN}]$ and $Y_j = [y_{j1}, y_{j2}, \ldots, y_{jN}]$.

17.4 SFDC FRAMEWORK

Our SFDC framework is built on a two-layer network architecture and consists of two phases: learning and data collection. During the learning phase that aims to analyze the collected data, we need to consider how best to create clusters within the network so as to make the most of the utility of spatial correlation between nodes in the SSIM assessment. Here, we will first introduce the clustering procedure performed for the cluster construction and the data-driven CH selection, which are very important basis for our data collection framework and determine the efficiency of the subsequent node scheduling scheme within cluster. Finally, we will describe the details of data collection phase.

17.4.1 CLUSTER CONSTRUCTION

Due to the prevalence of spatial correlation in environmental phenomena, the over-deployed sensor nodes used to sense environment will be highly correlated. If we group nodes with similar sense quality into the same clusters, it will be more efficient at exploiting the correlation of data. Naturally, we consider the physical distance as one of the metrics of the close relationship between any two nodes. Many mathematical models of spatial correlated sensor networks data have been established in Refs. [510–512], where the variogram (also called semivariance) that reveals the average square difference between the readings of sensor nodes is used to characterize spatial correlation in data.

Given a two-dimensional stationary process $S(x,y)$, where (x,y) is the geographic coordinate, the variogram is denoted as follows:

$$\gamma = \frac{1}{2}E[(S(x_1,y_1) - S(x_2,y_2))^2]. \tag{17.9}$$

For isotropic random processes [513,514], the variogram depends only on the Euclidean distance

$$r = \sqrt{(x_1 - x_2)^2 + (y_1 - y_2)^2}$$

between two nodes. The experiment results in Ref. [510] show that the variogram increases quickly when the distance grows. For large distance, the process exhibits a remarkably feature of independent and identically distribution process, where the variogram does not change with distance.

Therefore, the distance is a good indicator of correlation between nodes. However, we believe that it is insufficient in real world. The reason is that the phenomenon under observation is often influenced by multiple sources such as room air conditioners, heaters, and walls. Even though two nodes are physically close, the correlation between them may be less obvious.

We thus consider another statistic named Pearson's coefficient [515,516] as the supplementary metric to characterize spatial correlation in data. The Pearson's coefficient indicates the degree of linear correlation between two variables x and y, giving a range of $[-1,1]$, where 1 is total positive correlation, 0 is no correlation, and -1 is total negative correlation. Here, each individual sensor node can be regarded as a variable to record the change of observation in the monitored area where the sensor is located. Consequently, the formula for Pearson's coefficient when applied to two sensor variables is

$$\rho_{xy} = \frac{\sum_{i=1}^{m}(x_i - \bar{x})(y_i - \bar{y})}{\sqrt{\sum_{i=1}^{m}(x_i - \bar{x})^2}\sqrt{\sum_{i=1}^{m}(y_i - \bar{y})^2}}, \tag{17.10}$$

where m is the number of reading samples, and $\bar{x}$ and $\bar{y}$ are the average of reading samples for each sensor node. The Pearson's coefficient can identify how much one variable is linearly associated with another. In other words, the areas sensed by two nodes with larger value of the coefficient have more similar linear relation in terms of sensor readings. A known Pearson's distance is defined as

$$d_{xy} = 1 - \rho_{xy}. \tag{17.11}$$

From the fact that the Pearson's correlation coefficient falls between $[-1,1]$, it is seen that the Pearson's distance lies in $[0,2]$.

We address the cluster construction problem by using the Euclidean distance as the primary metric and the Pearson's distance between node readings as the secondary metric, where the correlation of sensor nodes is depicted in terms of the variance and the linear relation of their readings in the past. Given a node set G, we need to partition G into N disjoint clusters C such that sensor nodes within the same cluster are correlated. The details of the cluster formation applied in the sink are described in Algorithm 17.4.1.

We define the threshold max_dis as a tunable parameter that roughly determines the diameter of a cluster and the number of clusters. The predefined parameter changes inversely with the number of clusters or the percent of nodes being CHs, which means that it will be set to a small value when the percent of CHs in the network is large and vice versa. In LEACH [439], the authors indicate that there exists an optimal percentage α of nodes that should be CHs for a suitable energy dissipation consideration, i.e., $\alpha = 0.05$. In general, the predefined parameter max_dis is less than the communication radius of sensor node, where we only consider the senor nodes with the same communication ability for the homogeneous network. In Algorithm 17.4.1, we first put the nodes close to each other in terms of spatial distance together, and then sort them in order of the Pearson's distance. Only the nodes within the communication range R_{CH} of the CH join in the new cluster. The algorithm outputs a set of clusters in which all nodes are covered. The CH selection process is described in the following section.

17.4.2 CH SELECTION

Prior to the CH selection, we need to build a data transmission model between member nodes and CH. Normally, all nodes including the CH in the cluster sense the environment, and then member nodes report their readings to the CH via a hop distance.

Let $X_{M\times N}$ as described in Equation 17.7 represents the data set received by the CH, where M and N denote the discrete time domain and the number of sensor nodes in the cluster, respectively. Consider a node scheduling method, only $n \leq N$ sensor nodes are in a working state. In the case, the CH only receives $M \times n$ readings from the working node. In order to measure the SSIM index between the original data set X and the later data set Y, we need to find a way to fill the $M \times (N-n)$ remainder readings in Y. A polished model is designed for that, where all missing readings are replaced by the readings from CH without any increase in traffic. To reduce the error generated by this replacement, the mean squared deviations (MSD) on the readings between the CH and other nodes must be minimized. Based on the principle, the node with the minimal MSD (Min_MSD) will be selected as the CH. As member nodes in the cluster can receive and send the readings with each other, each node makes an independent decision based on the MSD value as to whether to be the CH.

A Min_MSD-based distributed CH selection algorithm named Min_MSD approach is described in Algorithm 17.4.1.

Algorithm 17.4.1 works by combining two main operations: cluster formation and CH selection. Two metrics (Euclidean and Pearson's distance) are used to group together nodes with high spatial correlation to form a cluster, and CH selection aims to minimize the structure distortion generated by our data transmission model between member nodes and CH. First, the geographically close nodes are roughly divided into a list of nodes S, where the node with the minimal MSD is elected as the CH. Second, the nodes in list S are sorted according to the increasing order of the Pearson's distance between them and the elected CH. Finally, the nodes both with low Pearson's distance value and within the communication radius of the CH have a better chance of joining the new cluster. It is important to emphasize that a greedy strategy is adopted in Algorithm 17.4.1 to ensure that all nodes are covered.

17.4.3 NODES SCHEDULING SCHEME BASED ON THE SSIM INDEX

After the cluster construction, the member nodes send all readings to the single CH node. In this case, the CH creates continuous images over time, where each image is composed of the readings sensed by all nodes at an instant. For a nodes scheduling scheme, only a subset of nodes is selected to work and sample, while the rest of sensor nodes go to sleep to save energy. Based on our data transmission model, the readings of unselected sensor nodes are replaced by the CH readings from the same time. The replaced data set denoted as $Y_{M\times N}$ can provide a good SSIM index over the monitored area during the subsequent data collection phase.

Algorithm 17.4.2 Cluster formation and cluster head selection

Input: a node set G

Output: a set of clusters covering the set G and the cluster head for each cluster

Set all nodes in G as uncovered
while G is not null **do**
    Randomly pick up the node υ from G
    Pick up all the nodes distance from less than the threshold max_dis and put them into a list of S
    for every node i in the list S **do**
        $MSD(i) = \frac{1}{M}\sum_{j=1}^{M}\sum_{k\in S, k\neq i}(x_{ji}-x_{jk})^2$
    end for
    Cluster head $CH = \{i \mid \min(MSD(i))\}$
    Calculate the Pearson's distance of each node in S and CH
    Sort the nodes in S according to the increasing order of the Pearson's distance value
    Create a new cluster C including only CH
    while S is not null **do**
        Pick up the next node s from S in turn
        if s is within the communication radius R_{CH} of the CH **then**
            Put s into the cluster C
        end if
    end while
    Output the cluster C and cluster head CH
    Remove all nodes in C from G
end while

The main goal of our scheduling scheme is to enable the nodes with greater contribution to the SSIM index to enter active mode. Since the SSIM index is a way to measure the image structure quality from three different aspects: mean distortion, variance distortion, and loss correlation, it is hard to calculate each individual node's contribution to the overall structure of collected data. From the property (3) of the SSIM index, however, we know that the maximal value 1 is achieved only when the image signals being compared are identical. In other words, the closer they are to each other, the closer SSIM is to 1. The increase in the deviation between the distorted image and the reference image inevitably leads to a decrease in the SSIM index. Based on the replacement principle as described above, our nodes scheduling scheme try to sort the member nodes in decreasing order of the deviation between the readings of them and the CH, and add the sensor node one by one into the active node set until the predefined structure fidelity is met, such that the SSIM index will undergo the largest increase as the addition of the candidate node.

We present a heuristic algorithm to tackle the problem of active nodes selection. Initially, the active node subset A only includes the CH node while all elements in the member node set Q are unselected. At this point, the CH generates a data sequence viewed as the first image with the largest distortion

$$Y_i^1 = [x_{iC}, x_{iC}, \ldots, x_{iC}],$$

where x_{iC} denotes the reading from the CH at the time i and all readings of member nodes are replaced by x_{iC}. Let x_{ij} be the reading of the jth member node at the time i, and the reference image is represented by

$$X_i = [x_{i1}, x_{i2}, \ldots, x_{ij}, \ldots, x_{i(N-1)}, x_{iC}].$$

The node j among the set Q is selected into the active set A if and only if x_{ij} is with the biggest difference with x_{iC}. Then, the node j is removed from the set Q and the second distorted image is given by

$$Y_i^2 = [x_{iC}, x_{iC}, \ldots, x_{ij}, \ldots, x_{iC}, x_{iC}].$$

The value of $\mathrm{SSIM}(X_i, Y_i^2)$ can be calculated through Equation 17.6. So repeatedly until a minimal subset is built while meeting the predefined threshold. Y_i^N is equal to X_i, where all nodes will be incorporated into the active node set A when the threshold is set to 1.

Here, we define the contribution rate of node to the SSIM.

Definition 17.4.1 *Let Y_i^k be the nodes data sequence after inserting the reading x_{ij} of the jth node at the time i into Y_i^{k-1}, the contribution rate of node j to the structure of reference node data sequence X_i is defined as*

$$\alpha_j = \mathrm{SSIM}(X_i, Y_i^k) - \mathrm{SSIM}(X_i, Y_i^{k-1}), \tag{17.12}$$

where

$$\mathrm{SSIM}(X_i, Y_i^k) = \frac{4\mu_{X_i}\mu_{Y_i^k}\sigma_{X_iY_i^k}}{(\mu_{X_i}^2 + \mu_{Y_i^k}^2)(\sigma_{X_i}^2 + \sigma_{Y_i^k}^2)}$$

and

$$\mathrm{SSIM}(X_i, Y_i^{k-1}) = \frac{4\mu_{X_i}\mu_{Y_i^{k-1}}\sigma_{X_iY_i^{k-1}}}{(\mu_{X_i}^2 + \mu_{Y_i^{k-1}}^2)(\sigma_{X_i}^2 + \sigma_{Y_i^{k-1}}^2)}.$$

In order to statistically assess the node contribution rate, the deviation and SSIM index can be extended over multiple time instances. We still use X to denote the data sequences of all nodes as described in Equation 17.13. The CH generates the first data matrix Y^1 represented by Equation 17.14, where the elements for each row are identical and equal to x_{iC} $(1 \le i \le M)$.

$$X = \begin{bmatrix} X_1 \\ X_2 \\ \vdots \\ X_M \end{bmatrix} = \begin{bmatrix} x_{11} & x_{12} & \cdots & x_{1(N-1)} & x_{1C} \\ x_{21} & x_{22} & \cdots & x_{2(N-1)} & x_{2C} \\ \vdots & \vdots & \ddots & \vdots & \vdots \\ x_{M1} & x_{M2} & \cdots & x_{M(N-1)} & x_{MC} \end{bmatrix} \tag{17.13}$$

$$Y^1 = \begin{bmatrix} Y_1^1 \\ Y_2^1 \\ \vdots \\ Y_M^1 \end{bmatrix} = \begin{bmatrix} x_{1C} & x_{1C} & \cdots & x_{1C} & x_{1C} \\ x_{2C} & x_{2C} & \cdots & x_{2C} & x_{2C} \\ \vdots & \vdots & \ddots & \vdots & \vdots \\ x_{MC} & x_{MC} & \cdots & x_{MC} & x_{MC} \end{bmatrix}_{M \times N} \tag{17.14}$$

If the MSD of a candidate node and the CH node is maximal among all such nodes, then the descriptive power of the active nodes set will maximally grow if the candidate is added into it. The values of MSD for all member nodes and the CH are known to the CH based on the CH selection procedure, without any increase in the computing load. We can easily obtain the value of $\mathrm{SSIM}(X, Y)$ defined in Equation 17.8 to evaluate the fidelity of the data from active nodes set. We named the active nodes selection as the *Max_MSD* approach. The detailed algorithm for finding the set of active nodes with a minimal number of nodes in a cluster is presented in Algorithm 17.4.3:

Algorithm 17.4.3 Active nodes selection

Input: predefined fidelity threshold δ
Output: a set of active nodes A
1: $A = \{\text{CH node}\}, \bar{A} = \{\text{all member nodes}\}$
2: $k = 1$
3: **while** $\text{SSIM}(X, Y^k) < \delta$ **do**
4: Candidate node $j = \{i | \max(MSD(i, CH)), i \in \bar{A}\}$
5: Put j into A and remove from $\bar{A}$
6: $k = k + 1$
7: **end while**
8: Output the node set A

17.4.4 DATA COLLECTION

Once clusters are formed and active sensors serving as samplers to collect data are selected, no-sampler nodes enter the energy-efficient sleep mode without much degradation of structure fidelity. Note that with a fixed number of sampler nodes determined throughout the data collection phase, it is possible that the fidelity is not always superior to the preset threshold. In essence, the phenomenon that WSN is sensing is in a constant state of flux and continual change. Even an anomaly may occur in a particular time slot, e.g., a sudden flood, forest fire, or earthquake involved in the event detection problem. More likely, the intracluster nodes relationships may imperceptibly change due to the slight and gradual changes in environment. Such observation on temperature and humidity is clearly affected by seasonal changes. These changes can lead to the negative effects on the performance of our data collection framework. The first is that the nodes in a cluster no longer have a close spatial correlation. Furthermore, sensor nodes have a different sort for the data deviation compared to previous data, which would further affect the quality of CH and sampler nodes selection.

We define the global parameter denoted by T_d as the period of data collection and T_l (i.e., M) as the period of learning. The setting of T_d is mainly for the consideration of two aspects. On the one hand, the sampler nodes consume more energy compared to the sleeping nodes, since they carry out the task of sensing and report the readings. The CH node has additional responsibility for collecting the data from all sampler nodes in the cluster and forwarding them to the sink. In this case, energy dissipation is not balanced especially for the large number of sampler nodes. Moreover, the spatial correlation of nodes may change with the change of sensor readings. As a result, the nodes scheduling scheme fails to keep stability in a long run. Consequently, the value of T_d should be small enough to balance the energy consumption and maintain the correlation between different nodes. On the other hand, too small value of T_d cannot result in a substantial saving in energy consumption. Its value is expected to be much large than the learning period T_l. Thus, the data collection period should be dynamically adjusted to maintain the stability of correlation structure in the cluster in response to environmental changes.

In order to deal with the dynamics, a feasible solution to adaptively adjust the data collection period is that the CH node broadcast a "relearning" message to all member nodes when the drastic deterioration in sensor readings happen. In our nodes scheduling scheme, what we are most concerned is any trend changes occurred in the monitored field. Therefore, we define a metric to approximately measure the cluster trend based on the sensor time sequences.

Definition 17.4.2 *Two sensor nodes are isotonic if there is $m_1/m > t$ for their time sequences $\{x_1, x_2, \ldots, x_m\}$ and $\{y_1, y_2, \ldots, y_m\}$, where m_1 is the number of pairs (x_i, y_i) in the time sequences that satisfy $(x_i - x_{i-1})(y_i - y_{i-1}) \geq 0$, $2 \leq i \leq m$.*

Definition 17.4.3 *The cluster trend is deteriorating if any two nodes in the cluster are not isotonic.*

According to Definition 17.4.2, the isotonicity means that the value of a senor will increase/decrease with the value of another correlated senor increase/decrease in most of the time. As a more precise indicator to highlight the impact of sensor readings' change on data structure, the isotonicity is different from the former Pearson's coefficient, which roughly classifies the sensor nodes from the perspective of linear relation and thus more suitable to be used at the time of cluster launch. The value of t depends on the requirement of the application at hand. Once the trend deterioration based on Definition 17.4.3 is identified, the CH broadcasts a "relearning" message that means terminating the data collection procedure. After receiving the message, awakened no-sampler and working sampler nodes start a new learning phase only including selection of CH and active nodes. If the trend deterioration occurs during the learning period, the CH broadcasts a increased value of T_l to all member nodes to guarantee the accuracy of the video fidelity index. Moreover, if the percentage of cluster required relearning is more than a value (i.e., 50%), the sink broadcasts a similar relearning message to all sensor nodes via the hierarchical routing. The complete learning process will be performed in the whole network.

Our method by adjusting the data collection period based on the cluster deterioration avoids the structure index instability resulting from the change of the internode relationships, thus ensuring the adaptation of the data collection system without human intervention.

17.4.5 ENERGY CONSUMPTION

Unlike other models using the radio distance in calculating the energy cost, we analyze the energy consumption from the perspective of the number of working/sleeping sensor nodes in unit of cluster on the learning and data collection phases, since our approach aims to energy saving by reducing the number of nodes that have no need to be active. First, all member nodes send the readings to the CH node during the learning phase and thus there is no sleeping node (or energy conservation). The energy cost in a cluster with N nodes during the learning phase is

$$E_l = T_l N e, \tag{17.15}$$

where e is the energy cost on transmitting a unit data between two nodes.

During the data collection phase, the energy cost in the cluster is

$$E_d = T_d N_a e, \tag{17.16}$$

where N_a is the number of active nodes.

Comparing the energy consumption without any nodes scheduling, the percentage on energy saving within a cluster during a cycle (i.e., learning and data collection) is

$$P_E = 1 - \frac{T_l N e + T_d N_a e}{(T_l + T_d) N e} = \frac{T_d (N - N_a)}{(T_l + T_d) N} = \frac{T_d}{T_l + T_d} \frac{N - N_a}{N}. \tag{17.17}$$

Equation 17.17 indicates that a good energy conservation can be achieved by extending the data collection period T_d and reducing the number of the working nodes N_a. T_d and N_a depend on the sensed environment and the predefined fidelity threshold, respectively. We will evaluate them in the following simulations.

17.5 PERFORMANCE EVALUATION

17.5.1 REAL DATA SET

The experiment data set plays an important role in the simulation and implementation of the SFDC framework. Thus, we used the real-world data set from the Intel Berkeley Research lab [517] as the evaluation object. The data set, including humidity, temperature, light, and voltage values, is collected from 54 Mica2Dot sensor nodes deployed in the lab layout for a period of 1 month. In the publicly available data set, we selected the temperature readings sampled every 31 s for all nodes to perform a data analysis. Since 2 sensor nodes are without data available for a long period of observation in the data set, only 52 nodes are used in our experiment. As shown in Figure 17.1, the location information of all nodes is fixed at configuration time and hence the distance between nodes can be calculated. The X- and Y-axes in Figure 17.1 represent the coordinates of sensors in meters relative to the lower left corner of the lab.

17.5.1.1 Correctness of Clustering with SFDC

To group together nodes with high spatial correlation to form a cluster, Algorithm 17.4.1 is designed to heuristically obtain a rough approximation on correlation based on two indexes of geographical distance and Pearson's distance. The network is partitioned to several subareas in which the distance of any two nodes is smaller than max_dis. As the growth in sensor network scale, max_dis will be set to be a larger value such that an optimal number of clusters is achieved. For a fixed-scale sensor network, the smaller max_dis leads to the more number of clusters. The optimal number of clusters by minimizing the communication cost of the network presented in Ref. [439] can be used as a reference.

Another parameter needed to be supplied is the communication radius R_{CH} of CH nodes, and it depends on the application as well as the sensing hardware of sensor node. In general, the CHs have a larger communication radius than member nodes since they receive all data from member nodes and perform long-range transmissions to the remote base station. We assume that all CH nodes have the same value of R_{CH} to guarantee the expedite communication within their clusters.

As an example, Figure 17.1 illustrates the output of one of our clustering experiments with the default parameters $max\_dis = 15$, $R_{CH} = 25$, $M = 50$, and $N = 52$, where the sensor network is organized into six clusters. All nodes marked with the same sequence number belong to the same cluster, and the CH nodes are marked in red. It can be seen that the nodes adjacent to each other are grouped into one cluster and member nodes are close to the CHs in term of the Pearson's distance.

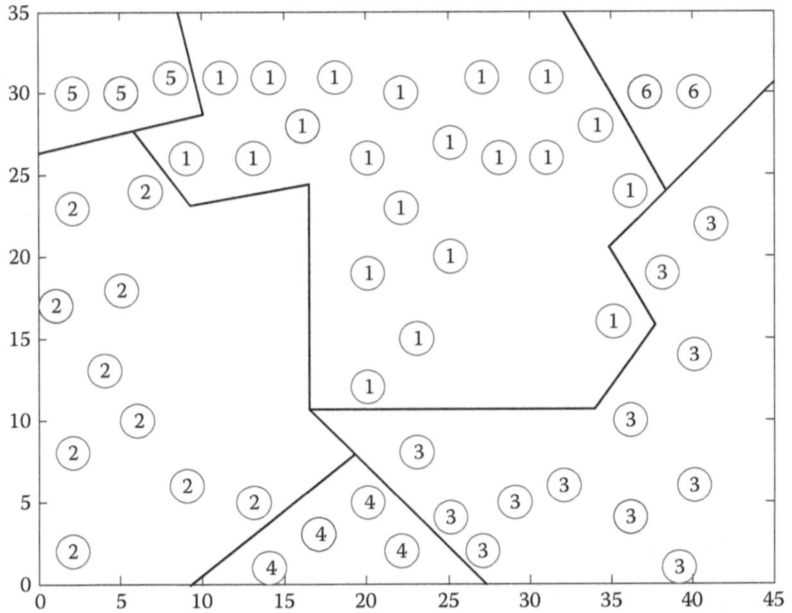

Figure 17.1 A clustering map is produced by implementing Algorithm 17.4.2.

Because the Pearson's distance based on the historical readings is time varying, the clustering map is not constant at different data collection phase. And hence it is reasonable to make the geographical distance that is constant as the primary metric without compromising the Pearson's distance.

17.5.1.2 Fidelity without the Dynamical Adjustment of T_d

In this section, we evaluate the fidelity without the dynamical adjustment of T_d during data collection by following the clustering map in Figure 17.1, and try to get some qualitative relations between the fidelity, the number of nodes, and the SNR. By default, the predefined fidelity threshold δ is set to 90%.

Figure 17.2 shows the sleeping node ratio (SNR) and nodes number for each cluster. SNR is defined by the following equation:

$$\text{SNR} = 1 - \frac{\text{Number of active nodes}}{\text{Total number of nodes}}, \tag{17.18}$$

which describes the ratio of the sleeping node for each cluster. In other words, the larger SNR, the more idle nodes that means the more energy conservation. Intuitively, when the number of nodes in the cluster decreases significantly, the SNR also appears to noteworthy decline. As we see in Figure 17.2, the SNR is more than 60% for the nodes in Cluster 1. We also note that the worst SNR 0 is obtained by Cluster 6 with the minimum number of nodes. Therefore, our SFDC framework is more suitable for the large-scale dense WSNs in order to achieve a better effect on energy saving. The conclusion is experimentally verified by running Algorithm 17.4.1 multiple times.

Figure 17.3 shows the fidelity curves for 6 clusters with different sizes for 100 time slots (31 seconds per time slot). For Clusters 1 and 2 with more number of nodes, the fidelity curves show a decline and instability. On the other hand, for Clusters 3 through 6 with fewer nodes, the fidelities are superior to the threshold throughout the data collection phase. Especially for Clusters 4 through 6 in which all nodes are active, it can be found that the fidelities during data collection are always 100%. It happens because the structure fidelity is less or not susceptible to the reading changes from the observed area where the less sleeping nodes are located. Obviously, the increase in the number of sleeping nodes leads to the performance degradation (structure distortion), which is amplified when the number and density of node deployment are increased.

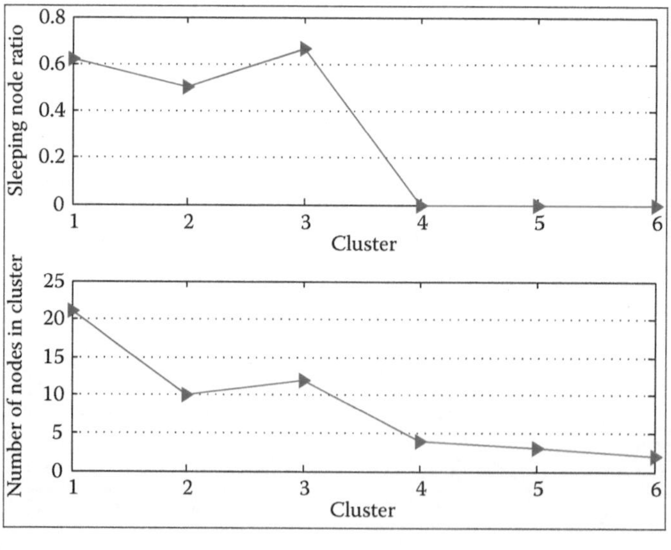

Figure 17.2 SNR versus nodes number.

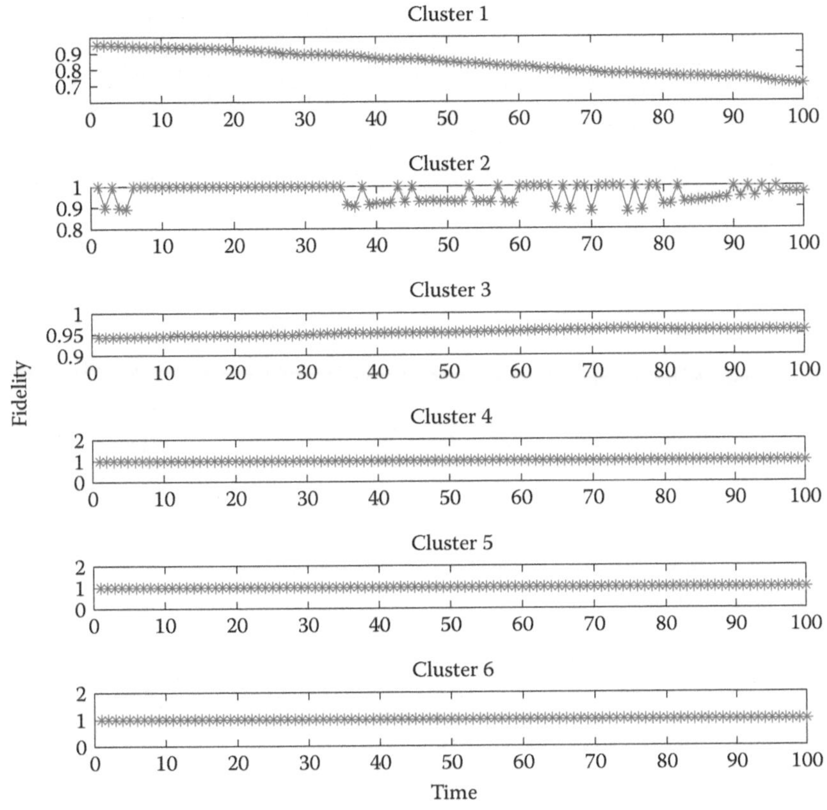

Figure 17.3 Fidelity curves without the dynamical adjustment of data collection period.

Therefore, we note that there is a trade-off between the fidelity in the data structure and the SNR. This raises the necessity to perform the dynamical adjustment of T_d to achieve large SNR while guaranteeing the low distortion.

17.5.1.3 Correctness of CH and Active Nodes Selection

The preceding discussion describes that our data collection framework benefits the large-scale dense WSNs in terms of SNR. To exploiting the potential of SFDC, we use all 52 sensor nodes as a single cluster to perform the following simulations.

For the purpose of comparison, we introduce a variant of our *Min_MSD* approach and *random* approach, where the CH node is randomly selected in the cluster. When the fidelity threshold is set to 90%, we compare the SNR for *Min_MSD* and *random* approaches through 20 learning periods. As shown in Figure 17.4, the *Min_MSD* presents better SNR than using the *random* approach throughout all learning periods. Based on the design in Algorithm 17.4.1, the result can be easily explained by the fact that the smaller deviation produces the smaller distortion in our data transmission model.

In previous evaluation, both of *Min_MSD* and random approaches used the same active nodes selection scheme as described in Algorithm 17.4.3 to derive the SNR. For simplicity, we also name the random active node selection the *random* approach. Figure 17.5 shows the effect of active nodes selection using the *Max_MSD* and *random* approaches on SNR. It is obvious that the greater SNR is achieved by our *Max_MSD* approach since each selected active node can eliminate the most difference between the current and original active nodes set. Another result viewed from Figures 17.4 and 17.5 is that the performance degradation of *random* approach shows a distinct advantage provided

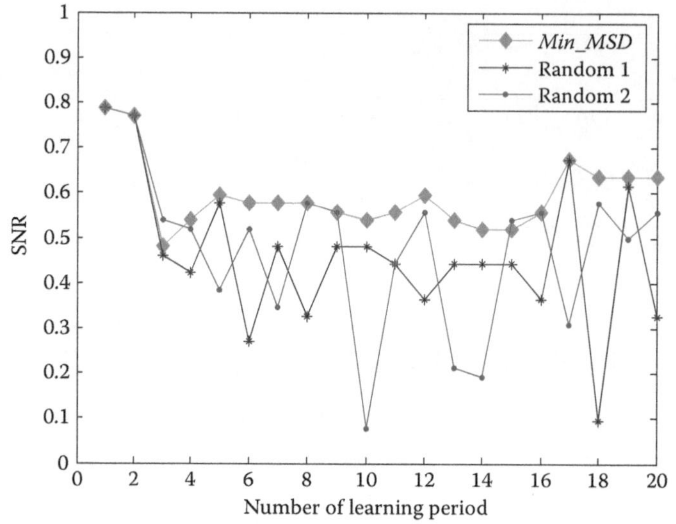

Figure 17.4 Effect of CH selection using the *Min_MSD* and *random* approaches on SNR.

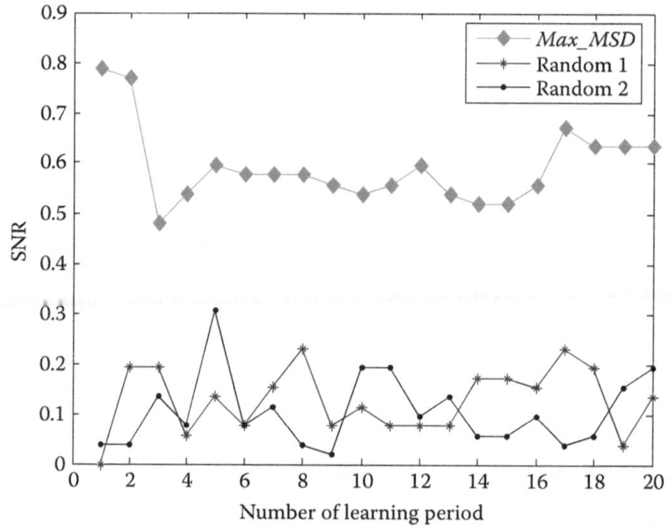

Figure 17.5 Effect of active nodes selection using the *Max_MSD* and *random* approaches on SNR.

by *Max_MSD* in terms of SNR, compared to that of the *Min_MSD*. The reason is that more active nodes with maximal MSD are selected instead of completely randomized nodes.

17.5.2 NODE CONTRIBUTION RATE

In this evaluation, we try to examine the node contribution rate α to structure fidelity in order to provide a way to determine the fidelity threshold in practice. Figure 17.6 shows the structure fidelity when all 52 nodes are added into the active nodes set one by one. It can be observed that the fidelity monotonically increases with the active node ratio (ANR), which is the ratio of active nodes to all nodes. Consequently, the fidelity is optimal (100%) when the ANR is equal to 1. This is no surprise because more selected active nodes means smaller distortion. Based on Definition 17.4.1, we calculate the contribution rate for different number of nodes. The result is given in Table 17.1.

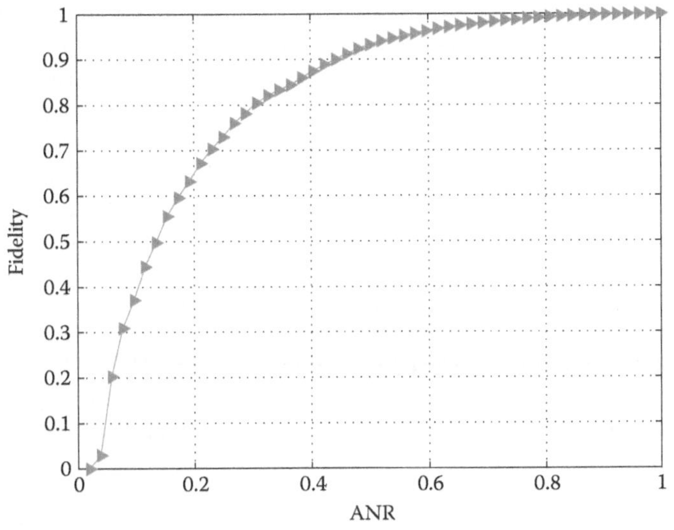

Figure 17.6 Structure fidelity versus ANR.

Table 17.1
Nodes Contribution Rate with Different Active Node Rate

Number of Nodes	1–22	23–32	33–42	43–52
Nodes contribution rate	0.9056	0.0642	0.0246	0.0056
ANR	0.4231	0.1923	0.1923	0.1923

The initial 22 active nodes with 42.31% of ANR contribute 90.56% of fidelity, while the subsequent three active node sets only provide 6.42%, 2.46%, and 0.56% of the whole fidelity, respectively. For the evaluated scenarios, therefore, 61% of the nodes need to work when the target fidelity is set to a high value of 96%. That means, it is possible to save 39% of energy on nodes dissipation while only losing 4% of fidelity.

Based on the above analysis, we can draw the conclusion that the spatial redundancy in the experimental scenario is high since last 30 nodes representing about 58% of all nodes account for less than 10% of fidelity. Thus, a way is introduced for evaluating the level of nodes redundancy. For the WSNs with high spatial redundancy, the satisfying energy saving can be achieved while the expected threshold is met.

17.5.2.1 Effect of Adaptive Data Collection

The principal remaining problem is that the underlying physical phenomenon is complex and not constant. The only information available is what can be learned from all reported readings. Then, the adaptive data collection will be executed in response to the spatial correlation changes. Once the CH detects that a cluster should be relearned, it asks all sensor nodes in the corresponding cluster to collect the reading simultaneously. In the worst case, the sink can re-cluster the whole network when most clusters trend in the network is deteriorating.

We experimentally test our adaptive data collection scheme to observe the changing fidelity. The settings of parameters are as follows: $t = 0.4$, $T_l = 50$, $T_d = 4$, and $T_l = 200$. Based on the results in the previous section, the SNR is 39% when the fidelity threshold is set to 96%. By evaluating the isotonicity between nodes to judge the cluster trend, we get the result shown in Figure 17.7. The fidelity during the learning phase keeps a constant value of 1. In two periods of data collection, the fidelity curves show a clear decline. However, after each round of data collection, the cluster starts a relearning process when the CH detected the deterioration of cluster trend. Consequently, the only suitable candidate is elected as the new CH by executing Algorithm 17.4.1. Afterward, an updated working schedule is created by the CH, thus avoiding the continued decline in the fidelity. As we can see in Figure 17.7, the result is verified that the structure fidelity with the adaptive data collection is acceptable.

According to Equation 17.17, 31.2% of energy for the network with 52 sensor nodes is saved. We can set the more strict value of t or shorten the period of data collection to enhance the fidelity, however, that will not achieve the significant reductions in energy. The trade-off between the fidelity and the energy saving depends on the tolerance of distortion in the particular sensor applications. Note that there are several outliers in the second T_d, where much lower fidelity scores than they should supply are given. In fact, most of these significant outliers correspond to the sensor readings with large global motions at some instants. An outlier detection approach is beyond the scope of this section, which is primarily aimed at the structural distortion.

17.5.3 SYNTHETIC DATA

To verify our data collection framework for the large-scale WSNs, we adopted the environmental model in Ref. [518] for synthetic data generation that provides a possibility for the realism of simulated data. Two sets of sensor nodes each with the number of 500 and 1000 are scattered randomly in a square field, in which an unknown number of diffusion heat sources move randomly at a set of positions and data were collected for every time instant at each sensor position. Modeling of the temperature variations due to heat conduction can be obtained by using the following partial differential equation:

$$\frac{\partial V(X,t)}{\partial(t)} - k\Delta V(X,t) = Q, \tag{17.19}$$

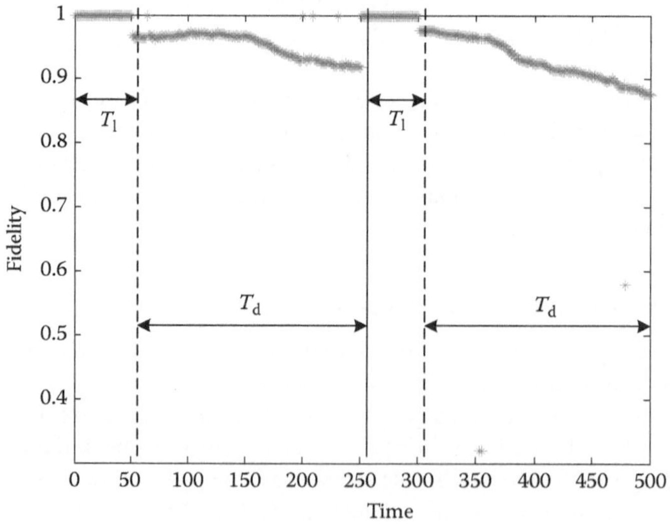

Figure 17.7 The changing fidelity during the data collection.

where $V(X,t)$ is the temperature value at the coordinate point $X(x_1,x_2)$ and time t, k is the diffusion coefficient for heat propagation,

$$\Delta V(X,t) = \sum_{i=1}^{2} \frac{\partial^2 V(X,t)}{\partial^2 x_i}$$

and Q is the heat source in Joule.

By setting the following parameters in MATLAB® simulation: $k = 0.1\,\text{m}^2$/instant, 3 heat sources of $Q = 200\,\text{W}$ with 20 random positions and 20 m radius, 2000 observations for every position (or sensor node) were collected. Figure 17.8 shows the scenario of 1000 sensor nodes with triangular meshes, in which the red circle indicates the position of head resource at a time. Since the correctness that CH and active nodes selected in the learning phase of SFDC is validated in Section 17.5.1.3, more attention is paid to the performance of ANR and adaptive data collection in the case when a large number of sensor nodes are used in the simulation.

17.5.3.1 Node Contribution and ANR

We conduct simulations for two scenarios: Scenario 1 with 500 nodes and Scenario 2 with 1000 nodes deployed over an area of size 800 m × 800 m. Figure 17.9 shows how the structure fidelity change over ANR and the corresponding statistic is given in Table 17.2. As we can see, the results that the fidelity monotonically increases with ANR are similar to those in Figure 17.6. However, it is obvious that there are differences between the two curves in Figure 17.9. For Scenario 1, the initial 183 active nodes accounting for 36.6% of 500 nodes contribute 90.03% of the whole fidelity, while for Scenario 2, the initial 174 active nodes only accounting for 17.4% of 1000 nodes also provide 90.01% of the whole fidelity. On the other hand, we can see from the last column in Table 17.2 that 35% of the nodes for Scenario 2 and 18% of the nodes for Scenario 1 do not really need to work when 1% of distortion is permitted. Because of the higher node redundancy, we conclude that the large-scale dense WSNs have an advantage in the SNR by using our data collection approach.

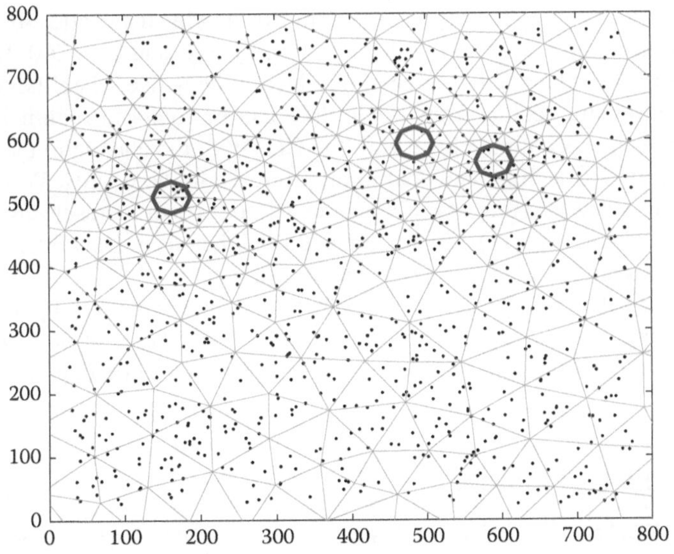

Figure 17.8 Random scenario with 3 heat sources and 1000 sensor nodes.

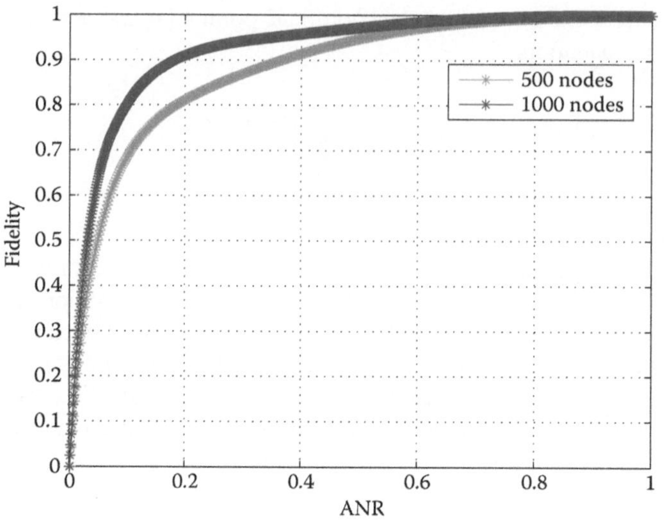

Figure 17.9 Structure fidelity versus ANR.

Table 17.2
Nodes Contribution Rate with Different Active Node Rate

	Number of Active Nodes	**1–183**	**184–300**	**301–410**	**411–500**
500 nodes	Nodes contribution rate	90.03%	7.09%	2.35%	0.53%
	ANR	36.6%	23.4%	22%	18%
	Number of Active Nodes	1–174	175–358	359–650	651–1000
1000 nodes	Nodes contribution rate	90.01%	6.05%	3.02%	0.92%
	ANR	17.4%	18.4%	29.2%	35%

17.5.3.2 Effect of Adaptive Data Collection

Figure 17.10 shows the changing fidelity for Scenario 2 during the data collection by using the same parameter settings in Figure 17.7, i.e., $t = 0.4$, $T_l = 50$, $T_d = 4$, and $T_l = 200$. During the learning time of T_l, about 35.8% of the nodes can be determined to be active when the fidelity threshold is set to 96%. Then, a fidelity decline from 1% to 96% can be found in Figure 17.10 because only 35.8% of the member nodes communicate with the CH. With the change in the spatial correlation of nodes over time, there is a slight and continuing decrease in the fidelity. When the CH detected that the cluster trend is deteriorating, the CH and active nodes will be reselected to still maintain a high level of fidelity. Compared to Figure 17.7, in addition, there is not any fidelity outlier appearing in Figure 17.10 because of the ideal property of synthetic data.

We also calculated the energy saving for both of the two scenarios by Equation 17.17 and obtained the following results: $P_E = 32.8\%$ for Scenario 1 and $P_E = 51.36\%$ for Scenario 2. The main reason for more energy saving in Scenario 2 with a denser node deployment is that a larger SNR is achieved.

17.6 CONCLUDING REMARKS

Our cluster-based SFDC framework takes advantage of the SSIM index to perform the energy-aware wake-up/sleep nodes scheduling scheme by which the spatial correlation is leveraged to save even more energy. Unlike other typical clustering techniques considering the spatial correlation, both of

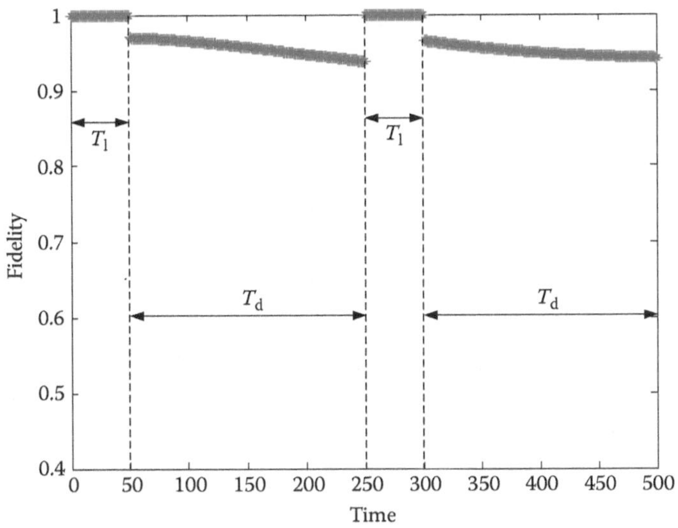

Figure 17.10 The changing fidelity for the scenario with 1000 nodes during the data collection.

the Euclidean distance and the Pearson's distance are used to be two metrics of the correlation of sensor nodes in our clustering algorithm. With the node scheduling scheme, only a part of nodes in the cluster are required to work for sampling, while the rest of nodes switch to the sleep mode for energy saving. Since the active node selections are based on the features of sampling data changing over time, we dynamically adjust the data collection period against the cluster deterioration to avoid the structure index instability resulted from the change of the internode relationships. Simulation results show that SFDC is more suitable for the large-scale dense WSNs in order to achieve a better effect on energy saving. Our clustering and node selection algorithms have remarkable advantages in the SNR when compared to the random CHs and active nodes selection. We also discuss the node contribution rate and provide a way to determine the fidelity threshold in practice. Finally, the adaptive data collection scheme is proved to be an effective technique in response to the spatial correlation changes.

As a future work, we intend to explore the temporal correlation to further reduce the energy consumption without the decline of structural fidelity. And the data prediction and compression techniques can be exploited and used to achieve the objective. In addition, we will consider the problems of sensor failure and calibration into our data collection framework in future work.

18 Trade-Off between Network Lifetime and Utility in Wireless Sensor Networks

We consider the challenge to simultaneously achieve the maximal lifetime and the maximal total utility in a multihop wireless sensor network (WSN) communication system, for there is usually a contradiction between these two objectives but a trade-off exists between them. In order to guarantee bandwidth allocation among individual sensor node while to extend the network lifetime to the maximum, we propose herein a network utility maximization (NUM) model, which is essentially a cross-layer constrained maximization problem. The retransmission scheme is suggested to be in a hop-by-hop manner to guarantee reliable communication and the Media Access Control (MAC) constraints are therefore addressed as channel access time constraints. By using a dual decomposition method, we establish two subproblems: one concerns the rate allocation issue and the other one concerns the energy conservation issue. By studying these two subproblems, we propose a general approach to characterize the trade-off from a cross-layer point of view.

We propose some key theorems and algorithms to achieve the solution of the best trade-off, which are theoretically analyzed and then evaluated through simulations. Being sharply different from other existing works, where the network optimization models only account for either network lifetime or network utility without transmission reliability being incorporated, our approaches (models and algorithms) present a unified framework to consider these three key elements simultaneously in the network optimization model and the solving procedures. This point brings our algorithms and solutions to be promising for implementation in a cross-layer manner in real WSN communication systems.

18.1 INTRODUCTION

WSN information and communication systems [7] have received strong research interests in recent years due to their wide applications of detecting and conveying information. Generally, WSN information and communication systems have an ad hoc topology with each node being able to relay the data toward the sink. Normally, the performance of a WSN system not only depends on the amount of gathered data but also relies on the accuracy of information obtained by the network [558] and the latency of data transmissions as well in most cases. It is seen that both the accuracy and the latency are strongly relevant to the data sending rates that the sensor nodes can be allowed for, which are subject to the condition of the total amount of resource (bandwidth) in the wireless links [538,566]. But it remains a challenge how to dynamically allocate the limited resource (bandwidth) of a WSN information and communication system among its sensor nodes toward the maximal network lifetime and the satisfactory application performance.

The NUM theory, that is, pioneered by Kelly [104,105] and followed up by, for example, [112, 364,532], has been widely applied in resource allocation and performance optimization for both wire-lined networks, including the Internet and wireless networks. The NUM model is formulated as an optimization problem, where the objective function to be maximized is the summation of the utility function of all users, and the constraint is that the aggregate rate of users is less than the capacity of the corresponding link.

In Ref. [555], in wireless networks, Tan et al. provided a unified framework for resource allocation by using the NUM theory and solved the NUM problem using the Karush–Kuhn–Tucker condition. The authors in Ref. [560] propose a novel transmission scheme for jointly optimal allocation of the base station (BS) broadcasting power and time sharing among the wireless nodes, which maximizes the overall network throughput, under the constraint of average transmit power and maximum transmit power at the BS. The paper [529] provides a model that focuses on the resource prediction issues, together with the resource estimation and reservation issue for Internet of Things systems.

Within the WSN information and communication system area, many research efforts have been directed to rate allocation issue (see, for example, [531,539,541,545,548,552,553,557,561]). The sparse energy in sensor node comes to be a handicap that holds back the application scope of WSN information and communication systems. To be exact, if a sensor node transmits data packets at a higher data rate, more energy will be consumed, which brings the network lifetime to be shorter. To tackle the above problem, a number of energy aware or energy-efficient algorithms have recently been proposed in the literature (for example, [431,536,551]). To maximize the network lifetime, distributed routing algorithms and cross-layer design techniques have also been proposed (see, for example, [543,544,550]). In Ref. [532], Nama et al. studied the utility and lifetime trade-off issue by introducing a lifetime-penalty function to penalize large values of the inverse of the network system lifetime. However, they do not offer a distributed algorithm that can be implemented in distinct network layers.

Most models presented in the previous papers often assume fixed source rates, which is impractical in engineering. In addition, how to ensure the reliability of end-to-end data transmission has not been considered yet, which is a very important requirement for a lossy link. Since the application performance is often related to the reliable data rates in WSN information and communication systems, it is very important to guarantee the data to be delivered reliably in WSN information and communication systems [566]. There are discussions on data delivery reliability issue in WSN information and communication systems arising from various aspects. However, most papers are dealing with this issue from the transport layer [528,533,534,554,559], which are aimed at reducing either the probability of loss or the error by adopting a certain retransmission mechanism. To achieve communication reliability, we herein adopt a hop-by-hop retransmission mechanism as introduced in Ref. [566], which utilizes the idea in the approach proposed by Deb et al. [534].

In WSN information and communication systems, an additional important resource constraint is the MAC constraint. In a wireless network, because multiple frequencies or codes can be available for transmission, parallel communications can be performed by using those orthogonal channels within the neighborhoods. In the wireless MAC layer, a wireless network will utilize a time-division policy to access the channel [530,542], in the manner that the physical channel is accessible only by a single user at each time instant at any place in space. In WSN information and communication systems, there are other constraints, for instance, the "Clique constraints" [535,564], the relaxed "MAC constraints" [540,556], and the constraints being put on the capacity limitation [549]. It is appropriate to consider the MAC constrain Ref. [128] when links do not have significant interference to each other at the time instance, or when all links are performing transmission by various frequencies or by various codes. In Ref. [565], the MAC constraint is seen to be strictly stronger than link constraint in all cases. Following [540,556], we assume herein that different links, use different transmission channels, so one only needs to consider the MAC constraints [128,565].

The main contributions presented herein are summarized as follows:

- The existing discussions are solely aimed at maximizing the total utility for WSN information and communication systems. With deviation to those studies, the present approach will take the utility maximization into account together with another key objective, that is, the network lifetime. We are going to address the utility and lifetime trade-off issue for resource-constrained WSN communication system.
- Our present work differs from other previous works in that the MAC constraints in multichannel WSN information and communication systems are all considered, while other works only consider the simple link capacity constraint.

- The issue of bandwidth allocation and congestion control in WSN information and communication systems is forced to be attacked in a cross-layer designing framework due to the fact that a trade-off exists between the whole network lifetime and the total utility of the network system. Note that the network utility will determine the quality of service (QoS). To facilitate analysis, we cast these two conflicting objectives into a unified weighted objective with the aid of a trade-off parameter and subsequently we are able to formulate the problem into a constrained convex programming NUM model. Then this trade-off model is decomposed into two subproblems by using the dual decomposition method: the rate control problem and the network lifetime maximization problem, both of which interact through the dual variables. By using the gradient projection method, we obtain a partially distributed algorithm that solves the separated two subproblems and the dual problem. In the solutions, we obtain the global information being outputted from the globally exchanging manipulations among sensor nodes at each iteration within the whole network, which comes out the network lifetime. The proposed algorithm so far is not completely distributed, which may restrict its application scope in terms of scalability with the scale of the WSNs. We next propose a fully distributed algorithm, in which we use the subgradient solving method. All the algorithms are ready to be implemented in WSNs.

  We list all nomenclatures in Table 18.1.

The remaining of this chapter is organized as follows: In Section 18.2, we present the NUM model and the fair rate allocation problem. In Section 18.3, we formulate the network utility and lifetime trade-off issue into a NUM model, then transform it into a separable and convex optimization mathematical model, and finally propose a partially distributed algorithm to solve the targeted problem. In Section 18.4, we propose a new model to achieve a fully distributed algorithm. Section 18.5 is about the convergence properties of the distributed algorithm. Section 18.6 presents numerical results to demonstrate the network performance. Finally, we conclude the whole chapter in Section 18.7 and discuss the issue for future work.

18.2 NUM AND SYSTEM MODEL

Let us consider a general L link communication network. Every link has capacity c_l bps, and there are S sources. The transmission rate is denoted by x_s bps for $s \in S$. If each source s only sends one flow using a fixed set $L(s)$ of links in its route path, and each source (flow) is related to a utility

Table 18.1
Nomenclature

S	The set of sensor nodes in the wireless sensor network
L	The set of logical links in the wireless sensor network
$L(s)$	The fixed subset of links L used by sensor node s in its route path
$S(l)$	The subset of sensor nodes using link l
c_l	The capacity of link l
$\mathbf{c}$	The capacity of links in the wireless sensor network
e_s	The initial energy of sensor node s
$\mathbf{e}$	The vector of initial sensor node energy
m_s	The minimum data rate for sensor node s
M_s	The maximum data rate for sensor node s
$A(s)$	The collection of all nodes in the path of sensor node s
β	A trade-off parameter
δ	A mapping parameter
p_e	The link error probability

function $U_s(x_s)$. At each link l, there are a number of sources (flows), which set is denoted by $S(l)$, competing for the link capacity. The original NUM model for a wired network was proposed to solve the following distributed end-to-end rate allocation problem

$$\begin{aligned} &\max \sum_s U_s(x_s) \\ &\text{s.t.} \sum_{s\in L(s)} x_s \le c_l, \quad \forall l. \end{aligned} \tag{18.1}$$

The above problem can be solved in a distributed way as suggested by Ref. [112]. However, it is noted that in the above model, only the data packet flow is considered, but the impact of the ACK packet flow in performance is not considered. In the recent paper [364], for a two-way flow scenario, Ge and Tan establish a novel NUM model.

Now let us consider a static multihop WSN information and communication system, which consists of a set of sensor nodes $S = \{1,2,\ldots,S,S+1\}$, where the former s nodes are the normal sensor nodes while the last one is the sink node. The sink node is collecting data through every sensor node. In the network, each sensor node transmits its gathered data packets to its peer nodes. These transmissions form the end-to-end flows, traversing a number of hops starting from a source to its destination. The sensor nodes are powered by nonrechargeable and irreplaceable batteries, and they are fulfilling the tasks of sensing, collecting, and transmitting data.

The WSN information and communication system can be modeled as a graph $G = (V,E)$, where V is the set N of nodes, and E is the set of all un-directional links. The sensor nodes send their collected data to the sink through a set $L = \{1,2,\ldots,L\}$ of links. Each link has a capacity c_l, $l \in L$. A wireless link is deemed to be existed between a pair of nodes, if direct communications happen between these peer-to-peer nodes that are within each other's transmission range. In reality, although wireless channels are in nature time varying, we can take an averaged capacity for each link, which is determined by physical layer transmission scheme. Therefore, it can be assumed that each link has a fixed capacity. Further, we assume that the specific wireless links are alive long enough so that the response of the congestion control scheme can be taken.

18.2.1 MAC CONSTRAINTS

In terms of the wireless MAC protocol, at a time instant through a given channel, only one peer-to-peer node pair can communicate with each other to avoid collision. The simple "MAC constraints" can mathematically be expressed [537,556] as

$$\sum_{s\in S(j)} x_s \left(\frac{1}{c_{l_i(j,s)}} + \frac{1}{c_{l_o(j,s)}} \right) \le 1, \quad \forall s \in N, \tag{18.2}$$

where $S(j)$ is the set of flows traversing Node j. The components $l_i(j,s)$ and $l_o(j,s)$ denote the input and output links for Flow s at Node j, respectively. Flows either entering into or departing from a link generate the corresponding $1/c_l$ terms in the above expression. However, the above "ideal" time constraints are not precise [537,556] in real networks as there is usually resource loss in a wireless network due to a number of reasons, for example, the channel attenuation. A feasible MAC condition is attained if the time constraints are relaxed by replacing the term "1" by a positive parameter that is less than 1 [537]. To attain the simplicity of implementation, we simply adopt formula 18.2 as the MAC constraints in later expositions.

We define the following incident matrix $\mathbf{G} \in \mathbf{R}_+^{(S+1)\times L}$, whose element (s,l) is specified by

$$G_{sl} = \begin{cases} 1, & \text{when Link } l \text{ is incident to Node } s, \\ 0, & \text{otherwise.} \end{cases} \tag{18.3}$$

Let $A_{ls} = 1/c_l$, if the flow emitted by sensor node s uses Link l and 0 otherwise. Similarly, one specifies the matrix $\mathbf{A} \in \mathbf{R}_+^{L\times S}$, with its element (s,l) being defined by

$$A_{ls} = \begin{cases} \frac{1}{c_l}, & \text{if the flow emitted by Sensor } s \text{ uses Link } l \\ 0, & \text{otherwise} \end{cases} \tag{18.4}$$

We use **G** and **A** to reduce the time constraint Equation 18.2 into the following formulation:

$$\mathbf{GAx} \leq \mathbf{I}_N, \tag{18.5}$$

where $\mathbf{x} = [x_1, x_2, \ldots, x_S]^T$ denotes the data rate vector and $\mathbf{I}_N$ is the identity matrix. The matrix $\mathbf{G} \cdot \mathbf{A}$ is termed as the generalized routing matrix, which is denoted by **F**. That is, $\mathbf{F} = \mathbf{G} \cdot \mathbf{A}$. Therefore, the above routing expression can be further rewritten as

$$\mathbf{Fx} \leq \mathbf{I}_N. \tag{18.6}$$

18.2.2 NETWORK UTILITY MAXIMIZATION

In a WSN information and communication system, each sensor node s can be associated with a utility function $U_s(x_s)$, where the variable x_s is the transmission rate and must satisfy $m_s \leq x_s \leq M_s$. The components m_s and M_s are the minimal and maximal transmission rates, respectively, which are required by Source s to satisfy $m_s \geq 0$ and $M_s < \infty$. We introduce a parameter ω_s as a weight factor to represent the level of importance of different users in the total utility summation. In particular, those nodes nearer the sink have heavier task to relay the packets from other nodes, and they are deemed to be more important than those further away from the sink. A larger value of weight are thus put for those nearby nodes in utility summation. Hence, the basic NUM framework at hand is then described as follows:

$$\begin{aligned} P1: &\max_{\mathbf{x}} \sum_{s=1}^{S} \omega_s U_s(x_s) \\ \text{s.t.} &\begin{cases} \mathbf{Fx} \leq \mathbf{I}_N, \\ m_s \leq x_s \leq M_s, \quad \forall s \in S \end{cases} \end{aligned} \tag{18.7}$$

Using the formula 18.2 and by substituting it into the first constraint in Equation 18.7, we obtain the equivalent expression as follows:

$$\begin{aligned} P1: &\max_{\mathbf{x}} \sum_{s=1}^{S} \omega_s U_s(x_s) \\ \text{s.t.} &\begin{cases} \sum\limits_{s \in S(j)} x_s \left(\frac{1}{c_{l_i(j,s)}} + \frac{1}{c_{l_o(j,s)}} \right) \leq 1, \quad \forall s \in N, \\ m_s \leq x_s \leq M_s, \quad \forall s \in S. \end{cases} \end{aligned} \tag{18.8}$$

18.2.3 NETWORK LIFETIME MAXIMIZATION

This section is dedicated to a discussion of energy consumption model. In a typical WSN information and communication system, sensor nodes spend much more energy than the sink node, and as such we will only focus on the energy dissipated in the sensor nodes. For a specific node, all processes including sensing, transmitting, and receiving data consume energy [462]; however, communications spend the mostly large portion of energy. Therefore, energy consumed by the tasks other than communication can be ignored. We denote the amount of energy consumed per bit in sensing activities by ε_{s}, in transmitting activities by $\varepsilon_{\mathrm{sl}}^t$, and in the activities of receiving data by ε_{r}. Note that, for reliable communication, $\varepsilon_{\mathrm{sl}}^t$ also includes the radiated energy, which is the energy consumption for transmitting one unit data over link l. We have

$$\varepsilon_{\mathrm{sl}}^t = \rho + \sigma d_{\mathrm{sl}}^n, \quad 2 \leq n \leq 4.$$

Here, ρ is the portion of energy, which is used to run the transmitter circuitry, and σ is the distance-related coefficient of the transmit amplifier. Both ρ and σ are related to the physical layer and the environment circumstances. The component d_{sl} is the logic distance from sensor node s to link l, and n is the path loss factor.

There are usually packet losses in the transmission process in a wireless channel, which are mainly caused by link error together with congestion. The channel condition in physical layer and the environment where networks are deployed and so on will lead to packet loss through transmission errors; while just like in wire-lined network if congestion happens, packet loss also occurs. If there is no route to the destination available, a packet may be dropped at the source. In the situation, where the buffer that is storing and queuing the pending packets are full, the incoming packets may be also dropped.

At an intermediate host, the packets may also be dropped if the link to the next hop was unavailable due to being broken. To consider the consequence of packet loss and to ensure reliable transmission, we use a scheme in the hop-by-hop retransmission manner herein. For simplicity, all links are assumed to have the same probability p_e, which is the so-called link error probability, presenting the chance of a packet getting lost or being received by mistakes. Based on this probability, the following formula gives the averaged number of transmission $\overline{n}$, that is, required to deliver a packet safely over each hop:

$$\overline{n} = (1-p_e)+2p_e(1-p_e)+3{p_e}^2(1-p_e)+\cdots = \frac{1}{1-p_e}.$$

It is assumed that every node has the identical power dissipation features, which enables us to define an energy consumption matrix $\mathbf{E} = [E_{sn}] \in \mathbf{R}_+^{S\times S}$, where its element is specified by

$$E_{sn} = \begin{cases} \varepsilon_s + \frac{\varepsilon_{sl}^t}{1-p_e}, & \text{if } s = n, \\ \frac{(\varepsilon_r+\varepsilon_{sl}^t)\varepsilon_{sl}^t}{1-p_e}, & \text{if Node } s \text{ is an relay of Node } n, \\ 0, & \text{other.} \end{cases} \tag{18.9}$$

Consider the case, where each source emits only one unique flow to the corresponding sink transiting a number of shops through the network, the above energy consumption matrix is a square one. The average power, which is dissipated at node s, is then obtained as

$$P_s = \sum_{n\in N(s)} E_{sn}x_n, \tag{18.10}$$

where $N(s)$ denotes a set of sensor nodes, whose elements are all the sensor nodes acting as the role of a relay for others plus Node s itself.

The initial energy of Node s is denoted by e_s and the vector $\mathbf{e}$ is constructed from these initial energy components. For Node s, its lifetime is defined as $t_s = {}^{e_s}/_{P_s}$. As every node may have different lifetime, we define the time as the lifetime of the whole WSN, at which the first sensor node becomes "died," that is, it runs out of its energy. In other words, the lifetime of the whole WSN is defined as $T_{net} = \min\{t_s|s = 1,2,\ldots,S\}$.

Therefore, if we denote $q = 1/T_{net}$, which is the inverse portion of lifetime of the network, we have the following maximum problem for obtaining the whole lifetime of the network:

$$\begin{aligned} &P2: \max\ T_{net} \\ &\text{s.t.} \begin{cases} \mathbf{Fx} \le \mathbf{I}_N, \\ \mathbf{Ex} \le \mathbf{e}q, \\ m_s \le x_s \le M_s, \quad \forall s \in S. \end{cases} \end{aligned} \tag{18.11}$$

It should be noted that, in the above maximum network lifetime model, the first constraint is that the summation of flow rates at the links should be capped by the link capacity while the second one is that the power dissipation at each node should be less than its initial energy. However, the fact that the objective function in the above formulation is not strictly convex in the variables, the dual function is subsequently not differentiable, brings a very difficulty in solving the above problem. We next, therefore, transform the above problem into a convex optimization problem by using a method similar to that used in Refs. [176,538].

We propose the lifetime-penalty function $F(q) = q^2$. Besides the property of being strictly convex, this function is continuously differentiable and increasing with $q \geq 0$. Since to minimize q can be realized through to minimize q^2, then we have the following equivalent network lifetime maximization problem:

$$P2: \min q^2 \quad \text{s.t.} \begin{cases} \mathbf{Fx} \leq \mathbf{I}_N, \\ \mathbf{Ex} \leq \mathbf{e}q, \\ m_s \leq x_s \leq M_s, \quad \forall s \in S. \end{cases} \tag{18.12}$$

We will work on the above model P2 together with the model P1 in the later development to solve the problem of trade-offing between utility and lifetime.

18.3 PARTIALLY DISTRIBUTED ALGORITHM FROM DUALITY DECOMPOSITION

We are now ready to propose a novel model by considering fair rate allocation and MAC time constraints jointly into the utility–lifetime trade-off problem. In order to achieve the fair bandwidth allocation among individual nodes and to maximize the whole network lifetime, we introduce two parameters: $\beta \in [0,1]$ and δ. The former helps us in combining the two objective functions into a single function, while the latter brings the two objective functions into the same order in magnitude.

Given the above analyses on the introduced constraints, the trade-off problem at hand is then formulated as follows:

$$P3: \max_{\mathbf{x},q} \beta \sum_{s=1}^{S} \omega_s U_s(x_s) - (1-\beta)\delta q^2 \quad \text{s.t.} \begin{cases} \mathbf{Fx} \leq \mathbf{I}_N, \\ \mathbf{Ex} \leq \mathbf{e}q, \\ m_s \leq x_s \leq M_s, \quad \forall s \in S. \end{cases} \tag{18.13}$$

Further, the above utility and lifetime maximization problem can be equivalent to the following expression:

$$P3: \max_{\mathbf{x},q} \beta \sum_{s=1}^{S} \omega_s U_s(x_s) - (1-\beta)\delta q^2 \quad \text{s.t.} \begin{cases} \sum_{s \in S(j)} x_s \left(\frac{1}{c_{l_i(j,s)}} + \frac{1}{c_{l_o(j,s)}} \right) \leq 1, \quad \forall s \in N, \\ \sum_{s=1}^{S} E_{ks} x_s \leq e_k q, \quad \forall k \in S, \\ m_s \leq x_s \leq M_s, \quad \forall s \in S. \end{cases} \tag{18.14}$$

On the above basis, by using the duality approach, we achieve a partially distributed algorithm as outlined hereafter.

18.3.1 DUALITY PROBLEM

The problem (Equation 18.14) is a convex optimization problem, where we are thus able to solve it by using the duality decomposition approach. Equation 18.15 formulates the Lagrangian duality function, which corresponds to the primal problem described by Equation 18.14, where we have introduced two Lagrange multipliers: λ and μ.

$$\begin{aligned} D(\lambda,\mu) &= \max_{\mathbf{x},q} \beta \sum_{s=1}^{S} \omega_s U_s(x_s) - (1-\beta)\delta q^2 - \lambda^T(\mathbf{Fx} - \mathbf{I}_N) - \mu^T(\mathbf{Ex} - \mathbf{e}q) \\ &= \max_{\mathbf{x},q} \beta \sum_{s=1}^{S} \omega_s U_s(x_s) - (1-\beta)\delta q^2 - \sum_{j=1}^{N} \lambda_j \left(\sum_{s \in S(j)} x_s \left(\frac{1}{c_{l_i(j,s)}} + \frac{1}{c_{l_o(j,s)}} \right) - 1 \right) \end{aligned}$$

$$
\begin{aligned}
&-\sum_{k=1}^{S}\mu_k\left(\sum_{s=1}^{S}E_{ks}x_s-e_kq\right)\\
&=\max_{\mathbf{x},q}\sum_{s=1}^{S}\left\{\beta\omega_s U_s(x_s)-\left(\sum_{j\in A(s)}\left(\frac{1}{c_{l_i}(j,s)}+\frac{1}{c_{l_o}(j,s)}\right)\lambda_j+\sum_{k=1}^{S}\mu_k E_{ks}\right)x_s\right\}\\
&+\sum_{k=1}^{S}\mu_k e_k q-(1-\beta)\delta q^2+\sum_{j=1}^{N}\lambda_j\\
&=\sum_{s=1}^{S}\max_{x_s}\left\{\beta\omega_s U_s(x_s)-\left(\sum_{j\in A(s)}\left(\frac{1}{c_{l_i}(j,s)}+\frac{1}{c_{l_o}(j,s)}\right)\lambda_j+\sum_{k=1}^{S}\mu_k E_{ks}\right)x_s\right\}\\
&+\max_{q}\left(\sum_{k=1}^{S}\mu_k e_k q-(1-\beta)\delta q^2\right)+\sum_{j=1}^{N}\lambda_j\\
&=D_1(\lambda,\mu)+D_2(\lambda,\mu)+\sum_{j=1}^{N}\lambda_j
\end{aligned}
\tag{18.15}
$$

For the above two subproblems, one can separate the two objective functions. They are strictly concave ones about their individual variables. Therefore, one can solve the above two subproblems to obtain the unique analytical solution on the variables x_s and q. They are given by

$$
x_s=\left[U_s'^{-1}\left(\frac{\sum_{j\in A(s)}\left(\left(\frac{1}{c_{l_i}(j,s)}+\frac{1}{c_{l_o}(j,s)}\right)\lambda_j\right)+\sum_{k=1}^{S}\mu_k E_{ks}}{\beta\omega_s}\right)\right]_{m_s}^{M_s},\quad \forall s\in S,
\tag{18.16}
$$

$$
q=\left[\frac{\sum_{k=1}^{S}\mu_k e_k}{2(1-\beta)\delta}\right]^{+},
\tag{18.17}
$$

where we have denoted $[\bullet]_m^M=\max\{\min\{\bullet,M\},m\}$, and $[\cdot]^+$ denotes the projection onto the range R^+. Given that, in the primal problem, the objective function is strictly concave on the variables $\mathbf{x}$ and q, one can use the gradient projection algorithm to solve it. One also notes that the primal variables $\mathbf{x}$ and q are implied in the duality variables through Equations 18.16 and 18.17. One further verifies that

$$\nabla_\lambda D=\mathbf{I}_N-\mathbf{Fx}$$

and

$$\nabla_\mu D=\mathbf{e}q-\mathbf{Ex}$$

are the partial gradients of the dual function (Equation 18.15) regarding the dual variables λ and μ at point (λ,μ), respectively.

Then, using the gradient projection algorithm, the dual problem (Equation 18.15) can be further solved to obtain

$$\lambda(t+1)=[\lambda(t)-\gamma(I_N-\mathbf{Fx}(t))]^+, \tag{18.18}$$

$$\mu(t+1)=[\mu(t)-\gamma(\mathbf{e}q(t)-\mathbf{Ex}(t))]^+, \tag{18.19}$$

where $\gamma>0$ is a constant scalar step size.

Moreover, with λ_j denoting the node congestion price and μ_k denoting the node lifetime price, the duality variables have implication of resource-pricing mechanism. The algorithm described by Equations 18.18 and 18.19 can be understood as the law of supply and demand similar to Ref. [538]. One observes from Equation 18.16 that the node congestion price decreases during the tth iteration if the total fraction of time spent at a node is capped by 1, and the reduced node congestion price leads to a larger source rate for all sources using this node as a relay and vice versa, thus leading to an efficient node resource utilization. At the same way, if the actual average power dissipation $(\mathbf{Ex}(t))_k$ is less than the maximum averaged power dissipated by $e_k q$, the node lifetime price will decrease during the tth iteration. Therefore, larger source rates for all sources are only obtained at the cost of the reduced node lifetime price.

18.3.2 PARTIALLY DISTRIBUTED IMPLEMENTATION

Now we can outline the proposed partially distributed algorithm as Algorithm 18.3.1.

Algorithm 18.3.1 A partially distributed algorithm for the source rate iteration and the network inverse-lifetime iteration.

Step 1: Let $t \leftarrow 0$, to put the initial values to the variables: $\left(\lambda^{(0)}, \mu^{(0)}\right)$, $\varepsilon \geq 0$ and to organize the initial information $(U_s(\cdot), e_s, E_{ks})$ for every node;

Step 2: The node sending rate is updated on the basis of the aggregate prices of the node by

$$x_s(t+1) = \left[(U_s)'^{-1} \left(\frac{\sum_{j \in A(s)} \left(\left(\frac{1}{c_{l_i(j,s)}} + \frac{1}{c_{l_o(j,s)}} \right) \lambda_j(t) \right) + \sum_{k=1}^{S} \mu_k(t) E_{ks}}{\beta \omega_s} \right) \right]_{m_s}^{M_s}, \quad \forall s \in S. \tag{18.20}$$

Then, node s communicates the new rate $x_s(t+1)$ to all nodes along its path;

Step 3: The inverse of lifetime is updated by

$$q(t+1) = \left[\frac{\sum_{k=1}^{S} \mu_k(t) e_k}{2(1-\beta)\delta} \right]^+, \tag{18.21}$$

and then broadcasts to all the sensor nodes in the sensor network;

Step 4: Each node j updates its congestion price λ_j according to

$$\lambda_j(t+1) = \left[\lambda_j(t) - \gamma \left(1 - \sum_{s \in S(j)} x_s(t) \left(\frac{1}{c_{l_i(j,s)}} + \frac{1}{c_{l_o(j,s)}} \right) \right) \right]^+, \quad \forall j \in N. \tag{18.22}$$

and then this information is broadcasted to all the relaying nodes of node s;

Step 5: Each sensor node k updates its energy price μ_k in terms of

$$\mu_k(t+1) = \left[\mu_k(t) - \gamma \left(e_k q(t) - \sum_{s=1}^{S} E_{ks} x_s(t) \right) \right]^+, \quad \forall k \in S \tag{18.23}$$

and then node s broadcasts this piece of information to all other relaying nodes;

Step 6: If $\|D(\lambda(t+1), \mu(t+1)) - D(\lambda(t), \mu(t+1))\| \leq \varepsilon$, stop; otherwise set $t = t+1$, turn to Step 2.

In this proposed algorithm, at each iteration instant t, at the sink node the inverse of lifetime q is adjusted by Equations 18.20 and 18.21, while every sensor node updates its values of x_s locally, correspondingly. Based on Equations 18.22 and 18.23, the values of the Lagrange multipliers λ and μ are also iterated dynamically, respectively.

The mostly notable feature of this algorithm is that the individual variable corresponding to each node is not necessarily circulated through the whole network, but only the aggregate variables are needed to be passed through the network. Therefore, the required updates can be piggybacked on the data packets to be directed to the sink, and then they are broadcasted from the sink to all sensor nodes. These manipulations can be realized by acknowledgment packets in TCP [111,538].

However, the algorithm unfortunately needs the feedback from the sink about the inverse of network lifetime updates $q(t)$. This brings a certain amount of communications to the network, which then generate communication loads. Thus, the partially distributed Algorithm 18.3.1 is not efficient to our most satisfaction. An alternative algorithm will be offered in the next section, which is fully distributed.

18.4 FULLY DISTRIBUTED ALGORITHMS

This section is dedicated to seek a completely distributed algorithm.

In this section, we firstly present the combined single optimization primal problem, which is decomposable. Then for achieving fully distributed implementation, we use dual decomposition to separate the primal problem into several subproblems, individually for each node. To ensure that the primal problem keeps decomposable characteristics, we use the similar techniques in Ref. [547]. We propose a method to completely decentralize the problem by introducing additional variables, which are corresponding to the upper bound on the inverse lifetime of each node. This novel trade-off model can be reshaped into the following convex optimization problem:

$$P4: \max \sum_{s=1}^{S} \left(\beta \omega_s U_s(x_s) - (1-\beta)\delta q_s^2\right),$$
$$s.t. \begin{cases} \sum_{s \in S(j)} x_s \left(\frac{1}{c_{l_i}(j,s)} + \frac{1}{c_{l_0}(j,s)} \right) \leq 1, \quad \forall j \in N, \\ \sum_{s=1}^{S} E_{ks} x_s \leq e_k q_k, \quad \forall k \in S, \\ q_s \leq q_j, \quad \forall s \in S, j \in N_s, \\ m_s \leq x_s \leq M_s, \quad \forall s \in S. \end{cases} \tag{18.24}$$

Here, the constraints are the MAC constraints and energy conservation constraints. In addition, an objective function is considered, that is, quadratic and strictly convex regarding to its variable q_i. Furthermore, there are additional constraint that enforces all q_s to be equal. We assume the Slater's condition for constraint qualification is satisfied and hence strong duality holds for this problem. Thus, we can solve the primal problem (Equation 18.24) via the dual theory.

To ensure that the duality function is differentiable, by considering in all the coupling constrains in problem (Equation 18.24), for the first three constraints concerning the three prices: the congestion price, the energy price, and the inconsistent coordination price, we are hereby introducing the Lagrangian multiplier vector:

$$\lambda \geq 0, \quad \mu \geq 0, \quad v \geq 0.$$

The Lagrangian duality function associated with the above primal problem is given by Equation 18.26. Thus the dual function can be evaluated separately in the variables corresponding to each node $s \in S$. From above, the duality problem is then formulated into

$$\min_{\lambda \geq \mathbf{0}, \mu \geq \mathbf{0}, v \geq \mathbf{0}} D(\lambda, \mu, v) \tag{18.25}$$

$$
\begin{aligned}
D(\lambda,\mu,v) &= \max_{\mathbf{x},\mathbf{q}} \sum_{s=1}^{S} \left(\beta\omega_s U_s(x_s) - (1-\beta)\delta q_s^2\right) - \sum_{j=1}^{S+1} \lambda_j \left(\sum_{s\in S(j)} x_s \left(\frac{1}{c_{l_i(j,s)}} + \frac{1}{c_{l_o(j,s)}} \right) - 1 \right) \\
&\quad - \sum_{k=1}^{S} \mu_k \left(\sum_{s=1}^{S} E_{ks} x_s - e_k q_k \right) - \sum_{s=1}^{S} \sum_{j\in N_s} v_{sj} (q_s - q_j) \\
&= \max_{\mathbf{x},\mathbf{q}} \sum_{s=1}^{S} \left\{ \beta\omega_s U_s(x_s) - (1-\beta)\delta q_s^2 - \left(\sum_{j\in A(s)} \lambda_j \left(\frac{1}{c_{l_i(j,s)}} + \frac{1}{c_{l_o(j,s)}} \right) + \sum_{k=1}^{S} \mu_k E_{ks} \right) x_s \right. \\
&\quad \left. + \mu_s e_s q_s - q_s \sum_{j\in N_s} (v_{sj} - v_{js}) \right\} + \sum_{j=1}^{S+1} \lambda_j. \qquad (18.26)
\end{aligned}
$$

Similarly, if the Slater's condition of the constraint is satisfied, then the strong duality holds. As in Refs. [543,547], we can use a subgradient algorithm to solve the duality problem, which method is designed into Algorithm 18.4.1 as a fully distributive algorithm.

18.4.1 SUBGRADIENT-BASED ALGORITHM

To solve the duality problem (Equation 18.25), we first consider the Lagrangian duality function (Equation 18.26). Given the Lagrangian is separable, at each source node s, the maximization of Lagrangian over (x, q) can be performed in a parallel way. During the tth iteration of the subgradient algorithm, given $\lambda^{(t)}, \mu^{(t)}, v^{(t)}$, each sensor node s solves the following convex program with variables x_s and q_s.

$$
\begin{aligned}
&\max \quad \beta\omega_s U_s(x_s) - (1-\beta)\delta q_s^2 + \mu_s^{(t)} e_s q_s - q_s \sum_{j\in N_s} \left(v_{sj}^{(t)} - v_{js}^{(t)} \right) \\
&\quad - \left(\sum_{j\in A(s)} \lambda_j^{(t)} \left(\frac{1}{c_{l_i}(j,s)} + \frac{1}{c_{l_o}(j,s)} \right) + \sum_{k=1}^{S} \mu_k^{(t)} E_{ks} \right) x_s \qquad (18.27) \\
&s.t. \quad m_s \le x_s \le M_s, q_s \ge 0.
\end{aligned}
$$

Given a convex function $f : R^n \to R$, a vector $d \in R^n$ is a subgradient of f at a point $u \in R^n$ if

$$f(v) \ge f(u) + (v-u)^T d, \quad \forall v \in R^n.$$

Let $x_s^{(t)}$ and $q_s^{(t)}$ denote the optimal solutions of the above problem. One checks that during the tth iteration the subgradient of the duality function (Equation 18.26) with respect to the dual variable λ at point (λ, μ, v), denoted by h, is given by

$$h_j^{(t)} = 1 - \sum_{s\in S(j)} x_s^{(t)} \left(\frac{1}{c_{l_i}(j,s)} + \frac{1}{c_{l_o}(j,s)} \right), \quad \forall j \in N. \qquad (18.28)$$

Similarly, the subgradient with respect to μ and $\mathbf{v}$ at point (λ, μ, v), denoted by d and g, are given by

$$d_k^{(t)} = e_k q_k^{(t)} - \sum_{s=1}^{S} E_{ks} x_s^{(t)}, \quad \forall k \in S \qquad (18.29)$$

$$g_{kj}^{(t)} = q_j^{(t)} - q_k^{(t)}, \quad \forall k \in S, j \in N_k \qquad (18.30)$$

The subproblem (Equation 18.27) can be solved in each iteration within a given set of feasible duality variables. Using these obtained solutions, we are able to further compute the subgradients of the duality function (Equation 18.26) and to evaluate a new set of dual variables for the next iteration. Using the subgradient, which corresponds to the duality variables at the tth iteration, the duality variables are updated at the $(t+1)^{\text{th}}$ iteration in the following way:

$$\lambda_j^{(t+1)} = \left[\lambda_j^{(t)} - \gamma^{(\mathrm{t})} h_j^{(t)}\right]^+, \quad \forall j \in N. \tag{18.31}$$

$$\mu_k^{(t+1)} = \left[\mu_k^{(t)} - \gamma^{(t)} d_k^{(t)}\right]^+, \quad \forall k \in S. \tag{18.32}$$

$$v_{sj}^{(t+1)} = \left[v_{sj}^{(t+1)} - \gamma^{(t)} f_{sj}^{(t)}\right]^+, \quad \forall s \in S,\ j \in N_{\mathrm{s}}. \tag{18.33}$$

In the above, the component $\gamma^{(\mathrm{t})}$ is a positive scalar step size, which satisfies the conditions in Theorem 18.5.1.

18.4.2 IMPLEMENTATIONS

The duality variables $(\lambda_j,\ \mu_k,\ v_{kj})$ and the subgradients $(h_j,\ d_k,\ g_{kj})$ can be explained by an intuitive economic implication. λ_j represents the price to use the node time ratio, μ_k represents the price to use the node's energy, and v_{kj} is the inconsistent coordination price between q_j and its neighbor q_k for sensor node k. The subgradient h_j represents the excess time ratio at node k, the subgradient d_k represents the excess energy at node k, and g_{kj} as the inconsistent coordination difference at node k with its neighbor j. Based on the updates of the duality variable as specified by Equations 18.31 through 18.33 at each time instant, one sees that the node congestion and energy prices decrease if the total node time ratio is below 1 or the total power dissipation is less than the initial energy, respectively, while the inconsistent coordination difference will disappear when the sensor node's lifetime is consistent with its neighbors.

Based on the above analyses, it is seen that with the help of the proposed subgradient approach, we are able to attain a fully distributed solution of Equation 18.24. Steered from that, a remarkable duality decomposition to the problem then enables us to evaluate the total utility and lifetime in a decoupled way. This is the mostly desirable characteristic of the algorithms presented herein.

18.5 ANALYSES TO THE CONVERGENCE OF THE DISTRIBUTED ALGORITHMS

If the assumption of strong duality holds, we now discuss the convergence property of the two distributed algorithms, which are proposed in the previous sections. Let us suppose the following assumptions hold:

- The utility function $U_{\mathrm{s}}(x_{\mathrm{s}})$ is increasing with its variable x_{s}, strictly concave on its variable x_{s}, and twice continuously differentiable on its variable x_{s}. Further, there is a parameter α_n such that $-U_{\mathrm{s}}''(x_{\mathrm{s}}) \geq {}^1/_{\alpha_n} > 0$.
- The lifetime-penalty function $F(q)$ is also increasing with q, strictly convex on q, and twice continuously differentiable with respect to q, $F'(q) < \infty$ and there is a parameter ϕ such that $F''(q) > {}^1/_{\phi} > 0$.

Note that the proposed Algorithm 18.3.1 is a gradient projection. By following a similar convergency analysis of the distributed gradient algorithm [259], we find that there is a positive number K, for all $\gamma < K$; Algorithm 18.3.1 converges to the desired global optimal solution.

The previously derived fully distributed Algorithm 18.4.1 is a subgradient algorithm. Based on the mathematical results given in Ref. [546], we can verify that if the step size $\gamma^{(\mathrm{t})}$ in 18.34 through 18.36 satisfies $\gamma^{(\mathrm{t})} \to 0$, when $t \to \infty$ and $\sum_{t=1}^{\infty} \gamma^{(\mathrm{t})} = \infty$, then the duality variables $\left(\lambda^{(t)},\ \mu^{(t)},\ v^{(t)}\right)$ in

Algorithm 18.4.1 A fully distributed algorithm for iterations of congestion price, energy price, and the inconsistent coordination price.

Step 1: Let $t \leftarrow 0$, to put the initial value with $\left(\lambda^{(0)}, \mu^{(0)}, \nu^{(0)}\right)$, $\varepsilon \geq 0$, and the local information $(U_s(\cdot), e_s, E_{ks})$.

Step 2: Each sensor node s locally solves the following convex programming problem 18.27 to update the variables $q_s^{(t)}$ and $x_s^{(t)}$, and then broadcasts them to the nodes that are on the path of node s. Note in the above problem, each node s only needs the values of the local duality variables, which are computed by itself and its neighbors; hence, solving this problem can be performed by a distribution manner at each sources.

Step 3: The node j updates its congestion price in terms of

$$\lambda_j^{(t+1)} = \left[\lambda_j^{(t)} - \gamma^{(t)}\left(1 - \sum_{s \in S(j)} x_s^{(t)}\left(\frac{1}{c_{l_i}(j,s)} + \frac{1}{c_{l_o}(j,s)}\right)\right)\right]^+, \quad \forall j \in N. \tag{18.34}$$

Then this congestion price is broadcasted from node s to those neighbor nodes that are using node s to relay their data packets.

Step 4: Each sensor node k updates its energy price by

$$\mu_k^{(t+1)} = \left[\mu_k^{(t)} - \gamma^{(t)}\left(e_k q_k^{(t)} - \sum_{s=1}^{S} E_{ks} x_s^{(t)}\right)\right]^+, \quad \forall k \in S. \tag{18.35}$$

and then this piece of information is broadcasted to the nodes that are using node k to relay their data packets. Note that each node k can update its μ_k by its local information, where q_k is the solution of the problem (Equation 18.27) for a given (λ, μ, ν).

Step 5: Each sensor node k updates its inconsistent coordination price by

$$\nu_{kj}^{(t+1)} = \left[\nu_{kj}^{(t+1)} - \gamma^{(t)}\left(q_j^{(t)} - q_k^{(t)}\right)\right]^+, \quad \forall k \in S, \quad j \in N_k. \tag{18.36}$$

and then node k broadcasts this piece of information to all other relaying nodes. Note that each node k can only use its local information to update ν_{kj}. The components q_j and q_k are the solution to the problem (Equation 18.27) for a given (λ, μ, ν). The component $\gamma^{(t)}$ in Steps 3 through 5 is a positive scalar step size.

Step 6: If $\left\| D(\lambda^{(t+1)}, \mu^{(t+1)}, \nu^{(t+1)}) - D(\lambda^{(t)}, \mu^{(t)}, \nu^{(t)}) \right\| \leq \varepsilon$, stop; otherwise, let $t = t+1$, turn to Step 2.

Algorithm 18.4.1 converges to the optimal dual solution $(\lambda^*, \mu^*, \nu^*)$ that solves the corresponding dual problem.

Define

$$w(t) = (\mathbf{x}(t), \mathbf{q}(t)),$$

where $\mathbf{x}(t)$ and $\mathbf{q}(t)$ are obtained as the solution to Equation 18.25 by Algorithm 18.4.1. Let W^* be the set of the optimal solution to the problem. Besides, we define

$$d(w(t), W^*) = \min_{w^* \in W^*} \|w(t) - w^*\|,$$

where $\|\cdot\|$ denotes the Euclidian distance. Then, we can obtain the following convergence results for our distributed Algorithm.

Theorem 18.5.1

We have the following two statements:

- On the condition that we find a positive constant K such that the step size γ in Algorithm 18.3.1 satisfies $\gamma < K$, if we put any initial value to the variables to have $x(0) \geq 0$, $q(0) \geq 0$, $\lambda(0) \geq 0$, and $\mu \geq 0$, based on them and starting from them, the time series $(\mathbf{x}(t), q(t), \lambda(t), \mu(t))$, which is computed from the proposed Algorithm 18.3.1, is a primal-dual optimal solution;
- If the step size $\gamma^{(t)}$ in Algorithm 18.4.1 satisfies $\gamma^{(t)} \to 0$, when $t \to \infty$, and

$$\sum_{t=1}^{\infty} \gamma^{(t)} = \infty.$$

Then, there holds

$$\lim_{t \to \infty} d\left(w(t), W^*\right) = 0.$$

■

18.6 NUMERICAL STUDIES AND PERFORMANCE ANALYSES

Let us consider a WSN information and communication system in the same network topology and the same scenario as in Ref. [563], which is shown in Figure 18.1 to present the numerical results for the two proposed algorithms. This WSN has six sensor nodes, which are numbered by $1, 2, 3, 4, 5, 6$, and one sink node. They are distributed randomly over a square area with the size of $100\,\text{m} \times 100\,\text{m}$. There are seven links indexed by $a - g$, through which the six sensor nodes are sending their data to the sink node. The sink node only receives data from the sensor nodes but does not perform sending data duty. We assume that, each sensor node can automatically find its route to transmit the packets to the sink, the first important task to us is thus to find the suitable sending rate for each node. Of course, the mechanism of finding a route for each sensor node is not trivial, but it is out of the scope of our present focus.

Let us assume that the utility function $U(\cdot)$ is a logarithmic function. We set the weight factor ω_s for each sensor node to be 1. It is rationale to assume the sink node has enough energy for all

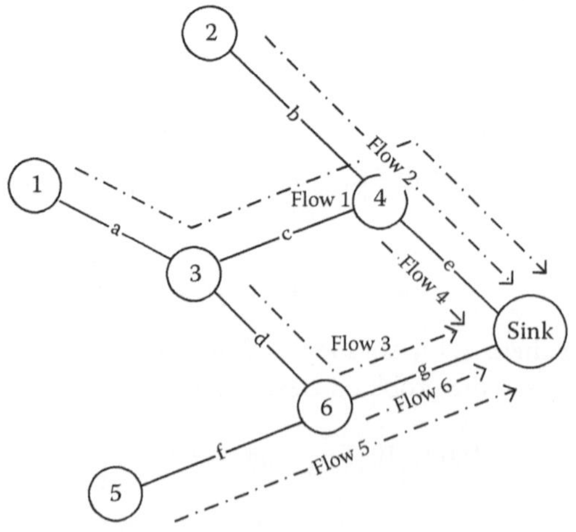

Figure 18.1 A WSN system for simulations.

communications. Therefore, our concerns are focused on the energy that is dissipated in these nodes for the purpose of numerical studies. We list all related parameters in the simulation as follows:

$$m_s = 0.1 \text{ Mbps}, M_s = 2.0 \text{ Mbps}, \varepsilon_s = 50 \text{ nJ/bit}, \varepsilon_r = 50 \text{ nJ/bit}, \rho = 50 \text{ nJ/bit}, n = 4,$$
$$\sigma = 0.0013 \text{ nJ/bit/m}^4, \delta = 1.3454 \times 10^8, \mathbf{e} = [30;28;25;22;26;20] \text{ (KJ)}, p_e = 0.01,$$
$$\mathbf{c} = [2; 3; 2.5; 3; 4; 2.5; 4] \text{ (Mbps)}.$$

18.6.1 NUMERICAL RESULTS ON CONVERGENCE AND THE TRADE-OFF OBTAINED BY ALGORITHM 18.1

18.6.1.1 Convergence

After the initial point $(\lambda(0), \mu(0))$ has been randomly chosen, one firstly obtains the convergence results of sources rates, normalized lifetime, and Lagrangian duality variables as displayed from Figures 18.2 through 18.6. After setting $\beta = 0.6$ and collecting the outputted node sending rate values after iterations, we plot them into Figures 18.2 and 18.3 with $\gamma = 0.02$ and $\gamma = 0.05$, respectively. At the beginning of the iterations, the node sending rates of all sensor nodes change sharply and dynamically, and afterward converge to the desired optimal solution very quickly. This is consistent with the stability of the algorithms, which is also we are expecting. From Figures 18.2 and 18.3, one also observes that, a larger step size γ will lead to a faster approaching process to the optimum value. But this produces larger oscillations as a side effect. Besides, we can find the rates of Nodes 1, 2, 3, and 5 are relatively small and the rates of Node 4 and Node 6 are relatively large. Note that Nodes 4 and 6 are the bottleneck ones as they need to relay data from other sensor nodes, so they need larger bandwidth.

The normalized network lifetime corresponding to the computed optimal lifetime value normalized with regard to the minimum node lifetime at each iteration is shown in Figure 18.4. After about 1800 iterations, the computed network lifetime is very close to the optimal solution to the linear program with high-level accuracy.

The corresponding duality congestion and energy price iterations are shown in Figures 18.5 and 18.6. All the duality node price updates except nodes 4 and 6 all go to zero very quickly. Since nodes

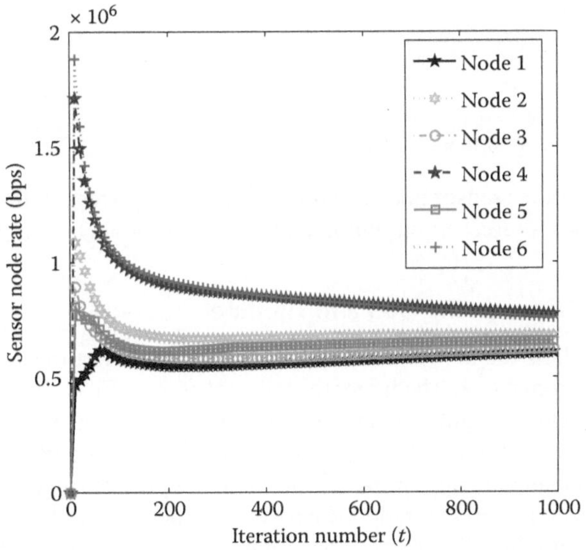

Figure 18.2 Sending rates of sensor nodes with step size $\gamma = 0.02$ outputted from Algorithm 18.3.1.

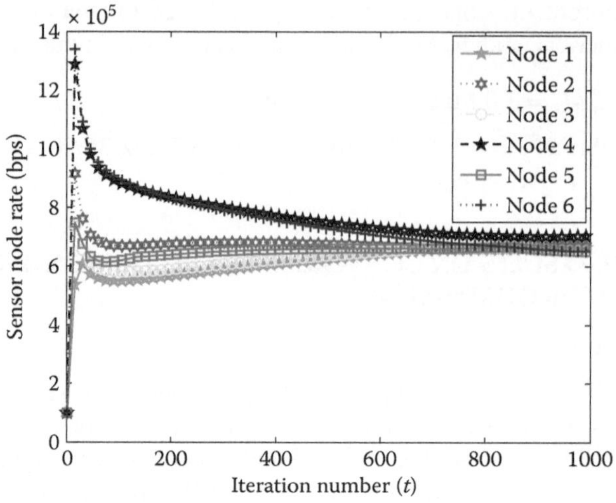

Figure 18.3 Sending rates of sensor nodes with step size $\gamma = 0.05$ outputted from Algorithm 18.3.1.

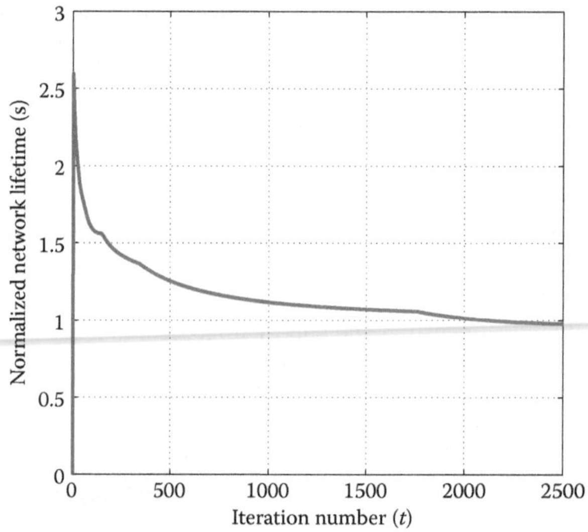

Figure 18.4 The evolution of normalized network lifetime.

4 and 6 relay traffic for other sensor nodes, and thus they have correspondingly a higher source rate representing bottleneck resources in our numerical model, as displayed in Figures 18.2 and 18.3.

18.6.1.2 Trade-Off between the QoS and Lifetime

Figures 18.7 and 18.8 show the network throughput and network lifetime curve as the system trade-off parameter β varies from 0 to 1. The larger the value of β is, more weight is put on the increase of network throughput and less weight on extending the network lifetime. Figures 18.7 and 18.8 clearly illustrate that the desired trade-off between the network throughput and the lifetime maximization can be achieved by choosing the appropriate β based on the performance requirement.

The impact of the factor β on the trade-off between the network utility and the network lifetime is outputted from the partially distributed Algorithm 18.3.1, and then is plotted into Figure 18.9. Note by network utility, we mean the aggregate total utility as a summation to the individual utility of each node in the whole network.

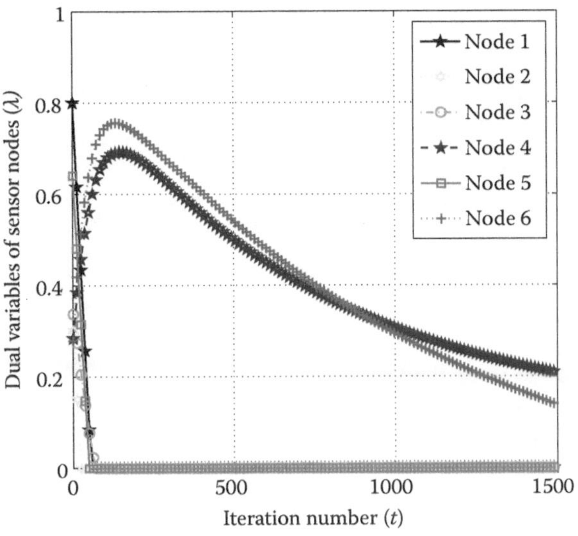

Figure 18.5 Congestion price, λ.

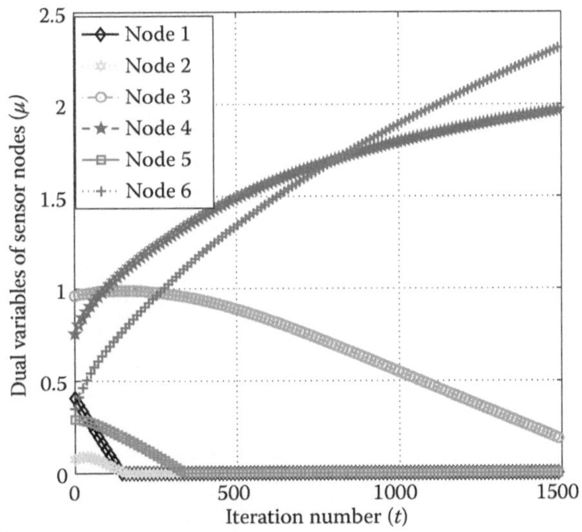

Figure 18.6 Energy price, μ.

In one hand, the model in the situation of $\beta = 1$ comes out to a pure NUM problem without network lifetime being taken into account. Therefore, theoretically in this model, all nodes are allowed to spend as much energy as they want. This is of course the trivial situation in the models presented herein.

On the other hand, if we let $\beta = 0$, then the original maximization problem will be in another trivial formulation to maximize the network lifetime alone. Being directed by Figure 18.9 and on the basis of the service (application) requirements, the system designer is guided to choose the suitable value of β to achieve the maximized network lifetime.

In summary, when we come to configuring a WSN, Figures 18.7 through 18.9 are very informative and helpful in that the trade-off choices between the QoS (performance) and the lifetime costs

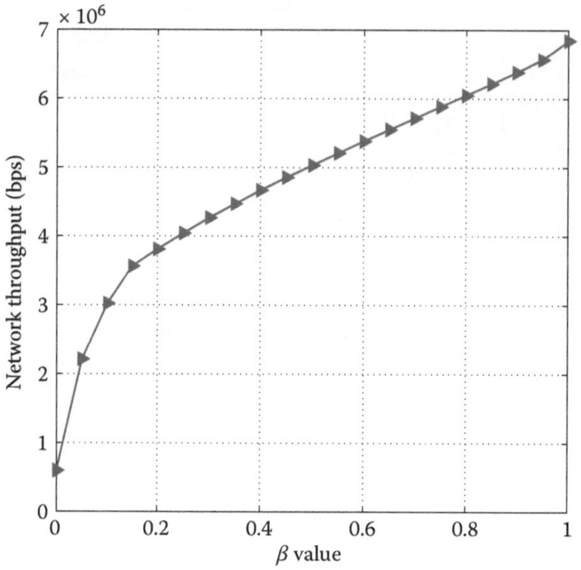

Figure 18.7 The convergence of network throughput with respect to β.

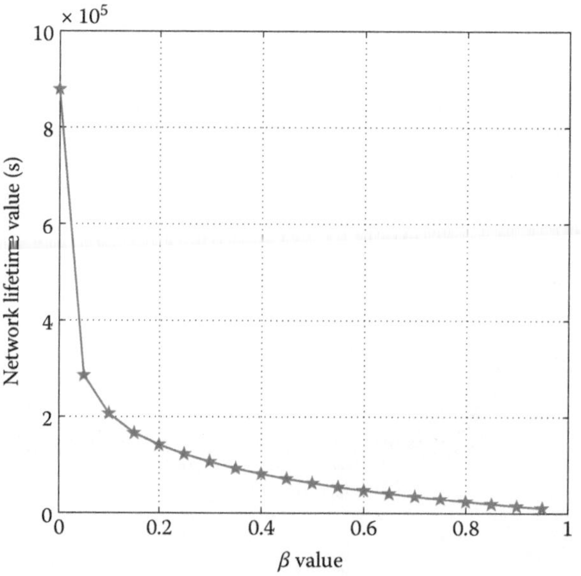

Figure 18.8 The convergence of network lifetime with respect to β.

in an energy limited WSN can be made in a systematic way. The key among these choices are the parameters β.

18.6.2 NUMERICAL CONVERGENCE RESULTS OF ALGORITHM 18.4.1

18.6.2.1 Convergence Property

Figure 18.10 shows the convergence results of the fully distributed algorithm, namely Algorithm 18.4.1, which are derived previously with step size $\gamma^{(t)} = \max\{1/t+1, 0.02\}$ at the tth instant. The

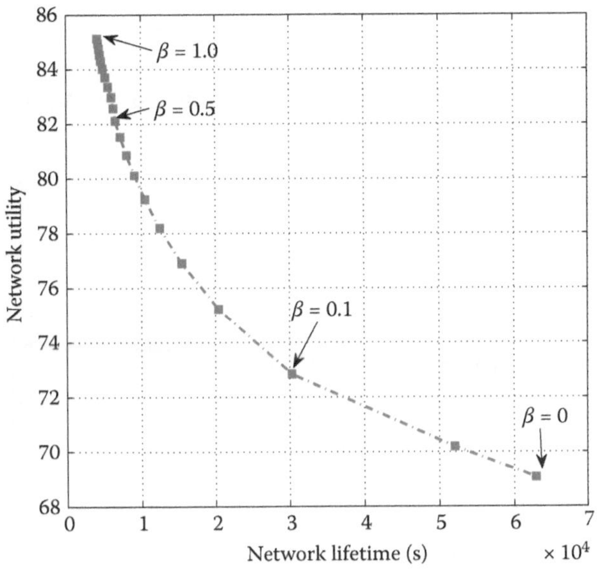

Figure 18.9 Network utility and lifetime with variations of β.

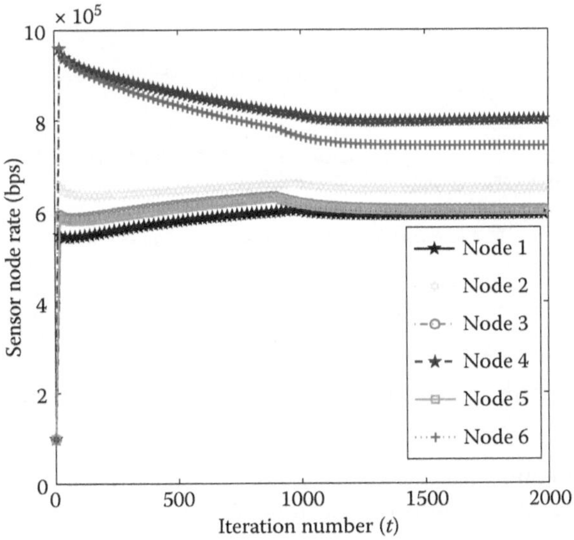

Figure 18.10 The convergence of node sending rates outputted from Algorithm 18.4.1.

convergence is slower for smaller values of step size. Besides, the completely distributed algorithm converges to the optimal solution after a substantially large number of iterations. As we can observe from Figure 18.10, the convergence of the fully distributed algorithm is much slower than that of the partially distributed algorithm. Thus, Algorithm 18.4.1 has a trade-off between the amount of communication during each iteration and the number of iterations with comparison to Algorithm 18.3.1.

The comparisons between the lifetime and node sending rate being computed from Algorithm 18.3.1 under different step size and Algorithm 18.4.1 at 1000th iteration are displayed in Table 18.2. From this table, the lifetime of sensor Node 6 has the lowest value, so the lifetime of Node 6

Table 18.2
Lifetime and Rate for Each Sensor Node Computed by Algorithm 18.3.1 and 18.4.1

Node	Algorithm 18.3.1, $\gamma = 0.02$		Algorithm 18.3.1, $\gamma = 0.05$		Algorithm 18.4.1, $\gamma^{(t)} = \max\{\frac{1}{t+1}, 0.02\}$	
	Rate ($\times 10^5$ bps)	Lifetime ($\times 10^4$ s)	Rate ($\times 10^5$ bps)	Lifetime ($\times 10^4$ s)	Rate ($\times 10^5$ bps)	Lifetime ($\times 10^4$ s)
1	6.1422	4.7545	6.5856	4.3906	6.0013	4.8181
2	6.6659	2.1412	6.6792	2.1261	6.5934	2.1537
3	6.3335	1.4297	6.5771	1.3426	6.1827	1.4525
4	7.8323	1.0531	7.1990	1.0654	8.1115	1.0531
5	6.4701	2.0172	6.5771	1.9669	6.1476	2.1044
6	7.5475	0.9251	6.5882	0.9562	7.6479	0.9452

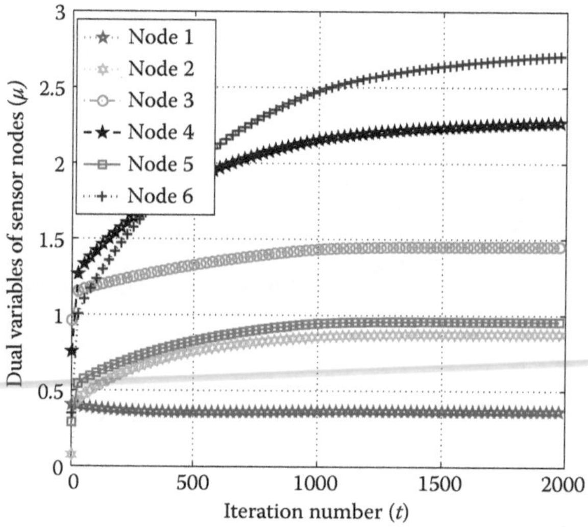

Figure 18.11 The convergence of node energy priceμ of sensor nodes in Algorithm 18.4.1.

denotes the network lifetime. The network lifetime computed from both Algorithms 18.3.1 and 18.4.1 approaches to the minimal sensor node lifetime very well.

Figure 18.11 plots the updated results on the duality sensor node energy price λ, which illustrates the convergence property of the iterations to the optimal values. Due to space limitations, from the figure, we omitted all the other two duality variables μ and $\mathbf{v}$ updates, which are all converging to their optimums within a few hundred iterations. From Figure 18.11, we observe that sensor nodes 4 and 6 have larger dual prices than other sensor nodes and thus represent bottleneck. In addition, corresponding to Figure 18.11, we see that the source rates allocated to sensor nodes 4 and 6 are higher than other sensor nodes as displayed by Figure 18.10.

18.6.2.2 Impact of the Trade-Off Parameter on the Network Properties

Figures 18.13 and 18.14 show the sensor node rate and lifetime as the weight β changes. Without surprise, the sensor node's sending rates increase with the parameter β increases. Subsequently, the lifetime of each sensor node decreases. When $\beta = 1$, the problem at hand reduces to a pure NUM

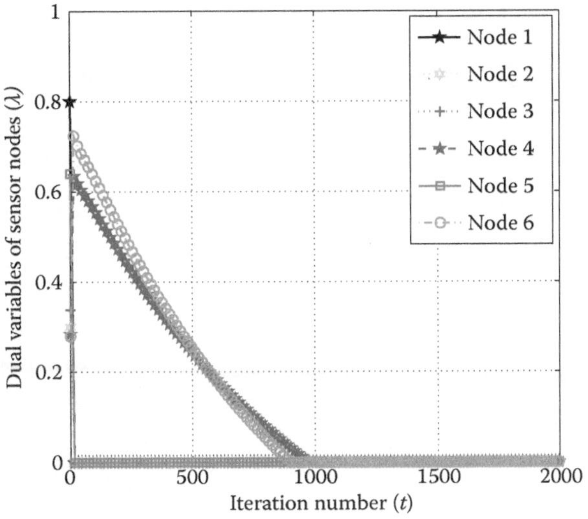

Figure 18.12 The convergence of node energy priceλ of sensor nodes in Algorithm 18.4.1.

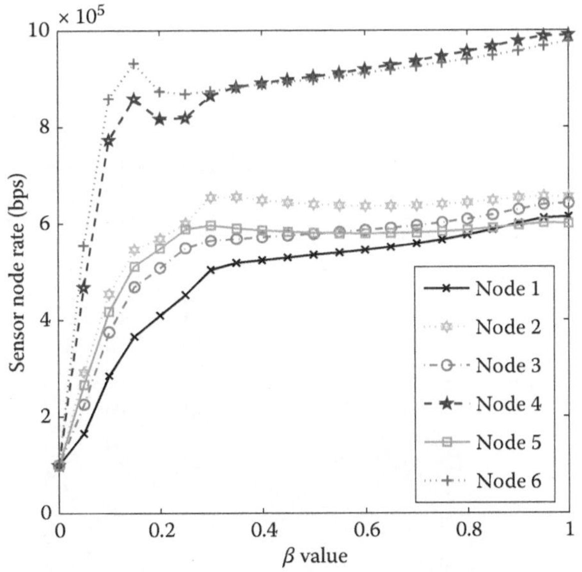

Figure 18.13 The node sending rates as a function of β in Algorithm 18.4.1.

problem by controlling the sensor node sending rates. In this case, each node is permitted to spend energy to its extreme, to maximize the node utility and then to maximize the total utility. This will shorten the node lifetime and thus the whole network lifetime as the consequence. Contrarily, the case of $\beta = 0$ gives us only a network lifetime maximization problem alone, under which each sensor node obtains a lower data rate in order to prolong the network lifetime. The network sending rate curve, where the parameter β is changing from 0 to 1, is plotted into Figure 18.15. For enough large β, the node sending rate will be maximized, while when the variable β is sufficiently small, the network lifetime can be maximized.

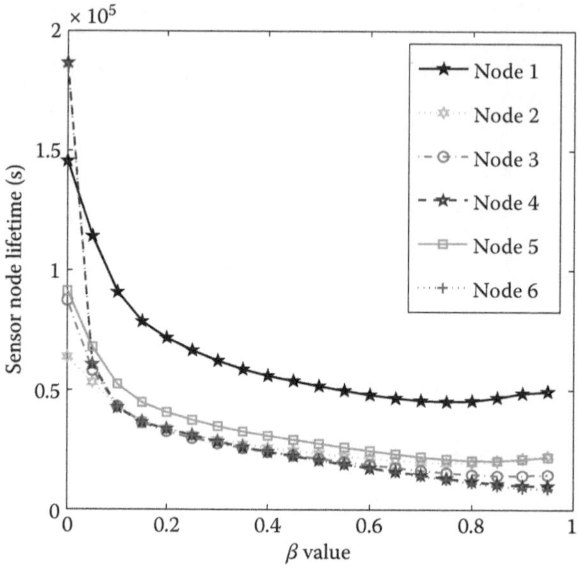

Figure 18.14 The node lifetime as a function of β in Algorithm 18.4.1.

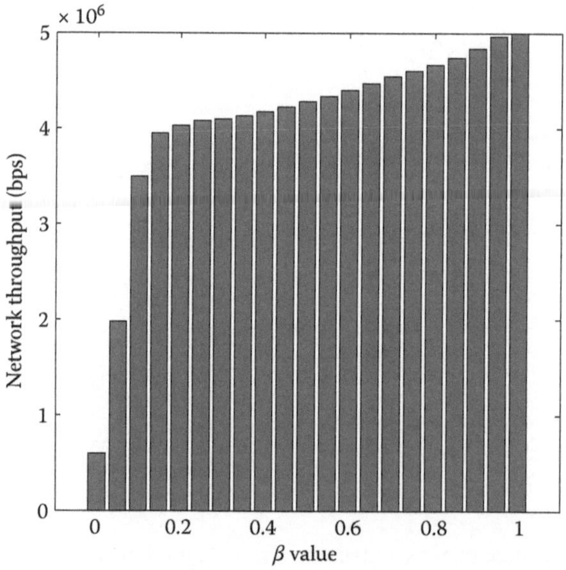

Figure 18.15 Network throughput with respect to β in Algorithm 18.4.1.

Figure 18.16 gives us the curve of utility and lifetime trade-off in a WSN, regarding the parameter β variations. One observes from this figure that in general, the network lifetime is inversely proportional to a scale of the trade-off parameter asymptotically. This suggests useful guideline on the selection of β when we configure a WSN. Further, one can always find a more suitable operating point on the curve, rather than in the below of the curve. Hence, operations on the trade-off curve are always the best choices.

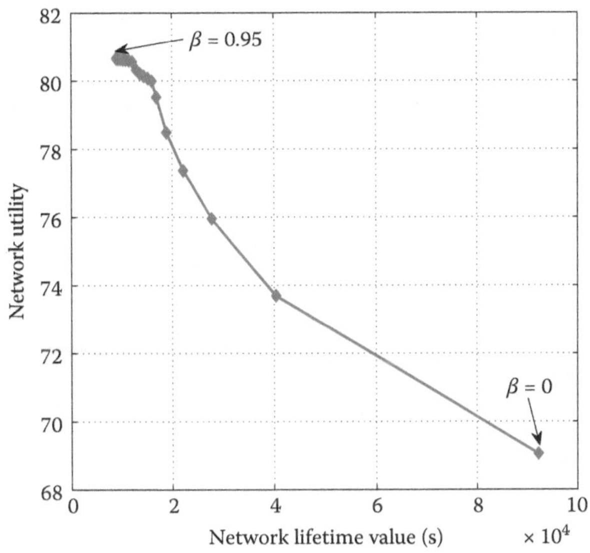

Figure 18.16 The curve of trading off between utility and lifetime with respect to β in Algorithm 18.4.1.

18.6.3 NETWORK PROPERTIES UNDER VARIOUS NONUNIFORM LINK ERROR PROBABILITIES

In the previous section, it is assumed that, every link has the same error probability 0.1. However, this may not cover all the situations in practice. Generally, the link error probability of a given link is determined by many factors, such as the power level at the receiver, the interference degree in the surrounding areas, and others. Therefore, the error probabilities of links are dynamically variable [566]. Subsequently, it is necessary to study how the link error probabilities affect the trade-off between node's sending rate and lifetime.

We firstly consider the example used in Ref. [566], where the link error probability p_e is given by $0.01 + 0.1 \times n$, where n is the total neighboring node number of the receiver. The lifetime and node sending rates under various error distributions are outlined in Tables 18.3 and 18.4, respectively. Observing from these two tables, we find that the rate allocation obtained in the situation, where the error probability of a given link is given by $0.01 + 0.1 \times n$, changes more sharply than the previous one with the constant link error probabilities. This implies that the effect of link error on performance is important to consider when we are aimed at less energy consumption and reliable communication in WSN information and communication systems.

The trade-off curve outputted from Algorithm 18.3.1, under the assumption of the error probability in all links taking the fixed value of 0.01, is also shown in Figure 18.17. Given another algorithm also has the similar trend, we just study Algorithm 18.3.1 in this instance. Observing from these two curves, we find that the nonuniform link error probability curve is below another, and also they have different trade-off trends if using the same value of β. Hence, our experimental results are consistent with [566]. Therefore, one should take link error probability into account when modeling the data transmission and energy consumption behavior. Otherwise, we are not able to obtain the actual optimal solution.

Next, we consider a particular situation, where except the link g all link error probabilities are the same value. For link g, we set its error probability as 0.9, while for other links, we keep the error probability to be 0.01 as before. Figures 18.18 and 18.19 plot the lifetime and the node sending rate under different error distributions, respectively. From these two figures, we observe that some changes in the link error rate probability tend to cause great changes to the rate allocation and

Table 18.3
Lifetime and Sending Rate of Sensor Nodes under Various Link Error Situations: The Outcomes of Algorithm 18.3.1

Node	No Link Error		Uniform Link Error $p_e = 0.01$		Nonuniform Link Error $p_e = 0.1 + 0.01 \times n$	
	Rate ($\times 10^5$ bps)	Lifetime ($\times 10^4$ s)	Rate ($\times 10^5$ bps)	Lifetime ($\times 10^4$ s)	Rate ($\times 10^5$ bps)	Lifetime ($\times 10^4$ s)
1	6.0834	4.7778	6.1422	4.7545	5.7844	4.0798
2	6.6295	2.1581	6.6659	2.1412	6.9448	1.5438
3	6.3022	1.4410	6.3335	1.4297	6.0057	1.1070
4	7.9945	1.0616	7.8323	1.0531	8.0091	0.8728
5	6.4102	2.0334	6.4701	2.0172	6.6978	1.4560
6	7.7046	0.9325	7.5475	0.9251	7.5433	0.7716

Table 18.4
Lifetime and Sending Rate of Sensor Nodes under Various Link Error Situations: The Outcomes of Algorithm 18.4.1

Node	No Link Error		Uniform Link Error $p_e = 0.01$		Nonuniform Link Error $p_e = 0.1 + 0.01 \times n$	
	Rate ($\times 10^5$ bps)	Lifetime ($\times 10^4$ s)	Rate ($\times 10^5$ bps)	Lifetime ($\times 10^4$ s)	Rate ($\times 10^5$ bps)	Lifetime ($\times 10^4$ s)
1	5.9638	4.8736	6.0013	4.8181	5.2700	1.1780
2	6.5465	2.1055	6.5934	2.1537	6.0807	1.7631
3	6.3566	1.4504	6.1827	1.4525	5.6877	1.1946
4	8.2741	1.0577	8.1115	1.0531	8.9647	0.8964
5	6.2460	2.0869	6.1476	2.1044	5.5423	1.7595
6	7.8215	0.9322	7.6479	0.9452	8.1843	0.8088

lifetime in the whole network. Therefore, we draw the conclusion that it is important to account for the link error probability factor when modeling the energy consumption behavior of WSNs.

18.7 CONCLUDING REMARKS

For WSN information and communication systems with the MAC constraints, we have formulated the trade-off between the two conflicting metrics, namely, the utility and the lifetime, into an optimization model with certain constraint. We then apply the Lagrangian dual decomposition approach to obtain the desired solution. To ensure the end-to-end communication reliability, we consider the intrinsic energy efficiency. Instead of the traditional end-to-end retransmission scheme, we use a hop-by-hop schedule to yield the reliable transmissions. Subsequently, when modeling the trade-off between the whole lifetime and the whole utility maximization, we obtain two subproblems: one for rate allocation and the other for energy conservation issue. By studying these two subproblems, we are able to propose a workable approach to address the trade-off for designing in a cross-layer manner. The crossed layers in the design include the MAC layer for rate adjusting in terms of the

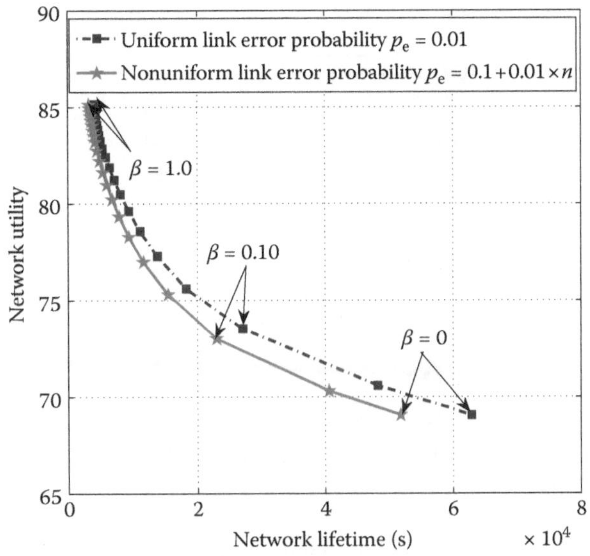

Figure 18.17 The impact of trade-off parameter β on the trade-off between network lifetime and utility under different link error probabilities.

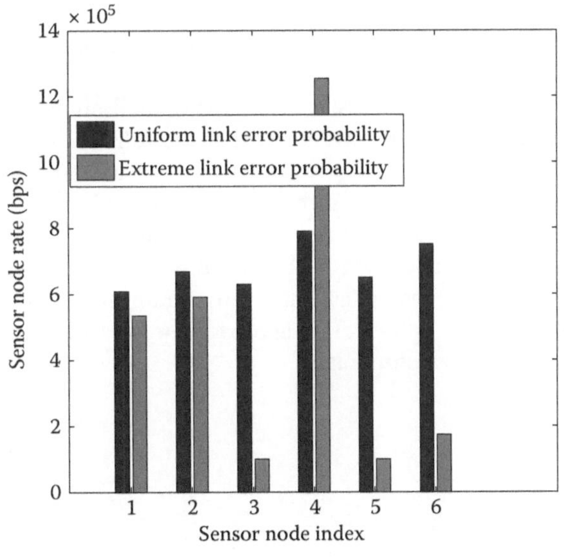

Figure 18.18 The impact of link error probabilities on the rate allocation of sensor nodes outputted from Algorithm 18.3.1.

MAC constraint, the link layer for accounting for the link error issue, and the network layer for achieving the maximal lifetime.

We propose a number of key results to achieve the best trade-off, which are theoretically analyzed and then evaluated through simulations. Both theoretical findings and simulation studies verified our theoretical model and algorithms. With striking contrast to other existing expositions, where the network optimization models only account for either network lifetime or network utility without transmission reliability being incorporated, our approach (models and algorithms) presents a unified framework to incorporate these three key elements in the network optimization model and the

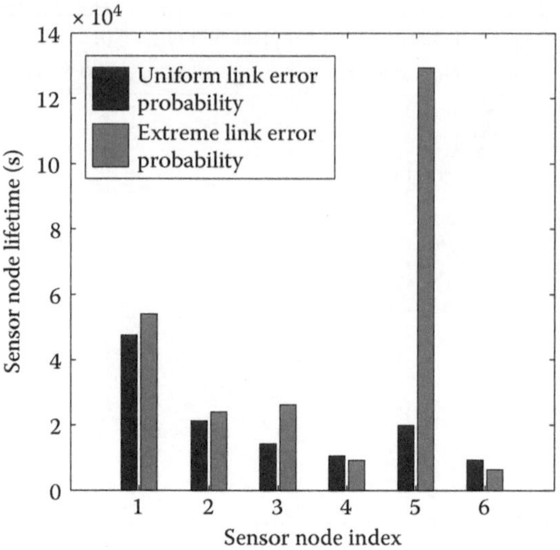

Figure 18.19 Lifetime of sensor node under the uniform and extreme link error probabilities by Algorithm 18.3.1.

resulted solving procedures. In particular, we are able to achieve the NUM with the trade-off of network lifetime while with transmission reliability being guaranteed. This brings our approach to be promising for implementation in a cross-layer manner in real WSN information and communication systems.

The communication overhead is one issue in the distributed algorithm to achieve the desired trade-off. Future work would be worthwhile to look at the refinement technique in reducing the communication overhead. One also needs to consider those issues hurting the performance of the algorithms in practice, and achieves a balance between optimality and efficiency. For instance, dynamically routing needs to be studied to see how it can be incorporated into the obtained utility and performance optimization approach in the future work. The theoretical assumption is also needed to be lifted, that sensor nodes update their information in a synchronized manner. Future work is thus called for to study the optimization approach for the situation, where sensor nodes and the sink node are working in an asynchronous way.

19 Resource Allocation among Real-Time Multimedia Users in Wireless Networks: Approximate NUM Model and Solution

Resource allocation among multimedia users to meet the various quality of service (QoS) requirements in a wireless network is still a challenge. So far, satisfactory results have only been developed along the line of network utility maximization (NUM) approach for the resource allocation among elastic traffic in a wireless network, which is eased by the concave utility function of such traffic. However, adaptive multimedia real-time users have nonconcave utility functions, which means the traditional convex optimization methods are not suitable. Facing the above challenge, we describe the attacked resource allocation problem into a NUM model, in which the objective function is formulated by the nonconcave utility functions. Rather than to solve this nonconcave optimization problem, we propose an approximate concave model for the original model, which is easier to solve. We proposed a gradient-type greedy iteration method to obtain the solution for the approximate model. Then, we use the obtained solution to the approximate model as our starting point of iteration to solve the original problem (OP). We also provide a flexible admission control protocol, based on the solution we obtained, from the optimal algorithm that can adapt to systems with different requirement. The performance of the proposed method is further evaluated via simulation results, which demonstrate that our algorithm can be adapted to all kinds of inelastic flows and the solution can be used as a guideline for admission control protocol design in practice.

19.1 INTRODUCTION

19.1.1 MOTIVATION

In this chapter, we attack the problem of resource allocation in wireless networks with competing multimedia users, with the complex issue coming from the fact that the resource (bandwidth, timeslot, or frequency) of the networks is very scarce and limited while the multimedia users have various QoS requirements to occupy the resource in operation in a competing manner. The resource allocated to users may be guided by many factors such as user's satisfaction and fairness. In this chapter, we focus on the user's satisfaction as our guideline for bandwidth allocation. The wireless multimedia network has developed rapidly in recent years.

There are many different kinds of multimedia services (e.g., data, audio, and video), each of them has a different bandwidth requirement. Providing QoS to different users or applications according to their bandwidth requirement is an important task for multimedia network. Unlike the wired network, bandwidth is a major bottleneck for wireless network, and it is much more easy to be affected by many factors such as channel quality and interference by other devices. So bandwidth allocation for wireless network has become much more challenge than in wired network.

For nowadays multimedia users or applications, there are mainly three kinds of traffic, including:

1. Non-real-time traffic for file transmission or e-mail, which can adapt their bandwidth to various network load since they can tolerate certain delays,
2. Adaptive real-time traffic for online video watching, which must acquire at least a certain amount of bandwidth to guarantee the basic requirement and more bandwidth after the full satisfaction level reached will become useless,
3. Hard real-time traffic for real-time video conference, which has a hard requirement for a certain amount of bandwidth. It is useless when the bandwidth allocated to the users (applications) is less than their preferable amount of bandwidth. Also, there is no need to allocate more bandwidth to them as long as their basic requirement is acquired.

Usually, one may use a utility function to reflect the degree of user's satisfaction level relating to the bandwidth they acquired. A utility function $U(r)$ is a nondecreasing function with respect to the amount of allocated resource r. The user's satisfaction will become higher because of the more resource it has been allocated. The basic model for our problem is shown in Equation 19.1, and our goal is to maximize the total utility function, which means to achieve the highest total satisfaction, which can be seen in Figure 19.1.

$$\begin{aligned} \max \quad & \sum_{i=1}^{n} U_i(r_i) \\ \text{s.t.} \quad & \sum_{i=1}^{n} r_i \leq C \\ & r_i \geq 0 \end{aligned} \tag{19.1}$$

The expression of a utility function is varied, and it depends on the type of the traffic and user's specific requirement of the bandwidth. The adaptive real-time traffic refers to the traffic that requires a preferable amount of bandwidth. In order to avoid congestion, it can tolerate a certain flexibility of bandwidth in a certain range. The bandwidth requirement of the adaptive real-time traffic is often described by the sigmoid function with respect to the bandwidth resource, which is shown in Figure 19.2. It can be interpreted as follows. When the bandwidth allocated to the user is unacceptable

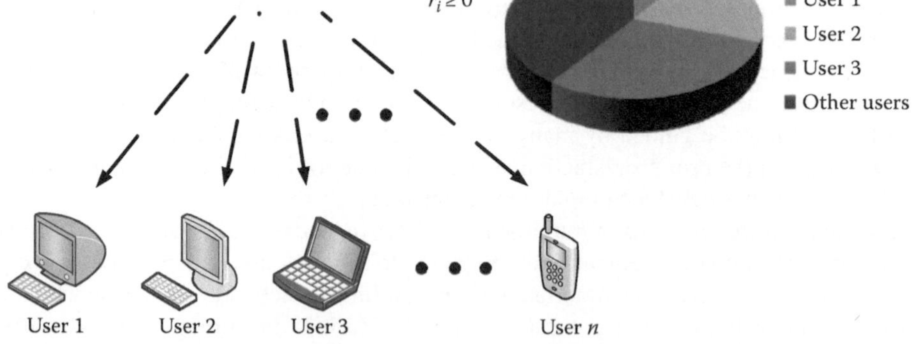

Figure 19.1 The basic NUM problem for resource allocation in wireless networks.

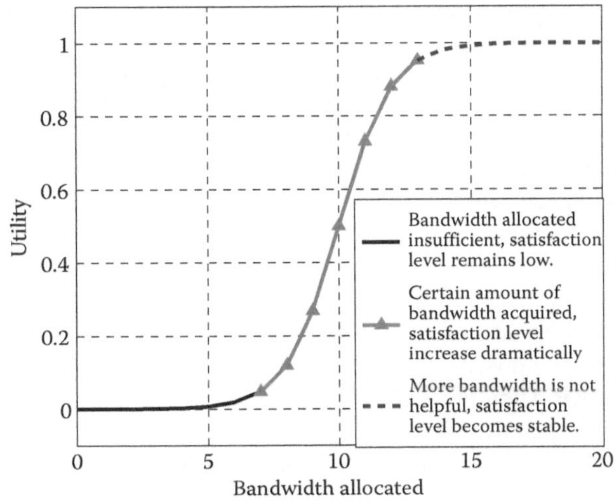

Figure 19.2 Sigmoid function for adaptive real-time traffic bandwidth requirements.

insufficient, it is useless for the real-time traffic so the user's satisfaction level remains very low. As the bandwidth allocated gradually increase to a certain level, the user's satisfaction level will increase dramatically. If the allocated bandwidth has exceeded the preferable amount of the user, which means more bandwidth is not helpful, then the satisfaction level will keep stable.

In order to guarantee the QoS in future wireless networks, it is desirable to implement certain resource allocation mechanism to allocate the limited resource (e.g., bandwidth) among all the users fairly and efficiently. Utility functions have been widely applied by many adaptive bandwidth allocation schemes for QoS provisioning in wireless networks. Kelly et al. [104,105] have presented the classic utility maximization model, which is used as the basic model in many later research. In [567], the authors propose an adaptive bandwidth reservation scheme to provide QoS guarantees for multimedia traffic in wireless networks. Hande et al. [568] have proposed an optimization framework and optimality conditions for the adaptive real-time traffic flows. Nasser et al. [569] described an adaptive bandwidth allocation framework, which can adjust the bandwidth of ongoing calls during their stay in the cell whenever there are resource fluctuations in wireless networks.

In [570], a bandwidth adaptation scheme is developed for wireless networks to guarantee the upper bound of the call degradation probability. In [617,626], Kuo and Liao proposed a resource allocation algorithm, which achieves utility maximization for both hard real-time traffic and non-real-time traffic, and then extends the discussion to the adaptive real-time traffic to give a suboptimal solution to the utility maximization problem. Babayev and Bell [571] have proposed a gradient iteration method to solve the similar nonconcave optimization problem. Lu [572] has proposed a discrete iteration method to solve the utility maximization problem. In [573], Ferragut and Paganini proposed a network resource allocation scheme for multiple connections considering both fairness and stability.

Although numerous research have been done in this field, bandwidth allocation to adaptive real-time traffic is still a challenge. Since unlike the other types of traffic, the utility function of adaptive real-time traffic is a part convex and part concave function, which means the traditional optimization algorithm is not suitable to solve it. The traditional greedy algorithm, which only considers, the current most increment step, is also not suitable, because the increase rate of the sigmoid function is "slow–fast–slow" and the slope of the "fast" part can be different due to different types of traffic. So the flow slow increasing rate at the current moment and may dramatically increase when a certain amount of bandwidth is acquired, which means allocating the resource to the current most increment flow may not achieve the total utility maximization. In summary, a new method is needed to solve the problem.

19.1.2 MAIN CONTRIBUTIONS AND NOVELTY

In order to solve the problem, first we present an approximate utility function, which is concave throughout its domain and similar to the original function. Then a gradient type greedy iteration method has been proposed to solve the approximate problem (AP), which is different from the traditional greedy algorithm. Then, unlike the traditional greedy algorithm, which will only take the most advantages step in the current moment.

We creatively define the crossing point for each flows, so that the later increasing rate can be foreseen even when the current increasing rate for that flow is very low. The algorithm can be seen as a farsight greedy method, which means both the most advantage step in the current moment and the later increase rate will be taken into consideration. Combining these two main factors, a most potential traffic flows will be chosen out to allocate the bandwidth each time and finally, we obtained our optimal solution to AP. Then the optimal solution of AP will be taken as the starting point of our final iteration to solve the OP.

Besides, based on the solution we obtained, we also present a fast suboptimal admission control protocol for the system to decide whether a new coming flow can access in a quick time. It is a flexible protocol that can make a trade-off between time and resource consuming with the total utility the system achieved. Finally, we test our algorithm, and the result shows that our algorithm can adapt to different sigmoid functions or even mixed with multiple different kinds of sigmoid functions and the fast suboptimal admission control protocol also show great effect when dealing with new coming flows.

The rest of the chapter is organized as follows: In Section 19.2, theoretical analysis and an algorithm for solving the utility maximization problem are proposed. In Section 19.3, the admission control protocol is presented based on the solution of the utility maximization problem. We verify our results via simulations in Section 19.4. Finally, the conclusion is given in Section 19.5.

19.2 THE BANDWIDTH ALLOCATION ALGORITHM

19.2.1 SYSTEM MODEL AND PROBLEM DESCRIPTION

Under the circumstance of multimedia wireless network, different users have different bandwidth requirements. To provide suitable QoS support for users based on their bandwidth requirement in a bandwidth-limited environment, a basic utility model (function) for traffic flows is needed to reflect the satisfaction level of the end users. A utility function is defined as a curve mapping the amount of the bandwidth received by the application to the performance as perceived by the end users. A utility function is monotonically nondecreasing, which means the more bandwidth allocated to the users will not lead to a degrade of the application performance. Adaptive real-time traffic refers to the applications that have flexible bandwidth requirements. Typical examples for adaptive real-time traffic are multimedia service and video on demand. The utility function for adaptive real-time traffic is assumed to be a sigmoid function.

There are several different kinds of sigmoid functions, which are shown below:

$$U(r) = \frac{1}{1+e^{(-ar+b)}} \tag{19.2}$$

$$U(r) = \begin{cases} qe^{p(r-r_p)} & r < r_p \\ 1-(1-q)e^{-p(r-r_p)} & r \geq r_p \end{cases} \tag{19.3}$$

$$U(r) = 1 - e^{-\frac{kr^2}{m+r}} \tag{19.4}$$

The above three sigmoid functions are shown in Figure 19.3. Each of the sigmoid function has its own properties.

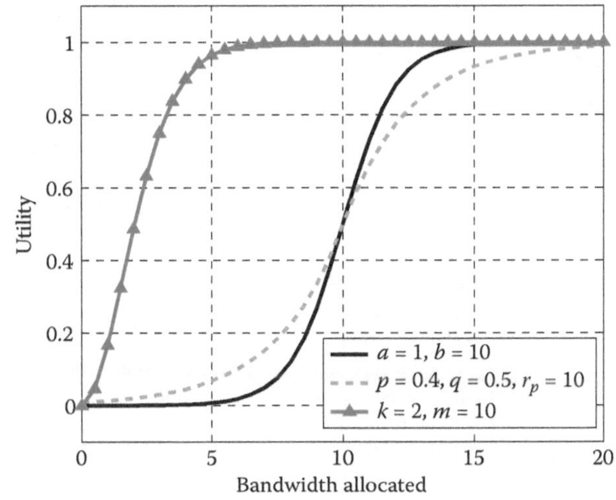

Figure 19.3 Sigmoid functions for modeling the users' utility.

For Equation 19.2, the slope is determined by the parameter a and the preferable bandwidth is b/a and the utility when acquiring the preferable amount of bandwidth is always 0.5 and it is a central symmetric function at point $(b/a, U(b/a))$.

Equation 19.3 is a piecewise function, where r is the bandwidth allocated to the users, which may also refer to time slots or radio frequency occupied by this user. The component r_p denotes the preferable amount of resource for the adaptive real-time traffic. Component q is the channel quality parameter, which represents the ratio of actual amount of resource received by the user to the amount of resource allocated by the base station, which ranges from [0, 1]. The parameter p determines the slope of the utility function, and it characterizes the flexibility level to bandwidth requirement of the flow. When p increases, the shape of the sigmoid function becomes closer to a unit-step function, which is very similar to a hard real-time traffic.

For Equation 19.4, k and m are two positive parameters, which determine the shape of the utility function, and ensure that when the maximum bandwidth is acquired, the achieved utility is approximately equal to 1. More details can be seen in [572]. We presented a model-free iteration method to solve the OP, which means our algorithm can adapt to any sigmoid-type functions. But in order to make it clear to readers and without losing generality, we will use Equation 19.3 as our basic system model.

19.2.2 APPROXIMATE MODEL AND SOLUTION

We list all the notations for variables used throughout this chapter into Table 19.1.

Normally, one may attempt to use the primal-dual method to solve the problem under the assumption that the utility functions are concave. If not, it is either impossible to define a dual function or the marginal value of the dual function is not equal to the primal function, which means a duality gap is present. Unfortunately, under the circumstance of adaptive real-time traffic, the utility functions always present like a sigmoid shape and it is not concave through its domain, which means the traditional convex optimization method is not suitable for our problem. In our algorithm, we will firstly create an approximate model from the OP, which is concave through its domain. So it is easy to obtain its optimal solution through various mathematical methods. Then, the solution of the approximate model is used as a starting point for the further iteration.

Table 19.1
Nomenclature

Symbols	Explanation
C	The total amount of the bandwidth resource available to allocate
r_i	The amount of resource allocated to flow i
q_i	The channel quality of flow i
$U(*)$	The sigmoid utility function
$u(*)$	The derivation of $U(*)$
r_p	The preferable amount of resource for the elastic traffic
r_i^p	The preferable amount of resource for flow i
$AU(*)$	The approximate sigmoid utility function
$au(*)$	The derivation of the approximate sigmoid utility function
CP^i	The crossing point set of flow i
cp_B^A	Flow B crossing flow A at $x = cp_B^A$

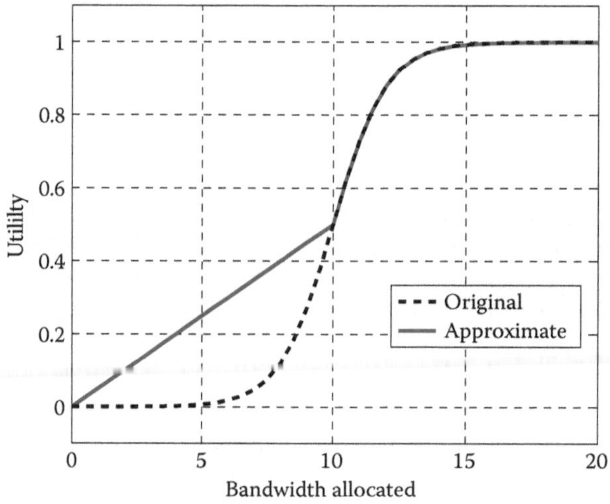

Figure 19.4 Approximate utility function.

Since $U(r)$ is convex when $r < r_p$ and concave when $r \geq r_p$, we can create the approximate function by using a line to connect the two ends of the convex part, which is shown in Figure 19.4, and our approximate model is

$$AU(r) = \begin{cases} \frac{U(r_p)}{r_p} r = \frac{q}{r_p} r & r < r_p \\ U(r) = 1 - (1-q)e^{-p(r-r_p)} & r \geq r_p. \end{cases} \tag{19.5}$$

Since $AU(r)$ is concave through its domain, it is easy to obtain the optimal solution by primal-dual method, that is

$$\begin{aligned} \min \sum_{i=1}^{n} -AU_i(r_i) \\ \text{s.t.} \sum_{i=1}^{n} r_i = C \\ r_i \geq 0. \end{aligned}$$

The Lagrangian is

$$L(r,\lambda,v) = \sum_{i=1}^{n} -AU_i(r_i) + \sum_{i=1}^{n} \lambda_i(-r_i) + v(\sum_{i=i}^{n} r_i - C)$$
$$= -vC + \sum_{i=1}^{n} [-AU_i(r_i) + vr_i - \lambda_i r_i].$$

The dual function is

$$g(\lambda,v) = -vC + \inf_{x}[\sum_{i=1}^{n} -AU_i(r_i) - \lambda_i r_i + vr_i]$$
$$= -vC + \sum_{i=1}^{n} \inf_{x}[-AU_i(r_i) - \lambda_i r_i + vr_i].$$

So, the dual problem is

$$\begin{aligned} &\max\ g(\lambda,v) \\ &\text{s.t.}\ \ \lambda_i \geq 0 \\ &\qquad v \in R. \end{aligned}$$

We present a farsight greedy iteration method to solve the upon optimization problem. Since the utility functions for different users can be obtained beforehand, we first sort the flows by $\frac{U_i(r_i^p)}{r_i^p}$ (the average increment through 0 to r_i^p). Then, we have

$$\frac{U_1(r_1^p)}{r_1^p} \geq \frac{U_2(r_2^p)}{r_2^p} \geq \cdots \geq \frac{U_n(r_n^p)}{r_n^p}.$$

If

$$\frac{U_i(r_i^p)}{r_i^p} = \frac{U_j(r_j^p)}{r_j^p},$$

we choose the flow with a higher increase rate at r_p in front, that is

$$u_i(r_i^p) > u_j(r_j^p) \rightarrow flow\ i\ in\ front\ of\ flow\ j,$$

where $u_i(*)$ is the derivation of $U_i(*)$ and all the positions described (first, second, last, front, before, behind, adjacent, etc.) are refer to this order. The main idea of this algorithm is to choose a small parameter Δr, so we add this Δr to the flow, which has the most unit increment

$$k = \arg\max_{i} \left\{ \frac{AU_i(r_i + \Delta r) - AU_i(r_i)}{\Delta r} \right\}$$

each time and decrease the total resource $C = C - \Delta r$. Since the smaller Δr we choose the more accurate result we will get, the so-called "optimal solution" in the later text is revelant to this parameter. Before we present the rest of the algorithm, we give the definition of crossing point and equilibrium point.

Definition 19.2.1 ***(Crossing Point):*** *For Flows A and B (A is in front of B), if they satisfies*

$$AU_A(r) = AU_B(r) \tag{19.6}$$

and

$$au_A(r) < au_B(r), \tag{19.7}$$

then we say Flow A has a crossing point at $x = r$ by Flow B, denoted as cp_B^A.

The crossing point can reflect the later increasing rate of the flow, which means the flow having a low increasing rate at the current moment may have a dramatic increase when acquiring a certain amount of bandwidth. For example, in Figure 19.5, there are 8-unit total bandwidth to allocate to Flow A and B. The starting point is $(0,0)$.

Traditional greedy algorithm will choose the flow, which has the most increasing rate at the current moment, which is Flow A. So the dramatic increase in Flow B cannot be foreseen since the current resource allocated to Flow B is 0 and the result will be (8, 0). But if we take the crossing point into consideration, we will know the fact that allocating 8-unit bandwidth to Flow B can achieve a higher utility even though the starting increasing rate of Flow B is lower than that of Flow A. So the optimal result will be $(0,8)$. This example shows that focusing on the current most increment flow cannot obtain the global optimal solution. A global greedy algorithm can be designed only if we take the crossing point into consideration.

Since we use Equation 19.3 as our system model, we can solve Equations 19.6 and 19.7 to find the crossing points between two flows, that is:

1. Flow B's concave part crosses Flow A's segment part:

$$\begin{cases} \frac{q_A}{r_A^p} r = 1-(1-q_B)e^{-p_B(r-r_B^p)} \\ \frac{q}{r_A^p} < (1-q_B)p_B e^{-p_B(r-r_B^p)} \end{cases} \tag{19.8}$$

2. Flow B's concave part crosses Flow A's concave part:

$$\begin{cases} 1-(1-q_A)e^{-p_A(r-r_A^p)} = 1-(1-q_B)e^{-p_B(r-r_B^p)} \\ (1-q_A)p_A e^{-p_A(r-r_A^p)} < (1-q_B)p_B e^{-p_B(r-r_B^p)} \end{cases} \tag{19.9}$$

Since Equation 19.8 is not easy to solve directly, we can use an iteration method to find the point that satisfies Equation 19.7. For Equation 19.9, since the parameter p reflects the slope of the function, if Flow B crosses Flow A, it must satisfy $p_A < p_B$. So the solution is

$$r = \frac{\ln\left(\frac{1-q_A}{1-q_B}\right) + (p_A r_A^p - p_B r_B^p)}{p_A - p_B},$$

$$r < \frac{\ln\left(\frac{(1-q_A)p_A}{(1-q_B)p_B}\right) + (p_A r_A^p - p_B r_B^p)}{p_A - p_B}.$$

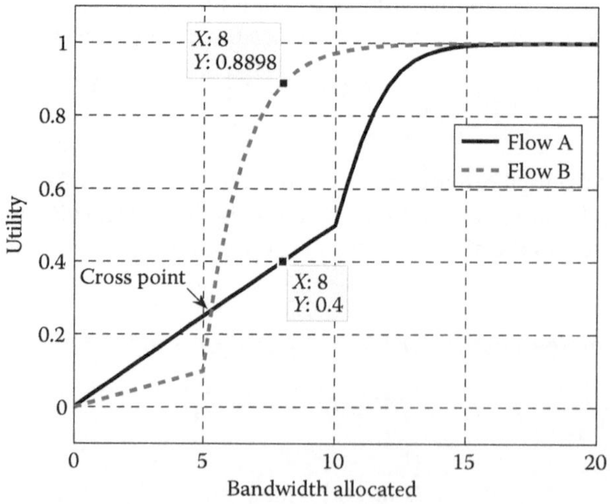

Figure 19.5 Definition of crossing point.

For any sigmoid function, which is difficult to solve Equations 19.6 and 19.7 directly, we can use the iteration methods

$$AU_A(r) > AU_B(r)$$

and

$$AU_A(r+\Delta r) < AU_B(r+\Delta r)$$

to find all the crossing points between them.

Now we present some properties for the crossing points between two flows and by using these properties, we can create a high-efficiency method to find all the crossing points avoiding to compare every each two flows. For two adjacent flows i and j (i is in front of j), we have:

Property 1: If Flow j has crossing point on Flow i, then Flow j may have crossing point on Flow $i-1$.

Property 2: If Flow j has no crossing point on Flow i:

1. If Flow i has no crossing point on Flow $i-1$, then Flow j has no crossing point on any Flow before i.
2. If Flow i has crossing point on Flow $i-1$, then Flow j may have crossing point on Flow $i-1$.

Property 3: Flow j has crossing point on Flow i only when Flow j's concave part cross Flow i.

Following these properties, we can create the flowchart to find all the crossing points, which is shown in Figure 19.6. As we mentioned before, the crossing point can reflect the later increasing rate for those flows, which have not allocated a certain amount of bandwidth. So during the iteration, when one flow (Flow k) reaches a crossing point and $r_k > r_{cross}$ (the bandwidth allocated to the flow which crosses Flow k), it means there is another flow that can reach the same utility when allocating the same bandwidth, but has a higher increasing rate in the further iteration and that means the

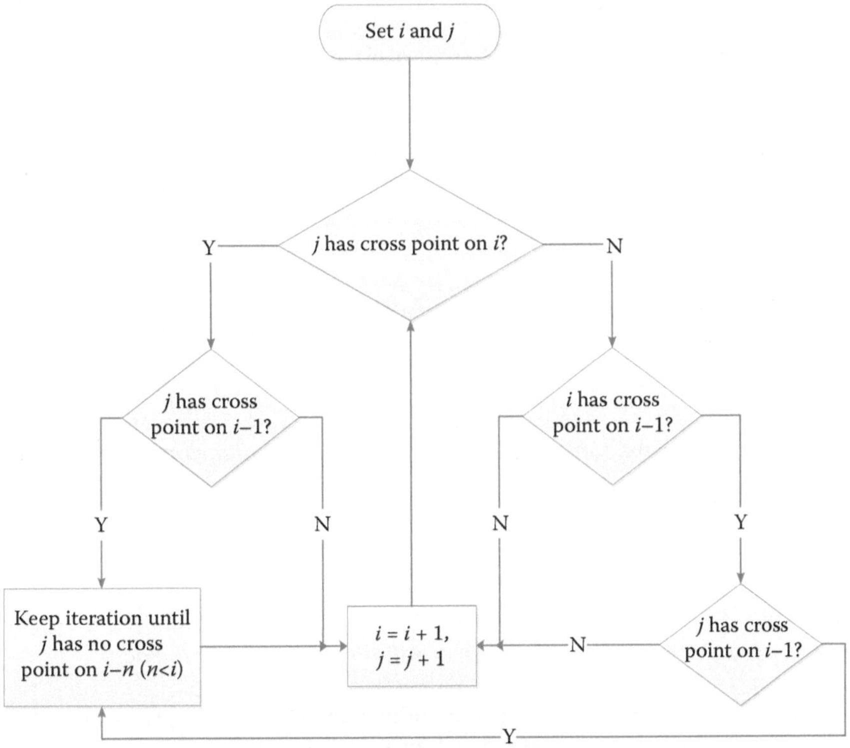

Figure 19.6 Flowchart for finding all the crossing points.

current flow will have to switch the bandwidth with the flow that has crossed it in order to acquire a higher increasing rate. (Noting that there may still have a flow that cross the flow, which crosses Flow k, so this switch will not stop until no crossing point exists.) Based on the above statement, we present Algorithm 19.2.2 to find the solution for the AP model.

Algorithm 19.2.2 Find the solution for the AP model

Input: $r_i = 0 \ \forall i$

Output: AR^*(The solution for the AP model)

```
Sort the flows by U_i(r_i^P)/r_i^P in decreasing order
for all flows do
    Find all the crossing points on them
end for
while C ≥ Δr do
    k = arg max_i{(AU_i(r_i+Δr) − AU_i(r_i))/Δr} r_k = r_k + Δr, C = C − Δr
    if r_k reaches the crossing point AND r_k > r_cross then
        Switch the bandwidth to the cross flow until no crossing point exists.
    end if
end while
Return AR*
```

Proposition 19.2.1 *For a solution obtained from the above algorithm, there is at most one flow, we say the kth flow that the resource allocated is in the interval of* $(0, r_k^P)$.

Proof: As we have sorted the flows by $U_i(r_i^P)/r_i^P$, so not until the flow before it (we say $k-1$) has allocated the resource larger than r_{k-1}^P that Flow k can become the most increment flow. (Since the line segment part's increasing rate is fixed, so the current flow will continue to be the most increment flow until it acquires the preferable amount of bandwidth.) When allocating to Flow k, if there exists any crossing point, the bandwidth switch to that flow is also larger than the preferable amount, because the crossing point only appears in the concave part (Property 3 for crossing point). If the remaining resource is larger than r_k^P, then r_k will allocated to r_k^P and the iteration continues, whereas if the remaining resource is less than r_k^P, then all the remaining resource is allocated to Flow k and the iteration terminates.

Definition 19.2.2 ***(Equilibrium point):*** *For a solution* $R = \{r_1, r_2, \ldots, r_n\}$, *if for any* $i,\ j \in n$ *and* $r_i \neq 0,\ r_j \neq 0$, *it satisfies*

$$u_i(r_i) = u_j(r_j)$$

or for any two flows i and j, sharing Δr *from i to j will only lead a decrease of total utility. We say R has reached the equilibrium point. Since the accuracy of our algorithm is related to* Δr, *we can rewrite it by:*

$$\left|u_i(r_i) - u_j(r_j)\right| < \mathcal{O}(\Delta r),$$

where $\mathcal{O}(\Delta r)$ *is the high-order infinitesimal of* Δr.

Proposition 19.2.2 *For all r, if either* $r_i = 0$ *or* $r_i \geq r_i^P$ *and equilibrium point reached, then the solution to the approximate model is the optimal solution to the original model.*

Proof: For $AR = \{r_1, r_2, \ldots, r_k, 0, \ldots, 0\}$ is the optimal solution to the approximate model, since it reaches the equilibrium point and for all i, $r_i = 0$ or $r_i \geq r_i^P$ holds. Then, we have

$$AU(AR) = U(AR).$$

Since *AU(*)* and *U(*)* is the same when $r_i > r_i^p$, thus

$$\sum_{i=1}^{n} AU(r_i) = \sum_{i=1}^{n} U(r_i).$$

Further, we have

$$AU(*) \geq U(*)$$

through its domain, so $\sum_{i=1}^{n} AU(r_i)$ gives the upper bound for $\sum_{i=1}^{n} U(r_i)$. So it is naturally to realize that $\sum_{i=1}^{n} U(r_i)$ reaches its maximization when $r_i = 0$ or $r_i \geq r_i^p$ for all i.

Proposition 19.2.3 *Assume* $AR = \{r_1, r_2, \ldots, r_k\}$ *(*r_k *is the last allocated flow when the algorithm end) is the solution obtained from the above algorithm, then we have* $au_1(r_1) = au_2(r_2) = \cdots = au_{k-1}(r_{k-1})$*, if* $r_i \neq 0$.

Proof: When the algorithm begins, it will allocate the resource to the first flow. Based on the criterion

$$k = \arg\max_i \left\{ \frac{AU_i(r_i + \Delta r) - AU_i(r_i)}{\Delta r} \right\},$$

it will not start to allocate the second flow until

$$au_1(r_1) = \frac{AU_2(r_2^p)}{r_2^p}.$$

So if the kth flow is starting to allocate, it must satisfy that

$$au_1(r_1) = au_2(r_2) = \cdots = au_{k-1}(r_{k-1}) = \frac{AU_k(r_k^p)}{r_k^p}.$$

Since k is the last flow when the algorithm stops, the remaining resource allocated to k must be less than r_k^p. If it is the first time to allocate bandwidth to Flow k, then all the remaining resource will allocate to Flow k and

$$au_1(r_1) = au_2(r_2) = \cdots = au_{k-1}(r_{k-1}) = \frac{AU_k(r_k^p)}{r_k^p} = au_k(r_k)$$

holds. If k has already allocated a certain amount of bandwidth, which is larger than r_k^p, then choosing k as the most increment flow means the former $k-1$ flows have reached the equilibrium. That is

$$au_1(r_1) = au_2(r_2) = \cdots = au_{k-1}(r_{k-1}),$$

which concludes the proof.

19.2.3 HEURISTIC RESOURCE ALLOCATION ALGORITHM

Our final goal is to find the optimal solution for the original model based on the solution we obtained from the approximate model. Before we propose our algorithm, several lemmas should be taken as the guideline for the algorithm.

Lemma 19.2.1 *For an optimal solution* R^**, if* $r_i > 0$ *then* $u_i(r_i) = v$*, where* v *is the Lagrange multiplier. (Herein, we assume the sigmoid function is differentiable in the definition domain, or differentiable except the point* $(r_p, U(r_p))$*.)*

Proof: We can write the original utility maximization problem 19.1 into the Lagrange multiplier form

$$L(r,\lambda,v) = \sum_{i=1}^{n} -U_i(r_i) - \lambda_i r_i + v\left(\sum_{i=1}^{n} r_i - C\right).$$

Following the Karush–Kuhn–Tucker condition [601], we have

$$\nabla_r L(r,\lambda,v) = -u_i(r_i) - \lambda_i + v = 0,$$
$$\lambda_i r_i = 0, \lambda_i \geq 0.$$

Observing from the above, for any $r_i \neq 0$, we have $\lambda_i = 0$, which leads to $u_i(r_i) = v$. This concludes the proof.

Lemma 19.2.2 [625]: *For an optimal solution R^*, there is at most one flow whose $u'(r_i) > 0$.*

Now we will propose the final iteration algorithm to find the solution to the OP. By Proposition 19.2.2, we have

$$au_1(r_1) = au_2(r_2) = \cdots = au_{k-1}(r_{k-1})$$

if $r_i \neq 0$. If $r_k \geq r_k^p$ and the equilibrium point is reached, the solution of the AP is the solution to the final solution (Proposition 19.2.2) and no more work need to be done. If $r_k < r_k^p$ or no equilibrium point is reached, we have

$$u_1(r_1) = u_2(r_2) = \cdots = u_{k-1}(r_{k-1}) \neq u_k(r_k)$$

for the OP. The next step is to keep iteration until the equilibrium point is reached. The following two cases should be taken into consideration:

Case 1. Flows $i \in [1, k-1]$ share their bandwidth to Flow k until they reach the equilibrium point.
Case 2. Flow k share its bandwidth to Flows $i \in [1, k-1]$ until it reaches the equilibrium point.

The one that has a higher total utility will be our final solution.

For Case 1, we will find one Flow $m \in [1, k-1]$ each time, which has the least decrease rate by the criterion

$$m = \arg\min_i \left\{ \frac{U_i(r_i) - U_i(r_i - \Delta r)}{\Delta r} \right\},$$

then $r_k = r_k + \Delta r$ and $r_m = r_m - \Delta r$ (if there exits any crossing point on Flow k, switch the bandwidth to the cross flow and the cross flow will replace Flow k to finish the iteration) until

$$\frac{U_k(r_k + \Delta r) - U_k(r_k)}{\Delta r} < \frac{U_m(r_m) - U_m(r_m - \Delta r)}{\Delta r},$$

showing that any more resources allocate to Flow k will only lead to a decrease of the total utility, which means the equilibrium point is reached.

For Case 2, we will find the flow, which has the most increase rate by

$$m = \arg\max_i \left\{ \frac{U_i(r_i + \Delta r) - U_i(r_i)}{\Delta r} \right\},$$

then $r_k = r_k - \Delta r$ and $r_m = r_m + \Delta r$ (if there exists any crossing point, switch the bandwidth to the cross flow to finish the iteration) until

$$\frac{U_m(r_m + \Delta r) - U_m(r_m)}{\Delta r} < \frac{U_k(r_k) - U_k(r_k - \Delta r)}{\Delta r},$$

which also means the equilibrium point is reached. Our algorithm will be shown in Algorithm 19.2.3.

Algorithm 19.2.3 Find solutions for the OP

Input: AR^*
Output: R^*(The solution for the OP)

Let $R = AR^*$
$m = \arg\min_i \left\{ \frac{U_i(r_i) - U_i(r_i - \Delta r)}{\Delta r} \right\}$
while $r_k < r_k^p$ OR $\frac{U_k(r_k+\Delta r)-U_k(r_k)}{\Delta r} < \frac{U_m(r_m)-U_m(r_m-\Delta r)}{\Delta r}$ **do**
  $m = \arg\min_i \left\{ \frac{U_i(r_i) - U_i(r_i - \Delta r)}{\Delta r} \right\}$
  $r_k = r_k + \Delta r$
  $r_m = r_m - \Delta r$
end while
Let $R_1 = R$ and $R = AR^*$
$m = \arg\max_i \left\{ \frac{U_i(r_i+\Delta r) - U_i(r_i)}{\Delta r} \right\}$
while $r_k > 0$ AND $\frac{U_m(r_m+\Delta r)-U_m(r_m)}{\Delta r} < \frac{U_k(r_k)-U_k(r_k-\Delta r)}{\Delta r}$ **do**
  $m = \arg\max_i \left\{ \frac{U_i(r_i+\Delta r) - U_i(r_i)}{\Delta r} \right\}$
  $r_k = r_k - \Delta r$
  $r_m = r_m + \Delta r$
end while
Let $R_2 = R$
if $\sum_{i=1}^{n} U_i(r_i^1) > \sum_{i=1}^{n} U_i(r_i^2)$ **then**
  $R^* = R_1$
end if
$R^* = R_2$
Return R^*

Proposition 19.2.4 *The solution R^* obtained from the above algorithm satisfies Lemma 19.2.1.*

Proof: Since the algorithm will not terminate until it reaches the equilibrium point, so it's obvious that R^* satisfies Lemma 19.2.1.

Proposition 19.2.5 *The solution R^* has at most one flow that $u'(r) > 0$.*

Proof: If the solution R^* is obtained from Case 1, each time one flow will be selected to share its resource to Flow k. Assume one of the flows, say m, and share its resource until $r_m = r_m^p$. Since the utility function is convex when $r_m < r_m^p$, the decrease of this flow will become smaller when r_m keep decreasing and the increment of Flow k will keep increasing until $r_k = r_k^p$. So each time Flow m will be selected as the least decreases flow to share its resource and no equilibrium will be reach until $r_m = 0$ or $r_k \geq r_k^p$. For Case 2, either Flow k shares its resource to other flows until the equilibrium point is reached or r_k decreases to 0. So the other flows will always keep $r > r_p$ and $r_k \in [0, r_k^p)$. So the solution R^* will always satisfy Lemma 19.2.2.

19.3 FAST SUBOPTIMAL ADMISSION CONTROL PROTOCOL

Based on the result from our algorithm, we know that the optimal solution will be obtained either in Case 1 or Case 2. Each time a new flow is about to come in, the system can re-sort the current flows by $U_i(r_i^p)/r_i^p$ in increasing order and run the above algorithm again to calculate the new allocation scheme. If the new allocation scheme satisfies Case 1, which means all the current flows share their bandwidth to the last flow in new order, that is, the new coming flow is allowed to access. Otherwise, the last flow in new order (may not the new coming flow) will have to share its bandwidth to others to achieve a higher total utility, which means one flow will be denied to access.

In nowadays wireless network facility, one AP access may connect to hundreds of devices, which means to run the whole algorithm each time when a new flow is about to come is a waste of time and resource, which is not sensible in real device. So we proposed a fast suboptimal admission control protocol to deal with the trade-off between time, resource consuming, and the total utility achieved in order to satisfy different system requirements. The basic idea of the protocol is to share the bandwidth from the current flows to the new coming flow until the equilibrium point is reached. The system can accept as much flows as possible under the condition that the total utility decrease is within a certain limit by the criterion

$$\frac{\sum U(R) - \sum U(R_{\text{New}})}{\sum U(R)} < Threshold, \tag{19.10}$$

where $\sum U(R)$ is the original total utility and $\sum U(R_{\text{New}})$ is the new total utility when equilibrium point is reached. This criterion makes the system has some redundancy to tolerate certain degree of utility decrease, but allowing more flows to access, which can adapt to different system requirements by setting the parameter "Threshold" (T). Once the new flow access, the new allocation scheme R_{New} can be written as

$$R_{\text{New}} = \{R_{\text{Current}}^{\text{New}}, r_{\text{New}}\},$$

where $R_{\text{Current}}^{\text{New}}$ is the current flows' new allocation scheme after they share their bandwidth to the new coming flow and r_{New} is the bandwidth acquired by the new coming flow. So the new total utility is

$$\sum U(R_{\text{New}}) = \sum U(R_{\text{Current}}^{\text{New}}) + U_{\text{New}}(r_{\text{New}}),$$

where $\sum U(R_{\text{Current}}^{\text{New}})$ is the new total utility of the current flows and $U(r_{\text{New}})$ is the utility of the new coming flow.

Once the "Threshold" is set, Equation 19.10 can be rewritten as

$$(1-T)\sum U(R) < \sum U(R_{\text{Current}}^{\text{New}}) + U_{\text{New}}(r_{\text{New}}).$$

Since $\sum U(R)$ and $T$are fixed and the sigmoid function is a monotonically increasing function, we can calculate the relationship between the new total utility of the current flows and the bandwidth shared to the new coming flow, which is

$$U_{\text{New}}^{-1}[(1-T)\sum U(R) - \sum U(R_{\text{Current}}^{\text{New}})] < r_{\text{New}}, \tag{19.11}$$

where $U^{-1}(*)$ is the inverse function of $U(*)$.

Since we use Equation 19.3 as our basic sigmoid function and we consider the equilibrium point satisfying Case 1. Assuming that we have n current flows, then we have

$$r_{\text{New}} > r_{\text{New}}^{p} - \frac{1}{p_{\text{New}}} \ln\left[\frac{1-(1-T)\sum U(R) + \sum U(R_{\text{Current}}^{\text{New}})}{1-q_{\text{New}}}\right] \tag{19.12}$$

and

$$\begin{aligned} r_{\text{New}} &= \sum_{i=1}^{n} \left(r_i^{\text{Before}} - r_i^{\text{Current}}\right) \\ &= C - \sum_{i=1}^{n} r_i^{\text{Current}}, \end{aligned} \tag{19.13}$$

where r_i^{Before} is the allocation scheme of each current flows before the new coming flow access and r_i^{Current} is the allocation scheme of each current flows after the new coming flow access and C is the total bandwidth. We use Equations 19.11 and 19.13 as the basic principle for the fast suboptimal admission control protocol. Let us use Q to simplify Equation 19.12 by

$$Q = (1-T)\sum U(R) - \sum U(R_{\text{Current}}^{\text{New}})$$

$$= (1-T)\sum_{i=1}^{n} U_i(r_i^{\text{Before}}) - \sum_{i=1}^{n} U_i(r_i^{\text{Current}}),$$

then we have

$$r_{\text{New}} > r_{\text{New}}^{p} - \frac{1}{p_{\text{New}}} \ln\left(\frac{1-Q}{1-q_{\text{New}}}\right).$$

Since

$$0 < q_{\text{New}} < 1,$$

r_{New} has solution when $Q < 1$. When we use Equation 19.2 as the basic model, will be changed to

$$r_{\text{New}} > \frac{b - \ln(\frac{1-Q}{Q})}{a} \quad Q > 0$$

or

$$r_{\text{New}} < \frac{b - \ln(\frac{1-Q}{Q})}{a} \quad Q < 0,$$

when

$$Q = (1-T)\sum U(R) - \sum U(R_{\text{Current}}^{\text{New}}).$$

r_{New} has solution when $(1-Q)/Q > 0$. So, we have

$$r_{\text{New}} > \frac{b - \ln(\frac{1-Q}{Q})}{a},$$

when

$$0 < Q < 1.$$

Also, when we use Equation 19.4 as the basic model, Equation 19.12 will become

$$kr_{\text{New}}^2 + r_{\text{New}} \ln(1-Q) + m\ln(1-Q) > 0$$

and

$$Q = (1-T)\sum U(R) - \sum U(R_{\text{Current}}^{\text{New}}).$$

r_{New} has two solutions when

$$\ln^2(1-Q) - 4km\ln(1-Q) > 0$$

and

$$1 - Q > 0.$$

So we have

$$0 < r_{\text{New}} < \frac{-\ln(1-Q) - \sqrt{\ln^2(1-Q) - 4km\ln(1-Q)}}{2k}$$

or

$$r_{\text{New}} > \frac{-\ln(1-Q) + \sqrt{\ln^2(1-Q) - 4km\ln(1-Q)}}{2k},$$

when

$$Q < 1 - e^{4km}.$$

Or

$$r_{\text{New}} > 0,$$

when

$$Q \geq 1 - e^{4km}.$$

So our final goal is to calculate the new allocation scheme when the new coming flow is about to access. If the decrease of new total utility is higher than the threshold we set, the new coming flow will be denied to access, otherwise the system will allow the new coming flow to access based on the new allocation scheme. Based on the principle above, we present a gradient-type iteration method to solve the problem. First, the bandwidth allocated to the new coming flow will be set as 0. Each time the least decrease flow in the current flows will be chosen out by the criterion

$$k = \arg\min_i \left\{ \frac{U_i(r_i) - U_i(r_i - \Delta r)}{\Delta r} \right\} \quad i \in current\ flows.$$

Then

$$r_{\text{New}} = r_{\text{New}} + \Delta r,$$

$$r_k = r_k - \Delta r.$$

Until

$$\frac{U_{\text{New}}(r_{\text{New}} + \Delta r) - U_{\text{New}}(r_{\text{New}})}{\Delta r} < \frac{U_k(r_k) - U_k(r_k - \Delta r)}{\Delta r},$$

which means the new allocation scheme including the new coming flow has reached the equilibrium point. Then we have

$$u_1(r_1) = u_2(r_2) = \cdots = u_n(r_n) = u_{\text{New}}(r_{\text{New}}),$$

for any $r_i \neq 0$. Note that the equilibrium point will reach both in Case 1 or 2, which is similar to Algorithm 19.2.3, and we will not terminate the algorithm until the new coming flow's bandwidth is higher than its preferable amount of bandwidth ($r_{\text{New}} > r^p_{\text{New}}$). After the iteration stops, we obtain the new allocation scheme (R_{New}). We calculate the decrease rate of the total utility and compare it with "Threshold" to decide whether the new coming flow can access. The details of the fast suboptimal admission control algorithm are shown in Algorithm 19.3.4.

Algorithm 19.3.4 Fast Suboptimal Admission Control

Input: $R = \{r_1, r_2, \ldots, r_n\}$ and $Flow_{\text{New}}$

Output: $R_{\text{New}} = \{r_1, r_2, \ldots, r_n, r_{\text{New}}\}$

```
1: Let r_New = 0
2: k = arg min_i{ (U_i(r_i) − U_i(r_i − Δr)) / Δr } i ∈ current flows
3: while
   r_New < r^p_New OR (U_New(r_New+Δr) − U_New(r_New))/Δr < (U_k(r_k) − U_k(r_k − Δr))/Δr do
4:    k = arg min_i{ (U_i(r_i) − U_i(r_i − Δr)) / Δr } i ∈ current flows
5:    r_k = r_k − Δr, r_New = r_New + Δr
6:    if (ΣU(R) − ΣU(R_New)) / ΣU(R) < Threshold then
7:        Access is approved
8:    end if
9: end while
10: Access is denied
11: Return R_New
```

Algorithm 19.3.4 can be used to decide whether to let the new coming flow access in quick time without using too much time and resource, but cannot guarantee the new allocation scheme has the highest total utility. So a timer is needed to recalculate the allocation scheme by Algorithms 19.2.2 and 19.2.3 once in a certain time (e.g., once for 1 h or once for 10 new flows come in). This timer can be set as a trade-off factor to decide whether time and resource consumption or the total utility is more important in the current system. We simulate it in Example 1 to make it more clear to the readers.

19.4 SIMULATION RESULTS

In this section, we will give two examples focus on different points to testify our algorithm. For Example 1, we focus on the bandwidth allocation algorithm and the admission control protocol and Example 2 will focus on the crossing point and illustrate that the algorithm we proposed can deal with any sigmoid shape utility functions.

Example 1:

The system model is

$$U_i(r_i) = \frac{1}{1+e^{(-a_i r_i + b_i)}}.$$

In this model, we assume the total available bandwidth C is 18, and there are five users with different requirements. The users' relevant parameter is

$$\begin{pmatrix} a_1 & a_2 & a_3 & a_4 & a_5 \\ b_1 & b_2 & b_3 & b_4 & b_5 \end{pmatrix} = \begin{pmatrix} 1 & 3 & 2 & 3 & 6 \\ 3 & 9 & 8 & 6 & 12 \end{pmatrix}.$$

First, we sort the users as the order we mentioned before

$$\begin{pmatrix} a_1 & a_2 & a_3 & a_4 & a_5 \\ b_1 & b_2 & b_3 & b_4 & b_5 \end{pmatrix} = \begin{pmatrix} 6 & 3 & 3 & 1 & 2 \\ 12 & 6 & 9 & 3 & 8 \end{pmatrix}.$$

and Δr is chosen to be 0.01.

The solutions for the approximate model are shown in Table 19.2, and the final solution for the OP is shown in Table 19.3 and Figure 19.7.

From the result, we can see that when there are one to four flows, the solution to AP model satisfied Proposition 19.2.2 so the solution to AP model is the final solution to the OP. When there are five flows to be allocated, the solution to AP model of Flow 5 is at the interval of $(0, r_5^p]$ and a further iteration is needed to proceed. The final result shows that Case 1 can achieve a higher total utility. So the final solution is $R = \{2.54, 2.83, 3.83, 3.81, 4.99\}$.

Now, we assume there are two new flows ($a = 0.5, b = 1.5$ and $a = 1, b = 3.5$) about to come in and the threshold is 5%. The fast admission control protocol is proceeded to decide whether they can access. For the first new flow ($a = 0.5, b = 1.5$), the result is shown in Figure 19.8. From the result, we can see that the access of the new coming flow when it reaches the equilibrium point will

Table 19.2
Solution to the AP Model (Example 1)

1 Flow	18.00	0	0	0	0
2 Flows	6.74	11.26	0	0	0
3 Flows	4.29	6.36	7.35	0	0
4 Flows	2.94	3.64	4.64	6.78	0
5 Flows	2.64	3.03	4.03	4.76	3.54

Table 19.3
Final Solution (Example 1)

						Utility
1 Flow	18.00	0	0	0	0	1.000
2 Flows	6.74	11.26	0	0	0	2.000
3 Flows	4.29	6.36	7.35	0	0	3.000
4 Flows	2.94	3.64	4.64	6.78	0	3.960
5 Flows	2.54	2.83	3.83	3.81	4.99	4.380

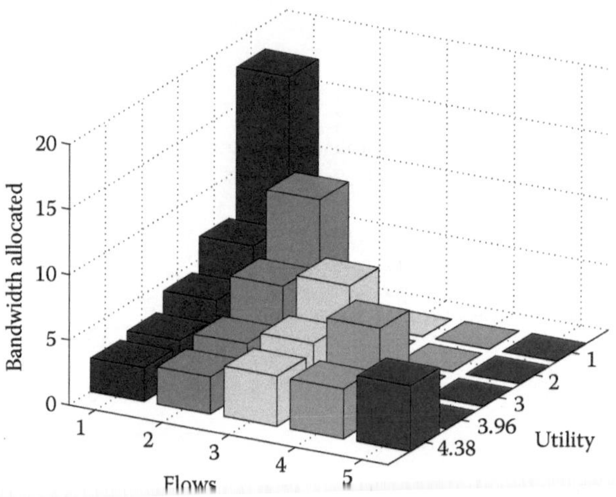

Figure 19.7 Final solution for Example 1.

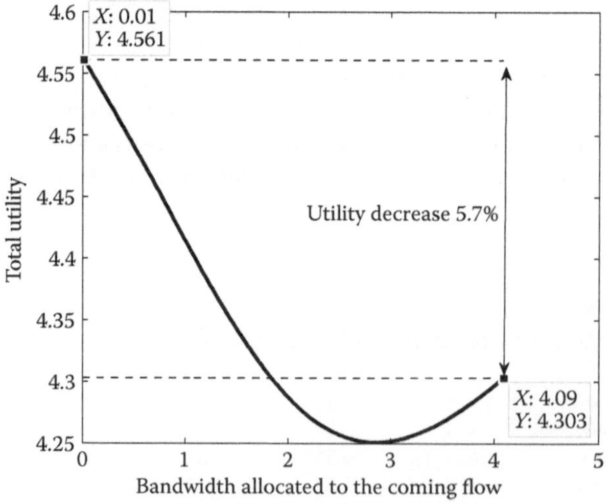

Figure 19.8 Total utility related to the bandwidth allocated to the first new flow.

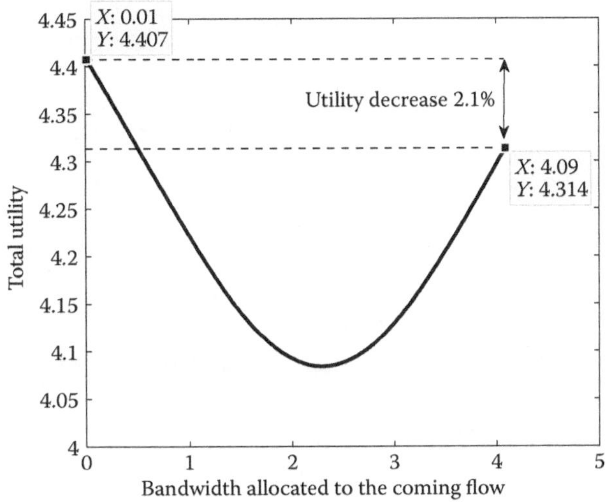

Figure 19.9 Total utility related to the bandwidth allocated to the second new flow.

cause a 5.7% decrease in the total utility, which is higher than the threshold, that is, this flow will be denied to access. For the second new flow ($a = 1, b = 3.5$), the result is shown in Figure 19.9. The decrease of the total utility is 2.1% when the second flow access, so the second coming flow will be allowed to access and the new allocation scheme is $R = \{2.51, 2.76, 3.76, 0.00, 4.48, 4.09\}$, with total utility $= 4.3136$.

Since the second new flow has access to the system, the current allocation scheme may not achieve the highest total utility. After a certain period of time, the system will reproceed Algorithm 19.2.2 and Algorithm 19.2.3 to adjust the allocation scheme. The allocation scheme after adjustment is $R = \{2.57, 2.88, 3.88, 4.09, 0, 4.58\}$, with total utility $= 4.3303$.

Example 2:

Assume there are six users with different bandwidth requirement and the total bandwidth C is 60, 70, 80, we use different sigmoid functions to reflect their requirement. For users 1, 2, 3, 5, we use Equation 19.2, which is

$$U(r) = \frac{1}{1 + e^{(-ar+b)}}$$

as the sigmoid function and the relevant parameter is

$$\begin{pmatrix} a_1 & a_2 & a_3 & a_4 \\ b_1 & b_2 & b_3 & b_4 \end{pmatrix} = \begin{pmatrix} 1 & 0.5 & 2 & 2 \\ 10 & 5 & 24 & 26 \end{pmatrix}.$$

While Equation 19.3

$$U(r) = \begin{cases} qe^{p(r-r_p)} & r < r_p \\ 1 - (1-q)e^{-p(r-r_p)} & r \geq r_p \end{cases}$$

is applied to users 4, 6 and

$$\begin{pmatrix} p_1 & p_2 \\ q_1 & q_2 \\ r_{p1} & r_{p2} \end{pmatrix} = \begin{pmatrix} 0.7 & 0.3 \\ 0.2 & 0.1 \\ 5 & 5 \end{pmatrix}.$$

The approximate model is shown in Figure 19.10. All the crossing points are listed in Table 19.4. The solution of the AP problem is shown in Table 19.5, and the final solution of the OP problem is shown in Table 19.6 and Figure 19.11.

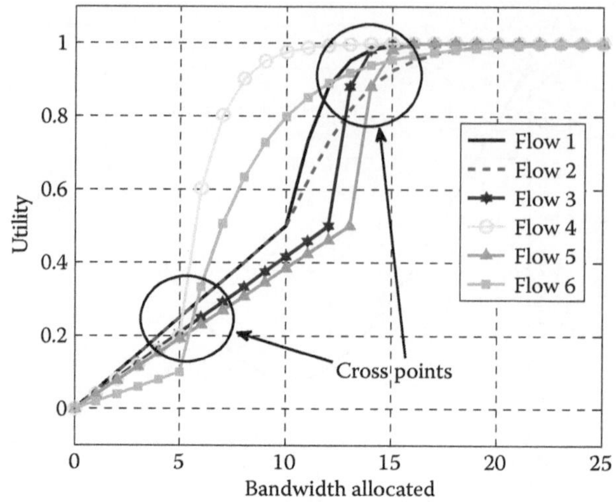

Figure 19.10 AP model for Example 2.

Table 19.4
Crossing Point Table (Example 2)

	Crossing Points (Coordinate/Cross Flow ID)			
Flow 1	5.11/4	5.79/6		
Flow 2	5.11/4	5.79/6	12.67/3	14.01/5
Flow 3	5.02/4	5.53/6		
Flow 4	No Crossing Point			
Flow 5	5.43/6			
Flow 6	12.14/1	13.26/3	14.45/5	

Table 19.5
Solution to the AP Model (Example 2)

Bandwidth	Flow 1	Flow 2	Flow 3	Flow 4	Flow 5	Flow 6
60	13.18	14.78	13.96	8.83	9.26	0
70	13.87	16.27	14.29	9.76	15.29	0.53
80	13.87	16.27	14.29	9.76	15.29	10.52

Table 19.6
Final Solution (Example 2)

Bandwidth	Flow 1	Flow 2	Flow 3	Flow 4	Flow 5	Flow 6	Utility
60	12.17	12.28	13.49	7.58	14.49	0	4.4494
70	12.34	12.75	13.56	7.77	14.56	9.03	5.242
80	13.51	15.49	14.11	9.26	15.11	12.52	5.7468

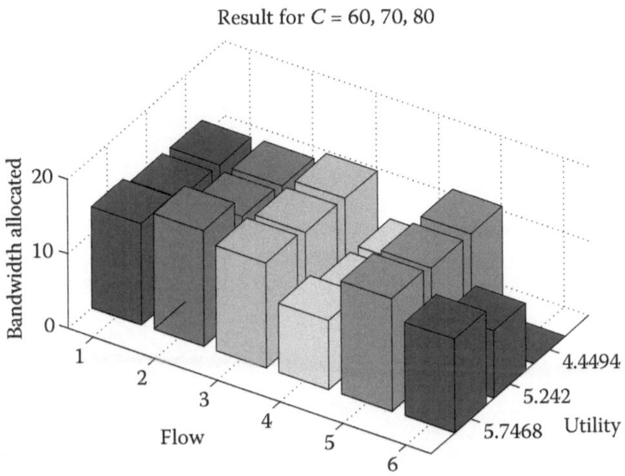

Figure 19.11 Final solution for Example 2.

19.5 CONCLUSION

We study the utility maximization problem for adaptive real-time traffic for multimedia wireless networks. An approximate system model and a heuristic iteration method have thus been proposed to solve the nonconvex optimization problem for bandwidth (resource) allocation of a wireless network among the competing real-time multimedia flows. We thereby establish an admission control protocol, which can be used in multimedia wireless networks for resource management and protocol design. Our theoretical findings are verified by numerical examples and simulation results, which have shown that our algorithm is efficient and effective in dealing with complex sigmoid-type utility maximization problem for wireless multimedia networks and can satisfy the various QoS requirements of users based on our proposed admission control criterion for various wireless systems. Since the non-real-time traffic and hard real-time traffic are the special cases of adaptive real-time traffic, the presented approach can be extended to cover the situation that all types of traffic flows, including the non-real-time traffic, the hard real-time traffic, and the adaptive real-time traffic, are appearing simultaneously in the network and that will be the worthwhile in future work.

20 Resource Allocation for Hard QoS Traffic and Elastic Traffic in Wireless Networks

This chapter studies the wireless resource allocation problem in the downlink of a wireless network with a central control system, such as a cellular base station. We attempt to maximize the total utility of all users. We consider two common types of traffic: hard quality of service (QoS) traffic and best-effort traffic. Three allocation algorithms are presented for these two types of traffic, namely

- The Hard Quality of Service (HQ) allocation for hard QoS traffic
- The elastic allocation for best-effort traffic
- The mixed allocation for the coexistence of both types of traffic.

A number of key theorems as the general design guidelines for utility-based resource allocation in wireless networks are given. The performance of the proposed schemes is validated via simulations. The results presented in this chapter tell us that optimal wireless resource allocation needs to account for the traffic demand, total available resource, and wireless channel quality, rather than solely be dependent on channel quality or traffic type as assumed in most existing work.

20.1 INTRODUCTION

Resource allocation is a challenging research topic in wireless networks, given the fact that in such networks, radio resource is limited, and the channel quality of each user may vary with time, while the users' service requirements are of diversity. There have been enormous efforts being directed to this topic, to name a few, see [603–612,616,623]. Resource allocation in wireless networks can be performed in the following manner:

- Centralized radio resource allocation scheme: a central entity like a base station makes channel assignment and power allocation decisions, and communicates these decisions to the sensor nodes.
- Cluster-based radio resource allocation schemes: a wireless sensor network is usually designed into the clustered architecture. The sensor nodes in the network are divided into small groups called clusters where each cluster is controlled by a central entity called cluster head (CH). In the cluster-based radio resource allocation scheme, the CH indicates the cluster members about spectrum sharing and power allocation.
- Distributed resource allocation schemes: each wireless communication node makes its transmission decisions in an autonomous manner.

The schemes are designed on the basis of performance optimization criteria that include energy efficiency, throughput maximization, QoS assurance, interference avoidance, fairness and priority consideration, and hand-off reduction. Particularly, being sharply different from the wire-lined communication networks, e.g., the Internet, where centralized control is impossible, in wireless networks, resource allocation and congestion control can be realized in a centralized manner. Note that in wireless networks, the base station has centralized controlling capability.

Assuming that the channel conditions at the wireless links and the total amount of available resource are known to the centralized system, the system may allocate resource to users either according to some performance metrics such as throughput and fairness [603,623] or according to the types of traffic [604]. Usually, the types of traffic are categorized as hard QoS traffic, soft QoS traffic, and elastic traffic in terms of their flexibility degree to the supplied bandwidth. Throughput and fairness are seen to be conflicting performance metrics. If we just want to maximize system throughput, more resource will be allocated to the users with better channel conditions. The radio resource will subsequently be occupied by a small number of users, leading to unfairness. On the other hand, to provide fairness to all users, the system tends to allocate more resource to the users in worse channel conditions so as to compensate for their shares. As a result, the system throughput may be degraded dramatically. The studies in Refs. [605,606] show that the system can behave either throughput oriented or fairness oriented by adjusting certain parameters. However, determining and justifying the value of these parameters is left unsolved.

In the approach presented in [625,626], user satisfaction is used as a criterion for resource allocation to avoid the above throughput–fairness dilemma. Since it is difficult to fully satisfy the diversity of demands of all users, one is forced to turn to maximize the total degree of user satisfaction. The degree of user satisfaction under a given amount of resource is described by the utility function $UT(rs)$, a nondecreasing function with respect to the given amount of resource rs. This suggests that if more resource is allocated to the user, the user is more satisfied.

The marginal utility function defined by

$$ut(rs) = \frac{dUT(rs)}{drs}$$

is the derivative of the utility function $UT(rs)$ with respect to the given amount of resource rs. The exact expression of a utility function relies on traffic types and the feeling of the user. It can be obtained by studying the behavior and feeling of users. This is lying in the scopes of psychologists and economists. One usually focuses on maximizing the total utility for a given set of utility functions on the assumption that the utility functions are known.

For the relatively early studies of utility-based resource management in wireless networks, we mention the following. Schemes in Refs. [607,611] are the examples that are based on the utility functions of different objectives for wireless networks. In Ref. [607], a utility-based power control scheme with respect to channel quality is proposed, in which users with higher SIR values have higher utilities and thus are more likely to transmit packets. Therefore, this utility-based power control scheme can fully take advantage of wireless medium, and thus, the transmission power can be conserved.

The work in Ref. [608] proposes a utility-based bandwidth allocation scheme, which can adapt to channel conditions and guarantee the minimum utility requested by each user. In Refs. [609,610], the authors designed a utility-based fair allocation scheme to ensure the same utility value for each user. However, this may not be an efficient way of using wireless resource by driving users with different traffic demands to achieve an identical level of satisfaction. Traffic with worse channel condition, which is difficult to be satisfied, tends to consume most of the system resource. This causes another kind of unfairness. In Ref. [611], a utility-based scheduler together with a forward error correction (FEC) and an automatic repeat request (ARQ) scheme is proposed. That work gives lagging users more resource and thus results in a similar performance level (i.e., fixed utility value) for each user. The work in Refs. [618,619] is applicable to multihop wireless networks.

Utility functions have also been widely used in Internet pricing and congestion control [104]. Many bandwidth pricing schemes have been proposed for wireless networks (e.g., [612,616]). The typical approach is to set a price to radio resource and to allocate tokens to users. The objective is then to maximize the social welfare through a bidding process. These kinds of bidding schemes, though useful for Internet pricing and congestion control, may not be practical for wireless networks. In wireless environments, the types of traffic, the number of users, and channel conditions are all dynamically time varying. It would be impractical and expensive to implement a wireless bidding process because the users would have to keep exchanging control messages for real-time bidding, and the control protocols of the wireless system need to be modified to accommodate this process. Still further, the complexity and efficiency of wireless bidding are very difficult to control.

In this chapter, we study the wireless resource allocation problem in the downlink of a wireless network with a central control system, such as a cellular base station. We attempt to maximize the total utility of all users. We consider two common types of traffic: hard QoS traffic and best-effort traffic. Three allocation algorithms are presented for these two types of traffic, namely

- The HQ allocation for hard QoS traffic,
- The elastic allocation for best-effort traffic
- The mixed allocation for the coexistence of both types of traffic.

These three allocation schemes are all polynomial time solutions and proved to be optimal under certain conditions, and in any case, the difference between the total utilities obtained by the solutions herein and the optimal utility are bounded. Some theorems as the general design guidelines for utility-based resource allocation in wireless networks are given. The performance of the proposed schemes is validated via simulations. The results presented in this chapter tell us that optimal wireless resource allocation needs to account for the traffic demand, total available resource, and wireless channel quality, rather than solely be dependent on channel quality or traffic type as assumed in most existing work.

20.2 RESOURCE ALLOCATION AMONG HARD QoS TRAFFIC AND BEST-EFFORT TRAFFIC: NETWORK UTILITY MAXIMIZATION (NUM) APPROACH

20.2.1 MODEL AND PROBLEM STATEMENT

Suppose that there are n users served by a base station. Let rs_{tt} denotes the total amount of radio resource available at the base station and rs_i the amount of resource to be allocated to user i. Due to the data loss in a wireless link that is related to its channel quality, users within the same kind of traffic category may not feel the same way even the same amount of resource is given to them.

Let ql_i denotes the channel quality of user i, $0 \leq ql_i \leq 1$, and $i = 1, 2, \ldots, n$. A small value of ql_i indicates the poor channel quality. Given an amount of resource rs_i and channel quality ql_i, the essential amount of resource that is used by user i is given by

$$\theta_i = rs_i \times ql_i.$$

Let $T(i)$ denotes the type of traffic of user i. The utility function of user i is expressed by

$$UT_i(rs_i) = UT_{T(i)}(rs_i \times ql_i),$$

where $UT_{T(i)}(\bullet)$ is the utility function of traffic $T(i)$, and $UT_i(\bullet)$ is the utility function for the type of traffic described by $UT_{T(i)}(\bullet)$ but taking into account the channel quality of user i. Assuming that the marginal utility function of $UT_{T(i)}(\bullet)$ is $ut_{T(i)}(\bullet)$, the marginal utility function of $UT_i(\bullet)$ is then

$$\frac{\mathrm{d}UT_{T(i)}(rs_i \times ql_i)}{\mathrm{d}rs_i} = ql_i \times ut_{T(i)}(rs_i \times ql_i)$$

Our objective is to maximize the aggregate utility of all users subject to the restriction of total available resources. The problem at hand is thus formulated into the following optimization model:

$$max_{rs_i} \sum_{i=1}^{n} UT_i(rs_i),$$

$$s.t. \sum_{i=1}^{n} rs_i \leq rs_{tt},$$

$$\forall r_i \geq 0,$$

where rs_{tt} denotes the total resource that is available for sharing among users at the base station. Note that mathematically the optimal allocation may not be unique.

We define the term of optimal allocation for n users with the total available resource rs_{tt} in the following manner.

Definition 20.2.1 *A resource allocation* $RS^* = \{rs_1, rs_2, \ldots, rs_n\}$ *for n users is an optimal allocation if for all feasible allocations* $RS_a = \{rs_1^{'}, rs_2^{'}, \ldots, rs_n^{'}\}$, $UT(RS^*) = \Sigma_{i=1}^{n} UT_i(r_i)$, *and* $UT(RS_a) = \Sigma_{j=1}^{n} UT_j(rs_j^{'})$, *such that* $UT(RS^*) \geq UT(RS_a)$.

Note that for an allocation $RS^* = \{rs_1, rs_2, \ldots, rs_n\}$, if it satisfies $\Sigma_{i=1}^{n} rs_i \leq rs_{tt}$, then this allocation is called a feasible allocation.

Definition 20.2.2 *A unit-step utility function* $UT_{step}(rs)$ *refers to a utility function whose* $ut_{step}(r_M) = \infty$ *if* $rs = rs_M$, *and* $ut_{step}(rs) = 0$, *otherwise, where* $ut_{step}(rs) = \frac{dUT_{step}(rs)}{drs}$.

Definition 20.2.3 *A concave utility function* $UT_{concave}(rs)$ *refers to a utility function whose* $ut_{concave}(rs) > 0$ *and* $ut_{concave}^{'}(rs) < 0$ *for all rs, where* $ut_{concave}(rs) = \frac{dUT_{concave}(rs)}{drs}$ *and* $ut_{concave}^{'}(rs) = \frac{dut_{concave}(rs)}{drs}$.

By definition, a unit-step function is a discrete function, and a concave utility function is a nondecreasing and continuous function with respect to resource *rs*. Figures 20.2 and 20.3 plot the marginal utility functions for the two types of traffic shown in Figure 20.1. More terminology used in this chapter is defined as follows.

Definition 20.2.4 $RS = \{rs_1, rs_2, \ldots, rs_n\}$ *is a full allocation if* $\Sigma_{\forall rs_i \in R} rs_i = rs_{tt}$.

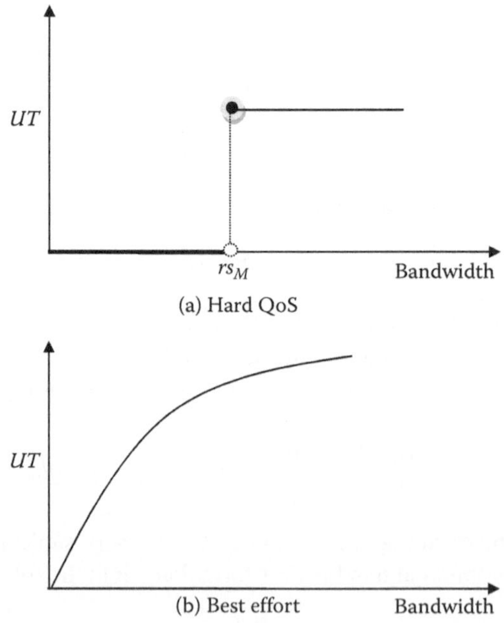

Figure 20.1 The utility function of two types of traffic.

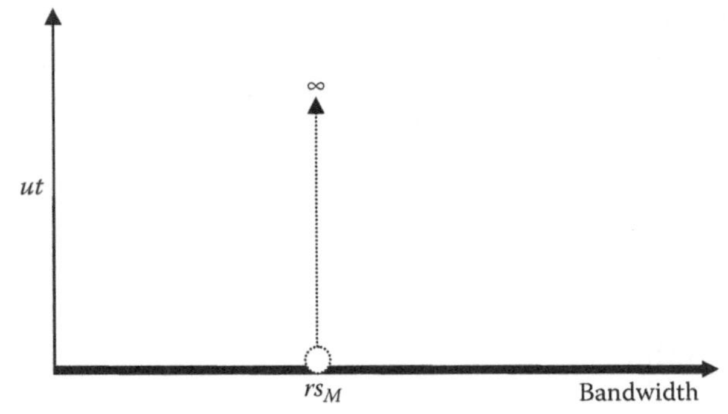

Figure 20.2 The marginal utility function of hard QoS traffic.

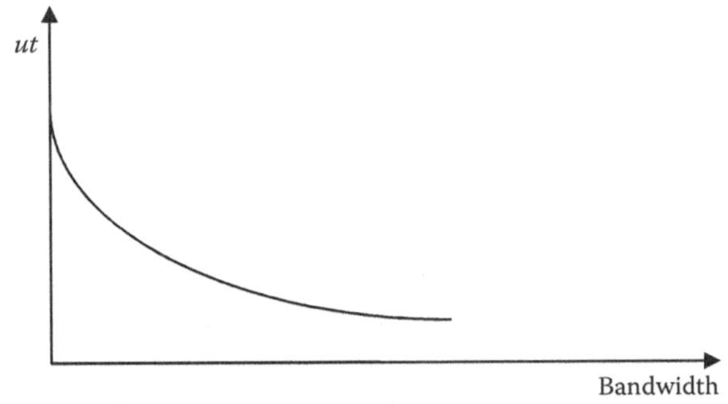

Figure 20.3 The marginal utility function of elastic traffic.

Definition 20.2.5 *For $RS = \{rs_1, rs_2, \ldots, rs_n\}$, all users i with $rs_i > 0$ (i.e., those who are allocated resources) are referred to as allocated users, and all users j with $rs_j = 0$ (i.e., those who are not allocated resources) are unallocated users.*

Definition 20.2.6 *$RS = \{rs_1, rs_2, \ldots, rs_n\}$ is marginally fair if it satisfies the following two conditions:*

1. Each allocated user i (i.e., with $rs_i > 0$) in RS must have the same marginal utility value.

2. For each unallocated user j (i.e., with $rs_j = 0$) in RS, its marginal utility value $ut_j(0)$ cannot exceed $ut_i(rs_i)$, which is the marginal utility of each allocated user i.

Definition 20.2.7 *For a marginally fair allocation $RS = \{rs_1, rs_2, \ldots, rs_n\}$, the marginal utility value of each allocated user in RS is equal and is referred to as the allocated marginal utility $ut_m(RS)$. Thus, for each allocated user i in RS, $ut_i(rs_i) = ut_m(RS)$, and for each unallocated user j, $ut_j(0) < ut_m(RS)$.*

20.2.2 HQ ALLOCATION FOR HARD QoS TRAFFIC

For a wireless communication system, suppose that there are n users in the queue, and all are in hard QoS traffic. Let rs_{rsd} denotes the residual resource in the system. The resource allocation algorithm designed for users whose utility functions are all unit-step functions is referred to as the HQ allocation, and the output is denoted by $RS_{HQ} = \{rs_1, rs_2, \ldots, rs_n\}$. Given the total available resource in the system rs_{tt}, the channel quality ql_i and utility function $UT_{T(i)}(\bullet)$ for all user i, RS_{HQ} can be obtained as follows.

The utility function for user i with hard QoS traffic is described by $UT_{T(i)}(rs) = UT_{M_i} \times f_u(ql_i \times rs - rs_{M_i})$, where $f_u(\bullet)$ is the unit-step function, ql_i is the channel quality of this user, M_i is the kind of QoS traffic, rs_{M_i} is the preferred amount of resource to be allocated.

The allocation rule of the **HQ allocation** is to assign resources to users in the descending order of $\frac{UT_{M_i} \times ql_i}{rs_{M_i}}$, subject to $\sum_{i=1}^{k} rs_i \leq rs_{tt}$, where k is the largest value satisfying this constraint, as illustrated in Figure 20.4. The allocation problem for users with arbitrary unit-step utility functions is found to be NP-complete. The performance of the HQ allocation can be proved to be close to the optimum. When the utility functions of all users are identical, the HQ allocation algorithm is found to yield the desired optimal solution.

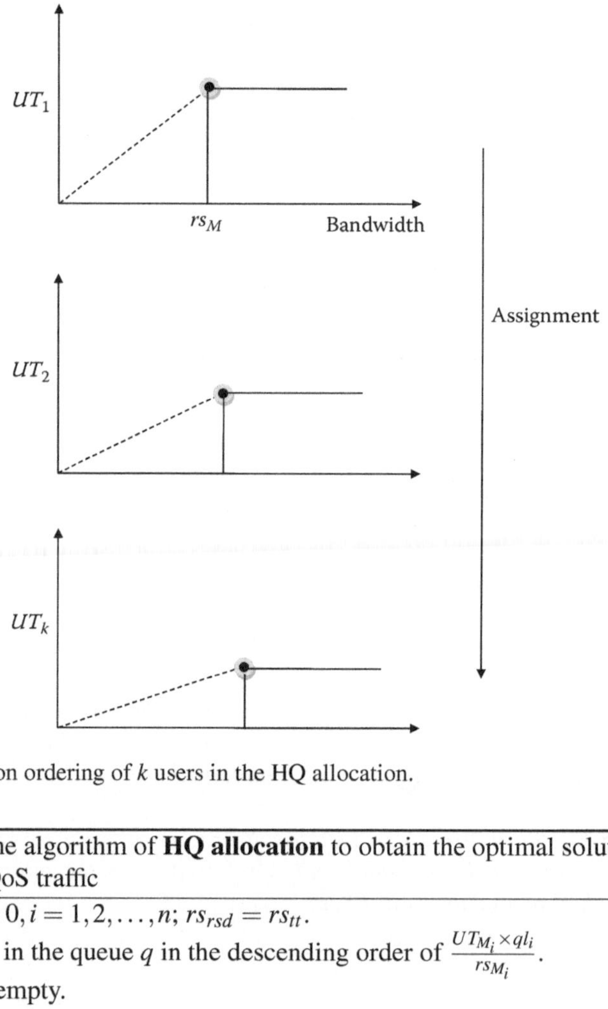

Figure 20.4 Allocation ordering of k users in the HQ allocation.

Algorithm 20.2.1 The algorithm of **HQ allocation** to obtain the optimal solution for resource allocation among hard QoS traffic

---

1. Initialize $rs_i = 0, i = 1, 2, \ldots, n$; $rs_{rsd} = rs_{tt}$.
2. Sort all users i in the queue q in the descending order of $\frac{UT_{M_i} \times ql_i}{rs_{M_i}}$.
3. while q is not empty.
4. fetch out user i who is now at the head of the queue q.
5. if $rs_{rsd} > \frac{rs_{M_i}}{ql_i}$, then

$$rs_i = \frac{rs_{M_i}}{ql_i};$$

$$rs_{rsd} = rs_{rsd} - rs_i.$$

6. endif
7. endwhile

Lemma 20.2.1

The allocation problem is NP-complete if the utility functions for all users are arbitrary unit-step function.

Proof: This problem can be reduced from the 0/1 knapsack problem, an NP-complete problem. Consider a knapsack with capacity $cp > 0$ and n items. Each item has a value of $vl_i > 0$ and a weight of $wt_i > 0$. The problem is then to find a selection of items that maximize $\sum_{i=1}^{n} \delta_i \times vl_i$ subject to $\sum_{i=1}^{n} \delta_i \times wt_i \leq cp$, where $\delta_i = 1$ if the item is selected, and 0, otherwise. Therefore, any instance of the knapsack can be reduced to an instance of our problem by substituting $rs_{tt} = cp$,

$$UT_i = vl_i \times f_{step}(ql_i \times rs - wt_i)$$

and $ql_i = 1$, for $i = 1, 2, \ldots, n$. Since an optimal solution to our problem is also a solution to the given knapsack problem, the knapsack problem is a special case of our problem, and it follows that our at-hand problem is NP-hard.

Next, it can be observed that our problem is an NP problem because any given solution, can be checked to be a feasible solution, and further, it is found that any given solution is bounded at a given utility value u in polynomial time. Since our problem is both in NP and NP-hard, it is, therefore, an NP-complete problem. ■

Theorem 20.2.2

One has

$$UT(RS_{HQ}) \geq UT(RS_{op}) - UT_{max},$$

where RS_{HQ} and RS_{op} are the proposed solution and the optimal solution, respectively, of this HQ allocation problem; $UT(x)$ is the total utility of all users for solution x; and $UT_{\max}$ is the maximum utility value taken over all users, i.e.,

$$UT_{\max} = \max_{1 \leq k \leq n} \{UT_{M_k}\}.$$

Proof: Let p denotes the original HQ allocation problem, which is NP-complete as proved in Lemma 20.2.2, and let p' denote the problem by relaxing the integrity constraint of problem p (i.e., each user in p' can be served fractionally). To better indicate the optimal solution to each problem, we let RS_{op}^{p} and $RS_{op}^{p'}$ represent the optimal solutions to problems p and p', respectively. Since p is a maximization problem, we have

$$UT(RS_{op}^{p'}) \geq UT(RS_{op}^{p}) \geq UT(RS_{HQ}^{p}),$$

where RS_{HQ}^{p} means RS_{HQ}.

The optimal solution to p', $RS_{op}^{p'}$, is obtained by sorting QoS users in the queue in the decreasing order of $\frac{UT_{M_i} \times ql_i}{rs_{M_i}}$ as in $UT(RS_{op}^{p})$. The difference between $RS_{op}^{p'}$ and RS_{HQ}^{p} is only in those users fractionally served in $RS_{op}^{p'}$. It follows that

$$UT(RS_{HQ}^{p}) + UT_x \geq UT(RS_{op}^{p'}),$$

where UT_x is the utility value of the unallocated user whose $\frac{UT_{M_i} \times ql_i}{rs_{M_i}}$ in RS_{HQ}^{p} is the largest. Thus, $UT_x \leq UT_{\max}$. Since

Algorithm 20.2.2 The algorithm of **elastic allocation** to obtain the optimal solution of resource allocation among the elastic traffic

1. For each user i, derive $ut_i^{-1}(ut)$, the inverted function of $ut_i(rs)$.
2. Derive $ut_\Sigma^{-1}(ut)$ by summing up $ut_i^{-1}(ut)$ over all users i, to obtain $ut_\Sigma^{-1}(ut) = \sum_i ut_i^{-1}(ut)$.
3. Find $ut_\Sigma(rs)$, the inverted function of $ut_\Sigma^{-1}(ut)$.
4. Find ut_{BE}, which is equal to $u_\Sigma(rs_{tt})$.
5. For all rs_i, $i = 1, 2, \ldots, n$.
6. If $ut_{BE} < ut_i(0)$, then $rs_i = ut_i^{-1}(ut_{BE})$.
7. Else $rs_i = 0$.
8. Endif
9. Endfor

$$UT(RS_{HQ}) + UT_{\max} \geq UT(RS_{HQ}^p) + UT_x > UT(RS_{op}^{p'}) \geq UT(RS_{op}^p),$$

subsequently we obtain

$$UT(RS_{HQ}^p) > UT(RS_{op}^p) - UT_{\max}.$$

■

Theorem 20.2.3

The HQ allocation solution RS_{HQ} is an optimal allocation if the unit-step utility functions $UT_{step}(rs)$ of all n users in the queue are identical.

Proof: Let UT_M denotes the unit-step utility function for all users. The targeted inequality in Theorem 20.2.2 can be rewritten as

$$n \times UT_M + UT_M > UT(RS_{op}^{p'}) \geq UT(RS_{op}^p) \geq n \times UT_M,$$

i.e.,

$$n + 1 > \frac{UT(RS_{op}^{p'})}{UT_M} \geq \frac{UT(RS_{op}^p)}{UT_M} \geq n.$$

Since both $UT(RS_{op}^p)/UT_M$ and n are integral,

$$UT(RS_{op}^p) = n \times UT_M.$$

Therefore, RS_{HQ} must be optimal. ■

Theorem 20.2.4

The time complexity of RS_{HQ} is $O(n \cdot \log\ n)$.

Proof: The time complexity of RS_{HQ} can be expressed by a function of the number of users in the network. Since the complexity of Step 2 (sorting) is $O(n \times \log\ n)$ and this iteration dominates the operation, the overall complexity of RS_{HQ} is, therefore, $O(n \times \log\ n)$. ■

20.2.3 ELASTIC ALLOCATION FOR THE BEST-EFFORT TRAFFIC

We next consider the best-effort traffic. The resource allocation algorithm for users with concave utility functions $UT_i(rs)$ is referred to as the elastic allocation, and the output is denoted by $RS_e =$

$\{rs_1, rs_2, \ldots, rs_n\}$. Given the total available resource rs_{tt}, the channel quality ql_i, and marginal utility function $ut_i(rs)$ for each user i, RS_e can be obtained by Algorithm 20.2.2.

The basic rule of the scheme **elastic allocation** is to

- Derive the aggregated utility function from the inverse functions of all users
- Calculate the allocated marginal utility from the aggregated utility function
- Determine rs_i for each user.

Serving as an example, Figure 20.5 illustrates the elastic allocation algorithm with two best-effort traffics.

Theorem 20.2.5

The elastic allocation R_e is optimal if the utility functions for all users are concave.

Proof: In the elastic allocation, (1) ut_{BE} is the marginal utility value at which point the total resource has been fully allocated and (2) for all allocated users i in R_e, $ut_{BE} < ut_i(0)$ and $ut_i(rs_i) = ut_{BE}$; for all unallocated users j, $ut_{BE} \geq ut_j(0)$. With concave utility functions, i.e., function with $ut(r) > 0$ for all r, it can be easily proved by contradiction that an optimal allocation for elastic users, i.e., R_e, must be full. Similarly, by using contradiction again, we can prove that R_e must also be marginally fair. Since all users utility functions are increasing, only one allocation can be both full and marginally fair. Therefore, if an allocation is full and marginally fair, it must be the desired allocation R_e. ■

Theorem 20.2.6

The time complexity of the algorithm elastic allocation is *O(n)*.

Proof: Since the operation at each step at most takes time *O(n)*, the time complexity of the elastic allocation algorithm is thus *O(n)*. ■

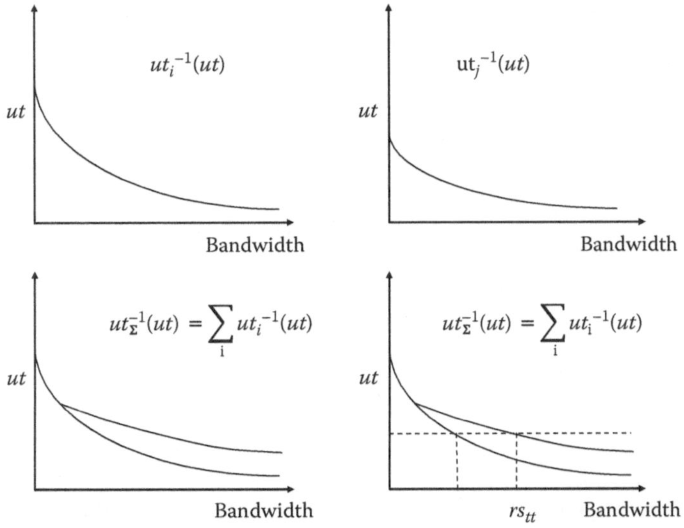

Figure 20.5 An example of the elastic allocation.

20.2.4 MIXTURE OF HARD QoS AND BEST-EFFORT TRAFFIC

Finally, we consider the coexistence of QoS and best-effort traffic in the system, which is referred to as mixed allocation and the output of which is denoted by

$$RS_m = \{rs_1, rs_1, \ldots, rs_n\}.$$

Let rs_{BE} denote the amount of residual resource to be given to best-effort traffic and ΔUT_i the utility gain by allocating resource rs_i to QoS user i. Other notations remain the same as in the HQ and the elastic allocations. Given the total available resource rs_{tt}, the channel quality ql_i, and marginal utility function $ut_i(rs)$ for each user i, RS_m can be obtained by the following procedure.

The allocation rule of this mixed allocation is to (1) allocate resource to the first k QoS users at the sorted queue and (2) then allocate the residual bandwidth (i.e., $rs_{tt} - rs_{QoS}$) to all best-effort users based on the elastic allocation. The value of k is determined based on the requirement that there is sufficient resource for this QoS user and the obtained utility gain ΣUT_k is positive (i.e., $rs_{BE} - rs_k$).

We have proved that the allocation problem for hard QoS traffic is NP-complete. It follows that the allocation problem for the coexistence of best-effort and hard QoS traffic is also NP-complete. Again, the proposed **mixed allocation** algorithm can achieve a performance metric lower bounded by $UT(RS_{op}) - UT_{\max}$. When all QoS users have an identical utility function, the mixed allocation is subsequently optimal.

Algorithm 20.2.3 The algorithm of **mixed allocation** to obtain the optimal solution to the resource allocation among mixture of hard QoS and elastic traffic

---

1. Initialize $rs_i = 0, i = 1, 2, \ldots, n$; and $rs_{BE} = rs_{tt}$.
2. Sort all QoS users i in the queue q in the descending order of $\frac{UT_{M_i} \times ql_i}{rs_{M_i}}$.
3. For each best-effort user j, calculate $ut_\Sigma^{-1}(ut)$ by summing up $ut_j^{-1}(ut)$ which is the inverted function of $ut_i(rs)$, i.e.,
$$ut_\Sigma^{-1}(ut) = \sum_j ut_j^{-1}(ut).$$
4. Find $ut_\Sigma(rs)$, the inverted function of $ut_\Sigma^{-1}(ut)$.
5. While the queue q is not empty,
6. then fetch out the QoS user i from the head of the queue.
7. if $rs_{BE} > \frac{rs_{M_i}}{ql_i}$
8. thn
$$rs_i = \frac{rs_{M_i}}{ql_i}.$$
9. Else
$$rs_i = 0;$$
$$\Delta UT_i = UT_{M_i} - \int_{rs_{BE}-rs_i}^{rs_{BE}} ut_\Sigma(rs)d(rs);$$
10. if $\Delta UT_i > 0$
11. $rs_{BE} = rs_{BE} - rs_i$; continue;
12. else
13. $rs_i = 0$; break;
14. endwhile
15. then for all unallocated best-effort user j, if $ut_\Sigma(rs_{BE}) < ut_j(0)$, then $rs_j = ut_j^{-1}(ut_\Sigma(rs_{BE}))$; else $rs_j = 0$.
16. Endif.

Theorem 20.2.7

For the mixed allocation problem, $UT(RS_m) \geq UT(RS_{op}) - UT_{\max}$, where RS_m is the proposed solution, RS_{op} is the optimal solution, $UT(x)$ is the total utility of all users for solution x, and $UT_{\max}$ is the maximum utility value taken over all users, i.e.,

$$UT_{\max} = \max_{1 \leq k \leq n} \{UT_{M_k}\}.$$

Proof: This theorem can be proved by relaxing the constraint in the original problem as in Theorem 20.2.2. Therefore, the difference in the total utility between RS_m and the relaxed problem is bounded by $UT_{\max}$. That is,

$$UT(RS_m) \geq UT(RS_{op}) - UT_{\max}.$$

■

Theorem 20.2.8

The mixed allocation RS_m is an optimal allocation for traffic mixed with identical unit-step functions and arbitrary concave functions.

Proof: Let us consider the first k QoS users in the sorted queue, where k is the maximum possible value satisfying $\sum_{i=1}^{k} rs_i \geq rs_{tt}$. The set of all possible optimal allocations is then $RS_0, RS_1, \ldots, RS_i, \ldots, RS_k$, where RS_i is the allocation in which the first i QoS users are allocated resource in an amount of $rs_j = rs_M / ql_j, j = 1, 2, \ldots, i$, and the residual bandwidth is all allocated to the best-effort users. The obtained utility gain $\Delta UT_i, i = 1, 2, \ldots, k$, is therefore expressed by

$$\Delta UT_i = UT(RS_i) - UT(RS_{i-1}) = UT_{M_i} - \int_{rs_{BE} - rs_i}^{rs_{BE}} ut_{\Sigma}(rs) d(rs).$$

Since QoS users are sorted in the decreasing order of their ql_i, this leads to that rs_i is allocated in the increasing order of ql_i resulting in $UT_i \geq UT_{i+1}$. Thus, the allocation RS_i in $[RS_0, RS_1, \ldots, RS_i, \ldots, RS_k]$ is the optimal allocation, where i is the largest value satisfying $\Delta UT_i > 0$.

■

20.3 NUMERICAL RESULTS

This section presents comprehensive numerical results that demonstrate the proposed algorithms: the algorithm of **HQ allocation** (namely Algorithm 20.2.1), the algorithm of **elastic allocation** (namely Algorithm 20.2.2), and the algorithm of **mixed allocation** (namely Algorithm 20.2.3). All the numerical results are outputted for the wireless communication system, in which there are 26 users competing for the limited total bandwidth at the base station. We consider the following three scenarios:

- Scenario 1: there are 26 users of hard QoS traffic
- Scenario 2: there are 26 users of best-effort traffic.
- Scenario 3: there are first 13 users of hard QoS traffic and the remaining 13 users are the best-effort traffic.

For the the hard QoS traffic user i, the utility function is defined by

$$UT_i(rs_i) = UT_M \times f_u(ql_i \times rs_i - rs_M), \tag{20.1}$$

where we choose the specific parameters $UT_M = 100$, $rs_M = 100$ while letting the channel quality parameters ql_i be varying randomly between the range of $(0,1)$. For the the best-effort traffic user i, the utility function is defined by

$$UT_i(rs_i) = \frac{ql_i \times rs_i}{ql_i \times rs_i + 1}, \tag{20.2}$$

where again we let the channel quality parameters ql_i be varying randomly between the range of $(0,1)$.

Viewing the fact that in any wireless communication system, the channel quality is usually dynamically varying due to a number of causes, in order to see the adaptability of the proposed three algorithms to the wireless communication environment: the algorithm of **HQ allocation**, the algorithm of *elastic allocation*, and the algorithm of *mixed allocation*, corresponding to the above three scenarios, we use a generator in MATLAB to randomly generate the channel quality parameter from the range of $(0,1)$. For each of the above three scenarios, we generate 10 sets of the channel quality parameters and then run the above three algorithms to yield the plots shown in Figures 20.6 through 20.35. Note that the numerical results shown in Figures 20.6 through 20.15 are the outputs of the *HQ allocation* algorithm for Scenario 1, the numerical results shown in Figures 20.16 through 20.25 are the outputs of the *elastic allocation* algorithm, and, finally, the numerical results shown in Figures 20.26 through 20.35 are the outputs of the *mixed allocation* algorithm. In summary, the connection between the used channel quality parameters and the numerical result plots is outlined in Table 20.1.

Next, we take a look at the fairness degree under the proposed three allocation schemes. Corresponding to the aforementioned Scenarios 1 through 3, the fairness index is computed by the following equation:

$$F(rs_1, rs_2, \ldots, rs_{26}) = \frac{(\sum_{i=1}^{26} rs_i)^2}{26 \times \sum_{i=1}^{26} (rs_i)^2},$$

where the resource allocation $(rs_1, rs_2, \ldots, rs_{26})$ is outputted by the algorithm of *HQ allocation* (namely Algorithm 20.2.1), the algorithm of *elastic allocation* (namely Algorithm 20.2.2), and the

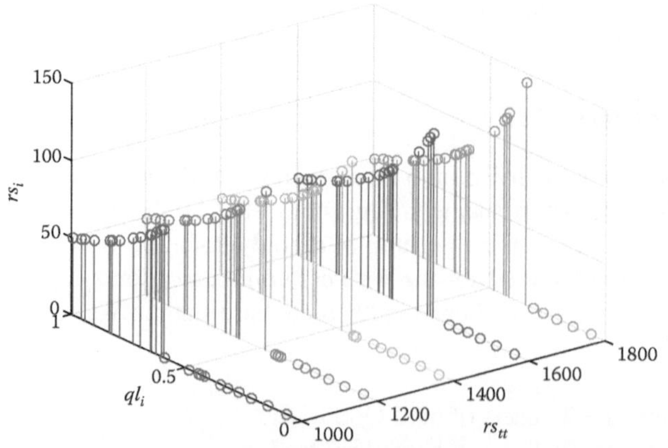

Figure 20.6 The optimal allocation results of hard QoS traffic based on Algorithm 20.2.1 under the channel quality parameter choice of SET_1 in Table 20.1. The three-dimensional coordinates are the resource allocation rs_i to user i, the channel quality of user i, and the total available resource rs_{tt} in the communication system.

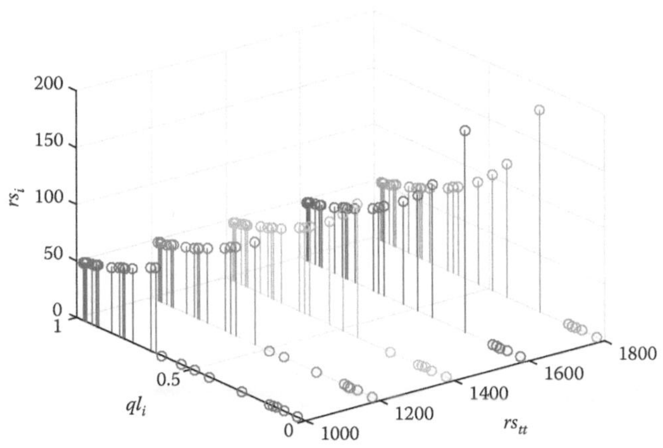

Figure 20.7 The optimal allocation results of hard QoS traffic based on Algorithm 20.2.1 under the channel quality parameter choice of SET_2 in Table 20.1. The three-dimensional coordinates are the resource allocation rs_i to user i, the channel quality of user i, and the total available resource rs_{tt} in the communication system.

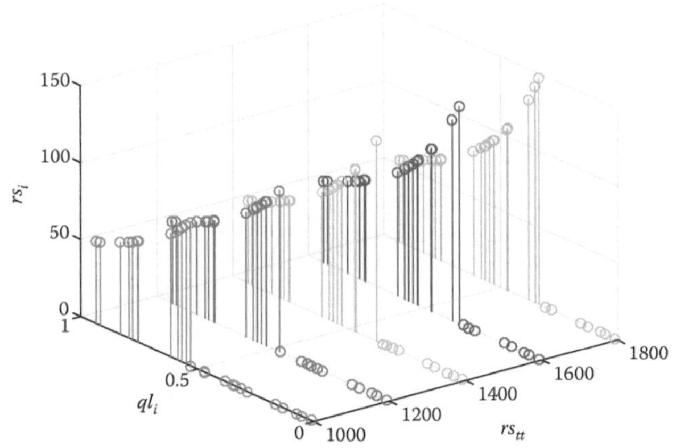

Figure 20.8 The optimal allocation results of hard QoS traffic based on Algorithm 20.2.1 under the channel quality parameter choice of SET_3 in Table 20.1. The three-dimensional coordinates are the resource allocation rs_i to user i, the channel quality of user i, and the total available resource rs_{tt} in the communication system.

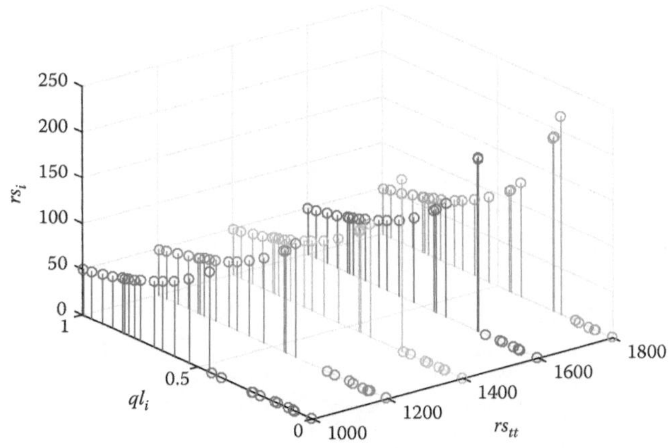

Figure 20.9 The optimal allocation results of hard QoS traffic based on Algorithm 20.2.1 under the channel quality parameter choice of SET_4 in Table 20.1. The three-dimensional coordinates are the resource allocation rs_i to user i, the channel quality of user i, and the total available resource rs_{tt} in the communication system.

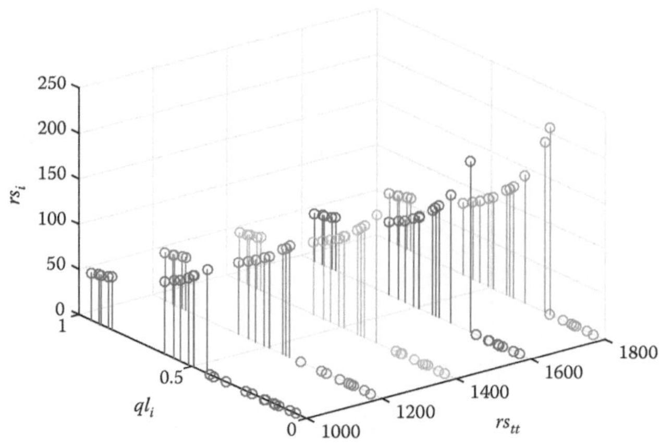

Figure 20.10 The optimal allocation results of hard QoS traffic based on Algorithm 20.2.1 under the channel quality parameter choice of SET_5 in Table 20.1. The three-dimensional coordinates are the resource allocation rs_i to user i, the channel quality of user i, and the total available resource rs_{tt} in the communication system.

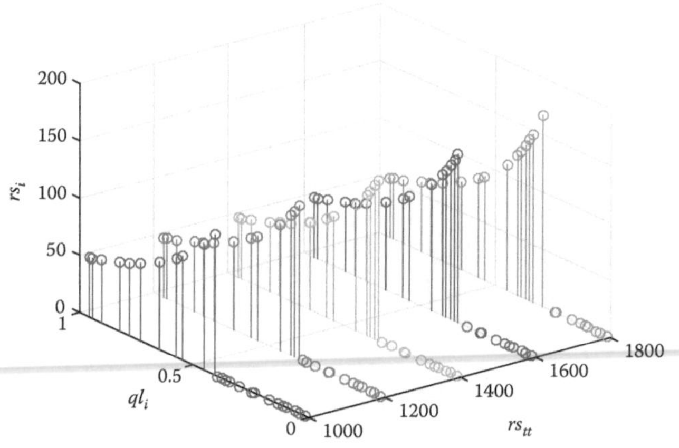

Figure 20.11 The optimal allocation results of hard QoS traffic based on Algorithm 20.2.1 under the channel quality parameter choice of SET_6 in Table 20.1. The three-dimensional coordinates are the resource allocation rs_i to user i, the channel quality of user i, and the total available resource rs_{tt} in the communication system.

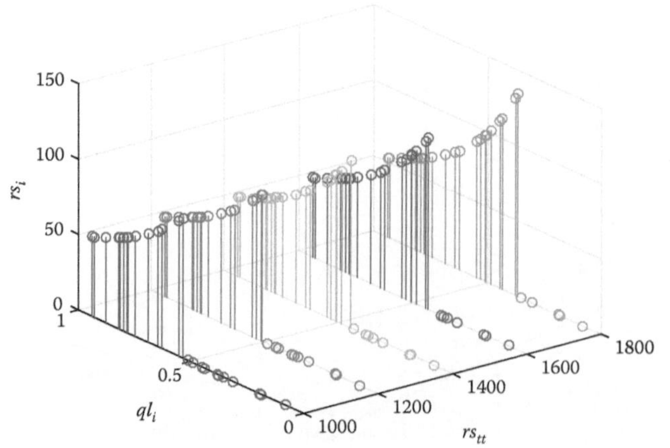

Figure 20.12 The optimal allocation results of hard QoS traffic based on Algorithm 20.2.1 under the channel quality parameter choice of SET_7 in Table 20.1. The three-dimensional coordinates are the resource allocation rs_i to user i, the channel quality of user i, and the total available resource rs_{tt} in the communication system.

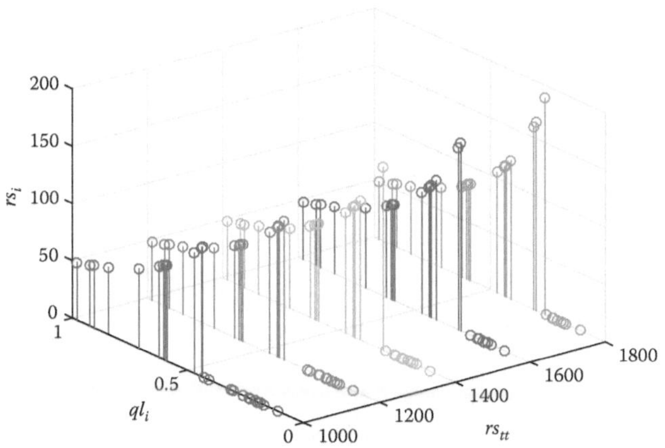

Figure 20.13 The optimal allocation results of hard QoS traffic based on Algorithm 20.2.1 under the channel quality parameter choice of SET_8 in Table 20.1. The three-dimensional coordinates are the resource allocation rs_i to user i, the channel quality of user i, and the total available resource rs_{tt} in the communication system.

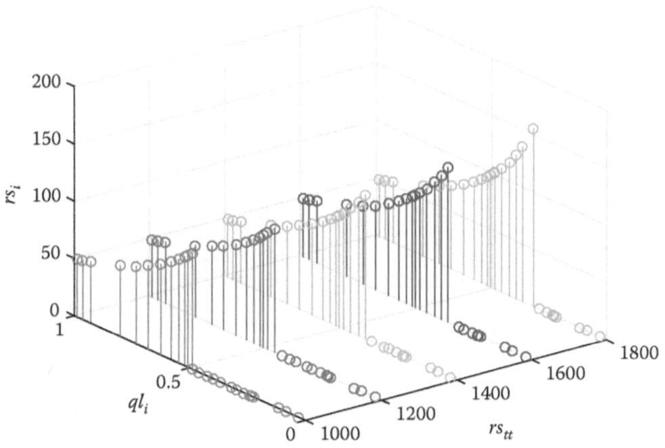

Figure 20.14 The optimal allocation results of hard QoS traffic based on Algorithm 20.2.1 under the channel quality parameter choice of SET_9 in Table 20.1. The three-dimensional coordinates are the resource allocation rs_i to user i, the channel quality of user i, and the total available resource rs_{tt} in the communication system.

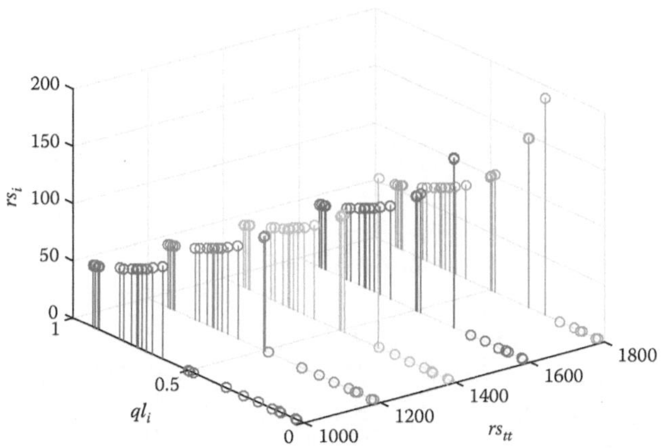

Figure 20.15 The optimal allocation results of hard QoS traffic based on Algorithm 20.2.1 under the channel quality parameter choice of SET_{10} in Table 20.1. The three-dimensional coordinates are the resource allocation rs_i to user i, the channel quality of user i, and the total available resource rs_{tt} in the communication system.

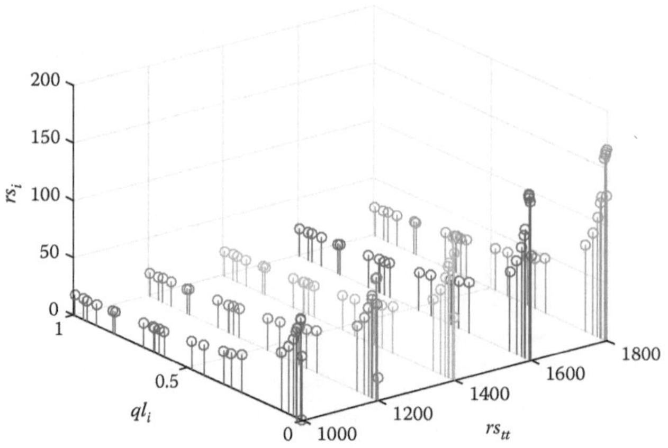

Figure 20.16 The optimal allocation results of elastic traffic based on Algorithm 20.2.2 under the channel quality parameter choice of SET_1 in Table 20.1. The three-dimensional coordinates are the resource allocation rs_i to user i, the channel quality of user i, and the total available resource rs_{tt} in the communication system.

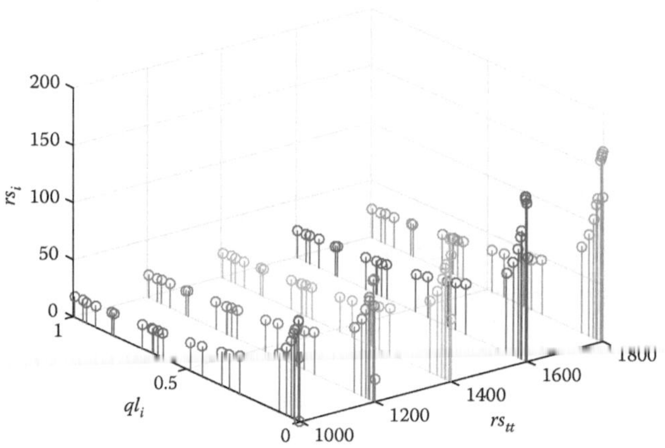

Figure 20.17 The optimal allocation results of elastic traffic based on Algorithm 20.2.2 under the channel quality parameter choice of SET_2 in Table 20.1. The three-dimensional coordinates are the resource allocation rs_i to user i, the channel quality of user i, and the total available resource rs_{tt} in the communication system.

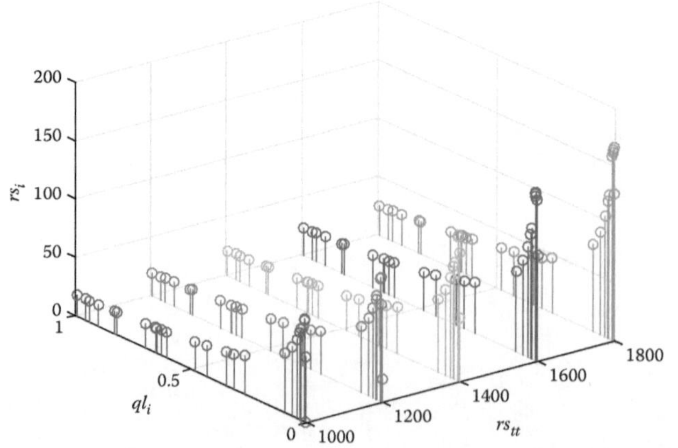

Figure 20.18 The optimal allocation results of elastic traffic based on Algorithm 20.2.2 under the channel quality parameter choice of SET_3 in Table 20.1. The three-dimensional coordinates are the resource allocation rs_i to user i, the channel quality of user i, and the total available resource rs_{tt} in the communication system.

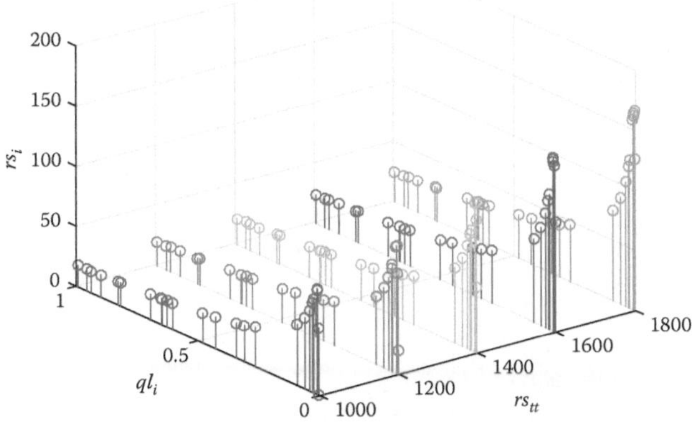

Figure 20.19 The optimal allocation results of elastic traffic based on Algorithm 20.2.2 under the channel quality parameter choice of SET_4 in Table 20.1. The three-dimensional coordinates are the resource allocation rs_i to user i, the channel quality of user i, and the total available resource rs_{tt} in the communication system.

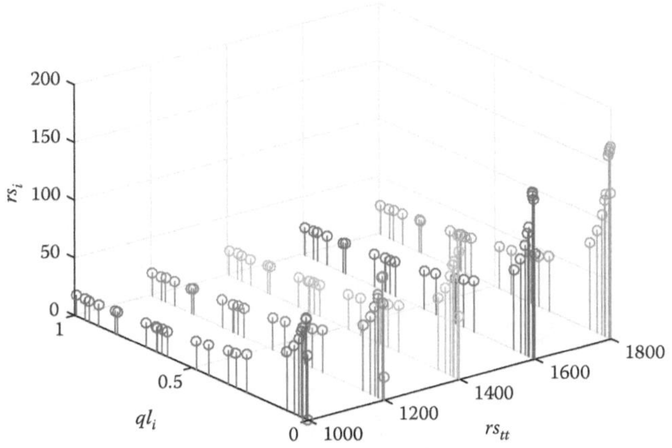

Figure 20.20 The optimal allocation results of elastic traffic based on Algorithm 20.2.2 under the channel quality parameter choice of SET_5 in Table 20.1. The three-dimensional coordinates are the resource allocation rs_i to user i, the channel quality of user i, and the total available resource rs_{tt} in the communication system.

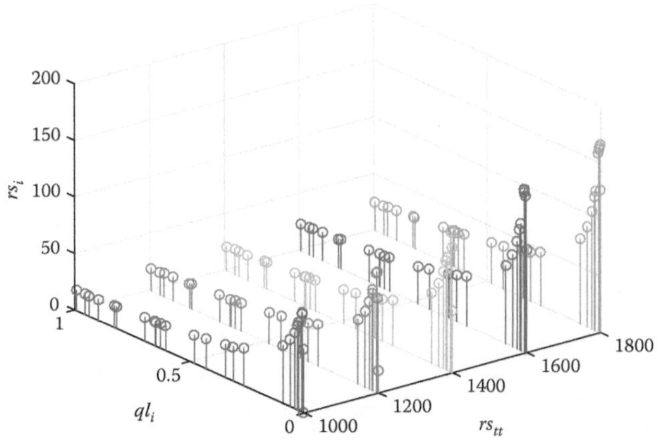

Figure 20.21 The optimal allocation results of elastic traffic based on Algorithm 20.2.2 under the channel quality parameter choice of SET_6 in Table 20.1. The three-dimensional coordinates are the resource allocation rs_i to user i, the channel quality of user i, and the total available resource rs_{tt} in the communication system.

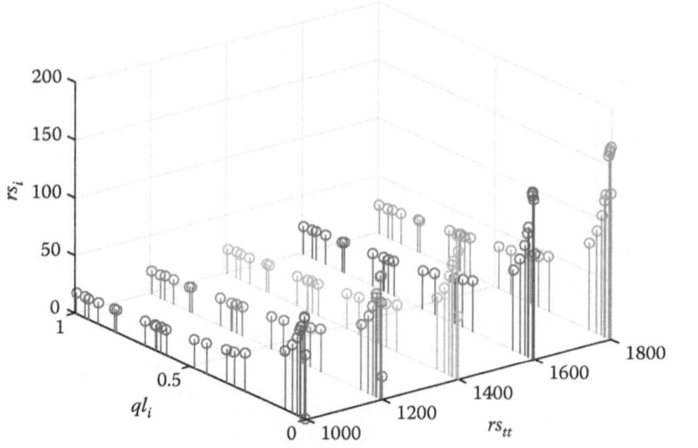

Figure 20.22 The optimal allocation results of elastic traffic based on Algorithm 20.2.2 under the channel quality parameter choice of SET_7 in Table 20.1. The three-dimensional coordinates are the resource allocation rs_i to user i, the channel quality of user i, and the total available resource rs_{tt} in the communication system.

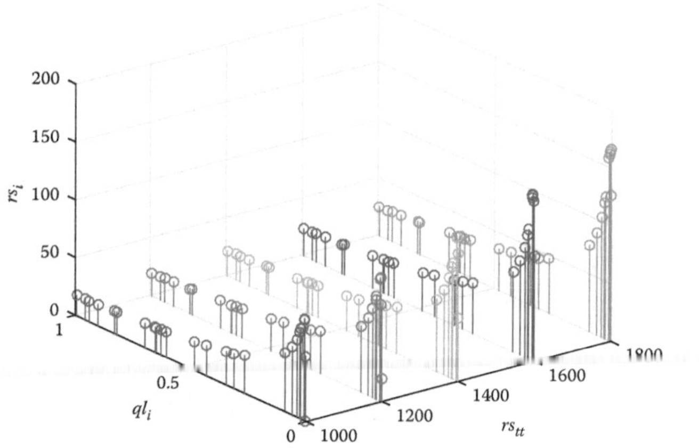

Figure 20.23 The optimal allocation results of elastic traffic based on Algorithm 20.2.2 under the channel quality parameter choice of SET_8 in Table 20.1. The three-dimensional coordinates are the resource allocation rs_i to user i, the channel quality of user i, and the total available resource rs_{tt} in the communication system.

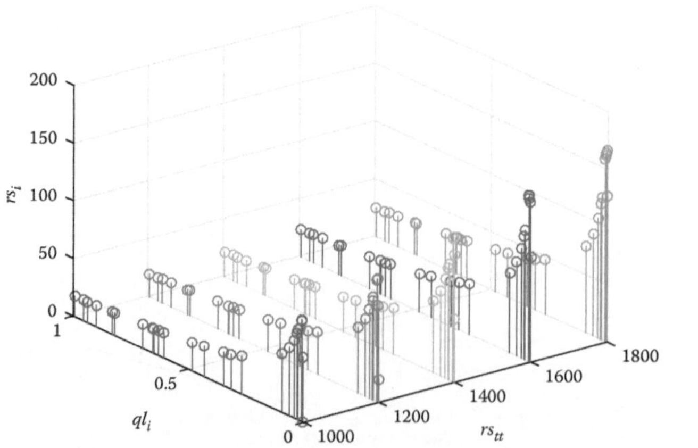

Figure 20.24 The optimal allocation results of elastic traffic based on Algorithm 20.2.2 under the channel quality parameter choice of SET_9 in Table 20.1. The three-dimensional coordinates are the resource allocation rs_i to user i, the channel quality of user i, and the total available resource rs_{tt} in the communication system.

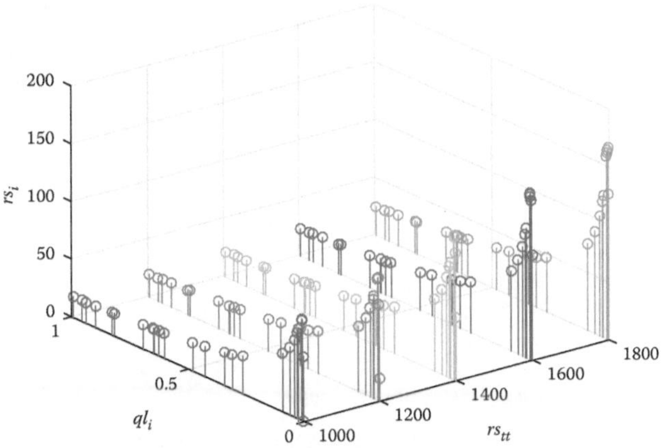

Figure 20.25 The optimal allocation results of elastic traffic based on Algorithm 20.2.2 under the channel quality parameter choice of SET_{10} in Table 20.1. The three-dimensional coordinates are the resource allocation rs_i to user i, the channel quality of user i, and the total available resource rs_{tt} in the communication system.

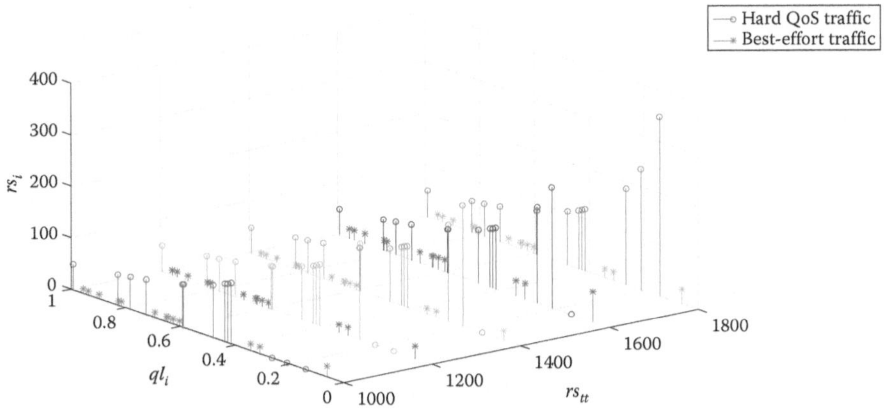

Figure 20.26 The optimal allocation results of mixture traffic based on Algorithm 20.2.3 under the channel quality parameter choice of SET_1 in Table 20.1. The three-dimensional coordinates are the resource allocation rs_i to user i, the channel quality of user i, and the total available resource rs_{tt} in the communication system.

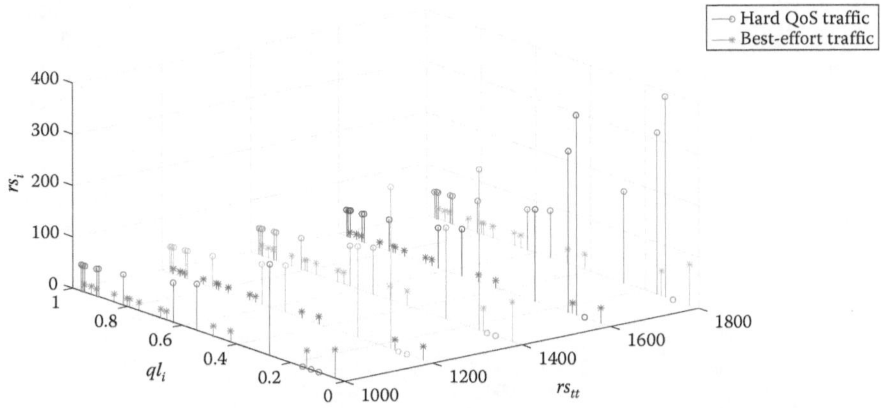

Figure 20.27 The optimal allocation results of mixture traffic based on Algorithm 20.2.3 under the channel quality parameter choice of SET_2 in Table 20.1. The three-dimensional coordinates are the resource allocation rs_i to user i, the channel quality of user i, and the total available resource rs_{tt} in the communication system.

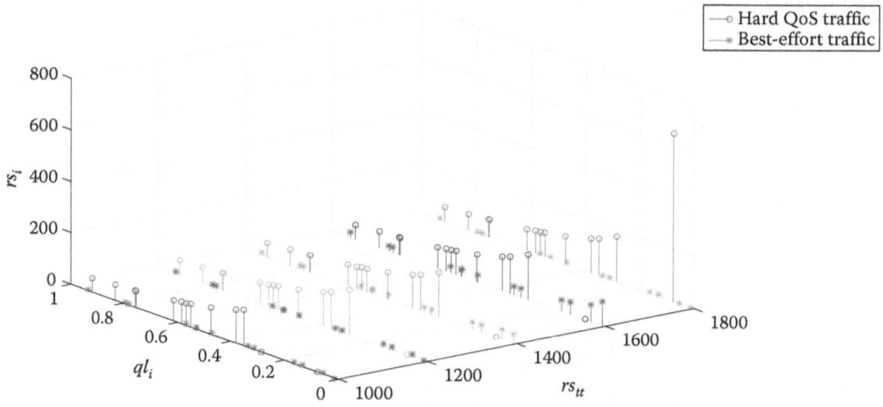

Figure 20.28 The optimal allocation results of mixture traffic based on Algorithm 20.2.3 under the channel quality parameter choice of SET_3 in Table 20.1. The three-dimensional coordinates are the resource allocation rs_i to user i, the channel quality of user i, and the total available resource rs_{tt} in the communication system.

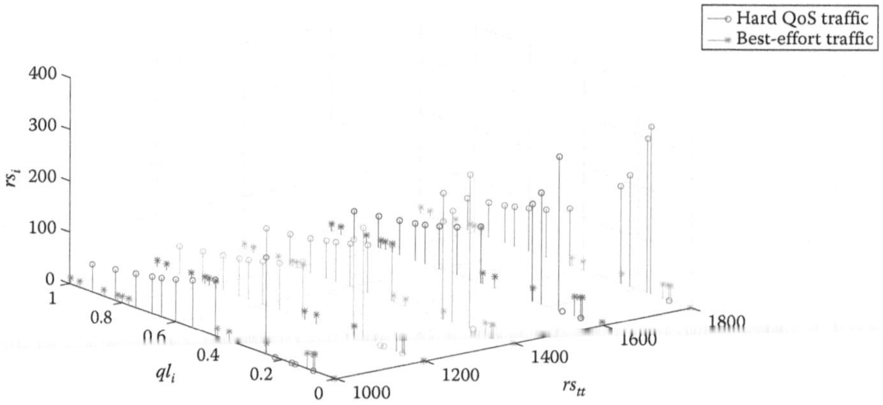

Figure 20.29 The optimal allocation results of mixture traffic based on Algorithm 20.2.3 under the channel quality parameter choice of SET_4 in Table 20.1. The three-dimensional coordinates are the resource allocation rs_i to user i, the channel quality of user i, and the total available resource rs_{tt} in the communication system.

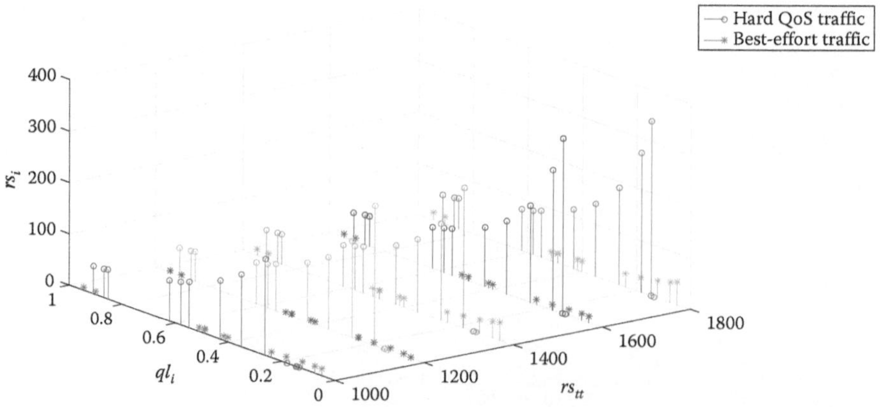

Figure 20.30 The optimal allocation results of mixture traffic based on Algorithm 20.2.3 under the channel quality parameter choice of SET_5 in Table 20.1. The three-dimensional coordinates are the resource allocation rs_i to user i, the channel quality of user i, and the total available resource rs_{tt} in the communication system.

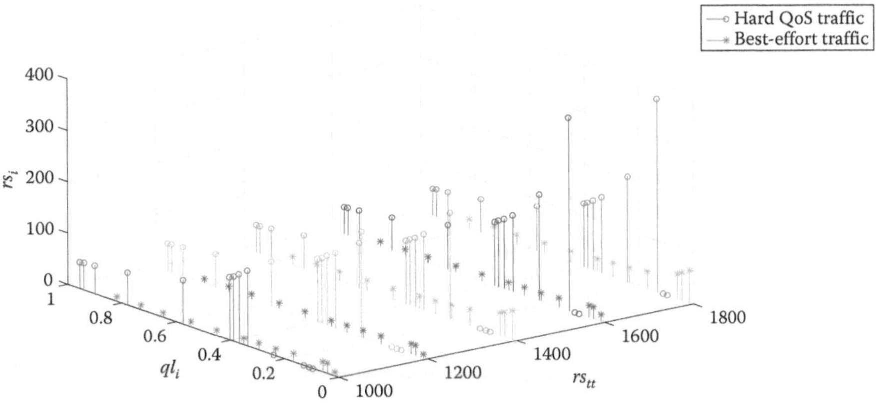

Figure 20.31 The optimal allocation results of mixture traffic based on Algorithm 20.2.3 under the channel quality parameter choice of SET_6 in Table 20.1. The three-dimensional coordinates are the resource allocation rs_i to user i, the channel quality of user i, and the total available resource rs_{tt} in the communication system.

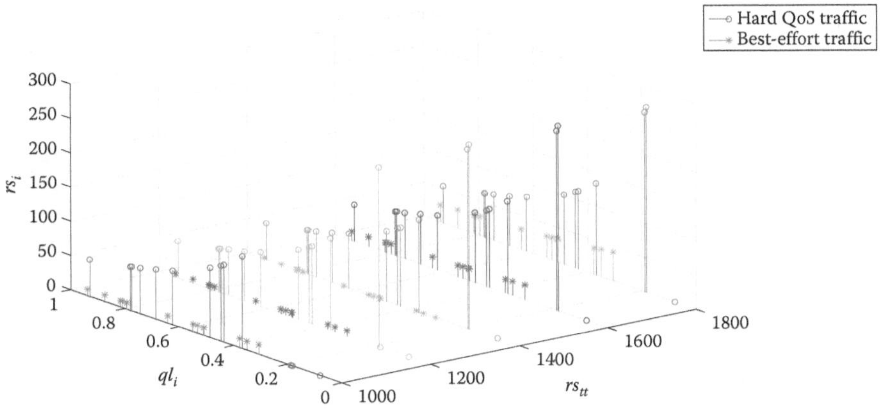

Figure 20.32 The optimal allocation results of mixture traffic based on Algorithm 20.2.3 under the channel quality parameter choice of SET_7 in Table 20.1. The three-dimensional coordinates are the resource allocation rs_i to user i, the channel quality of user i, and the total available resource rs_{tt} in the communication system.

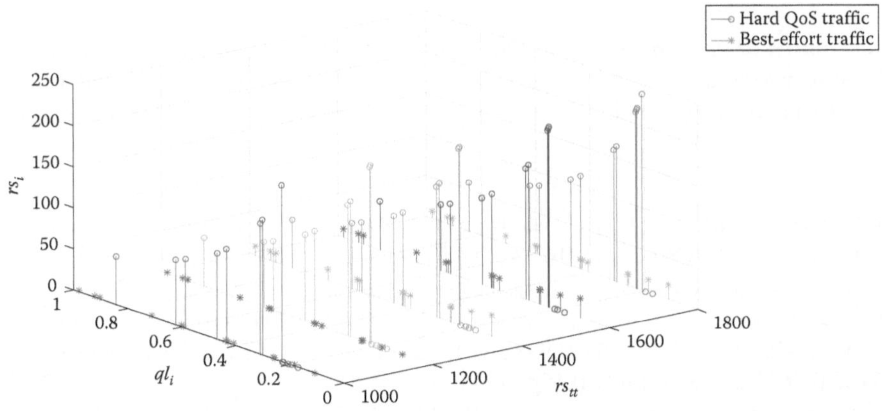

Figure 20.33 The optimal allocation results of mixture traffic based on Algorithm 20.2.3 under the channel quality parameter choice of SET_8 in Table 20.1. The three-dimensional coordinates are the resource allocation rs_i to user i, the channel quality of user i, and the total available resource rs_{tt} in the communication system.

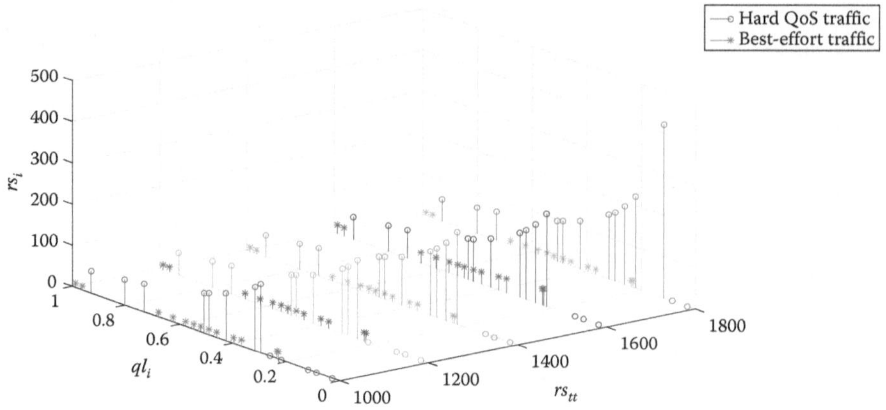

Figure 20.34 The optimal allocation results of mixture traffic based on Algorithm 20.2.3 under the channel quality parameter choice of SET_9 in Table 20.1. The three-dimensional coordinates are the resource allocation rs_i to user i, the channel quality of user i, and the total available resource rs_{tt} in the communication system.

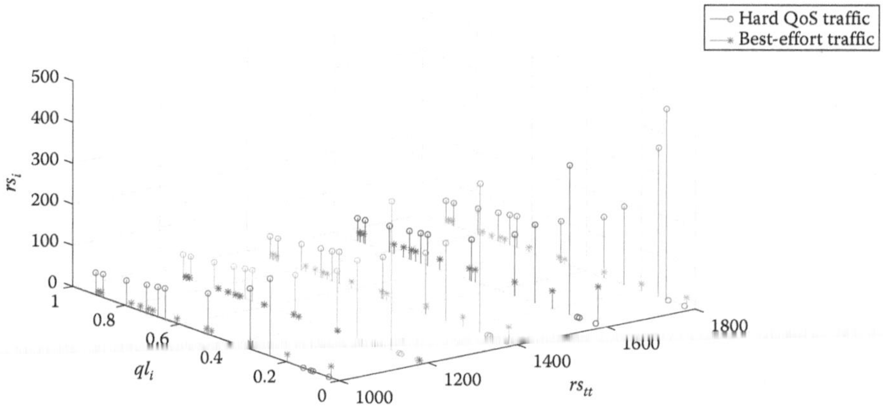

Figure 20.35 The optimal allocation results of mixture traffic based on Algorithm 20.2.3 under the channel quality parameter choice of SET_{10} in Table 20.1. The three-dimensional coordinates are the resource allocation rs_i to user i, the channel quality of user i, and the total available resource rs_{tt} in the communication system.

algorithm of *mixed allocation* (namely Algorithm 20.2.3). The numerical results of fairness index for the above scenarios are plotted in Figures 20.36 through 20.38, respectively, under the following channel quality parameters, which are randomly selected from $[0, 1]$ with uniform distribution:

- For Scenario 1, $[ql_i,\ i = 1, 2, \ldots, 26]$ = [0.0026 0.0091 0.0044 0.0300 0.0096 0.0421 0.0933 0.0262 0.4284 0.0108 0.2654 0.9932 0.4808 0.8278 0.6032 0.8178 0.9555 0.6504 0.6276 0.6466 0.0621 0.9382 0.8987 0.6949 0.3104 0.3435].
- For Scenario 2, $[ql_i,\ i = 1, 2, \ldots, 26]$ =[0.0026 0.0091 0.0044 0.0300 0.0096 0.0421 0.0933 0.0262 0.4284 0.0108 0.2654 0.9932 0.4808 0.8278 0.6032 0.8178 0.9555 0.6504 0.6276 0.6466 0.0621 0.9382 0.8987 0.6949 0.3104 0.3435].
- For Scenario 3, $[ql_i,\ i = 1, 2, \ldots, 26]$ =[0.0026 0.0091 0.0044 0.0300 0.0096 0.0421 0.0933 0.0262 0.4284 0.0108 0.2654 0.9932 0.4808 0.8278 0.6032 0.8178 0.9555 0.6504 0.6276 0.6466 0.0621 0.9382 0.8987 0.6949 0.3104 0.3435].

Table 20.1
Randomly Generated Channel Quality Parameters That Are Used in the Numerical Studies

The numerical result plots	The used channel quality parameters ($SET_j = [ql_i,\ i = 1,2,\ldots,26]$, $j = 1,2,\ldots,10$), which are randomly generated from $(0, 1)$ with uniform distribution
Figures 20.6, 20.16, and 20.26	SET_1 = [0.4160 0.4391 0.7844 0.8300 0.5896 0.1421 0.5933 0.7262 0.4284 0.2108 0.2654 0.9932 0.4808 0.8278 0.6032 0.8178 0.9555 0.6504 0.6276 0.6466 0.0621 0.9382 0.8987 0.6949 0.3104 0.3435]
Figures 20.7, 20.17, and 20.27	SET_2 = [0.8147 0.9058 0.1270 0.9134 0.6324 0.0975 0.2785 0.5469 0.9575 0.9649 0.1576 0.9706 0.9572 0.4854 0.8003 0.1419 0.4218 0.9157 0.7922 0.9595 0.6557 0.0357 0.8491 0.9340 0.6787 0.7577]
Figures 20.8, 20.18, and 20.28	SET_3 = [0.6160 0.4733 0.3517 0.8308 0.5853 0.5497 0.9172 0.2858 0.7572 0.7537 0.3804 0.5678 0.0759 0.0540 0.5308 0.7792 0.9340 0.1299 0.5688 0.4694 0.0119 0.3371 0.1622 0.7943 0.3112 0.5285]
Figures 20.9, 20.19, and 20.29	SET_4 = [0.1656 0.6020 0.2630 0.6541 0.6892 0.7482 0.4505 0.0838 0.2290 0.9133 0.1524 0.8258 0.5383 0.9961 0.0782 0.4427 0.1067 0.9619 0.0046 0.7749 0.8173 0.8687 0.0844 0.3998 0.2599 0.8001]
Figures 20.10, 20.20, and 20.30	SET_5 = [0.4314 0.9106 0.1818 0.2638 0.1455 0.1361 0.8693 0.5797 0.5499 0.1450 0.8530 0.6221 0.3510 0.5132 0.4018 0.0760 0.2399 0.1233 0.1839 0.2400 0.4173 0.0497 0.9027 0.9448 0.4909 0.4893]
Figures 20.11, 20.21, and 20.31	SET_6 = [0.3377 0.9001 0.3692 0.1112 0.7803 0.3897 0.2417 0.4039 0.0965 0.1320 0.9421 0.9561 0.5752 0.0598 0.2348 0.3532 0.8212 0.0154 0.0430 0.1690 0.6491 0.7317 0.6477 0.4509 0.5470 0.2963]
Figures 20.12, 20.22, and 20.32	SET_7 = [0.7447 0.1890 0.6868 0.1835 0.3685 0.6256 0.7802 0.0811 0.9294 0.7757 0.4868 0.4359 0.4468 0.3063 0.5085 0.5108 0.8176 0.7948 0.6443 0.3786 0.8116 0.5328 0.3507 0.9390 0.8759 0.5502]
Figures 20.13, 20.23, and 20.33	SET_8 = [0.6225 0.5870 0.2077 0.3012 0.4709 0.2305 0.8443 0.1948 0.2259 0.1707 0.2277 0.4357 0.3111 0.9234 0.4302 0.1848 0.9049 0.9797 0.4389 0.1111 0.2581 0.4087 0.5949 0.2622 0.6028 0.7112]
Figures 20.14, 20.24, and 20.34	SET_9 = [0.2217 0.1174 0.2967 0.3188 0.4242 0.5079 0.0855 0.2625 0.8010 0.0292 0.9289 0.7303 0.4886 0.5785 0.2373 0.4588 0.9631 0.5468 0.5211 0.2316 0.4889 0.6241 0.6791 0.3955 0.3674 0.9880]
Figures 20.15, 20.25, and 20.35	SET_{10} = [0.0377 0.8852 0.9133 0.7962 0.0987 0.2619 0.3354 0.6797 0.1366 0.7212 0.1068 0.6538 0.4942 0.7791 0.7150 0.9037 0.8909 0.3342 0.6987 0.1978 0.0305 0.7441 0.5000 0.4799 0.9047 0.6099]

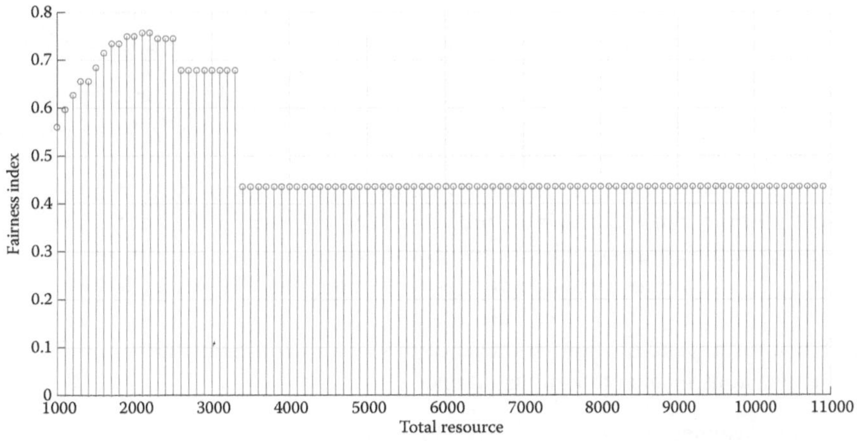

Figure 20.36 Fairness index of hard QoS allocation for Scenario 1.

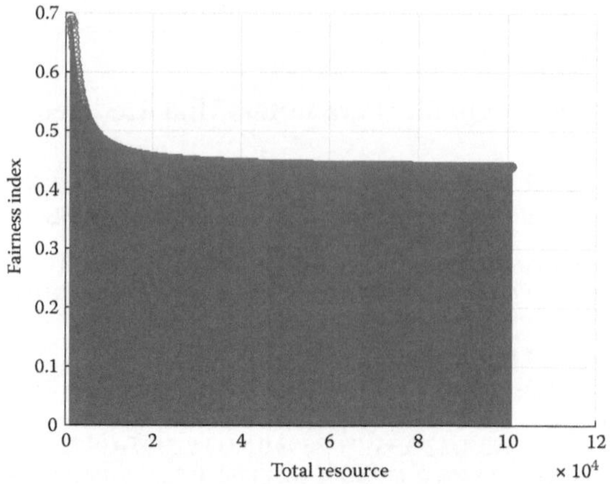

Figure 20.37 Fairness index of elastic allocation for Scenario 2.

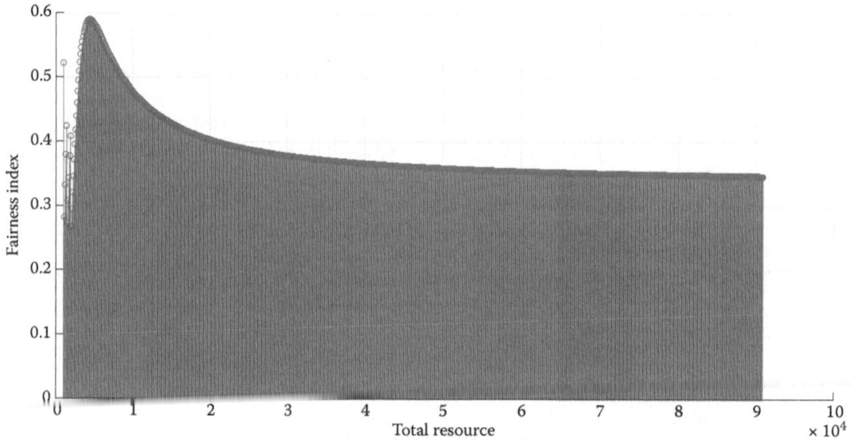

Figure 20.38 Fairness index of mixed allocation for mixture of hard QoS traffic and best-effort traffic for Scenario 3.

20.4 RADIO RESOURCE ALLOCATION IN WIRELESS NETWORKS: IMPLEMENTATION AND PERFORMANCE OPTIMIZATION

Resource allocation in wireless networks can be performed in the following three manners:

- Centralized radio resource allocation scheme: a central entity like a base station makes channel assignment and power allocation decisions, and communicates these decisions to the sensor nodes.
- Cluster-based radio resource allocation schemes: a wireless sensor network is usually designed into the clustered architecture. The sensor nodes in the network are divided into small groups called clusters where each cluster is controlled by a central entity called CH. In the cluster-based radio resource allocation scheme, the CH indicates the cluster members about spectrum sharing and power allocation.
- Distributed resource allocation schemes: each wireless communication node makes its transmission decisions in an autonomous manner.

The schemes are designed on the basis of performance optimization criteria that include energy efficiency, throughput maximization, QoS assurance, interference avoidance, fairness and priority consideration, and hand-off reduction.

21 Utility Optimization-Based Resource Allocation for Soft QoS Traffic

Optimization theory and nonlinear programming method have successfully been applied into wire-lined networks (e.g., the Internet) in developing efficient resource allocation and congestion control schemes. The resource (e.g., bandwidth) allocation in a communication network has been modeled into an optimization problem: the objective is to maximize the source aggregate utility subject to the network resource constraint. However, for wireless networks, allocating the resource among the soft quality of service (QoS) traffic remains an important design challenge. Mathematically, the most difficulty comes from the nonconcave utility function of soft QoS traffic in the network utility maximization (NUM) problem. Previous result on this problem has only been able to find its suboptimal solution. Facing this challenge, this chapter establishes some key theorems to find the optimal solution, and then present a complete algorithm called utility-based allocation for soft QoS (USQ) to obtain the desired optimal solution. The proposed theorems and algorithm act as designing guidelines for resource allocation of soft QoS traffic in a wireless network, which take into account the total available resource of network, the users traffic characteristics, and the users channel qualities. By numerical examples, we illustrate the explicit solution procedures.

21.1 INTRODUCTION

The utility optimization model has been widely adopted in the research of end-to-end congestion control [104,105,112,624] by viewing variant congestion control protocols as distributed algorithms to solve some basic NUM problems. Consider a communication network (e.g., the Internet) with L links, each with a fixed capacity of c_l bps, and S sources (i.e., end users or flows), each transmitting at a source rate of x_s bps. It is assumed that each source s emits only one flow, using a fixed set $L(s)$ of links in its path, and has a utility function $U_s(x_s)$. The utility function describes the degree of user satisfaction when allocated by a certain amount of bandwidth. Each link l is shared by a set $S(l)$ of sources. NUM is the following problem of maximizing the total utility of the network:

$$\sum_{s} U_s(x_s),$$

over the source rates x_s, subject to linear flow constraints

$$\sum_{s:l\in L(s)} x_s \leq c_l$$

for all links l. The solution strategy of the above NUM problem has successfully been developed [111] into distributed primal-dual algorithms for congestion controlling of elastic traffic in a wire-lined network (e.g., the Internet).

In telecommunication networks, QoS refers to several related aspects that allow the transport of traffic with special requirements. QoS is the ability to provide different priority to different applications or to guarantee a certain level of performance to a data flow. For example, a required bit rate, delay, jitter, packet dropping probability, and/or bit error rate may be guaranteed. QoS guarantees are important if the network capacity is insufficient, especially for real-time streaming multimedia applications such as voice over Internet protocol (IP), online games and Internet protocol Television (IP-TV), and in networks where the capacity is a limited resource, for example, in cellular data communication.

In order to guarantee the QoS in future wireless networks, it is desirable to implement certain resource allocation mechanism to allocate the limited resource (e.g., bandwidth) among all the users fairly and efficiently. The paper [627] is concerned with the bandwidth allocation problem for cooperative relay networks. The bandwidth allocation problem is therein formulated as a Nash bargaining problem, and then the bandwidth allocation algorithm is discussed using the subgradient method. For wireless mesh networks (WMNs), a solution to address channel assignment termed as the extended level-based channel assignment (ELCA) scheme is presented in [628]. In [632], the theoretically achievable average channel capacity (in the Shannon sense) per user of a hybrid cellular direct sequence/fast frequency hopping code-division multiple-access (DS/FFH-CDMA) system, operating in a Rayleigh fading environment, was examined. The paper [629] presents a performance analysis of dynamic channel allocation (DCA) based on the greedy approach (GA) for orthogonal, frequency-division, multiple-access downlink systems over Rayleigh fading channels. Relevant work also includes [631], where for mobile ad hoc networks, an energy-efficient routing approach is presented to study the resource allocation issue.

It is established that the wireless communication system may allocate its resource to users according to some performance metrics such as throughput and fairness [128,623,634] or according to the type of traffic [620,630]. Due to the difficulty in trading off throughput and fairness, the allocation problem is usually approached by focusing on users' satisfaction [128,626], which is specific to traffic type. The degree of user satisfaction can be described by the utility function of the traffic type under consideration. Different shapes of utility functions lead to optimal resource allocations that satisfy the existing definitions of fairness [622], such as max–min fairness [139], proportional fairness [105], and $(\alpha - n)$ proportional fairness [115]. And, utility functions can provide a new metric to define optimality of resource allocation efficiency.

Utility-based allocation schemes in wireless networks are proposed in [609,633]. The paper [625] studied the issue of resource allocation for two types of traffic, i.e., best effort and hard QoS traffic. For soft QoS traffic that demands a preferred amount of bandwidth with some flexibility for normal operation, the paper [626] proposed a utility-based resource allocation scheme. Unfortunately, the proposed algorithm is only suboptimal, and the authors pointed out that it was difficult to give an approach to obtain the optimal solution. The present chapter is to face this challenge and to obtain the optimal solution for this problem. We establish some key theorems to find the optimal solution to this problem; on this basis, we then present a complete algorithm called USQ to obtain the desired optimal solution. The proposed theorems and algorithm act as designing guidelines for resource allocation of soft QoS traffic in a wireless network, which take into account the total available resource of network, the users' traffic characteristics, and the users' channel qualities. By numerical examples, we illustrate the explicit solution procedures.

Our approach establishes that in a wireless network, the base station can optimally allocate the resource among the competing soft QoS traffic to maximize the total utility of all users. This allocation takes into account the users' utility characteristic, the users' channel conditions, and the total available resource of the system. The new developments like the adaptive modulation and coding (AMC) schemes [621] in wireless networks enable the upper layers to access user channel conditions. Therefore, our algorithm can be implemented at the base station in a wireless network to perform the bandwidth allocation function by using these new technologies.

The rest of this chapter is organized as follows: in Section 21.2, the theoretical analysis and the main results are presented. The USQ algorithm is proposed in Section 21.3. In Section 21.4, we then verify the performance of this algorithm via some numerical examples. Finally, Section 21.5 concludes this chapter.

21.2 PROBLEM DESCRIPTION AND OPTIMAL SOLUTION OF UTILITY MAXIMIZATION

A utility function is defined as a curve mapping the amount of bandwidth received by the application to the performance as perceived by the end user. Utility function is monotonically nondecreasing; in other words, more bandwidth allocation should not lead to degraded application performance. The key advantage of utility function is that it can inherently reflect the QoS requirements of the end user. The exact expression of a utility function may depend on the type of traffic and can be derived from some metrics that reflect the user's perception or the content quality.

21.2.1 UTILITY FUNCTION OF SOFT QOS TRAFFIC AND THE UTILITY MAXIMIZATION PROBLEM

Soft QoS traffic [602] refers to the applications that have flexible bandwidth requirements. In case of congestion, they can gracefully adjust their transmission rates to adapt to various network conditions. However, such applications have an intrinsic bandwidth requirement because the data generation rate is independent of the network congestion. Typical examples of soft QoS traffic are interactive multimedia services and video on demand. The utility function of soft QoS traffic is assumed to be a sigmoidal function (see, e.g., [626]) in NUM approach, which is a nonconcave function with a shape needed to be determined by a couple of parameters.

Without loss of generality, we study the following sigmoid utility function $U(r)$:

$$U(r) = \begin{cases} qe^{p(r-rc)} & r < rc, \\ 1-(1-q)e^{-p(r-rc)} & r \geq rc, \end{cases} \tag{21.1}$$

where r is the bandwidth being allocated to the user, which may also refer to time slots or radio frequency occupied by this user. The component rc denotes the preferable amount of resource for the soft QoS traffic. The component q is the channel quality parameter representing the ratio of actual amount of resource received by the user to the amount of resource allocated by the base station to the user, which is in the range of $[0,1]$. It is the utility value when $r = rc$. The parameter p determines the slope of the utility function, and it characterizes the flexibility level to bandwidth requirement of the flow. When p increases, the shape of the sigmoid utility function becomes closer to that of a unit-step function. Note that a unit-step function represents the utility function of hard QoS traffic. Hard QoS traffic refers to the applications with stringent bandwidth requirements. A call belonging to this type of traffic requires strict end-to-end performance guarantees and does not show any adaptive properties. For $q = 0.4, p = 0.12,$ and $rc = 40$, we depict a utility function in Figure 21.1.

The utility function $U(r)$ given by Equation 21.1 is a nondecreasing function with respect to the amount of allocated resource r: the more the resource is allocated, the more the user is satisfied. To see this feature, we take a look at its marginal utility function $u(r)$, which is the derivative of the utility function $U(r)$ with respect to r. Namely,

$$u(r) = U'(r) = \begin{cases} qpe^{p(r-rc)} & r < rc, \\ (1-q)pe^{-p(r-rc)} & r \geq rc. \end{cases} \tag{21.2}$$

By taking $q = 0.4, p = 0.12,$ and $rc = 40$, for example, the above marginal utility function $u(r)$ is depicted in Figure 21.2. The utility function given by Equation 21.1 is a nonconcave function with respect to r. This is seen from the following derivative of the marginal utility function:

$$u'(r) = U''(r) = \begin{cases} qp^2e^{p(r-rc)} & r < rc, \\ -p^2(1-q)e^{-p(r-rc)} & r \geq rc. \end{cases} \tag{21.3}$$

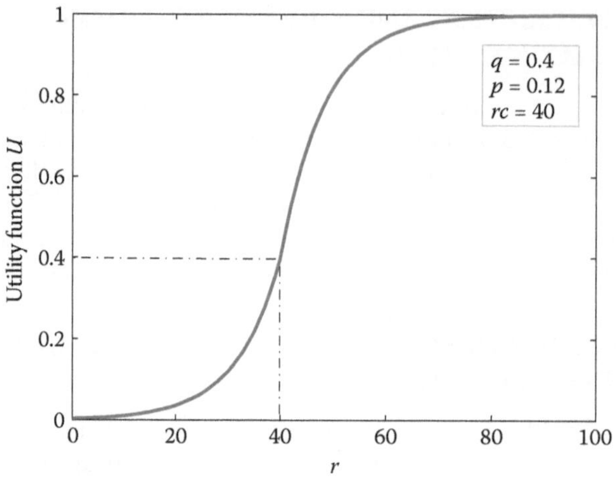

Figure 21.1 Sigmoid utility function.

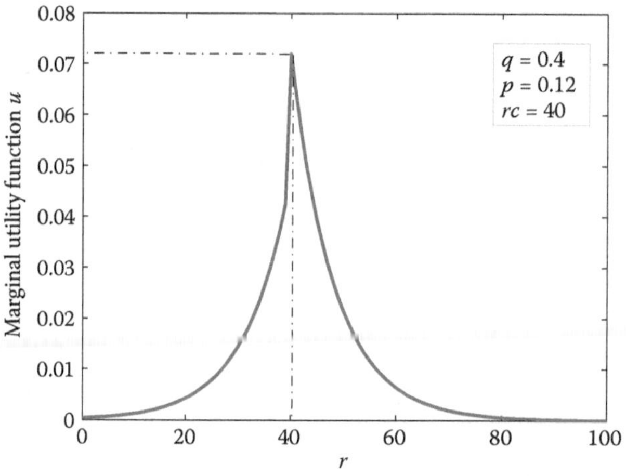

Figure 21.2 Marginal utility function.

Again by taking $q = 0.4, p = 0.12$, and $rc = 40$, the above function is depicted in Figure 21.3. Observing from Equation 21.3 and Figure 21.3, one notes that there is a special point $r = rc$. rc is referred to as the preferable amount of resource for the soft QoS traffic. It is the single inflexion point of the derivative of the marginal utility function satisfying $U''(rc) = 0$. We observe that if $r < rc$, $u'(r) > 0$; if $r \geq rc$, $u'(r) \leq 0$. It can be interpreted as follows: when the amount of bandwidth r is given insufficiently, it is unsatisfactory for user for real-time application; as the bandwidth value r approaches its desired value, the flow becomes gradually operational, and thus the marginal utility increases dramatically. Once the allocated amount of resource r exceeds rc too much, allocating more resource may not be helpful for operation, the user still runs around its normal operation, and therefore the marginal utility drops hence forth.

Let us consider a wireless network that consists of a base station and a set of soft QoS traffic (users) denoted by Γ. All the soft QoS traffic are competing for the total resource (e.g., bandwidth) valued at C at the base station in communications. Denotations of all the variables are given in Table 21.1. We depict this communication system in Figure 21.4.

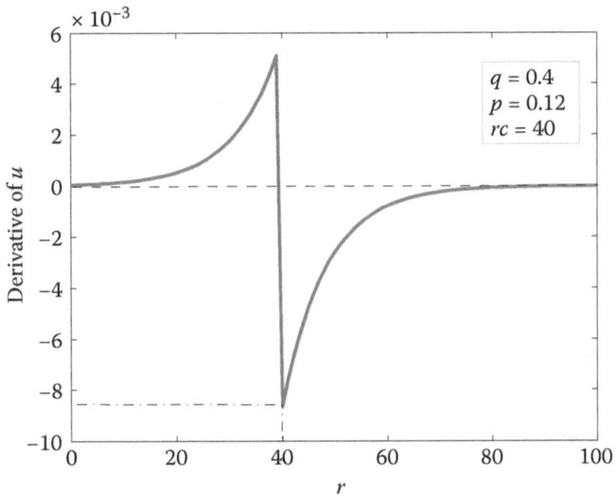

Figure 21.3 Derivative of marginal utility function.

Table 21.1
Nomenclature

Γ	The set of soft QoS traffic in a wireless network, $\lvert\Gamma\rvert = n$
C	The total amount of radio resource available at the base station
r_i	The amount of resource to be allocated to flow i
q_i	The channel quality of flow i
$U(\bullet)$	The sigmoid utility function
rc	The preferable amount of resource for the soft QoS traffic
$U_i(\bullet)$	The utility function of flow i
rc_i	The preferred value flow i
$u_i(\bullet)$	The marginal utility function of $U_i(\bullet)$
$u'_i(\bullet)$	The derivation of $u_i(\bullet)$

Assume that each flow belonging to Γ has a sigmoid utility function $U(\bullet)$ given by Equation 21.1. Let r_i denote the amount of resource allocated to flow i, $i \in \Gamma$, and q_i the channel quality of flow i, where $0 \leq q_i \leq 1$. The bigger the value q_i, the better the channel quality. Let p_i be the parameter determining the flexibility level to its preferred bandwidth value for the soft QoS traffic i. It is reasonable to assume that (1) the base station has the information of each user's parameters like q_i, p_i and (2) users under different channel conditions can transmit data at different rates.

Given the channel quality q_i of flow i, if it is allocated by the bandwidth value r_i at the base station, the amount of bandwidth it receives for operation is actually r_iq_i. Therefore, the actual utility function of flow i can be expressed as $U_i(\bullet) = U(r_iq_i)$, where $U(\bullet)$ is the utility function of the soft QoS traffic under consideration (formulated by 21.1) and $U_i(\bullet)$ is the utility function for the specific traffic described by $U(\bullet)$ but taking into account the channel quality parameter q_i and others. One sees that the actual beneficial utility value the user i "received" is less than the "allocated" utility value for a bandwidth allocation, i.e., $U_i(r_iq_i) \leq U_i(r_i)$. For flow i, the marginal utility function of $U_i(\bullet)$, denoted by $u_i(\bullet)$, is defined by $u_i(r_i) = \mathrm{d}U(r_iq_i)/\mathrm{d}r_i = q_i \times u(q_i \times r_i)$. Considering the channel condition of each flow may not be identical, the preferable amount of resource for flow i is denoted by rc_i. For each flow i, $i \in \Gamma$, $U_i(rc_i) = U(rc) = U_c$, and thus $rc_i = rc/q_i$.

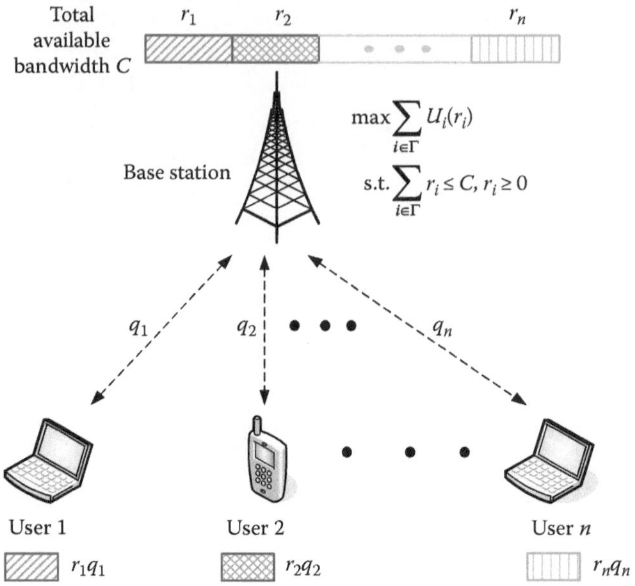

Figure 21.4 The network utility optimization for bandwidth allocation in a wireless network.

Now, we are ready to describe the targeted problem as follows (refer to Figure 21.4):

$$P1 : \max \sum_{i\in\Gamma} U_i(r_i) = \sum_{i\in\Gamma} U(r_i q_i)$$
$$\text{s.t.} : \sum_{i\in\Gamma} r_i \le C, r_i \ge 0, i \in \Gamma,$$

where

$$U(r_i q_i) = \begin{cases} q_i e^{p_i(r_i q_i - rc)} & r_i q_i < rc, \\ 1-(1-q_i)e^{-p_i(r_i q_i - rc)} & r_i q_i \ge rc. \end{cases} \tag{21.4}$$

21.2.2 OPTIMAL SOLUTION TO THE UTILITY MAXIMIZATION PROBLEM

Definition 21.2.1 *The solution $\mathscr{R} = \{r_i, i \in \Gamma\}$ to the above problem $P1$ is optimal if, for any other allocation solution $\mathscr{R}' = \{r_j', j \in \Gamma\}$, we have $U(\mathscr{R}) \ge U(\mathscr{R}')$, where*

$$U(\mathscr{R}) = \sum_{i\in\Gamma} U_i(r_i), U(\mathscr{R}') = \sum_{i\in\Gamma} U_i(r_i').$$

According to Lemma 2.2 in [626], the optimal solution of $P1$ is at most one flow's derivation of the marginal utility function which is positive. To obtain the optimal solution, we only need to consider the following *two classes*:

1. $\forall i \in \Gamma$, $u_i'(r_i) < 0$ if $r_i > 0$.
2. $\forall i \in \Gamma$, there exists unique flow i whose $u_i'(r_i) > 0$ if $r_i > 0$.

For the first case, in [626], the authors proposed a suboptimal algorithm. In the meantime, they pointed out that it was difficult to give an approach to obtain the optimal solution of the whole problem. The current chapter attacks this problem. Our procedure works according to the following mechanism: first to obtain the optimal solution that satisfies (ii), and then to compare it with the optimal solution that satisfies (i). As the result, the solution that achieves larger utility is the optimal one.

Let $\mathscr{R}_j^1$ and $\mathscr{R}_j^2$ denote the optimal solutions that allocate resource to a total of j flows and satisfy Classes (i) and (ii), respectively. Note that $\mathscr{R}_j^1$ and $\mathscr{R}_j^2$ are n-dimensional vectors in the form of $(r_1, r_2, \cdots, r_n)$, where $r_{j+1} = \cdots = r_n = 0$.

For simplicity, we assume that all flows in the queue are sorted in the decreasing order of their channel qualities.

Lemma 21.2.1

For $\mathscr{R}_j^2$, we have $u_j^{'}(r_j) > 0$.

Proof: Assume that $u_j^{'}(r_j) < 0$, then there exists an $i < j$, $u_i^{'}(r_i) > 0$. From

$$u_j^{'}(r_j) = q_j^2 u^{'}(r_j q_j) < 0,$$

we have $r_j q_j > rc$. By Lemma 2.3 in [626],

$$q_i r_i = \theta_i \geq \theta_j = q_j r_j,$$

so

$$r_i \geq \frac{q_j r_j}{q_i} > \frac{rc}{q_i} = rc_i,$$

hence $u_i^{'}(r_i) < 0$, it is a contradiction. ■

By Lemma 21.2.1, we find that in $\mathscr{R}_j^2, j = 1, 2, \cdots, n$, inequality $r_i > rc_i$ holds for each allocated flows $i, i = 1, 2, \cdots, j-1$ and $r_j < rc_j$.

Let

$$\hat{u}_i(x_i) = u_i(x_i + rc_i), \quad i = 1, 2, \cdots, j-1,$$

then $\hat{u}_i(x_i)$ is a decreasing function with respect to x_i, and $r_i = x_i + rc_i$ for $i = 1, 2, \cdots, j-1$ and $u_j(r_j)$ is increasing. Let $\hat{u}_i^{-1}(\bullet)$ be the inverse function of $\hat{u}_i(\bullet), i = 1, 2, \cdots, j-1$ and $u_j^{-1}(\bullet)$ be the inverse function of $u_j(\bullet)$.

To find the optimal solution of $P1$, we first establish the sufficiency and necessity condition of optimal solution. Based on the optimization theory, we have the following result.

Theorem 21.2.1

Let a set of bandwidth allocation $\vec{r} = (r_1, r_2, \cdots, r_n)$ and

$$L(r, \lambda, \lambda_1, \cdots, \lambda_n) = \sum_{i=1}^{n}(-U_i(r_i)) - \lambda(C - \sum_{i=1}^{n} r_i) - \sum_{i=1}^{n} \lambda_i r_i.$$

The bandwidth allocation $\vec{r}$ is an optimal solution to $P1$ if and only if there exists Lagrange multiplier $(\lambda, \lambda_1, \cdots, \lambda_n) \geq 0$ such that

$$u_i(r_i) = \lambda > 0, for \;\; r_i > 0, \quad i = 1, \cdots, n, \quad C = \sum_{i=1}^{n} r_i. \tag{21.5}$$

■

Proof: $P1$ is equivalent to the following form:

$$\min \sum_{i \in \Gamma} (-U_i(r_i))$$
$$h(r) = C - \sum_{i \in \Gamma} r_i \geq 0,$$
$$r_i \geq 0, i \in \Gamma.$$

Let $I = \{i | r_i = 0\}$, because at least one $r_{i_0} > 0$, so gradients

$$\nabla_{\vec{r}} h(\vec{r}) = (-1, \cdots, -1)^T,$$
$$\nabla_{\vec{r}} r_i = (0, \cdots, 0, 1, 0, \cdots, 0)^T, i \in I,$$

(where the ith component is 1, and the others are 0) are linear independent. If $\vec{r}$ is an optimal solution of $P1$, by the Karush–Kuhn–Tucker condition [144], we have

$$-u_i(r_i) + \lambda - \lambda_i = 0,$$
$$\lambda h(r) = 0, \lambda_i r_i = 0,$$
$$\lambda, \lambda_i \geq 0, i = 1, \cdots, n.$$

If $r_i > 0$, then $\lambda_i = 0$, hence $u_i(r_i) = \lambda$. Since $U_i(\cdot), i \in \Gamma$ are sigmoid utility functions, hence $u_i(\cdot) > 0$, and $\lambda_i \geq 0$, so $\lambda > 0$, hence $h(\vec{r}) = 0$, that is, $C = \sum_{i \in \Gamma} r_i$.

Since $\hat{u}_i(x_i)(i = 1, \cdots, j)$ is decreasing for class (1), and $\hat{u}_i(x_i)(i = 1, \cdots, j-1)$ is decreasing and $u_j(r_j)$ is increasing for class (2), then the solution of $\lambda = u_i(r_i)$ is always unique. Then, if $\vec{r}$ satisfies (1), it must be optimal. Hence, we obtain our desirable result.

Now, we can consider the solution of $P1$.

Case 1: if $r_j > rc_j$, by Theorem 21.2.1, we have

$$\hat{u}_1(x_1) = \cdots = \hat{u}_j(x_j) = \lambda > 0.$$

Let $rr_j = C - \sum_{i=1}^{j} rc_i$, then

$$rr_j = \sum_{i=1}^{j} \hat{u}_i^{-1}(\lambda). \tag{21.6}$$

Denote its solution by $\lambda = k_0$. Then,

$$r_i = rc_i + \hat{u}_i^{-1}(k_0), i = 1, \cdots, j;$$
$$r_i = 0, i = j+1, \cdots, n. \tag{21.7}$$

Case 2: if $r_j < rc_j$, set $rr_{j-1} = C - \sum_{i=1}^{j-1} rc_i$, then

$$rr_{j-1} = \sum_{i=1}^{j-1} \hat{u}_i^{-1}(\lambda) + u_j^{-1}(\lambda). \tag{21.8}$$

Denote its solution by $\lambda = k_1$. Then,

$$\begin{aligned} r_i = rc_i + \hat{u}_i^{-1}(k_1), i = 1, \cdots, j-1, \\ r_j = u_i^{-1}(k_1), \\ r_i = 0, i = j+1, \cdots, n. \end{aligned} \tag{21.9}$$

Note that $U_i(r_i), i = 1, 2, \cdots, n$ all are increasing, so $P1$ always has an optimal solution. Hence, at least one of Equations 21.6 and 21.8 has a solution. In Equation 21.8, we have $\hat{u}_i(x_i)$ is a decreasing function with respect to x_i for $i = 1, \cdots, j-1$, by adding up them as $\sum_{i=1}^{j-1} \hat{u}_i$ (as shown in Figure 21.5), at the same coordinate system, we can obtain the plots of u_j and $\sum_{i=1}^{j-1} \hat{u}_i$. Figure 21.5 illustrates the procedures of how to construct Equations 21.6 and 21.8, which work as follows: Figure 21.5a shows the plots of sigmoid function $U_i(\cdot)$ $(i = 1, 2, \ldots, j-1)$, and the plots of their derivations are shown in Figure 21.5b. Along rc_i, we can divide them into two parts; in each part, they are all monotonous, parallel moving the last part to exactly intersect y-axis, which is shown in Figure 21.5d, and then adding up all of them to obtain the right part Figure 21.5e. Combining this and the left plot of derivation of $U_j(\cdot)$, we have the fifth subfigure. At the same system, we show them in Figure 21.5f.

If $rr_{j-1} = a + c$, i.e., the resource is allocated among the first j users, and $r_j < rc_j$, then by Equation 21.8, we have solution (21.9).

If $rr_{j-1} = b$, *i.e.*, the resource is allocated only among the first $j-1$ users and $r_i \geq rc_i, i = 1, \cdots, j-1$, then by Equation 21.6, we have solution (21.7).

Based on the above discussions, we have the following main result.

Theorem 21.2.2

The component $\mathscr{R}_j^2$ is the one and only optimal solution that satisfies class (ii), and its component must satisfy (21.9). ■

Proof: Since

$$\begin{aligned} \sum_{i=1}^{n} r_i &= \sum_{i=1}^{j-1} (rc_i + \hat{u}_i^{-1}(k_1)) + u_j^{-1}(k_1) \\ &= \sum_{i=1}^{j-1} rc_i + \sum_{i=1}^{j-1} (\hat{u}_i^{-1}(k_1)) + u_j^{-1}(k_1) = C, \end{aligned}$$

For $i = 1, 2, \cdots, j-1$, we have

$$u_i(r_i) = u_i(rc_i + \hat{u}_i^{-1}(k_1)) = \hat{u}_i(\hat{u}_i^{-1}(k_1)) = k_1$$

and

$$u_j(r_j) = u_j(u_j^{-1}(k_1)) = k_1.$$

For $i = 1, \cdots, n$, if $r_i > 0$, set $\lambda = k_1$, $\lambda_i = 0$;

if $r_i = 0$, set $\lambda_i = \lambda - u_i(0) = k_1 - u_i(0) > 0$.

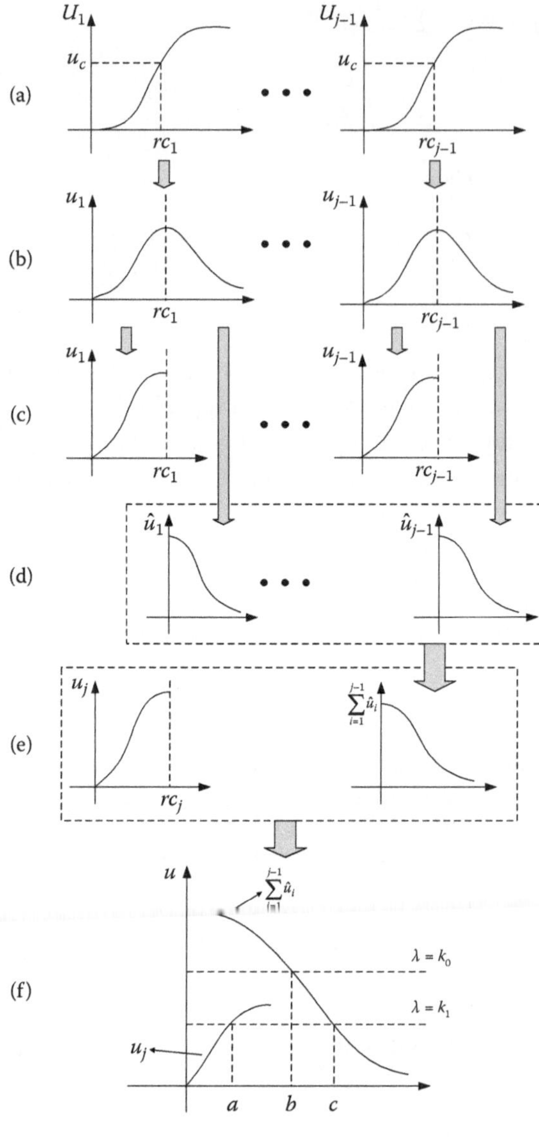

Figure 21.5 The procedure of constructing Equations 21.6 and 21.8.

That is, there exists $(\lambda, \lambda_1, \cdots, \lambda_n) \geq 0$ such that

$$-u_i(r_i) + \lambda - \lambda_i = 0,$$
$$\lambda_i r_i = 0, i = 1, \ldots, n,$$
$$C = \sum_{i=1}^{n} r_i.$$

By Theorem 21.2.1, (21.9) is an optimal solution in which a total of j flows are allocated by resource. By the proof of Theorem 21.2.1, we know that the solution of $\lambda = u_i(r_i)$ is unique. Hence, we obtain the desired result.

The above theorem is our new finding that enables us to find the desired optimal solution by using both the conditions (21.7) and (21.9). In contrast, the paper [626] is only able to find a sub-optimal solution using condition (21.7) alone. Based on Theorem 21.2.2, the explicit procedures for obtaining the desired optimal solution are detailed in the next section.

21.3 ALGORITHM USQ TO OBTAIN THE OPTIMAL SOLUTION

Suppose that there are n flows sorted in the decreasing order of their channel qualities in the system, and all flows have the same utility function: sigmoid utility function $U(\cdot)$. To find the optimal solution, the following resource allocation algorithm, termed by USQ, is designed.

USQ Algorithm

1. Initialize $r_i = 0, i = 1, 2, \cdots, n, rr_0 = C$, and $U(\mathscr{R}_{sigmoid}) = 0$.
2. Sort all flows i in descending order of q_i, and store them in the queue.
3. For $j = 1$ to n
   a. Solve Equations 21.6 and 21.8.
   Case 1: If (21.6) has a solution but (21.8) does not have solution. Denote this solution by $u = k_0$. Then $r_i = rc_i + \hat{u}_i^{-1}(k_0)$, for $i = 1, \cdots, j$, and $r_i = 0$, for $i = j+1, \cdots, n$. Let $\mathscr{R}_j = (r_1, \cdots, r_n)$.
   Case 2: If (21.8) has a solution but (21.6) does not have solution. Denote this solution by $u = k_1$. Then $r_i = rc_i + \hat{u}_i^{-1}(k_1)$, for $i = 1, \cdots, j-1$, $r_j = u_i(k_1)^{-1}$ and $r_i = 0$ for $i = j+1, \cdots, n$. Let $\mathscr{R}_j = (r_1, \cdots, r_n)$.
   Case 3: If both (21.6) and (21.8) have solution. Calculate $U(\mathscr{R}_j^1), U(\mathscr{R}_j^2)$, let $\mathscr{R}_j = argmax\{U(\mathscr{R}_j^1), U(\mathscr{R}_j^2)\}$.
   b. Calculate $U(\mathscr{R}_j)$, if $U(\mathscr{R}_j) > U(\mathscr{R}_{sigmoid})$, $\mathscr{R}_{sigmoid} = \mathscr{R}_j, U(\mathscr{R}_{sigmoid}) = U(\mathscr{R}_j)$.
4. Return $\mathscr{R}_{sigmoid}$ and $U(\mathscr{R}_{sigmoid})$.

Note that the algorithm given in [626] only solves Equation 21.6 repeatedly, and then the obtained solution is certainly suboptimal.

21.4 NUMERICAL EXAMPLES

In this section, we give two examples to demonstrate the explicit techniques of USQ and the solution procedures.

Example 1:

We assume that there are 10 soft QoS traffic in a wireless network, in which sigmoid utility functions are given [622] by

$$U_i(r_i) = \frac{1}{1 + e^{-a_i r_i + b_i}}, \quad i = 1, 2, \ldots, 10,$$

where

$$(a_1, a_2, \ldots, a_{10}) = (1, 2, 1, 3, 2, 4, 1, 5, 3, 6),$$

$$(b_1, b_2, \ldots, b_{10}) = (5, 10, 3, 9, 8, 16, 16, 30, 6, 12).$$

It should be noted that the above sigmoid utility function can be transformed into the form of (21.1). The transformation details are, however, omitted here. For the total capacity at the base station $C = 58$, we run the proposed algorithm USQ. The optimization problem at hand is, therefore, described into

$$\max \sum_{i=1}^{10} \frac{1}{1+e^{-a_i r_i + b_i}}$$

$$\text{s.t.}: \sum_{i=1}^{10} r_i \leq 58, r_i \geq 0.$$

Mathematical manipulations arrive that

$$\frac{dU_i(r_i)}{dr_i} = u_i(r_i) = \frac{a_i e^{-a_i r_i + b_i}}{(1+e^{-a_i r_i + b_i})^2},$$

$$\frac{d^2U_i(r_i)}{dr_i^2} = u_i'(r_i) = \frac{a_i^2 e^{-a_i r_i + b_i}(1+e^{-a_i r_i + b_i})(e^{-a_i r_i + b_i} - 1)}{(1+e^{-a_i r_i + b_i})^4}.$$

From

$$\frac{d^2U_i(r_i)}{dr_i^2} = 0,$$

we obtain that

$$rc_i = \frac{b_i}{a_i}.$$

Furthermore,

$$\hat{u}_i(r_i) = u_i(r_i + rc_i) = \frac{a_i e^{-a_i r_i}}{(1+e^{-a_i r_i})^2},$$

$$\hat{u}_i^{-1}(r_i) = \frac{1}{a_i} \ln \frac{a_i - 2r_i + \sqrt{a_i^2 - 4a_i r_i}}{2r_i},$$

$$u_i^{-1}(r_i) = \frac{1}{a_i}[\ln \frac{a_i - 2r_i - \sqrt{a_i^2 - 4a_i r_i}}{2r_i} + b_i].$$

When $C = 58$, we run Algorithm USQ. For the j $(j = 1,2,\ldots,10)$ allocated flows, we solve the Equations 21.6 and 21.8, and then we obtain the corresponding allocated resource to every flow; we further compute the corresponding utility of the system, in which the allocated resource corresponding to larger utility is the optimal solution for the j allocated flows. The results of this procedure are listed in Table 21.2. Figure 21.6 plots the procedure and optimal solution to this utility optimization problem. Hence, the resulted optimal solution of the system is

$$(5.8277, 5.9948, 3.8277, 3.8308, 4.9948, 4.7059, 16.8277, 6.6143, 2.8308, 2.5450).$$

Example 2:

In a wireless network, we assume that the total resource is C. The considered soft QoS traffic has the following sigmoid utility function:

$$U(r) = \begin{cases} qe^{0.2(r-9)} & r < 9, \\ 1-(1-q)e^{-0.2(r-9)} & r \geq 9. \end{cases}$$

Assume that there are five users in the system with its sigmoid utility function being $U_i(r_i)$ $(i = 1,2,3,4,5)$. Set the channel quality of each user as follows: $q_1 = 0.9, q_2 = 0.6, q_3 = 0.5, q_4 = 0.3$, and $q_5 = 0.1$. Let $p_i = 0.2$, for $i = 1,2,3,4,5$. By the analysis in Section 1.2, we know that $U(r)$ and $U_i(r_i)$ have the following relations: $U_i(r_i) = U(q_i r_i)$, where q_i is the channel quality of flow i. So,

$$U_i(r_i) = \begin{cases} q_i e^{0.2(q_i r_i - 9)} & q_i r_i < 9, \\ 1-(1-q_i)e^{-0.2(q_i r_i - 9)} & q_i r_i \geq 9, \end{cases}$$

Table 21.2
Results of Running Algorithm USQ When $C = 58$

	Solution of Equation 21.6	Solution of Equation 21.8	Optimal Solution	Utility
$j=1$	9.6027×10^{-24}	No solution	(58, 0, 0, 0, 0, 0, 0, 0, 0, 0)	1.0000
$j=2$	1.5956×10^{-14}	No solution	(36.7690, 21.2310, 0, 0, 0, 0, 0, 0, 0, 0)	2.0000
$j=3$	1.7495×10^{-8}	No solution	(22.8614, 14.2773, 20.8614, 0, 0, 0, 0, 0, 0, 0)	3.0000
$j=4$	4.6933×10^{-7}	3.7795×10^{-9}	(19.5720, 12.6326, 17.5720, 8.2235, 0, 0, 0, 0, 0, 0)	4.0000
$j=5$	1.5383×10^{-5}	9.8938×10^{-8}	(16.0822, 10.8877, 14.0822, 7.0603, 9.8877, 0, 0, 0, 0, 0)	4.9999
$j=6$	1.1209×10^{-4}	2.0484×10^{-5}	(14.0960, 9.8946, 12.0960, 6.3982, 8.8946, 6.6206, 0, 0, 0, 0)	5.9996
$j=7$	2.5651×10^{-2}	1.6195×10^{-3}	(8.6098, 7.1651, 6.6098, 4.5815, 6.1651, 5.2591, 19.6098, 0, 0, 0)	6.8799
$j=8$	9.7807×10^{-2}	7.2471×10^{-2}	(7.0920, 6.4560, 5.0920, 4.1182, 5.4560, 4.9151, 18.0920, 6.7788, 0, 0)	7.4885
$j=9$	1.5649×10^{-1}	1.1463×10^{-1}	(6.4229, 6.1845, 4.4229, 3.9466, 5.1845, 4.7894, 17.4229, 6.6797, 2.9466, 0)	8.0627
$j=10$	2.1163×10^{-1}	1.8775×10^{-1}	(5.8277, 5.9948, 3.8277, 3.8308, 4.9948, 4.7059, 16.8277, 6.6143, 2.8308, 2.5450)	8.5572

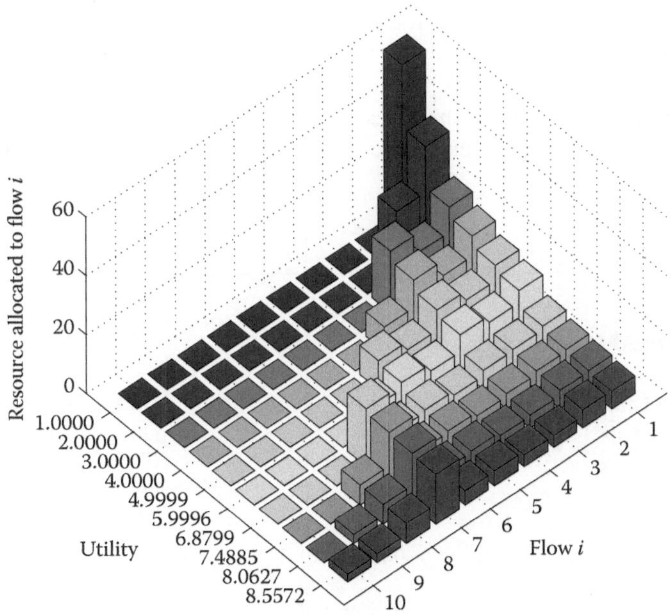

Figure 21.6 Utility optimized bandwidth allocation in Example 1: procedures and optimal solution.

for $i = 1,2,3,4,5$ and $rc_1 = 10, rc_2 = 15, rc_3 = 18, rc_4 = 30, rc_5 = 90$. Therefore, the optimization problem can be described as follows:

$$\max \sum_{i=1}^{5} U_i(r_i),$$

$$\text{s.t.} : \sum_{i=1}^{5} r_i \leq C, r_i \geq 0.$$

When $C = 200$, we run Algorithm USQ. For the j $(j = 1,2,\ldots,5)$ allocated flows, we solve Equations 21.6 and 21.8, and then we obtain the corresponding allocated resource to every flow; we then compute the corresponding utility of the system, in which the allocated resource corresponding to larger utility is the optimal solution for the j allocated flows. The results of this procedure are listed in Table 21.3. Hence, the resulted optimal solution of the system is

$$(22.9436, 42.5890, 51.5150, 82.9524, 0).$$

Figure 21.7 plots the procedure and optimal solution to this utility optimization problem.

It is interesting to note that, as displayed in Example 1, the optimal solution outputted from our algorithm is in accordance with the suboptimal solution outputted by the algorithm in [626], while as displayed in Example 2, the optimal solution outputted from our algorithm has deviations from the suboptimal solution outputted by the algorithm in [626]. From these, we know that the optimal solution may be obtained in either class (1) or class (2). Therefore, our algorithm inherits an advantage over that in [626] that one can always achieve the optimal solution by using our algorithm.

21.5 CONCLUSION

Because of the nonconcavity of utility function of soft QoS traffic, it is a challenge to allocate the limited resource among the soft QoS traffic in wireless networks. The authors in [626] gave an algorithm to find its suboptimal solution. In order to find the optimal solution, in this chapter, we first establish some key theorems; on this basis, we then present a complete algorithm called utility-based allocation for soft QoS (USQ) to obtain the desired optimal solution. The proposed theorems and algorithm act as designing guidelines for resource allocation of soft QoS traffic in a wireless network, which take into account the total available resource of network, the users' traffic

Table 21.3
Results of Running Algorithm USQ When C = 200

	Solution of Equation 21.6	Solution of Equation 21.8	Optimal Solution	Utility
$j=1$	2.5258×10^{-17}	No solution	$(200, 0, 0, 0, 0)$	1.0000
$j=2$	0.1093×10^{-6}	2.6424×10^{28}	$(76.7306, 123.2694, 0, 0, 0)$	2.0000
$j=3$	0.5437×10^{-4}	0.3119×10^{-19}	$(42.2349, 71.5259, 86.2393, 0, 0)$	2.9987
$j=4$	0.1752×10^{-2}	0.5299×10^{-8}	$(22.9436, 42.5890, 51.5150, 82.9524, 0)$	3.9290
$j=5$	0.1713×10^{-1}	0.2569×10^{-6}	$(10.2754, 23.5866, 28.7122, 44.9477, 92.4782)$	3.4488

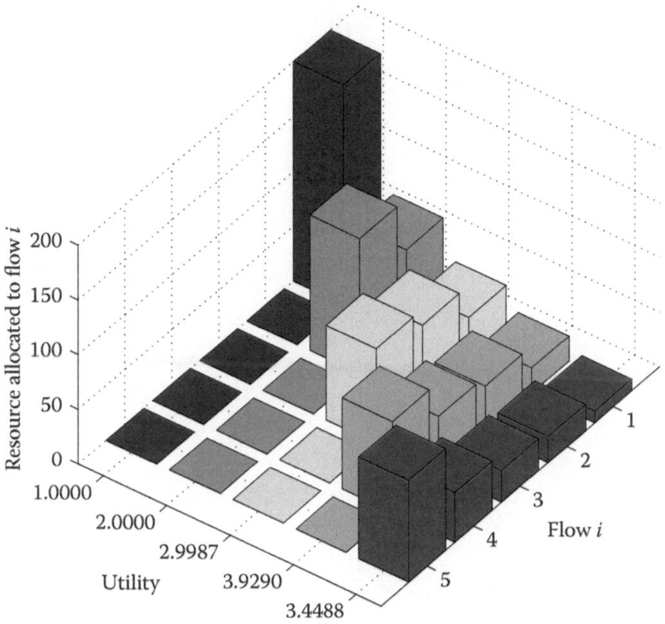

Figure 21.7 Utility optimized bandwidth allocation in Example 2: procedures and optimal solution.

characteristics, and the users' channel qualities. By numerical examples, we illustrate the explicit solution procedures. Future research would be interesting to study the scenario where best-effort, hard QoS, and soft QoS traffic coexist in a wireless communication system. A unified framework for resource allocation for this scenario is hopefully developed by following the optimization approach proposed in this chapter.

References

1. P. Parry, Overlooking 3G, *IEEE Potential*, vol. 21, no. 4, pp. 6–9, 2002.
2. F. Abrishamkar and E. Biglieri, An overview of wireless communications, *1994 IEEE Military Communications Conference (MILCOM '94)*, Fort Monmouth, NJ, vol. 3, pp. 900–905, Oct 1994.
3. L. Aravamudhan, S. Faccin, R. Mononen, B. Patil, Y. Saifullah, S. Sharma, and S. Sreemanthula, Getting to know wireless networks and technology, *Inform. IT*, July 4, 2003. http://www.informit.com/articles/printerfriendly/98132.
4. A. A. Huurdeman, *The Worldwide History of Telecommunications*, John Wiley & Sons, Hoboken, NJ, 2003.
5. United States Patent and Trademark Office, Statement of Use, s/n 75799629, US Patent and Trademark Office Trademark Status and Document Retrieval, August 23, 2005, retrieved 21 September, 2014, first used the Certification Mark August 1999.
6. ABC News, CSIRO wins legal battle over Wi-Fi patent, 6:23 am, April 02, 2012. http://www.abc.net.au/news/2012-04-01/csiro-receives-payment-for-wifi-technology/3925814.
7. I. F. Akyildiz, X. Wang, and W. Wang, Wireless mesh networks: A survey, *Computer Networks*, vol. 47, no. 4, pp. 445–487, 2005.
8. F. Akyildiz and I. H. Kasimoglu, Wireless sensor and actor networks: Research challenges, *Ad Hoc Networks*, vol. 2, no. 4, pp. 351–367, 2004.
9. W. Dargie and C. Poellabauer, *Fundamentals of Wireless Sensor Networks: Theory and Practice*, John Wiley and Sons, Chichester, 2010.
10. K. Sohraby, D. Minoli, and T. Znati, *Wireless Sensor Networks: Technology, Protocols, and Applications*, John Wiley and Sons, Hoboken, NJ, 2007.
11. C. K. Toh, *Ad Hoc Mobile Wireless Networks: Protocols and Systems*, Prentice Hall Publishers, Upper Saddle River, NJ, 2002.
12. C. S. R. Murthy and B. S. Manoj, *Ad Hoc Wireless Networks: Architectures and Protocols*, Prentice Hall PTR, Upper Saddle River, NJ, 2004.
13. K. H. Almotairi and X. S. Shen, A distributed multi-channel MAC protocol for ad hoc wireless networks, *IEEE Transactions on Mobile Computing*, vol. 14, no. 1, pp. 1–13, 2014.
14. J. Qi, J. Zollner, J. Robert, L. Stadelmeier, and N. Loghin, Redundancy on demand–Extending the coverage area of terrestrial broadcast via broadband networks, *IEEE Transactions on Broadcasting*, vol. 61, no. 3, pp. 337–345, 2015.
15. T. Arthanayake, Interference aspects of off-shore terrestrial microwave networks, *Radio & Electronic Engineer*, vol. 50, no. 8, pp. 390–396, 1980.
16. S. W. Smith, *The Scientist and Engineer's Guide to Digital Signal Processing*, copyright 1997–1998 by Steven W. Smith. www.DSPguide.com.
17. J. M. Blackledge, *Digital Signal Processing: Mathematical and Computational Methods, Software Development and Applications*, Horwood Publishing, Chichester, 2003, ISBN 1-898563-48-9.
18. J. E. Padgett, C. G. Gunther, and T. Hattori, Overview of wireless personal communications, *IEEE Communications Magazine*, vol. 33, no. 1, pp. 28–41, 1995.
19. A. L. I. Semeia, *Wireless Network Performance Analysis for Adaptive Bandwidth Resource Allocations*, PhD Thesis, Stevens Institute of Technology, Hoboken, NJ, 2003.
20. C. Weber and G. Bahadur, Wireless networking security, TechNet Microsoft. https://technet.microsoft.com/en-us/library/bb457019.aspx.
21. G. Pierobon, A. Zanella, and A. Salloum, Contention-TDMA protocol: Performance evaluation, *IEEE Transactions on Vehicular Technology*, vol. 51, no. 4, pp. 781–788, 2002.

22. R. Nelson and L. Kleinrock, Spatial TDMA—A collision-free multihop channel access protocol, *IEEE Transactions on Communications*, vol. 33, no. 9, pp. 934–944, 1985.
23. S. Hara and R. Prasad, Overview of multicarrier CDMA, *IEEE Communications Magazine*, vol. 35, no. 12, pp. 126–133, 1997.
24. A. Viterbi, CDMA: Principles of spread spectrum communication, *Communications Technology,* vol. 9, no. 1, pp. 155–213, 1995.
25. N. Lu, *Utility-Based Bandwidth Adaptation for QoS Provisioning in Multimedia Wireless Networks*, PhD Thesis, Queen Mary, University of London, London, April 2007.
26. H. G. Myung, J. Lim, and D. J. Goodman, Single carrier FDMA for uplink wireless transmission, *IEEE Vehicular Technology Magazine*, vol. 1, no. 3, pp. 30–38, 2006.
27. M. D. Nisar, H. Nottensteiner, and T. Hindelang, On performance limits of DFT-spread OFDM systems, in *Sixteenth IST Mobile Summit*, Budapest, Hungary, July 2007.
28. H. G. Myung, J. Lim, and D. J. Goodman, Peak-to-average power ratio of single carrier FDMA signals with pulse shaping, *The 17th Annual IEEE International Symposium on Personal, Indoor and Mobile Radio Communications (PIMRC '06)*, Helsinki, Finland, September 2006.
29. D. Falconer, S. L. Ariyavisitakul, A. Benyamin-Seeyar, and B. Eidson, Frequency domain equalization for single-carrier broadband wireless systems, *IEEE Communications Magazine*, vol. 40, no. 4, pp. 58–66, 2002.
30. N. R. Prasad, GSM evolution towards third generation UMTIS/IMT2000, in *Proceedings of the IEEE International Conference on Personal Wireless Communication*, Jaipur, India, pp. 50–54, February 1999.
31. Z. Zvonar, P. Jung, and K. Kammerlander, *GSM: Evolution towards 3rd Generation Systems*, Kluwer Academic Publishers, Boston, MA, 1999.
32. N. Narang and S. Kasera, *2G Mobile Networks: GSM and HSCSD*, Tata McGraw-Hill Education, India, 2006.
33. P. Decker and B. Walke, A general packet radio service proposed for GSM, *ETSI SMG Workshop GSM in a Future Competitive Environment,* Helsinki, Finland, pp. 1–20, October 13, 1993.
34. B. Walke, The roots of GPRS: The first system for mobile packet-based global internet access, *IEEE Wireless Communications*, vol. 20, no. 5, pp. 12–23, 2013.
35. Wireless Moves, Evolved EDGE—The new kid on the block, *Thoughts on the Evolution of Wireless Networks and Mobile Web 2.0.* http://mobilesociety.typepad.com/mobile_life/2006/11/evolved_edge_th.html.
36. M. Decina and E. Scace, CCITT recommendations on the ISDN: A review, *CCITT Red Book*, vol. 4, no. 3, pp. 320–325, 1986.
37. R. Aaron and R. Wyndrum, Future trends, *IEEE Communications Magazine* (AT&T Bell Laboratories), vol. 24, no. 3, pp. 38–43, March 1986.
38. A. Tarantola, The next generation of DSL can pump 1 Gbps through copper phone lines, *Gizmodo*, December 18, 2013. http://gizmodo.com/the-next-generation-of-dsl-can-pump-1gbps-through-coppe-1484256467.
39. C. Huynh, J. Lee, and C. Nguyen, Multi-band radio-frequency integrated circuits for multiband and multimode wireless communication, radar and sensing systems in harsh environments, in *Proceedings of 2014 IEEE International Conference on Acoustics, Speech and Signal Processing (ICASSP)*, Florence, Italy, doi: 10.1109/ICASSP.2014.6853704.
40. T. Ojanpera and R. Prasad, An overview of third-generation wireless personal communications: A European perspective, *IEEE Personal Communications*, vol. 5, no. 6, pp. 59–65, 1998.
41. N. Padovan, M. Ryan, and L. Godara, An overview of third generation mobile communications systems: IMT-2000, in *Proceedings of the IEEE Region 10 International Conference on Global Connectivity in Energy, Computer, Communication and Control (TENCON '98)*, New Delhi, India, vol. 2, pp. 360–364, December 1998.

42. P. Chaudhury, W. Mohr, and S. Onoe, The 3GPP proposal for IMT-2000, *IEEE Communications Magazine*, vol. 37, no. 12, pp. 72–81, 1999.
43. D. N. Knisely, S. Kumar, S. Laha, and S. Nanda, Evolution of wireless data services: IS-95 to CDMA2000, *IEEE Communications Magazine*, vol. 36, no. 10, pp. 140–149, 1998.
44. B. Li, D. Xie, S. Cheng, J. Chen, P. Zhang, W. Zhu, and B. Li, Recent advances on TD-SCDMA in China, *IEEE Communications Magazine*, vol. 43, no. 1, pp. 30–37, 2005.
45. H. H. Chen, *The Next Generation CDMA Technologies*, John Wiley and Sons, Chichester, 2007.
46. C. Blanchard, Security for the third generation (3G) mobile system, *Information Security Technical Report*, vol. 5, no. 3, pp. 55–65, Sept. 2000.
47. C. E. Perkins, D. B. Johnson, and Jari Arkko, Mobility Support in IPv6, *Internet Engineering Task Force* (IETF), Request for Comments: 6275, July 2011. https://tools.ietf.org/html/rfc6275.
48. I. Batisti and J. Stojanovski, Quality of service in 3G network, *Hrvatska znanstvena bibliografija i MZOS-Svibor*, 2010.
49. N. C. Wang, J. W. Jiang, and Y. F. Huang, RSVP extensions for real-time services in heterogeneous wireless networks, *Computer Communications*, vol. 30, no. 10, pp. 2248–2257, 2007.
50. L. Zhang, S. Berson, S. Herzog, S. Jamin, and R Braden, Resource Reservation Protocol (RSVP)—Version 1 functional specification, *IETF RFC*, vol. 40, no. 5, pp. 116–127, 1997.
51. K. Murota and NTT DoCoMo, Mobile communications trends in Japan and DoCoMos activities towards 21st century, in *Proceedings of the 4th ACTS Mobile Communications Summit*, Sorrento, June 1999.
52. F. Xia, N. Y. Asabere, A. M. Ahmed, J. Li, and X. Kong, Mobile multimedia recommendation in smart communities: A survey, *IEEE Access*, vol. 1, pp. 606–624, 2003.
53. V. Mohanan and R. Budiarto, Wireless communication technologies for vehicular nodes: A survey, *International Journal of Mobile Computing & Multimedia Communications*, vol. 5, no. 2, pp. 58–77, 2013.
54. S. Frattasi, H. Fathi. F. H. P. Fitzek, R. Prasad, and M. D. Katz, Defining 4G technology from the user's perspective, *IEEE Network*, vol. 20, no. 1, pp. 35–41, 2006.
55. R. D. Leonardis, C. Sansotta, B. Testagrossa, M. Ferlazzo, and G. Vermiglio, Wired and wireless network solution for the integrated management of data and images, *La Radiologia Medica*, vol. 104, no. 3, pp. 194–202, 2002.
56. K. Mohanta and D. V. Khanaa, 4G technology, *International Journal of Engineering & Computer Science*, vol. 168, no. 4, pp. 113–118, 1984.
57. I. Bose, Fourth generation wireless systems: Requirements and challenges for the next frontier, *Communications of AIS*, vol. 17, pp. 693–713, 2006.
58. S. Devi, Quality of services in 4G networks, *International Journal of Research in Computer and Communication Technology*, vol. 2, no. 9, pp. 747–753, 2013.
59. P. Rengaraju, C. H. Lung, and A. Srinivasan, QoS and protection of relay nodes in 4G wireless networks using network coding, in *Proceedings of 2013 9th International Wireless Communications and Mobile Computing Conference (IWCMC)*, Sardinia, Italy, vol. 1–5, pp. 282–287, July 2013.
60. D. Kim, S. Lee, H. Lee, and H. Yoon, An effective hotspot cell management scheme using adaptive handover time in 4G mobile networks, in *Proceedings of TENCON 2005, IEEE Region 10*, Melbourne, QLD, vol. 21–24, pp. 1–6, November 2005.
61. R. Maallawi, N. Agoulmine, B. Radier, and T. B. Meriem, A comprehensive survey on offload techniques and management in wireless access and core networks, *IEEE Communications Surveys & Tutorials*, vol. 17, no. 3, pp. 1582–1604, 2014.

62. J. M. Pereira, Fourth generation: Now, it is personal!, in *Proceedings of the 11th IEEE International Symposium on Personal, Indoor and Mobile Radio Communications (PIMRC '00)*, London, vol. 2, pp. 1009–1016, September 2000.
63. J. Z. Sun, J. Sauvola, and D. Howie, Features in future: 4G visions from a technical perspective, in *Proceedings of IEEE Global Telecommunications Conference (GLOBECOM '01)*, San Antonio, TX, vol. 6, pp. 3533–3537, November 2001.
64. Y. Chen, L. Duan, and Q. Zhang, Financial analysis of 4G network deployment, in *Proceedings of 2015 IEEE Conference on Computer Communications (INFOCOM)*, Kowloon, Hong Kong, pp. 1607–1615, April 26–May 1, 2015.
65. Q. Li, G. Li, W. Lee, M. Lee, D. Mazzarese, B. Clerckx, and Z. Li, MIMO techniques in WiMAX and LTE: A feature overview, *IEEE Communications Magazine*, vol. 48, no. 5, pp. 86–92, 2010.
66. E. Hossain, IEEE802.16/WiMAX-based broadband wireless networks: Protocol engineering, applications, and services, in *Proceedings of Fifth Annual Conference on Communication Networks and Services Research (CNSR '07)*, Frederlcton, NB, 2007.
67. L. M. Gavrilovska and V. M. Atanasovski, Resource management in wireless heterogeneous networks (WHNs), in *Proceedings of the 9th International Conference on Telecommunication in Modern Satellite, Cable, and Broadcasting Services (TELSIKS '09)*, Nis, Serbia, 2009.
68. China Internet Network Information Center (CINIC), The 36rd report on the statistics of internet developing status in China, July 2015. http://cnnic.cn/hlwfzyj/hlwxzbg/hlwtjbg/201507/P020150723549500667087.pdf
69. T. S. Rappaport, S. Shu, R. Mayzus, Z. Hang, Y. Azar, K. Wang, G. N. Wong, J. K. Schulz, M. Samimi, and F. Gutierrez, Millimeter wave mobile communications for 5G cellular: It will work!, *IEEE Access*, vol. 1, pp. 335–349, 2013.
70. T. S. Rappaport, F. Gutierrez, E. Ben-Dor, J. N. Murdock, Y. Qiao, and J. I. Tamir, Broadband millimeter-wave propagation measurements and models using adaptive-beam antennas for outdoor urban cellular communications, *IEEE Transactions on Antennas and Propagation*, vol. 61, no. 4, pp. 1850–1859, 2013.
71. T. S. Rappaport, R. W Heath, J. Robert Daniels, and J. Murdock, *Millimeter Wave Wireless Communications*, Prentice Hall, Upper Saddle River, NJ, 2014.
72. B. Kouassi, I. Ghauri, and L. Deneire, Reciprocity-based cognitive transmissions using a MU massive MIMO approach, in *Proceedings of IEEE International Conference on Communications (ICC)*, Budapest, Hungary, 2013.
73. T. L. Marzetta, Noncooperative cellular wireless with unlimited numbers of base station antennas, *IEEE Transactions on Wireless Communications*, vol. 9, no. 11, pp. 3590–3600, 2010.
74. J. Hoydis, S. T. Brink, M. Debbah, Massive MIMO in the UL/DL of cellular networks: How many antennas do we need?, *IEEE Journal on Selected Areas in Communications*, vol. 31, no. 2. pp. 160–171, 2013.
75. F. Rusek, D. Persson, K. L. Buon, E. G. Larsson, T. L. Marzetta, O. Edfors, and F. Tufvesson, Scaling up MIMO: Opportunities and challenges with very large arrays, *IEEE Signal Processing Magazine*, vol. 30, no. 1, pp. 40–60, 2013.
76. E. Bastug, M. Bennis, and M. Debbah, Living on the edge: The role of proactive caching in 5G wireless networks, *IEEE Communications Magazine*, vol. 52, no. 8, pp. 82–89, 2014.
77. E. Bastug, M. Bennis, M. Kountouris, and M. Debbah, Cache-enabled small cell networks: Modeling and tradeoffs, *EURASIP Journal on Wireless Communications and Networking*, vol. 41, 2015. http://www.jwcn.eurasipjournals.com/content/pdf/s13638-015-0250-4.pdf.
78. IEEE Communication Magazine, Communications, caching, and computing for content-centric mobile networks, Feature Topic: Call For Papers. http://www.comsoc.org/commag/cfp/communications-caching-and-computing-content-centric-mobile-networks.

79. D. Gesbert, S. Hanly, H. Huang, S. Shamai, O. Simeone, and W. Yu, Multi-cell MIMO cooperative networks: A new look at interference, *IEEE Journal on Selected Areas in Communications*, vol. 28, no. 9. pp. 1380–1408, 2010.
80. E. Bjornson and E. Jorswieck, Optimal resource allocation in coordinated multi-cell systems, *Foundations and Trends in Communications and Information Theory*, vol. 9, no. 2–3, pp. 113–381, 2013.
81. R. Baldemair, E. Dahlman, G. Fodor, G. Mildh, S. Parkvall, Y. Selen, H. Tullberg, and K. Balachandran, Evolving wireless communications: Addressing the challenges and expectations of the future, *IEEE Vehicular Technology Magazine*, vol. 8, no. 1, pp. 24–30, 2013.
82. A. Gani, X. Li, L. Yang, O. Zakaria, and N. B. Anuar, Multi-bandwidth data path design for 5G wireless mobile internets, *WSEAS Transactions on Information Science and Applications*, vol. 6, no. 2, pp. 159–168, 2009.
83. C. Liang and F. R. Yu, Wireless network virtualization: A survey, some research issues and challenges, *IEEE Communications Surveys & Tutorials*, vol. 17, no. 1, pp. 358–380, 2015.
84. C. I. Badoi, N. Prasad, V. Croitoru, and R. Prasad, 5G based cognitive radio, *Wireless Personal Communications*, vol. 57, no. 3, pp. 441–464, 2013.
85. L. S. Cardoso, M. Maso, M. Kobayashi, and M. L. Debbah, Orthogonal LTE two-tier cellular networks, in *Procceedings of 2011 IEEE International Conference on Communications (ICC)*, Kyoto, pp. 1–5, June 2011.
86. S. Akhtar, Evolution of technologies, standards, and deployment of 2G-5G networks, *Encyclopedia of Multimedia Technology and Networking*, 2nd ed., IGI Global, Hershey, PA, pp. 522–532, 2005. doi: 10.4018/978-1-60566-014-1.ch070.
87. A. Tzanakaki, M. P. Anastasopoulos, and D. Simeonidou, Evaluation of converged networks for 5G infrastructures, in *Proceedings of the 7th International Workshop on Reliable Networks Design and Modeling (RNDM)*, Munich, Germany, pp. 1–6, October 5–7, 2015.
88. A. Tzanakaki, M. Anastasopoulos, K. Georgakilas, G. Landi, G. Bernini, N. Ciulli, J. F. Riera, E. Escalona, J. A. Garcia-Espin, X. Hesselbach, S. Figuerola, S. Peng, R. Nejabati, D. Simeonidou, D. Parniewicz, B. Belter, and M. J. Rodriguez, Planning of dynamic virtual optical cloud structures: The GEYSERS approach, *IEEE Communication Magazine*, vol. 52, no. 1, pp. 26–34, 2014.
89. A. Tzanakaki, M. P. Anastasopoulos, G. S. Zervas, B. R. Rofoee, R. Nejabati, and D. Simeonidou, Virtualization of heterogeneous wireless-optical network and IT support and mobile services, *IEEE Communication Magazine*, vol. 51, no. 8, pp. 155–161, 2013.
90. S. H. Yeganeh, A. Tootoonchian, and Y. Ganjali, On scalability of software-defined networking, *IEEE Communication Magazine*, vol. 51, no. 2, pp. 136–141, 2013.
91. 3GPP TS 36.213, Technical specification group radio access network: Evolved universal terrestrial radio access (E-UTRA): Physical layer procedures, URL: https://portal.3gpp.org/desktopmodules/Specifications/SpecificationDetails.aspx?specificationId=2427.
92. G. S. Zervas, J. Triay, N. Amaya, Y. Qin, C. Cervello-Pastor, and D. Simeonidou, Time shared optical network (TSON): A novel metro-architecture for flexible multi-granular services, *Optical Express*, vol. 19, no. 26, pp. 1–3, 2011.
93. C. Perera, A. Zaslavsky, P. Christen and D. Georgakopoulos, Context aware computing for the internet of things: A survey, *IEEE Communications Surveys & Tutorials*, vol. 16, no. 1, pp. 414–454, 2014.
94. H. Sundmaeker, P. Guillemin, P. Friess, and S. Woelffle, Vision and challenges for realising the internet of things, European Commission Information Society and Media, Technical Report, March 2010. http://www.internet-of-things-research.eu/pdf/IoT Clusterbook March 2010.pdf.
95. A. Zaslavsky, C. Perera, and D. Georgakopoulos, Sensing as a service and big data, in *Proceedings of International Conference on Advances in Cloud Computing (ACC-2012)*, Bangalore, India, pp. 21–29, July 2012.

96. C. Associati, The evolution of internet of things, Casaleggio Associati Technical Report, February 2011. http://www.casaleggio.it/pubblicazioni/Focus internet of things v1.81.
97. European Commission, Internet of things in 2020 road map for the future, Technical Report Working Group RFID of the ETP EPOSS, May 2008. http://ec.europa.eu/information society/policy/rfid/ documents/iotprague2009.pdf.
98. NGMN Alliance, *5G White Paper*, Next Generation Mobile Networks Ltd, 17 February 2015.
99. S. Shakkottai and R. Srikant, Network optimization and control, *Foundations and Trends in Networking*, vol. 2, no. 3, pp. 271–379, 2007.
100. S. Athuraliya, V. H. Li, S. H. Low, and Q. Yin, REM: Active queue management, *IEEE Network*, vol. 15, no. 3, pp. 48–53, 2001.
101. M. Carugi and D. McDysan, Service requirements for layer 3 provider provisioned virtual private networks (PPVPNs), Requirest for Comments (RFC) 4031, April 2005.
102. J. Eriksson, M. Faloutsos, and S. Krishnamurthy, Justice: Flexible and enforceable per-source bandwidth allocation, in *2005 IFIP Networking Conference*, Waterloo, CA, May 2–6, 2005.
103. J. Feigenbaum and S. Shenker, Distributed algorithmic mechanism design: Recent results and future directions, in *Sixth International Workshop on Discrete Algorithms and Methods for Mobile Computing and Communications (Dial'M 2002)*, Atlanta, September 2002.
104. F. P. Kelly, Charging and rate control for elastic traffic, *European Transactions on Telecommunications*, vol. 8, no. 1, pp. 33–37, 1997.
105. F. P. Kelly, A. Maulloo, and D. Tan, Rate control for communication networks: Shadow prices, proportional fairness and stability, *Journal of Operations Research Society*, vol. 49, no. 3, pp. 237–252, 1998.
106. A. Kozakiewicz and K. Malinowski, Network traffic routing using effective bandwidth theory, *European Transactions on Telecommunications*, vol. 20, no. 7, pp. 660–667, 2009.
107. H. T. Kung, R. Morris, T. Charuhas, and D. Lin, Use of link-by-link flow control in maximizing ATM network performance: Simulation results, in *Proceedings of IEEE Hot Interconnects Symposium'93*, Palo Alto, CA, August 5–6, 1993.
108. H. T. Kung and S. Y. Wang, TCP trunking: Design, implementation, and performance, in *Proceedings of IEEE International Conference on Network Protocols'99*, Toronto, ON, pp. 222–231, October 1999.
109. P. Lassila, A. Penttinen, and J. Virtamo, Dimensioning of data networks: A flow-level perspective, *European Transactions on Telecommunications*, vol. 20, no. 6, pp. 549–563, 2009.
110. D. C. Lay, *Linear Algebra and Its Applications*, 2nd ed., Addison-Wesley, Reading, MA, 1996.
111. S. H. Low, A duality model of TCP and queue management algorithms, *IEEE/ACM Transactions on Networking*, vol. 11, no. 4, pp. 525–536, 2003.
112. S. H. Low and D. E. Lapsley, Optimization flow control, I: Basic algorithm and convergence, *IEEE/ACM Transactions on Networking*, vol. 7, no. 6, pp. 861–874, 1999.
113. N. Maxemchuk, I. Ouveysi, and M. Zukerman, A quantitative measure for telecommunications networks topology design, *IEEE/ACM Transactions on Networking*, vol. 13, no. 4, pp. 731–742, 2005.
114. M. Minoux, *Mathematical Programming: Theory and Algorithms*, Wiley, Chichester, 1986.
115. J. Mo and J. Walrand, Fair end-to-end window-based congestion control, *IEEE/ACM Transactions on Networking*, vol. 8, no. 5, pp. 556–567, 2000.
116. N. Nisan and A. Ronen, Algorithmic mechanism design, *Games and Economic Behavior*, vol. 35, no. 1–2, pp. 166–196, 2001.
117. J. Nocedal and S. Wright, *Numerical Optimization*, Springer, New York, NY, 1999.
118. R. Pletka, A. Kind, M. Waldvogel, and S. Mannal, Closed-loop congestion control for mixed responsive and non-responsive traffic, in *Proceedings of IEEE GLOBECOM 2003*, San Francisco, CA, pp. 4180–4186, December 2003.

119. M. Welzl, M. Scharf, and B. Briscoe, Open research issues in internet congestion control, September 2010. Internet Draft, http://tools.ietf.org/html/draft-irtf-iccrg-welzl-congestion-control-open-research-08.
120. C. R. Rao and S. K. Mitra, *Generalized Inverse of Matrices and its Applications*, John Wiley, New York, NY, 1971.
121. M. Rodríguez-Pérez, S. Herrería-Alonso, M. Fernández-Veiga, and C. López-García, The persistent congestion problem of FAST-TCP: Analysis and solutions, *European Transactions on Telecommunications*, vol. 21, no. 6, pp. 504–518, 2010.
122. M. Rodríguez-Pérez, S. Herrería-Alonso, M. Fernández-Veiga, and C. López-García, Common problems in delay-based congestion control algorithms: A gallery of solutions, *European Transactions on Telecommunications*, vol. 22, no. 4, pp. 168–178, 2011.
123. N. Semret, R. R.-F. Liao, A. T. Campbell, and A. A. Lazar, Pricing, provisioning and peering: Dynamic markets for differentiated Internet services and implications for network interconnections, *IEEE Journal on Selected Areas in Communications*, vol. 18, no. 12, pp. 2499–2513, 2000.
124. L. Tan, H. Wang, and M. Zukerman, Adaptive bandwidth allocation for metropolitan and wide area networks, *IEEE Communications Letters*, vol. 9, no. 6, pp. 561–563, 2005.
125. L. Tan and X. Wang, On IP traffic matrix estimation, in *Proceedings of 16th International Conference on Computer Communications and Networks (ICCCN 2007)*, Honolulu, HI, August 13–16, 2007.
126. L. Tan and X. Wang, A novel method to estimate IP traffic matrix, *IEEE Communications Letters*, vol. 11, no. 11, pp. 907–909, 2007.
127. L. Tan, W. Zhang, G. Peng, and G. Chen, Stability of TCP/RED systems in AQM routers, *IEEE Transactions on Automatic Control*, vol. 51, no. 8, pp. 1393–1398, 2006.
128. L. Tan, X. Zhang, L. L. H. Andrew, S. Chan, and M. Zukerman, Price-based max-min fair rate allocation in wireless multi-hop networks, *IEEE Communications Letters*, vol. 10, no. 1, pp. 31–33, 2006.
129. A. Tang, J. Wang, and S. H. Low, Counter-intuitive throughput behaviors in networks under end-to-end control, *IEEE/ACM Transactions on Networking*, vol. 14, no. 2, pp. 355–368, 2006.
130. Y. Vardi, Network tomography: Estimating source-destination traffic intensities from link data, *Journal of the American Statistical Association*, vol. 91, no. 433, pp. 365–377, 1996.
131. W. Wang and B. Li, Market-driven bandwidth allocation in selfish overlay networks, in *Proceedings of IEEE INFOCOM*, Miami, FL, vol. 3, pp. 2578–2589, March 13–17, 2005.
132. Y. Yi and M. Chiang, Stochastic network utility maximization, *European Transactions on Telecommunications*, vol. 19, no. 4, pp. 421–442, 2008.
133. G. Yu, Y. Jiang, L. Xu, and G. Y. Li, Multi-objective energy-efficient resource allocation for multi-RAT heterogeneous networks, *IEEE Journal on Selected Areas in Communications*, vol. 33, no. 10, pp. 2118–2127, 2015.
134. S. Navaratnapajah, A. Saeed, M. Dianati, M. Dianati, and M. A. Imran, Energy efficiency in heterogeneous wireless access networks, *IEEE Wireless Communications*, vol. 20, no. 5, pp. 37–43, 2013.
135. J. Geanakoplos, Arrow-Debreu model of general equilibrium, *The New Palgrave: A Dictionary of Economics*, vol. 1, pp. 116–124, 1987.
136. V. Bhm and H. Haller, Demand theory, *The New Palgrave: A Dictionary of Economics*, vol. 1, pp. 785–792, 1987.
137. R. L. Keeney and H. Raiffa, *Decisions with Multiple Objectives*, Cambridge University Press, Cambridge, 1993, ISBN 0-521-44185-4.
138. P. Hurley, J. Y. Le Boudec, and P. Thiran, A note on the fairness of additive increase and multiplicative decrease, in *Proceedings of ITC-16*, Edinburgh, Scotland, pp. 467–478, June 1999.

139. D. Bertsekas and R. Gailager, *Data Networks*, Prentice-Hall, Englewood Cliffs, NJ, 1992.
140. E. Hahne, Round-robin scheduling for max-min fairness in data network, *IEEE Journal of Selected Areas in Communications*, vol. 9, no. 7, pp. 1024–1039, 1991.
141. R. Jain, Congestion control and traffic management in ATM networks: Recent advances and a survey, *Computer Networks ISDN System*, vol. 28, no. 13, pp. 1723–1738, 1996.
142. A. Charny, *An Algorithm for Rate Allocation in a Packet-Switching Network with Feedback*, Master's Thesis, Massachusetts Institute of Technology, Cambridge, MA, 1994.
143. A. Mayer, Y. Ofek, and M. Yung, Approximating max-min fair rates via distributed local scheduling with partial information, in *Proceedings of IEEE lnfocom' 96*, San Francisco, CA, pp. 926–936, 1996.
144. M. Avriel, *Nonlinear Programming*, Prentice-Hall, Englewood Cliffs, NJ, 1976.
145. J. Rawls, *A Theory of Justice*, Oxford University Press, Oxford, 1999.
146. J. S. Mill, *Utilitarianism in on Liberty and Other Essays,* Chapter 5, ed. J. Gray, Oxford University Press, Oxford, 1991.
147. D. Bertsekas, *Nonlinear Programming*, Athena Scientific, Belmont, MA, 1995.
148. F. Kelly, Fairness and stability of end-to-end congestion control, *European Journal of Control*, vol. 9, no. 2, pp. 159–176, 2003.
149. G. Vinnicombe, On the stability of networks operating TCP-likc congestion control, in *Proceedings of IFAC World Congress*, Barcelona, Spain, 2002, University of Cambridge Technical Report CUED/F-INFENG/TR. 398. http://www.eng.cam.ac.uk/ gv.
150. R. Johari and D. Tan, End-to-end congestion control far the internet: Delays and stability, *IEEE/ACM Transactions on Networking*, vol. 9, no. 6, pp. 818–832, 2001.
151. G. Vinnicombe, On the stability of end-to-end congestion control for the internet, University of Cambridge Technical Report CUED/FINFENG/TR 398, 2001. http://www.eng.cam.ac.uk/ gv.
152. G. Vinnicombe, Robust congestion control for the internet, University of Cambridge Technical Report, 2002. http://www.eng.cam.ac.uk/ gv.
153. S. Kunniyur and R. Srikant, Analysis and design of an adaptive virtual queue algorithm for active queue management, in *Proceedings of ACM Sigcomm*, San Diego, CA, pp. 123–134, 2001.
154. S. Kunniyur and R. Srikant, A time-scale decomposition approach 10 adaptive ECN marking, *IEEE Transaction on Automatic Control*, vol. 47, no. 6, pp. 882–894, 2002.
155. S. Kunniyur and R. Srikant, Note on the stability of the AVQ scheme, in *Proceedings of Conference on Information Sciences and Systems*, Princeton, NJ, 2002.
156. S. Kunniyur and R. Srikant, Designing AVQ parameters for a general topology network, in *Proceedings of Asian Control Conference*, Singapore, 2002.
157. H. Yaiche, R. R. Mazumdar, and C. Rosenberg, A game-theoretic framework for bandwidth allocation and pricing in broadband network, *IEEE/ACM Transaction on Networking*, vol. 8, no. 5, pp. 667–678, 2000.
158. F. Paganini, J. Doyle, and S. Low, Scalable laws for stable network congestion control, in *Proceedings of IEEE Conference on Decision arid Control*, Orlando, FL, 2001.
159. F. Paganini, Z. Wang, J. Doyle, and S. Low, A new TCP/AQM for stable operation in fast networks, in *Proceedings of IEEE Infocom*, San Francisco, CA, 2003.
160. J. Wen and M. Arcak, A unifying passivity framework for network flow control, in *Proceedings of IEEE Infocom*, San Francisco, CA, 2003.
161. S. Liu, T. Basar, and R. Srikant, Controlling the internet: A survey and some new results, in *Proceedings of the 42nd IEEE Conference on Decision and Control*, Maui, HI, pp. 3048–3057, 2003.
162. H. W. Kuhn and A. W. Tucker, Nonlinear programming, in *Proceedings of 2nd Berkeley Symposium*, University of California Press, Berkeley, CA, pp. 481–492, 1951.

163. S. Boyd and L. Vandenberghe, *Convex Optimization*, Cambridge University Press, Cambridge, 2004.
164. L. Tan, P. Yang, W. Zhang, and F. Ge, On utility-optimized router-level bandwidth allocation, *Transactions on Emerging Telecommunications Technologies*, vol. 24, no. 3, pp. 303–316, 2013. doi: 10.1002/ett.2540.
165. J. Wang, *A Theoretical Study of Internet Congestion Control: Equilibrium and Dynamics*, PhD Thesis, California Institute of Technology, Pasadena, CA, 2005.
166. S. H. Low, F. Paganini, and J. C. Doyle, Internet congestion control, *IEEE Control Systems Magazine*, vol. 22, no. 1, pp. 28–43, 2002.
167. H. Lawrence, *Introduction to Data Multicasting*, Althos Publishing, Fuquay Varina, NC, 2008.
168. V. Jacobson and M. Karels, Congestion avoidance and control, *ACM SIGCOMM Computer Communication Review*, vol. 25, no. 1, pp. 157–187, 1988. doi: 10.1145/205447.205462
169. L. S. Brakmo, S. W. Omalley, and L. L. Peterson, TCP Vegas: New techniques for congestion detection and avoidance, in *Proceedings of 1994 SIGCOMM*, London, 1994.
170. S. Floyd, Highspeed TCP for large congestion windows, IETF, Internet Draft, 2002. http://www.ietf.org/internet-drafts/draft-floyd-tcp-highspeed-00.txt.
171. T. Kelly, Scalable TCP: Improving performance in highspeed wide area networks, *ACM SIGCOMM Computer Communication Review*, vol. 33, no. 2, pp. 83–91, 2003.
172. C. Jin, D. Wei, and S.H. Low, FAST TCP: Motivation, architecture, algorithms, performance, in *Proceedings of IEEE INFOCOM 2004*, Hong Kong, China, vol. 4, pp. 2490–2501, March 7–11, 2004.
173. A. Lakshmikantha, C. L. Beck, and R. Srikant, Robustness of real and virtual queue-based active queue management schemes, *IEEE/ACM Transactions on Networking*, vol. 13, no. 1, pp. 81–92, 2005.
174. S. H. Low, L. L. Peterson, and L. Wang, Understanding Vegas: A duality model, *Journal of the ACM*, vol. 49, no. 2, pp. 207–235, 2002.
175. L. Tan, C. Yuan, and M.Zukerman, A price-based internet congestion control scheme, *IEEE Communications Letters*, vol. 12, no. 4, pp. 331–333, 2008.
176. C. Yuan, L. Tan, L. L. H. Andrew, W. Zhang, and M. Zukerman, A generalized FAST TCP scheme, *Computer Communications*, vol. 31, no. 14, pp. 3242–3249, 2008.
177. D. E. Comer, *Internetworking with TCP/IP*, 5th ed., Prentice Hall, Upper Saddle River, NJ, 2005. ISBN 978-0131876712.
178. S. Floyd and V. Jacobson, Random early detection (RED) gateways for congestion avoidance, *IEEE/ACM Transactions on Networking*, vol. 1, no. 4, pp. 397–413, 1993. doi: 10.1109/90.251892.
179. W. C. Feng, D. D. Kandlur, D. Saha, and K. G. Shin, BLUE: A new class of active queue management algorithms, Computer Science Technical Report (University of Michigan) (CSE-TR-387-99), June 8, 2013.
180. I. F. Akyildiz, W. Su, Y. Sankarasubramaniam, and E. Cayirci, Wireless sensor networks: A survey, *Computer Networks*, vol. 38, no.4, pp. 393–422, 2002.
181. V. Kumar, S. Jain, and S. Tiwari, Energy efficient clustering algorithms in wireless sensor networks: A survey, *IJCSI International Journal of Computer Science Issues*, vol. 8, no. 5, pp. 246–253, 2011.
182. L. Tan, Y. Yang, C. Lin, N. Xiong, and M. Zukerman, Scalable parameter tuning for AVQ, *IEEE Communications Letters*, vol. 9, no.1, pp. 90–92, 2005.
183. T. Alpcan and T. Basar, A game-theoretic framework far congestion control in general topology networks, in *Proceedings of 41th IEEE Conference on Dccirion and Control*, Las Vegas, NV, December 2002.

184. M. Chiang, S. H. Low, A. R. Calderbank, and J. C. Doyle, Layering as optimization decomposition: A mathematical theory of network architectures, in *Proceedings of the IEEE*, vol. 95, no. 1, pp. 255–312, 2007.
185. D. Palomar and M. Chiang, A tutorial to decompositon methods for network utility maximization, *IEEE Journal of Selected Areas in Communications*, vol. 24, no. 8, pp. 1439–1450, 2006.
186. D. Palomar and M. Chiang, Alternative decompositions for distributed maximization of network utility: Framework and applications, in *Proceedings of IEEE INFOCOM*, Barcelona, Spain, April 2006.
187. R. Srikant, *The Mathematics of Internet Congestion Control*, Birkhäuse, Boston, MA, 2004.
188. R. Gopalakrishnan, S. D. Nixon, and J. R. Marden, Stable utility design for distributed resource allocation, in *Proceedings of 53rd IEEE Conference on Decision and Control*, Los Angeles, CA, December 15–17, 2014.
189. H. L. Chen, T. Roughgarden, and G. Valiant, Designing network protocols for good equilibria, *SIAM Journal on Computing*, vol. 39, no. 5, pp. 1799–1832, 2010.
190. J. R. Marden and A. Wierman, Overcoming the limitations of utility design for multiagent systems, *IEEE Transactions on Automatic Control*, vol. 58, no. 6, pp. 1402–1415, 2013.
191. S. V. Hanly and D. N. C. Tse, Power control and capacity of spread spectrum wireless networks, *Automatica*, vol. 35, no. 12, pp. 1987–2012, 1999.
192. R. D. Yates, A framework for uplink power control in cellular radio systems, *IEEE Journal on Selected Areas in Communications*, vol. 13, no. 7, pp. 1341–1348, 1995.
193. C. W. Tan, Wireless network optimization by Perron-Frobenius theory, in *Proceedings of Conference on Information Sciences & Systems*, Princeton, NJ, pp. 1–6, 2014.
194. J. T. Wen and M. Arcak, A unifying passivity framework for network flow control, *IEEE Transcations on Automatic Control*, vol. 49, no. 2, pp. 162–174, 2004.
195. G. Raina and D. Wischik, Buffer sizes for large multiplexers: TCP queuing theory and stability analysis, in *Proceedings of Next Generation Internet Networks*, Rome, Italy, pp. 173–180, 2005.
196. S. Ramanathan, A unified framework and algorithms for channel assignment in wireless networks, *Wireless Networks*, vol. 5, no. 2, pp. 81–94, 1999.
197. P. Ranjan, R. J. La, and E. H. Abed, Characterization of global stability conditions with an arbitrary communication delay, *IEEE/ACM Transactions on Networking*, vol. 14, no. 2, pp. 94–107, 2006.
198. Y. Li, M. Chiang, A. R. Calderbank, and S. Diggavi, Optimal delay-rate-reliability tradeoff in networks with composite links, in *Proceedings of IEEE INFOCOM*, Barcelona, Spain, pp. 526–534, May 2007.
199. C. Yuen and P. Marbach, Price-based rate control in random access networks, *IEEE/ACM Transactions on Networking*, vol. 13, no. 5, pp. 1027–1040, 2005.
200. J. Ponsajapan and S. H. Low, Reverse engineering TCP/IP-like networks using delay-sensitive utility functions, in *Proceedings of IEEE INFOCOM*, Anchorage, AK, May 2007.
201. A. L. Stolyar, On the asymptotic optimality of the gradient scheduling algorithm for multi-user throughput allocation, *Operations Research*, vol. 53, no. 1, pp. 12–25, 2005.
202. A. L. Stolyar, Maximizing queueing network utility subject to statbility: Greedy primal-dual algorithm, *Queueing Systems*, vol. 50, no. 4, pp. 401–457, 2005.
203. X. Lin, N. B. Shroff, and R. Srikant, A tutorial on cross-layer design in wireless networks, *IEEE Journal of Selected Areas in Communications*, vol. 24, no. 8, pp. 1452–1463, 2006.
204. D. S. Lun, N. Ratnakar, M. Mdard, R. Koetter, D. R. Karger, T. Ho, and E. Ahmed, Minimum-cost multicast over coded packet networks, *IEEE Transactions on Information Theory*, vol. 52, no. 6, pp. 2608–2623, 2006.

205. L. Chen, S. H. Low, and J. C. Doyle, Joint TCP congestion control and medium access control, in *Proceedings of IEEE INFOCOM*, Miami, FL, vol. 3, no. 7, pp. 2212–2222, 2005.
206. K. Kar, S. Sarkar, and L. Tassiulas, Achieving proportional fairness using local information in Aloha networks, *IEEE Transactions of Automatic Control*, vol. 49, no. 10, pp. 1858–1862, 2004.
207. J. W. Lee, M. Chiang, and R. A. Calderbank, Jointly optimal congestion and contention control in wireless ad hoc networks, *IEEE Communication Letters*, vol. 10, no. 3, pp. 216–218, 2006.
208. X. Wang and K. Kar, Cross-layer rate optimization for proportional fairness in multihop wireless networks with random access, *IEEE Journal on Selected Areas in*, vol. 24, no. 8, pp. 1548–1559, 2006.
209. J. Zhang and D. Zheng, A stochastic primal-dual algorithm for joint flow control and MAC design in multihop wireless networks, in *Proceedings of CISS*, Princeton, NJ, March 2006.
210. J. Zhang, D. Zheng, and M. Chiang, Impacts of stochastic noisy feedback in network utility maximization, in *Proceedings of IEEE INFOCOM*, Barcelona, Spain, May 2007.
211. M. Chiang, Balancing transport and physical layer in wireless multihop networks: Jointly optimal congestion control and power control, *IEEE Journal on Selected Areas in Communication*, vol. 23, no. 1, pp. 104–116, 2005.
212. M. Chiang and J. Bell, Balancing supply and demand of bandwidth in wireless cellular networks: Utility maximization over powers and rates, in *Proceedings of IEEE INFOCOM*, Hong Kong, China, vol. 4, pp. 2800–2811, 2004.
213. J. W. Lee, M. Chiang, and R. A. Calderbank, Price-based distributed algorithm for optimal rate-reliability tradeoff in network utility maximization, *IEEE Journal on Selected Areas in Communication*, vol. 24, no. 5, pp. 962–976, May 2006.
214. M. Andrews, Joint optimization of scheduling and congestion control in communication networks, in *Proceedings of CISS*, Princeton, NJ, March 2006.
215. L. Bui, A. Eryilmaz, R. Srikant, and X. Wu, Joint congestion control and distributed scheduling in multi-hop wireless networks with a node exclusive interference model, in *Proceedings of IEEE INFOCOM*, Barcelona, Spain, April 2006.
216. A. Eryilmaz and R. Srikant, Fair resource allocation in wireless networks using queue-length-based scheduling and congestion control, in *Proceedings of IEEE INFOCOM*, Miami, FL, vol. 3, pp. 1794–1803, March 13–17, 2005.
217. P. Marbach and Y. Lu, Active queue management and scheduling for wireless networks: The single cell case, in *Proceedings of CISS*, Princeton, NJ, March 2006.
218. B. Johansson, P. Soldata, and M. Johansson, Mathematical decomposition techniques for distributed cross-layer optimization of data networks, *IEEE Journal on Selected Areas in Communications*, vol. 24, no. 8, pp. 1535–1547, 2006.
219. M. J. Neely, E. Modiano, and C. E. Rohrs, Dynamic power allocation and routing time varying wireless networks, *IEEE Journal on Selected Areas in Communications*, vol. 23, no. 1, pp. 89–103, 2005.
220. L. Xiao, M. Johansson, and S. Boyd, Joint routing and resource allocation via dual decomposition, *IEEE Transactions on Communications*, vol. 52, no. 7, pp. 1136–1144, 2004.
221. L. Chen, S. H. Low, M. Chiang, and J. C. Doyle, Joint optimal congestion control, routing, and scheduling in wireless ad hoc networks, in *Proceedings of IEEE INFOCOM*, Barcelona, Spain, April 2006.
222. A. Eryilmaz and R. Srikant, Joint congestion control, routing and MAC for stability and fairness in wireless networks, *IEEE Journal on Selected Areas in Communications*, vol. 24, no. 8, pp. 1514–1524, 2006.
223. X. Lin and N. B. Shroff, Joint rate control and scheduling in multi-hop wireless networks, in *Proceedings of IEEE CDC*, Nassau, Bahamas, vol. 2, pp. 1484–1489, December 14–17, 2004.

224. X. Lin and N. B. Shroff, The impact of imperfect scheduling on cross-layer rate control in wireless networks, *IEEE/ACM Transactions on Network*, vol. 14, no. 2, pp. 302–315, 2006.
225. M. J. Neely, E. Modiano, and C. P. Li, Fairness and optimal stochastic control for heterogeneous networks, in *Proceedings of IEEE INFOCOM*, Miami, FL, vol. 3, pp. 1723–1734, March 13–17, 2005.
226. R. L. Cruz and A. Santhanam, Optimal routing, link scheduling, and power control in multihop wireless networks, in *Proceedings of IEEE INFOCOM*, San Francisco, CA, vol. 1, pp. 702–711, March 30–April 3, 2003.
227. Y. Xi and E. Yeh, Node-based distributed optimal control of wireless networks, in *Proceedings of CISS*, Princeton, NJ, March 2006.
228. J. Huang, Z. Li, M. Chiang, and A. K. Katsaggelos, Pricing-based rate control and joint packet scheduling for multi-user wireless uplink video streaming, in *Proceedings of IEEE Packet Video Workshop*, Seattle, WA, April 2006.
229. W. Yu and J. Yuan, Joint source coding, routing, and resource allocation for wireless sensor networks, in *Proceedings of IEEE ICC*, Seoul, South Korea, May 2005.
230. J. He, M. Chiang, and J. Rexford, TCP/IP interaction based on congestion prices: Stability and optimality, in *Proceedings of IEEE ICC*, Istanbul, Turkey, June 2006.
231. J. Wang, L. Li, S. H. Low, and J. C. Doyle, Cross-layer optimization in TCP/IP networks, *IEEE/ACM Transactions on Networking*, vol. 13, no. 3, pp. 582–268, 2005.
232. C. S. Chang and Z. Liu, A bandwidth sharing theory for a large number of HTTP-like connections, *IEEE/ACM Transactions on Networking*, vol. 12, no. 5, pp. 952–962, 2004.
233. H. Han, S. Shakkottai, C. V. Hollot, R. Srikant, and D. Towsley, Overlay TCP for multi-path routing and congestion control, in *Proceedings of IMA Workshop Measurement and Modeling of the Internet*, January 2004.
234. K. Kar, S. Sarkar, and L. Tassiulas, Optimization based rate control for multipath sessions, in *Proceedings of Int. Teletraffic Congress*, Anchorage, AK, December 2001.
235. F. P. Kelly and T. Voice, Stability of end-to-end algorithms for joint routing and rate control, *Computer Communication Review*, vol. 35, no. 2, pp. 5–12, 2005.
236. P. Key, L. Massoulie, and D. Towsley, Combining multipath routing and congestion control for robustness, in *Proceedings of CISS*, Princeton, NJ, March 2006.
237. F. Paganini, Congestion control with adaptive multipath routing based on optimization, in *Proceedings of CISS*, Princeton, NJ, March 2006.
238. H. Nama, M. Chiang, and N. Mandayam, Utility lifetime tradeoff in self regulating wireless sensor networks: A cross-layer design approach, in *Proceedings of IEEE ICC*, Istanbul, Turkey, June 2006.
239. J. Zhu, S. Chen, B. Bensaou, and K. L. Hung, Tradeoff between lifetime and rate allocation in wireless sensor networks: A cross layer approach, in *Proceedings of the 26th IEEE International Conference on Computer Communications (INFOCOM 2007)*, Anchorage, AK, pp. 267–275, May 6–12, 2007.
240. P. Jayachandran and T. Abdelzaher, Bandwidth allocation for elastic real-time flows in multihop wireless networks based on network utility maximization, in *Proceedings of the 28th International Conference on Distributed Computing Systems (ICDCS '08)*, Beijing, China, pp. 849–857, 2008.
241. A. Ferragut, J. Garciia, and F. Paganini, Network utility maximization for overcoming inefficiency in multi-rate wireless networks, in *Proceedings of the 8th International Symposium on Modeling and Optimization in Mobile, Ad Hoc and Wireless Networks (WiOpt)*, Avignon, France, pp. 393–401, May 31–June 4, 2010.
242. Y. Liang, H. V. Poor, and L. Ying, Secure communications over wireless broadcast networks: Stability and utility maximization, *IEEE Transactions on Information Forensics and Security*, vol. 6, no. 3, pp. 682–692, 2011.

243. E. Liu, Q. Zhang, and K. K. Leung, Clique-based utility maximization in wireless mesh networks, *IEEE Transactions on Wireless Communications*, vol. 10, no. 3, pp. 948–957, 2011.
244. F. P. Kelly, Mathematical modelling of the internet, in *Proceedings of 4th International Congress on Industrial Applied Mathematics*, Edinburgh, UK, July 1999. http://www.statslab.cam.ac. uk/ frank/mmi.html.
245. S. H. Low, A duality model of TCP flow controls, in *Proceedings of ITC Specialist Seminar on IP Traffic Measurement, Modeling and Management*, New Orleans, LA, September 18–20, 2000. http://netlab.caltech.edu.
246. T. V. Lakshman and U. Madhow, The performance of TCP/IP for networks with high bandwidth-delay products and random loss, *IEEE/ACM Transactions on Networking*, vol. 5, no. 3, pp. 336–350, 1997.
247. M. Mathis, J. Semke, J. Mahdavi, and T. Ott, The macroscopic behavior of the TCP congestion avoidance algorithm, *ACM Computer Communication Review*, vol. 27, no. 3, pp. 67–82, 1997.
248. B. Radunovic and J. Y. L. Boudec, A unified framework for max-min and min-max fairness with applications, in *Proceedings of 40th Annual Allerton Conference*, Allerton, IL, October 2002.
249. L. Georgadis, P. Georgatsos, K. Floros, and S. Sartzetakis, Lexicographically optimal balanced networks, in *INFOCOM'01*, Anchorage, AK, vol. 2, no. 6, pp. 818–829, 2001.
250. Y. Hou, H. Tzeng, and S. Panwar, A generalized max-min rate allocation policy and its distributed implementation using the ABR flow control mechanism, in *INFOCOM '98*, San Francisco, CA, vol. 3, pp. 1366–1375, 1998.
251. H. Tzeng and K. Siu, On max-min fair congestion control for multicast ABR service in ATM, *IEEE Journal on Selected Areas in Communications,* vol. 15, no. 3, pp. 545–556, 1997.
252. D. Rubenstein, J. Kurose, and D. Towsley, The impact of multicast layering on network fairness, *IEEE/ACM Transactions on Networking,* vol. 10, no. 2, pp. 169–182, 2002.
253. A. Mas-Colell, M. Whinston, and J. Green. *Microeconomic Theory*. Oxford University Press, 1995.
254. J. Ros and W. Tsai, A theory of convergence order of maxmin rate allocation and an optimal protocol, in *INFOCOM'01*, Anchorage, AK, vol. 2, pp. 717–726, 2001.
255. P. Marbach, Priority service and max-min fairness, in *INFOCOM'02*, New York, 2002.
256. S. Sarkar and L. Tassiulas, Fair allocation of discrete bandwidth layers in multicast networks, in *INFOCOM'00*, Tel Aviv, Israel, vol. 3, no. 23, pp. 1491–1500, 2000.
257. Z. Cao and E. Zegura, Utility max-min: An application-oriented bandwidth allocation scheme, in *INFOCOM'99*, New York, pp. 793–801, 1999.
258. S. Li, W. Sun, and C. Hua, Fair resource allocation and stability for communication networks with multipath routing, *International Journal of Systems Science*, vol. 45, no. 11, pp. 2342–2353, 2014.
259. P. Bender, P. Black, M. Grob, R. Padovani, N. Sindhushyana, and A. Viterbi, A bandwidth efficient high speed wireless data service for nomadic users, *IEEE Communications Magazine*, vol. 38, no. 7, pp. 70–77, 2000.
260. T. Bu, L. Li, and R. Ramjee, Generalized proportional fair scheduling in third generation wireless data networks, in *Proceedings of the 25th IEEE International Conference on Computer Communications (INFOCOM 2006)*, Barcelona, Spain, pp. 1–12, April 2006.
261. M. S. Bazaraa and J. J. Jarvis, *Linear Programming and Network Flows,* John Wiley & Sons, Hoboken, NJ, 1976, ISBN: 0-471-06015-1.
262. G. E. Bolton and A. Ockenfels, How do efficiency and equity trade-off when a majority rules? Working paper, University of Cologne, 2003. http://ockenfels.uni-koeln.de/download/papers/fair-vs-eff.pdf.
263. T. Bonald and L. Massoulie, Impact of fairness on internet performance, in *Proceedings of ACM Sigmetrics'01,* Guilin, China, pp. 82–91, June 2001.

264. M. Butler and H. P. Williams, Fairness versus efficiency in charging for the use of common facilities, *Journal of Operational Research Society*, vol. 53, no. 12, pp. 1324–1329, 2002.
265. A. Charny, D. D. Clark, and R. Jain, Congestion control with explicit rate indication, in *Proceedings of IEEE International Conference on Communications (ICC'95)*, Seattle, WA, pp. 18–22, June 1995.
266. F. Davik, M. Yilmaz, S. Gjessing, and N. Uzún, IEEE 802.17 resilient packet ring tutorial, *IEEE Communications Magazine*, vol. 42, no. 3, pp. 112–118, 2004.
267. V. Gambiroza, P. Yuan, L. Balzano, Y. Liu, S. Sheafor, and E. Knightly, Design, analysis, and implementation of DVSR: A fair high-performance protocol for packet rings, *IEEE/ACM Transactions on Networking*, vol. 12, no. 1, pp. 85–102, 2004.
268. W. Güth, H. Kliemt, and A. Ockenfels, Fairness versus efficiency: An experimental study of (mutual) gift giving, *Journal of Economic Behaviour and Organization*, vol. 50, pp. 465–475, 2003.
269. E. L. Hahne, A. K. Choudhury, and N. F. Maxemchuk, DQDB networks with and without bandwidth balancing, *IEEE Transactions on Communications*, vol. 40, no. 7, pp. 1192–1204, 1992.
270. C. N. Hawkins, J. Green, M. Sharma, and K. Vasani, *Resilient Packet Rings for Metro Networks*, August 2001. http://www.rpralliance.org/.
271. S. L. Hsu, Fairness versus efficiency in environmental law, *Ecology Law Quarterly* , vol. 31, no. 2, pp. 303–401, 2004.
272. T. G. Kolda, R. M. Lewis, and V. Torczon, Optimization by direct search: New perspectives on some classical and modern methods, *SIAM Review* , vol. 45, no. 3, pp. 385–482, 2003.
273. S. Kunniyur and R. Srikant, End-to-end congestion control: utility functions, random losses and ECN marks, *IEEE/ACM Transactions on Networking*, vol. 11, no. 5, pp. 689–702, 2003.
274. H. Luo, S. Lu, V. Bharghavan, J. Cheng, and G. Zhong, A packet scheduling approach to QoS support in multihop wireless networks, *ACM Journal of Mobile Networks and Applications (MONET), Special Issue on QoS in Heterogeneous Wireless Networks* , vol. 9, no. 3, pp. 193–206, 2004.
275. M. A. Mammadov and R. Orsi, H_∞ synthesis via a nonsmooth, nonconvex optimization approach, *Pacific Journal of Optimization*, vol. 1, no. 2, pp. 405–420, 2005.
276. M. A. Mammadov, A. M. Rubinov, and J. Yearwood, Dynamical systems described by relational elasticities with applications to global optimization. in *Continuous Optimisation: Current Trends and Applications*, eds V. Jeyakumar and A. Rubinov, Springer, New York, pp. 365–387, 2005.
277. M. A. Mammadov, J. Yearwood, and L. Aliyeva, Multi label classification and drug-reaction associations using global optimization techniques, in *Proceedings of The Sixth International Conference on Optimization: Techniques and Applications (ICOTA6)*, Ballarat, VIC, December 2004.
278. L. Massoulie and J. Roberts, Bandwidth sharing: Objectives and algorithms, in *Proceedings of INFOCOM'99*, New York, pp. 1395–1403, March 1999.
279. A. Neumaier, O. Shcherbina, W. Huyer, and T. Vinko, A comparison of complete global optimization solvers. http://citeseer.ifi.unizh.ch/642030.html.
280. J. Pinter, *LGO—A Model Development System for Continuous Global Optimization*, User's Guide (Revised edition), Pinter Consulting Services, Halifax, NS, Canada, 2003.
281. P. M. Pardalos and J. B. Rosen, Methods for global concave minimization: A bibliographic survey, *SIAM Review* vol. 28, no. 3, pp. 367–379, 1986.
282. A. M. Rubinov and X. Yang. *Lagrange-Type Functions in Constrained Non-Convex Optimization*, Kluwer Academic, Springer, New York, vol. 115, no. 4, 2003.
283. R. Srinivasan and A. Somani, On achieving fairness and efficiency in high-speed shared medium access, *IEEE/ACM Transactions on Networking*, vol. 11, no. 1, pp. 111–124, 2003.

284. A. Tang, J. Wang, and S. H. Low, Is fair allocation always inefficient? in *Proceedings of IEEE INFOCOM*, Hong Kong, China, pp. 35–45, 2004.
285. B. Vandalore, S. Fahmy, R. Jain, R. Goyal, and M. Goyal, A definition of general weighted fairness and its support in explicit rate switch algorithms, in *Proceedings of Sixth International Conference on Network Protocols (ICNP)*, Austin, TX, pp. 22–30, October 1998.
286. B. Vandalore, S. Fahmy, R. Jain, R. Goyal, and M. Goyal, General weighted fairness and its support in explicit rate switch algorithms, *Computer Communications*, vol. 23, no. 2, pp. 149–161, 2000.
287. M. Zukerman and P. Potter, The DQDB protocol and its performance under overload traffic conditions, *Computer Network and ISDN Systems*, vol. 20, no. 1–5, pp. 261–270, 1990.
288. M. Zukerman, L. Tan, H. Wang, and I. Ouveysi, Efficiency-fairness tradeoff in telecommunications networks, *IEEE Communications Letters*, vol. 9, no. 7, pp. 643–645, 2005.
289. M. Zukerman, M. Mammadov, L. Tan, I. Ouveysi, and L. Andrew, To be fair or efficient or bit of both, *Computers and Operations Research*, vol. 35, no. 12, pp. 3787–3806, 2008.
290. C. Jin, D. Wei, and S. H. Low, FAST TCP for high-speed long-distance networks, Internet draft draft-jwl-tcp-fast-01.txt. http://netlab.caltech.edu/pub/papers/draft-jwl-tcp-fast-01.txt.
291. F. Paganini, Z. Wang, J. C. Doyle, and S. H. Low, Congestion control for high performance, stability, and fairness in general networks, *IEEE/ACM Transactions on Networking*, vol. 13, no. 1, pp. 43–56, 2005.
292. M. Allman, V. Paxson, and W. Stevens, TCP congestion control, *IETF RFC2581*, April 1999.
293. M. Mathis, J. Mahdavi, S. Floyd, and A. Romanow, TCP selective acknowledgment options, *RFC 2018*, ftp://ftp.isi.edu/ in-notes/rfc2018.txt, October 1996.
294. V. Jacobson, R. Braden, and D. Borman, TCP extensions for high performance, *RFC 1323*, ftp://ftp.isi.edu/in-notes/ rfc1323.txt, May 1992.
295. J. Hoe, Improving the startup behavior of a congestion control scheme for TCP, in *ACM Sigcomm96*, August 1996. http://www. acm.org/sigcomm/sigcomm96/program.html.
296. R. Jain, A delay-based approach for congestion avoidance in interconnected heterogeneous computer networks, *ACM Computer Communication Review*, vol. 19, no. 5, pp. 56–71, 1989.
297. Z. Wang and J. Crowcroft, Eliminating periodic packet losses in the 4.3-Tahoe BSD TCP congestion control algorithm, *ACM Computer Communications Review*, vol. 22, pp. 9–16, 1992.
298. L. Tan, W. Zhang, and C. Yuan, On parameter tuning for FAST TCP, *IEEE Communications Letters*, vol. 11, no. 5, pp. 458–460, 2007.
299. A. Jalali, R. Padovani, and R. Pankaj, Data throughput of CDMA HDR: A high efficiency-high data rate personal communication wireless system, in *Proceedings of IEEE Vehicular Technology Conference*, Tokyo, Japan, May 2000.
300. P. Viswanath, D. Tse, and R. Laroia, Opportunistic beam-forming using dumb antennas, *IEEE Transactions on Information Theory*, vol. 48, pp. 1277–1294, 2002.
301. S. Borst, User-level performance of channel-aware scheduling algorithms in wireless data networks, in *Proceedings of INFOCOM*, San Francisco, CA, April 2003.
302. M. Andrews, K. Kumaran, K. Ramanan, A. L. Stolyar, R. Vijayakumar, and P. Whiting, Providing quality of service over a shared wireless link, *IEEE Communications Magazine*, vol. 39, no. 2, pp. 150–154, 2001.
303. S. Shakkottai and A. Stolyar, Scheduling algorithms for a mixture of real-time and non-real-time data in HDR, in *Proceedings of ITC-17*, Salvador da Bahia, Brazil, pp. 793–804, September 2001.
304. S. Shakkottai and A. L. Stolyar, Scheduling for multiple flows sharing a time-varying channel: The exponential rule, *Analytic Methods in Applied Probability*, vol. 207, pp. 185–202, 2002.
305. A. Stolyar and K. Ramanan, Largest weighted delay first scheduling: Large deviations and optimality, *Annals of Applied Probability*, vol. 11, no. 1, pp. 1–48, 2001.

306. K. Norlund, T. Ottosson, and A. Brunstrom, Fairness measures for best effort traffic in wireless networks, in *Proceedings of PIMRC*, Barcelona, Spain, 2004.
307. C. Westphal, Monitoring proportional fairness in CDMA2000 high data rate networks, in *Proceedings of Globecom*, Dallas, TX, 2004.
308. S. Keshav, *An Engineering Approach to Computer Networking*, Addisson Wesley, Longman Publishing Co., Inc. Boston, MA, 1997.
309. A. Sang, X. Wang, M. Madihian, and R. D. Gitlin, Coordinated load balancing, handoff/cell-site selection, and scheduling in multi-cell packet data systems, in *Proceedings of the 10th Annual International Conference on Mobile Computing and Networking*, Dallas, TX, pp. 302–314, 2004.
310. S. Das, H. Viswanathan, and G. Rittenhouse, Dynamic load balancing through coordinated scheduling in packet data systems, in *Proceedings of IEEE INFOCOM*, San Francisco, CA, 2003.
311. Y. Bejerano, S. J. Han, and L. E. Li, Fairness and load balancing in wireless lans using association control, in *Proceedings of the 10th Annual International Conference on Mobile Computing and Networking*, Philadelphia, PA, pp. 315–329, 2004.
312. P. J. Argibay-Losada, K. Nozhnina, A. Surez-Gonzlez, C. Lpez-Garcła, and M. Fernndez-Veiga, Loss-based proportional fairness in multihop wireless networks, *Wireless Networks*, vol. 20, no. 5, pp. 805–816, 2014.
313. T. Alpcan and T. Basar, A utility-based congestion control scheme for internet-style networks with delay, in *Proceedings of IEEE INFOCOM'03*, San Francisco, CA, pp. 2039–2048, March 2003.
314. L. S. Brakmo and L. L. Peterson, TCP Vegas: End-to-end congestion avoidance on a global internet, *IEEE Journal on Selected Areas in Communications*, vol. 13, no. 8, pp. 1465–1480, 1995.
315. E. W. Kamen, On the relationship between zero criteria for two variable polynomials and asymptotic stability of delay differential equation, *IEEE Transactions on Automatic Control*, vol. AC-25, no. 5, pp. 983–984, 1980.
316. S. H. Lehnigk, *Stability Theorems for Linear Motions*, Prentice-Hall, Englewood Cliffs, NJ, 1966.
317. M. Marden, *Geometry of Polynomials*, American Mathematical Society, Providence, RI, 1966.
318. A. Tang, K. Jacobsson, L. L. H. Andrew, and S. H. Low, An accurate link model and its application to stability analysis of FAST TCP, in *Proceedings of IEEE INFOCOM 2007*, Anchorage, Alaska, pp. 161–169, May 2007.
319. A. Tang, L. L. H. Andrew, K. Jacobsson, K. H. Johansson, S. H. Low, and H. Hjalmarsson, Window flow control: Macroscopic properties from microscopic factors, in *Proceedings of IEEE INFOCOM*, Phoenix, AZ, April 2008. (to appear). http://netlab.caltech.edu/ lachlan/abstract/ACanalysisTR.pdf.
320. Y. P. Tian and G. Chen, Stability of the primal-dual algorithm for congestion control, *International Journal of Control*, vol. 79, no. 6, pp. 662–676, 2006.
321. USC/ISI, Los Angeles, CA. The NS simulator and the documentation. http://www.isi.edu/ nsnam/ns/.
322. Z. Wang and F. Paganini, Global stability with time-delay of a primal-dual congestion control, in *Proceedings of 42nd IEEE Conference on Decision and Control (CDC'03)*, Maui, HI, pp. 3671–3676, December 2003.
323. M. Welzl and D. Paradimitriou, Open research issues in internet congestion control, Internet Draft. http://www.ietf.org/internet-drafts/draft-irtf-iccrg-welzl-congestion-control-open-research-00.txt.

324. J. Y. Choi, K. Koo, J. S. Lee, and S. H. Low, Global stability of FAST TCP in single-link single-source network, in *Proceedings of 44th IEEE Conference on Decision and Control (CDC)*, Seville, Spain, December 12–15, 2005.
325. T. Cui and L. Andrew, FAST TCP simulator module for ns-2, version 1.1. http://www.cubinlab.ee.mu.oz.au/ns2fasttcp.
326. G. F. Franklin, J. D. Powell, and A. Emami-Naeini, *Feedback Control of Dynamic Systems*, 3rd ed., Addison-Wesley, Reading, MA, 1995.
327. S. H. Low, F. Paganini, J. Wang, and J. C. Doyle, Linear stability of TCP/RED and a scalable control, *Computer Networks Journal*, vol. 43, no. 5, pp. 633–647, 2003.
328. A. Tang, K. Jacobsson, L. L. H. Andrew, and S. H. Low, Linear stability analysis of FAST TCP using a new accurate link model, in *Proceedings of 44th Annual Allerton Conference on Communication, Control and Computing*, Monticello, IL, September 2006.
329. J. Wang, A. Tang, and S. H. Low, Local stability of FAST TCP, in *Proceedings of 43th IEEE Conference on Decision and Control (CDC)*, Paradise Island, Bahamas, December 2004.
330. J. Wang, D. X. Wei, and S. H. Low, Modeling and stability of FAST TCP, in *Proceedings of IEEE INFOCOM 2005*, Miami, FL, pp. 938–948, March 2005.
331. M. Christiansen, K. Jeffay, D. Ott, and F. D. Smith, Tuning RED for web traffic, *IEEE/ACM Transactions on Networking*, vol. 9, no. 3, pp. 249–264, 2001.
332. L. Tan, Y. Yang, W. Zhang, and M. Zukerman, On control gain selection in dynamic-RED, *IEEE Communications Letters*, vol. 9, no. 1, pp. 81–83, 2005.
333. B. Wydrowski and M. Zukerman, QoS in best-effort networks, *IEEE Communications Magazine*, vol. 40, no. 12, pp. 44–49, 2002.
334. C. Jin, D. X. Wei, S. H. Low, G. Buhrmaster, J. Bunn, D. H. Choe, R. L. A. Cottrell, J. C. Doyle, W. Feng, O. Martin, H. Newman, F. Paganini, S. Ravot, and S. Singh, FAST TCP: From theory to experiments, *IEEE Network*, vol. 19, no. 1, pp. 4–11, 2005.
335. C. A. Desoer and Y. T. Yang, On the generalized Nyquist stability criterion, *IEEE Transactions on Automatic Control*, vol. 25, no. 2, pp. 187–196, 1979.
336. G. H. Hardy, J. E. Littlewood, and G. Pólya, *Inequalities*, Cambridge University Press, Cambridge, 1934.
337. B. Wydrowski and M. Zukerman, MaxNet: A congestion control architecture, *IEEE Communications Letters*, vol. 6, no. 11, pp. 512–514, 2002.
338. R. Jain, D. Chiu, W. Hawe, A quantitative measure of fairness and discrimination for resource allocation in shared computer systems, Technical Report DEC-TR-301, Digital Equipment Corporation, September 1984.
339. R. Mazumdar, L. G. Mason, and C. Douligeris, Fairness in network optimal flow control: Optimality of product forms, *IEEE Transactions on Communications*, vol. 39, no. 5, pp. 775–782, 1991.
340. D. M. Chiu and R. Jain, Analysis of the increase and decrease algorithms for congestion avoidance in computer networks, *Computer Networks and ISDN Systems*, vol. 17, no. 1, pp. 1–14, 1989.
341. S. Chan and M. Zukerman, Is max-min fairness achievable in the presence of insubordinate users? *IEEE Communications Letters*, vol. 6, no. 3, pp. 120–122, 2002.
342. B. Vandalore, S. Fahmy, R. Jain, R. Goyal, and M. Goyal, General weighted fairness and its support in explicit rate switch algorithms, *Computer Communications*, vol. 23, no. 2, pp. 149–161, 2000.
343. I. Cidon and Y. Ofek, Metaring-a full-duplex ring with fairness and spatial reuse, *IEEE Transacation on Communication*, vol. COM-41, no. 1, pp. 110–120, 1993.
344. Y. Ofek, Overview of the metaring architecture, *Computer Networks and ISDN Systems*, vol. 26, no. 6–8, pp. 817–830, 1994.

345. G. Anastasi, L. Lenzini, M. La Porta, and Y. Ofek, Dynamic max-min fairness in ring networks, *Cluster Computing*, vol. 3, no. 3, pp. 215–230, 2000.
346. G. Anastasi, L. Lenzini, and Y. Ofek, Tradeoff between the cycle complexity and the fairness of ring networks, *Microprocessors and Microsystems*, vol. 25, no. 1, pp. 41–59, 2001.
347. J. L. Boudec, Rate adaptation, congestion control and fairness: A tutorial. http://ica1www.epfl.ch/PS_files/LEB3132.pdf.
348. S. Hegde, D. Lapsley, B. Wydrowski, J. Lindheim, D. Wei, C. Jin, S. H. Low, and H. Newman, FAST TCP in high speed networks: An experimental study, in *Proceedings of GridNets*, San Jose, CA, October 2004.
349. L. A. Grieco and S. Mascolo, Performance evaluation and comparison of Westwood+, New Reno, and Vegas TCP congestion control, *ACM SIGCOMM Computer Communication Review*, vol. 34, no. 2, pp. 25–38.
350. A. Tang, J. Wang, S. Hegde, and S. H. Low, Equilibrium and fairness of networks shared by TCP Reno and Vegas/FAST, *Telecommunications Systems–Special Issue on High Speed Transport Protocols*, vol. 30, no. 4, pp. 417–439, 2005.
351. L. L. H. Andrew, L. Tan, T. Cui, and M. Zukerman, Fairness comparison of FAST TCP and TCP Vegas, in *Proceedings of 19th ITC*, Beijing, China, pp. 1375–1384, August 2005.
352. L. Tan, C. Yuan, and M. Zukerman, FAST TCP: Fairness and queuing issues, *IEEE Communications Letters*, vol. 9, no. 8, pp. 262–764, 2005.
353. J. Mo, R. J. La, V. Anantharam, and J. Walrand, Analysis and comparison of TCP Reno and Vegas, in *Proceedings of IEEE INFOCOM*, New York, vol. 3, pp. 1556–1563, March 1999.
354. K. Kurata, G. Hasegawa, and M. Murata, Fairness comparisons between TCP Reno and TCP Vegas for future deployment of TCP Vegas, in *Proceedings of INET 2000*, Yokohama, Japan, July 18–21, 2000. http://www.isoc.org/inet2000/cdproceedings/2d/2d_2.htm.
355. D.D. Luong and J. Bíró, Bandwidth sharing scheme of end-to-end congestion control protocols, *Periodica Polytechnica, Series Electrical Engineering*, vol. 45, no. 2, pp. 79–90, 2001.
356. Measurement and network analysis: AMP raw data query. http://watt.nlanr.net/Active/raw_data/cgi-bin/data_form.cgi.
357. L. Kalampoukas, A. Varma, and K. Ramakrishnan, Two-way TCP traffic over rate controlled channels (effects and analysis), *IEEEACM Transactions on Networking*, vol. 6, no. 6, pp. 729–743, 1998.
358. L. X. Zhang, S. Shenker, and D. D. Clark, Observations on the dynamics of a congestion control algorithm: The effects of two-way traffic, *SIGCOMM Computer Communication Review*, vol. 21, no. 4, pp. 133–147, 1991.
359. H. Balakrishnan and V. N. Padmanabhan, How network asymmetry affects TCP, *IEEE Communications Magazine*, vol. 39, no. 4, pp. 60–67, 2001.
360. K. T. Truong, S. Weber, and R. W. Heath, Transmission capacity of two way communication in wireless ad hoc networks, in *Proceedings of IEEE ICC 2009*, Dresden, Germany, pp. 1–5, 2009.
361. E. Lopez-Aguilera, M. Heusse, Y. Grunenberger, F. Rousseau, A. Duda, and J. Casademont, An asymmetric access point for solving the unfairness problem in WLANs, *IEEE Transactions on Mobile Computing*, vol. 7, no. 10, pp. 1213–1227, 2008.
362. T. V. Lakshman, U. Madhow, and B. Suter, TCP/IP performance with random loss and bidirectional congestion, *IEEE/ACM Transactions on Networking*, vol. 8, no. 5, pp. 541–555, 2000.
363. C. P. Fu and S. C. Liew, A remedy for performance degradation of TCP Vegas in asymmetric networks, *IEEE Communications Letters*, vol. 7, no. 1, pp. 42–44, 2003.
364. F. Ge and L. Tan, Network utility maximization in two-way flow scenario, *ACM SIGCOMM Computer Communication Review*, vol. 44, no. 2, pp. 14–19, 2014.

365. M. A. Marsan, A. Bianco, P. Giaccone, E. Leonardi, and F. Neri, Multicast traffic in input-queued switches: Optimal scheduling and maximum throughput, *IEEE/ACM Transactions on Networking*, vol. 11, no. 3, pp. 465–477, 2003.
366. L. Massoulie and J. Roberts, Bandwidth sharing: Objectives and algorithms, *IEEE/ACM Transactions on Networking*, vol. 10, no. 3, pp. 320–328, 2002.
367. W. H. Wang, M. Palaniswami, and S. H. Low, Application-oriented flow control: Fundamentals, algorithms and fairness, *IEEE/ACM Transactions on Networking*, vol. 14, no. 6, pp. 1282–1291, 2006.
368. I. Juva, Robust load balancing, in *Proceedings of GLOBECOM07*, Washington, DC, pp. 2708–2713, 2007.
369. I. Juva, Robust laod balancing in wireless networks, in *Proceedings of the EuroFGI Workshop on IP Qos and Traffic Control*, Lisbon, Portugal, pp. 127–134, 2007.
370. Y. Ohsita, T. Miyamura, S. Arakawa, S. Ata, E. Oki, K. Shiomoto, and M. Murata, Gradually reconfiguring virtual network topologies based on estimated traffic matrices, in *Proceedings of INFOCOM07*, Barcelona, Spain, pp. 2511–2515, 2007.
371. H. X. Nguyen and P. Thiran, The Boolean solution to the congested IP link location problem: Theory and practice, in *Proceedings of INFOCOM07*, Barcelona, Spain, pp. 2117–2125, 2007.
372. Q. Zhao, A. Kumar, J. Wang, and J. Xu, Data streaming algorithms for accurate and efficient measurement of traffic and flow matrices, *ACM SIGMETRICS Performance Evaluation Review*, vol. 33, no. 1, pp. 350–361, 2005.
373. G. Liang and B. Yu, Maximum pseudo likelihood estimation in network tomography, *IEEE Transactions on Signal Processing,* vol. 51, no. 8, pp. 2043–2053, 2003.
374. A. Soule, A. Nucci, R. Cruz, E. Leonardi, and N. Taft, How to identify and estimate the largest traffic matrix elements in a dynamic environment, *ACM SIGMETRICS Performance Evaluation Review,* vol. 32, no. 1, pp. 73–84, 2004.
375. Y. Zhang, M. Roughan, N. Duffield, and A. Greenberg, Fast accurate computation of large-scale IP traffic matrices from link loads, in *Proceedings of ACM SIGMETRICS 2003,* San Diego, CA, vol. 31, no. 1, pp. 206–217, 2003.
376. J. Cao, D. Davis, S. V. Wiel, and B. Yu, Time-varying network tomography, *Journal of the American Statistical Association*, vol. 95, no. 452, pp. 1063–1075, 2000.
377. A. Medina, N. Taft, K. Salamatian, S. Bhattacharyya, and C. Diot, Traffic matrix estimation: Existing techniques and new directions, in *Proceedings of ACM SIGCOMM 2002*, Pittsburgh, PA, August 2002.
378. C. Tebaldi and M. West, Bayesian inference on network traffic using link count data, *Journal of the American Statistical Association*, vol. 93, no. 442, pp. 557–576, 1998.
379. Y. Zhang, M. Roughan, C. Lund, and D. Donoho, An information-theoretic approach to traffic matrix estimation, in *Proceedings of ACM SIGCOMM 2003,* Karlsruhe, Germany, vol. 1, no. 1, pp. 326–327, August 2003.
380. Y. Zhang, M. Roughan, C. Lund, and D. Donoho, Estimating point-to-point and point-to-multipoint traffic matrices: An information-theoretic approach, *IEEE/ACM Transactions on Networking,* vol. 13, no. 5, pp. 947–960, 2005.
381. Y. Xiong, M. Vandenhoute, and H. Cankaya, Control architecture in optical burst switched WDM networks, *IEEE Journal on Selected Areas in Communications*, vol. 18, no. 10, pp. 1838–1851, 2000.
382. C. Qiao, Labeled optical burst switching for IP-over-WDM intergration, *IEEE Communications Magazine*, vol. 38, no. 9, pp. 104–114, 2000.
383. J. Turner, Terabit burst switching, *Journal of High Speed Networks*, vol. 8, no. 1, pp. 3–16, 1999.
384. C. Qiao and M. Yoo, Optical burst switching (OBS)—a new paradigm for an optical internet, *Journal of High Speed Networks*, vol. 8, no. 1, pp. 69–84, 1999.

385. X. Yu, Y. Chen, and C. Qiao, Study of traffic statistics of assembled burst traffic in optical burst switched networks, in *Proceedings of Opticomm*, pp. 149–159, 2002.
386. M. Yoo, C. Qiao, and S. Dixit, Optical burst switching for service differentiation in the next-generation optical internet, *IEEE Communications Magazine*, vol. 39, no. 2, pp. 98–104, 2001.
387. T. Battestilli and H. G. Perros, Optical burst switching for the next genetration internet, *IEEE Potentials*, vol. 23, no. 5 pp. 40–43, 2004.
388. M. Yoo, C. Qiao, and S. Dixit, QoS performance of optical burst switching in IP-over-WDM networks, *IEEE Journal on Selected Areas in Communications*, vol. 18, no. 10, pp. 2268–2278, 2000.
389. IEEE Standard 802.17: Resilient packet ring (draft version 1.0) *IEEE Wireless Communications*, Aug. 2002.
390. G. Aybay, M. O Connor, R. Vasani, and T. Wu, RPR alliance white paper, *Journal of Software*. http://www.rpralliance.org, October 2001.
391. A. Mekkittikuletal, Alladin proposal for IEEE Standard 802.17, *IEEE Transactions on Wireless Communications*, http://grouper.ieee.org/groups/802/17/documents/drafts/zz_draft_04.zip, November, 2001.
392. J. Kaoetal, Gandalf proposal for IEEE Standard 802.17. http://grouper.ieee.org/groups/802/17/documents/drafts/gandalf_04Plus.pdf, November 2002.
393. V. Gambiroza, Y. Liu, P. Yuan, and E. Knightly, High performance fair bandwidth allocation for resilient packet rings, in *Proceedings of 5th ITC Specialist Seminar on Traffic Engineering and Traffic Management,* Wurzburg, Germany, July 22–24, 2002.
394. L. Balzano, V. Gambiroza, Y. Liu, S. Shaefor, P. Yuan, and E. Knightly, Design, analysis, and implementation of DVSR: An enhanced protocol for packet rings, Rice University Technical Report. http://www.ece.rice.edu/ knightly, January 2002.
395. E. Knightly, L. Balzano, V. Gambiroza, Y. Liu, S. Shaefor, P. Yuan, and H. Zhang, Achieving high performance with Darwin's fairness algorithm, Rice University, March 2002. http://www.ece.rice.edu/ knightly.
396. A. W. Barnhart, Explicit rate performance evaluation, ATM Forum 94-0983R1, 1994.
397. L. Kalampoukas, A. Varma, and K. K. Ramakrishnan, An efficient rate allocation algorithm for ATM networks providing max-min fairness, in *Proceedings of the 6th IFIP International Conference on High Performance Networking,* Balearic Islands, Spain, pp. 143–154, September 1995.
398. K. Siu and T. Tzeng, Intelligent congestion control for ABR service in ATM networks, *Computer Communication Review*, vol. 24, no. 5, pp. 81–106, 1995.
399. F. Bonomi and K. W. Fendick, The rate-based flow control framework for the available bit rate ATM service, *IEEE Network Magazine*, vol. 9, no. 2, pp. 25–39, 1995.
400. Y. Afek, Y. Mansour, and Z. Ostfeld, Phantom: A simple and effective flow control scheme, in *Proceedings of ACM SIGCOMM 96 Conference*, Stanford University, Stanford, CA, pp. 169–182, August 1996.
401. D. H. K. Tsang, W. K. F. Wong, S. M. Jiiang, and E. Y. S. Liu, A fast switch algorithm for ABR traffic to achieve max-min fairness, in *Proceedings of IEEE International Zurich Seminar on Digital Communications,* pp. 19–23, 1996.
402. D. H. K. Tsang and W. K. F. Wong, A new rate-based switch algorithm for ABR traffic to achieve max-min fairness with analytical approximation and delay adjustment, in *Proceedings of IEEE INFOCOM'96*, San Francisco, CA, pp. 1174–1181, 1996.
403. D. Hughes, Fair share in the context of MCR, ATM Forum/AF-TM 94-0977, 1994.
404. Y. T. Hou, H. Tzeng, and S. S. Panwar, A simple ABR switch algorithm for the weighted max–min fairness policy, in *Proceedings of IEEE ATM97 Workshop*, Lisboa, Portugal, pp. 329–338, May 1997.

405. L. Tan, A. C. Pugh, and M. Yin, Rate-based congestion control in ATM switching networks using a recursive digital filter, *Control Engineering Practice (Special Issue on Control Methods for Telecommunication Networks)*, vol. 11, no. 10, pp. 1171–1181, 2003.
406. X. Zhang, K. G. Shin, D. Saha, and D. Kandlur, Scalable flow control for multicast ABR services, in *Proceedings of IEEE INFOCOM'99*, New York, pp. 837–846, March 1999.
407. J. Byers, M. Frumin, G. Horn, and M. Luby, FLID-DL: Congestion control for layered multicast, in *Proceedings of the 2nd International Workshop on Networked Group Communication (NGC2000)*, Stanford University, Palo Alto, CA, pp. 71–81, November 2000.
408. E. W. Kamen and B. S. Heck, *Fundamentals of Signals and Systems Using the Web and Matlab*, 2nd ed., Prentice Hall, Inc., Upper Saddle River, NJ, 2002.
409. S. Corson and J. Macker, Mobile ad hoc networking (MANET): Routing protocol performance issues and evaluation considerations, in *IETF RFC2501*, January 1999.
410. C. Wang, C. Wu, and G. Chen, P-MANET: Efficient power saving protocol for multihop mobile ad hoc networks, in *Proceedings of Third International Conference on Information Technology and Applications (ICITA'05),* Washington, DC, vol. 2, pp. 271–276, 2005.
411. R. Zheng and R. Kravets, On-demand power management for ad hoc networks, in *Proceedings of IEEE INFOCOM*, San Francisco, CA, vol. 1, pp. 481–491, April 2003.
412. S. Wu, Y. Tseng, and J. Sheu, Intelligent medium access for mobile ad hoc networks with busy tones and power control,' *IEEE Journal on Selected Areas in Communications*, vol. 18, no. 9, pp. 1647–1657, 2000.
413. J. Gomez and A. Campbell, A case for variable-range transmission power control in wireless ad hoc network, in *Proceedings of IEEE INFOCOM*, Hong Kong, China, vol. 2, pp 1425–1436, 2004.
414. J. Monks, V. Bharghavan, and W. Hwu, A power controlled multiple access protocol for wireless packet networks, in *Proceedings of IEEE INFOCOM*, Anchorage, AK, vol. 1, pp. 219–228, April 2001.
415. M. Zawodniok and S. Jagannathan, A distributed transmission power control protocol for mobile ad hoc networks, in *Proceedings of the IEEE WCNC,* Atlanta, GA, pp. 1915–1920, 2004.
416. D. B. Johnson and D. A. Maltz, Dynamic source routing in ad hoc wireless networks, *Mobile Computing*, eds T. Imielinski and H. Korth, Kluwer Academic Publishers, Big Island, HI, pp. 153–181, 1996.
417. C. Perkins, Ad-hoc on demand distance vector (AODV) routing, intemet-draft, draft-ieff-manet-aodv-00.txt, November 1997.
418. C. E. Perkins and E. M. Royer, Ad-hoc on-demand distance vector routing, in *Proceedings of the 2nd IEEE Workshop on Mobile Computing Systems and Applications*, New Orleans, LA, pp. 90–100, February 1999.
419. L. Junhai, Y. Danxia, X. Liu, and F. Mingyu, A survey of multicast routing protocols for mobile ad hoc networks, *IEEE Communications Surveys Tutorials*, vol. 11, no. 1, pp. 78–91, 2009.
420. J. Zhu and X. Wang, Model and protocol for energy-efficient routing over mobile ad hoc networks, *IEEE Transactions on Mobile Computing*, vol. 10, no. 11, pp. 1546–1557, Nov. 2011.
421. J.-H. Chang and L. Tassiulas, Energy conserving routing in wireless ad-hoc networks, in *Proceedings of Nineteenth Annual Joint Conference of the IEEE Computer and Communications Societies,* Tel Aviv, Israel, vol. 1, pp. 22–31, 2000.
422. I. Chlamtac, M. Conti, and J. J.-N. Liu, Mobile ad hoc networking: Imperatives and challenges, *IJCA Special Issue on MANETs*, no. 3, pp. 153–158, 2010.
423. S. Singh, M. Woo, and C. S. Mghavendra, Power-aware routing in mobile ad hoc networks, in *Proceedings of ACM/IEEE MobiCom98,* Chicago, IL, 1998.

424. G. Anastasia, M. Contib, and M. Di Francescoa, Energy conservation in wireless sensor networks: A survey, *Ad Hoc Networks*, vol 7, no. 3, pp. 537–568, 2009.
425. P. Jacquet, P. Muhlethaler, T. Clausen, A. Laouiti, A. Qayyum, and L. Viennot, Optimized link state routing protocol for ad hoc networks, in *IEEE International Proceedings of Multi Topic Conference, INMIC'01*, Lahore, Pakistan, 2001.
426. C. E. Perkins and P. Bhagwat, Highly dynamic destination-sequenced distance vector routing (DSDV) for mobile computers, in *Proceedings of ACM SIGCOMM '94,* London, UK, pp. 234–244, October 1994.
427. T. Camp, J. Boleng, and V. Davies, A survey of mobility models for ad hoc network research, in *Proceedings of Wireless Communications and Mobile Computing (WCMC'02),* vol. 2, no. 5, pp. 483–502, 2002.
428. P. Gupta and P. Kumar, Critical power for asymptotic connectivity in wireless networks, *Stochastic Analysis, Control, Optimization and Applications*, Boston, pp. 547–566, 1998.
429. K. Scott and N. Bambos, Routing and channel assignment for low power transmission in PCS, in *Proceedings of Fifth IEEE International Conference on Universal Personal Communication (ICUPC '96),* Cambridge, MA, October 1996.
430. S. Doshi, S. Bhandare, and T.X. Brown, An on-demand minimum energy routing protocol for a wireless ad hoc network, *ACM Mobile Computing and Comminications Review*, vol. 6, no. 3, pp. 50–66, 2002.
431. V. Rodoplu and T. Meng, Minimum energy mobile wireless networks, *IEEE Journal on Selected Areas in Communications*, vol. 17, no. 8, pp. 1333–1344, 1999.
432. C. K. Toh, H. Cobb, and D. Scott, Performance evaluation of battery-life-aware routing schemes for wireless ad hoc networks, in *Proceedings of IEEE International Conference on Communication (ICC '01),* Helsinki, Finland, June 2001.
433. A. Misra and S. Banerjee, MRPC: Maximizing network lifetime for reliable routing in wireless environments, in *Proceedings of IEEE Wireless Communication and Networking Conference (WCNC '02),* Orlando, FL, March 2002.
434. T. Rappaport, *Wireless Communications: Principles and Practice*. Prentice Hall, Upper Saddle River, NJ, 1996.
435. M. Bhardwaj, T. Garnett, and A. Chandrakasan, Upper bounds on the lifetime of sensor networks, in *Proceedings of IEEE International Conference on Communication,* Helsinki, Finland, vol. 3, pp. 785–790, June 2001.
436. I. F. Akyildiz, W. L. Su, Y. Sankarasubramaniam, and E. Cayirci, A survey on sensor networks, *IEEE Communications Magazine*, vol. 40, no. 8, pp. 102–114, 2002.
437. J. N. Al-Karaki and A. E. Kamal, A survey on routing protocol for wireless sensor networks, *IEEE Wireless Communications*, vol. 11, no. 6, pp. 6–28, 2004.
438. Y. Tang, M. T. Zhou, and X. Zhang, Overview of routing protocols in wireless sensor networks, *Journal of Software*, vol. 17, no. 3, pp. 410–421, 2006.
439. W. Heinzelman, A. Chanrakasan, and H. Balakrishnan, Energy-efficient communication protocol for wireless micro-sensor networks, in *Proceedings of the 33rd Hawaii International Conference on System Sciences*, Maui, HI, vol. 2, pp. 1–10, 2000.
440. W. Heinzelman, A. Chanrakasan, and H. Balakrishnan, An application-specific protocol architecture for wireless microsensor networks, *IEEE Transactions on Wireless Communications*, vol. 1, no. 4, pp. 660–670, 2002.
441. S. Lindsey and C. Raghavendra, PEGASIS: Power-efficient gathering in sensor information systems, in *Proceedings of IEEE Aerospace Conference*, Big Sky, MT, vol. 3, pp. 1125–1130, 2002.
442. A. Manjeshwar and D. P. Agarwal, TEEN: A routing protocol for enhanced efficiency in wireless sensor networks, in *Proceedings of 15th International Conference on Parallel and Distributed Processing Symposium*, San Francisco, CA, pp. 2009–2015, April 2002.

443. A. Manjeshwar and D. P. Agarwal, APTEEN: A hybrid protocol for efficient routing and comprehensive information retrieval in wireless sensor networks, in *Proceedings of 15th International Conference on Parallel and Distributed Processing Symposium*, Ft. Lauderdale, FL, pp. 195–202, 2002.
444. O. Younis and S. Fahmy, HEED: A hybrid, energy-efficient, distributed clustering approach for ad hoc sensor networks, *IEEE Transactions on Mobile Computing*, vol. 3, no. 4, pp. 660–669, 2004.
445. J. Yick, B. Mukherjee, and D. Ghosal, Wireless sensor network survey, *Computer Networks*, vol. 52, no. 12, pp. 2292–2330, 2008.
446. M. Tubaishat and S. Madria, Sensor networks: An overview, *IEEE Potentials*, vol. 22, no. 2, pp. 20–23, 2003.
447. O. Zytoune, M. El-aroussi, and D. Aboutajdine, A uniform balancing energy routing protocol for wireless sensor networks, *Wireless Personal Communications*, vol. 55, no. 2, pp. 147–161, 2010.
448. L. Tan, Y. Gong, and G. Chen, A balanced parallel clustering protocol for wireless sensor networks using K-means techniques, in *Proceedings of IEEE Proceedings of the Second International Conference on Sensor Technologies and Applications*, Cap Esterel, France, pp. 300–305, August 25–31, 2008.
449. Y. Gong, G. Chen, and L. Tan, A balanced serial K-means based clustering protocol for wireless sensor networks, in *Proceedings of the 4th IEEE International Conference on Wireless Communications, Networking and Mobile Computing,* Dalian, China, pp. 1–6, October 12–14, 2008.
450. A. Abdulla, H. Nishiyama, and N. Kato, Extending the lifetime of wireless sensor networks: A hybrid routing algorithm original research article, *Computer Communications*, vol. 35, no. 9, pp. 1056–1063, 2012.
451. A. F. Liu, J. Ren, X. Li, and Z. C. Xuemin, Design principles and improvement of cost function based energy aware routing algorithms for wireless sensor networks original research article, *Computer Networks*, vol. 56, no. 7, pp. 1951–1967, 2012.
452. L. Cobo, A. Quintero, and S. Pierre, Ant-based routing for wireless multimedia sensor networks using multiple QoS metrics original research article, *Computer Networks*, vol. 54, no. 17, pp. 2991–3010, 2010.
453. S. He, J. Chen, Y. Sun, D. Yau, and N. K. Yip, On optimal information capture by energy-constrained mobile sensors, *IEEE Transactions on Vehicular Technology*, vol. 59, no. 5, pp. 2472–2484, 2010.
454. X. Cao, J. Chen, C. Gao, and Y. Sun, An optimal control method for applications using wireless sensor/actuator networks, *Computers and Electrical Engineering*, vol. 35, no. 5, pp. 748–756, 2009.
455. J. Shlens, A tutorial on principal component analysis, December 2005. http://www.snl.salk.edu/ shlens/pca.pdf.
456. M.E. Wall et al., Singular value decomposition and principal component analysis, in D. P. Berrar, W. Dubitzky, and M. Granzow, *A Practical Approach to Microarray Data Analysis*, Kluwer Academic Publishers, Norwell, MA, vol. 2, no. 1, pp. 91–109, 2003.
457. H. Zha, C. Ding, M. Gu, X. F. He, and H. Simon, Spectral relaxation for K-means clustering, in *Advances in Neural Information Processing Systems 14 (NIPS'01)*, pp. 1057–1064, 2002.
458. C. Ding and X. F. He, K-means clustering via principal component analysis, in *Proceedings of ICML'2004*, Banff, Alberta, Canada, pp. 29, July 4–8, 2004.
459. S. P. Lloyd, Least squares quantization in PCM, *IEEE Transactions on Information Theory*, vol. 28, no. 2, pp. 128–137, 1982.
460. J. Ham and M. Kamber, *Data Mining: Concepts and Techniques*, 2nd ed., Morgan Kaufman Publishers, Burlington, MA, pp. 1–6, 2006.

461. W. R. Heinzelman, A. P. Chandrakasan, and H. Balakrishnan, Energy-efficient communication protocols for wireless microsensor networks, in *Proceedings of 33rd Hawaiian Int'l Conference System Sciences (HICSS 2000),* Maui, HI, pp. 3005–3014, January 2000.
462. W. R. Heinzelman, A. P. Chandrakasan, and H. Balakrishnan, An application-specific protocol architecture for wireless microsensor networks, *IEEE Transactions on Wireless Communications*, vol. 1, no. 4, pp. 660–670, 2002.
463. S. Lindsey and C. S. Raghavendra, PEGASIS: Power-efficient gathering in sensor information systems, in *Proceedings of IEEE Aerospace Conference,* Big Sky, MT, vol. 3, pp. 1125–1130, 2002.
464. W. R. Heinzelman, *Application-Specific Protocol Architecture for Wireless Networks*, PhD Thesis, Massachusetts Institute of Technology, June 2000.
465. S. Bandopadhyay, and E. J. Coyle, An energy-efficient hierarchical clustering algorithm for wireless sensor networks, in *Proceedings of IEEE INFOCOM 2003,* San Francisco, CA, vol. 3, pp. 1713–1723, 2003.
466. J. Kamimura, N. Wakamiya, and M. Murata, Energy-efficient clustering method for data gathering in sensor networks, in *Proceedings of the Annual International Conference on Broadband Networks,* October 2004.
467. V. Mhatre and C. Rosenberg, Design guidelines for wireless sensor networks: communication, clustering and aggregation, *Ad Hoc Networks Journal*, vol. 2, no. 1, pp. 45–63, 2004.
468. E. Hansen, J. Neander, M. Nolin, and M. Bj orkman, Energy efficient cluster formation for large sensor networks using a minimum separation distance, in *Proceedings of the Fifth Annual Mediterranean Ad Hoc Networking Workshop,* Lipari, Italy, June 2006.
469. G. Indranil, D. Riordan, and S. Sampalli, Cluster head selection using evolutionary computing in wireless sensor networks, in *Proceedings of Communication Networks and Services Research Conference,* Hangzhou, China, pp. 255–260, May 2005.
470. H. Chen and S. Megerian, Cluster sizing and head selection for efficient data aggregation and routing in sensor networks, in *Proceedings of Wireless Communications and Networking Conference, 2006 (WCNC 2006)*, Las Vegas, NV, vol. 4, pp. 2318–2323, 2006,
471. J. Hill, R. Szewczyk, A. Woo, S. Hollar, D. Culler, and K. Pister, System architecture directions for networked sensors, *Architectural Support for Programming Languages and Operating Systems*, vol. 35, no. 11, pp. 93–104, 2000.
472. F. Ye, H. Luo, J. Cheng, S. Lu, and L. Zhang, A two-tier data dissemination model for large-scale wireless sensor networks, in *Proceedings of IEEE/ACM MOBICOM,* 2002. http://citeseer.ist.psu.edu/ye02twotier.html.
473. S. Bandyopadhyay and E. J. Coyle, Minimizing communication costs in hierarchically clustered networks of wireless sensors, *Computer Networks,* vol. 44, no. 1, pp. 1–16, 2004.
474. C. Chiasserini, I. Chilamtac, P. Monti, and A. Nucci, An energy-efficient method for nodes assignment in cluster-based ad hoc networks, *Wireless Networks,* vol. 10, no. 3, pp. 223–231, 2004.
475. M. J. Handy, M. Hasse, and D. Timmerman, Low energy adaptive clustering hierarchy with deterministic cluster-head selection, in *Proceedings of the 4th International Workshop on Mobile and Wireless Communication Network,* Stockholm, Sweden, pp. 368–372, 2002.
476. W. Ye, J. Heihemann, and D. Estrin, An energy-efficient MAC protocol for wireless sensor networks, in *Proceedings of IEEE INFORCOM 2002,* New York, vol. 3, June 2002.
477. W. R. Heinzelman, A. Chandrakasan, and H. Balakrishnan, Energy-efficient communication protocol for wireless microsensor networks, in *Proceedings of the 33rd Annual Hawaii International Conference on System Sciences (HICSS)*, Maui, HI, vol. 2, pp. 1–10, January 2000.
478. J. N. Al-Karaki and A. E. Kamal, Routing techniques in wireless sensor networks: A survey, *IEEE Wireless Communications*, vol. 11, no. 6, pp. 6–28, 2004.

479. C. Intanagonwiwat, R. Govindan and D. Estrin, Directed difusion: A scalable and robust communication paradigm for sensor networks, in *Proceedings of ACM MobiCom*, Boston, MA, pp. 56–67, 2000.
480. S. Bandyopadhyay and E. Coyle, An energy efficient hierarchical clustering algorithm for wireless sensor networks, in *Proceedings of the 22nd Annual Joint Conference of the IEEE Computer and Communications Societies (Infocom'03)*, San Francisco, CA, vol. 3, pp. 1713–1723, April 2003.
481. Y. P. Chen, A. L. Liestman, and J. Liu, A hierarchical energy-efficient framework for data aggregation in wireless sensor networks, *IEEE Transactions on Vehicular Technology*, vol. 55, no. 3, pp. 789–796, 2006.
482. S. G. Foss and S. A. Zuyev, On a Voronoi aggregative process related to a bivariate Poisson process, *Advances in Applied Probability*, vol. 28, no. 4, pp. 965–981, 1996.
483. F. Xue and P. R. Kumar, The number of neighbors needed for connectivity of wireless networks, *Wireless Networks*, vol. 10, no. 2, pp. 169–181, 2004.
484. C. Zhu, C. Zheng, L. Shu, and G. Han, A survey on coverage and connectivity issues in wireless sensor networks, *Journal of Network Computer Applications*, vol. 35, no. 2, pp. 619–632, 2012.
485. P. Jiang, A new method for node fault detection in wireless sensor networks, *Sensors*, vol. 93, no. 10, pp. 2854–2857, 2008.
486. N. Xiong, A. V. Vasilakos, L. T. Yang, L. Song, Y. Pan, R. Kannan, and Y. Li, Comparative analysis of quality of service and memory usage for adaptive failure detectors in healthcare systems, *IEEE Journal on Selected Areas in Communication*, vol. 27, no. 4, pp. 495–509, 2009.
487. Z. Wang, L. Lu, and A. C. Bovik, Video quality assessment based on structural distortion measurement, *Signal Process Image Communication*, vol. 19, no. 2, pp. 121–132, 2004.
488. I. Epifanio, J. Gutierrez, and J. Malo, Linear transform for simultaneous diagonalization of covariance and perceptual metric matrix in image coding, *Pattern Recognition*, vol. 36, no. 8, pp. 1799–1811, 2003.
489. J. Liu, and P. Moulin, Information-theoretic analysis of interscale and intrascale dependencies between image wavelet coefficients, *IEEE Transactions Image Process.*, vol. 10, no. 11, pp. 1647–1658, 2001.
490. D. A. Silverstein and J. E. Farrell, The relationship between image fidelity and image quality, in *Proceedings of the International Conference on Image Processing*, Lausanne, Switzerland, vol. 1, pp. 881–884, September 16–19, 1996.
491. Z. Wang, A. C. Bovik, H. R. Sheikh, and E. P. Simoncelli, Image quality assessment: From error visibility to structural similarity, *IEEE Transactions on Image Processing*, vol. 13, no. 4, pp. 600–612, 2004.
492. D. Brunet, E. R. Vrscay, and Z. Wang, On the mathematical properties of the structural similarity index, *IEEE Transactions on Image Processing*, vol. 21, no. 4, pp. 1488–1499, 2012.
493. W. B. Heinzelman, A. P. Chandrakasan, and H. Balakrishnan, An application-specific protocol architecture for wireless microsensor networks, *IEEE Transactions on Wireless Communications*, vol. 21, no. 4, pp. 1488–1499, 2002.
494. V. Loscri, G. Morabito, and S. Marano, A two-levels hierarchy for low-energy adaptive clustering hierarchy (TL-LEACH), in *Proceedings of the IEEE Vehicular Technology Conference*, Dallas, TX, pp. 1809–1813, September 25–28, 2005.
495. D. Kumar, T. C. Aseri, and R. B. Patel, EEHC: Energy efficient heterogeneous clustered scheme for wireless sensor networks, *Computer Communications*, vol. 32, no. 4, pp. 662–667, 2009.
496. L. Badia, E. Fasolo, A. Paganini, and Zorzi, M. Data aggregation algorithms for sensor networks with geographic information awareness, in *Proceedings of the 9th International Sympo-*

sium on Wireless Personal Multimedia Communications (WPMC), San Diego, CA, September 2006.
497. S. Yoon and C. Shahabi, The clustered aggregation (CAG) technique leveraging spatial and temporal correlations in wireless sensor networks, *ACM Transactions on Sensor Networks (TOSN)*, vol. 3, no. 1, 2007.
498. L. Buttyan, P. Schaffer, and I. Vajda, CORA: Correlation-based resilient aggregation in sensor networks, *Ad Hoc Networks*, vol. 7, pp. 1035–1050, 2009.
499. Y. Ma, Y. Guo, X. Tian, and M. Ghanem, Distributed clustering-based aggregation algorithm for spatial correlated sensor networks, *IEEE Sensors Journal*, pp. 641–648, 2011.
500. B. Gedik, L. Liu, and P. S. Yu, ASAP: An adaptive sampling approach to data collection in sensor networks, *IEEE Transactions on Parallel Distributed Systems*, pp. 1766–1783, 2007.
501. J. Koshy, I. Wirjawan, R. Pandey, and Y. Ramin, Balancing computation and communication costs: The case for hybrid execution in sensor networks, *Ad Hoc Networks*, pp. 1185–1200, 2008.
502. M. Tahir, and R. Farrell, Optimal communication-computation tradeoff for wireless multimedia sensor network lifetime maximization, in *Proceedings of the IEEE Wireless Communications and Network Conferece*, Budapest, Hungary, pp. 1–6, April 5–8, 2009.
503. H. Jiang, S. Jin, and C. Wang, Prediction or not? An energy-efficient framework for clustering-based data collection in wireless sensor networks, *IEEE Transactions on Parallel and Distributed Systems*, vol. 22, pp. 1064–1071, 2011.
504. C. Carvalho, D. G. Gomes, N. Agoulmine, and J. N. de Souza, Improving prediction accuracy for WSN data reduction by applying multivariate spatio-temporal correlation, *Sensors*, vol. 11, pp. 10010–10037, 2011.
505. C. Carvalho, D. G. Gomes, J. N. de Souza, and N. Agoulmine, Multiple linear regression to improve prediction accuracy in WSN data reduction, in *Proceedings of the 7th Latin American Network Operations and Management Symposium (LANOMS)*, Quito, Ecuador, pp. 1–8, October 10–11, 2011.
506. M. A. Guvensan and A. G. Yavuz, On coverage issues in directional sensor networks: A survey, *Ad Hoc Networks*, vol. 9, no. 7, pp. 1238–1255, 2011.
507. J. M. Gil and Y. H. Han, A target coverage scheduling scheme based on genetic algorithms in directional sensor networks, *Sensors*, vol. 11, no. 2, pp. 1888–1906, 2011.
508. C. T. Cheng, C. K. Tse, and F. C. M. Lau, An energy-aware scheduling scheme for wireless sensor networks, *IEEE Transactions on Vehicular Technology*, vol. 59, no. 7, pp. 3427–3444, 2010.
509. P. Liaskovitis and C. Schurgers, Efficient sensing topology management for spatial monitoring with sensor networks, *Journal of Signal Processing Systems*, vol. 62, no. 3, pp. 415–426, 2011.
510. A. Jindal and K. Psounis, Modeling spatially correlated data in sensor networks, *ACM Transactions on Sensor Networks (TOSN)*, vol. 2, no. 4, pp. 466–499, 2006.
511. Y. Yu, D. Ganesan, L. Girod, D. Estrin, and R. Govindan, *Synthetic Data Generation to Support Irregular Sampling in Sensor Networks*, GeoSensor Networks, Portland, ME, USA, pp. 211–234, 2003.
512. M. Rahimi, R. Pon, W. J. Kaiser, G. S. Sukhatme, D. Estrin, and M. Srivastava, Adaptive sampling for environmental robotics, in *Proceedings of the International Conference on Robotics and Automation,* New Orleans, LA, vol. 4, pp. 3537–3544, April 26–May 1, 2004.
513. D. Marinucci and G. Peccati, *Random fields on the sphere: representation, limit theorems and cosmological applications*, Cambridge University Press, Cambridge, UK, 2011.
514. M. Haenggi, Stochastic geometry and random graphs for the analysis and design of wireless networks, *IEEE Journal on Selected Areas in Communications*, vol. 27, no. 7, pp. 1025–1028, 2009.

515. D. Freedman, *Statistical Models: Theory and Practice*, Cambridge University Press, Cambridge, UK, 2009.
516. P. Schaffer and I. Vajda, CORA: Correlation-based resilient aggregation in sensor networks, in *Proceedings of the 10th ACM Symposium on Modeling, Analysis, and Simulation of Wireless and Mobile Systems*, Chania, Crete Island, Greece, pp. 373–376, October 22–26, 2007.
517. P. Bodik, W. Hong, C. Guestrin, S. Madden, M. Paskin, and R. Thibaux, Intel lab data. http://db.csail.mit.edu/labdata/labdata.html.
518. Y. A. Borgne, M. Moussaid, and G. Bontempi, Simulation architecture for data processing algorithms in wireless sensor networks, in *20th International Conference on Advanced Information Networking and Applications (AINA)*, Vienna, Austria, pp. 5, 2006.
519. R. P. R. Alliance, An introduction to resilient packet ring technology, October 2001. http://www.rpralliance.org/.
520. D. Wang, K. K. Ramakrishnan, C. Kalmanek, R. Doverspike, and A. Smiljanic, Congestion control in resilient packet rings, in *Proceedings of the 12th IEEE International Conference on Network Protocol (ICNP'04)*, Berlin, Germany, pp. 108–117, October 5–8, 2004.
521. A. S. Tanenbaum, *Computer Networks*, 3rd ed., Prentice Hall, Inc., Upper Saddle River, NJ, 1996.
522. D. Tsiang and G. Suwala, The cisco SRP MAC layer protocol, *Internet RFC 2892*, August 2000.
523. P. Yuan, V. Gambiroza, and E. Knightly, The IEEE 802.17 media access protocol for high-speed metropolitan-area resilient packet rings, *IEEE Network*, vol. 18, no. 3, pp. 8–15, 2004.
524. A. Mekkittikuletal, Alladin proposal for IEEE standard 802.17, *IEEE Transactions on Wireless Communications*. http://grouper.ieee.org/groups/802/17/documents/drafts/zz_draft_04.zip, November, 2001.
525. J. Kaoetal, Darwin proposal for IEEE standard 802.17. http://www.ieee802.org/17/documents/presentations/jan2002/ib_fairdraft_02.pdf, January. 2002.
526. A. Charny, D. D. Clark, and R. Jain, Congestion control with explicit rate indication, in *Proceedings of IEEE International Conference on Communications*, Seattle, WA, pp. 1954–1963, June 1995.
527. L. Tan, H. Wang, and M. Zukerman, A distributed bandwidth fair allocation algorithm for RPR networks, in *Proceedings of 19th ITC*, Beijing University of Posts and Telecommunications Press, Beijing, China, pp. 1–10, August 29–September 2, 2005.
528. O. B. Akan and I. F. Akyildiz, Event-to-sink reliable transport inwireless sensor networks, *IEEE/ACM Transactions on Networking*, vol. 13, no. 5, pp. 1003–1016, 2005.
529. M. Aazam and E. N. Huh, Fog computing micro datacenter based dynamic resource estimation and pricing model for IoT, in *Proceedings of 2015 IEEE 29th International Conference on Advanced Information Networking and Applications (AINA)*, Gwangiu, South Korea, pp. 687–694, March 24–27, 2015.
530. P. Bender, P. Black, M. Grob, R. Padovani, N. Sindhushayana, and A. Viterbi, CDMA/HDR: A bandwidth efficient high speed wireless data service for nomadic users, *IEEE Communications Magazine*, vol. 38, no. 7, pp. 70–77, 2000.
531. L. Chen, J. Cheng, and Y. Tseng, Optimal path planning with spatial-temporal mobility modeling for individual-based emergency guiding, *IEEE Transactions on Systems, Man, and Cybernetics: Systems*, vol. 45, no. 12, pp. 1491–1501, 2015.
532. N. H. M. Chiang and N. Mandayam, Utility-lifetime trade-off in self-regulating wireless sensor networks: a cross-layer design approach, in *Proceedings of IEEE International Conference on Communications*, Istanbul, Turkey, vol. 8, 2006.
533. B. Deb, S. Bhatnagar, and B. Nath, ReInForM: Reliable information forwarding using multiple paths in sensor networks, in *Proceedings of the IEEE LCN*, Bonn/Konigswinter, Germany, 2003.

534. B. Deb, S. Bhatnagar, and B. Nath, Information assurance in sensor networks, in *Proceedings of the ACM WSNA,* Anchorage, AK, 2003.
535. Z. Fang and B. Bensaou, Fair bandwidth sharing algorithms based on game theory frameworks for wireless ad-hoc networks, in *Proceedings of IEEE INFOCOM 2004*, Hong Kong, China, pp. 1284–1295, 2004.
536. A. Goldsmith and S. Wicker, Design challenges for energy-constrainedad ad hoc wireless networks, *IEEE Transactions on Wireless Communications*, vol. 9, no. 4, pp. 8–27, 2002.
537. B. Hajek and G. Sasaki, Link scheduling in polynomial time, *IEEE Transactions on Information Theory*, vol. 34, no. 5, pp. 910–917, 1988.
538. N. Hithesh, M. Chiang, and N. Mandayam, Optimal utility-lifetime trade-off in wireless sensor networks: Characterization and distributed algorithms, in *Proceedings of IEEE ICC,* Princeton, NJ, 2006.
539. Y. T. Hou, Y. Shi, and H. D. Sherali, Rate allocation in wireless sensor networks with network lifetime requirement, in *Proceedings of ACM MobiHoc,* Roppongi Hills, Tokyo, Japan, pp. 67–77, 2004.
540. X. L. Huang and B. Bensaou, On max-min fairness and scheduling in wireless ad-hoc networks: Analytical framework and impmlementation, in *Proceedings of Workshop on Mobile Ad Hoc Networking and Computing (MobiHOC)*, pp. 221–231, 2001.
541. K. L. Hung, B. Bensaou, J. H. Zhu, and F. Nait-Abdesselam, Energy-aware fair routing in wireless sensor networks with maximum data collection, in *Proceedings of IEEE ICC,* Istanbul, Turkey, 2006.
542. IEEE Standard 802.11, Wireless LAN Medium Access Control (MAC) and Physical Layer (PHY) Specifications, IEEE Std. 802.11-1197, 1997.
543. R. Madan and S. Lall, Distributed algorithms for maximum lifetime routing in wireless sensor networks, *IEEE Transactions on Wireless Communication* vol. 5, no. 8, pp. 2185–2193, 2006.
544. R. Madan, S. Cui, S. Lall, and A. Goldsmith, Cross-layer design for lifetime maximization in interference-limited wireless sensor networks, in *Proceedings on IEEE INFOCOM,* Miami, FL, March 2005.
545. J. A. Maya, L. V. Rey Vega, and C. G. Galarza, Optimal resource allocation for detection of a Gaussian process using a MAC in WSNs, *IEEE Transactions on Signal Processing*, vol. 63, no. 8, pp. 2057–2069, 2015.
546. K. Miettinen, *Non-Linear Multi-Objective Optimization,* Kluwer Academic Publishers, Boston, 1999.
547. H. Nama, M. Chiang, and N. Mandayam, Optimal utility-lifetime trade-off in self-regulating wireless sensor networks: A distributed approach, in *Proceedings of the 40th Annual Conference on Information Sciences and Systems,* Princeton, NJ, pp. 789–794, 2006.
548. H. Park, J. Park, H. Kim, J. Jun, S. H. Son, T. Park, and J. Ko, ReLiSCE: utilizing resource-limited sensors for office activity context extraction, *IEEE Transactions on Systems, Man, and Cybernetics: Systems*, vol. 45, no. 8, pp. 1151–1164, 2015.
549. Y. Qiu and P. Marbach, Bandwidth allocation in wireless ad hoc networks: A price-based approach, in *Proceedings of IEEE INFOCOM 2003,* San Francisco, CA, vol. 2, pp. 797–807, 2003.
550. A. Sankar and Z. Liu, Maximum lifetime routing in wireless ad hoc networks, in *Proceedings of IEEE INFOCOM,* Hong Kong, China, March 2004.
551. R. C. Shah and J. M. Rabaey, Energy aware routing for low energy ad hoc sensor networks, in *Proceedings of IEEE WCNC,* Orlando, FL, March 2002.
552. V. Srinivasan, C. F. Chiasserini, P. S. Nuggehalli, and R. R. Rao, Optimal rate allocation for energy-efficient multipath routing in wireless ad hoc networks, *IEEE Transactions on Wireless Communications,* vol. 3, no. 3, pp. 891–899, 2004.

553. V. Srinivasan, P. Nuggehalli, C. F. Chiasserini, and R. Rao, Cooperatioin in wireless ad hoc networks, in *Proceedings of IEEE INFOCOMM,* San Francisco, CA, pp. 808–817, 2003.
554. F. Stann and J. Heidemann, RMST: Reliable data transport in sensor networks, in *Proceedings of the IEEE SNPA,* Anchorage, AK, 2003.
555. L. Tan, Z. Zhu, F. Ge, and N. Xiong, Utility maximization resource allocation in wireless networks: Methods and algorithms, *IEEE Transactions on Systems, Man and Cybernetics: Systems*, vol. 45, no. 7, pp. 1018–1034, 2015.
556. L. Tassiulas and S. Sarkar, Maxmin fair scheduling in wireless networks, in *Proceedings of IEEE INFOCOM 2002,* New York, NY, pp. 763–772, 2002.
557. K. Tei, R. Shimizu, Y. Fukazawa, and S. Honiden, Model-driven-development-based stepwise software development process for wireless sensor networks, *IEEE Transactions on Systems, Man, and Cybernetics: Systems*, vol. 45, no. 4, pp. 675–687, 2015.
558. S. Tilak, N. B. Abu-Ghazaleh, and W. Heinzelman, A taxonomy of wireless micro-sensor network models, *ACM SIGMOBILE Mobile Computing and Communication Review,* vol. 6, no. 2, pp. 28–36, 2002.
559. D. Tian and N. D. Georganas, Energy efficient routing with guaranteed delivery in wireless sensor networks, in *Proceedings of the IEEE WCNC,* New Orleans, LA, 2003.
560. Z. H. Velkov, I. Nikoloska, G. K. Karagiannidis and T. Q. Duong, Wireless networks with energy harvesting and power transfer: Joint power and time allocation, *IEEE Signal Processing Letters*, vol. 23, no. 1, pp. 50–54, 2015.
561. I. Vlasenko, I. Nikolaidis, and E. Stroulia, The smart-condo: Optimizing sensor placement for indoor localization, *IEEE Transactions on Systems, Man, and Cybernetics: Systems,* vol. 45, no. 3, pp. 436–453, 2015.
562. L. Xiao, M. Johansson, and S. Boyd, Simultaneous routing and resource allocation via dual decomposition, *IEEE Transactions on Communications*, vol. 52, no. 7, pp. 1136–1144, 2004.
563. W. Xu, Q. Shi, X. Wei, and Y. Wang, Distributed optimal rate-reliability-lifetime tradeoff in wireless sensor networks, *arXiv preprint*, arXiv:1303.3624, 2013.
564. Y. Xue, B. Li, and K. Nahrstedt, Price-based resource allocation in wireless ad hoc networks, in *Proceedings of International Workshop on Quality of Service (IWQoS)*, Berkeley, CA, pp. 79–96, June 2003.
565. Y. Yi and S. Shakkottai, Hop-by-hop congestion control over a wireless multi-hop network, in *Proceedings of IEEE INFOCOM 2004*, Hong Kong, China, pp. 2548–2558, 2004.
566. J. Zhu, K. L. Hung, B. Bensaou, and F. Nait-Abdesselam, Rate-lifetime tradeoff for reliable communication in wireless sensor networks, *Computer Networks,* vol. 52, no. 1, pp. 25–43, 2008.
567. C. Oliveira, An adaptive bandwidth reservation scheme for high-speed multimedia wireless networks, *IEEE Journal on Selected Areas in Communications,* vol. 16, no. 6, pp. 858–874, 1998.
568. P. Hande, S. Zhang, and M. Chiang, Distributed rate allocation for inelastic flows, *IEEE/ACM Transactions on Networking,* vol. 15, no. 6, pp. 1240–1253, 2007.
569. N. Nasser and H. Hassanein, Enabling seamless multimedia wireless access through QoS-based bandwidth adaptation, *Wireless Communications and Mobile Computing*, vol. 7, no. 1, pp. 53–67, 2007.
570. T. Kwon, Measurement-based call admission control for adaptive multimedia in wireless/mobile networks, *IEEE Wireless Communications and Networking Conference. (WCNC)*, New Orleans, LA, vol. 2, pp. 540–544, September 1999.
571. D. A. Babayev and G. I. Bell, An optimization problem with a separable non-convex objective function and a linear constraint, *Journal of Heuristics,* vol. 7, no. 2, pp. 169–184, 2001.

572. N. Lu, Utility-based bandwidth adaptation for QoS provisioning in multimedia wireless networks, Submitted for the degree of Doctor of Philosophy, Department of Electronic Engineering Queen Mary, University of London, UK, April 2007.
573. A. Ferragut and F. Paganini, Network resource allocation for users with multiple connections: Fairness and stability, *IEEE/ACM Transactions on Networking,* vol. 22, no. 2, pp. 349–362, 2014.
574. L. Xu, X. Shen, and M. Jon, Fair resource allocation with guaranteed statistical QoS for multimedia traffic in wideband CDMA cellular network, *IEEE Transactions on Mobile Computing.,* vol. 4, no. 2, pp. 166–177, 2005.
575. K. M. Ahn and S. Kim, Optimal bandwidth allocation for bandwidth adaptation in wireless multimedia networks, *Computers and Operations Research,* vol. 30, no. 13, pp. 1917–1929, 2003.
576. W. Wang, A multimedia quality-driven network resource management architecture for wireless sensor networks with stream authentication, *IEEE Transactions on Multimedia,* vol. 12, no. 5, pp. 439–447, 2010.
577. N. Muhammad, N. Sheneela, and G. Saira, A framework for real-time resource allocation in IP multimedia subsystem network, *International Journal of Computer Network and Information Security,* vol. 5, no. 3, pp. 32–39, 2013.
578. S. Yang, W. Song, and Z. Zhong, Resource allocation for aggregate multimedia and healthcare services over heterogeneous multi-hop wireless networks, *Wireless Personal Communications,* vol. 69, no. 1, pp. 229–251, 2012.
579. X. Sun, K. Piamrat, and C. Viho, QOE-based dynamic resource allocation for multimedia traffic in IEEE 802.11 wireless networks, presented at *the 2011 IEEE International Conference on Multimedia and Expo (ICME 2011),* Barcelona, Spain, July 2011.
580. B. Yang, Resource allocation methods in OFDMA distributed radio access networks, presented at *the 2011 4th IEEE International Conference on Broadband Network and Multimedia Technology (IC-BNMT),* Shenzhen, China, October 2011.
581. K. Shin, J. Y. Jung, and J. R. Lee, Resource allocation and pricing mechanisms for wireless multimedia service: Auction and bargaining models, *Journal of Internet Technology,* vol. 14, no. 7, pp. 1093–1103, 2013.
582. R. Devarajan, Energy-Aware resource allocation for cooperative cellular network using multi-objective optimization approach, *IEEE Transactions on Wireless Communications,* vol. 11, no. 5, pp. 1797–1807, 2012.
583. H. Park and M. Schaar, Fairness strategies for wireless resource allocation among autonomous multimedia users, *IEEE Transactions on Circuits and Systems for Video Technology,* vol. 20, no. 2, pp. 297–309, 2010.
584. Y. Li, Network resource allocation for competing multiple description transmissions, *IEEE Transactions on Communications,* vol. 58, no. 5, pp. 1493–1504, 2010.
585. Y. Zhang, Multihop packet delay bound violation modeling for resource allocation in video streaming over mesh networks, *IEEE Transactions on Multimedia,* vol. 12, no. 8, pp. 886–900, 2010.
586. K. Ivesic, M. Matijasevic, and L. S. Kapov, A framework for real-time resource allocation in IP multimedia subsystem network, presented at *2010 IEEE 21st International Symposium on Personal Indoor and Mobile Radio Communications (PIMRC),* Instanbul, Turkey, September 2010.
587. K. Ivesic, M. Matijasevic, and L. S. Kapov, Simulation based evaluation of dynamic resource allocation for adaptive multimedia services, presented at *the 7th International Conference on Network and Services Management (CNSM 2011),* Laxenburg, Austria, September 2011.

588. H. Zhao and J. Ilow, Adaptive resource allocation in OFDMA relay networks with wetwork coding, presented at *the 7th International Conference on Network and Services Management (CNSM 2011)*, Laxenburg, Austria, September 2011.
589. W. Gong, An optimal VM resource allocation for near-client-datacenter for multimedia cloud, presented at *2011 IEEE International Conference on Communications (ICC)*, Kyoto, Japan, June 2011.
590. A. Argyriou, D. Kosmanos, and L. Tassiulas, Joint time-domain resource partitioning, rate allocation, and video quality adaptation in heterogeneous cellular networks, *IEEE Transactions on Multimedia,* vol. 17, no. 5, pp. 687–699, 2015.
591. S. P. Chuah, Y. P. Tan, and Z. Chen, Rate and power allocation for joint coding and transmission in wireless video chat applications, *IEEE Transactions on Multimedia,* vol. 17, no. 5, pp. 736–745, 2015.
592. H. E. Egilmez and A. M. Tekalp, Distributed QoS architectures for multimedia streaming over software defined networks, *IEEE Transactions on Multimedia,* vol. 16, no. 6, pp. 1597–1609, 2014.
593. H. E. Egilmez, S. Civanlar, and A. M. Tekalp, An optimization framework for QoS-enabled adaptive video streaming over openflow networks, *IEEE Transactions on Multimedia,* vol. 15, no. 3, pp. 710–715, 2013.
594. H. Koraitim and S. Tohm, Resource allocation and connection admission control in satellite networks, *IEEE Journal on Selected Areas in Communication*, vol. 17, no. 2, pp. 360–372, 1999.
595. M. Chiang, Balancing transport and physical layers in wireless multihop networks: Jointly optimal congestion control and power control, *IEEE Journal on Selected Areas in Communication,* vol. 23, no. 1, pp. 104–116, 2005.
596. M. Ismail, Decentralized radio resource allocation for single-network and multi-homing services in cooperative heterogeneous wireless access medium, *IEEE Transactions on Wireless Communication,* vol. 11, no. 11, pp. 4085–4095, 2012.
597. L. Massoulie and J. W. Roberts, Bandwidth sharing and admission control for elastic traffic, *Telecommunication Systems,* vol. 15, no. 1, pp. 185–201, 2000.
598. X. Su, S. Chan, and J. H. Manton, Bandwidth allocation in wireless ad hoc networks: Challenges and prospects, *IEEE Communications Magazines,* vol. 48, no. 1, pp. 80–85, 2010.
599. L. Tan, P. Yang, W. Zhang, and F. Ge, On utility-optimised router-level bandwidth allocation, *Transactions on Emerging Telecommunications Technology,* vol. 24, no. 3, pp. 303–316, 2013.
600. D. P. Bertsekas, *Nonlinear Programming.* Athena Scientific, Belmont, MA, 1999.
601. D. P. Bertsekas, A. Nedic, and A. E. Ozdaglar, Min common max crossing duality: A simple geometric framework for convex optimization and minimax theory, Report LIDS-P-2536, Jan. 2002.
602. L. Tan, Z. Zhu, C. Yuan, and W. Zhang, A novel approach for bandwidth allocation among soft QoS traffic in wireless networks, *Transactions on Emerging Telecommunications Technologies*, vol. 25, no. 5, pp. 479–484, 2014.
603. D. Angelini and M. Zorzi, On the throughput and fairness performance of heterogeneous downlink packet traffic in a locally centralized CDMA/TDD system, in *Proceedings of IEEE VTC-Fall 2002*, Vancouver, BC, 2002.
604. Y. Cao, V. O. K. Li, and Z. Cao, Scheduling delay-sensitive and besteffort traffic in wireless networks, in *Proceedings of IEEE ICC 2003*, Anchorage, AK, vol. 3, pp. 2208–2212, 2003.
605. Z. Jiang and N. K. Shankaranarayana, Channel quality dependent scheduling for flexible wireless resource management, in *Proceedings of IEEE Globecom 2001*, San Antonio, TX, 2001.
606. W. T. Chen, K. C. Shih, and J. L. Chiang, Flexible packet scheduling for quality of service provisioning in wireless networks, in *Proceedings of ICPADS,* Taiwan, China, December 2002.

607. M. Xiao, N. B. Shroff, and E. K. P. Chong, A utility-based power control scheme in wireless cellular systems, *IEEE/ACM Transactions on Networking*, vol. 11, no. 2, pp. 210–221, 2003.
608. Y. Cao and V. O. K. Li, Utility-oriented adaptive QoS and bandwidth allocation in wireless networks, in *Proceedings of IEEE ICC 2002*, New York, vol. 5, pp. 3071–3075, 2002.
609. G. Bianchi and A. T. Campbell, A programmable MAC framework for utility-based adaptive quality of service support, *IEEE Journal on Selected Areas in Communications*, vol. 18, no. 2, pp. 244–255, 2000.
610. R. F. Liao and A. T. Campbell, A utility-based approach to quantitative adaptation in wireless packet networks, *ACM WINET*, vol. 7, no. 5, pp. 541–557, 2001.
611. X. Gao, T. Nandagopal, and V. Bharghavan, Achieving application level fairness through utility-based wireless fair scheduling, in *Proceedings of IEEE Globecom 2001*, San Antonio, TX, vol. 6, pp. 3257–3261, 2001.
612. V. A. Siris, B. Briscoe, and D. Songhurst, Economic models for resource control in wireless networks, in *Proceedings of IEEE PIMRC 2002*, Lisbon, Portugal, September 2002.
613. V. A. Siris, Resource control for elastic traffic in CDMA networks, in *Proceedings of IEEE MOBICOM 2002*, Pavilhao Altantico, Lisboa, Portugal, pp. 193–204, 2002.
614. V. A. Siris and C. Courcoubetis, Resource control for loss-sensitive traffic in CDMA networks, in *Proceedings of IEEE INFOCOM 2004*, Hong Kong, China, vol. 4, pp. 2790–2799, 2004.
615. P. Marbach and R. Berry, Downlink resource allocation and pricing for wireless networks, in *Proceedings of IEEE INFOCOM 2002*, New York, vol. 3, pp. 1470–1479, 2002.
616. P. Liu, R. Berry, M. L. Honig, and S. Jordan, Slow-rate utility-based resource allocation in wireless networks, in *Proceedings of IEEE Globecom*, Taipei, Taiwan, November 2002.
617. W. H. Kuo and W. J. Liao, Utility-based optimal resource allocation in wireless networks, in *Proceedings of IEEE Globecom 2005*, St. Louis, MO, vol. 6, no. 10, 3512, 2005.
618. K. D. Wu and W. Liao, Flow allocation in multi-hop wireless networks: A cross-layer approach, *IEEE Transactions on Wireless Communcations*, vol. 7, no. 1, pp. 269–276, 2008.
619. L. Chen, S. H. Low, and J. C. Doyle, Joint congestion control and media access control design for wireless ad hoc networks, in *Proceedings of IEEE Infocom*, Miami, FL, vol. 3, no. 7, pp. 2212–2222, March 2005.
620. Y. Cao, V. O. K. Li, and Z. Cao, Scheduling delay-sensitive and best-effort trffic in the wireless networks, in *Proceedings of IEEE ICC,* Anchorage, AK, vol. 3, pp. 2208–2212, 2003.
621. S. Catreux, V. Erceg, D. Gesbert, and R. W. Heath, Adaptive modulation and MIMO coding for broadband wireless data networks, *IEEE Communications on Magazine*, vol. 40, no. 6, pp. 108–115, 2002.
622. M. Chiang, Nonconvex optimization of communication systems, in *Advances in Mechanics and Mathematics, Special Volumn on Strang's 70th Birthday*, eds, D. Gao and H. Sherali, Springer, October 2008.
623. A. Furuskar, S. Parkvall, M. Persson, and M. Samuelsson, Performance of WCDMA high speed packet data, in *Proceedings of IEEE VTC-Fall 2002-Spring*, Birmingham, AL, vol. 3, pp. 1116–1120, 2002.
624. R. J. Gibbens and F. P. Kelly, Resource pricing and the evolution of congestion control, *Automatica*, vol. 35, no. 12, pp. 1969–1985, 1999.
625. W. H. Kuo and W. Liao, Utility-Based resource allocation in wireless networks, *IEEE Transactions on Wireless Communications*, vol. 6, no. 10, pp. 3600–3606, 2007.
626. W. H. Kuo and W. Liao, Utility-based radio resource allocation for QoS traffic in wireless networks, *IEEE Transactions on Wireless Communications*, vol. 7, no. 7, pp. 2714–2722, 2008.
627. K. Ma, Q. Han, C. Chen, and X. Guan, Bandwidth allocation for cooperative relay networks based on Nash bargaining solution, *International Journal of Communication Systems,* vol. 25, no. 8, pp. 1044–1058, 2012.

628. S. Mustafa, S. A. Madani, K. Bilal, K. Hayat, and S. U. Khan, Stable-path multi-channel routing with extended level channel assignment, *International Journal of Communication Systems,* vol. 25, no. 7, pp. 887–902, 2012.
629. E. Oh and C. Woo, Performance analysis of dynamic channel allocation based on the greedy approach for orthogonal frequency-division multiple access downlink systems, *International Journal of Communication Systems,* vol. 25, no. 7, pp. 953–961, 2012.
630. S. Shenker, Fundamental design issues for the future internet, *IEEE Journal on Selected Areas in Communications*, vol. 13, no. 7, pp. 1176–1188, 1995.
631. L. Tan, P. Yang, and S. Chan, Error-aware and energy efficient routing approach in MANETs, *International Journal of Communication Systems,* vol. 22, no. 1, pp. 37–51, 2009.
632. P. Varzakas, Channel capacity per user in a power and rate adaptive hybrid DS/FFH-CDMA cellular system over Rayleigh fading channels, *International Journal of Communication Systems,* vol. 25, no. 7, pp. 943–952, 2012.
633. K. D. Wu and W. Liao, Flow allocation in multi-hop wireless networks: A cross-layer approach, *IEEE Transactions on Wireless Communications*, vol. 7, no. 1, pp. 269–276, 2008.
634. X. Zhang, L. Tan, J. Li, S. Zhao, and H. H. Chen, Active-time based bandwidth allocation for multi-hop wireless ad hoc networks, in *Proceedings of IEEE ICC,* Istanbul, Turkey, pp. 3789–3794, June 2006.
635. S. Tang and L. Tan, Analysis of blocking probability of multi-class OBS with general burst size distribution, *IEEE Communications Letters*, vol. 20, no. 11, pp. 2153–2156, Nov. 2016.
636. F. Ge and L. Tan, Blocking performance approximation in flexi-grid networks, *Optical Fiber Technology*, vol. 32, pp. 58–65, 2016.
637. S. Tang and L. Tan, Erlangian approximation to finite time probability of blocking time of multi-class OBS nodes, *Photonic Network Communications*, vol. 30, no. 2, pp. 167–177, 2015.
638. L. Tan and M. Wu, Data reduction in wireless sensor networks: A hierarchical LMS prediction approach, *IEEE Sensors Journal*, vol. 16, no. 6, pp. 1708–1715, 2016.
639. M. Wu, L. Tan, and N. Xiong, Data prediction, compression, and recovery in clustered wireless sensor networks for environmental monitoring applications, *Information Sciences*, vol. 329, pp. 800–818, 2016.
640. S. Tang and L. Tan, Reward rate maximization and optimal transmission policy of EH device with temporal death in EH-WSNs, *IEEE Transactions on Wireless Communications*, vol. 16, no. 2, pp. 1157–1167, Feb. 2017.
641. L. Tan and S. Tang, Energy harvesting wireless sensor node with temporal death: Novel models and analyses, *IEEE/ACM Transactions on Networking*, vol. 25, no. 2, pp. 896–909, April 2017.
642. F. Ge, L. Tan, J. Sun, and M. Zukerman, Latency of FAST TCP for HTTP transactions, *IEEE Communications Letters*, vol. 15, no. 11, pp. 1259–1261, 2011.
643. F. Ge, S. Chan, L. L. H. Andrew, F. Li, L. Tan, and M. Zukerman, Performance effects of two-way FAST TCP, *Computer Networks*, vol. 55, pp. 2976–2984, 2011.
644. F. Ge, L. Tan, and R. A. Kennedy, Stability and throughput of FAST TCP traffic in bidirectional connections, *IET Communications*, vol. 4, no. 6, pp. 639–644, 2010.

Index

D

E

F

G

H

I

J

K

L

M

O

P

T

U

V

W